MAMMALIAN PROTEIN METABOLISM

Edited by

H. N. MUNRO

PHYSIOLOGICAL CHEMISTRY LABORATORIES
DEPARTMENT OF NUTRITION AND FOOD SCIENCE
MASSACHUSETTS INSTITUTE OF TECHNOLOGY
CAMBRIDGE, MASSACHUSETTS

VOLUME IV

ACADEMIC [illegible] New York and London

ACADEMIC PRESS, INC.
111 Fifth Avenue, New York, New York 10003

United Kingdom Edition published by
ACADEMIC PRESS, INC. (LONDON) LTD.
Berkeley Square House, London W1X 6BA

Library of Congress Catalog Card Number: 63-21397

PRINTED IN THE UNITED STATES OF AMERICA

List of Contributors

Numbers in parentheses indicate the pages on which the authors' contributions begin.

GEORGE F. CAHILL, JR., Elliott P. Joslin Research Laboratory, Department of Medicine, Harvard Medical School, Boston, Massachusetts (559)

DAVID H. ELWYN, Department of Surgery, Mt. Sinai School of Medicine, New York, New York (523)

G. FAUCONNEAU, Laboratoire d'Études des Métabolismes, C.R.Z.V., Theix, France (481)

JOEL H. KAPLAN, Departments of Oncology and Pathology, McArdle Memorial Laboratory, The Medical School, University of Wisconsin, Madison, Wisconsin (387)

FRANCIS T. KENNEY, Biology Division, Oak Ridge National Laboratory, Oak Ridge, Tennessee (131)

K. L. MANCHESTER, Department of Biochemistry, University College London, London, England (229)

M. C. MICHEL, Laboratoire de Physio-Pathologie de la Nutrition, C.R.Z.V., Theix, France (481)

H. N. MUNRO, Physiological Chemistry Laboratories, Department of Nutrition and Food Science, Massachusetts Institute of Technology, Cambridge, Massachusetts (3, 299)

OLIVER E. OWEN, Temple University School of Medicine, Philadelphia, Pennsylvania (559)

HENRY C. PITOT, Departments of Oncology and Pathology, McArdle Memorial Laboratory, The Medical School, University of Wisconsin, Madison, Wisconsin (387)

ROBERT T. SCHIMKE, Departments of Pharmacology and Biology, Stanford University School of Medicine, Palo Alto, California (177)

RICHARD J. WURTMAN, Department of Nutrition and Food Science, Massachusetts Institute of Technology, Cambridge, Massachusetts (445)

V. R. YOUNG, Department of Nutrition and Food Science, Massachusetts Institute of Technology, Cambridge, Massachusetts (585)

Preface

Nearly five years have elapsed since the late Dr. James Allison and I wrote the preface to Volumes I and II of this treatise. During the intervening period, there has been a steady accumulation of new observations in the field of mammalian protein metabolism, and it now seems timely to use some of this new knowledge to enlarge and add depth to the picture of protein metabolism provided by the earlier volumes. Accordingly, two additional volumes have been added to the treatise.

The first of these new volumes (Volume III) covers two aspects of protein metabolism not specifically considered in the earlier parts of the treatise. The first of these deals with changes in protein metabolism during evolution and during growth and development. The other part of Volume III provides a survey of methods appropriate to the study of protein metabolism in mammals.

The present volume is devoted to a single topic, the regulation of protein metabolism in mammals, which thus becomes the sixth part of the whole treatise. Regulation mechanisms are being increasingly studied, both at the subcellular level and at the level of the intact organ and of the whole animal. The sequence of chapters in this part of the book reflects this progression. The first two chapters deal with fundamental control mechanisms in the synthesis and turnover of proteins, followed by chapters on the mode of action of hormones on protein metabolism and on the role of free amino acid pools in regulation. Next, there is an account of the regulation of individual amino acid pathways, and a review of diurnal rhythms in protein metabolism. The final chapters deal with regulation of protein metabolism in four major organs and tissues of the body, namely the gastrointestinal tract, the liver, the kidney, and the musculature.

As in the case of the first three volumes, I have been fortunate in securing some of the principal authorities in the field of protein metabolism as authors of the specialist articles. In conformity with the practice in the earlier volumes, this part of the treatise is preceded by an introductory chapter providing a broad outline into which the succeeding chapters fit. It is hoped that this will allow the nonspecialist reader, and particularly the student, to orient himself before proceeding to the details of the specialized articles.

I am much indebted to friends and colleagues for helpful discussions of material I have used in writing these introductions to sections, and also for general advice about the assembly of the book. In particular, I should like to thank Dr. H. J. Sallach of the University of Wisconsin, Dr. L. M. Henderson of the University of Minnesota, and Dr. J. D. Finkelstein of Veterans Administration Hospital, Washington, for much help in evaluating selected areas of metabolic control.

June, 1970 H. N. Munro

Preface to Volumes I and II

Shortly after World War II, an international meeting was convened to consider nutritional deficiencies affecting populations in the post-war world. The agenda did not include protein deficiency. Since then, a revolution in outlook has taken place, so that in 1960 the Director of the Nutrition Division of the Food and Agricultural Division of the United Nations stated that, in the view of his organization, protein malnutrition is without doubt the most compelling nutritional problem in the underdeveloped countries today (M. Autret, 1961, *in* "Progress in Meeting Protein Needs of Infants and Preschool Children," National Academy of Sciences–National Research Council, Washington). In the more highly developed countries interest in the role of dietary proteins in the promotion of health has also quickened during the past few years.

The past decade has witnessed a revolution in a different aspect of protein studies. In less than ten years, the study of protein biosynthesis has progressed from a state of unproved and contradictory hypotheses to become the seat of intense and fruitful activity. When the dust of battle has settled, current work bids fair to resolving problems as varied as the nature of the inheritance mechanism and the genesis of the malignant cell.

These investigations into the importance of proteins in nutrition and into the mechanism of protein formation have each produced an abundant and specialized literature in which it is difficult for anyone other than the specialist to evaluate recent advances. In consequence, a barrier of noncommunication has arisen between experts in each field of inquiry, although in reality they are but opposite sides of the same coin: the doctor faced with a case of protein malnutrition is observing a natural experiment in the inhibition of protein synthesis. Other areas of protein metabolism present a similar rapid growth of knowledge. Changes in protein metabolism are now known to be involved in the action of many hormones. In the field of medicine, the physician observes an expanding series of errors in the metabolism of individual amino acids, and the surgeon continues to explore the significance and the repair of the large dissipation of body protein resulting from injury.

It is thus difficult nowadays for the research worker in any one branch of the subject to comprehend the advances which are taking place on all fronts in protein metabolism. The objective of this treatise is to provide an up-to-date account of current thought in all areas of protein metabolism which will meet the needs of specialists in nutrition, biochemistry, clinical chemistry, medicine, and indeed in all aspects of biology in which mammalian protein metabolism is studied. We are not aware that this has been done in recent years.

The contents of the treatise are presented in three parts. After a short historical account of early studies on protein metabolism, the first part provides a picture of the biochemical mechanisms involved in protein metabolism, the second part deals with the place of proteins in nutrition, the third part gives an account of protein metabolism in diseased states. Although written as a multi-author text, the book is designed to be read in continuity by those who have only a general knowledge of the outlines of protein metabolism, as presented in textbooks of general biochemistry. In order to emphasize this continuity, each main part—biochemical, nutritional, and pathological—is introduced by a short general survey of the area covered. It is hoped that, in this way, the treatise will serve as a source-book for the graduate student who requires an advanced survey of the whole subject, and, at the same time, will interest the specialist who needs an authoritative account of selected areas of protein metabolism.

As editors, we should like to say how fortunate we have been in securing as authors many of the leading investigators in the field of protein metabolism, and we are deeply conscious of the honor they do us in participating in the writing of this book. We should also like to acknowledge with gratitude the help and encouragement of many colleagues and friends who have advised us during the assembly of this book. In particular, it is a pleasure to thank Professor D. McKie of the Department of the History and Philosophy of Science at London University for the advice on the historical introduction; Dr. L. Fowden of University College, London, for information on the amino acids found in plants; Dr. J. G. Black of Unilever for providing us with a survey of literature on protein metabolism in the skin; and Dr. G. Leaf of the University of Glasgow for much general advice on protein metabolism. Dr. Renwick of the Genetics Department at Glasgow University kindly gave advice on some of the genetic points in abnormalities of protein biosynthesis discussed in the introductory material to Part III, and Mr. Robin Callander, Medical Artist at Glasgow University, provided a number of excellent illustrations. We are much indebted to these and many other friends for the help they have given us.

H. N. Munro

October, 1963 J. B. Allison

Table of Contents

Part VI

Regulation Mechanisms in Protein Metabolism

A GENERAL SURVEY OF MECHANISMS REGULATING PROTEIN METABOLISM IN MAMMALS

H. N. Munro

CHAPTER 31

HORMONAL REGULATION OF SYNTHESIS OF LIVER ENZYMES

Francis T. Kenney

CHAPTER 32

REGULATION OF PROTEIN DEGRADATION IN MAMMALIAN TISSUES

Robert T. Schimke

CHAPTER 33

SITES OF HORMONAL REGULATION OF PROTEIN METABOLISM

K. L. Manchester

CHAPTER 34

FREE AMINO ACID POOLS AND THEIR ROLE IN REGULATION

H. N. Munro

CHAPTER 35

THE REGULATION OF INTERMEDIARY AMINO ACID METABOLISM IN ANIMAL TISSUES

Joel H. Kaplan and Henry C. Pitot

CHAPTER 36

DIURNAL RHYTHMS IN MAMMALIAN PROTEIN METABOLISM

Richard J. Wurtman

CHAPTER 37

THE ROLE OF THE GASTROINTESTINAL TRACT IN THE REGULATION OF PROTEIN METABOLISM

G. Fauconneau and M. C. Michel

CHAPTER 38

THE ROLE OF THE LIVER IN REGULATION OF AMINO ACID AND PROTEIN METABOLSIM

David H. Elwyn

CHAPTER 39

THE ROLE OF THE KIDNEY IN THE REGULATION OF PROTEIN METABOLISM

George F. Cahill, Jr., and Oliver E. Owen

CHAPTER 40

THE ROLE OF SKELETAL AND CARDIAC MUSCLE IN THE REGULATION OF PROTEIN METABOLISM

V. R. Young

Contents of Other Volumes

Volume I

Volume II

Volume III

PART VI

Regulation Mechanisms in Protein Metabolism

A General Survey of Mechanisms Regulating Protein Metabolism in Mammals

H. N. Munro

Physiological Chemistry Laboratories,
Department of Nutrition and Food Science,
Massachusetts Institute of Technology,
Cambridge, Massachusetts

I. Introduction

The mammal is a metabolic machine equipped with control mechanisms that allow it both to adapt to changes in its own metabolic demands and also to survive in a variable and sometimes hostile environment. External stimuli to which it must be able to respond include heat and cold, rhythmic changes imposed by alternation of day and night, and the necessity to adjust to the intermittent intake of food. In addition, the animal itself provides internal metabolic variations, such as the menstrual cycle in the case of the adult woman, and the orderly sequence of events during growth of tissues followed by a plateau when maturity has been attained. All these external and internal factors demand metabolic responses, and in consequence there must be regulatory mechanisms with which to respond.

Regulation of protein metabolism is achieved both by mechanisms operating at a subcellular level (e.g., adaptive changes in enzyme activity) and also by coordinated actions between cells and tissues (e.g., secretion by an endocrine gland of a hormone in response to a stimulus with consequent changes in the metabolism of the target organ). The chapters making up this part of the treatise have been arranged to show this progression from underlying intracellular mechanisms to the more complex intercellular and interorgan relationships involved in regulation of protein metabolism in the whole animal. As an introduction to these specialized articles on regulation, this chapter provides a short integrated survey of the field of metabolic regulation, beginning with an account of how the regulatory mechanisms found in mammals may have arisen in the course of evolution.

II. Evolution of Regulatory Mechanisms

Since mammals have emerged as a consequence of millions of years of evolutionary history, it may be assumed that the mechanisms they possess for regulating their metabolism are those that have best allowed each species to adapt to its environment. It is therefore appropriate to look for metabolic regulatory mechanisms in mammals that are compatible with their evolutionary origin from the most primitive forms of life through various intermediate stages. The impact of evolution on protein metabolism has already been discussed in detail in Part IV of this treatise. Briefly, it seems likely that conditions on the primitive Earth were such that simple peptides and polynucleotides could have arisen by purely chemical means. These macromolecules may then have become associated in such a way that the presence of the polynucleotides determined the sequence of amino acids aggregating into peptides, and the peptides in turn regulated the formation of the polynucleotides. (For such a hypothetical scheme, see Fig. 3 of the introductory chapter to Part IV.) Interactions of this kind may thus have provided primitive self-replicating systems which would consequently be the forerunners of the sequence DNA → RNA → protein which is characteristic of control of protein formation in all living cells. Once this control of protein replication became established, further evolutionary development took place through chance variations in DNA nucleotide sequence caused by point mutations in single nucleotides and by duplication of sections of the DNA chain through breakage and reattachment of the fragments to other DNA chains. Organisms with this changed DNA survived if the resulting new proteins increased the efficiency of the organism in adapting to its surroundings, or at worst did nothing to diminish its competitiveness. This type of evolutionary change in protein structure has been relatively easy to follow by determining the amino acid sequence of a given protein in a series of organisms spread over a period of evolutionary history. It has also been possible to map with some certainty the changes in DNA nucleotide composition that must have taken place in order to produce these alterations in the amino acids making up the protein molecule. Such studies on the evolution of protein molecules are described in detail in Chapter 24 of Volume III.

In addition to causing progressive changes in the DNA of genes responsible for protein structure, evolution has also affected regulatory mechanisms encoded in the DNA of the organism, presumably also through alterations in nucleotide sequence brought about by point mutations and duplications. Perpetuation of these changes in regulatory control must have been determined on the same basis as that of changes in structural genes, namely in terms of their effect on adaptation of the organism to its environment. For example, it has been established for bacteria that survival of a species can depend on its success in regulating metabolic pathways. Thus, Baich and Johnson (1968) have demonstrated that a strain of *Escherichia coli* lacking the capacity to regulate proline synthesis is unable to multiply as rapidly as the wild-type *E. coli.* This occurs because there is no feedback inhibition of proline production by the mutant, so that it continues to waste energy on synthesis of this amino acid even when an adequate supply has been provided or has accumulated as the result of synthesis. Derepression of the tryptophan pathway in a mutant of *Bacillus subtilis* leads to an even more marked decrease in survival potential of this organism when compared with the wild type (Zamenhof and Eichhorn, 1967). In fact, uncontrolled production of metabolites is rare in nature, and it can be concluded that regulation of metabolic pathways must therefore be a significant determinant in the competitive survival of all organisms and thus of evolutionary success or failure.

In order to achieve a comprehensive view of metabolic control in mammals, it is necessary to appreciate the extensive explorations that have been made of control mechanisms occurring in other organisms, particularly bacteria. Studies on bacteria have demonstrated that control resides both at the chromosome level and at the enzyme level. In response to changing demands for a specific enzyme, the synthesis of the appropriate messenger RNA by polymerase acting on the bacterial chromosome can be repressed or derepressed. The details of such regulation of DNA expression have been elegantly assembled by Jacob and Monod into the operon hypothesis. Second, bacterial studies have provided much insight into the nature of feedback control of enzyme function, such as inhibition of the first enzyme in a reaction sequence by the end product of the series of reactions. Finally, the bacterial cell illustrates control of metabolism through changes in permeability to exogenous metabolites.

The transition from the bacterial cell to the animal cell involves considerable differences in the genetic apparatus, and consequently more complexity of chromosomal control may be anticipated. First, there is

TABLE I
AMOUNTS OF DNA PER CELL OR PER NUCLEUS FOR VARIOUS FORMS OF LIFE[a]

Group	Organism	Grams of DNA $\times 10^{-12}$ per cell or nucleus
Bacteria	*Escherichia coli*	0.005
Fungi	Yeast	0.02
	Neurospora	0.04
	Aspergillus	0.04
Fish	Shad	1
	Trout	5
	Shark	6
Dipnoans	Lungfish	100
Amphibians	Frog	15
	Amphiuma	168
Reptiles	Alligator	5
	Turtle	5
Birds	All species	2 to 4
Mammals	All species	7
Plants	Maize	5
	Pea	10
	Tulipa kaufmanniana	94
	Trillium erectum	120

[a] From Munro Volume III, Introduction to Part IV, Table I.

a general tendency for the DNA content per cell to increase as one progresses up the evolutionary scale from the bacterial cell to the cells of higher animals and plants (Table I). This suggests greater genetic complexity due not only to additional structural genes, but also to more subtle control mechanisms. Second, there is the emergence of a well-defined nucleus. The prokaryotic bacterial cell has no formal nucleus but contains a single chromosome made up of a circle of naked DNA. In the eukaryotic cells of all higher organisms, however, the DNA is almost exclusively present in a discrete nucleus and is associated there with histones and other proteins. The new structural features of the nucleus consequently add further points at which control mechanisms can occur. The only nonnuclear DNA occurs in the mitochondrion, where it forms a loop, thus being similar to bacterial DNA (Sinclair *et al.*, 1967).

Passing next from the unicellular (protozoan) animal to the multicellular (metazoan) animal, two additional control mechanisms become superimposed on those already required by the single animal cell. First, there has to be some coordination of metabolic control in different cells. This is achieved in part by secretion of hormones that affect the metabolism of their target organs. Second, the multicellular animal usually shows differentiation of tissues into many types of specialized cells. Since the mature animal arises from a single cell, the fertilized ovum, its DNA must be programmed to provide the orderly appearance of new cell types at the appropriate stages in development, and each of these new tissues must thereafter be provided with the instructions needed to grow to mature body size.

These various aspects of the evolution of control mechanisms in mammals will now be considered in more detail.

A. Adaptation of Bacterial Cells to Changes in Environment

Control of enzyme activity through environmental change was first recognized during the study of biosynthetic pathways in bacteria, particularly pathways for the formation of amino acids. The subject has recently been reviewed by Atkinson (1966, 1969), Umbarger (1969), and Datta (1969). One of the earliest studies was that of Abelson, who demonstrated in 1954 that isoleucine is synthesized from threonine by *Escherichia coli* and that this synthetic route can be inhibited by growing the cells in the presence of isoleucine. The underlying mechanism was elucidated by Umbarger (1956), who showed that threonine dehydrase, the enzyme responsible for the first step of the pathway, is strongly inhibited by isoleucine (Fig. 1). Thus, the end product (isoleucine) becomes the regulator of an enzyme (threonine dehydrase) involved in its own biosynthesis. It will be noted that the step subject to regulation is quite remote from the end product, and that the reaction catalyzed by threonine dehydrase is essentially irreversible. This picture is characteristic of numerous other metabolic pathways in which the phenomenon of *feedback control* is exhibited.

Changes in enzyme activity due to specific compounds in the intracellular or extracellular environment can occur from two causes. First, it can arise from direct action on the activity of enzyme molecules already present in the cell. This is termed *feedback inhibition.* Alternatively, it can be due to a change in the number of enzyme molecules in the cell. This can occur through reduced enzyme formation resulting from accumulation of an inhibitory end product (*repression*). Alternatively, we can have increased rapidity of enzyme formation through the presence of its substrate (*enzyme induction*). These effects on enzyme function (feedback inhibition) and enzyme synthesis (repression and induction) will now be separately considered in the context of control mechanisms.

1. *Feedback Inhibition*

This inhibition is illustrated by reference to the pathways of biosynthesis of branched-chain amino acids in *E. coli.*, illustrated in Fig. 1. These pathways demonstrate some of the common features of feedback inhibition. Thus, the site of inhibition is sometimes located at the initial step; for example, the formation of α-oxobutyrate from threonine is inhibited by L-isoleucine (Umbarger, 1956). Another example is the first step in the biosynthesis of histidine, where the reaction of ATP with 5-phosphoribosyl 1-pyrophosphate is inhibited by the end product L-histidine (R. G. Martin, 1963). A second common site of feedback inhibition is the first reaction after the branch-point in a series of reactions, again illustrated in Fig. 1, where the step from α-oxoisovalerate to β-carboxy-β-hydroxyisovalerate is inhibited by L-leucine. This has the advantage of partitioning the products between different pathways.

In some instances, two enzymes can catalyze the same reaction in the bacterial cell. Usually their sensitivities to inhibitors differ; for example, the formation of α-oxobutyrate from threonine is catalyzed by a second enzyme that is not subject to isoleucine feedback inhibition but is regulated by the amount of glucose in the medium (Umbarger and Brown, 1957). This is advantageous because threonine catabolism also yields energy through a series of steps beginning with α-oxobutyrate but diverging in subsequent reactions from the isoleucine-valine pathway. If the levels of isoleucine were the only determinant in regulating threonine catabolism, inhibition by accumulation of this amino acid could starve the cell of energy available from threonine degradation. Thus, a second enzyme with similar catalytic properties but insensitive to isoleucine feedback inhibition allows some of the threonine carbon to flow through the energy-yielding pathway. Two or more proteins with similar catalytic properties but differing in structure are termed *isozymes* (isoenzymes). E. R. Stadtman (1968) and Datta (1969) have recently reviewed a number of examples of the role of these multiple enzymes in the regulation of metabolic pathways in bacteria. Stadtman's article appears as part of a symposium edited by Vesell (1968) in which the functions of isozymes in higher organisms are also described.

The mechanism of the inhibitory action of end products of a reaction has to some extent been elucidated. Usually, the end product having such an inhibitory effect does not resemble or interact with either the substrate or the product of the reaction involved. Since the inhibitor molecule is not a steric analog of the substrate, its action is said to be an *allosteric effect.* This has led to the general proposition (Monod *et al.*, 1963) that the activity of an enzyme can be regulated by a

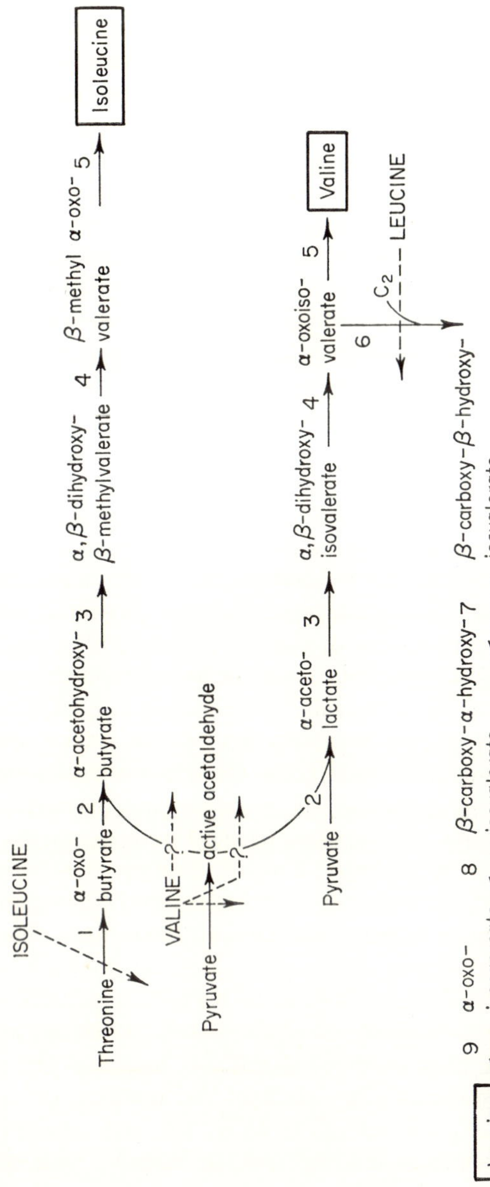

FIG. 1. Control points of feedback inhibition in the synthesis of the branched-chain amino acids in *Escherichia coli*. The products are shown in boxes. Reactions 2, 3, 4, and 5 are catalyzed by the same enzymes for both pathways, and the transaminase for reaction 5 also catalyzes reaction 9 in the synthesis of leucine. Sites of feedback inhibition are indicated by broken arrows. (From Fry, 1964.)

substance of low molecular weight through binding of this agent (allosteric effector) to a specific point in the enzyme protein (allosteric site) with consequent change in the conformation of the protein molecule and in its catalytic properties (allosteric transition). The allosteric effector may have an inhibitory or a stimulant action. The cause of allosteric changes in enzyme activity seems frequently to be connected with the subunit structure common to proteins. Some enzyme actions are known to depend on cooperation between several active sites on groups of protein subunits (protomers), and the dissociation or even partial dislocation of these catalytic sites caused by the conformational change due to the presence of the allosteric effector may be sufficient to induce loss of activity through separation of these cooperating sites (Monod *et al.*, 1963). The most thorough study of allosteric effects on protein structure and function has been made on the aspartate transcarbamylase of *Escherichia coli.* As shown in Fig. 2, this enzyme initiates the pathway of pyrimidine biosynthesis by catalyzing the reaction of aspartic acid with carbamyl phosphate to yield ureidosuccinic acid which eventually is transformed to the pyrimidine bases. An end product of this pathway, cytidine triphosphate (CTP), inhibits the activity of aspartate transcarbamylase, whereas ATP is an activator of this enzyme. A series of studies by Gerhart, Schachman, and Changeux (Gerhart and Schachman, 1965, 1968; Changeux *et al.*, 1968; Changeux and Rubin, 1968) have shown that the enzyme is made up of four protomers. At different locations on each protomer polypeptide chain, there is one regulatory site that binds CTP and one-half of the catalytic site that binds the substrates for reaction. The complete tetrameric structure of the enzyme is thus necessary in order to obtain cooperative effects between the catalytic sites on adjacent protomers. Attachment of CTP to the protomer appears to cause distortions of its polypeptide chain sufficient to prevent cooperation of the substrate binding site with similar sites on other protomers and in consequence the enzyme activity of the tetramer is inhibited.

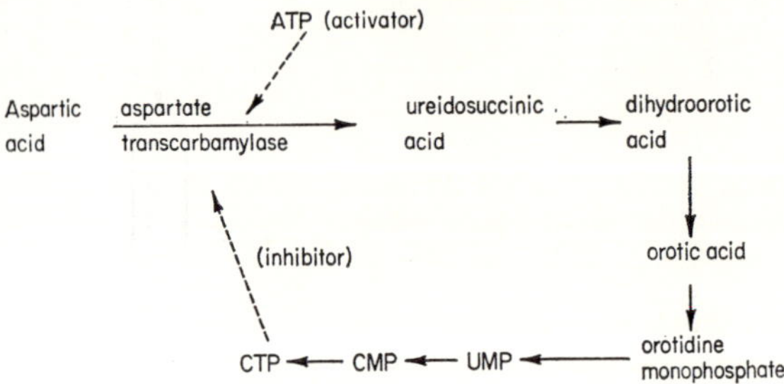

FIG. 2. Feedback control of the pathway of pyrimidine biosynthesis.

2. Repression and Induction

In contrast to the changes in enzyme activity due to feedback inhibition, repression and induction involve alterations in the *amount* of enzyme protein in the cell. When metabolites cause a reduction in protein formation, they result in *repression,* and when increased formation of the enzyme is produced, the reaction is termed *induction.*

Repression of enzyme synthesis is frequently due to the action of end products in a reaction chain and tends to involve the same enzymes as those subject to direct feedback inhibition. For example, isoleucine causes feedback inhibition of threonine degradation (Fig. 1), but it also represses synthesis of this enzyme (Umbarger and Brown, 1958). Enzymes subject to repression differ from those affected by feedback inhibition in one respect, namely that several related enzymes can often be simultaneously repressed by the same inhibitor; for example, a series of enzymes in the histidine biosynthesis pathway of *Salmonella typhimurium* are repressed simultaneously by free histidine (Ames and Garry, 1959). In contrast, feedback inhibition is generally specific to one enzyme or certain non-sequential enzymes in a chain (Fig. 1).

Increased synthesis of enzymes can arise from two causes. First, it can occur from *derepression,* when an end product that normally represses enzyme formation is reduced in amount. Thus arginine-requiring mutants of *E. coli* produce large amounts of ornithine transcarbamylase in arginine-poor media, whereas they repress this enzyme in arginine-rich media (Gorini and Maas, 1957). Second, enzyme synthesis can be stimulated by metabolites present in the medium. This represents *enzyme induction,* the best-known example being the series of enzymes formed in *E. coli* for the utilization of lactose in the medium.

Both repression and induction involve changes in the rate of protein formation and in consequence represent aspects of the control of protein synthesis. In the case of the bacterial cell, this has led to the hypothesis of the regulation of protein formation by the operon unit in the genetic apparatus. This will now be considered.

3. Regulation by the Operon of Enzyme Induction and Repression in the Bacterial Cell

The following account of the present status of the operon hypothesis is mainly taken from Beckwith (1967), who has reassessed the theory of F. Jacob and Monod (1961) in the light of subsequent studies and has concluded that the validity of the original hypothesis is largely substantiated by more recent evidence.

When a bacterial cell, such as *E. coli,* has become accustomed to growing on a medium in which the main carbon source is glucose and is then suddenly transferred to a medium containing lactose as the sugar, lactose utilization is made possible through two events involving synthesis of specific proteins. First, a protein (permease) responsible for lactose transportation into the cell makes its appearance; second, there is a rapid synthesis of β-galactosidase, an enzyme that catalyzes the hydrolysis of lactose to glucose and galactose. Synthesis of these two proteins is regulated by separate genes, named *y* for the permease and *z* for the β-galactosidase. In addition, exposure to lactose also induces accumulation of another enzyme, thiogalactoside transacetylase, although it is not apparently essential for utilization of lactose (Fox *et al.* 1966). This last enzyme is regulated by a gene designated *a*. In the bacterial chromosome, these genes are known to lie together in the sequence *z*-*y*-*a* (Fig. 3).

These three contiguous genes are able to code for a single messenger RNA strand that acts as a template for all three proteins. The signal

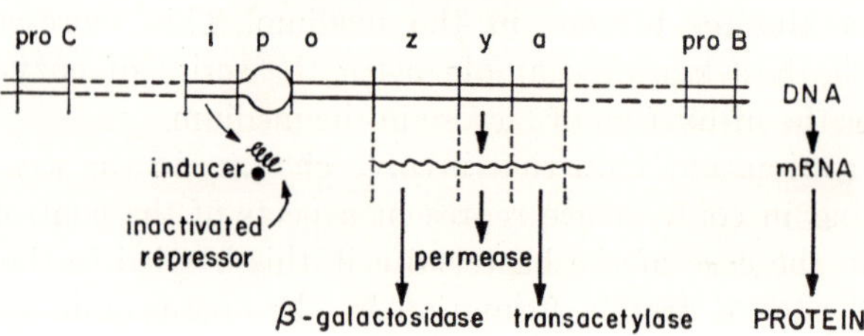

FIG. 3. The control of expression of the *lac* operon. The *lac* region and two of the markers surrounding it (genes involved in proline biosynthesis) on the chomosome are represented as regions of double-stranded DNA. In this scheme, the promoter is the site of initiation of RNA synthesis. The repressor binds to the operator and prevents movement of the RNA polymerase along the DNA, and in consequence the DNA strands undergo separation and the RNA polymerase begins to move along the DNA. An inducer alters the repressor so that it can no longer interact with the operator in this way. Pro C and pro B indicate two genes involved in the biosynthesis of proline. (From Beckwith, 1967, with inversion of the *o* and *p* sites on the basis of the findings of Ippen *et al.*, 1968.) Reproduced from *Science* **156,** 597. (May 5, 1967), copyright by the American Association for the Advancement of Science.

(Wade and Robinson, 1966; Ben-Hamida and Schlessinger, 1966; Natori and Mizuno, 1967). Furthermore, the free nucleotide pool of *Bacillus subtilis* has been examined in detail by Nierlich and Vielmetter (1968) and does not exhibit the nucleoside 3-phosphates that are diagnostic of the action of ribonuclease I activity. In consequence, ribonuclease I can be eliminated as an obligatory degradative enzyme, though it may play some role in the later stages of degradation by cleaving oligonucleotides produced by other nucleases that might otherwise inhibit the continued action of these nucleases (D. Mizuno and Anraku, 1967). It has similarly been also suggested that polynucleotide phosphorylase may not be essential for RNA turnover, since Ben-Hamida and Schlessinger (1966) found normal degradation of RNA in several bacterial mutants believed to lack this enzyme. However, Natori and Mizuno (1967) state that, in the case of at least one of these mutants, polynucleotide phosphorylase activity is present but is masked by rapid removal of the diphosphates through the action of another enzymes. It thus seems likely that, as originally suggested by Sekiguchi and Cohen (1963), the degradation of bacterial messenger and other forms of RNA is due to the cooperative actions of polynucleotide phosphorylase and ribonuclease II with the eventual production of nucleoside 5-phosphates that can if necesary be reincorporated into the nucleic acids of the cell.

From these studies it can be concluded that, during exponential growth, only messenger RNA undergoes degradation through the actions of ribonuclease II and perhaps also polynucleotide phosphorylase, both of which yield nucleoside 5-phosphates that can directly reenter the precursor pool. Since the bacterial messenger RNA has a half-life of 1–2 minutes, it is in rapid equilibrium with the pools of RNA precursors, which thus show the kinetics of a subdivided pool (Levinthal *et al.*, 1963; Nierlich and Vielmetter, 1968). On the basis of studies with digestion of polyuridylic acid by ribonuclease II and polynucleotide phosphorylase, Castles and Singer (1969) suggested that messenger RNA is protected against degradation by the ribosomes of polysomes. Messenger turnover is, of course, a desirable feature in a system which has to be ready to adapt its enzymes to varying external substrates. Under conditions of nutritional deficiency, other types of RNA and also cell protein begin to break down, thus providing a larger supply of nucleotides and also free amino acids. This offers an evolutionary advantage to the cell, which can use these products to form new messenger RNA and enzymes if an alternative nutrient source should become available.

When the cells of various bacteria are deprived of a required amino acid, there is evidence of interference with RNA synthesis, since uptake of labeled precursors into RNA falls to a few percent of its normal value (Sands and Roberts, 1952; Pardee and Prestidge, 1956; Gros and Gros, 1958; Stent and Brenner, 1961; Neidhardt, 1964). Messenger RNA is not necessarily affected (Lavallé and De Hauwer, 1968). The mechanism in *Escherichia coli* responsible for this inhibition of RNA synthesis is regulated by a genetic locus called RC. Maximum incorporation of precursors into RNA appears to depend on having all the aminoacyl-tRNA species charged with amino acids, since the same effect as amino acid starvation can be produced if one amino acid-activating enzyme is inactivated, an effect that can be produced in bacterial mutants having an amino acid-activating enzyme that is sensitive to a rise in the temperature of the incubation medium (Neidhardt, 1966). The intracellular relationship between pools of aminoacyl-tRNA and RNA synthesis also appears to account for the well-known overproduction of RNA in bacteria treated with chloramphenicol and some other antibiotics. These compounds inhibit bacterial protein synthesis; in consequence there is accumulation of charged tRNA, and thus RNA is stimulated to overproduce (Eidlic and Neidhardt, 1965; Ezekiel and Elkins, 1968; Ezekiel and Brockman, 1968).

The significant factor in shutting off RNA synthesis during amino acid starvation appears to be the accumulation of uncharged tRNA in the cell, since substitution of an amino acid analog that can charge tRNA but not enter into protein synthesis will relieve the block in RNA formation. For example, RNA synthesis in a valine-requiring mutant of *E. coli* can be reduced by valine starvation and restored by the valine analog chlorobutyrate which charges valyl-tRNA but does not participate in protein synthesis (Williams and Freundlich, 1969). Stent and Brenner (1961) and also Kurland and Maaløe (1962) postulated that uncharged tRNA may repress the activity of RNA polymerase. However, direct measurement of the polymerase *in vitro* has shown at most only a twofold difference in activity in the presence of charged versus uncharged tRNA (Tissières *et al.*, 1963; Bremer *et al.*, 1966), whereas the effects of amino acid starvation on incorporation of precursors into RNA are much more dramatic. Cashel, Gallant, and collaborators (Gallant and Cashel, 1967; Cashel and Gallant, 1968, 1969; Gallant and Harada, 1969; Cashel, 1969) have claimed that amino acid starvation acts by inhibiting the synthesis of nucleoside triphosphates from nucleoside monophosphates, and they concluded that the dependence of RNA synthesis on amino acid supply is mediated through general control of phosphorylation reactions by the RC gene. This theory has been challenged by Bagnara and Finch (1968) and by Edlin and Stent (1969), who failed to find changes in the triphosphate pools of amino acid-starved bacteria. However, the more widespread action of RC gene control on cell metabolism is supported by the recent finding that amino acid supply (Sokawa *et al.*, 1968) and charging of tRNA (Williams and Freundlich, 1969) can influence lipid synthesis in *E. coli* strains having stringent needs for exogenous amino acids. Other theories of amino acid regulation of RNA synthesis have also been proposed (e.g., Sykes, 1966; D. W. Morris and DeMoss, 1966; Stent, 1966; Delgarno and Gros, 1968; Nierlich, 1968; Ezekiel and Blumenthal, 1968). Edlin and Broda (1968) have provided a detailed review of the evidence up to 1968.

B. Evolution of Regulation

As discussed above, there is good reason to accept the Jacob-Monod model of the functioning of the *lac* operon. However, Pollock (1967) has assembled evidence that the classical polycistronic operon may be only a stage in the evolution of control mechanisms. For example, in the case of arginine biosynthesis by *Escherichia coli,* the eight genes involved are scattered around the bacterial chromosome in several clusters (Vogel *et al.,* 1963). Presumably cross-over has separated out what was once a coordinated, contiguous operon. This, of course, could still function as a single operon if, in agreement with study of the *lac* operon, a protein from a regulatory *i* gene can control synthesis of mRNA in noncontiguous cistrons by passing from the regulatory gene to the dispersed structural genes in the operon. However, attempts to examine the proposition in the arginine biosynthetic pathway of *E. coli* have led to somewhat equivocal results (Vogel and Vogel, 1967). To the situation in which the original sequential operon has been disseminated throughout the genome, the term *regulon* is applied. Although lack of knowledge of the evolutionary relationships between bacterial species limits such explorations, there have nevertheless been some speculations about the origin and development of the operon. Horowitz (1965) suggests that polycistronic operons first originated in living organisms with only the gene controlling synthesis of the final enzyme in the sequence. By means of duplication of this gene and subsequent mutation of the replicates, the genes for antecedent enzymes in the sequence were created. If this view is correct, then different enzyme proteins in a single operon should show some homology of amino acid sequence. This has not so far been tested. In this way, classical operons of the *lac* type could have arisen and then become disseminated to different chromosomal sites (regulon) by crossing-over. This evolutionary history is pictured diagrammatically in Fig. 4.

The passage from nonnucleated (prokaryote) to nucleated (eukaryote) forms of life offers additional points of control, since the DNA not only

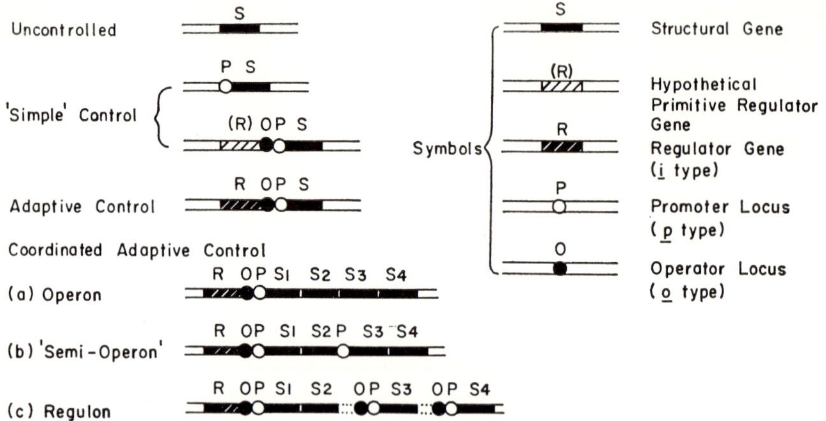

FIG. 4. Evolution of control systems for enzyme biosynthesis in bacteria. Note that *o* and *p* sites are the reverse of those shown in Fig. 3, since they are based on the original Jacob-Monod model. (From Pollock, 1967.)

is segregated in a nucleus, but is divided up into several chromosomes, where it is bound to histones and other proteins. The most primitive nucleated organisms are the fungi, and it is therefore of interest to compare metabolic pathways in these organisms with the corresponding pathways in bacteria. Thus, in the bacterium *Salmonella typhimurium* the 10 genes involved in histidine synthesis are arrayed as a classical compact operon (Ames *et al.*, 1967), whereas the same series in the mold *Neurospora crassa* are widely separated (Ahmed *et al.*, 1964). The interesting question here is whether the synthesis of the *Neurospora* histidine enzymes is regulated through a single regulator site as in the case of the corresponding *Salmonella* operon. This common control would be demonstrated by isolating a mutant of *Neurospora* in which all enzymes were increased in activity to a similar extent, thus suggesting loss of a single regulator. If this can be established, it would mean that the operon concept may also apply to eukaryote chromosomes containing histones as well as DNA, and that one should look within the nucleus for the regulatory protein made by an inhibitory gene.

Evolution of regulation in fungi may not be limited to nuclear mechanisms. There is also evidence that regulation of metabolic pathways in *Neurospora* and other eukaryote microorganisms is sometimes achieved by cytoplasmic coordination of the enzymes involved. For example, the synthesis of tryptophan by *Escherichia coli* is dependent on five reactions, for which the structural genes are localized contiguously within a small segment of the bacterial chromosome and are under the control of a regulator gene (Matsushiro *et al.*, 1965). In *E. coli*, the terminal enzyme of the pathway, tryptophan synthetase, consists of a heteropolymeric protein made up of pairs of two dissimilar subunits, α and β, which catalyze a rather complex reaction (Yanovsky, 1967; Crawford *et al.*, 1968). The first two reactions of this pathway are also catalyzed by a heteropolymeric enzyme (Ito and Yanovsky, 1966), and reactions 3 and 4 are promoted by a single enzyme (Creichton and Yanovsky, 1966). These cooperative reactions between enzymes in the pathway are also seen in the mold *Neurospora crassa.* Here the genes for the four enzymes are unlinked, but the cytoplasm contains a multienzyme complex carrying out reactions 1, 3, and 4 of the pathway (DeMoss and Wegman, 1965; Gaertner and DeMoss, 1969). Pollock (1967) suggested that such cytoplasmic association of enzymes involved in linked reactions in eukaryotes may have superseded the operon of the bacterial cell as the coordinator of regulation. However, this concept of the multienzyme complex is not exclusively confined to the eukaryote cell. For example, two multifunctional proteins are described in *Escherichia coli,* each of which possesses both aspartokinase and homoserine dehydrogenase activity and are involved in the biosynthesis of lysine, methionine, threonine, and the branched-chain amino acids, end products that can exert feedback and repressor regulation of the enzyme complex (Richaud and Cohen, 1968).

Although the association of several catalytic properties in a single multifunction enzyme complex is thus not peculiar to eukaryote life, nevertheless regulation among animals has to some extent undergone evolution through increasing complexity in enzyme structure. For example, the lamprey has a hemoglobin consisting of a single peptide chain, and uptake of oxygen by the blood of this species at different oxygen tensions follows a simple exponential curve. On the other hand, animals higher in the evolutionary scale have hemoglobin molecules consisting of more than one peptide chain, and as a result they exhibit a sigmoid curve of oxygen uptake that has certain physiological advantages in regulating oxygen acceptance and discharge in the body (see Chapter 24 in Volume III for further discussion). Evolution is also seen at work in the case of the dehydrogenases (Kaplan, 1965). Lactic dehydrogenase occurs in several isoenzymic forms representing the possible tetramers of two discrete types of subunit, called M (muscle) and H (heart) by Kaplan. The lactic dehydrogenase of heart muscle chiefly consists of tetramers of 4 H subunits, and is sensitive to substrate inhibition by pyruvate, whereas skeletal muscle contains mainly a dehydrogenase made up of 4 M subunits, and is little influenced by pyruvate concentration. This difference correlates with the aerobic removal of pyruvate by heart muscle through the tricarboxylic acid cycle, so that suppression of lactic dehydrogenase by excess pyruvate would be appropriate in this tissue; on the other hand, skeletal muscle often has to work under anaerobic conditions and must be able to transform excess pyruvate into lactate at all levels of pyruvate. It is therefore of interest that the lamprey, which is relatively anaerobic, has only M-type lactic dehydrogenase in all its tissues. Finally, evolution of function is seen in the case of glutamic dehydrogenase. In the mammal, this enzyme exists in at least three states, polymer, monomer x, and monomer y (Tomkins *et al.*, 1963, 1965). The monomer x form catalyzes primarily the dehydrogenation of glutamic acid, whereas monomer y also attacks alanine, valine, leucine, isoleucine, and methionine. The equilibrium between monomer x and monomer y is shifted in favor of the latter by estrogenic hormones and GTP, and in consequence the glutamate dehydrogenating activity of the enzyme is reduced and its reaction with the monocarboxylic amino acids is increased. ADP has the opposite action, and consequently favors dehydrogenation of glutamate but not of the other amino acids. Comparison of glutamic dehydrogenase activity in various species shows that the amphibian enzyme differs in several respects from the mammalian enzyme. A recent compilation by Balinsky (1970) suggests that the amphibian enzyme occurs only in one molecular form which catalyzes dehydrogenation of both glutamic acid and the other amino acids, and that ADP enhances enzymic attack not only on glutamic acid, but also on some of the other amino acids. These findings suggest that this key enzyme in the metabolism of amino acids is subject to different forms of regulation in the amphibian and the mammal, the latter possibly showing a finer sensitivity to regulators through its capacity to vary in substrate specificity.

These evolutionary alterations in enzymes affect regulation of metabolism in two ways. First, there are changes in specificity (e.g., glutamic dehydrogenase) and sensitivity to substrate concentration (e.g., lactic dehydrogenase) which alter the fate of metabolites. Secondly, the presence of different isozymes in different tissues occurs as part of the development of an animal during its differentiation.

In multicellular organisms, hormones provide a further means of coordinating or regulating metabolic activities in cells and tissues. Hormones occur throughout multicellular animal species; for example, extensive study of the action of molting hormones has provided insight into the mechanism of hormone action in insects (Karlson, 1968). There is some evidence that hormonal regulation of enzyme activity has increased in sophistication as animals have evolved. Thus the catabolic enzyme tryptophan pyrrolase found in the liver of the mammal undergoes an increase in activity after the administration of tryptophan or after the injection of adrenocortical hormones. Administration of tryptophan to amphibians also increases activity in this liver enzyme (Spiegel and Spiegel, 1964), but the amphibian liver enzyme is not affected by administration of corticosteroids (Spiegel, 1961; Chan and Cohen, 1964). Fish show even less responsiveness, since neither corticosteroids nor tryptophan affect the tryptophan pyrrolase content of their livers (J. N. Brown and Dodgen, 1968). An analogous situation occurs in the case of tyrosine metabolism. Administration of either tyrosine or corticosteroids increases tyrosine transaminase levels in the liver of the mammal, but not in the livers of amphibians or of lower species (Spiegel, 1961; Chan and Cohen, 1964). It is also interesting to note that the enzyme methylating norepinephrine to epinephrine in the adrenal medulla of mammals is dependent for its activity on the presence of adrenocortical hormones, whereas the corresponding enzyme in the frog appears not to be under adrenocortical control (Wurtman *et al.*, 1968). It would thus appear that the evolution of the mammals has imposed hormonal regulation on enzymes not so regulated in lower vertebrates, and has thus added further sensitivity to control of metabolic pathways.

III. Intracellular Regulation of Protein Metabolism in Mammals

As in the case of the bacterial cell, regulation is achieved in the mammalian cell through a combination of (a) control over the formation of its proteins and (b) changes in the activities of its enzymes. Each of these aspects of regulation will now be discussed.

A. Regulation of Protein Synthesis and Degradation

In the mammalian cell, regulation of protein synthesis is exercised at both the nuclear and cytoplasmic levels. The cell structures involved in protein synthesis in a typical protein-secreting cell, such as the hepatocyte, are shown in Fig. 5, where an attempt has been made to quantitate the relative amounts of DNA and RNA in different locations in the average liver cell of the rat.

1. The Nucleus and Regulation

In many of the studies to be described below, it has been necessary to separate nuclei from other cell constituents. The preparation of pure mammalian nuclei uncontaminated by cytoplasmic constituents is both difficult and controversial. The evaluation of nuclear purity by enzymic and other criteria is discussed by Roodyn (1963). Early methods of making nuclei, such as sedimentation at low centrifugal speeds from 0.25 *M* sucrose homogenates (Schneider, 1948), yielded preparations that were heavily contaminated with unbroken whole cells and cell debris.

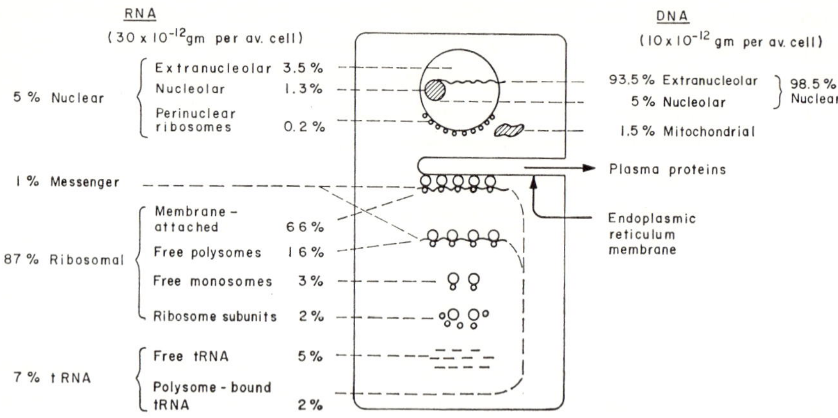

FIG. 5. The distribution in rat liver cells of RNA and DNA at different intracellular locations. The distribution of DNA between whole nucleus and mitochondria is taken from the data of Schneider and Kuff (1965), and the proportion recovered from isolated nucleoli is that reported in discussion at the end of the paper by Busch *et al.* (1966). The total RNA content of nuclei is for nuclei as isolated by Blobel and Potter (1966a). The nucleolar and extranucleolar distribution of nuclear RNA is taken from Muramatsu *et al.* (1963), and the value for ribosomal RNA adhering to the nuclear envelope (perinuclear RNA) comes from Whittle *et al.* (1968). The estimate of the amount of messenger RNA in the cytoplasm is based on the ratio of 90 nucleotides of messenger RNA per ribosome in a polysome, and the revised molecular weight of 2.4×10^6 (Loening, 1968) to 2.6×10^6 (McConkey and Hopkins, 1969) daltons for the ribosomal RNA in one 80 S ribosome. The relative amounts of membrane-attached and free polysomes are those given by Blobel and Potter (1967a); free monosomes are estimates by Enwonwu and Munro (1969). The amount of free subunit RNA is based on data from Munro *et al.* (1964). The amount of tRNA bound to polysomes was made on the basis that two tRNA molecules (each 25,000 daltons) are present on each ribosome (with RNA 2.4×10^6 daltons), so that tRNA represents 2% of the total RNA in the polysome. Hirsch (1967) has estimated by isotope dilution that tRNA accounts for 7% of the RNA in the liver cell; consequently, 5%, or three-fourths, must be free in the cell sap, and one-fourth is in polysome-bound form.

Chauveau *et al.* (1956) introduced the principle of sedimenting nuclei from homogenates made with 2.1 or 2.2 *M* sucrose. Since endoplasmic reticulum and most other cytoplasmic components have a lower density than the sucrose, they tend to float on centrifugation, whereas the nuclei are denser and form a pellet at the bottom of the tube. However, other dense particles, especially free ribosomes, also tend to sediment under these conditions and most users of this procedure now make the tissue homogenate in lower concentrations of sucrose (0.25–0.88 *M*) and layer it over 2.2 *M* sucrose, through which the pure nuclei sediment (e.g., Maggio *et al.*, 1963; Widnell and Tata, 1964). The yield by this procedure is much reduced and Blobel and Potter (1966a) have accordingly modified the two-layer method so that the upper layer is adjusted to 1.6 *M* sucrose and the lower layer is made with 2.3 *M* sucrose; on prolonged centrifugation, almost quantitative recovery of nuclei is obtained, whereas the endoplasmic reticulum and mitochondria float upward in the 1.6 *M* sucrose layer. Such nuclei are still contaminated, since the nucleus has a double membrane with ribosomes adhering to the outer layer (Fig. 5), and it is generally conceded that these perinuclear ribosomes should not be regarded as part of the nucleus. Penman and his colleagues (Penman, 1966; Holtzman *et al.*, 1966) describe a procedure for isolation of HeLa cell nuclei in which these tissue-culture cells are suspended in hypotonic solution, followed by mechanical shearing. The nuclear pellet is centrifuged down, then it is treated with a mixture of Tween 40 and sodium deoxycholate, which removes cytoplasmic contaminants and also the outer nuclear membrane. Such nuclei are almost devoid of the 18 S RNA characteristic of the smaller ribosomal subunit. Blobel and Potter (1966a) find that this detergent treatment causes complete fragmentation of liver nuclei; in their experience, another detergent, Triton X-100, selectively removes the outer nuclear membrane and its adherent perinuclear ribosomes. The outer nuclear membrane is also removed when the nuclei are prepared by the citric acid procedure (Umaña and Dounce, 1964). These technical details are of significance in studying the forms of RNA found within the nucleus. Johnston *et al.* (1968) describe a method of separating diploid and tetraploid liver nuclei by means of zonal centrifugation. For the purpose of preserving nucleolar structure, nuclei should be isolated in media containing Mg^{2+} ions (Busch *et al.*, 1967). Finally, it is sometimes desirable to prepare nuclei with minimal loss of their soluble contents, and for this purpose some form of the Behrens technique using nonaqueous solvents can be employed (Kay *et al.*, 1956; Okazaki *et al.*, 1968).

The unique feature of the nucleus is that it contains almost all the cellular complement of DNA in which is deposited the inherited information for the amino acid sequences of cell proteins, and also the mechanisms controlling expression of the messenger RNA templates appropriate to each cell as it differentiates and subsequently for regulating the amounts of these messenger and other species of RNA in response to variations in the environment of the cell. Although chromosome number in mammals varies from 17 in the diploid nucleus of the creeping vole to 78 in the case of the dog (Atkin *et al.*, 1965), the total amount of DNA per *diploid* nucleus is considered to be essentially constant at 7×10^{-12} gm from one mammal to another and also from one tissue to another in the same animal (Vendrely, 1955). The latter observation means that differences between tissues is achieved by suppressing much of the information contained in the structural genes of a cell so that only those proteins appropriate to the tissue (e.g., serum albumin, but not myosin, in liver) will be expressed. Even in specialized cells, the information is only dormant and has not been deleted from the genome. This was demonstrated by Gurdon and Uehlinger (1966) when they transferred the nuclei from the intestinal epithelium of the frog to an enucleated frog ovum and thereafter were able to obtain growth of a differentiating embryo from this synthetic cell. This experiment suggests also that cytoplasmic factors regulate the expression of messenger species appropriate to the type of cell.

a. Chromosome Structure and Function. The DNA of the nucleus is present in the form of chromosomes. Structural studies on chromosomes have recently been reviewed by Ris (1967). Each chromosome represents a large amount of genetic information since it contains many times the DNA of the bacterial genome. When inactive chromosomes (e.g., those in the nucleated red cells of birds) are unfolded, they are seen to be made up of fibers some 250 Å thick. These fibers can be further disrupted into fibrils 60–100 Å thick and composed of a central double helix of DNA with a protein coat, the DNA molecule being much folded on itself within the fibril. Chromosomes that are actively synthesizing RNA have been positively identified in only two special cases, namely as puffs and as lampbrush chromosomes. The puff is a swelling seen on the giant chromosomes of certain insects during periods of gene activity, as when the insect pupa is treated with the hormone ecdysone and undergoes molting. The puff is the seat of active RNA synthesis, and it is thought that this includes messenger RNA since the hormone causes the enzyme DOPA decarboxylase to appear at the same time (Sekeris, 1967). The other recognizable form of active chromosome, the lampbrush chromosome, occurs prominently in oocytes, where it forms a loop of chromosomal material (Callan, 1963). Histochemical and electron microscopic evidence shows that this chromosome loop contains a strand of DNA with associated RNA and protein (Ris, 1967; O. L. Miller and Beatty, 1969). It has been suggested by Callan (1967) that the lampbrush chromosome loop consists of copies (slave genes) of a master gene which is kept out of contact with RNA polymerase in order to preserve the authentic gene for cell division. Whitehouse (1967) has provided a cycloid model whereby the loop of slave copies can dissociate from the master gene during meiosis. There is evidence that these extra gene copies may be labile and undergoing extensive renewal or remodeling. Considerable variation in the number of copies of the genome for ribosomal RNA is known to occur during oogenesis and embryogenesis; thus, in the case of *Xenopus laevis,* hybridization experiments show that the number of gene-copies of rDNA increases several hundredfold during oogenesis, but by gastrulation the number of gene copies has declined sharply (D. D. Brown, 1968). A basic question is whether this type of adaptation of the number of slave copies is confined to ribosomal RNA synthesis or is a general mechanism for regulating mRNA production in cells by varying the amount of DNA template. In the case of the puffs on insect chromosomes it has been demonstrated (Pelling, 1966) that activity is associated with increases in the local DNA in steps of 1, 2, 4, 8, etc., suggesting creation of extra complete copies of the genetic information. Pelc considers from study of thymidine incorporation that part of the DNA of mammalian cells undergoes continued renewal even in cells not undergoing division or increase of DNA due to polyploidy (Stroun *et al.,* 1967; Pelc, 1968). If this can be confirmed, it implies that the process of gene replication may go on continuously, and thus provides a means of regulating the amount of DNA available for transcription by RNA polymerase. The further claim (Roels, 1966; Pelc, 1968) that measurable and even large changes in the DNA content of diploid nuclei can accompany changes in the metabolic activity of cells such as those of the thyroid gland has been generally disputed and must await independent confirmation. In line with the concept of metabolic DNA, J. Paul and Hunter (1968) find that synthesis of hemoglobin in bone marrow in response to erythropoietin requires prior synthesis of DNA not associated with cell multiplication. In conclusion, it remains an interesting but unproved possibility that one means of regulating the amount of protein in a cell is to alter the number of gene copies available for transcription.

The function of chromosomes requires that they must carry out both DNA and RNA synthesis while in the form known as chromatin, in which they exist between cell divisions. Bonner *et al.* (1968) have recently summarized the conclusions to be drawn from their studies on isolated plant and animal chromatins. These preparations contain about equal parts of DNA and histones, usually a small proportion of nonhistone protein, and finally some RNA. When incubated with ribonucleoside triphosphates, the isolated chromosomes exhibit some RNA polymerase activity, but it is not certain whether this enzyme is an integral component of the chromosome within the intact nucleus. If exogenous RNA polymerase is added to the chromosome preparation, much more RNA can be synthesized. This capacity of chromatin to act as a template is, however, much less than that of the same amount of DNA separated from its chromosomal protein. The suppression of DNA template activity in chromosomes varies from tissue to tissue. In tissues actively making RNA, such as liver or the growing cotyledons of the pea, the chromosomal preparations exhibit some 20–30% of the activity of DNA extracted from the same cells. On the other hand, the nucleated erythrocyte of the duck makes essentially no RNA, and its chromatin has almost no capacity to act as a template for RNA polymerase (Marushige and Bonner, 1966). It has been demonstrated by means of hybridization experiments (J. Paul and Gilmour, 1966) that the RNA species made by chromosomes from different tissues differ from one another. Since the total DNA of different cells is considered to be the same, this implies that each tissue allows only part of its DNA to be transcribed to RNA, and that the DNA available for this purpose differs from tissue to tissue. The histone fraction of the chromosomes appears to be responsible for masking DNA template capacity; preparations of DNA that still contain nonhistone proteins show full template activity comparable to that of purified DNA, and this activity can be reduced again by addition of histones to the preparation (Marushige and Bonner, 1966).

As might be guessed, much effort has gone into the study of histones as potential regulators of genetic information (see review by Stellwagen and Cole, 1969). Histones are basic proteins that have been fractionated into three major types, either lysine rich, moderately lysine rich, or arginine rich (Bonner *et al.*, 1968; Kinkade and Cole, 1966; Stellwagen and Cole, 1968a; Hnilica, 1967). Subclasses of these major species have been described, and are said to vary from one animal to another and from one tissue to another in the same species (Hnilica *et al.*, 1966a; Stellwagen and Cole, 1968b; Bustin and Cole, 1968). Attempts to demonstrate binding of different types of histones to specific regions of DNA do not suggest in general that they possess the high degree of selectivity necessary to turn off individual cistrons (Bonner *et al.*, 1968). However, certain classes of cistrons may be specifically suppressed by individual histones. Georgiev (1968) presents evidence to show that lysine-rich histones may specifically inhibit repeating DNA cistrons; this action may be linked to the function of lysine-rich histones in the tight packing of the deoxyribonucleoprotein into the chromosomal fibrils. This would endow the lysine-rich histones with regulation of the number of copies available for repeating DNA sequences. Histones exhibit chemical modifications that may alter their capacity to bind to DNA, and thus to repress transcription of genes. Isolated nuclei can methylate the ϵ-terminal amino group of lysine in histones (Allfrey *et al.*, 1964; Sekeris *et al.*, 1967), but methylation does not appear to be related to template activity for RNA polymerase (Tidwell *et al.*, 1968). A better correlation between chromosome template activity and acetylation of histones can be made. Histones can be acetylated independently of the synthesis of the remainder of the histone molecule. B. G. T. Pogo *et al.* (1966) have found that fractions of chromatin that are more active as templates for RNA polymerase have a larger proportion of acetyl groups on their histones, notably on the arginine-rich histones. Allfrey (1968) has recently reviewed the striking correlations between intensity of acetylation of arginine-rich histones in a tissue and gene transcription regulation; thus acetylation and gene activation go hand in hand in regenerating liver, steroid stimulation of liver RNA synthesis, and phytohemagglutinin stimulation of lymphocyte metabolism. A similar case can be made out for phosphorylation. The biologically more active fractions of chromosomes contain a larger proportion of phosphorylated serine and threonine residues on their histones (Allfrey *et al.*, 1966; Stevely and Stocken, 1968). In this connection, Langan (1967) has isolated nuclear phosphoproteins that bind to histones and modify their inhibitory action on RNA polymerase. Recently, Langan (1968a) describes nuclear phosphorylation of histones of the lysine-rich class through a kinase that transfers phosphate from ATP to the serine residues of the protein; this reaction is much accelerated by cyclic AMP, thus suggesting a means by which hormones releasing this nucleotide may influence RNA synthesis (Langan, 1968b). Acetylation and phosphorylation may alter the charge on the histone and thus its DNA-binding capacity.

In addition to histones, isolated chromosomes contain other proteins. The importance of these in gene transcription is suggested by the findings of Marushige *et al.* (1968) that (a) DNA-nonhistone protein complexes are as good substrates as DNA itself for transcription by RNA polymerase, (b) nonhistone proteins are abundant in active chromosomes but not in inert chromosomes. A series of acidic proteins have been fractionated by T. Y. Wang and Johns (1968), who also cite the many investigators who have surmised that these acidic proteins may be involved in gene regulation. Recently, J. Paul and Gilmour (1968) have found that, whereas addition of histones to DNA inhibits the capacity of the DNA to act as a template for RNA polymerase, the presence of nonhistone (acidic) chromosomal protein in the medium reduces the inhibitory effect of the histones. This suggests that the acidic proteins may play a role in regulating gene repression by histones. On the other hand, Bonner *et al.* (1968) have isolated a short-chain species of RNA rich in dihydropyrimidines from chromosomes. When tissue DNA and histones are mixed in the presence of chromosomal RNA from the same tissue, the resulting template transcribes RNA species that are tissue specific; removal of the chromosomal RNA leads to a synthetic chromosome that lacks this specificity. The existence of chromosomal RNA has been denied (Loeb, 1967) and confirmed (Prestayko and Busch, 1968) and ever claimed to be present as large amounts of ribosomal RNA (Maio and Schildkraut, 1967), the latter observation being made under the special circumstance of synchronous cell division at metaphase.

In summary, the following picture can be suggested. From the studies described above it can be concluded that histones probably act as nonspecific repressors of gene expression, except that a specific class of cistrons, those with repeating copies, may be specifically inhibited by lysine-rich histones. The general inhibitory action of histones can be rapidly modified by acetylation or phosphorylation, which alter the charges on various histones and thus the tightness of their binding to DNA. Hormones that cause release of cyclic AMP may stimulate histone phosphorylation. This allows coarse control of template activity which Allfrey (1968) likens to pulling out a file drawer before searching the files for a particular piece of information. Fine control (that is, the selection of a specific cistron for mRNA synthesis) may eventually reside in the acidic nuclear proteins or specific types of nuclear RNA which are sufficiently discriminating to select templates for RNA synthesis appropriate to the activities of the cell.

One of the questions raised by the action of histones and other nuclear proteins in regulating genetic expression is whether changes in regulation involve synthesis of new chromosomal protein. The bulk of histones in various mammalian tissues show the same metabolic stability as the tissue DNA (Byvoet, 1966; Piha *et al.*, 1966; Ohly *et al.*, 1967). Individual histone fractions do not seem to vary in rate of turnover (Dick and Johns, 1968), although Bustos-Valdes *et al.* (1968) consider that histones easily extractable with acid from rat liver nuclei have a higher incorporation of leucine-^{14}C than histones that are less easily extracted. Furthermore, Chalkley and Maurer (1965) have published data suggesting that lysine-rich histones are synthesized only during cell division, whereas arginine-rich histones are made at other times in the cell cycle. The question of the *site* of chromosomal protein synthesis is also uncertain. There seems no doubt that amino acids can be incorporated under *in vitro* conditions into isolated nuclei, even when adhering cytoplasmic ribosomes have been removed with detergents (Gallwitz and Mueller, 1969a; Zimmerman *et al.*, 1969). Furthermore, it has been claimed that isolated nucleoli (Birnsteil and Hyde, 1963; Maggio, 1966; Izawa and Kawashima, 1969; Zimmerman *et al.*, 1969) and chromatin (Sekeris *et al.*, 1966a,b; T. Y. Wang and Patel, 1967) can synthesize protein (the chromatin apparently by a system directly utilizing DNA as the template for amino acid incorporation). The study of Zimmerman *et al.* (1969) on *in vitro* incorporation by isolated HeLa nuclei was particularly detailed and showed that the main *in vitro* incorporation by nuclei occurred into nucleolar protein, where it was essentially represented by two peaks of radioactivity, one of which behaved like a histone. Reid *et al.* (1968) have also claimed that histones are made *in vitro* by isolated nuclei. This intranuclear synthesis may account for the formation of *some* of the histones in nucleoli needed for incorporation into nascent ribosomes. However, it seems unlikely that histone synthesis for formation of *chromosomes* involves nuclear protein synthesis, since Robbins and Borun (1967) and Gallwitz and Mueller (1969b) both found that histone synthesis occurs on cytoplasmic polysomes at the time when DNA is being made prior to cell division. There is good evidence that such a translocation of protein from cytoplasm to nucleus can take place. L. Goldstein and Prescott (1968) have observed rapid passage of protein in both directions between the nucleus and cytoplasm of *Amoeba proteus*. Berendes (1968) has shown by cytochemical techniques that swelling of puffs in insect chromosomes involves accumulation of acidic protein, which is translocated from elsewhere in the cell. In conclusion, it would appear that synthesis of chromosomal proteins probably takes place in the cytoplasm, and that the capacity of nuclei to incorporate labeled amino acids *in vitro* may represent some other function, perhaps analogous to the synthesis of *structural* proteins by mitochondria whose *enzyme* complement is made in the microsomes and transferred to the mitochondria.

b. The Nucleolus. Studies of the structure and function of this organelle up to the end of 1965 have been exhaustively reviewed in the Proceedings of the "International Symposium on the Nucleolus" (Vincent and Miller, 1966). These bodies occur in the nuclei of plant and animal cells, being absent only from mature spermatids and spermatozoa (Shea and Leblond, 1966) and at some of the early stages of embryonic development (D. D. Brown, 1966). A quantitative survey of the number and size of nucleoli in most tissues of the mouse has been published by Shea and Leblond (1966). In diploid cells, the number of nucleoli varies between 1 and 6, whereas tetraploid hepatocyte nuclei show up to 11 nucleoli. Shea and Leblond conclude that there are 6 nucleolar organizers per diploid set of mouse chromosomes, but that not all these organizers may be expressed, thus resulting in cells showing fewer than the maximum number of nucleoli. Mean nucleolar size was found by Shea and Leblond to vary over a threefold range in different mouse tissues, being generally correlated with the size of the nucleus. Nucleolar size is known to change under a variety of circumstances. Thus, liver cell nucleoli increase in size during the cell regeneration which follows partial hepatectomy (Kaufmann *et al.*, 1967) and after administration of thioacetamide (Kleinfeld, 1966), and can either increase or decrease with nutritional state (Lagerstedt, 1949; Stenram, 1958a; Nadal and Zajdela, 1967). The change in hepatocyte nucleolar size in response to dietary alterations can occur very rapidly. Lagerstedt (1949) found that the nucleoli of fasting rats were small, but they doubled their size within 3 hours of feeding a meal containing protein. These changes in nucleolar volume with amino acid supply are accompanied by corresponding increments in nucleolar RNA content (Stenram, 1958b; Munro *et al.*, 1965). Stenram (1958a) has shown that diet-induced changes in nucleoli are not detectable in other tissues or in liver tumors.

Bernhard (1966) has reviewed recent ultrastructural studies on the nucleolus (Fig. 6). This organelle shows a reticular network of RNA-containing fibrils and also other areas containing RNA consisting of granules similar in size to cytoplasmic ribosomes. In electron micrographs of nucleoli of normal liver cells, fibrillar areas represent about one-fourth the abundance of granular areas (Smetana *et al.*, 1966). Unuma *et al.* (1968) found by autoradiography that labeled uridine-^{3}H first appears in the fibrillar areas and subsequently in the granular, thus suggesting that ribosomal RNA precursor is first formed in the fibrillar material. The location of DNA in the nucleolus is of some interest. Surrounding the nucleolus is a double layer of deoxyribonucleoprotein, the nucleolus-associated chromatin, from which filaments of chromatin pass deep into the nucleolus and presumably act as DNA templates for nucleolar RNA synthesis. Chromatin around or within the liver nucleolus accounts for about one-third of the bulk of the nucleolus; most of this chromatin is perinucleolar (Unuma *et al.*, 1967). About 5% of the total DNA of rat liver cells appears to be associated with the nucleoli (see discussion at end of paper by Busch *et al.*, 1966); but only some 0.6% of this nucleolar DNA consists of ribosomal cistrons, according to the results of hybridization experiments with ribosomal RNA (Steele, 1968). The function of the major part of the nucleolar DNA is thus unknown. Nucleolus-associated chromatin is said to be especially abundant in cells with a high rate of protein synthesis, but in such instances it also adheres to the inner surface of the nuclear envelope and may not be a specific feature of the nucleolus. Since the nucleolus can change in size as discussed above, it is permissible to ask whether the DNA associated with this organelle shares in the increase or decrease. In the case of kidney cells in tissue culture, Granboulan and Granboulan (1964) have shown that nucleolar DNA incorporates thymidine actively. This thymidine incorporation might correspond to the metabolically active DNA of resting cells discussed above, and in consequence, thymidine incorporated by nucleoli could represent gene copying in the manner discussed earlier.

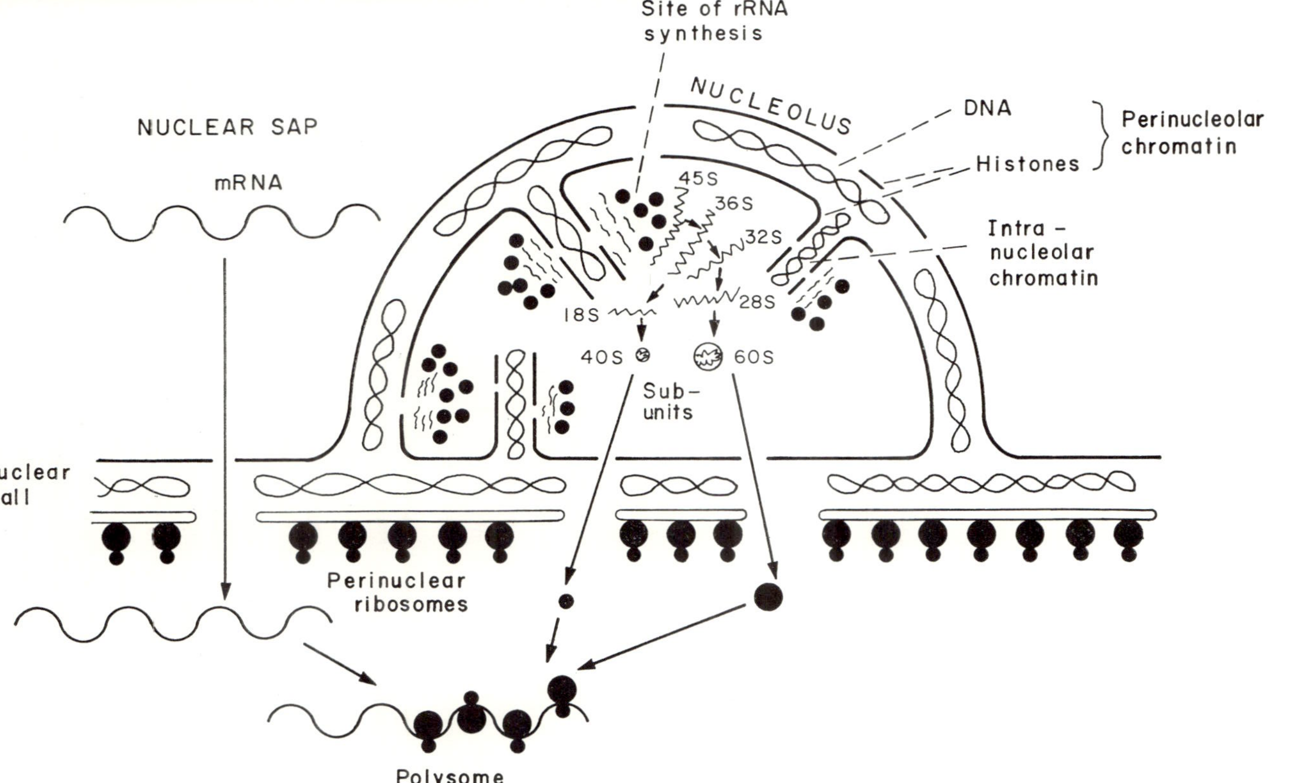

FIG. 6. Functional diagram of the nucleolus. The structural features shown are based on the review article by Bernhard (1966). The intermediate stages in ribosomal RNA formation are modified from Penman (1967). The figure shows the two ribosome subunits leaving separately through the nucleolar pores into the cytoplasm, where they form polyribosomes with messenger RNA coming from the chromosome via the nuclear sap. An alternative view is that the smaller ribosome subunit passes into the nuclear sap, joins with a strand of messenger RNA, and then leaves the nucleus together with the messenger RNA strand to form polysomes with the large subunits in the cytoplasm.

It would be interesting to know whether incorporation of thymidine into hepatocyte nucleolar DNA is influenced by factors causing rapid alterations in nucleolar size and metabolic activity, such as changes in the amino acid supply to the liver appear to induce. Finally, it has been pointed out by Bernhard (1966) that there are gaps in the perinuclear sheath of chromatin, and that, as often happens, when the nucleolus impinges on the nuclear wall these gaps correspond to similar pores seen in the nuclear membrane, so that the passage of nucleolar contents (ribosomal subunits) directly into the cytoplasm is an unproved possibility. This picture of nucleolar structure can be affected by many factors. In particular, nucleoli become disrupted following administration of various antimetabolites such as actinomycin D (Harris *et al.*, 1968; Kume and Chiga, 1967), ethionine (Shinozuka *et al.*, 1968), aflatoxin (Floyd *et al.*, 1968), proflavine and acid orange (Reynolds and Montgomery, 1967) and a number of carcinogenic agents (Reddy and Svoboda, 1968; Harris *et al.*, 1968; Floyd *et al.*, 1968; Kume and Chiga, 1968). These agents usually interfere with ribosomal RNA synthesis, not an unexpected consequence of nucleolar disintegration.

The function of nucleoli is to synthesize ribosomes. This is most conclusively demonstrated by the absence of new ribosome formation in embryos from the mutant of *Xenopus laevis* that is deficient in nucleolar organizer (D. D. Brown, 1966). These studies also dispel the belief that tRNA is made in nucleoli, since in the anucleolar mutant there is continuing synthesis and accumulation of this species of RNA. Regarding the proteins of the nucleolus, two important questions arise in connection with ribosome formation. First, do nucleoli contain protein species that are found also in ribosomes? Nucleoli have both basic and acidic proteins (Busch, 1965; Grogan *et al.*, 1966; Grogan and Busch, 1967), and extensive comparisons by gel electrophoresis and immunological precipitation leads to the conclusion that many of the proteins found in chicken liver nuclei (Lindsay, 1966) and starfish oocyte nucleoli (Vincent *et al.*, 1966; Mundell, 1967) are also present in the cytoplasmic ribosomes of these cells. This implies that assembly of the ribosome occurs within the nucleolus, and indeed, Warner and Soeiro (1967) and Izawa and Kawashima (1969) have isolated nucleolar particles consisting of ribosomal proteins associated with the 45 S and 32 S precursors of ribosomal RNA, implying that maturation of ribosomal RNA occurs within the ribosomal subunit. Liau and Perry (1969) have observed that the nucleolar particles containing precursor ribosomal RNA (45 S, 36 S, 32 S) are larger than mature ribosomal subunits containing 28 S RNA; consequently, maturation of the particle involves a reduction in protein content as well as RNA size.

Since ribosomes show continuous turnover (Chapter 32), there must be a steady outflow of protein from the nucleoli to the cytoplasm. Consequently, unlike chromosomal histones and other chromosomal proteins, there is an obvious requirement to provide a continuous supply of proteins to the nucleolus even when the cell is not undergoing division. It is thus no surprise that the turnover of nucleolar histones in plant cells (Birnstiel and Flamm, 1964) and in animal cells (Hnilica *et al.*, 1966b) is more rapid than the rate of renewal of histones in other parts of the nucleus. We thus come to the question, where in the cell are these ribosomal proteins manufactured? Birnstiel and Hyde (1963), Izawa and Kawashima (1969), Zimmerman *et al.* (1969), and Maggio (1966) have observed that isolated nucleoli can incorporate labeled amino acids under *in vitro* conditions. As discussed earlier, Zimmerman *et al.* (1969) found that one of the two proteins made *in vitro* was a histone. A disturbing feature of systems showing incorporation of amino acids into protein by isolated nucleoli is that uptake occurs briskly in the absence of an energy source and in the presence of high concentrations of ribonuclease. This urges caution in equating these *in vitro* findings with the site of synthesis of nucleolar proteins in the intact animal. In view of the ease with which proteins can pass from their site of biosynthesis on microsomes into mitochondria (Work *et al.*, 1968), there is no inherent reason why nucleoloribosomal proteins should not be made in the cytoplasm, and thereafter be transferred to the nucleus. Indeed, transfer of protein from cytoplasm to nucleus has been demonstrated by autoradiography in the case of hemoglobin in the nucleated erythrocyte of the chick (Kabat, 1968) and for *Amoeba proteus* (L. Goldstein and Prescott, 1968). Furthermore, the cytoplasm appears to contain polysomes making the structural proteins of ribosomes, since Ogata and his colleagues (Ogata *et al.*, 1967; Terao *et al.*, 1968) describe a cell-free polysome system that can incorporate cytoplasmic amino acids into basic proteins migrating on acrylamide gels in the same locations as the structural proteins of the ribosomes.

In this connection it would appear from separation studies on gels that at least 24 discrete species of proteins are present in mammalian ribosomes, and in *E. coli* ribosomes some 29 structural proteins have been identified (Low and Wool, 1967). In the case of the smaller subunit of the bacterial ribosome, there is evidence to show that all of the 13 to 15 proteins isolated from this particle are chemically distinct and probably have no peptide sequences in common (Moore *et al.*, 1968; Fogel and Sypherd, 1968). This finding implies that synthesis of each structural protein is controlled by a separate cistron. The cistrons active in different tissues of the same animal appear to be identical, since the proteins extracted from the ribosomes of various tissues exhibit a similar pattern on acrylamide gels (Low and Wool, 1967; Bielka *et al.*, 1967). It may be noted that ribosomal RNA from different tissues of the same animal has also been found to show the same base composition (Dingman and Sporn, 1962). The spectrum of ribosomal proteins is, however, distinguishably different from one mammalian species to another (Low and Wool, 1967; Bielka *et al.*, 1967; Keller *et al.*, 1968). Species variation in ribosomal RNA among different mammals has not so far been examined, but Loening (1968) has concluded that ribosomal RNA undergoes progressive evolution in size from primitive to higher organisms of both the plant and animal kingdoms.

c. *Synthesis of RNA by the Nucleus.* Recent investigations into the species of RNA formed by the mammalian cell have been reviewed by Penman (1967) and by Darnell (1968). There is substantial evidence to show that ribosomal RNA is first formed as a molecule with 45 S sedimentation (Fig. 7). This matures into the 28 S and 18 S species found in the ribosome subunits through a series of intermediate stages (45 S → 41 S → 36 S → 32 S → 28 S, and 20 S → 18 S), with accompanying methylation of the ribose. In addition, the separation of 20 S and 18 S from the precursor appears to involve further methylation with formation of dimethylaminopurine (Zimmerman, 1968). In the 60 S ribosome subunit there are two additional smaller RNA species, 5 S and 7 S RNA, both inserted in the course of ribosome synthesis in the nucleolus; their purpose in relation to ribosome structure and function is unknown (Darnell, 1968). Whereas the 7 S RNA appears to be formed as part of the 28 S RNA, the origin of the 5 S RNA is uncertain (Darnell, 1968); Perry and Kelley (1968b) have recently shown that 5 S RNA continues to be made when 28 S and 18 S RNA are inhibited by actinomycin in low doses. The site of synthesis of 4 S tRNA within the nucleolus is also uncertain except that, as pointed out above, it is still

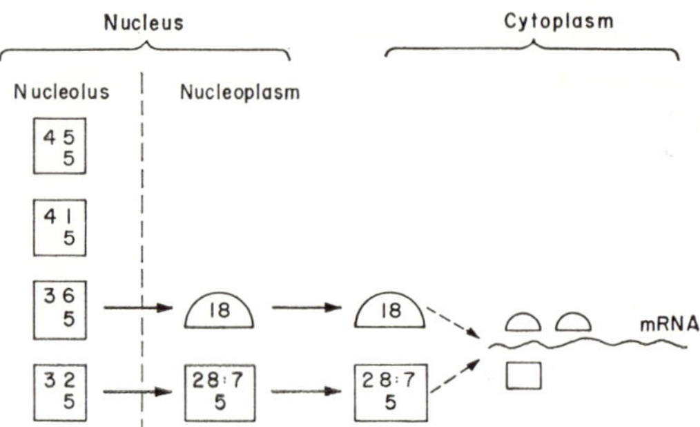

FIG. 7. Maturation of ribosomal RNA in mammalian cells. (From Darnell, 1968.)

made in *Xenopus levis* mutants lacking nucleoli. Finally, the messenger RNA of animal cells continues to be elusive. The presence of polysomes in the cytoplasm is good evidence of animal messenger RNA, as we shall see when polysomes making specific proteins are discussed below. Examination of the nucleus for mRNA has revealed that there are RNA species in the nucleoplasm that vary from 20 S to 100 S or more in size (Scherrer *et al.*, 1966; Attardi *et al.*, 1966; Penman *et al.*, 1968). The polydisperse nature of these molecules has led to the name "heterodisperse nuclear RNA" (HnRNA). These molecules have a DNA-like nucleotide composition, undergo rapid labeling, and have a rapid turnover rate, as many messengers have. However, they do not appear to be predominantly messenger RNA or a precursor of mRNA, since much of the HnRNA appears to undergo breakdown without leaving the nucleus. This is confirmed by hybridization studies which show in the case of L-cells (Shearer and McCarthy, 1967) and of liver cells (Drews *et al.*, 1968) that about 70–80% of the nuclear RNA is not found in the cytoplasm and thus does not leave the nucleus. The remaining 20–30% presumably includes messenger RNA destined for the cytoplasm. The HnRNA is not polydisperse through arising from different genes. Edström and Daneholt (1967) have identified heterogeneous RNA arising from single loci in the giant salivary chromosomes of insects. Careful quantitative studies on the relative amounts of different species of RNA made in HeLa cell nuclei (Soeiro *et al.*, 1968) show that HnRNA is a major product.

The DNA-directed synthesis of these RNA species by the nucleus is performed by RNA polymerase (nucleoside triphosphate:RNA nucleotidyl transferase) and this enzyme is consequently a primary site for regulation. RNA polymerase purified from *E. coli* consists of several nonidentical subunits (see Geiduschek and Haselkorn, 1969, for review). One subunit (σ) can be dissociated from the enzyme on phosphocellulose columns and is also dissociated after initiation of new RNA chains. Our knowledge of the function of RNA polymerase in genetic transcription has been reviewed by several authors (e.g., Hayes, 1967; G. S. Martin and Tocchini-Valentini, 1967; Chambon, 1968; Geiduschek and Haselkorn, 1969). The two strands of DNA are complementary and also antiparallel. Studies on synthesis of bacterial and viral RNA suggest that only one DNA strand is transcribed at a time. The experiments of A. Goldstein *et al.* (1965) favor growth from 5′ to 3′ in the polynucleotide chain. The enzyme carrying out this transcription was first described by S. B. Weiss and Gladstone (1959) and has been demonstrated in bacterial, plant, and animal cells, but much of the mechanism of enzyme action has been worked out using purified bacterial enzyme. To function, the enzyme requires DNA as template, all four ribonucleotide triphosphates, and Mg^{2+} or Mn^{2+} ions. *In vitro* studies with purified bacterial polymerase show that enzymic activity proceeds in a series of separable steps (Anthony *et al.*, 1966).

Association: DNA + enzyme ⇌ DNA-enzyme (1)

Initiation: DNA-enzyme + purine nucleoside triphosphate →
DNA-enzyme-purine nucleotide (2)

Polymerization: DNA-enzyme-purine nucleotide + nucleoside triphosphates →
DNA-enzyme-polyribonucleotide + pyrophosphate (3)

Termination: DNA-enzyme-polyribonucleotide → DNA + enzyme + RNA (4)

The first step is the binding of enzyme to a finite number of sites on the DNA (Richardson, 1966). Binding to DNA is accompanied by dissociation of the large polymerase aggregates (19 S to 24 S) to yield a protein of lower molecular weight (13 S) which has more catalytic activity than the large aggregate (Smith *et al.*, 1967; Pettijohn and Kamiyama, 1967). High ionic concentrations of monovalent salts—KCl, NH_4Cl, $(NH_4)_2SO_4$—prevent the enzyme from binding to DNA (Anthony *et al.*, 1966; Fuchs *et al.*, 1967).

The second step, chain initiation, requires specifically the presence of relatively high concentrations of purine nucleoside triphosphates (ATP, GTP). When the enzyme, DNA, and purine nucleoside triphosphate are mixed together, the complex DNA-enzyme-nucleotide formed is now resistant to subsequent addition of salt solutions of high ionic strength (Anthony *et al.*, 1966; Fuchs *et al.*, 1967). The relative amounts of ATP and GTP bound to DNA in the presence of polymerase varies considerably according to the nature of the DNA template. With *Micrococcus lysodeikticus* DNA as primer, the initiating nucleotide is predominantly GTP (Maitra and Hurwitz, 1965; Anthony *et al.*, 1966). With T4 bacteriophage DNA as the template for bacterial polymerase, ATP is the predominant nucleotide incorporated at the 5′-terminus though a significant amount of GTP is also incorporated as the initial nucleotide (Bremer and Bruner, 1968). It has recently been found with T4 DNA as template that large RNA molecules are initiated with ATP and shorter molecules with GTP (Mueller and Bremer, 1969; Bremer and Mueller, 1969).

In the third stage of enzyme action, polymerization, there is no differential effect of one nucleoside triphosphate over another (Anthony *et al.*, 1966) and incorporation of nucleotides is stimulated by high salt concentrations and by spermidine (Fuchs *et al.*, 1967). Since this stimulant action of salt concentration on chain elongation is not observed with single-stranded DNA as template, it has been concluded by Fuchs *et al.* (1967) that the salt causes separation ("melting") of the strands in the double-stranded DNA. The stimulant action of inorganic monovalent salts occurs maximally between 0.2 and 0.4 *M* and is to some extent dependent on the cation used (So *et al.*, 1967; Fuchs *et al.*, 1967).

The final step, chain termination, appears to occur when the enzyme reaches certain specific sequences in the DNA template (Chambon, 1968). In this respect, the presence of ribosomes, which are natural acceptors of mRNA, can accelerate both RNA transcription and also chain release (Shin and Moldave, 1966). Recently, a protein subunit (σ) has been separated from pure bacterial polymerase, which is required for initiation. After initiation, it is released from the polymerase–DNA complex and can be used for another round of initiation (Burgess *et al.*, 1969; Travers and Burgess, 1969).

The role of bivalent cations has been explored with purified bacterial RNA polymerase. Activity is dependent on the presence of Mg^{2+}, Mn^{2+}, or Co^{2+} (Fox and Weiss, 1964; Furth *et al.*, 1962; Fuchs *et al.*, 1967). The preferred metal ion for *in vitro* systems seems to vary with the bacterial source of enzyme, with the DNA template and with the substrate (Fox and Weiss, 1964; Hurwitz, 1959; Furth *et al.*, 1962; Fuchs *et al.*, 1967; Kerjean *et al.*, 1967). In the presence of 0.13 *M* NH_4Cl, the metal ions Mn^{2+} and Co^{2+} at 8 m*M* concentrations support considerable *E. coli* enzyme activity for several hours, whereas activity with 8 m*M* Mg^{2+} becomes plateaued after a short period of incubation (Fig. 8). The enzyme can be restarted by adding very high concentrations of Mg^{2+} (70 m*M*), or on addition of spermidine or by raising the concentration of monovalent salts (e.g., to 0.4 *M*), but not by adding Mn^{2+} or Co^{2+} (Fuchs *et al.*, 1967).

These various studies on bacterial RNA polymerase activity in different media should be evaluated in relation to the ionic conditions present within the bacterial cell. Calculations based on reported elementary composition of bacteria (Luria, 1960) show that the Mg content is about 0.5% of dry bacterial mass, whereas Mn content is only 0.002%, corresponding, respectively, to 40 m*M* Mg^{2+} and 0.1 m*M* Mn^{2+} on a wet weight basis. Such data suggest that Mg^{2+} is the physiologically active bivalent cation, since Mn^{2+} concentration is too low to be effective unless it is locally concentrated at the site of polymerase activity. Furthermore, the interior of *E. coli* provides a K^+ concentration of 0.2 *M* (Solomon, 1962) and a total salt concentration calculated to be 0.36 *M* (Fuchs *et al.*, 1967). In the presence of 40 m*M* Mg^{2+}, these salt concentrations will ensure continued enzyme activity without the plateau shown for 8 m*M* Mg^{2+} in Fig. 8, where salt concentration was only 0.13 *M*. It may be thought that the high intrabacterial salt concentration would

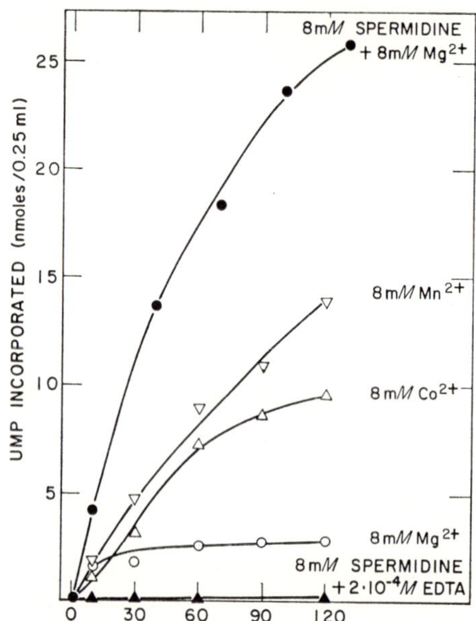

FIG. 8. Effect of divalent cations on the activity of bacterial RNA polymerase. The reaction mixtures contained UTP-^{3}H and other unlabeled nucleotide triphosphates, purified RNA polymerase, calf-thymus DNA template, 0.13 *M* NH_4Cl, to which Mg^{2+} acetate, $CoCl_2$, $MnCl_2$, spermidine phosphate, or EDTA was added. (From Fuchs *et al.*, 1967.)

be inimical to binding of the polymerase to the DNA template. However, salt-induced dissociation of enzyme and DNA may be prevented in the living cell by the continuous presence of purine nucleotide triphosphates in high concentration which promote the formation of the salt-resistant DNA-enzyme-purine nucleotide complex. Indeed, one way in which activity of the enzyme may be regulated could be through purine nucleotide supply, the formation of the DNA-enzyme complex being determined by the levels of nucleotide triphosphates in the cell. Finally, high intracellular salt concentration will favor dissociation of polymerase into 13 S subunits (Richardson, 1966; Pettijohn and Kamiyama, 1967), which have a higher activity and DNA-binding capacity than the 21 S aggregate enzyme (Smith *et al.*, 1967).

These somewhat technical aspects of recent studies on bacterial RNA polymerase are significant when the corresponding mammalian enzyme is considered. The *in vitro* investigation of RNA polymerase activity in extracts from nucleated cells has often been carried out using a deoxyribonucleoprotein aggregate enzyme prepared by a technique first described by S. B. Weiss (1960). In his original procedure, the nuclei are first separated and then lysed in hypotonic solution. The nuclear solids are then extracted with 0.4 M KCl, and the final gelatinous residue remaining ("aggregate enzyme") is used as the active preparation containing both DNA template and enzyme. It is obvious that the treatment with 0.4 M KCl, a known extractant of histones (Huang *et al.*, 1964), is likely to alter the properties of the chromosomal DNA template. Many recent investigators have therefore omitted this step (e.g., Widnell and Tata, 1964, 1966; Ramuz *et al.*, 1965). When lysed liver nuclei are incubated with the four nucleoside triphosphates in the presence of 5 mM Mg^{2+}, synthesis of RNA occurs but ceases after 10–15 minutes, whereas synthesis of RNA in the presence of 4 mM Mn^{2+} and 0.4 M $(NH_4)_2SO_4$ proceeds undiminished for at least an hour (Widnell and Tata, 1966). The stimulant action of $(NH_4)_2SO_4$ is much less if Mg^{2+} is substituted for Mn^{2+}, and KCl, NaCl, and NH_4Cl in combination with Mn^{2+} are less effective than $(NH_4)_2SO_4$ and Mn^{2+}. The picture bears a considerable resemblance to the findings obtained with the bacterial enzyme, which also shows (Fig. 8) an early plateau of activity with Mg^{2+} as cofactor, but sustained synthesis when suspended in a high salt concentration or with Mn^{2+} substituted for Mg^{2+}.

Once more we may ask what relationship the *in vitro* findings with mammalian RNA polymerase have to ionic conditions present within the nucleus of the mammalian cell. The total tissue content of Mg is equivalent to no more than 20 mM in liver and muscle and is somewhat less in other soft tissues (Wacker and Vallee, 1964); free ionic Mg^{2+} may be considerably less than the total Mg. On the basis of studies made on liver, there does not appear to be preferential concentration of Mg^{2+} in the nucleus (Thiers and Vallee, 1957; Okazaki *et al.*, 1968). The Mn content of mammalian tissues is much less than that of Mg, being equivalent to about 0.05 mM in liver and other glandular tissues (Fore and Morton, 1952). However, measurement of the Mn content of very impure preparations of liver nuclei showed that these had twice the cytoplasmic concentration of this metal (Thiers and Vallee, 1957), a finding which suggests that purified nuclei may well have very high concentrations of Mn. As regards monovalent ions, the concentration of K^+ in nuclei appears to be similar to that in cytoplasm (Okazaki *et al.*, 1968). Summarizing the picture, Okazaki and Farber (1969) have calculated the concentrations of several elements present in liver nuclei isolated in a nonaqueous medium and in liver cytoplasm. In the nucleus and cytoplasm, the respective concentrations are Ca, 1.9 and 1.1 mM; Mg, 12 and 11 mM; K, 110 and 98 mM; and Na, 46 and 5 mM. Since some of these metals are bound to macromolecules, such as ribosomes, these figures represent upper values for the free ion concentrations.

Clearly, further information about intranuclear ionic conditions is needed before one can realistically relate *in vitro* findings with polymerase to *in vivo* conditions of RNA synthesis.

The nature of the product formed under different ionic conditions has also been explored (Widnell and Tata, 1966; Blackburn and Klemperer, 1967; Nair *et al.*, 1967; Chambon *et al.*, 1968). Whereas the RNA synthesized in the presence of Mg^{2+} is rich in guanylic and cytidylic acids and sediments rather slowly on a sucrose gradient, the RNA formed in the presence of Mn^{2+} and $(NH_4)_2SO_4$ is richer in adenylic and uridylic acids and has a polydisperse distribution on sucrose gradients. It is presumed from these findings that Mg^{2+} favors synthesis of ribosomal RNA, and that Mn^{2+} and high salt concentration strip histones from the DNA and consequently result in synthesis of a product that is DNA-like in composition. A. O. Pogo *et al.* (1967) have attempted by autoradiography to localize the RNA formed under these two types of ionic conditions. They concluded that intact nuclei incubated in the presence of Mg^{2+} synthesize mainly ribosomal-type RNA located in the nucleolus, whereas addition of Mn^{2+} and 0.04 *M* $(NH_4)_2SO_4$ to the incubation medium causes RNA to be formed in the extranucleolar parts of the nucleus. Maul and Hamilton (1967) have arrived at the same conclusions by a similar approach. This distinction between the ionic conditions necessary for RNA synthesis by the nucleolus and by the chromosomes is, however, not absolute, since we (S. T. Jacob *et al.*, 1968a) have demonstrated that the polymerase of isolated nucleoli can respond to addition of Mn^{2+} and ammonium sulfate, provided the concentration of the latter is 0.4 *M*; no stimulation of nucleolar polymerase was observed at 0.04 *M* ammonium sulfate, the level at which autoradiography shows the polymerase in the extranucleolar chromatin to be activated in the presence of Mn^{2+}. This suggests that polymerase activity in the nucleolus may differ from that in the rest of the nucleus in being relatively insensitive to ionic conditions in the incubation medium.

Other evidence suggests the presence of more than one form of polymerase in mammalian nuclei. First, human lymphocytes incubated briefly in tissue culture medium show incorporation of precursors into a polydisperse type of RNA; there is almost no Mg^{2+}-activated polymerase in such cells, although Mn^{2+} ammonium sulfate-activated incorporation can be demonstrated (S. T. Jacob *et al.*, 1969a). It is therefore significant that resting lymphocytes show a virtual absence of nucleoli from their nuclei. A distinction between the polymerizing activity of liver nuclei incubated in Mg^{2+} and in Mn^{2+} combined with salt has also emerged from studies made with the inhibitor α-amanitin, a toxin isolated from the toadstool *Amanita phalloides*. When mice are injected with this compound, nuclei prepared from their livers show no significant change in polymerase activity when incubated with Mg^{2+}, but there is an extensive depression of polymerase activity in the presence of Mn^{2+} and ammonium sulfate (Stripe and Fiume, 1967).

These various observations with whole nuclei and with isolated nucleoli provide suggestive evidence that the nucleolus may contain a form of RNA polymerase that is especially dependent on Mg^{2+} and is relatively insensitive to ammonium sulfate. Because amanitin does not affect the polymerase activity of liver nuclei incubated in the presence of Mg^{2+}, it can be surmised that the nucleolar enzyme may be resistant to this inhibitor. However, the experiments reviewed above were performed with unresolved nuclear preparations containing both the polymerase and its deoxyribonucleoprotein template, and in consequence we cannot attribute the observed effects with certainty to differences in the properties of nucleolar and extranucleolar polymerase; the sensitivity of the deoxyribonucleoprotein template to salts and to inhibitors could be different in the nucleolus from that in other parts of the nucleus. Some of the uncertainties in experiments on mammalian RNA polymerase would be eliminated if the enzyme and its template of chromatin could be recovered separately from the aggregate preparation, since it might then be possible to distinguish effects of salts, hormones, and other factors on the DNA template from those on the enzyme itself. Several authors have obtained partial release of polymerase from animal cell nuclei (Furth and Ho, 1963, 1965; Ballard and Wiliams-Ashman, 1966; Ramuz *et al.*, 1965; Stout and Mans, 1967, 1968; D. D. Cunningham and Steiner, 1967; D. D. Cunningham *et al.*, 1969; Seifart and Sekeris, 1967; M. L. Goldberg *et al.*, 1969). Recently, we have achieved virtually complete extraction of RNA polymerase from whole liver nuclei by a procedure that probably destroys the nucleolar enzyme, leaving only extranucleolar enzyme in the extract (S. T. Jacob *et al.*, 1968b). We have also prepared soluble polymerase from isolated nucleoli and have compared its properties with those of the extranucleolar enzyme (S. T. Jacob *et al.*, 1969a). The results suggest that the liver nucleus contains nucleolar and extranucleolar polymerase that differ significantly in their ionic requirements for optimal activity and in their sensitivity to α-amanitin, shown here to be a specific inhibitor of mammalian polymerase protein. The extranucleolar enzyme is much more dependent than the nucleolar enzyme on Mn^{2+} and ammonium sulfate in the incubation medium. In the presence of ammonium sulfate, the extranucleolar polymerase is also subject to inhibition by amanitin, to which the nucleolar enzyme is resistant under the same conditions of incubation. It is suggested that the extranucleolar enzyme undergoes a salt-dependent change in configuration that renders it more susceptible to activation by Mn^{2+} and to inhibition by amanitin. The nucleolar enzyme is either resistant to configurational change in the same concentration of ammonium sulfate or alternatively differs structurally from the extranucleolar polymerase so that it responds differently to divalent cations and lacks a binding site for the inhibitor. Recently, Roeder and Rutter (1969) have isolated two distinct polymerase activities from rat liver nuclei, one of which appeared to be associated with the nucleolus and to have a greater preference for Mg^{2+} ions.

Inhibitors of RNA synthesis have made a significant contribution to the study of regulation of protein synthesis. A considerable number of compounds have been identified as having inhibitory properties, but not all have been fully investigated. Several have a direct action on RNA polymerization evidenced by the effects of *in vitro* addition to the enzyme. Antibiotics that inhibit the polymerase by binding to the DNA template include actinomycin D (Reich and Goldberg, 1964), miracil D (Weinstein *et al.*, 1965, 1967; Nicholson and Peacocke, 1966), nogalomycin (Bhuyan and Smith, 1965), chromomycin A_3 (Hartmann *et al.*, 1964; Ward *et al.*, 1964; Kaziro and Kamiyama, 1967), echinomycin, daunomycin, mithramycin, olivomycin (Ward *et al.*, 1964), aflatoxin (De Recondo *et al.*, 1966; Gelboin *et al.*, 1966; Sporn *et al.*, 1966; Moulé and Frayssinet, 1968), ethidium bromide (Waring, 1964, 1965), proflavine (Nicholson and Peacocke, 1966), nitrogen mustard (Ruddon and Johnson, 1968), acetylaminofluorene (Troll *et al.*, 1968), and probably also 4-nitroquinoline-*N*-oxide (J. S. Paul *et al.*, 1967). In each case tested, there is evidence that the antibiotic can inhibit RNA synthesis in both animal and bacterial cells, a finding to be expected since the DNA template, not the enzyme, is the site of inhibitory action. On the other hand, antibiotics that react with the enzyme molecule itself appear to be more species selective. Thus, the rifamycins inhibit bacterial polymerase, but not mammalian polymerase (Wehrli *et al.*, 1968a; S. T. Jacob *et al.*, 1968b). They bind stoichiometrically to the bacterial enzyme molecule (Wehrli *et al.*, 1968b) and appear to interfere with the initiation reaction in RNA synthesis (Sippel and Hartmann, 1968; Umezawa *et al.*, 1968). The mechanism of inhibition of the polymerase has not been identified (Di Mauro *et al.*, 1969; Travers and Burgess, 1969). Similarly, streptovaricin also acts as an inhibitor by preventing chain initiation by the polymerase, and again is effective only against bacterial polymerase, not against animal polymerase (S. Mizuno *et al.*, 1968). In the case of both streptovaricin and rifamycin, *E. coli* mutants are described with an RNA polymerase that is resistant to these antibiotics (Nita *et al.*, 1968; Ezekiel and Hutchins, 1968; Tocchini-

Valentini *et al.,* 1968). The resistance of these mutant enzymes and of mammalian RNA polymerase to inhibition indicates that the structure of the polymerase protein in the susceptible wild-type bacterial enzyme is critical for interaction with the inhibitor. Polyethylene sulfonate also inhibits bacterial RNA polymerase through a mechanism involving the enzyme itself, though the precise mode of action is not known (Chambon *et al.,* 1967). It has recently been demonstrated by us (S. T. Jacob *et al.,* 1969a) that α-amanitin, a peptide obtained from the toadstool *Amanita phalloides,* is a specific inhibitor which interacts with the mammalian polymerase enzyme, and is without effect on bacterial RNA polymerase. Amanitin is thus the mammalian counterpart of streptovaricin and rifamycin; however, it does not interfere only with chain initiation, as do the two bacterial inhibitors.

RNA polymerase activity can also be retarded through changes in available substrate. Thus two analogs of ATP, formycin triphosphate (Ikehara *et al.,* 1968) and tubercidin triphosphate (Nishimura *et al.,* 1966), have been found to retard synthesis of RNA when used as substrate in place of ATP in the polymerase reaction; the product is, of course, a fraudulent RNA in which the adenine is replaced by the analog. The exotoxin of *Bacillus thuringiensis* also appears to inhibit RNA synthesis by competing with ATP (Šebesta and Horská, 1968), and the inhibitory action of cordycepin, an adenosine analog that is phosphorylated in the cell, may also be due to this competitive mechanism (Siev *et al.,* 1969). Presumably analogs of other bases such as 8-azaguanine that become incorporated into cell RNA through intermediate nucleoside triphosphates (Grünberger *et al.,* 1966; Webb, 1967) have a similar retarding effect on RNA synthesis. The rapid inhibition of cytidine-^{3}H incorporation into liver RNA following administration of fluorouracil (Stenram and Willén, 1967) may also represent competition for substrate, since fluorouracil is known to become incorporated into the RNA of yeast cells exposed to this analog (DeKloet and Strijkert, 1966). Benzimidazole ribosides are also believed to reduce RNA synthesis through precursor pool depletion, since they are purine riboside antagonists (Sirlin and Loening, 1968). Reversible inhibition of RNA synthesis in mammalian cell cultures follows addition of mercaptopyridethyl-benzimidazole, an analog of adenine (Bucknall and Carter, 1967). Treated cells show no loss of cell-free polymerase activity (Summers and Mueller, 1968), as would be expected of a substrate competitor. However, the addition of adenosine to the cells does not reverse the inhibition (Bucknall and Carter, 1967). It is also reasonable to anticipate that factors reducing the supply of nucleoside triphosphates within cells will affect RNA synthesis adversely. This is believed to be the cause of the reduced incorporation of precursors into liver RNA after administration of ethionine, since the well-known depletion of liver ATP content by this methionine analog precedes its effect on RNA formation (Villa-Treviño *et al.*, 1966; Okazaki *et al.*, 1968; G. A. Stewart and Farber, 1968). RNA synthesis is also depressed by dinitrophenol, which might be attributed to the action of dinitrophenol as an uncoupler of oxidative phosphorylation of ATP in mitochondria. However, this effect occurs in bacteria (Woese *et al.*, 1963), where the mode of action of dinitrophenol is less certain. Furthermore, at low doses this drug has a selective action on bacterial RNA formation without parallel actions on the synthesis of protein and DNA (Simon *et al.*, 1966) and in anaerobic yeast cells, dinitrophenol and azide reduce RNA and protein formation without changing the ATP content of the cell (Jarett and Hendler, 1967). The action of dinitrophenol on RNA formation thus remains to be resolved; it does not have a direct action on RNA polymerase *in vitro* (Jarett and Hendler, 1967).

Mueller and his colleagues have examined the possibility that RNA synthesis can be inhibited by interfering with synthesis of the enzyme. Two analogs of morphine, levorphanol and levallorphan, inhibit RNA synthesis in both bacterial and mammalian cells. Although mammalian cells treated with these compounds show reduced RNA polymerase activity, direct addition of these inhibitors to the polymerase does not affect *in vitro* enzyme activity. Noteboom and Mueller (1966) suggest that these drugs may interfere with polymerase synthesis, since there is a similar rapid depression of RNA polymerase activity in tissue culture cells treated with cycloheximide or puromycin, two known inhibitors of protein synthesis (Summers *et al.*, 1966). The action of cycloheximide and puromycin on polymerase can be prevented by including in the culture medium mercaptopyridethyl-benzimidazole, a compound which is known to inhibit RNA synthesis without affecting polymerase activity. Summers and Mueller (1968) concluded that demonstration of the action of cycloheximide and puromycin on polymerase depends on continued RNA synthesis, and they suggested that these two antibiotics reduce the polymerase content of the cells because RNA synthesis in mammalian cells involves utilization of a critical, rate-limiting protein. Penman (1967) has pointed out that the effects of cycloheximide and puromycin are demonstrably different. In cultured cells, it takes about an hour to retard ribosomal RNA synthesis, first affecting the conversion of 45 S to subsequent products; the most likely reason for this slow action is lack of ribosomal protein. Puromycin acts more rapidly, probably because it results in the flooding of the nucleus with incomplete protein chains. The alkaloids vincristine and vinblastine cause rapid parallel inhibition of RNA and protein synthesis in sensitive leukemic leukocytes, but it has not been possible to determine whether the action on protein formation is responsible for the reduced RNA formation (Cline, 1968).

Actinomycin D and occasionally some of the other inhibitors have been widely used to suppress RNA synthesis in the course of studying regulation of protein biosynthesis by mammalian cells. The selective action of such inhibitors should be recognized. Actinomycin D at low doses prevents the formation of ribosomal RNA by inhibiting synthesis of its 45 S precursor (Penman *et al.*, 1968), whereas 4 S, 5 S, and heterodisperse RNA continue to be transcribed at normal or near normal levels (Perry *et al.*, 1964; S. J. Martin and Brown, 1967; Perry and Kelley, 1968b). This selective action on ribosomal RNA appears to be shared by other inhibitors such as aflatoxin (Lafarge *et al.*, 1966), cordycepin (Siev *et al.*, 1969), fluorouracil (DeKloet and Strijkert, 1966), benzimidazole riboside (Sirlin and Loening, 1968), and vinblastine (Wagner and Roizman, 1968). In order to inhibit synthesis of the other RNA species, much larger doses of actinomycin have to be used. In the case of liver RNA synthesis, the selective action on ribosomal RNA may occur because actinomycin is preferentially deposited in the chromatin surrounding the nucleoli (Ro *et al.*, 1966). It should also be noted that, following injection of actinomycin into intact animals, there are wide differences in uptake by different tissues, and that inhibition of uridine incorporation by the RNA of these tissues is correlated with the levels of the drug reached in each tissue (Schwartz *et al.*, 1968a,b). Consequently, dosages of actinomycin that are appropriate for suppressing ribosomal or total RNA synthesis in the liver are not necessarily optimal for these effects in other tissues. The period over which the drug remains inhibitory may also be shorter than suspected; Korner (1964) found that growth hormone could stimulate incorporation of label into rat liver RNA 12 hours after administration of a large dose of actinomycin D. Finally, it should be noted that evidence from bacterial studies has led to the suggestion that levels of actinomycin sufficient to suppress messenger RNA synthesis for one protein may not inhibit the messengers for all proteins made by the cell, and may even stimulate formation of some proteins (Pollock, 1963; Kadowaki *et al.*, 1966). However, Moses and Sharp (1966) point out that the data can also be interpreted in terms of different rates of decay of the messenger RNA for individual proteins following suppression of RNA synthesis by actinomycin. The applicability of this conclusion to mammalian cells will become evident when we consider later the stability of messenger RNA in the cytoplasm of the mammalian cell. It is nevertheless noteworthy that

8-azaguanine can inhibit induction by cortisone of some liver enzymes (Kvam and Parks, 1960), but not induction of liver tyrosine transaminase (Levitan and Webb, 1968). Presumably, in the presence of the inhibitor, messengers are made which in most cases fail to code for functional proteins, but sometimes the product is biologically active even if some constituent amino acids are not correct.

d. Fate of Nuclear RNA. The preceding description of RNA species formed in the nucleus leaves unspecified the mechanism by which they are transferred to the cytoplasm. In the case of ribosomal RNA, this becomes a question of what path to the cytoplasm is taken by the ribosomal subunits assembled in the nucleolus. Formation of tRNA appears to be preceded by synthesis of a slightly larger unmethylated precursor which undergoes maturation to the 4 S form (Darnell, 1968). The most intriguing problem is, however, the relationship of heterodisperse nuclear RNA to messenger RNA and the mechanism that determines the appearance of the latter but not the former in the cytoplasm. An early suggestion (Scherrer and Marcaud, 1965) was that, just as rRNA and tRNA are fashioned from larger precursor RNA species, so messenger RNA is formed from the heterodisperse nuclear RNA. This suggestion has been challenged by Attardi *et al.* (1966) on the basis that the cytoplasmic RNA is unrelated to the HnRNA kinetically or by base composition. Both Attardi and Attardi (1967) and Penman *et al.* (1968) describe a heterodisperse RNA in the cytoplasm that differs from polyribosome-associated messenger RNA in location and in kinetics. Both groups of investigators agree that this RNA does not have the characteristics of nuclear heterodisperse RNA, and Attardi and Attardi (1967) suggest that it may be secreted by the mitochondria, and may provide a template for synthesis of cell membrane structural proteins, which is the reputed function of the DNA of mitochondria. The data of Penman *et al.* (1968) would not seem to support this thesis, since, unlike the Attardis, they do not find this heterogeneous RNA specifically associated with membranes.

Despite these uncertainties about the nature and function of heterodisperse nuclear and cytoplasmic RNA, and their relationship to messenger RNA, it is still undoubted that messenger RNA is formed in the nucleus and passes to the cytoplasm. Several pieces of evidence suggest that a carrier may be involved in the exit of nonribosomal RNA from the nucleus. L. Goldstein and Prescott (1968) have used nuclear transplantation in *Amoeba proteus,* a unicellular organism, to demonstrate that protein migrates rapidly both out of and into the nucleus; they consider that their evidence is against such protein being histone in nature. Several authors have identified nuclear protein particles of some 30 to 45 S size associated with rapidly labeled RNA (Spirin, 1964; Samarina *et al.,* 1966; Parsons and McCarty, 1968; Köhler and Arends, 1968a,b; Monneron and Moulé, 1968). Georgiev and his colleagues (Samarina *et al.,* 1968; Georgiev, 1968), using ribonuclease inhibitor to preserve RNA associated with such particles, have isolated from nuclei units of such material varying from 30 S to 200 S in size. They named the basic 30 S unit of protein an "informofer," and suggested that several of these might be strung out along a large strand of DNA-like RNA in the nucleus, like polyribosomes in the cytoplasm. This interpretation is supported by Molnár (1969). Georgiev (1968) suggested that these informofers function in transporting the DNA-like RNA from the chromosomes to the cytoplasm, and may also be involved in cleavage of newly formed RNA. On the basis of hybridization studies, he concluded that almost all the DNA-like RNA of the nucleus is attached to informofers, and consequently that some of the heterodisperse nuclear RNA is present in this form, even though it does not leave the nucleus. In the cytoplasm, similar RNA-protein complexes have been observed (Spirin, 1966) and can be released by disrupting the polysomes with EDTA (Henshaw, 1968; Perry and Kelley, 1968a; Temmerman and Lebleu, 1969), or by the absence of Mg^{2+} (Armentraut and Weisberger, 1968), or by dissolving the microsomal membranes with deoxycholate (Samec *et al.,* 1968). It has been suggested by Spirin (1966) that this represents passage into the cytoplasm of a protein-mRNA complex, and that the protein accompanying the mRNA may regulate protein synthesis at the translational level in the cytoplasm. Henshaw (1968) believes that his recovery of the RNA-protein complex from the polysomes lends support to this theory. This is an interesting hypothesis, since we shall see later that there is evidence that membrane-attached polysomes are coded to make proteins for secretion, whereas polysomes lying free in the cytoplasm make a different set of proteins. It is thus puzzling to understand how the messenger secreted from the nucleus is directed toward either the ribosomes attached to membranes or to the free ribosomes according to the type of protein it is coded to make. The presence of a protein accompanying the mRNA would allow some sort of specificity in directing the template toward the appropriate set of ribosomes. Finally, in evaluating all these studies on RNA-protein complexes, it is as well to remember the demonstration by Girard and Baltimore (1966) that artifactual association of RNA and protein can occur on gradients.

Recently, the passage of information to the cytoplasm has taken a new direction with the claim from several laboratories that DNA other than mitochondrial DNA is a normal constituent of the cytoplasm of the eukaryote cell. A special type of DNA that incorporates thymidine actively has been described in liver microsomes by Bond *et al.* (1969) and by Schneider and Kuff (1969), and Bell (1969) has identified metabolically active DNA in the general cytoplasm from chick embryo preparations. It is implied that this DNA is the form in which the message is carried from the nucleus, a theory which implies that transcription by RNA polymerase must occur in the cytoplasm. Attempts to demonstrate RNA synthesis by preparations of cell cytoplasm, and particularly microsomes, have failed to provide evidence of the presence of an enzyme capable of utilizing DNA as a template on which to form heteropolynucleotides (Baltus *et al.*, 1965; Wykes and Smellie, 1966; Schneider and Kuff, 1969). The status of messenger DNA must therefore await further work.

2. The Cytoplasmic Mechanism for Protein Synthesis

As discussed above, messenger RNA, tRNA, and ribosomes are formed in the nucleus but perform their function of synthesizing proteins in the cytoplasm. Here, these nuclear products are assembled into polyribosomes which interact with cytoplasmic components (free amino acids, activating and other enzymes involved in protein synthesis, and nucleotide cofactors) in order to fabricate proteins. In this section we shall first summarize our knowledge of polysome structure and intracellular location, then discuss the mechanism of protein synthesis, finally dealing with direct inhibitors of protein synthesis.

a. Polysome Structure and Intracellular Location. In 1963, several laboratories demonstrated by electron microscopy and by sucrose gradient separation that ribosomes and messenger RNA are arranged within cells in aggregates. These observations have been followed up by surveys of many tissues for the presence and functional activity of such polyribosomes (polysomes). In addition to numerous papers showing the presence of polysome aggregates in rat liver and in bacterial cells, these investigations have demonstrated polysomes in primate liver (Means and Baker, 1969) in reticulocytes (e.g., Hori *et al.*, 1967), in spleen cells (La Via *et al.*, 1966; Talal and Kaltreider, 1968), in skeletal muscle cells (Heywood *et al.*, 1967), in cardiac muscle (Earl and Morgan, 1968), in brain tissue (e.g., Appel *et al.*, 1967; Vesco and Giuditta, 1968), in human placental tissue (Laga *et al.*, 1969), in uterine muscle (Teng and Hamilton, 1967), in the cells of the anterior pituitary lobe (Adiga *et al.*, 1968; Gospodarowicz and Laporte, 1968), in the lens of the eye (Schoenmakers *et al.*, 1967), in hepatoma cells (Plagemann, 1968a), in the mammary gland (Gaye and Denamur, 1968), in collagen-secreting fibroblasts (e.g., B. Goldberg and Green, 1967), in HeLa cells in culture (e.g., Borun *et al.*, 1967b), in Chang liver cells in culture (Eliasson *et al.*, 1967), in Chinese hamster cells in culture (Steward *et al.*, 1968), in virus-infected L-cells in culture (Prevec and Graham, 1966), in chick embryo cells (Humphreys and Bell, 1967), in slime molds (Mittermayer *et al.*, 1966), in yeast (L. Marcus *et al.*, 1963), in protozoans (Whitson *et al.*, 1966), in mitochondria (Küntzel and Noll, 1967), in germinating plant seeds (A. Marcus *et al.*, 1966; Jachymczyk and Cherry, 1968), in plant roots (C. Y. Lin and Key, 1966), in plant leaves (Lyttleton, 1967), in chloroplasts (M. F. Clark, 1964; Mehta *et al.*, 1968), in plants infected with tobacco mosaic virus (Kiho, 1968), and in other biological materials.

A characteristic sucrose density separation of polysome classes in rat liver is shown in Fig. 9. The profile exhibits separate peaks corresponding to monosomes (80 S), disomes (120 S), trisomes, tetrasomes, pentasomes, and thereafter the separation becomes indefinite as the profile merges into higher polysomes. On specially prepared gradients it is also possible to identify two ribosome subunit peaks (60 S and 40 S) at the top of the gradient; there is sometimes an additional identifiable peak at 105 S, which may represent a messenger strand with a monosome and one subunit attached to it (Pfuderer *et al.*, 1965). The disome peak in liver cell preparations is preferentially increased under some conditions

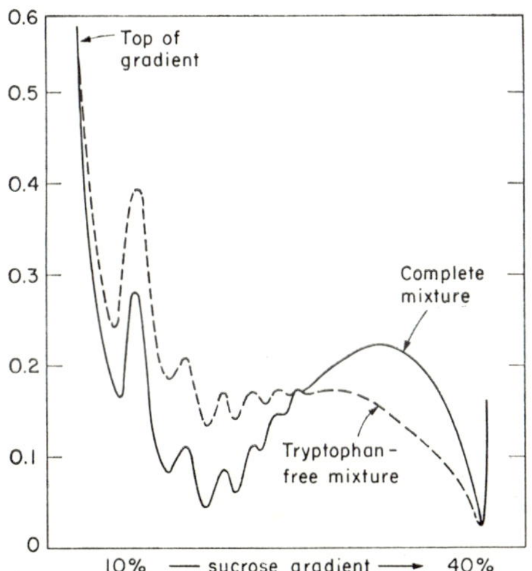

FIG. 9. Sucrose density gradients of liver polysomes prepared by homogenizing rat liver and then removing the nuclei, mitochondria, and ferritin, the latter by means of a specific antiserum. Deoxycholate was added to the postmitochondrial supernatant, which was then put on a gradient of sucrose. One profile shows the relative abundance of different classes of polysomes in rats fed a nutritionally complete mixture of amino acids; the other profile is of the polysomes of rats given a similar mixture with tryptophan deleted. (From Drysdale and Munro, 1967.)

in which monosomes accumulate; this is probably due to physical aggregation of monosomes that have neither peptide chains nor messenger attached to them. The phenomenon has been shown to occur only with the monomers of some species but not of others (Reader and Stanners, 1967), and may depend on the presence of free SH groups on the monomers (J. H. C. Wang and Matheson, 1966). The identity of aggregate size for each peak can be confirmed by electron microscopy. In the case of bacterial polysomes, Oppenheim *et al.* (1968) have provided gradient profiles and elegant micrographs of the particles in each constituent peak over a range beginning with the subunits and progressing in size up to aggregates of 25 ribosomes attached to each messenger strand.

Both in intact tissue (Behnke, 1963; Andersson-Cedergren and Karlsson, 1967; P. Weiss and Grover, 1968) and after isolation (Pfuderer and Swartzendruber, 1966), electron microscopy reveals a helical structure made up of a messenger strand at a pitch of some 20°, with from 4 to 6 ribosomes attached for each complete turn. Wettstein *et al.* (1963) first defined the dimensions of the polysome units. Each ribosome has a diameter of some 200 Å, and from the center of one ribosome to the next on the polysome, the messenger extends for 90 nucleotides, or 30 codons. Since the ribosomes are usually packed closely, the size of the polysome aggregate is a measure of the length of the messenger strand. Thus, a peptide chain containing about 560 amino acids (e.g., serum albumin) would be coded for by a messenger that can carry 19 ribosomes. This relationship of aggregate size to messenger dimensions has been confirmed for several proteins, either by identifying the protein attached to one class of polysome, or by incubating separated polysomes of a given size with a cell-free system in order to incorporate radioactivity into the nascent peptide chains and then identify the product. In one or other way, it has been concluded that histones are made on small polysomes (Borun *et al.*, 1967b), myosin is made on larger polysomes of 50–60 ribosomes per strand (Heywood *et al.*, 1967), the polycistronic messenger of tobacco mosaic virus requires 60–80 ribosomes (Kiho, 1968), but there is dispute as to whether the polysomes for collagen synthesis carry about 100 ribosomes (Kretsinger *et al.*, 1964) or are much smaller and possibly of various lengths (B. Goldberg and Green, 1967; Speakman, 1968). Furthermore, Ragnotti *et al.* (1969) have found that nascent microsomal cytochrome *c* reductase is attached to liver polysomes of varying sizes.

Changes in the frequency of different classes of polysomes have been repeatedly used to study control of protein synthesis in mammalian cells. In such experiments, it is usually desirable to obtain a polysome profile that represents as closely as possible the size-distribution of the polysomes in the intact cell. Ideally, this demands recovery of polysomes quantitatively and without degradation. The earliest of these methods, the procedure of Wettstein *et al.* (1963) for obtaining C-ribosomes, has been extensively used to study changes in polysome profile. In this method, the cell is gently homogenized and the homogenate is then spun for 10 minutes at 10,000 *g* to obtain a postmitochondrial supernatant fraction. This fraction is then treated with deoxycholate to liberate membrane-bound polysomes, and the resulting solution is sedimented centrifugally on a discontinuous gradient through a layer of 0.5 *M* sucrose into a 2 *M* sucrose, where the C-ribosomes form a pellet. The polysomes are finally resolved on a linear gradient. This procedure provides a selected polysome profile that is not representative of the profile in the intact cell for two reasons. First, the centrifugation conditions for obtaining a postmitochondrial supernatant fraction were devised at a time when cells were exhaustively homogenized to prepare microsomes and these conditions of homogenization shear the endoplasmic reticulum of the intact cell into quite small microsomal vesicles. The gentler disruption of the mammalian cell necessary to obtain intact polysomes results in larger fragments of endoplasmic reticulum, many of which sediment with the mitochondria. Howell *et al.* (1964) drew attention to the large loss of membrane-attached ribosomes in the nuclear and mitochondrial debris discarded under such conditions, and concluded that it represented a different spectrum of particles since only monosomes and disomes were recovered when deoxycholate was added to this fraction. Blobel and Potter (1966b) and Lawford *et al.* (1966) have demonstrated that this result is due to ribonuclease present in the microsome-mitochondrial fraction; if liver cell sap containing ribonuclease inhibitor is added to the microsome-mitochondrial fraction before treatment with deoxycholate, a satisfactory polysome spectrum is now obtained. In order to achieve a complete recovery of polysome species from the cell, the nuclei and debris should first be briefly spun off from the homogenate, and then the postnuclear supernatant fraction should be treated with deoxycholate. Blobel and Potter (1967c) recommend adding liver cell sap to the homogenizing medium in order to eliminate ribonuclease action, and we confirm that this causes a small improvement in the polysome patterns obtained; it should, however, be remembered that liver cell sap contains ferritin, which can add to the absorption of the monosome peak. The second source of selection of polysomes is due to spinning them through a discontinuous sucrose gradient to harvest the C-ribosomes. This procedure was adopted by Wettstein *et al.* (1963) because ferritin molecules present in tissues such as liver tend to contaminate the monosome peak in polysome gradients unless steps are taken to remove the ferritin, which absorbs ultraviolet light strongly because of its high iron content (Drysdale and Munro, 1965). By using the discontinuous sucrose gradient, Wettstein *et al.* (1963) were able to keep most of the ferritin suspended in the upper layer of sucrose, but they also lost the ribosome subunits and most of the monosomes. In order to overcome ferritin contamination without loss of monosomes, Drysdale and Munro (1967) added antiferritin serum to the tissue homogenate before spinning down the debris, and were then able to apply the deoxycholate-treated supernatant preparation directly to a linear sucrose gradient for resolution of the polysome profile (Fig. 9).

Other methods of cell disruption and of ribonuclease inhibition have been proposed. Many cells, including most bacteria, are more difficult to break than liver cells. Flessel *et al.* (1967) recommend lysozyme to disrupt bacteria gently, while Godson and Sinsheimer (1967) describe a new detergent Brij-58 for cell dissolution following lysozyme treatment. In the case of *Azotobacter vinelandii,* Oppenheim *et al.* (1968) prefer osmotic shock, which they contrast with several other methods of cell disruption. They point out that even several passages of the isolated bacterial polysomes through a fine pipette is sufficient to disaggregate them through shearing forces. For suspensions of mammalian cells, such as tissue culture cells or lymphocytes, the detergent Nonidet P-40 liberates cytoplasmic polysomes without the necessity of homogenization (Borun *et al.*, 1967a). In order to inhibit ribonuclease more effectively, Haschemeyer and Gross (1967) add yeast RNA to tissue homogenates. Rowley and Morris (1966) recommend adding heparin, a known ribonuclease inhibitor; heparin dissociates free ribosomes but not ribosomes in polysome array (A. O. A. Miller, 1968). The use of bentonite to adsorb ribonuclease is undesirable since it may selectively precipitate monosomes from the polysome profile (Tester and Dure, 1966), although a specially purified bentonite is said not to distort the polysome profile (Watts and Mathias, 1967). As mentioned above, Blobel and Potter (1967c) add extra liver cell sap at the time of homogenization; we have also obtained somewhat fewer monosomes by this procedure. Earl and Morgan (1968) found with cardiac muscle that the inclusion of EDTA and NH_4Cl in the homogenizing medium reduced nuclease activity and yielded polysomes that incorporated amino acids into protein more vigorously. We have confirmed the action of EDTA for human placenta (Laga *et al.,* 1969). Finally, it should be pointed out that monosomes tend to precipitate from polysome preparations unless the profile is spun quickly through the sucrose gradient after harvesting (Pronczuk *et al.*, 1969). A similar phenomenon in the case of lysed reticulocytes has been attributed to continuing peptide bond formation and ribosome migration even in the cold (Danon and Cividalli, 1968). These various precautions emphasize that study of polysome profiles should be carried out under carefully controlled conditions in order to obtain meaningful differences.

The study of polysomes is further complicated by the finding that some are attached to the membranes of the endoplasmic reticulum, whereas others lie free within the cytoplasm. Several authors have described procedures for separating free and membrane-attached polysomes, such as the procedure of Blobel and Potter (1967a,b,c). This involves preparation of the postmitochondrial fraction from the tissue homogenate, followed by centrifugation through a discontinuous gradient. The free ribosomes pass through 2 *M* sucrose and form a pellet at the bottom of the centrifuge tube, whereas the membrane fraction is held at the interface. It can be recovered and treated with deoxycholate in the presence of the ribonuclease inhibitor of cell sap, and the polysomes thus released can then be harvested. The total yields of ribosomes obtained by these procedures have been extensively tested by Blobel and Potter (1967a,b,c). We are not satisfied, however, that the conditions worked out for the centrifugation of free polyribosomes avoid the criticism of the original C-ribosome procedure of Wettstein *et al.* (1963), namely that, in order to avoid contamination with UV-absorbing ferritin particles, the free polysomes are passed through 2 *M* sucrose and may thus lose monosome particles with sedimentation properties similar to these of ferritin which is being excluded. Monosome abundance is a particularly sensitive indicator of alterations in cell polysome content, since loss of polysomes is usually accompanied by accumulation of monosomes. We have therefore devised the simple technique of treating the tissue homogenate with ferritin antiserum, followed by preparation of the postmitochondrial supernatant. Half of this is put directly onto a sucrose gradient, to give a profile of monosomes and other free ribosomes in the upper half of the gradient, with a large mass of undissolved microsomes at the bottom of the gradient (Fig. 10). The other half of the postmitochondrial supernatant is treated with deoxycholate to dissolve the microsomal membranes, and on a sucrose gradient yields the profile of the combined free and membrane-attached ribosomes (Fig. 10). It will be seen from these gradients that the monosome peak is much larger for the total ribosome population than for the free ribosomes, implying that many monosomes remain firmly attached to the endoplasmic reticulum membrane, and do not become dislodged when the end of translation of a messenger RNA has been reached. This picture differs from that presented for liver by Rizzo and Webb (1968), who found by the technique of Blobel and Potter that most of the monomers occur

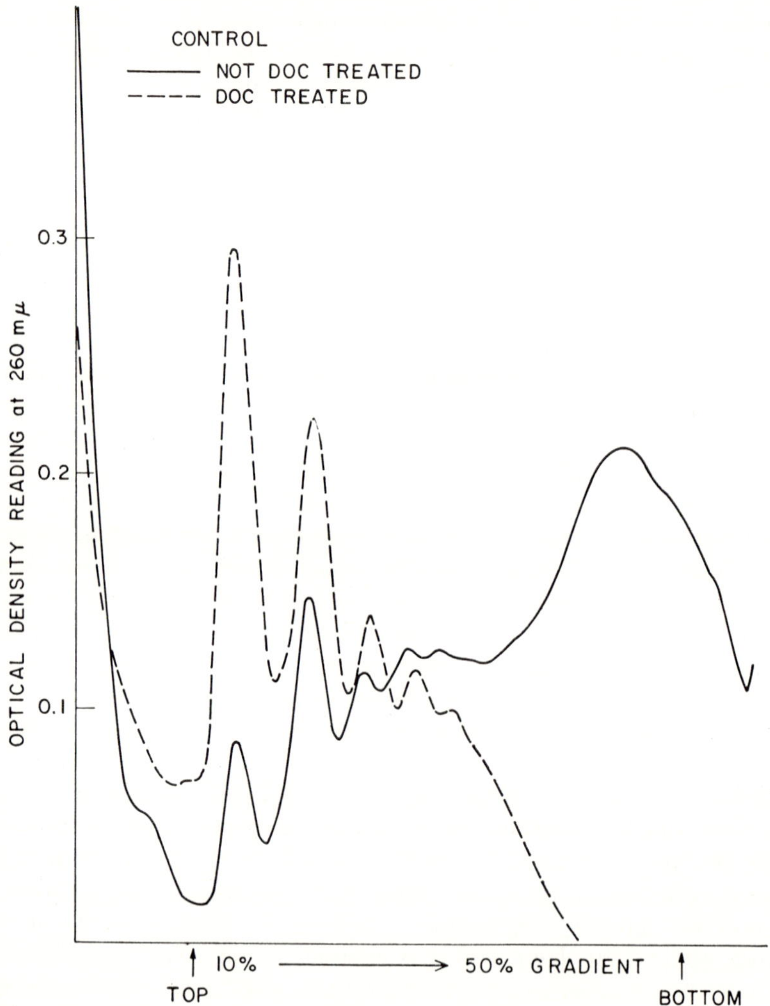

FIG. 10. Profiles on a sucrose gradient of free and total polysomes of the livers of adrenalectomized rats. The free ribosome pattern was obtained by putting the untreated postmitochondrial fraction of the liver on the sucrose gradient. The total ribosome fraction was prepared by putting the deoxycholate-treated postmitochondrial fraction on a similar gradient. The small abundance of polysomes is characteristic of fasting adrenalectomized rats. (From Enwonwu and Munro, 1969.)

unattached to membranes. On the other hand, it is supported by studies of free and membrane-attached polysomes in spleen, in which Talal and Kaltreider (1968) found many membrane-attached monosomes.

These conclusions agree with other evidence of a stable attachment of ribosomes to endoplasmic reticulum membrane. By applying a pyrophosphate solution to liver microsomes, Goswami *et al.* (1962) separated off an RNA species with the composition of the RNA in the smaller ribosome subunit, leaving RNA attached to the membrane which was richer in guanylic and adenylic acids. This was erroneously identified as a new species of membrane RNA, but in reality has a composition similar to the 28 S RNA of the larger mammalian ribosome subunit. In a more comprehensive study, Sabatini *et al.* (1966) showed that treatment of liver microsomes with increasing concentrations of EDTA first causes detachment of the smaller (40 S) ribosomal subunit from the larger subunit, followed at higher concentrations of EDTA by separation of the larger (60 S) subunit from the membrane. It has been suggested that the larger subunit is attached to the membrane by means of the nascent peptide chain (Chefurka and Hayashi, 1966), or by divalent cations (Sabatini *et al.,* 1966), but neither explanation is acceptable to Bennett and Hallinan (1968). The firm binding of the ribosomes to the membrane of the endoplasmic reticulum is also consistent with incorporation of precursors into the ribosomes. Within 1 or 2 hours after administration of a precursor of RNA, the specific activity of the RNA in the bound polysomes of the liver rises much more slowly than the specific activity of the free polysomes of that organ (Hallinan and Munro, 1965; Murty and Hallinan, 1968), but several hours later the specific activities of the two ribosome populations are similar (Bouvet and Moulé, 1964). This suggests that the newly synthesized ribosomes appear first in the pool of free ribosomes or their subunits, and over a period of a few hours equilibrate with the ribosomes attached to membranes. A similar conclusion from labeling studies on kidney has been drawn by Malt and LeMaitre (1967). It has been periodically reported (Sachs, 1958; Goswami *et al.,* 1962; Petrovic *et al.,* 1965; Bergeron-Bouvet and Moulé, 1966; Shapot and Pitot, 1966; Rodionova and Shapot, 1966; Tata, 1967a) that microsomal membranes contain an unusual RNA component; some authors claim that it is of low molecular weight (Gardner and Hoagland, 1968; King and Fitschen, 1968; Lèbre *et al.,* 1969) and is specifically associated with the smooth agranular membranes (King and Fitchen, 1968). Such membrane RNA has been attributed to degradation of ribosomal RNA during extraction from microsomes (Hoagland *et al.,* 1968) and to the presence of some residual ribosomes in smooth membrane preparations (Hallinan and Munro, 1964; Munns and Hallinan, 1968). In consequence, membrane RNA should be regarded as an unproved entity, probably produced as an artifact during membrane isolation.

The presence of two populations of ribosomes, membrane-attached and free, led Birbeck and Mercer (1961) to make the eminently sensible suggestion that those attached to membranes are engaged in making proteins for export by way of the membrane channels (serum proteins in the case of liver), whereas ribosomes lying free in the cytoplasm make proteins that are retained within the cell. With the emergence of techniques for separating the two populations of polysomes, it has been possible to test this hypothesis by identifying the proteins made by each species of polysomes. From evidence obtained both by *in vivo* and *in vitro* incorporation of labeled amino acids, membrane-bound polysomes would appear to be the exclusive site of serum albumin synthesis in the liver cell (Redman, 1968; Takagi and Ogata, 1968; Hicks *et al.*, 1969), a finding which confirms several earlier studies with microsomes (e.g., Campbell *et al.*, 1968). Ganoza and Williams (1969) have demonstrated that other serum proteins are also made by membrane-bound liver polysomes, and Hallinan *et al.* (1968) have identified membrane-bound polysomes as the exclusive site at which the insertion of carbohydrate onto glycoprotein apopolypeptides is initiated. In contrast to these secreted proteins, Ganoza and Williams (1969) have shown that liver cell sap proteins in general are made by free polysomes, and Hicks *et al.* (1969) have identified free polysomes as the preferential site of ferritin synthesis in the liver cell. However, Andrews and Tata (1968) have shown that proteins made by brain microsomes are not secreted through the microsomal membrane. A further discrepancy has been disclosed by Ragnotti *et al.* (1969), who have shown that both membrane-attached and free polysomes prepared from rat liver are able to incorporate labeled amino acids into the cytochrome *c* reductase found in liver microsomes. Although this is not a secreted protein, it is nevertheless located in the membranes of the endoplasmic reticulum. Some of these contradictions in the evidence obtained from cell-free systems arise because such systems are presently very inefficient in making proteins; the emergence of an *in vitro* system making and releasing appreciable amounts of protein would allow more rigorous evaluation of the relationship of free and bound polysomes to protein synthesis. If it is confirmed that different polysome populations specialize in the synthesis of different proteins, we are left with the problem of how the appropriate messenger reaches each group of ribosomes.

Some attention has also been paid to the mechanism by which proteins are secreted through endoplasmic reticulum membrane. It has been demonstrated that proteolytic enzymes fail to digest some of the nascent peptide chain on the ribosome (L. I. Malkin and Rich, 1967), and it has been concluded that the growing end of the peptide is buried within the ribosome and emerges from the apex of the large subunit. If this view is correct, in the case of membrane-attached ribosomes the emerging peptide would erupt at the point of attachment of the ribosome to the membrane and presumably would pass through. It has been suggested by Eylar (1965) that all secreted proteins are glycoproteins; while this is largely true, serum albumin is an important exception, since pure albumin is free of carbohydrate (T. Peters, 1969). The process of secretion would thus not appear to depend on initiation of a polysaccharide chain.

The function of the membranes of the endoplasmic reticulum appears to be that of segregating proteins that are subsequently secreted from the cell. This has been elegantly demonstrated by autoradiography in the case of pancreatic exocrine cells. Caro and Palade (1964) have shown that rough endoplasmic reticulum at one pole of the cell forms the proteins which then pass through the smooth reticulum of the Golgi region and eventually collect in the zymogen granules at the apex of the cell, from which they are discharged into the pancreatic juice. In rat liver, the proportions of rough and smooth reticulum are 60% and 40%, respectively (Loud, 1968). Studies of the quantities of RNA and of phospholipid in the liver cells of various mammals (Munro and Downie, 1964) suggest that larger species may have greater proportions of smooth membrane and smaller amounts of rough membrane with attached ribosomes (see also Chapter 25 in Volume III).

In addition to polysomes and intact monosomes, the cytoplasm of mammalian cells contains ribosomal elements in the form of free subunits. These have been identified on the basis of particle size, of contained RNA species, and of ribosomal protein content, in liver cells (Henshaw *et al.*, 1965; Wunner *et al.*, 1966), in reticulocytes (Bishop, 1966), in plasma tumor cells (Kedes *et al.*, 1966), in HeLa cells (Girard *et al.*, 1965; Joklik and Becker, 1965), in KB tumor cells (Ristow and Köhler, 1966), and in ascites tumor cells (Hogan and Korner, 1968). In the case of the liver, ribosomal subunits can be recovered as a postmicrosomal pellet on prolonged centrifugation of the conventional preparation of cell sap (Munro *et al.*, 1964), thus indicating that the customary cell sap preparation contains some ribosomal RNA, as well as tRNA. The abundance of this postmicrosomal pellet and the relative proportions of 60 S and 40 S subunits contained in it vary according to the previous diet of the animal (Munro *et al.*, 1964; Munro, 1968). The postmicrosomal pellet also contains high concentrations of several amino acid-activating enzymes and can incorporate methionine and leucine into an acid-precipitable product by an ATP-dependent reaction that is resistant to ribonuclease (Hird *et al.*, 1964). It is not known whether these properties of the postmicrosomal fraction are related to the function of its contained ribosomal subunits, nor is it known whether they have any biological importance.

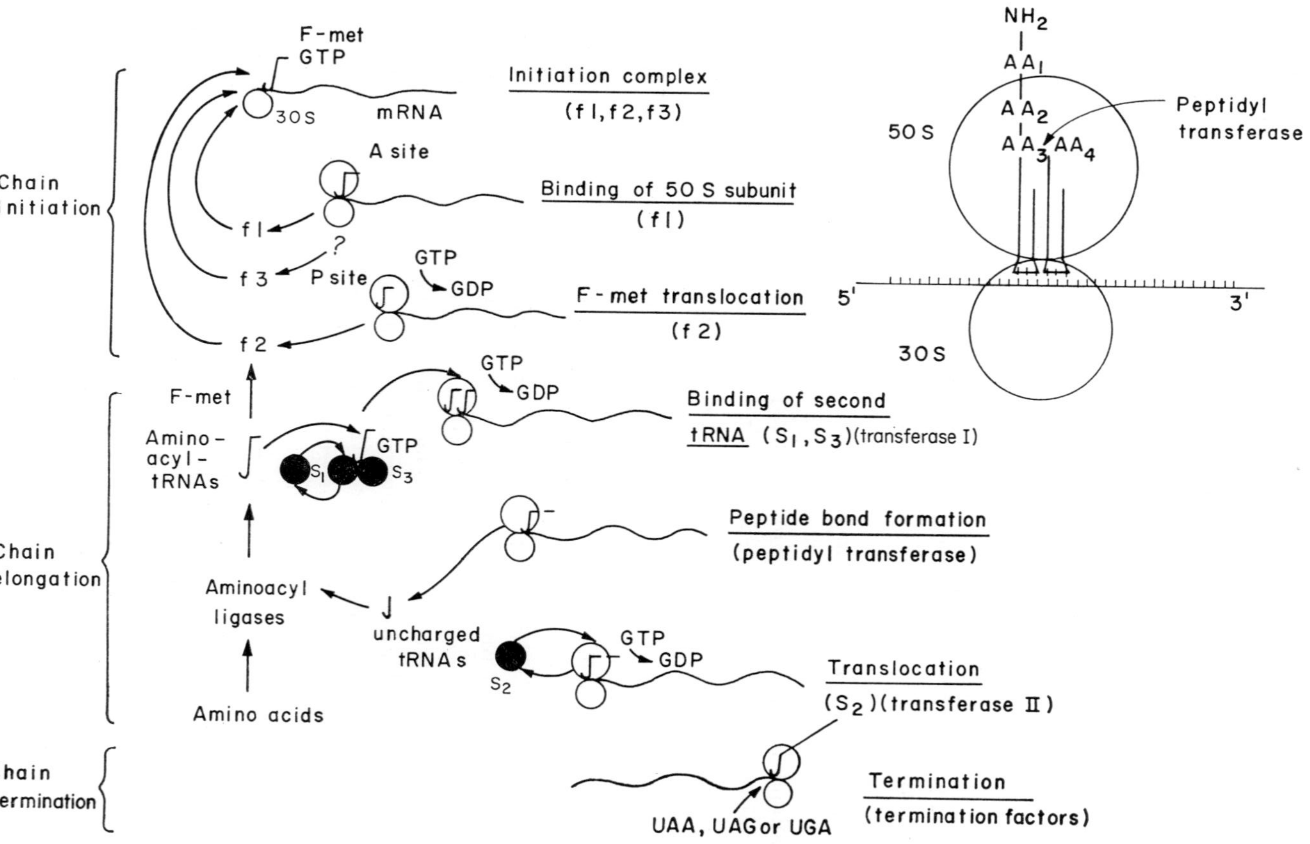

FIG. 11. (For legend see opposite page.)

Labeling studies show that ribosomal RNA first enters the cytoplasm as subunits, then passes into the polysomes, and finally appears in the free ribosome fraction (Girard *et al.*, 1965; Joklik and Becker, 1965; Ristow and Kohler, 1966; Hogan and Korner, 1968; Kabat and Rich, 1969). In the case of bacteria, it has been concluded that ribosomes present on polysomes undergo dissociation into subunits when they are released from the messenger at the end of each round of translation, and that the subunits reassociate again along with messenger for initiation of a new peptide chain (Schlessinger *et al.*, 1967; Kaempfer *et al.*, 1968). The free 30 S subunit of *E. coli* is known to bind the initiator tRNA for formylmethionine, followed by binding of the 50 S subunit (Nomura and Lowry, 1967). In the case of mammalian cells, the presence of a monosome population has frequently been demonstrated, and in many instances it has been possible to show that the monosomes are in equilibrium with the polysomes, as, for example, when reticulocyte polysomes break down when tryptophan is removed from the incubation medium and reaggregate when tryptophan is restored (Hori *et al.*, 1967). Colombo *et al.* (1968) consider that ribosome subunit formation is an obligatory step in reassembly of monosomes into polysomes. They have found that fluoride can prevent dissociation of monosomes into subunits, and that the ensuing depletion of the subunit pool is followed by cessation of protein synthesis in the reticulocyte. Removal of fluoride is followed by the reappearance of subunits and then by restoration of protein synthesis. In the case of bacteria, Subramanian *et al.* (1968) have suggested that one of the protein factors required for peptide chain initiation may regulate breakdown of monosomes into subunits in order to prime them for reutilization. However, Kabat and Rich (1969) have concluded from a study of ^{32}P-labeling of ribosomes in embryonic skeletal muscle that single ribosomes do not participate in the subunit-polysome cycle. They can offer no evidence as to the function of free ribosomes in cells.

b. Mechanism of Peptide Bond Formation. The mechanism of protein synthesis is most completely understood for bacteria (for review, see Lengyel and Söll, 1969). On the other hand, the evidence for protein formation in mammalian cells is defective at several points. Figure 11

FIG. 11. Diagram to illustrate the mechanism of protein synthesis in bacterial and mammalian cells. The factors initiating peptide chains are understood only in the case of the bacterial system. For this illustration, the initiation scheme proposed by Hershey *et al.* (1969) has been chosen. The mechanism of subsequent elongation of peptide chains depends on enzymes differently named for the bacterial system (S_1, S_2, S_3) and for the mammalian system (transferase I and transferase II, or translocase). Chain termination is not well understood. The upper right-hand diagram shows the positions of messenger RNA and the two tRNA molecules on the ribosome.

is an attempt to present the current status of our knowledge of protein synthesis in bacteria and its counterpart in mammalian cells, naming the enzymes separately for each system. The first step in protein formation in both bacterial and metazoan cells is activation of the carboxyl residues of the amino acids that are to be used to form peptides. This process is generally conceived as having two steps, first the formation of an aminoacyl adenylate, followed by the reaction of this adenylate with tRNA to yield the aminoacyl-tRNA. Loftfield and Eigner (1968, 1969) emphasize that the reactions are not divided in this way and that the process should be considered as involving amino acid, ATP, enzyme, general base, and tRNA, and the products as being aminoacyl-tRNA, AMP, pyrophosphate, free enzyme, and protonated base. The enzymes involved are often called amino acid-activating enzymes but are more correctly designated aminoacyl-tRNA ligases or aminoacyl-tRNA synthetases. Each such enzyme is specific for a given amino acid and its tRNA. These enzymes are highly specific and do not readily misplace amino acids. For example, Baldwin and Berg (1966) prepared pure isoleucyl-tRNA synthetase and challenged it with valine; although valine-enzyme-AMP complex was formed, it dissociated when the tRNA specific for isoleucine was added to the system. Thus no spurious valine-carrying isoleucyl-tRNA was formed. This provides a safeguard against errors in protein structure due to substitution of one natural amino acid for another. A few exceptional cases of partial mischarging of tRNA by aminoacyl ligase from special sources have since been recorded (Arca *et al.*, 1967; Holten and Jacobson, 1969). In addition, some synthetic analogs of amino acids are activated. For example, aminoacyl synthetases prepared from *Neurospora* can charge the transfer RNA acceptors for methionine and for phenylalanine with the analogs ethionine and *p*-fluorophenylalanine, respectively (Shearn and Horowitz, 1969). On the whole, however, errors in protein structure due to erroneous insertion of the wrong amino acids attached to tRNA is generally conceded to be a rare event.

For many amino acids, there is more than one tRNA species. These isoaccepting tRNA species can arise in two ways. First, they can occur through *degeneracy* of the code, so that different isoaccepting tRNA's can have different anticodons corresponding to dissimilar codons on the messenger RNA strand. This degeneracy of the triplet code for amino acids is possible because mathematically the four nucleotide bases of the code can be arranged in 64 different triplets. These possible triplets have now been equated with individual amino acids out of the 20 found in proteins and are listed in Table I of Chapter 24 in Volume III of this treatise. Those amino acids that are present in proteins in highest concentration (e.g., leucine) show the most extensive degeneracy, whereas the least abundant amino acid, tryptophan, has only one codon. Second, isoaccepting tRNA's can have identical codons, but may still differ in other aspects of their structure so that they can be separated chromatographically. These have been termed *redundant* tRNA's (Bergquist *et al.*, 1968) and presumably arise by duplication of the tRNA gene, followed by mutations that cause changes in the secondary structure of the tRNA copies without affecting either the anticodon or the capacity for recognition of the aminoacyl-tRNA ligase (Jukes, 1967). Muench and Safille (1968) list 56 separate tRNA species chromotographically isolated from *E. coli,* including five species of tryptophanyl-tRNA, for which only one codon is believed to exist. In a mouse tumor, Yang and Novelli (1968) also observed more tRNA species than there are codons for lysine, aspartic acid, tyrosine, and methionine.

The question arises, can all the tRNA species for one amino acid be charged by a single aminoacyl synthetase? Some of the evidence favors a single enzyme capable of charging different tRNA species for the same amino acid. Thus purified seryl-tRNA synthetase from *E. coli* was reacted with three tRNA species for serine, two of which were for codons with totally unrelated sequences; in spite of this, all three tRNA species were equally well charged by the synthetase (Sundharadas *et al.*, 1968). The lack of importance of the anticodon for recognition of the tRNA by its ligase is further emphasized by the studies of Carbon and Curry (1968). By partial deamination with nitrous acid, they transformed the anticodon CCU of glycyl-tRNA into UCU, which nevertheless continued to be charged with glycine. Presumably portions of tRNA molecule other than the anticodon determine its specificity for charging with a given amino acid. This property must be shared by some isoaccepting tRNA's, as demonstrated by the studies on serine cited above. Nevertheless, there is evidence from studies of leucyl and alanyl-tRNA charging by ligases from rabbit tissues (Strehler *et al.*, 1967), leucyl-tRNA activation by *E. coli* enzymes (Yu, 1966) and leucyl-tRNA charging by plant aminoacyl-tRNA ligases (M. B. Anderson and Cherry, 1969) that a single enzyme may not charge all isoaccepting tRNA species.

Once the tRNA has been charged with an amino acid, it can direct that amino acid to its proper site in the peptide chain identified by the appropriate codon on the messenger strand. Consequently, two tRNA species with different anticodons for the same amino acid will go to different points in the messenger sequence; this has been verified for insertion of leucine into hemoglobin by two different leucyl-tRNA's (Weisblum *et al.*, 1965). The occurrence of multiple forms of tRNA for carrying a single amino acid suggests a means by which individual proteins might be suppressed. If, say, one of the tRNA's for leucine is not made by a cell or is not recognized by the ligase, all proteins having this codon for leucine in their messenger RNA's will fail to be completed. Evidence relating to this type of control will be discussed later in this chapter.

The aminoacyl-tRNA's are now ready to carry out protein synthesis on the ribosome, which provides structural support for the series of enzymic reactions occurring in peptide bond formation and is also a carrier of the enzymes involved. The intact ribosome has two loci to which tRNA binds, namely an A (aminoacyl) site to which the incoming charged tRNA adheres, and a P (peptidyl) site into which the tRNA migrates once it has acquired the growing peptide chain. According to the Bretscher (1968) model of peptide chain elongation, there is continuous movement of the larger subunit on the smaller subunit and messenger. In the case of bacterial protein synthesis, it has been established that the initiator tRNA undergoes reactions distinct from those of tRNA during subsequent peptide chain elongation. Bacterial peptide chain formation appears to begin with the assembly of an initiating complex consisting of the 30 S ribosomal subunit, messenger RNA, formylmethionyl-tRNA, three protein initiation factors (f-1, f-2, and f-3), and GTP (Nomura and Lowry, 1967). The formylmethionyl-tRNA binds to the AUG initiator codon. The next event is the completion of the 70 S ribosome through binding of a 50 S subunit to the initiating complex. Coincident with this formation of the 70 S ribosome-messenger complex, Hershey *et al.* (1969) find that there is release of the factor f-1, which is now free to fertilize another initiating complex. At this point, the formylmethionyl-tRNA is still at the A site on the ribosome. It is then translocated to the P site by a reaction involving initiating factor f-2 and energy from hydrolysis of GTP (Kolakofsky *et al.*, 1968). Hershey *et al.* (1969) think that f-2, like f-1 before it, now separates from the ribosome since neither factor can be recovered from the 70 S ribosome, but Mangiarotti (1969) claims that failure to find these factors in 70 S ribosomes is due to technical losses in preparation. Presumably f-3, which serves especially in the binding of messenger to the 30 S ribosome (Iwasaki *et al.*, 1968), is also discharged at the end of initiation.

A second series of protein factors are responsible for adding subsequent aminoacyl-tRNA species during chain elongation. Three such factors have been identified in bacteria, named S_1, S_2, and S_3 in the nomenclature of Lengyel (Ono *et al.*, 1968). The S_1 and S_3 factors begin by complexing with GTP. This complex then reacts with any aminoacyl-tRNA (except formylmethionyl-tRNA), the product formed being a complex of S_3,

GTP, and aminoacyl-tRNA, with release of S_1. This complex then binds to the ribosome so that the incoming aminoacyl-tRNA is inserted on the ribosome at the A site, with cleavage of GTP. The 50 S subunit of the ribosome contains peptidyl transferase (peptide synthetase), an enzyme that now catalyzes peptide bond formation between the free carboxyl group of the formylmethionine (or of the growing peptide chain) on the P site and the amino group of the entering aminoacyl-tRNA on the A site. This is followed by translocation of the newly made peptidyl-tRNA from the A site to the P site, with concurrent movement of the messenger in the same direction, and elimination of the uncharged tRNA from the P site. This movement is catalyzed by the enzyme S_2, with energy derived from breakdown of GTP. Finally, on reaching the codons UAA, UAG, or UGA, the peptide chain is terminated and released with the aid of termination factors (Capecchi, 1967; Scolnick *et al.*, 1968). The formylmethionine inserted at the start of peptide chain synthesis is then excised from the finished protein molecule by a deformylating enzyme and an aminopeptidase (Livingston and Leder, 1969).

The picture of protein synthesis in mammalian cells is much less complete. The nature of chain initiation is disputed. Formylmethionyl-tRNA is present in mitochondria, but not in other parts of mammalian cells (Galper and Darnell, 1969; Reboud, 1969). Arnstein and Rahamimoff (1968) suggest that free aminoacyl tRNA species themselves may be initiators, and Mosteller *et al.* (1968) propose initiation by deacylated tRNA. However, Smith and Marcker (1970) provide convincing evidence of unformylate methionine as initiator, and Prichard *et al.* (1970) describe three protein initiating factors. Mammalian chain elongation also seems to be similar to that of bacteria (Moldave *et al.*, 1968). The mammalian binding enzyme is named transferase I; it forms a complex with aminoacyl-tRNA. This complex integrates with the ribosome at the A site in the presence of GTP. Moldave considers that this binding function of GTP does not require its hydrolysis. However, several authors (e.g., McKeehan *et al.*, 1969) have shown that GTP hydrolysis is an essential subsequent step before the bound aminoacyl-tRNA will form a peptide bond in either the bacterial or mammalian systems. Under the influence of peptidyl transferase (a constituent enzyme of the ribosome), the incoming amino acid on the A site now accepts the growing peptide chain from the peptidyl-tRNA on the P site, while the tRNA on the P site leaves the ribosome. Thus, in a random population of ribosomes, the growing peptide chain can be on the P site or the A site, depending on whether this transfer has been made. In practice, the majority of peptide chains in rat liver ribosomes are on the A site, suggesting that the action of peptidyl transferase is very rapid, whereas the subsequent reaction (translocation from the A to the P site) is slower. This last event requires the enzyme transferase II (translocase) and energy derived from the breakdown of GTP. This movement of the peptidyl-tRNA to the P site leaves the A site vacant for the next amino-acyl-tRNA complex to bind and thus results in repetition of the cycle of peptide transfer and translocation once more.

A large array of compounds has been described that can inhibit protein synthesis at one or other of its stages. A few of these inhibitors, such as puromycin and cycloheximide, are extensively employed to demonstrate whether or not protein synthesis is involved in various biological phenomena, whereas many other inhibitors have been used only rarely and are not commercially available. A partial list of inhibitors of protein synthesis will now be presented according to their presumed site of action. In some instances, the site of action is not known; these compounds will be listed at the end. In some other cases, the action on protein synthesis is secondary to inhibition of RNA formation; for example, nitrogen mustards inhibit uptake of amino acids into liver proteins, but the effective dose has to be substantial enough to suppress a large part of liver RNA synthesis (C. M. Clark *et al.*, 1958). These indirect inhibitors of protein synthesis will not be discussed here.

First, protein synthesis can be inhibited through the use of analogs of its substrate, the amino acids (see review by Richmond, 1962). Many of these analogs are now available, and a proportion can be activated by the aminoacyl ligases and become incorporated into the cell protein. For example, *p*-fluorophenylalanine can replace phenylalanine in bacterial alkaline phosphatase without affecting enzyme activity (Richmond, 1963), whereas replacement of histidine with triazolealanine results in phosphatase subunits that are unable to aggregate (Schlesinger and Schlesinger, 1967). With some amino acid analogs, it has been demonstrated that the analog not only replaces the natural amino acid in protein synthesis, but it also retards the rate of protein synthesis. It cannot, however, be assumed that the site of retardation is exclusively limited to amino acid activation. For example, β-2-thienylalanine, the structural analog of phenylalanine, reduces the rate of uptake of glycine-^{14}C into the protein of liver slices incubated with the inhibitor. This action can be prevented by adding phenylalanine to the medium at the time of inhibitor addition, but not if the phenylalanine is added subsequently (Munro and Clark, 1958). This finding suggests that the main competition between the natural amino acid and its analog probably occurs during transport into the liver cell, and that the subsequent toxic action of thienylalanine on liver protein synthesis may be unrelated to competition with phenylalanine for activation. Competitive inhibition of activation can be demonstrated using analogs that cannot undergo carboxyl activation. For example, amines (e.g., cysteamine), amino alcohols (e.g., isoleucinol), notably their aminoalkyl adenylates (e.g., isoleucinol-AMP), and amino acid esters (e.g., leucine ethyl ester) are known to act as specific competitive inhibitors in the activation of the corresponding amino acids in cell-free systems prepared from microorganisms (Cassio *et al.*, 1967; Shiftlet and Bucovaz, 1967; Owens and Bell, 1967; Owens and Blum, 1967).

Inhibitors can reduce the capacity of tRNA to accept amino acids by causing structural changes in this nucleic acid. Acceptor activity is reduced by the ready formation of complexes between tRNA and proflavin (Wérenne *et al.*, 1966; Weinstein and Finkelstein, 1967), with N-cyclohexyl-N'-β-(4-methylmorpholinium), ethyl carbodiimide p-toluene-sulfonate (Girshovich *et al.*, 1966), ethylenimine (Reid, 1968) and perhaps also with steffimycin (Reusser, 1969b). Actions limited to individual aminoacyl-tRNA species have also been described. The antibiotic borrelidin has been found to prevent specifically the formation of threonyl-tRNA through a specific inhibitory action on bacterial threonyl-tRNA-ligase (Hütter *et al.*, 1966; Poralla and Zähner, 1968; G. Nass *et al.*, 1969), and Okamoto (1967) has described factors extracted from *Bacillus stearothermophilus* that inhibit the attachment of valine and phenylalanine to tRNA.

Inhibitors are known for each of the subsequent stages of protein synthesis on ribosomes. As regards initiation, pactamycin is believed to inhibit bacterial protein synthesis by preventing the initial stages in peptide chain formation (Cohen *et al.*, 1969). The triphenylmethane dye aurintricarboxylate has a specific inhibitory action on the complexing of messenger RNA for the f2 virus to bacterial ribosomes (Grollman and Stewart, 1968). Initiation of peptide chains in animal cells appears to be sensitive to cycloheximide, an action that will be discussed when the main effect of this inhibitor is described.

The binding of tRNA to the A site on the ribosome is subject to inhibition of various kinds. The tetracyclines, which inhibit protein synthesis in both bacterial and mammalian cells (Greenberger, 1967), prevent the binding of aminoacyl-tRNA to the A site (Gottesman, 1967). This inhibitory action includes the binding of formylmethionyl-tRNA to the A site on the 30 S ribosome during peptide chain initiation in bacteria (Sarkar and Thach, 1968). It seems likely that the antibiotics of the PA 114 group (Ennis, 1966) and kasugamycin (N. Tanaka *et al.*, 1966) also inhibit bacterial protein synthesis by preventing the attachment of aminoacyl-tRNA to ribosomes. Misreading of the code by the bound tRNA occurs in the presence of streptomycin (Kaji, 1967) and kanamycin (Masukawa *et al.*, 1968). As pointed out earlier, the GTP analog guanylylmethylene diphosphonate permits binding of aminoacyl-tRNA to ribosomes, but does not undergo hydrolysis, so that subsequent steps are blocked (Hershey and Monro, 1966).

The next reaction is formation of the peptide bond, a reaction catalyzed by peptidyl transferase present in the 50 S ribosome subunit, and this step is subject to inhibition by a number of compounds. Identification of the site of action of these inhibitors has been greatly facilitated by using 50 S subunits treated with ethanol or methanol which activates the enzyme and allows it to form peptide bonds between formylmethionyl-tRNA and puromycin. Using this simplified analog of protein synthesis, Monro *et al.* (1969a) have shown that peptidyl transferase in bacterial ribosomes is inhibited by chloramphenicol, lincomycin, streptogramin A, carbomycin, and spiramycin, whereas anisomycin acts only on the enzyme of animal ribosomes, and sparsomycin, amicetin, and gougerotin inhibit both. The action of sparsomycin has been examined in greater detail by Monro *et al.* (1969b), who find, in agreement with Herner *et al.* (1969), that it has a unique action in binding the peptidyl-tRNA to the 50 S subunit. This may be connected with the finding of Baliga *et al.* (1969) that GTP hydrolysis in a mammalian cell-free protein-synthesizing system is not significantly inhibited by sparsomycin, although aminoacyl-^{14}C transfer is severely reduced; this suggests uncoupling of GTP hydrolysis by the inhibitor, another unusual feature of this antibiotic. Although chloramphenicol is generally considered to inhibit protein synthesis only in bacteria, periodic claims are made of an action on mammalian systems. Zelkowitz *et al.* (1968) refute the claim that protein synthesis in reticulocytes is sensitive to this drug. Nevertheless, there appears to be substantial evidence that antibody synthesis in the spleen is inhibited by low doses of chloramphenicol (Talal and Exum, 1966; Weisberger and Daniel, 1969), and that HeLa cells in tissue culture can no longer synthesize mitochondrial cytochromes in the presence of chloramphenicol in the medium (Firkin and Linnane, 1968).

After peptide bond formation, the peptidyl-tRNA is translocated from the A site to the P site with concomitant hydrolysis of GTP. This reaction, which is catalyzed by transferase II in animal cells, appears to be one site of inhibition by cycloheximide (Actidione) added in large amounts to mammalian cell-free systems for protein synthesis (Munro *et al.*, 1968; Baliga *et al.*, 1969). Using a cell-free system in which polysome reaggregation was dependent on addition of amino acids to the medium, Baliga *et al.* (1969) found that the reaggregation could be arrested by very low concentrations of cycloheximide, thus suggesting that this antibiotic may also inhibit some phase of chain initiation in mammalian protein synthesis in addition to its effect on chain elongation. These authors also showed that the inhibitory action of cycloheximide on chain elongation could be diminished considerably by adding sulfhydryl reagents to the medium, apparently because transferase II is a sulfhydryl-dependent enzyme; on the other hand, sulfhydryl addition did not prevent the action of cycloheximide on polysome reaggregation, emphasizing that the two actions of the inhibitor are different. Streptovitacin A, which is structurally related to cycloheximide, has a similar inhibitory action on transferase II that can be prevented by addition of sulfhydryl reagents to the medium. Emetine, a drug used in the treatment of amebic dysentery, was thought by Grollman (1966) to owe its inhibitory action on protein synthesis to its structural relationship to cycloheximide, but Baliga *et al.* (1969) found that the action of this antibiotic was not affected by the sulfhydryl content of the medium. Other aspects of emetine action, notably its failure to prevent nascent peptide release from ribosomes by puromycin, differentiate it from that of cycloheximide (Grollman, 1968). Tubulosine is another antibiotic that has structural similarities to cycloheximide and may act in a similar manner (Grollman, 1967), though this requires confirmation. The steroid antibiotic fusidic acid inhibits breakdown of GTP in presence of transferase II prepared from reticulocytes (M. Malkin and Lipmann, 1969) and the corresponding translocase of bacteria (N. Tanaka *et al.*, 1968, 1969). Bottromycin A_2 is thought to inhibit translocation in bacteria but fails to affect hydrolysis of GTP (Y. C. Lin *et al.*, 1968), thus suggesting that the action of this antibiotic differs from that of cycloheximide and fusidic acid. Finally, diphtheria toxin inhibits protein synthesis in mammalian cells through inactivation of transferase II. In the presence of nicotinamide adenine dinucleotide (NAD), the toxin catalyzes the transfer of adenine-diphosphate ribose from NAD to the enzyme, thereby inactivating its capacity for translocation and for GTP hydrolysis (Goor *et al.*, 1967; Raeburn *et al.*, 1968; Honjo *et al.*, 1968; Gill *et al.*, 1969; Collier and Cole, 1969). Although diphtheria toxin does not have a similar action on translocation in bacteria, Goto *et al.* (1968) have found that low doses of the toxin inhibit binding of messenger RNA to bacterial ribosomes.

Once peptidyl-tRNA has been translocated from the A site to the P site, it can react with yet another inhibitor, puromycin. Due to its structural similarity to aminoacyl-tRNA, puromycin enters the A site on the ribosome and forms a peptide bond with the peptidyl-tRNA on the P site (Traut and Monro, 1964). Since the resulting peptidyl-puromycin lacks most of the polynucleotide portion of aminoacyl-tRNA, it does not continue to adhere to the ribosome and the peptidyl-puromycin is released from the ribosome. In order to cause maximal peptide chain release, the peptidyl-tRNA, which is mostly on the A site in harvested ribosomes, has then to be translocated to the P site, and the enzyme peptidyl transferase must be active in order to form the peptide bond to puromycin. Consequently, inhibitors of either of these enzymes can prevent the release of nascent peptide chains by puromycin. For example, Baliga *et al.* (1969) showed that *in vitro* release of nascent peptide from liver polysomes by puromycin treatment could be inhibited by cycloheximide, but this effect of cycloheximide did not occur if the sulfhydryl content of the medium was raised and translocase activity was thus protected against cycloheximide. Fusidic acid inhibits puromycin peptide release by a similar mechanism (N. Tanaka *et al.*, 1968).

Finally, a number of inhibitors of protein synthesis have been discovered without as yet identifying the site of their action. These include the mikamycins (Yamaguchi *et al.*, 1966), nivalenol (Ueno *et al.*, 1968), homogentisic acid (Raghupathy *et al.*, 1968), the phenanthrene alkaloids tylophorine, tylocrebrine, and cryptopleurine (Donaldson *et al.*, 1968; Haslam *et al.*, 1968), nucleocidin (Florini *et al.*, 1966), erythromycin (Taubman *et al.*, 1966; K. Tanaka *et al.*, 1966), and berninamycin (Reusser, 1969a).

3. *Regulation of the Protein Content of Mammalian Cells*

Regulation of the protein content of a cell can take two forms, namely qualitative and quantitative. Qualitative regulation is exercised when a protein species appears or disappears in a cell. In terms of the genetic potential of the animal, information about protein structure encoded in the genome of the species is withheld until it is needed, when it is then expressed as a new messenger RNA species in the cytoplasm and there becomes active in protein synthesis. For example, embryonic development, with its creation of specialized cells from undifferentiated cells, represents a series of regulatory acts in which the genetic potential of the fertilized ovum to make certain proteins becomes active in some cells but remains repressed in others. The converse of this can occur with the onset of malignant change, which is frequently accompanied by loss of the capacity to form the specialized proteins of the parent cell, a change that may indicate the closing down of transcription of certain genetic loci as the cancer cell returns to a less differentiated form. Hormone treatment can also cause the appearance of a new protein. For example, when hens first begin to lay eggs, three new proteins appear in the liver and are transported to the oviduct for incorporation into the egg. The synthesis of these proteins by the liver depends on new messenger species induced by estrogens in the hen; induction by estrogen administration can be prevented by prior administration of actinomycin D (Greengard *et al.*, 1965). From this description, it can be deduced that qualitative regulation of protein synthesis is usually brought about by a nuclear event involving the appearance or nonappearance of messenger RNA.

Quantitative regulation of the protein content of a cell is achieved by much more varied mechanisms. To begin with, the amount of a protein can be altered either by a change in rate of its synthesis or of its degradation. In consequence, accumulation of a protein can represent either increased synthesis or reduced breakdown. The regulation of *breakdown* is dealt with in detail in Chapter 32. When we come to consider how *synthesis* is regulated quantitatively, it is possible to conceive of two major methods. Either the number of cells in a tissue making the protein could be varied or the amount of protein formed by each cell could be regulated. The first of these control mechanisms is but one example of the all-or-none principle best exemplified by the control of the contraction of a skeletal muscle by bringing in more or fewer muscle fibers, each of which contracts to its maximal extent. This has been claimed as the means by which rate of synthesis of albumin, fibrinogen (Hamashima *et al.*, 1964), prothrombin (Barnhart, 1960; Barnhart and Anderson, 1961), transferrin (Lane, 1967, 1968), and haptoglobin (J. H. Peters and Alper, 1966) are regulated in the liver. In each case the evidence is essentially similar. An antibody is prepared against the specific tissue protein under investigation, and the antibody is then coupled to a fluorescent dye. When histological sections of the tissue are made and are treated with the antibody, those cells that contain the protein (antigen) fluoresce. In all the instances cited above, only some of the cells in the liver showed such fluorescence, suggesting that not all were actively synthesizing the protein; further, appropriate stimulation of the tissue to form more of the protein resulted in a greater proportion of the cells becoming fluorescent. We have used this technique to study the synthesis of serum albumin by the livers of rats receiving different amounts of dietary protein (Chandrasakharam *et al.*, 1967), the results being illustrated in Fig. 2 in the introductory chapter to Section V in Volume III of this treatise. Rats receiving a protein-free diet showed almost no cells that fluoresced with an antibody to rat albumin, those on a stock diet showed 10% of fluorescent cells, while rats on a diet rich in protein had fluorescence in 60% of cells. Consequently, these results would appear to support the thesis that amino acid supply regulates albumin synthesis by varying the number of liver cells engaged in formation of this protein. However, it is also possible that cells reacting with the antiserum do so because of albumin present in the cavity of the endoplasmic reticulum, and that the cells of the liver may have a distended reticulum more frequently when the rate of synthesis is increased. Indeed, it is quite possible that, a few seconds later, these cells would not longer fluoresce because of discharge of the reticulum to the exterior and a different set of cells would show fluorescence. Consequently, the number of fluorescent cells seen at any one time might simply reflect the frequency of filling and emptying of these channels. It is noteworthy that all such immunofluorescent studies have been carried out on cell proteins that are secreted through the endoplasmic reticulum and are subject to the same criticism. There is, however, one finding that makes this interpretation less likely. J. H. Peters and Alper (1966) found that binucleate liver cells showed antibody to haptoglobin more frequently than other liver cells when synthesis of this protein was stimulated by turpentine injections. This suggests that some cells (here binucleate) may be *selectively* stimulated to greater protein synthesis. These observations thus demonstrate that the all-or-nothing response theory of cell protein synthesis to a stimulus still requires a definitive experiment.

The alternative mechanism for regulating rate of protein synthesis, namely by more or less uniform change in the synthetic rate in all cells of the tissue, is generally assumed to be the method operating under most conditions. Any point in the chain of events leading to formation of a protein molecule can serve as the rate-limiting step regulating the amount of the protein synthesized. Protein may be made more rapidly because more messenger RNA passes into the cytoplasm or because the rate of breakdown of messenger RNA is retarded. It can also occur because of an increased supply of amino acids or even of one amino acid, and the rate of protein synthesis can also be determined by energy supply. These and other factors regulating rate of protein synthesis in mammalian cells will now be considered in more detail.

a. Variations in Amount of Messenger RNA. In cells that are synthesizing many different proteins, it is reasonable to assume that the relative quantities of different proteins made will depend on the amount of each species of messenger RNA in the cytoplasm. Indeed, there is evidence that the relative rates of synthesis of the β-chain of hemoglobin A and the δ-chain of hemoglobin A_2 is determined solely by the relative abundance of the messengers for these chains within the same reticulocyte (Kazazian and Itano, 1968). The quantity of a given messenger can be varied by changing its rate of synthesis in the nucleus and transport into the cytoplasm, or by regulating its rate of decay in the cytoplasm.

Increased formation of messenger is considered to account for the many instances in which hormones stimulate the rate of formation of given proteins in the cells of the target organs, but fail to do so when the animal is pretreated with actinomycin D. The classical example of this is the increase in tryptophan pyrrolase level in the liver of the rat following administration of corticosteroids, a response that is obliterated by prior treatment with 700 μg actinomycin D per kilogram of body weight (Greengard *et al.*, 1963). The effect of a single dose of steroid on pyrrolase level in the liver is limited to a few hours; from this it may be deduced that the survival of messenger for this protein is brief. On the other hand, some mammalian messengers are known to survive for long periods in the cytoplasm. The hemoglobin messenger of mammalian reticulocytes is one such example. Another nonnucleated cell is the blood platelet, which continues to incorporate ^{14}C-labeled amino acids into its contractile protein even after 72 hours of *in vitro* incubation (Booyse and Rafelson, 1967). Thus, the messenger for this protein probably survives for the life-span of the platelet, which is 3–8 days. Similarly, the emergence of the mature fiber cell in the lens of the eye is accompanied by loss of nuclear RNA synthesis, but the cell continues to make lens protein, using stable messenger previously laid down in the cytoplasm (J. A. Stewart and Papaconstantinou, 1967).

Long-lived messenger is, however, not confined to such specialized cells. In the liver, Wilson and Hoagland (1967) have shown that there are both short-lived and long-lived families of polysomes after administration of actinomycin D, a finding compatible with similar studies by T. Staehelin *et al.* (1963). Furthermore, Wilson *et al.* (1967) found that starvation preferentially removed the short-lived messenger, whereas the messenger for albumin synthesis persisted. We have observed that loss of liver RNA due to protein depletion does not impair the capacity of the liver to make ferritin protein in response to an iron load (Drysdale *et al.*, 1967), thus suggesting that the ferritin messenger is also very stable. The presence of both short- and long-lived messengers in the same cell accounts for changes in the relative amounts of proteins made in the same organ under different conditions. For example, Rothschild *et al.* (1969a) perfused the livers of fasting or fed rabbits and measured the amounts of plasma albumin synthesized under these conditions. In the case of the *fasting* rabbit, the liver made 18 mg of albumin over a 2.5-hour period, and this increased to 49 mg when tryptophan was added to the perfusion fluid. The livers of *fed* rabbits formed 33 mg of albumin in a 2.5-hour period, but addition of tryptophan to the medium failed to stimulate a further output of albumin. The authors suggest that the liver of the fasting rabbit has lost some of the short-lived templates so that there are fewer templates competing with the albumin messenger for available ribosomes and the other components involved in protein synthesis; in consequence, the albumin messenger can be used more efficiently under the stimulus of tryptophan. Another possible example is provided by Moscona *et al.* (1968), who found that steroid-induced glutamine synthetase activity in the retina of the chick embryo was not inhibited if large doses of actinomycin were used late in the induction phenomenon, but that low doses of this antibiotic prevented further increase in enzyme activity. It is possible that low doses of actinomycin allow formation of some short-lived messengers which compete with the glutamine synthetase messenger for ribosomes, etc., whereas at high doses of the inhibitor these transient messengers are eliminated by decay and the long-lived glutamine synthetase messenger is more effectively translated.

In using actinomycin D for studies of messenger formation and survival, it is important to recognize that low doses selectively inhibit synthesis of ribosomal RNA, and that higher doses are needed to prevent messenger synthesis. If a small dose is administered, it is conceivable that the continued secretion of messenger RNA will result in an imbalance between messenger and ribosomes in the cytoplasm. Indeed, we have observed that such a dose of actinomycin can change the polysome spectrum in the liver by reducing the number of monosomes and increasing the polysomes, presumably because excess messenger is now competing for all available free ribosomes (Enwonwu and Munro, 1969). Even with the larger doses needed to shut off synthesis of messenger RNA, there is also some evidence that the action of actinomycin on different messenger species may be selective. In the case of bacteria, actinomycin treatment can reduce the levels of some proteins and not others, suggesting a selective action (Pollock, 1963; Kadowaki *et al.*, 1966), but this may also be interpreted as differing rates of decay of the messenger in the cell (Moses and Sharp, 1966). A similar difficulty in interpretation haunts the use of such inhibitors in mammalian studies. Sarma *et al.* (1969) found that actinomycin caused disaggregation mainly of polysomes free in the cytoplasm of the liver cell, but not of those attached to membranes. This could mean either that actinomycin selectively inhibits synthesis of certain messengers, or else that the templates for secreted proteins are more stable, and, in consequence, cessation of synthesis of all new messenger RNA does not have an early effect on the number of polysomes containing stable messengers, whereas polysomes containing short-lived messengers are not maintained. Second, Korner (1964) has provided evidence from studies of labeling of cytoplasmic and of nuclear RNA on gradients that growth hormone can still cause the appearance of newly labeled RNA in rats treated with enough actinomycin to suppress all RNA labeling in control animals. Again, this need not necessarily mean that actinomycin has a differential action on different messenger species. It is more probable that the free actinomycin is present only briefly in the liver nuclei, and that the hormone acts later and causes exposure of DNA that was previously protected by histones, etc., against the action of the antibiotic.

Finally, the observation that some messengers are short-lived and others are more stable implies a differential rate of breakdown. Stability does not appear to depend on the messenger being in polysome form in the cytoplasm. Functional messenger is known to persist more or less intact when the polysomes of HeLa cells disaggregate during cell division (Hodge *et al.*, 1969) and when polysomes from cultures of Chang liver cells (Eliasson *et al.*, 1967) and from cultured mouse cells (Chen *et al.*, 1968) break down following transfer of the cells to an amino acid-deficient medium. The evidence for this conclusion is that addition of large doses of actinomycin D to the medium while the polysomes are disaggregated does not prevent reaggregation when cell division has been completed, or when amino acids are restored to the medium, respectively. Disaggregation of the polysomes of hepatoma cells with phenethyl alcohol is also reversed after several hours by removal of the inhibitor in presence of actinomycin (Plagemann, 1968b). It seems more likely that stability or instability is an inherent feature of the messenger species. Thus, in the disease thalassemia, the production of the β-chain of hemoglobin falls off markedly as the erythroid cell matures (Braverman and Bank, 1969). One explanation is that the β-chain mRNA in this disease is abnormally susceptible to nucleases and undergoes premature degradation compared with normal β-messenger.

It should be added that actinomycin has been claimed both to shorten and to lengthen the half-life of messenger RNA (Acs *et al.*, 1963; Kennell, 1964; Kennell, 1964; Trakatellis *et al.*, 1964a,b; Chanarenne, 1965). Much of this evidence is rather indirect and should be viewed with suspicion. Nevertheless, as discussed in Chapter 31, changes in rate of breakdown of messenger RNA by antibiotics would explain some otherwise puzzling aspects of induction phenomena, notably when the administration of the inhibitor at a strategic time increases the capacity of the cell to make the induced protein. Finally, it should be pointed out that some antibiotics stimulate adrenocortical secretion. This occurs in the case of actinomycin D (Lippe and Szego, 1965), cycloheximide (Jondorf *et al.*, 1966), and 8-azaguanine (Levitan and Webb, 1968).

b. Ribosomes. While the rate of protein synthesis is not generally consided to be limited on account of availability of ribosomes, it is noteworthy that many stimuli that increase the rate of RNA synthesis in mammalian cells do so with an accompanying major increase in rate of ribosomal RNA formation. For example, this occurs in the liver within 3 hours of administering hydrocortisone to rats (S. T. Jacob *et al.*, 1969b); the increased rate of ribosomal RNA formation coincides with the time of maximal induction of tryptophan pyrrolase and tyrosine transaminase in the cytoplasm. Inspection of Fig. 5 shows that the free ribosome population of the average liver cell represents only 3% of the total cell RNA, and ribosome subunits account for a further 2%. Thus at any one time only some 5% of the total ribosome population is not employed in polysome formation, and presumably provides the pool from which new polysomes can be synthesized. It is also likely that newly formed ribosomes and their subunits will first enter this pool. Since the replacement of ribosomes in the liver of the normal rat occurs at a rate of some 1% per hour (Hirsch and Hiatt, 1966; Enwonwu and Munro, 1970), a change in rate of ribosome production could double or halve the pool of free ribosomes and subunits in a few hours. Availability of free ribosomes might thus become a general mechanism for regulating rate of protein synthesis. The experiments of Rothschild *et al.* (1969a) cited above may perhaps exemplify ribosome availability as a factor limiting the rate of protein synthesis. It will be recollected that, when the perfused liver of a fasting rabbit was stimulated by infusion of tryptophan, it achieved a greater rate of albumin production than the liver of a fed rabbit. This was attributed to the loss of unstable mRNA templates from the liver of the fasting rabbit, thus leaving fewer messengers to compete with the stable albumin messenger for the pool of free ribosomes.

An aspect of the abundance of ribosomes and other RNA-containing components involved in protein synthesis is that mammals of increasing mature body size show a progressive reduction in the RNA and ribosome content of their tissues. As shown in Table II, the RNA content of the liver cell and of the liver microsomes diminishes steadily from the mouse to the cow, and at the same time the rate of liver protein synthesis, as indicated by albumin renewal, decreases in parallel. This suggests that the rate of protein synthesis in different species is regulated through reduction in the abundance of the apparatus for synthesis rather than by change in the output per unit. Other examples and further implications of changes in metabolic intensity in relation to body size are provided in Chapter 25, Section IV in Volume III of this treatise.

There is evidence that the ribosome itself can vary in capacity to translate the message. Wool has found that skeletal muscle ribosomes obtained from diabetic rats are less efficient than those from normal

TABLE II
PLASMA ALBUMIN TURNOVER, LIVER CELL COMPOSITION PER MILLIGRAM OF DNA AND COMPOSITION OF LIVER MICROSOMES OF DIFFERENT MAMMALS ARRANGED IN ASCENDING ORDER OF WEIGHT[a]

Species	Half-life of plasma albumin (days)	Whole liver composition			Microsome fraction (% of total dry matter)	
		RNA/DNA	Phospholipid/DNA	Protein/DNA	RNA	Phospholipid
Mouse	1.2	4.45	14.2	86	9.6	24.2
Rat	2.5	3.05	9.8	60	7.3	22.9
Rabbit	5.7	2.44	11.6	77	6.2	29.2
Dog	8.2	1.69	9.4	70	—	—
Cow	20.7	1.29	8.7	47	5.9	31.3

[a] From Munro and Downie (1964).

rats. T. E. Martin and Wool (1968) dissociated the ribosomes into 60 S and 40 S subunits by means of KCl treatment and found that, on reassociation, the ribosomes were still capable of translating a polyuridylic acid messenger under *in vitro* conditions, but the reconstituted ribosomes from the diabetic animals were less efficient. This difference was traced to a defect in the 60 S subunit obtained from the diabetic rats, as shown by reacting polyuridylic acid with reconstituted ribosomes containing hybrid subunits. A similar defect in the ability of liver ribosomes to translate messenger RNA has been claimed to occur in hypophysectomized rats, and to be reversed by growth hormone administration (M. Staehelin, 1965; Garren *et al.*, 1967; Korner and Gumbley, 1966; Herrlich and Lang, 1967; Korner, 1969). In contrast, Sells and his colleagues find that polysomes prepared from the livers of normal and of hypophysectomized rats show similar protein-synthesizing capacities when the size-classes of the two sets of polysomes are similar (Brewer *et al.*, 1969) and, moreover, that ribosomes prepared by reassociation in all possible combinations from the subunits of each group show identical capacities to translate a synthetic messenger (Foster and Sells, 1969). Sells draws the conclusion that cell-free systems obtained from hypophysectomized rats often undergo damage during preparation because of the high level of active ribonuclease in the livers of such animals.

c. Regulation through Supply of Aminoacyl-tRNA. The provision of aminoacyl-tRNA is an obvious potential site of regulation in protein synthesis. Baliga *et al.* (1968) describe a cell-free protein-synthesizing system from liver that is dependent on addition of free amino acids. Each of the twenty amino acids requires to be provided in order to obtain incorporation of leucine-^{14}C; aggregation of polysomes in the system is also dependent on all amino acids being present.

Several studies show that protein synthesis can be rate-limited by scarcity of a single aminoacyl-tRNA. W. F. Anderson (1969) has recently studied the translation of a synthetic random-ordered poly AG messenger by *E. coli* ribosomes, using *E. coli* tRNA as the amino acid source. The synthetic messenger coded for arginine, glutamic acid, lysine, and glycine. The speed of translation was found to be rate-limited by the amount of a subfraction of arginyl-tRNA present in the *E. coli* tRNA. However, most work on the effect of amino acid supply on rate of translation has been done on hemoglobin synthesis by reticulocytes. Hunt *et al.* (1969a) measured the rate of synthesis of the α- and β-chains of hemoglobin when intact reticulocytes were incubated in a complete amino acid mixture, and observed that each α-chain took 140 seconds to be made, whereas the β-chain needed 210 seconds. These investigators suggest that differences in the amino acid composition of the two chains may make translation of one messenger strand more susceptible than the other to availability of specific aminoacyl-tRNA species. For example, threonine occurs much more frequently in the α-chain than in the β-chain; this may account for the subsequent findings of Hunt *et al.* (1969b) that incubation of reticulocytes in a threonine-deficient medium reduces the rate of translation of the α-chain much more than that of the β-chain, so that the rates of formation of the two chains now become equal. W. F. Anderson and Gilbert (1969) have confirmed this selective action on synthesis of the α-chain in a cell-free protein-synthesizing system prepared from rabbit reticulocytes. When supplied with unfractionated aminoacyl-tRNA, this system made equal amounts of the α- and β-chains of hemoglobin, but α-chain production was selectively reduced when a group of aminoacyl-tRNA species were deleted from the tRNA mixture added to the system.

A special example of regulation of hemoglobin synthesis through amino acid availability concerns the supply of tryptophan. If rabbit reticulocytes are incubated in a medium lacking tryptophan, their polysomes disaggregate, whereas deficiency of other amino acids such as valine does not have this effect (Hori *et al.*, 1967). As pointed out by these authors, tryptophan occurs only near the N-terminal ends of the α- and β-chains of rabbit hemoglobin, namely at position 14 in the α-chain and at positions 15 and 37 in the β-chain, so that completion of each chain beyond these points does not require tryptophanyl-tRNA. Consequently, ribosomes translating later sections of these messengers are able to complete each chain and roll off the messenger even when tryptophan is not available, whereas ribosomes present on the messenger strands before the point of tryptophan insertion cannot advance if there is a lack of this amino acid; in consequence, unemployed free ribosomes accumulate and the polysomes become disaggregated. On the other hand, valine occurs in many positions throughout both hemoglobin chains, so that deletion of this amino acid from the medium prevents extension of the peptide chain at many points; in consequence, the ribosomes are immobilized along the whole messenger. Ribosomes, stuck because of lack of tryptophanyl-tRNA, can be mobilized and move on to the next codon if the growing peptide chain is discharged by low doses of puromycin (Freedman *et al.*, 1968). Using a different approach, Hunt *et al.* (1968) have confirmed the retarding action of tryptophan deficiency on ribosome movement along the α- and β-chain hemoglobin messengers. Another and more complex example of the effect of nonavailability of one aminoacyl-tRNA on polysomes is the use of *O*-methylthreonine, a competitive analog of isoleucine which prevents the entry of isoleucine into positions 10, 15, and 55 in the α-chain and position 112 in the β-chain of rabbit reticulocyte hemoglobin (Hori and Rabinovitz, 1968). This analog should thus produce a proximal rate-limiting block in the α-chain messenger and a distal block in the β-chain messenger; Kazazian and Freedman (1968) have in fact observed clusters of very small and very large polysomes in reticulocytes treated with *O*-methylthreonine.

The genetic code shows extensive degeneracy, since most of the twenty amino acids are represented by more than one tRNA species corresponding to different codons. This has prompted several investigators to look for evidence that a protein or group of proteins can be deleted from a cell by suppressing one tRNA species, such as one of the six isoaccepting tRNA's for leucine. All messenger RNA species containing the codon for this particular leucyl-tRNA would then cease to be translated. A striking but artificial example of this phenomenon is seen in the case of hemoglobin synthesis by reticulocytes obtained from newborn infants with hemolytic disease. Such reticulocytes contain messenger RNA not only for the α- and β-chains of adult hemoglobin, but also for the γ-chains of fetal hemoglobin. Only the γ-chain of human hemoglobin contains isoleucine, so that incubation of these reticulocytes in an isoleucine-free medium permits the α- and β-chains to be translated uninterrupted, whereas translation of the γ-chain is suppressed for lack of isoleucyl-tRNA (Honig *et al.*, 1969). It is also possible to show that at least one species of tRNA has been completely deleted from the spectrum of tRNA species found in cell sap, although the animal cell can make that specific tRNA. Formylmethionyl-tRNA is not present in the cell sap of animal cells, although this tRNA species is found in the mitochondria of the same cells (Galper and Darnell, 1969; Reboud, 1969). Does the DNA responsible for the presence of formylmethionyl-tRNA in animal mitochondria reside in the cell nucleus or is it mitochondrial DNA? The latter is almost certainly true, since in the case of rat liver, mitochondrial tRNA species have been shown to hybridize better to mitochondrial DNA than do cytoplasmic tRNA species (M. M. K. Nass and Buck, 1969), thus confirming earlier evidence of a different set of tRNA species in mitochondria (Barnett and Brown, 1967). In spite of this example of selective deletion of tRNA, it is difficult to believe that a control mechanism involving deletion of one or other of the isoaccepting tRNA species can enjoy widespread use in determining the appearance or nonappearance of proteins in the cells of higher organisms. The complete genetic code appears to have been established very early in evolution and must thus have been disseminated throughout the nucleotide sequence of many proteins essential to the function of all cells; it would thus seem unlikely that in the course of evolution an isoaccepting tRNA could be selected to regulate the translation of a specialized protein in a higher organism which would not also affect translation of some other protein essential to cell function. Nevertheless, there have been extensive searches for meaningful variations in the relative abundance of tRNA species in different tissues, and especially in malignant cells where loss of specific proteins is common. Three types of variation have been encountered. First, the relative abundance of different tRNA species varies moderately in normal tissues (Taylor *et al.*, 1968) and extensive differences have been noted in tumor cells (e.g., Mushinski and Potter, 1969) and in embryonic tissue (Lee and Ingram, 1967). We do not know whether such variations in availability of specific tRNA species will selectively alter the rate of synthesis of some proteins more than others, as we saw earlier in the case of the rates of formation of the α- and β-chains of hemoglobin. Second, certain tRNA species may be completely absent, especially in tumor cells (Goldman *et al.*, 1969; Mushinski and Potter, 1969); whether this completely prevents synthesis of certain proteins by these cells is unknown. Finally, new acceptor species can appear, especially on chromatograms of tRNA from tumor cells (Taylor *et al.*, 1967, 1968; Gonano and Chiarugi, 1969) and from virus-infected cells (e.g., Kano-Sueoka and Sueoka, 1966; Subak-Sharpe *et al.*, 1966). These modified or new tRNA species may show altered function, such as the decreased capacity of prolyl-tRNA from phage-infected *E. coli* to bind to ribosomes (Hung and Overby, 1968). Presumably such aberrant tRNA species could also be less efficiently recognized by the ligases, though in the case of viruses this would select against their survival. It is still too early to try to assess the regulatory significance of these differences in the quantity and quality of different tRNA species.

In order to function in protein synthesis, the tRNA must be charged with the appropriate amino acid. Studies on both animal tissues (Strehler *et al.*, 1967; Ceccarini *et al.*, 1967) and plant cells (M. B. Anderson and Cherry, 1969) show that these tissues can contain more than one aminoacyl-tRNA ligase for each amino acid, that not all isoaccepting tRNA species are charged by the different ligases for a single amino acid, and that the abundance of the different ligases for a single amino acid can vary from tissue to tissue. The last two observations imply that tissues lacking one of the set of ligases for, say, leucine would be unable to charge all the isoaccepting tRNA's for that amino acid. This would have the same effect as deleting that particular tRNA from the spectrum made by the tissue.

Changes in the activities of aminoacyl ligases have been described in rat liver and muscle in response to alterations in dietary protein intake (Gaetani *et al.*, 1964) and in the seminal vesicle of the rat following administration of testosterone (Tóth and Mányai, 1968). The effects of such changes in enzyme activity on the availability of aminoacyl-tRNA have not been established. There is also a need for more information about the degree of saturation of different tRNA species with amino acids. This information is required because it is commonly assumed that deficiency of an amino acid will produce effects on protein synthesis through the failure of tRNA species carrying that amino acid to remain fully charged. As discussed in Chapter 34, Section III,A,2, free amino acid concentrations can vary widely and still be compatible with maximal protein deposition in the growing animal, which suggests that some tRNA species may remain fully charged over a wide range of free amino acid concentrations. On the other hand, protein synthesis in the liver of the normal animal is especially sensitive to the availability in the diet of tryptophan, but not of other amino acids (Fleck *et al.*, 1965; Wunner *et al.*, 1966; Pronczuk *et al.*, 1968; Sidransky *et al.*, 1968). Some recent experiments (Allen *et al.*, 1969) on rats tube-fed various amino acid mixtures show that in fact omission of several of the essential amino acids from the mixture had little or no effect on the charging of tRNA with that amino acid, whereas omission of tryptophan from the administered mixture reduced tryptophan charging of tRNA to one-third its normal level. Since tryptophan is the least frequent amino acid in the average liver protein, availability of charged tryptophanyl-tRNA could then be rate-limiting for protein synthesis. In circumstances under which the supply of tryptophan was too low to permit complete charging of its tRNA, retardation of protein synthesis and disaggregation of polysomes would occur for the same reasons as those given above in the case of reticulocytes. A more extensive discussion of the relationship of amino acid supply to tissue protein synthesis is provided in Chapter 34, Section III,A,1.

d. Energy Supply. High energy phosphate bonds are utilized at several stages in protein synthesis: in the substrate for RNA polymerase, for amino acid activation by the ligases, for turnover of 3-terminal nucleotides of tRNA, and in the GTP requirement for binding and translocation of tRNA on the ribosomes. Although there is evidence to show that a severe inadequacy of cellular energy supply can interfere with protein synthesis, in no case has it been demonstrated how this occurs. Thus lack of oxygen or treatment with dinitrophenol causes the polysomes of soybean roots to undergo rapid disaggregation; restoration of aerobic conditions reverses this polysome dissociation (C. Y. Lin and Key, 1967), but the mechanism of this adverse effect of cutting off energy supply is not known. Incubation with fluoride has been used to cause dissaggregation of polysomes in reticulocytes (Marks *et al.*, 1965; Conconi *et al.*, 1966; Dreyfus and Schapira, 1966), an effect that has been attributed to interference by the fluoride with generation of ATP (Marks *et al.*, 1965). However, Colombo *et al.* (1968) consider that fluoride acts as an inhibitor of protein synthesis by preventing dissociation of free ribosomes into their subunits, an obligatory step in the reassociation of subunits with the messenger. This site of action of fluoride is supported by the data of Hunt *et al.* (1968) on inhibition of hemoglobin synthesis. Wittmann *et al.* (1969) have shown that the liver polysomes undergo aggregation when fasting rats are fed glucose; however, the time needed for this response, more than 6 hours, suggests that this action is indirect rather than a direct consequence of making more energy available to the liver.

Perhaps the most striking studies are those involving rats treated with ethionine, a well-known agent for trapping ATP as *S*-adenosylethionine at a rate faster than the liver cell can synthesize it, so that the intrahepatic concentration of ATP can be reduced to 30% or less of its initial level within 2 hours of administration of this methionine analog (Farber *et al.*, 1964; Okazaki *et al.*, 1968; G. A. Stewart and Farber, 1968; Oler *et al.*, 1969). In female rats, this severe reduction in liver ATP concentration is paralleled by an extensive decrease in the protein-synthesizing capacity of liver ribosomes and by almost total disaggregation of liver polysomes (Villa-Treviño *et al.*, 1963, 1964). The mechanism by which ethionine exerts this adverse effect on liver protein synthesis is not clear, except that RNA formation is known not to be involved (G. A. Stewart and Farber, 1967). A direct action of ATP supply on protein synthesis seems improbable since male rats respond to ethionine treatment with an equally dramatic and rapid fall in ATP concentration, but there is no change in the capacity of their liver ribosomes to synthesize protein and only a slight alteration in their liver polysome profiles (Oler *et al.,* 1969). Indeed, these data show that liver protein synthesis is relatively insensitive to ATP concentration in the liver, since the male rat can tolerate an 80% reduction without seriously interfering with the mechanism of protein synthesis. This conclusion is compatible with the observation that, when a rat with a liver containing abundant glycogen is fed a meal of protein or amino acids, the glycogen rapidly disappears from the liver and the ATP concentration falls from 1.65 m*M* to 0.45 m*M* (C. M. Clark *et al.,* 1960). Despite this adverse effect of an amino acid influx on liver levels of ATP, liver protein synthesis is quickly stimulated by feeding a meal of protein alone to a rat (C. M. Clark *et al.,* 1957), which is good news for carnivores who might otherwise fail to make liver proteins after a meal of meat. It implies that the primary impact of a meal on liver protein synthesis is due to the amino acids absorbed from the intestine. The differing roles of amino acid supply and energy supply in regulating protein metabolism are discussed in detail in Chapter 10, Section III,A, in Volume I of this treatise.

Finally, polysomes have been found to undergo disaggregation in cells treated with agents that act primarily on the cell membrane. This might be assumed to be due to loss of metabolites such as ATP through the damaged cell membrane, but there is no evidence to support this contention. Thus, phenethyl alcohol induces a loss of polysomes from rat hepatoma cells in tissue culture (Plagemann, 1968a) and is thought to act primarily through changes in cell membrane characteristics (Plagemann, 1968b), but the consequences of such cell membrane changes have not been established. Freedman *et al.* (1967) have treated reticulocytes with *n*-butanol, which resulted in reversible disaggregation of the reticulocyte polysomes. They were unable to find any escape of metabolites from the butanol-treated cells, and concluded that the reticulocyte membrane plays a part in polysome formation and integrity. It has been shown that about one-sixth of the ribosomes in rabbit reticulocytes are bound to the cell membrane and turn over at a different rate from the free reticulocyte ribosomes (Burka *et al.,* 1967a; Schreml and Burka, 1968; Kruh, 1968). However, butanol treatment does not affect these attached ribosomes (Burka *et al.,* 1967b). Finally, cyclic AMP appears to stimulate synthesis of the enzyme tryptophanase, but not other enzymes of *E. coli,* through an effect at the level of translation of the messenger (Pastan and Perlman, 1969).

e. Post-translational Factors Regulating Protein Formation. After its release from the ribosome, the peptide chain may have to undergo further changes before becoming a functionally complete protein. Often the change involves aggregation of several peptide subunits into a polymer. Furthermore, the peptide or polymer may not become functional until it acquires a prosthetic group. Alternatively, the released peptide chain may undergo alterations in chemical structure, such as the modification of an amino acid or cleavage of the chain. Presumably all such reactions are subject to regulation affecting the final yield of the completed protein; in many cases, it is probable that accumulation of an intermediate will cause feedback inhibition of protein synthesis by the appropriate polysomes, as seems to be the case with the mechanism for regulating hemoglobin production discussed below. The relevance of these various post-translational changes to control of the amounts of individual proteins formed will now be considered.

It is well known that many functionally active proteins are made up of peptide subunits, which may be similar or dissimilar. When the subunits differ in structure, two messengers are usually involved. The only known exception to this is the subunit structure of insulin. This protein is made on its polysome as a single polypeptide chain, from which a segment of some 30 amino acid residues is then excised in order to liberate the A and B subunit chains of insulin, which then combine through disulfide bonds (see review by Steiner, 1969). In some cases, it is known that such a polycistronic messenger is not involved in the formation of dissimilar subunits. For example, the predominant hemoglobin of adult subjects consists of two α and two β chains. From genetic evidence (London *et al.,* 1967), it is believed that these two chains are coded for by completely independent cistrons on the chromosomes, and in consequence, the two peptides are made on different messenger RNA's. How is the rate of synthesis on these two messengers coordinated so that approximately equal amounts of each chain are synthesized? Colombo and Baglioni (1966) first suggested that free β chains are needed to aid in the release of nascent α chains from the polysomes of the reticulocyte, thus coordinating synthesis of α and β chains. This hypothesis is disproved by the finding of large amounts of free α chains in the reticulocytes of β-thalassemia cases, in whom synthesis of β chains is severely reduced (Bargellesi *et al.,* 1967; Bank *et al.,* 1968). From later evidence, Baglioni and Campana (1967) deduced that the reverse may be true, the free α chains binding to the nascent β chains. However, such a mechanism cannot be a significant factor in β chain translation, since hemoglobin H, composed of four β chains, occurs in patients with α-thalassemia who suffer from severely defective α-chain synthesis (London *et al.*, 1967). A similar mechanism has been suggested in the case of synthesis of immunoglobulins, which consist of two light and two heavy peptide chains. In cells making immunoglobulins, the cytoplasm is found to contain free light chains but not free heavy chains, and Askonas and Williamson (1966) have proposed that the free light chains bind to the nascent heavy chains on the polysomes and control their release, thus ensuring balanced production of light and heavy chains. However, a later search for the half-molecules of immunoglobulins (one light and one heavy chain) demanded by this hypothesis was unsuccessful (Askonas and Williamson, 1968). Thus, well-authenticated examples of mechanisms for coordinating the synthesis of two dissimilar subunits of a protein have yet to be revealed.

The role of the prosthetic group as a site of regulation can be illustrated by the relationship of heme availability to globin formation in the reticulocyte. Figure 12, taken in slightly modified form from a recent review by London *et al.* (1967), suggests diagrammatically how the synthesis of heme and of globin may be coordinated in the immature red cell. There is a good deal of evidence to support their proposed mechanism. It has been repeatedly demonstrated that both the synthesis of globin from labeled amino acids and aggregation of polysomes are stimulated by adding hemin to the medium used for incubation of intact reticulocytes (Bruns and London, 1964; Grayzel *et al.*, 1966; Waxman

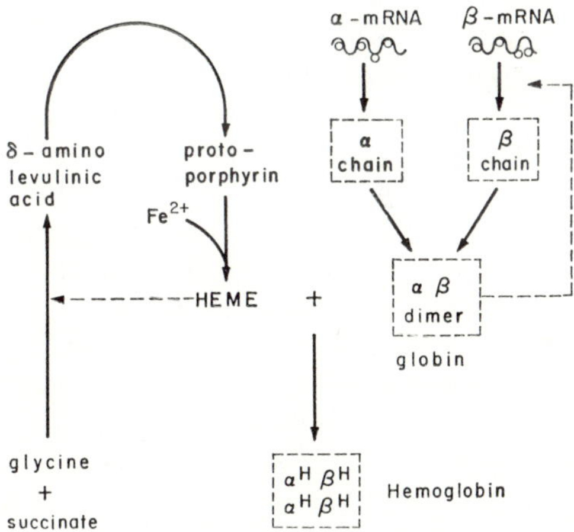

FIG. 12. Regulation of the biosynthesis of hemoglobin. The dotted lines represent feedback inhibition of the activity of δ-aminolevulinic acid synthetase by heme and inhibition of globin synthesis by the $\alpha\beta$ dimer. (Modified from London *et al.*, 1967.)

and Rabinovitz, 1966; Tavill *et al.*, 1968a) or to cell-free systems made from lysed reticulocytes (Adamson *et al.*, 1968, 1969; Maxwell and Rabinovitz, 1969). Several investigators have explored possible mechanisms by which hemin could cause this increase in globin formation. Hemin does not appear to bind in significant quantities to nascent peptides that are still attached to the reticulocyte ribosomes, but it binds to free $\alpha\beta$ dimers formed from the released α and β chains (Felicetti *et al.*, 1966; A. J. Morris and Liang, 1968). The product of this nonenzymic reaction between heme and the free dimer is tetrameric hemoglobin. If reticulocytes are prepared from iron-deficient or anemic animals, they are found to contain significant amounts of free α chains and $\alpha\beta$ dimers (Felicetti *et al.*, 1966; Zucker and Schulman, 1967; Shaeffer, 1967; Tavill *et al.*, 1968a), which presumably have accumulated for lack of sufficient heme to complete the formation of hemoglobin. Thus, the accumulation of free monomeric and dimeric subunits appears to result in feedback inhibition of α- and β-chain formation. There is some direct experimental support for this view. Maxwell and Rabinovitz (1969) provide evidence that soluble factors, possibly the soluble free subunits, accumulate in a cell-free system made without hemin addition from reticulocytes, and eventually inhibit protein synthesis by the reticulocyte polysomes; addition of hemin to the system stimulates globin synthesis by removing the soluble inhibitors, presumably as hemoglobin. Blum and co-workers (Kruh and Blum, 1968; Blum *et al.*, 1969) consider that globin synthesis is regulated through feedback inhibition by α-hemoglobin (α-chains combined with heme), rather than by the accumulation of free α or $\alpha\beta$ chains themselves.

The study of hemoglobin formation is also interesting for the light it sheds on the regulation of heme synthesis (Fig. 12). If globin formation is blocked by incubating reticulocytes with an inhibitor of protein synthesis, such as puromycin or cycloheximide, there is a parallel inhibition of heme synthesis from glycine, but not of heme synthesis from α-aminolevulinic acid (Grayzel *et al.*, 1967). It appears that cessation of globin formation causes accumulation of unused heme in the reticulocyte, and this exerts feedback inhibition on the formation of δ-aminolevulinic acid, an intermediate in the pathway of heme synthesis (London *et al.*, 1967). Thus, the rates of formation of heme and of globin in the reticulocyte are coordinated through the level of heme in the cell. A control mechanism similar to that illustrated in Fig. 12 has been postulated by Levere and Granick (1967) to account for the stimulation of hemoglobin formation that occurs on treating chick embryo blastoderms with δ-aminolevulinic acid.

Another example of coordinate synthesis of a protein and its prosthetic group is the formation of the liver enzyme tryptophan pyrrolase (oxygenase) which consists of an apoenzyme and a heme prosthetic group (T. Tanaka and Knox, 1959). Figure 13 summarizes the details of regulation of enzyme activity described below. Tryptophan pyrrolase activity in the liver can be increased in two ways. First, administration of corticosteroid hormones promotes synthesis of more messenger RNA for the enzyme; this mechanism can therefore be blocked by treatment with actinomycin D (Greengard *et al.*, 1963). Second, the amount of enzyme can be increased by giving the enzyme substrate, tryptophan; this occurs through a mechanism that does not require more messenger to be secreted (Greengard *et al.*, 1963), but involves a reduction in the rate of breakdown of enzyme protein so that enzyme accumulates as synthesis proceeds (Schimke *et al.*, 1965). On closer analysis, the increase in enzyme activity due to tryptophan administration appears to involve three steps:

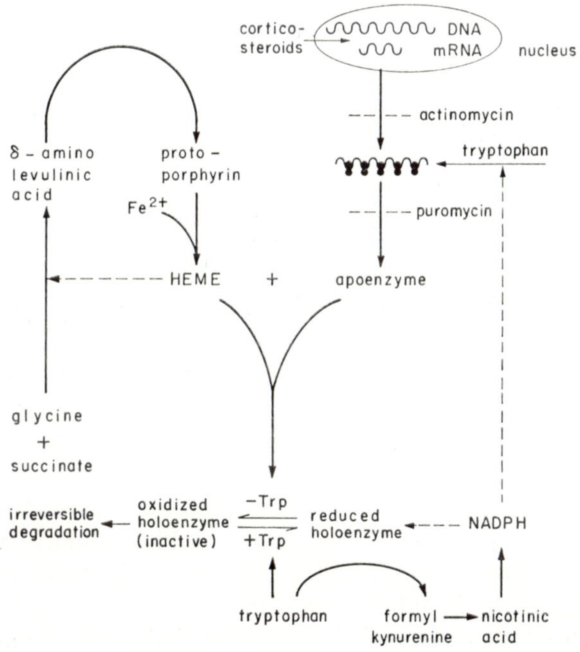

FIG. 13. Diagram of regulation of synthesis of tryptophan pyrrolase (oxygenase) in the liver cell. The diagram incorporates the evidence for mechanisms provided by Greengard *et al.* (1963), Knox and Piras (1966), Marver *et al.* (1966a), and Cho-Chung and Pitot (1968). The dotted lines from heme and from NADPH represent feedback inhibition. There is some doubt about the form of the enzyme that is irreversibly degraded. The stimulation of enzyme synthesis by tryptophan may not be a major regulator.

(1) tryptophan promotes the conjugation of heme with the apoenzyme to form the holoenzyme (Knox and Piras, 1967; Piras and Knox, 1967); (2) after conjugation, the holoenzyme exists in equilibrium between an active (reduced) form and an inactive (oxidized) form, and the active form is favored by tryptophan (Knox and Piras, 1966); (3) tryptophan can also stimulate moderately the translation of the messenger template for the apoenzyme (Knox and Piras, 1967; Cho-Chung and Pitot, 1968), an action that might be due to the general aggregation and stimulation of polysomes following tryptophan administration (see Fig. 9 and also Section III,A,3,c). Whether the form of enzyme susceptible to irreversible degradation is the apoenzyme or the oxidized holoenzyme is uncertain on present evidence. The cell mechanism controlling oxidation-reduction regulation has been the subject of further study. Whereas Knox and Piras (1966) use ascorbic acid in their homogenate system for transforming one form of holoenzyme to the other, Chytil (1968; Julian and Chytil, 1969) seems to favor xanthine oxidase as a source of H_2O_2 for regulating the equilibrium between inactive and active enzyme; Chytil (1968) links such a regulatory mechanism to the pathway of degradation of cyclic AMP. If this is so, it might explain why glucagon, epinephrine, and insulin induce tryptophan pyrrolase activity in the liver. However, Chytil's identification of the nature of xanthine oxidase regulation is still imprecise. Cho-Chung and Pitot (1968) find that NADPH, an end product of the pathway of tryptophan metabolism, causes direct feedback inhibition of the enzyme and also inhibits the stimulus of translation by tryptophan.

It will be noted that heme pigments are required to obtain enzyme activity, and indeed Marver *et al.* (1966a) have examined the role of heme availability in the regulation mechanism, and also the regulation of heme synthesis in relation to its requirement for production of the holoenzyme. Since administration of tryptophan causes increased binding of heme to the apoenzyme, the free heme concentration falls and derepresses δ-aminolevulinic acid synthetase so that more heme is made by means of derepression of the enzyme making δ-aminolevulinic acid. There is accordingly coordinated regulation of the two constituents of the holoenzyme. On the other hand, cortisol administration results mostly in apoenzyme formation, and thus does not induce the δ-aminolevulinic acid synthetase. The regulation of tryptophan pyrrolase by hormones is discussed at length in Chapter 31, Section II,A, and the mode of action of tryptophan in retarding degradation of the enzyme is dealt with in Chapter 32, Section IV,B,1.

Stabilization at the stage of aggregation has also been suggested as the means by which the ferritin content of the liver and of intestinal mucosa is regulated (see review by Munro and Drysdale, 1970). Ferritin consists of an apoferritin shell of twenty identical subunits surrounding a core of iron. A few hours after parenteral administration of iron, there is accumulation of ferritin in the liver. If a pulse dose of a labeled amino acid is given, the deposition of extra ferritin is seen to be accompanied by retention of ^{14}C in the apoferritin (Fineberg and Greenberg, 1955; Drysdale and Munro, 1966). Administration of actinomycin D failed to inhibit this response (Drysdale and Munro, 1966), and it was concluded that a cytoplasmic regulation mechanism is involved. Iron also stabilizes ferritin after its formation (Drysdale and Munro, 1966). If the liver ferritin of a rat is first labeled by injecting leucine-^{14}C, and then the animal is given repeated doses of iron over a period of several days, the rate of loss of label is retarded, indicating that the iron has reduced protein breakdown. Munro and Drysdale (1970) have therefore postulated that the presence of iron may decide the fate of free apoferritin subunits. As shown in Fig. 14, the precursor subunit of ferritin is considered to have a short half-life unless it combines with iron to make ferritin. The administration of iron would therefore provide stabilization of nascent subunits which, if radioactive from a pulse dose of leucine-^{14}C, would give the impression of increased translation of the ferritin template. Similarly, iron administration is thought to reduce the chances of ferritin dissociating and subunits reentering the pool, so that the half-life of prelabeled ferritin protein becomes prolonged. In HeLa cells, addition of iron to the medium stimulated ferritin protein synthesis by a mechanism resistant to actinomycin treatment but removal of iron from the cell with a chelating agent failed

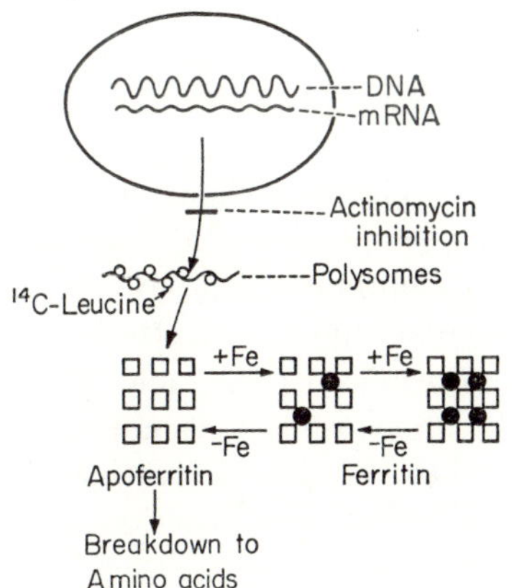

FIG. 14. Diagram of a mechanism proposed for regulation by iron of the biosynthesis and accumulation of liver ferritin protein. (From Munro and Drysdale, 1970.)

to accelerate ferritin protein breakdown (Chu and Fineberg, 1969). However, bleeding of rats sufficient to cause extensive loss of iron from the liver does accelerate liver ferritin breakdown (Munro and Drysdale, 1970). Obviously, more work has to be done to determine whether the model shown in Fig. 14 can be sustained.

Finally, the amino acid residues in completed peptide chain can undergo chemical modifications of several kinds. A number of proteins contain amino acids not included in the genetic code for protein synthesis. In each case, however, the aberrant amino acid is a derivative of one of the twenty amino acids represented in the code. Thus, collagen contains hydroxyproline and hydroxylysine (see Chapter 7, Volume I), the actin and myosin of muscle contain 3-methylhistidine (Trayer *et al.*, 1968), and the peptide chains of elastin are cross-linked by desmosine and isodemosine, derived from the lysine residues in the primary structure of elastin (Partridge *et al.*, 1964). Pyrrolidonecarboxylic acid derived from glutamic acid is found in an N-terminal position in the β-peptides of fibrinogen (Mross and Doolittle, 1967) and in the heavy chain of immunoglobulins (Wilkinson *et al.*, 1966); fibrinogen also contains sulfated tyrosine (Bettelheim, 1954). Phosphoproteins, such as casein and phosvitin, contain their phosphorus atoms as phosphoserine (Lipmann and Levene, 1932; Burnett and Kennedy, 1954), and the histones of chromosomes show phosphorylation of their serine and threonine residues (Allfrey *et al.*, 1966). These proteins also exhibit methylation of the ϵ-terminal amino group of lysine (Allfrey *et al.*, 1964) and acetylation (B. G. T. Pogo *et al.*, 1966); ϵ-*N*-acetyllysyl residues may also account for the acetyl groups found in fowl hemoglobin (Moss and Thompson, 1969). Finally, thyroglobulin contains mono- and diiodotyrosine, triiodothyronine, and thyroxine, all derived from tyrosine. It is generally conceded that the derived amino acid is not likely to be activated by the aminoacyl-tRNA ligases, so that the modification in chemical structure must take place at a subsequent stage in protein synthesis. There is evidence that the chemical change in amino acid structure takes place after formation of the peptide chain in the case of hydroxyproline and hydroxylysine formation in collagen (Bhatnagar *et al.*, 1967), formation of desmosine and isodesmosine in elastin (Partridge *et al.*, 1964), sulfation of tyrosine in fibrinogen (Segal and Mologne, 1959), acetylation of histones (Allfrey, 1968), phosphorylation of histones (Langan, 1968b), and formation of the iodotyrosine derivatives in thyroglobulin (Seed and Goldberg, 1965). Whether these reactions of amino acid side chains regulate the rate at which the protein is made is uncertain, but in the case of the hydroxylation of the proline and lysine residues in collagen, this step would appear to be an obligatory requirement for extrusion of the collagen from the fibroblast (Juva *et al.*, 1966).

f. Regulation of Secreted Proteins. Proteins that are secreted from cells provide special aspects additional to the general picture of regulation presented above. The product leaves the cell of origin, and accordingly regulation of the amount made must depend on information about its extracellular concentration. This information has to be transmitted to the mechanism making that particular protein without causing a general change in the rates of synthesis of other proteins by the cell. This is presumably achieved by a change in the amount of messenger specific for the protein. If the secreted protein is a hormone, its action on the target organ provides the basis for a feedback mechanism of control over its own synthesis. For example, thyrotropin secreted by the pituitary gland is controlled by the levels of thyroid hormones in the circulation, whereas ACTH secretion from the same pituitary gland is regulated by corticosteroid levels. In each case, a negative feedback loop is activated, so that secretion by the target organ (thyroid gland, adrenal cortex) specifically suppresses release of the appropriate pituitary polypetide hormone (Kendall, 1962). Regulation thus becomes a question of how the levels of thyroid hormones or of corticosteroids in the plasma act on the hypothalamus to diminish the rate of synthesis and of release of the pituitary hormones. A possible example of a positive feedback mechanism is provided by the recent thesis of Felig *et al.* (1969b) that insulin secretion from the β-cells of the pancreas is under the control of the free amino acid levels in the systemic plasma. As documented in more detail in Chapter 34, Section III,B, obese subjects show excessive levels of plasma free amino acids and of plasma insulin. It is postulated that uptake of amino acids into muscle is impeded in these subjects because of insulin resistance associated with obesity, and that the resulting chronic hyperaminoacidemia stimulates the β-cells to secrete a persistently high level of insulin. It has been inferred from this that one of several factors normally regulating the secretion of insulin is variation in the levels of free amino acids in the plasma.

The secretion of nonhormonal proteins is also subject to regulation mechanisms. For example, the level of albumin in the plasma can be regulated by changes in rate of synthesis in the liver and of degradation at unidentified sites in the body. These are subject to independent control; rate of synthesis is much more rapidly responsive to nutrient intake (Rothschild *et al.*, 1968a, 1969a; Kirsch *et al.*, 1968), whereas degradation shows delayed responses. Thus, Kirsch *et al.* (1968) examined albumin metabolism in rats subjected to 12 days of depletion on a protein-free diet, followed by refeeding on a normal diet. The rate of albumin synthesis by the liver fell steadily during the period of depletion and rebounded rapidly when protein was reintroduced into the diet. On the other hand, catabolism of albumin responded more sluggishly to the dietary changes and appeared to be related to the level of albumin circulating at the time. Albumin synthesis and degradation also react to changes in plasma albumin level from causes other than nutritional factors. It has been noted that depressed levels of albumin in the plasma are often associated with increased plasma concentrations of globulins (Bjørneboe and Schwartz, 1959), or after infusion of dextran (Carbone *et al.*, 1955), and it has been suggested that albumin synthesis may be regulated by the colloid osmotic pressure of the plasma or of the extracellular fluid within the liver (Rothschild *et al.*, 1966). Attempts to demonstrate such a mechanism by infusing additional albumin into animals (Rothschild *et al.*, 1968b) or human subjects (Tavill *et al.*, 1968b) have yielded negative or equivocal results, partly because albumin catabolism accelerates and removes the excess albumin. However, Rothschild *et al.* (1969b) have tested the thesis of osmotic control of albumin synthesis on excised rabbit livers perfused with rabbit blood modified to provide differences in albumin content and/or osmotic pressure. When the albumin content of the perfusate was reduced, albumin synthesis was accelerated, whereas synthesis was depressed by raising the albumin content of the blood above the physiological range, or by adding sucrose to it. These results support the osmotic regulation hypothesis and agree with similar studies by Tracht *et al.* (1967) on perfused rat liver, but disagree with the findings of Katz *et al.* (1967), also on perfused rat liver. The nature of osmotic control over rate of albumin synthesis is unknown. The levels of other plasma proteins are also subject to regulation of both synthesis and degradation, though less is known about them. Synthesis of fibrinogen can be much increased as a result of defibrination of the blood (Barnhart, 1967), but excessive levels of fibrinogen do not appear to inhibit synthesis, the excess being removed by increased catabolism (Atencio *et al.*, 1969).

Many secreted proteins contain groups other than amino acid residues, commonly carbohydrate. The carbohydrate groups of glycoproteins are inserted by enzymes present in the membranes of the endoplasmic reticulum and in the plasma membrane of the cell surface (Hagopian *et al.*, 1968; Hagopian and Eylar, 1969). Plasma glycoprotein is thought to be synthesized by formation of the peptide chains on membrane-attached ribosomes in the liver. Some of the sugar residues are inserted on the end of this peptide while it is still attached to the ribosome, whereas the terminal sialic acid residue is implanted in the molecule by enzymes present in the smooth vesicles as the glycoprotein passes along toward the exterior of the liver cell (Lawford and Schachter, 1966; Simkin and Jamieson, 1967; Li *et al.*, 1968). A similar sequential addition of carbohydrate residues to the peptide chain occurs in the biosynthesis of thyroglobulin by membrane-attached ribosomes in the thyroid gland (R. G. Spiro and Spiro, 1966; M. J. Spiro and Spiro, 1968; Cheftel and Bouchilloux, 1968; Cheftel *et al.,* 1968). No doubt there are mechanisms that coordinate and regulate synthesis of the peptide and carbohydrate portions of glycoproteins.

Proteins with lipid components occur in the plasma. The liver has been identified as the source of plasma α-lipoprotein (Radding *et al.,* 1958) and the liver and intestine for β-lipoprotein (Windmueller and Levy, 1967, 1968). The function of the latter appears to be to carry triglycerides from both liver and intestine, so that patients with abetalipoproteinemia show an inability to transport triglycerides from these organs to the peripheral tissues. Administration of orotic acid to rats has been found to cause inhibition of production of the β-lipoprotein in the liver but not the intestine (Windmueller, 1964; Windmueller and Levy, 1967, 1968). Consequently, triglycerides accumulate in the liver, while plasma concentrations of triglycerides and of β-lipoprotein fall to low levels, but transport of fat as lipoprotein from the intestine to the rest of the body by way of the intestinal lymphatics is not interrupted. It appears that the suppression of β-lipoprotein secretion from the liver is due to changes in the adenine nucleotide pool of that organ, and can be reversed by adenine administration (von Euler *et al.,* 1963; Windmueller, 1964). These responses to orotic acid administration appear to be specific to the rat, since neither triglyceride transport nor adenine nucleotide pools in the livers of the mouse, chick, or monkey are affected by orotic acid administration (Valli *et al.,* 1968). The details of the relationship between adenine nucleotide metabolism and lipoprotein secretion are unknown. The role of triglyceride in regulating the formation of the apoprotein moiety has provoked some investigation. Possible models of lipoprotein synthesis are discussed by Trams and Brown (1966). Perfusion studies with rat liver (Ruderman *et al.,* 1968) showed that addition of fatty acids to the perfusate resulted in increased release of triglyceride bound to carrier protein that had been synthesized *de novo.* Addition of puromycin to such a perfusion system caused immediate termination of protein formation, and output of labeled fatty acids as triglycerides also ceased sharply thereafter (Buckley *et al.,* 1968). Puromycin does not, however, appear to block triglyceride transport in chylomicrons from the intestine (Redgrave and Zilversmit, 1969). These studies on perfused livers and also observations made on rats receiving orotic acid suggest an obligatory relationship between apolipoprotein formation and triglyceride transport from the liver, in which the fat plays a part in regulating the synthesis of the apolipoprotein. It would be a useful confirmation of this relationship to quantitate the amounts of triglycerides and of β-lipoprotein turned over per day; if the thesis of obligatory apolipoprotein formation for triglyceride transport is correct, the amount of apoprotein formed and degraded should be sufficient to account for the transport of all the triglyceride. In this connection, it may be noted that in patients with hyperbetalipoproteinemia, the rate of degradation of the apoprotein is reduced and β-lipoprotein thus accumulates in the plasma (Langer *et al.,* 1970).

B. Regulation of Amino Acid Metabolism

In addition to protein synthesis, free amino acids follow many other metabolic pathways, as summarized in Fig. 15. These pathways involve enzymes subject to control mechanisms basically similar to those identified in the regulation of bacterial metabolism. As discussed in Section II,A of this chapter, regulation of enzyme activity in the bacterial cell has been shown to occur through two principal mechanisms, namely (a) by varying the number of enzyme molecules through induction and repression at the gene level, and (b) by regulating the activity of the enzyme itself through feedback inhibition, commonly by an allosteric change in a key enzyme of a metabolic pathway caused by an end product of the pathway. These two basic principles of regulation are represented in the control of amino acid metabolism in mammals, which is discussed in detail in Chapter 35.

Mechanisms available for regulating the *quantities* of enzymes in mammalian cells have already been discussed (Section III,B,3). The *direct control* of catalytic activity remains to be discussed. The activities of mammalian enzymes are directly affected by concentrations of substrates, cofactors, and inhibitors acting on their active centers. Enzyme activity is also regulated by feedback inhibition or stimulation through sites other than the active center; these occur not only through allosteric changes, but also from homosteric inhibition (McElroy *et al.,* 1967) and by elastoplasticity, in which a metabolite structurally modifies the enzyme protein by reacting with an amino acid residue (Grisolia, 1968). Finally, multiple forms of enzymes (isozymes) are found in the pathways of mammalian amino acid metabolism. In the case of several of the transaminases, such as tyrosine aminotransferase (J. E. Miller and Litwak, 1969; Koler *et al.,* 1969), aspartate aminotransferase (Waksman and Rendon, 1968), and branched-chain amino acid transaminase (Aki *et al.,* 1967), the transaminase isozyme found in the mitochondria has been shown to differ from that in the cell sap.

Many of the enzymes involved in mammalian amino acid metabolism catalyze reactions similar to those found in bacteria, but, in view of evolutionary changes in protein structure, it is not surprising that

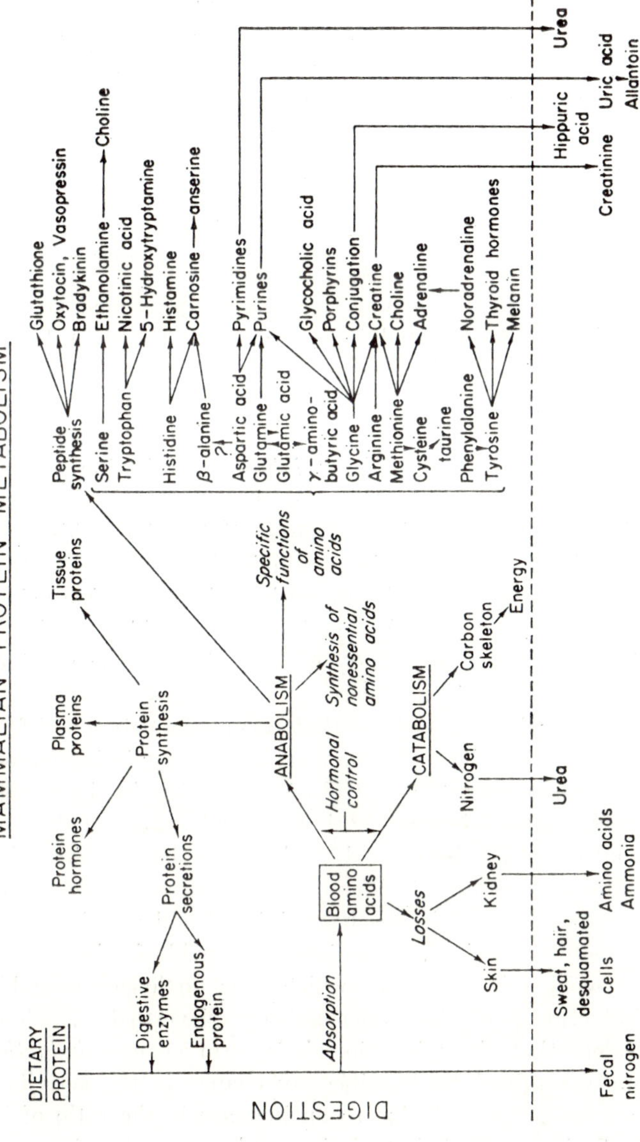

FIG. 15. Diagram showing the main pathways of mammalian protein metabolism. (Reprinted from Volume I page 32.)

allosteric and other feedback regulators differ. For example, aspartate transcarbamylase has been examined in normal and malignant mammalian tissues for feedback control by nucleotides (Heise and Görlich, 1966), but the responses differ from those obtained with the bacterial enzyme (Fig. 2). Another example is the so-called phosphorylated pathway of serine biosynthesis, which shows differences in the sites of feedback control in bacteria and mammals. This pathway takes the following form:

Fructose diphosphate → 3-phosphoglycerate → phosphohydroxypyruvate → phosphoserine → serine

Feedback inhibition of the enzyme regulating the second step in this pathway has been demonstrated in bacteria (Pizer, 1963) but not in cultured human cells in which serine inhibits the final step, namely cleavage of phosphoserine by a specific phosphatase (Pizer, 1964). There is, however, some evidence that, although the earlier enzymes in the pathway are not subject to *feedback inhibition,* addition of serine to the culture medium causes *repression* of their synthesis (Pizer, 1964). The regulation of the synthesis of nonessential amino acids by mammalian cells in tissue culture has been examined systematically by Eagle *et al.* (1965), who report that the formation of some amino acids, such as serine, is suppressed when that amino acid is added to the culture medium, whereas addition of other amino acids, such as alanine, does not inhibit biosynthesis in its pathway.

While these examples demonstrate that enzyme induction and feedback inhibition are represented among the mechanisms regulating enzyme activity in mammalian cells, the control of metabolism in the mammalian cell is also achieved by other means that take advantage of the complexities of animal structure, both at the subcellular level and at the level of the whole animal. Important differences between control of metabolism in the mammalian cell and in the bacterial cell will now be considered. Some of the features of the mammalian cell provide further refinements in regulation of protein synthesis, which thus no longer resembles the bacterial induction-repression mechanism. Nevertheless, Greengard (1967) considers that the microbiologist's term *enzyme induction* can still be usefully retained to describe any change in the activity of an enzyme in a mammalian cell due to an increase in the ratio of its rate of synthesis to its rate of degradation. Other aspects of mammalian control mechanisms described below involve regulation of the enzyme protein to perform its catalytic function.

1. The chromosomal apparatus is encapsulated within the nucleus. As discussed in Section III,A,1 of this chapter, the eukaryote nucleus

synthesizes appreciable amounts of HnRNA, which apparently does not pass into the cytoplasm. It is not presently known whether messenger RNA represents a part of this RNA that is selectively transferred to the cytoplasm. In any case, the presence of a nuclear envelope allows segregation of some of the RNA formed, and also requires the use of transport mechanisms of the informosome type to carry the message into the cytoplasm. Such processes are obvious sites for control.

2. Many animal cells secrete proteins that are made by polysomes attached to membranous secretory channels (Section III,A,2). Some unknown mechanism must direct the messenger for such proteins to the appropriate site for secretion.

3. Protein synthesis occurs in the cell cytoplasm and is subject to regulatory factors present in the cytoplasm. These occur both at the stage of translation, such as availability of tRNA species, and also through post-translational modification of the protein, such as attachment of a prosthetic group.

4. Unlike the bacterial cell during logarithmic growth, the mammalian cell shows significant degradation of proteins, and in consequence enzyme concentration can be regulated through variations in rate of breakdown of the enzyme. This type of regulation is discussed at length in Chapter 32.

5. Location is also a means by which mammalian cells provide additional regulation mechanisms for enzymes. This is well known in the case of enzymes related to the tricarboxylic acid cycle, some of which occur both within and outside the mitochondria and are thus subject to different environmental factors. This is also the case for tyrosine aminotransferase. The cell sap and mitochondrial forms of this enzyme are different species of protein (J. E. Miller and Litwack, 1969; Koler *et al.*, 1969). Both transferases are inhibited by excess of glutamate; however, this reaction product does not accumulate in the mitochondria, where it is rapidly transformed to α-ketoglutarate. A liver mitochondrial transaminase has been described for leucine which differs from the cell sap enzyme in its response to lack of protein in the diet (McFarlane and von Holt, 1969) and there is no doubt that future investigators of enzyme regulation will have to pay more attention to multiple forms of enzymes and the effects of location on enzyme function. A different example of the importance of intracellular enzyme location is provided by the synthesis of lactose in the mammary gland. Synthesis of this disaccharide requires two proteins, one being a galactosyltransferase, the other α-lactalbumin, a major protein of milk. The lactalbumin acts by specifying the transfer of the galactose to glucose as the preferred acceptor, so that lactose is the only product of reaction (Brew *et al.*,

1968). It now appears that lactalbumin is made on polysomes attached to the endoplasmic reticulum of the mammary gland, and thereafter this protein passes along the channels of the reticulum until it meets the transferase in the Golgi region, where lactose biosynthesis takes place, and thereafter the whole enzymic complex with lactose is secreted into the milk (Brew, 1969). Location of enzymes can be under genetic control. This was observed by Ganschow and Paigen (1967) when studying β-glucuronidase, which is present in the lysosomes and also in the endoplasmic reticulum membranes of mouse liver. A mutant mouse strain was found in which the enzyme was absent from the endoplasmic reticulum but not from the lysosomes, and this was demonstrated to depend on a gene other than that for the glucuronidase protein messenger.

6. In multicellular organisms, metabolism in the cells of one tissue can be regulated by metabolic products coming from another tissue. Hormones, which represent a specialized form of such product, will be discussed in the next section. In addition, however, there is a continuous exchange of metabolites between tissues, and these can cause regulatory changes in tissues receiving these metabolites. For example, administration of phlorizin causes rapid breakdown of carcass protein and in consequence, additional amounts of amino acids pass into the blood and the levels of free threonine and serine in the liver rise extensively; some time later, the serine-threonine dehydratase of the liver increases and then the concentrations of these amino acids decline (Bojanowska and Williamson, 1968). A striking example of such indirect actions is illustrated by the rapid effect of insulin in suppressing gluconeogenesis in the liver. This appears to be due indirectly to the antilipolytic action of insulin, which prevents release of fatty acids from the fat depots (Friedmann *et al.*, 1967). It has been found that free fatty acids have a stimulant effect on gluconeogenesis (Williamson *et al.*, 1966). This type of indirect effect makes interpretation of hormonal and other actions on whole animals difficult, and in the past has led to many oversimplifications. Even within a single organ, different types of cells may cooperate in producing a reaction. Thus the synthesis of antibodies by the spleen seems to require the interaction of at least two types of cell (Mosier and Coppleson, 1968). Agarwal *et al.* (1969) have recently performed studies on the induction of tryptophan pyrrolase by cortisol administration, in which they claim that blockade of the reticuloendothelial (Kupffer) cells of the liver by administration of Thorotrast prevents the response from occurring in the hepatocytes.

7. As discussed in detail in Chapter 33, hormones act as regulators of metabolism in the cells of the target organ, though it has usually proved difficult to identify the primary changes caused by each hormone.

The following points represent some of the problems facing the investigator of hormonal action at the subcellular level.

a. Each hormone has some degree of specificity regarding the tissues it influences, and this has led to the search in the target organs for receptor molecules that bind the hormones specifically. For example, Toft *et al.* (1967) have extracted from rat uterus a soluble protein which binds estrogens but not other steroids. A second smaller estrogen-binding protein has been obtained from uterine nuclei, and the evidence suggests that the cell-sap binding factor enters the nucleus and either transfers the steroid to the nuclear protein or becomes the nuclear protein carrier through dissociation (Jensen *et al.*, 1968; Shyamala and Gorski, 1969; Korenman and Rao, 1968).

b. The primary effect(s) of the hormone on metabolism in the target cell has been extensively sought. In the case of hormones such as estrogens, which appear to concentrate in target cell nuclei, it is reasonable to expect that this will result in a change in RNA synthesis. Indeed, Dahmus and Bonner (1965) observed that rat liver chromatin becomes a better template for bacterial RNA polymerase if the rat is pretreated with cortisone, and it has been demonstrated that arginine-rich histones prepared from liver (Sluyser, 1966) and from calf thymus (Sluyser, 1969) bind cortisol tenaciously. Since histones may play a part in regulating template exposure to the polymerase (Section III,A,1), this binding effect may appear to indicate the primary cause of the well-known increase in liver RNA synthesis after giving corticosteroids (Chapter 31). However, costicosteroids have the opposite action on thymus RNA synthesis, which is extensively *diminished* by the steroid (Makman *et al.*, 1968). The specificity of hormone responses in the liver and thymus gland is therefore likely to reside in some feature of each target organ other than the arginine-rich histones.

c. Some striking examples of the appearance of new cell proteins following hormone administration, such as induction of the enzyme DOPA decarboxylase by administering molting hormone to insect pupae (Karlson, 1968), have tended to overshadow the fact that a major product of hormonal stimulation of RNA synthesis is usually ribosomal RNA, which has been observed as an early and even major product of estrogen action on the uterus (Hamilton *et al.*, 1968), of cortisol action on liver (Drews and Brawerman, 1967; S. T. Jacob *et al.*, 1969b), of testosterone action on prostate (Liao and Stumpf, 1968), and of triiodothyronine and growth hormone action on liver cells (Wyatt and Tata, 1968). As pointed out earlier, the efficient utilization of messenger may be enhanced by the supply of other RNA species, such as ribosomal RNA. This could explain the finding of Marver *et al.* (1966b) that the injection of cortisol into adrenalectomized rats does not induce δ-aminolevulinic acid synthetase in their livers, but the presence of corticosteroids is necessary in order to permit allylisopropylacetamide to function as an inducer of this enzyme.

d. The expression of increased protein synthesis as a result of hormonal action can involve regulation of other major cell components, such as formation of membranes of the endoplasmic reticulum (Tata, 1967b, 1968). It is not known whether these ancillary effects represent separate regulatory actions of the hormone, or in what way they are coordinated with the increase in protein synthesis.

e. As discussed in Chapter 34, an increased inflow of amino acids, even nonmetabolizable model amino acids, usually accompanies the stimulant action of hormones on protein synthesis in specific target organs such as the uterus, but there is little to support the view that this influx is a trigger mechanism in producing the increased formation of protein. However, it is possible that protein synthesis in the liver may benefit from increased flow of amino acids from the carcass after corticosteroid administration. In the case of the inflow of amino acids into the uterus following estrogens, this would appear to be due to release of cyclic AMP (Griffin and Szego, 1968), an action that may be an independent effect of the hormone from the well-known changes in RNA formation.

f. The initial rapid stimulation of protein synthesis in the liver following thyroid hormone administration is said to be dependent on a primary mitochondrial effect and independent of increased synthesis of RNA by the nucleus (Sokoloff *et al.*, 1968).

g. Hormones may act directly in regulating the activities of enzymes involved in amino acid metabolism. Glutamic dehydrogenase activity is depressed by *in vitro* addition of estrogens (Tomkins *et al.*, 1963, 1965), an action whose physiological significance has yet to be evaluated.

IV. Regulation of Protein Metabolism in the Whole Animal

The mammal consists of a series of organs and tissues with differing regulatory mechanisms, which are described in the later chapters of this volume dealing with regulation of protein metabolism in the gastrointestinal tract (Chapter 37), in the liver (Chapter 38), in the kidney (Chapter 39), and in skeletal and cardiac muscle (Chapter 40). In consequence, regulation of protein metabolism in the whole animal is made up of the sum total of the regulatory responses in these individual parts of the body. Chapter 35 describes in detail regulation mechanisms affecting individual pathways of amino acid metabolism. Here, we shall consider some general features of protein metabolism in the intact animal, dealing first with the general regulation of the pathways of protein metabolism in the whole animal, then with the contribution of individual tissues to the regulation of protein metabolism, followed finally by the overall action of hormones on protein metabolism.

A. General Regulation of Metabolic Pathways

The major reactions undergone by free amino acids within the body of the mammal are displayed in Fig. 15. Although this presents a diverse and complicated picture, for descriptive purposes the reactions undergone by free amino acids can be grouped into three categories: (a) synthesis of proteins; (b) synthesis of small molecules such as purines, creatine, etc.; (c) degradative reactions in which the carbon skeleton is removed as CO_2 or deposited in the tissues as carbohydrate or fat, and the nitrogen forms urea. When one of these routes of disposal of free amino acids alters, there is frequently a reciprocal compensatory change in other pathways. This is notably seen in the relationship between intensity of body protein synthesis and rate of amino acid degradation. Thus, administration of puromycin along with phenylalanine-^{14}C to rats results in a reduced uptake of the phenylalanine into tissue proteins, but causes an increase in the plasma levels of most free amino acids, and greater degradation of the labeled amino acid to $^{14}CO_2$ (Godin, 1967b). This diversion of amino acids from protein synthesis to degradation has also been observed in isolated perfused livers on addition of puromycin (John and Miller, 1966) or aflatoxin (John and Miller, 1969) to the perfusate. The opposite type of change is produced by inhibiting amino acid catabolism with the drug hydrazine. This interferes with the function of pyridoxal phosphate in transamination (McCormick and Snell, 1961). Animals treated with hydrazine show a reduction in amino acid degradation, an increase in the levels of free amino acids in the blood and tissues, and increased protein synthesis in the liver (Amenta and Johnston, 1963).

These various routes of disposal of amino acids can be investigated in the whole animal by administering ^{14}C-labeled amino acids and following their fate. This not only allows the relative magnitude of different pathways to be quantitated, but also provides information about regulatory responses in metabolism. In the older literature, there are a number of observations showing the percentage of $^{14}CO_2$ formed, and in some cases the proportion of ^{14}C deposited in tissues as protein and in other forms, a few hours after administration of ^{14}C-labeled leucine (Borsook *et al.*, 1950; Ul-Hassan and Greenberg, 1952), histidine (Borsook *et al.*, 1950; Lim-Sylianco and Berg, 1969), lysine (Borsook *et al.*, 1950), glycine (Borsook *et al.*, 1950), tyrosine (Winnick *et al.*, 1948; Dalgliesh and Tabechian, 1956; Godin and Gajda, 1967), phenylalanine (Dalgliesh and Tabechian, 1956; Godin, 1967a), tryptophan (Henderson and Hankes, 1956; Mathur *et al.,* 1964; Hankes *et al.,* 1967a; Madras and Sourkes, 1968), valine (H. M. Cunningham, 1967), and methionine (Edwards *et al.,* 1960). Most of these studies involved injection of the labeled amino acid, and may not be representative of the disposal of amino acids under physiological conditions, especially when the dose of amino acid administered was large. In the past few years, several investigators have provided comprehensive pictures of the fate of an oral dose of a ^{14}C-labeled amino acid fed under defined conditions. For example, Soliman and King (1969) describe the corporeal distribution of uniformly labeled histidine-^{14}C at various times up to 6 hours after feeding a diet containing this and other unlabeled amino acids to rats. At 6 hours after administering a meal containing a well-balanced mixture of amino acids ("corrected" diet), the percentages of absorbed radioactivity recovered (1) in body protein, (2) in acid-soluble form, and (3) in CO_2, excreta, and tissue fat, were 47%, 37%, and 14%, respectively, of the administered histidine-^{14}C. When the rats were first conditioned to a similar diet with a much reduced histidine content ("histidine-imbalanced" diet), a larger percentage of the histidine radioactivity was retained as body protein and less was lost as CO_2, etc. Thus, the least abundant essential amino acid in the imbalanced diet was used more efficiently, indicating adaptation of its catabolism.

Harper (1968) has emphasized that the metabolic pathways for degradation of amino acids show adaptation in relation to intake. He has compared the behavior of different rat liver enzymes toward increasing levels of protein in the diet from none up to very large amounts. In the case of glutamic–pyruvic and glutamic–oxaloacetic transaminases, the hepatic enzyme levels rise in proportion to the protein content of the diet, whereas threonine dehydratase level does not increase with increasing protein intakes until an adequate amount of protein in the diet has been reached, at which point the activity of this liver enzyme rises sharply. This has prompted Harper to conclude that enzymes involved in the metabolism of nonessential amino acids undergo adaptation related to the amount of protein consumed, whereas enzymes for catabolism of essential amino acids adapt to dietary protein level as it relates to the needs of the body.

This is an attractive concept, and it is supported by the finding of McFarlane and von Holt (1969) that the overall oxidative degradation of leucine and phenylalanine, measured *in vivo* in rats fed on a 2% casein diet, shows a marked decrease, whereas oxidation of glutamate and alanine is unaffected by the low protein diet. Furthermore, they were able to correlate the lack of leucine oxidation by the whole animal with loss of leucine transaminase activity from the liver mitochondria of the protein-depleted rats; another leucine transaminase found in liver cell sap was not affected by the dietary deficiency of protein. Nevertheless, the concept requires further exploration. For example, it is necessary to take into account the behavior of degradative enzymes present in tissues other than the liver. Thus, Mimura *et al.* (1968) have found that the transaminases for the branched-chain amino acids in muscle, kidney, and liver do not decrease in activity when rats are transferred from a normal to a low intake of protein. However, this observation may not take account of possible leucine transaminase activity associates with mitochondria, which in the case of liver mitochondria is sensitive to protein deficiency in the diet (McFarlane and von Holt, 1969). As pointed out in an earlier section, mitochondrial enzyme species distinct from the corresponding enzyme in the cell sap have now been recognized for several of the key degradative reactions of amino acid metabolism, and in future it will be necessary to assay these separately. What is now required is a series of studies similar to those of McFarlane and von Holt (1969), in which the utilization of labeled amino acids under different dietary and other conditions is correlated with changes in the enzymes of amino acid metabolism *throughout the body.*

As would be expected, excessive intake of amino acids leads to adaptive increases in degradative pathways. For example, it has been demonstrated in the rat (Hankes *et al.*, 1967a) and in man (Hankes *et al.*, 1967b) that conversion of tryptophan-^{14}C to $^{14}CO_2$ is markedly increased when a loading dose of tryptophan is administered along with the labeled amino acid. This is presumably due to increased activity of tryptophan pyrrolase and other enzymes of the tryptophan degradative pathway. Daniel and Waisman (1969) have made a comprehensive study in rats of enzyme adaptation to excessive intake of methionine and have found that, as the rats became accustomed to the diet, enzymes involved in taurine formation from methionine decreased in activity, whereas enzymes channeling the excess methionine to yield pyruvate showed increases in activity. They describe cogently why increasing certain pathways is advantageous to the animal receiving a large intake of methionine. The effects of excessive intakes of amino acids are discussed in Chapter 13, in Volume II of this treatise.

Finally, attention should be drawn to the use of respiration pattern analysis in studying the metabolic pathways of amino acids in the whole animal. This technique involves determination of the specific activity of CO_2 after administration of ^{14}C-labeled compounds. From the time-course of $^{14}CO_2$ evolution, some deductions about the metabolic pathway can be made. For example, Kretchmar and Price (1969) describe a study of serine metabolism in which mice were injected with labeled serine, formate, or bicarbonate and the $^{14}CO_2$ was collected for 60 minutes thereafter. When unlabeled formate was given along with serine-3-^{14}C, the pattern of $^{14}CO_2$ output was altered in normal mice but not in mice receiving a folic acid antagonist. It was concluded from these and other experiments that 71% of the C-3 atom of serine is oxidized in fasting mice by way of formate, and 29% by other pathways. The applications of respiration pattern analysis in the study of intermediary metabolism are discussed by Tolbert (1963).

B. Metabolic Regulation in Individual Tissues

In the mammal, protein metabolism is achieved through a series of reactions that are both integrated and cooperative. First, the reactions are *integrated* because, as pointed out above, there is a continuous exchange of metabolites between tissues, and metabolic regulations are made in response to this flow of compounds arriving at each cell. The impact of free amino acids circulating in the plasma varies from tissue to tissue. One reason for this difference between the sensitivity of tissues is the degree to which protein turnover contributes to the internal free amino acid pool. In the case of the liver, recycling of amino acids from liver protein degradation provides half the free pool of the fed rat, and 90% of the pool of the fasting rat (Gan and Jeffay, 1967). Consequently, the endogenous supply of amino acids can profoundly modify changes imposed by the external supply. Nevertheless, liver metabolism appears to be quite sensitive to changes in amino acid supply (Chapters 34 and 38).

Second, the metabolism of amino acids in the whole animal is *cooperative* because not every tissue can perform the complete series of reactions occurring in the intact organism. Thus glutamine is not an essential amino acid for the mammal since the liver and intestine perform the synthesis of enough glutamine for the requirements of other tissues (Chapter 25, Section II,A, in Volume III). Similarly, the synthesis of tyrosine from phenylalanine is normally dependent on phenylalanine hydroxylase in the liver, though this reaction has recently been shown to occur also in the pancreas and kidney (Tourian *et al.*, 1969) and in the mammary gland (Jorgensen and Larson, 1968). It is notorious that urea synthesis is limited to the liver (Chapter 5, Volume I), Perfusion studies with labeled amino acids (Table III, Chapter 38) show that the liver is also required for the degradation to CO_2 of arginine, histidine, lysine, methionine, phenylalanine, threonine, tryptophan, and tyrosine, whereas isoleucine, leucine, valine, and the nonessential amino acids are degraded mainly in the peripheral tissues. This evidence receives some support from studies of enzyme distributions in different tissues. Tryptophan pyrrolase is usually considered to be present exclusively in the liver of the mammal; however, Yamamoto and Hayaishi (1967) have recently described its occurrence in rabbit intestinal wall, though its physiological importance here remains uncertain. The lack of degradation of methionine-^{14}C to $^{14}CO_2$ by hepatectomized rats (Table III, Chapter 38) is, however, misleading since other tissues contain enzymes involved in methionine degradation (Finkelstein, 1967). The reason for the discrepancy may be that the carbon skeleton of methionine forms α-oxobutyrate after donating the sulfur to form cysteine, and the liver may be necessary for the further metabolism of α-oxobutyrate to CO_2. The lack of degradation of lysine in hepatectomized rats may relate to the high concentrations in the liver of the enzyme lysine-ketoglutarate reductase (Hutzler and Dancis, 1968), which produces saccharopine, a likely intermediate in lysine breakdown (Grove *et al.,* 1969). In the case of the branched-chain amino acids, the extrahepatic degradation of these is well supported by the finding of high concentrations of a specific transaminase in heart, kidney, and skeletal muscle and in very much smaller concentrations in liver (Ichihara and Koyama, 1966).

These observations indicate that enzymes regulating the reactions of amino acid metabolism are distributed differently in various tissues. Consequently, an influx of amino acids into the plasma after a meal of protein or as a result of discharge from the tissues will be dealt with by increased catabolism taking place partly in the liver and partly in the peripheral tissues. Amino nitrogen resulting from peripheral degradative reactions is then transported to the liver mainly in the form of alanine (Felig *et al.*, 1969a), where it is excreted as urea. The magnitude of such cooperative tissue interchanges during the course of amino acid metabolism have not been sufficiently studied to allow us to assemble a satisfactory quantitative picture of the contribution of individual tissues to amino acid metabolism in the whole animal.

C. Role of Hormones in the Protein Metabolism of the Whole Animal

Hormones play an important role in the integration of adaptive changes in the protein metabolism of the whole animal. It is not possible here to review all the ways in which endocrine secretory activity is known to change the metabolism of amino acids, but some general features deserve comment.

1. Many hormones are continuously present in the blood, and thus are normal constituents of the medium surrounding the cells of the body.

Consequently, excision of an endocrine gland not only prevents the gland from responding to a stimulus, but also deprives the tissues of the normal circulating level of the hormone. In the case of the liver, the synthesis of RNA is influenced by corticosteroids, and therefore the ribosome population of the liver cell diminishes considerably in adrenalectomized rats (Enwonwu and Munro, 1969). This may explain why certain reactions involving the liver require the presence of corticosteroids even if they do not directly stimulate the reaction. An example of this permissive action of corticosteroids is seen in the case of induction of δ-aminolevulinic acid synthetase discussed in Section III,A.

2. Alterations in the blood levels of hormones cause changes in the metabolism of the target organ. The speed with which the secretory activity of an endocrine gland can significantly alter the blood level of its hormone is related to the half-life of the hormone in the body, which seems to be dependent on species size, as judged by the data for thyroxine turnover (see Chapter 25 in Volume III). In man, the half-lives of hormones range from 8 minutes for insulin (Tomasi *et al.*, 1967) through 1.5 hours for free cortisol (Forsham and Melman, 1968) to 16 hours for triiodothyronine and 6.7 days for thyroxine (Ingbar and Woeber, 1968).

3. The magnitude of the change in the overall protein metabolism of the animal produced by hormonal action depends on whether the hormone affects amino acid utilization in major metabolic tissues (liver, muscle, etc.) or whether its effects are restricted to lesser tissues (such as the action of physiological doses of estrogen on the uterus). In the case of hormones producing measurable changes in overall protein metabolism, the protein content of the body may either be increased by anabolic hormones (e.g., androgens, growth hormone) or decreased by catabolic hormones (e.g., adrenocorticoids, thyroid hormones) as discussed in detail in Chapter 10, Section IV, in Volume I.

4. Most hormones display some degree of specificity regarding the tissues which they influence. Consequently, hormones having a major effect on body protein metabolism generally alter the distribution of amino acid metabolism between tissues. Redistribution of body protein due to hormone action has already been discussed in Section IV of Chapter 10, and only a few features of that survey will be repeated here. For example, the androgens, which are anabolic in action, tend to cause a preferential increase in the protein content of skeletal muscle. On the other hand, the corticosteroids, which are catabolic, decrease the protein content of muscle but cause protein to accumulate in the liver. The latter effect also characterizes many responses of body protein metabolism to environmental factors. Thus, as discussed in Chapter 10,

the liver gains protein while the body as a whole is losing protein following general exposure to X-rays, in tumor-bearing rats, or in the case of the alloxan-diabetic rat. In each of these instances, it appears that this paradoxical effect is due to stimulation of adrenocortical secretion, and thus represents a general stress phenomenon. Nevertheless, stress applied to fasting adrenalectomized rats also stimulates protein synthesis in the liver; this appears to be mediated through insulin secretion (Leon and Chackerian, 1968).

TABLE III
CHANGES IN GROSS BODY COMPOSITION DUE TO ADMINISTRATION OF HORMONES TO RATS[a]

Measurement	Growth hormone treatment[b]	Testosterone treatment[c]	Cortisone treatment[d]
Duration of treatment (days)	21	23	14
Initial body weight (gm)	211	162	229
Weight change (gm)	+81	+16	−8
Body protein (gm)	+19	+5.2	−1.7
Body fat (gm)	−12	−4.1	+4.5
Body water (gm)	+70	+14.5	—
Mineral ash (gm)	+5	+0.3	—

[a] The data are expressed as the difference between the body composition of the hormone-treated group and of controls fed the same amount of diet.

[b] Data of Young (1945).

[c] Data of Korner and Young (1955). The change in body mineral content has been calculated from the difference between the change in body weight minus the change in other body constituents.

[d] Data of Silber and Porter (1953). The data represent the difference in body composition of two groups of adrenalectomized rats pair-fed a low protein diet and given either a small or a large daily dose of cortisone.

5. The action of hormones on protein metabolism has to be viewed as part of a general effect on metabolism. This is illustrated in Table III for three hormones, two of which (growth hormone, testosterone) are anabolic and cause a net gain in body protein while at the same time decreasing the fat content of the body as compared with control animals receiving the same food intake. The third hormone, cortisone, induces a substantial loss of body protein but a gain in body fat.

6. The relationship between hormones and amino acid metabolism is a two-way interaction. There is good reason to believe that the levels of free amino acids in the plasma act as regulators for the secretion of insulin, and possibly also for the secretion of glucagon, growth hormone, and corticosteroids. The detailed evidence is discussed in Chapter

34, Section III,B. Felig *et al.* (1969b) propose a feedback loop in which an increase in free amino acid levels in the plasma stimulates insulin secretion which in turn causes the amino acids to be rapidly deposited in muscle, thus lowering the plasma level of amino acids again. This may indeed be an important mechanism in the regulation of plasma amino acid concentration, since muscle acts as an extensive reservoir for free amino acids (Chapter 34, Section II,A, in this volume), particularly in the case of certain amino acids such as lysine (Chapter 40, Section IV).

These various relationships of hormones to the overall amino acid metabolism of the body illustrate how a comprehensive description of the regulation of protein metabolism can involve extensive exploration of regulatory mechanisms operating in various tissues. Consequently, the chapters in this volume have been arranged to provide an overview of the mechanisms involved in regulation of protein metabolism, first at the subcellular level, and second at the level of individual tissues and organs.

References

Abelson, P. H. (1954). *J. Biol. Chem.* **206,** 335.

Acs, G., Reich, E., and Valanju, S. (1963). *Biochim. Biophys. Acta* **76,** 68.

Adamson, S. D., Herbert, E., and Godchaux, W. (1968). *Arch. Biochem. Biophys.* **125,** 671.

Adamson, S. D., Herbert, E., and Kemp, S. F. (1969). *J. Mol. Biol.* **42,** 247.

Adiga, P. R., Hussa, R. O., Robertson, M. C., Hohl, H. R., and Winnick, T. (1968). *Proc. Natl. Acad. Sci. U.S.* **60,** 606.

Agarwal, M. K., Hoffman, W. W., and Rosen, F. (1969). *Biochim. Biophys. Acta* **177,** 250.

Ahmed, A., Case, M., and Giles, N. H. (1964). *Brookhaven Symp. Biol.* **17,** 53.

Aki, K., Ogawa, K., Shirai, A., and Ichihara, A. (1967). *J. Biochem.* (*Tokyo*) **62,** 610.

Allen, R. E., Raines, P. L., and Regen, D. M. (1969). *Biochim. Biophys. Acta* **190,** 323.

Allfrey, V. G. (1968). *In* "Regulatory Mechanisms for Protein Synthesis in Mammalian Cells" (A. San Pietro, M. R. Lamborg, and F. T. Kenney, eds.), p. 65. Academic Press, New York.

Allfrey, V. G., Faulkner, R., and Mirsky, A. E. (1964). *Proc. Natl. Acad. Sci. U.S.* **51,** 786.

Allfrey, V. G., Pogo, B. G. T., Pogo, A. O., Kleinsmith, L., and Mirsky, A. E. (1966). *In* "Histones" (A. V. S. de Rueck and J. Knight, eds.), p. 42. Churchill, London.

Amenta, J. S., and Johnston, E. H. (1963). *Lab. Invest.* **12,** 921.

Ames, B. N., and Garry, B. (1959). *Proc. Natl. Acad. Sci. U.S.* **45,** 1453.

Ames, B. N., Goldberger, R. F., Hartman, P. E., Martin, R. G., and Roth, J. R. (1967). *In* "Regulation of Nucleic Acid and Protein Biosynthesis" (V. V. Koningsberger and L. Bosch, eds.), p. 272. Elsevier, Amsterdam.

Anderson, M. B., and Cherry, J. H. (1969). *Proc. Natl. Acad. Sci. U.S.* **62,** 202.
Anderson, W. F. (1969). *Proc. Natl. Acad. Sci. U.S.* **62,** 566.
Anderson, W. F., and Gilbert, J. M. (1969). *Biochem. Biophys. Res. Commun.* **36,** 456.
Andersson-Cedergren, E., and Karlsson, U. (1967). *J. Ultrastruct. Res.* **19,** 409.
Andrews, T. M., and Tata, J. R. (1968). *Biochem. Biophys. Res. Commun.* **32,** 1050.
Anraku, Y. (1968). *J. Biol. Chem.* **243,** 3116, 3123, and 3128.
Anthony, D. D., Zeszotek, E., and Goldthwait, D. A. (1966). *Proc. Natl. Acad. Sci. U.S.* **56,** 1026.
Appel, S. H., Davis, W., and Scott, S. (1967). *Science* **157,** 836.
Arca, M., Frontali, L., Sapora, O., and Tecce, G. (1967). *Biochim. Biophys. Acta* **145,** 284.
Armentrout, S. A., and Weisberger, A. S. (1968). *Biochim. Biophys. Acta* **161,** 180.
Arnstein, H. H. V., and Rahamimoff, H. (1968). *Nature* **219,** 942.
Askonas, B. A., and Williamson, A. R. (1966). *Nature* **211,** 369.
Askonas, B. A., and Williamson, A. R. (1968). *Biochem. J.* **109,** 637.
Atencio, A. C., Joiner, K., and Reeve, E. B. (1969). *Am. J. Physiol.* **216,** 764.
Atkin, N. B., Mattinson, G., Becak, W., and Ohno, S. (1965). *Chromosoma* **17,** 1.
Atkinson, D. E. (1966). *Ann. Rev. Biochem.* **35,** Part I, 85.
Atkinson, D. E. (1969). *Ann. Rev. Microbiol.* **23,** 47.
Attardi, G., and Attardi, B. (1967). *Proc. Natl. Acad, Sci. U.S.* **58,** 1051.
Attardi, G., Parnas, H., Hwang, M. I. H., and Attardi, B. (1966). *J. Mol. Biol.* **20,** 145.
Baglioni, C., and Campana, T. (1967). *European J. Biochem.* **2,** 480.
Bagnara, A. S., and Finch, L. R. (1968). *Biochem. Biophys. Res. Commun.* **33,** 15.
Baich, A., and Johnson, M. (1968). *Nature* **218,** 464.
Baldwin, A. N., and Berg, P. (1966). *J. Biol. Chem.* **241,** 839.
Baliga, B. S., Pronczuk, A. W., and Munro, H. N. (1968). *J. Mol. Biol.* **34,** 199.
Baliga, B. S., Pronczuk, A. W., and Munro, H. N. (1969). *J. Biol. Chem.* **244,** 4480.
Balinsky, J. B. (1970). *In* "Comparative Biochemistry of Nitrogen Metabolism" (J. W. Campbell, ed.), Vol. 2. Academic Press, New York.
Ballard, P. L., and Williams-Ashman, H. G. (1966). *J. Biol. Chem.* **241,** 1602.
Baltus, E., Quertier, J., Ficq, A., and Brachet, J. (1965). *Biochim. Biophys. Acta* **95,** 408.
Bank, A., Braverman, A. S., O'Donnell, J. V., and Marks, P. A. (1968). *Blood* **31,** 226.
Bargellesi, A., Pontremoli, S., and Conconi, F. (1967). *European J. Biochem.* **1,** 73.
Barnett, W. E., and Brown, D. H. (1967). *Proc. Natl. Acad. Sci. U.S.* **57,** 452.
Barnhart, M. I. (1960). *Am. J. Physiol.* **199,** 360.
Barnhart, M. I. (1967). *In* "Blood Clotting Enzymology" (W. H. Seegers, ed.), p. 217. Academic Press, New York.
Barnhart, M. I., and Anderson, G. F. (1961). *Federation Proc.* **20,** 52.
Beckwith, J. R. (1967). *Science* **156,** 597.
Behnke, O. (1963). *Exptl. Cell Res.* **30,** 597.
Bell, E. (1969). *Nature* **224,** 326.
Ben-Hamida, F., and Schlessinger, D. (1966). *Biochim. Biophys. Acta* **119,** 183.
Bennett, J., and Hallinan, T. H. (1968). *Life Sci.* **7,** 553.
Berendes, H. D. (1968). *Chromosoma* **24,** 418.

Bergeron-Bouvet, C., and Moulé, Y. (1966). *Biochim. Biophys. Acta* **123,** 617.
Bergquist, P. L., Burns, D. J. W., and Plinston, C. A. (1968). *Biochemistry* **7,** 1751.
Bernhard, W. (1966). *Natl. Cancer Inst. Monograph* **23,** 13.
Bettelheim, F. R. (1954). *J. Am. Chem. Soc.* **76,** 2838.
Bhatnagar, R. S., Prockop, D. J., and Rosenbloom, J. (1967). *Science* **158,** 492.
Bhuyan, B. K., and Smith, C. G. (1965). *Proc. Natl. Acad. Sci. U.S.* **54,** 566.
Bielka, H., Welfle, H., and Schneiders, I. (1967). *Acta Biol. Med. Ger.* **19,** K13.
Birbeck, M. S. C., and Mercer, E. H. (1961). *Nature* **189,** 558.
Birnstiel, M. L., and Flamm, W. G. (1964). *Science* **145,** 1435.
Birnstiel, M. L., and Hyde, B. B. (1963). *J. Cell Biol.* **18,** 41.
Bishop, J. O. (1966). *J. Mol. Biol.* **17,** 285.
Bjørneboe, M., and Schwartz, M. (1959). *J. Exptl. Med.* **110,** 259.
Blackburn, K. J., and Klemperer, H. G. (1967). *Biochem. J.* **102,** 168.
Blobel, G., and Potter, V. R. (1966a). *Science* **154,** 1662.
Blobel, G., and Potter, V. R. (1966b). *Proc. Natl. Acad. Sci. U.S.* **55,** 1283.
Blobel, G., and Potter, V. R. (1967a). *J. Mol. Biol.* **26,** 279.
Blobel, G., and Potter, V. R. (1967b). *J. Mol. Biol.* **26,** 293.
Blobel, G., and Potter, V. R. (1967c). *J. Mol. Biol.* **28,** 539.
Blum, N., Maleknia, N., and Schapira, G. (1969). *Biochim. Biophys. Acta* **179,** 448.
Bojanowska, K., and Williamson, D. H. (1968). *Biochim. Biophys. Acta* **159,** 560.
Bond, H. E., Cooper, J. A., Courington, D. P., and Wood, J. S. (1969). *Science* **165,** 705.
Bonner, J., Dahmus, M. E., Fambrough, D., Huang, R. C. C., Marushige, K., and Tuan, D. Y. H. (1968). *Science* **159,** 47.
Booyse, F. M., and Rafelson, M. E. (1967). *Biochim. Biophys. Acta* **145,** 188.
Borsook, H., Deasy, C. L., Haagen-Smit, A. J., Keighley, C., and Lowy, P. H. (1950). *J. Biol. Chem.* **187,** 839.
Borun, T. W., Scharff, M. D., and Robbins, E. (1967a). *Biochim. Biophys. Acta* **149,** 302.
Borun, T. W., Scharff, M. D., and Robbins, E. (1967b). *Proc. Natl. Acad. Sci. U.S.* **58,** 1977.
Bouvet, C., and Moulé, Y. (1964). *Exptl. Cell Res.* **33,** 330.
Braverman, A. S., and Bank, A. (1969), *J. Mol. Biol.* **42,** 57.
Bremer, H., and Bruner, R. (1968). *Mol. Gen. Genet.* **101,** 6.
Bremer, H., and Mueller, K. (1969). *J. Mol. Biol.* **43,** 109.
Bremer, H., Yegian, C., and Konrad, M. (1966). *J. Mol. Biol.* **16,** 94.
Bretscher, M. (1968). *Nature* **218,** 675.
Brew, K. (1969). *Nature* **223,** 671.
Brew, K., Vanaman, T. C., and Hill, R. L. (1968). *Proc. Natl. Acad. Sci. U.S.* **59,** 491.
Brewer, E. N., Foster, L. B., and Sells, B. H. (1969). *J. Biol. Chem.* **244,** 1389.
Brown, D. D. (1966). *Natl. Cancer Inst. Monograph* **23,** 297.
Brown, D. D. (1968). *Current Topics Develop. Biol.* **2,** 47.
Brown, J. N., and Dodgen, C. L. (1968). *Biochim. Biophys. Acta* **165,** 463.
Bruns, G. P., and London, I. M. (1964). *Biochem. Biophys. Res. Commun.* **18,** 236.
Buckley, J. T., Delahunty, T. J., and Rubinstein, D. (1968). *Can. J. Biochem.* **46,** 341.
Bucknall, R. A., and Carter, S. B. (1967). *Nature* **213,** 1099.

Burgess, R. R., Travers, A. A., Dunn, J. J., and Bautz, E. K. F. (1969). *Nature* **221,** 43.

Burka, E. R., Schreml, W., and Kick, C. (1967a). *Biochem. Biophys. Res. Commun.* **26,** 334.

Burka, E. R., Schreml, W., and Kick, C. (1967b). *Biochemistry* **6,** 2840.

Burnett, G., and Kennedy, E. P. (1954). *J. Biol. Chem.* **211,** 969.

Busch, H. (1965). "Histones and Other Nuclear Proteins." Academic Press, New York.

Busch, H., Desjardins, R., and Grogan, D. E., Higashi, K., Jacob, S. T., Muramatsu, M., Ro, T. S., and Steele, W. J. (1966). *Natl. Cancer Inst. Monograph* **23,** 193.

Busch, H., Narayan, K. S., and Hamilton, J. (1967). *Exptl. Cell Res.* **47,** 329.

Bustin, M., and Cole, R. D. (1968). *J. Biol. Chem.* **243,** 4500.

Bustos-Valdes, S. E., Deisseroth, A., and Dounce, A. L. (1968). *Arch. Biochem. Biophys.* **126,** 848.

Byvoet, P. (1966). *J. Mol. Biol.* **17,** 311.

Callan, H. G. (1963). *Intern. Rev. Cytol.* **15,** 1.

Callan, H. G. (1967). *J. Cell Sci.* **2,** 1.

Campbell, P. N., Lawford, G. R., and Gonzalez-Cadavid, N. F. (1968). *Proc. 7th Intern. Congr. Biochem., Tokyo,* 1967 Symp. No. 2, p. 123. Sci. Council Japan, Tokyo.

Capecchi, M. R. (1967). *Proc. Natl. Acad. Sci. U.S.* **58,** 1144.

Carbon, J. A., and Curry, J. B. (1968). *Proc. Natl. Acad. Sci. U.S.* **59,** 467.

Carbone, J. V., Uzman, L. L., and Plough, I. C. (1955). *Proc. Soc. Exptl. Biol. Med.* **90,** 68.

Caro, L. G., and Palade, G. E. (1964). *J. Cell Biol.* **20,** 473.

Cashel, M. (1969). *J. Biol. Chem.* **244,** 3133.

Cashel, M., and Gallant, J. (1968). *J. Mol. Biol.* **34,** 317.

Cashel, M., and Gallant, J. (1969). *Nature* **221,** 838.

Cassio, D., Lemoine, F., Waller, J. P., Sandrin, E., and Boissonas, R. A. (1967). *Biochemistry* **6,** 827.

Castles, J. J., and Singer, M. F. (1969). *J. Mol. Biol.* **40,** 1.

Ceccarini, C., Maggio, R., and Barbata, G. (1967). *Proc. Natl. Acad. Sci. U.S.* **58,** 2235.

Chalkley, G. R., and Maurer, H. R. (1965). *Proc. Natl. Acad. Sci. U.S.* **54,** 498.

Chaloupka, J., and Kreckova, P. (1966). *Fol. Microbiol.* **11,** 82 and 89.

Chambon, P. (1968). *Bull. Soc. Chim. Biol.* **50,** 349.

Chambon, P., Ramuz, M., Mandel, P., and Doly, J. (1967). *Biochim. Biophys. Acta* **149,** 584.

Chambon, P., Ramuz, M., Mandel, P., and Doly, J. (1968). *Biochim. Biophys. Acta* **157,** 504.

Chan, S. K., and Cohen, P. P. (1964). *Arch. Biochem. Biophys.* **104,** 331 and 335.

Chandrasakharam, N., Fleck, A., and Munro, H. N. (1967). *J. Nutr.* **92,** 497.

Changeux, J. P., and Rubin, M. M. (1968). *Biochemistry* **7,** 553.

Changeux, J. P., Gerhart, J. C., and Schachman, H. K. (1968). *Biochemistry* **7,** 531.

Chantrenne, H. (1965). *Biochim. Biophys. Acta* **95,** 351.

Chauveau, J., Moulé, Y., and Rouiller, C. (1956). *Exptl. Cell Res.* **11,** 317.

Cheftel, C., and Bouchilloux, S. (1968). *Biochim. Biophys. Acta* **170,** 15.

Cheftel, C., Bouchilloux, S., and Chabaud, O. (1968). *Biochim. Biophys. Acta* **170,** 29.

Chefurka, W., and Hayashi, Y. (1966). *Biochem. Biophys. Res. Commun.* **24,** 633.

Chen, H. W., Hersh, R. T., and Kitos, P. A. (1968). *Exptl. Cell Res.* **52,** 490.

Cho-Chung, Y. S., and Pitot, H. C. (1968). *European J. Biochem.* **3,** 401.

Chu, L. L. H., and Fineberg, R. A. (1969). *J. Biol. Chem.* **244,** 3847.

Chytil, F. (1968). *J. Biol. Chem.* **243,** 893.

Clark, C. M., Naismith, D. J., and Munro, H. N. (1957). *Biochim. Biophys. Acta* **23,** 587.

Clark, C. M., Goodlad, G. A. J., and Munro, H. N. (1958). *Biochem. Pharmacol.* **1,** 213.

Clark, C. M., Goodlad, G. A. J., Chisholm, J., and Munro, H. N. (1960). *Nature* **186,** 719.

Clark, M. F. (1964). *Biochim. Biophys. Acta* **91,** 671.

Cline, M. J. (1968). *Brit. J. Haematol.* **14,** 21.

Cohen, L. B., Herner, A. E., and Goldberg, I. H. (1969). *Biochemistry* **8,** 1312.

Collier, R. J., and Cole, H. A. (1969). *Science* **164,** 1179.

Colombo, B., and Baglioni, C. (1966). *J. Mol. Biol.* **16,** 51.

Colombo, B., Vesco, C., and Baglioni, C. (1968). *Proc. Natl. Acad. Sci. U.S.* **61,** 651.

Conconi, F. M., Bank, A., and Marks, P. A. (1966). *J. Mol. Biol.* **19,** 525.

Corradino, R. A., and Wasserman, R. H. (1968). *Arch. Biochem. Biophys.* **126,** 957.

Crawford, I. P., Ito, J., and Hatanaka, M. (1968). *Ann. N.Y. Acad. Sci.* **151,** 171.

Creichton, T. E., and Yanovsky, C. (1966). *J. Biol. Chem.* **241,** 4616.

Cunningham, D. D., and Steiner, D. F. (1967). *Biochim. Biophys. Acta* **145,** 834.

Cunningham, D. D., Chou, S., and Steiner, D. F. (1969). *Biochim. Biophys. Acta* **171,** 67.

Cunningham, H. M. (1967). *J. Animal Sci.* **26,** 1332.

Dahmus, M. E., and Bonner, J. (1965). *Proc. Natl. Acad. Sci. U.S.* **54,** 1370.

Dalgliesh, C. E., and Tabechian, H. (1956). *Biochem. J.* **62,** 625.

Daniel, R. G., and Waisman, H. A. (1969). *J. Nutr.* **99,** 299.

Danon, D., and Cividalli, L. (1968). *Biochem. Biophys. Res. Commun.* **30,** 717.

Darnell, J. E., Jr. (1968). *Bacteriol. Rev.* **32,** 262.

Datta, P. (1969). *Science* **165,** 556.

DeKloet, S. R., and Strijkert, P. J. (1966). *Biochem. Biophys. Res. Commun.* **23,** 49.

Delgarno, L., and Gros, F. (1968). *Biochim. Biophys. Acta* **157,** 64.

DeMoss, J. A., and Wegman, J. (1965). *Proc. Natl. Acad. Sci. U.S.* **54,** 241.

De Recondo, A. M., Frayssinet, C., Lafarge, C., and LeBreton, E. (1966). *Biochim. Biophys. Acta* **119,** 322.

Dick, C., and Johns, E. W. (1968). *Biochem. J.* **109,** 15P.

Di Mauro, E., Snyder, L., Marino, P., Lamberti, A., Coppo, A., and Tocchini-Valentini, G. P. (1969). *Nature* **222,** 533.

Dingman, C. W., and Sporn, M. B. (1962). *Biochim. Biophys. Acta* **61,** 164.

Donaldson, G. R., Atkinson, M. R., and Murray, A. W. (1968). *Biochem. Biophys. Res. Commun.* **31,** 104.

Drews, J., and Brawerman, G. (1967). *J. Biol. Chem.* **242,** 801.

Drews, J., Brawerman, G., and Morris, H. P. (1968). *European J. Biochem.* **3,** 284.

Dreyfus, J. C., and Schapira, G. (1966). *Biochim. Biophys. Acta* **129,** 601.

Drysdale, J. W., and Munro, H. N. (1965). *Biochem. J.* **95,** 851.
Drysdale, J. W., and Munro, H. N. (1966). *J. Biol. Chem.* **241,** 3630.
Drysdale, J. W., and Munro, H. N. (1967). *Biochim. Biophys. Acta* **138,** 616.
Drysdale, J. W., Olafsdottir, E., and Munro, H. N. (1967). *J. Biol. Chem.* **243,** 552.
Eagle, H., Washington, C. L., and Levy, M. (1965). *J. Biol. Chem.* **240,** 3944.
Earl, D. C. N., and Morgan, H. E. (1968). *Arch. Biochem. Biophys.* **128,** 460.
Edlin, G., and Broda, P. (1968). *Bacteriol. Rev.* **32,** 206.
Edlin, G., and Stent, G. S. (1969). *Proc. Natl. Acad. Sci. U.S.* **62,** 475.
Edström, J. E., and Daneholt, B. (1967). *J. Mol. Biol.* **28,** 331.
Edwards, C. H., Gadsden, E. L., and Edwards, C. A. (1960). *J. Nutr.* **72,** 185.
Eidlic, L., and Neidhardt, F. C. (1965). *J. Bacteriol.* **89,** 706.
Eliasson, E., Bauer, G. E., and Hultin, T. (1967). *J. Cell Biol.* **33,** 287.
Ennis, H. L. (1966). *Mol. Pharmacol.* **2,** 444.
Enwonwu, C. O., and Munro, H. N. (1969). Unpublished observations.
Enwonwu, C. O., and Munro, H. N. (1970). *Arch. Biochem. Biophys.* (in press).
Eylar, E. H. (1965). *J. Theoret. Biol.* **10,** 89.
Ezekiel, D. H., and Blumenthal, A. (1968). *Biochim. Biophys. Acta* **161,** 494.
Ezekiel, D. H., and Brockman, H. (1968). *J. Mol. Biol.* **31,** 541.
Ezekiel, D. H., and Elkins, B. N. (1968). *Biochim. Biophys. Acta* **166,** 466.
Ezekiel, D. H., and Hutchins, J. E. (1968). *Nature* **220,** 276.
Farber, E., Shull, K. H., Villa-Treviño, S., Lombardi, B., and Thomas, M. (1964). *Nature* **203,** 34.
Felicetti, L., Colombo, B., and Baglioni, C. (1966). *Biochim. Biophys. Acta* **129,** 380.
Felig, P., Owen, O. E., Wahren, J., and Cahill, G. F., Jr. (1969a). *J. Clin. Invest.* **48,** 584.
Felig, P., Marliss, E., and Cahill, G. F., Jr. (1969b). *New Engl. J. Med.* **281,** 811.
Fineberg, R. A., and Greenberg, D. M. (1955). *J. Biol. Chem.* **214,** 107.
Finkelstein, J. D. (1967). *Arch. Biochem. Biophys.* **122,** 583.
Firkin, F. C., and Linnane, A. W. (1968). *Biochem. Biophys. Res. Commun.* **32,** 398.
Fleck, A., Shepherd, J., and Munro, H. N. (1965). *Science* **150,** 628.
Flessel, C. P., Ralph, P., and Rich, A. (1967). *Science* **158,** 658.
Florini, J. R., Bird, H. H., and Bell, P. H. (1966). *J. Biol. Chem.* **241,** 1091.
Floyd, L. R., Unuma, T., and Busch, H. (1968). *Exptl. Cell Res.* **51,** 423.
Fogel, S., and Sypherd, P. S. (1968). *Proc. Natl. Acad. Sci. U.S.* **59,** 1329.
Fore, H., and Morton, R. A. (1952). *Biochem. J.* **51,** 600.
Forsham, P. H., and Melmon, K. L. (1968). *In* "Textbook of Endocrinology" (R. H. Williams, ed.), p. 287. Saunders, Philadelphia, Pennsylvania.
Foster, L. B., and Sells, B. H. (1969). *Arch. Biochem. Biophys.* **132,** 561.
Fox, C. F., and Weiss, S. B. (1964). *J. Biol. Chem.* **239,** 175.
Fox, C. F., Beckwith, J. R., Epstein, W., and Singer, E. R. (1966). *J. Mol. Biol.* **19,** 576.
Freedman, M. L., Hori, M., and Rabinovitz, M. (1967). *Science* **157,** 323.
Freedman, M. L., Fisher, J. M., and Rabinovitz, M. (1968). *J. Mol. Biol.* **33,** 315.
Freundlich, M., Burns, R. O., and Umbarger, H. E. (1962). *Proc. Natl. Acad. Sci. U.S.* **48,** 1804.

Friedmann, B., Goodman, E. H., and Weinhouse, S. (1967). *J. Biol. Chem.* **242,** 3620.

Fry, B. A. (1964). *Proc. Nutr. Soc.* (*Engl. Scot.*) **23,** 170.

Fuchs, E., Millette, R. L., Zillig, W., and Walter, G. (1967). *European J. Biochem.* **3,** 183.

Furth, J. J., and Ho, P. (1963). *Biochem. Biophys. Res. Commun.* **13,** 100.

Furth, J. J., and Ho, P. (1965). *J. Biol. Chem.* **240,** 2602.

Furth, J. J., Hurwitz, J., and Anders, M. (1962). *J. Biol. Chem.* **237,** 2611.

Futai, M., Anraku, Y., and Mizuno, D. (1966). *Biochim. Biophys. Acta* **119,** 373.

Gaertner, F. H., and DeMoss, J. A. (1969). *J. Biol. Chem.* **244,** 2716.

Gaetani, S., Paolucci, A. M., Spadoni, M. A., and Tomassi, G. (1964). *J. Nutr.* **84,** 173.

Gallant, J., and Cashel, M. (1967). *J. Mol. Biol.* **25,** 545.

Gallant, J., and Harada, B. (1969). *J. Biol. Chem.* **244,** 3125.

Gallwitz, D., and Mueller, G. C. (1969a). *European J. Biochem.* **9,** 431.

Gallwitz, D., and Mueller, G. C. (1969b). *Science* **163,** 1351.

Galper, J. B., and Darnell, J. E., Jr. (1969). *Biochem. Biophys. Res. Commun.* **34,** 205.

Gan, J. C., and Jeffay, H. (1967). *Biochim. Biophys. Acta* **148,** 448.

Ganoza, M. C., and Williams, C. A. (1969). *Proc. Natl. Acad. Sci. U.S.* **63,** 1370.

Ganschow, R., and Paigen, K. (1967). *Proc. Natl. Acad. Sci. U.S.* **58,** 938.

Gardner, J. A. A., and Hoagland, M. B. (1968). *J. Biol. Chem.* **243,** 10.

Garren, L. D., Richardson, A. P., and Crocco, R. M. (1967). *J. Biol. Chem.* **242,** 650.

Gaye, P., and Denamur, R. (1968). *Bull. Soc. Chim. Biol.* **50,** 1273.

Geiduschek, E. P., and Haselkorn, R. (1969). *Ann. Rev. Biochem.* **38,** 647.

Gelboin, H. V., Wortham, J. S., Wilson, R. G., Friedman, M., and Wogan, G. N. (1966). *Science* **154,** 1205.

Georgiev, G. P. (1968). *In* "Regulatory Mechanisms for Protein Synthesis in Mammalian Cells" (A. San Pietro, M. R. Lamborg, and F. T. Kenney, eds.), p. 25. Academic Press, New York.

Gerhart, J. C., and Schachman, H. K. (1965). *Biochemistry* **4,** 1054.

Gerhart, J. C., and Schachman, H. K. (1968). *Biochemistry* **7,** 538.

Gilbert, W., and Müller-Hill, B. (1966). *Proc. Natl. Acad. Sci. U.S.* **56,** 1891.

Gill, D. M., Pappenheimer, A. M., Brown, K., and Kurnick, J. T. (1969). *J. Exptl. Med.* **129,** 1.

Girard, M., and Baltimore, D. (1966). *Proc. Natl. Acad. Sci. U.S.* **56,** 999.

Girard, M., Latham, H., Penman, S., and Darnell, J. E., Jr. (1965). *J. Mol. Biol.* **11,** 187.

Girshovich, A. S., Knorre, D. G., Nelidova, O. D., and Ovander, M. N. (1966). *Biochim. Biophys. Acta* **119,** 216.

Godin, C. (1967a). *Can. J. Biochem.* **45,** 715.

Godin, C. (1967b). *Can. J. Biochem.* **45,** 1961.

Godin, C., and Gajda, A. T. (1967). *Can. J. Biochem.* **45,** 745.

Godson, G. N., and Sinsheimer, R. L. (1967). *Biochim. Biophys. Acta* **149,** 476 and 489.

Goldberg, B., and Green, H. (1967). *J. Mol. Biol.* **26,** 1.

Goldberg, M. L., Moon, H. D., and Rosenau, W. (1969). *Biochim. Biophys. Acta* **171,** 192.

Goldman, M., Johnston, W. M., and Griffin, A. C. (1969). *Cancer Res.* **29,** 1051.
Goldstein, A., Kirschbaum, J. B., and Roman, A. (1965). *Proc. Natl. Acad. Sci. U.S.* **54,** 1669.
Goldstein, L., and Prescott, D. M. (1968). *J. Cell Biol.* **36,** 53.
Gonano, F., and Chiarugi, V. P. (1969). *Exptl. Mol. Pathol.* **10,** 99.
Goor, R. S., Pappenheimer, A. M., and Ames, E. (1967). *J. Exptl. Med.* **126,** 923.
Gorini, L., and Maas, W. K. (1957). *Biochim. Biophys. Acta* **25,** 208.
Gospodarowicz, D., and Laporte, J. (1968). *Biochim. Biophys. Acta* **169,** 212.
Goswami, P., Barr, G. C., and Munro, H. N. (1962). *Biochim. Biophys. Acta* **55,** 408.
Goto, N., Kato, I., and Sato, H. (1968). *Japan. J. Exptl. Med.* **38,** 185.
Gottesman, M. (1967). *J. Biol. Chem.* **242,** 5564.
Granboulan, N., and Granboulan, P. (1964). *Exptl. Cell Res.* **34,** 71.
Grayzel, A. I., Hörchner, P., and London, I. M. (1966). *Proc. Natl. Acad. Sci. U.S.* **55,** 650.
Grayzel, A. I., Fuhr, J. E., and London, I. M. (1967). *Biochem. Biophys. Res. Commun.* **28,** 705.
Greenberger, N. J. (1967). *Nature* **214,** 702.
Greengard, O. (1967). *Enzymol. Biol. Clin.* **8,** 81.
Greengard, O., Smith, M. A., and Acs, G. (1963). *J. Biol. Chem.* **238,** 1548.
Greengard, O., Gordon, M., Smith, M. A., and Acs, G. (1965). *J. Biol. Chem.* **239,** 2079.
Griffin, D. M., and Szego, C. M. (1968). *Life Sci.* **7,** 1017.
Grisolia, S. (1968). *Biochem. Biophys. Res. Commun.* **32,** 56.
Grogan, D. E., and Busch, H. (1967). *Biochemistry* **6,** 573.
Grogan, D. E., Desjardins, R., and Busch, H. (1966). *Cancer Res.* **26,** 775.
Grollman, A. P. (1966). *Proc. Natl. Acad. Sci. U.S.* **56,** 1867.
Grollman, A. P. (1967). *Science* **157,** 84.
Grollman, A. P. (1968). *J. Biol. Chem.* **243,** 4089.
Grollman, A. P., and Stewart, M. L. (1968). *Proc. Natl. Acad. Sci. U.S.* **61,** 719.
Gros, F., and Gros, F. (1958). *Exptl. Cell Res.* **14,** 104.
Gross, S. R. (1965). *Proc. Natl. Acad. Sci. U.S.* **54,** 1538.
Grove, T. A., Gilbertson, T. J., Hammerstedt, R. H., and Henderson, L. M. (1969). *Biochim. Biophys. Acta* **184,** 329.
Grünberger, D., Meissner, L., Holý, A., and Šorm, F. (1966). *Biochim. Biophys. Acta* **119,** 432.
Gurdon, J. B., and Uehlinger, V. (1966). *Nature* **210,** 1240.
Hagopian, A., and Eylar, E. H. (1969). *Arch. Biochem. Biophys.* **129,** 447 and 515.
Hagopian, A., Bosmann, H. B., and Eylar, E. H. (1968). *Arch. Biochem. Biophys.* **128,** 387.
Hallinan, T. H., and Munro, H. N. (1964). *Biochim. Biophys. Acta* **80,** 166.
Hallinan, T. H., and Munro, H. N. (1965). *Biochim. Biophys. Acta* **108,** 285.
Hallinan, T. H., Murty, C. N., and Grant, J. H. (1968). *Life Sci.* **7,** 225.
Hamashima, Y., Harter, J. G., and Coons, A. H. (1964). *J. Cell Biol.* **20,** 271.
Hamilton, T. H., Widnell, C. C., and Tata, J. R. (1968). *J. Biol. Chem.* **243,** 408.

Hankes, L. V., Brown, R. R., and Schmaeler, M. (1967a). *Proc. Soc. Exptl. Biol. Med.* **124,** 1212.

Hankes, L. V., Brown, R. R., Lippincott, S., and Schmaeler, M. (1967b). *J. Lab. Clin. Med.* **69,** 313.

Harper, A. E. (1968). *Am. J. Clin. Nutr.* **21,** 358.

Harris, C., Reddy, J., and Svoboda, D. (1968). *Exptl. Cell Res.* **51,** 268.

Hartmann, G., Goller, H., Koschel, K., Kersten, W., and Kersten, H. (1964). *Biochem. Z.* **341,** 126.

Haschemeyer, A. E. V., and Gross, J. (1967). *Biochim. Biophys. Acta* **145,** 76.

Haslam, J. M., Davey, P. J., Linnane, A. W., and Atkinson, M. R. (1968). *Biochem. Biophys. Res. Commun.* **33,** 368.

Hayes, D. (1967). *Ann. Rev. Microbiol.* **21,** 369.

Heise, E., and Görlich, M. (1966). *Acta Biol. Med. Ger.* **17,** 17.

Henderson, L. M., and Hankes, L. V. (1956). *J. Biol. Chem.* **222,** 1068.

Henshaw, E. C. (1968). *J. Mol. Biol.* **36,** 401.

Henshaw, E. C., Revel, M., and Hiatt, H. H. (1965). *J. Mol. Biol.* **14,** 241.

Herner, A. E., Goldberg, I. H., and Cohen, L. B. (1969). *Biochemistry* **8,** 1335.

Herrlich, P., and Lang, N. (1967). *Z. Physiol. Chem.* **348,** 1377.

Hershey, J. W. B., and Monro, R. E. (1966). *J. Mol. Biol.* **18,** 68.

Hershey, J. W. B., Dewey, K. F., and Thach, R. E. (1969). *Nature* **222,** 944.

Heywood, S. M., Dowben, R. M., and Rich, A. (1967). *Proc. Natl. Acad. Sci. U.S.* **57,** 1002.

Hicks, S. J., Drysdale, J. W., and Munro, H. N. (1969). *Science* **164,** 584.

Hird, H. J., McLean, E. J. T., and Munro, H. N. (1964). *Biochim. Biophys. Acta* **87,** 219.

Hirsch, C. A. (1967). *J. Biol. Chem.* **242,** 2822.

Hirsch, C. A., and Hiatt, H. H. (1966). *J. Biol. Chem.* **241,** 5936.

Hnilica, L. S. (1967). *Progr. Nucleic Acid Res. Mol. Biol.* **7,** 25.

Hnilica, L. S., Edwards, L. J., and Hey, A. E. (1966a). *Biochim. Biophys. Acta* **124,** 109.

Hnilica, L. S., Liau, M. C., and Hurlbert, R. B. (1966b). *Science* **152,** 521.

Hoagland, M. B., Wilson, S. H., and Quincey, R. V. (1968). *In* "Regulatory Mechanisms for Protein Synthesis in Mammalian Cells" (A. San Pietro, M. R. Lamborg, and F. T. Kenney, eds.), p. 179. Academic Press, New York.

Hodge, L. D., Robbins, E., and Scharff, M. D. (1969). *J. Cell Biol.* **40,** 497.

Hogan, B. L. M., and Korner, A. (1968). *Biochim. Biophys. Acta* **169,** 129, 139.

Holten, V. Z., and Jacobson, K. B. (1969). *Arch. Biochem. Biophys.* **129,** 283.

Holtzman, E., Smith, I., and Penman, S. (1966). *J. Mol. Biol.* **17,** 131.

Honig, G. R., Rowan, B. Q., and Mason, R. G. (1969). *J. Biol. Chem.* **244,** 2027.

Honjo, T., Nishizuka, Y., Hayaishi, O., and Kato, I. (1968). *J. Biol. Chem.* **243,** 3553.

Hori, M., and Rabinovitz, M. (1968). *Proc. Natl. Acad. Sci. U.S.* **59,** 1349.

Hori, M., Fisher, J. M., and Rabinovitz, M. (1967). *Science* **155,** 83.

Horowitz, N. H. (1965). *In* "Evolving Genes and Proteins" (V. Bryson and H. J. Vogel, eds.), p. 15. Academic Press, New York.

Howell, R. R., Loeb, J. N., and Tomkins, G. M. (1964). *Proc. Natl. Acad. Sci. U.S.* **52,** 1241.

Huang, R. C. C., Bonner, J., and Murray, K. (1964). *J. Mol. Biol.* **8,** 54.

Humphreys, T., and Bell, E. (1967). *Biochem. Biophys. Res. Commun.* **27,** 443.

Hung, P. P., and Overby, L. R. (1968). *J. Biol. Chem.* **243,** 5525.
Hunt, R. T., Hunter, A. R., and Munro, A. J. (1968). *J. Mol. Biol.* **36,** 31.
Hunt, R. T., Hunter, A. R., and Munro, A. J. (1969a). *J. Mol. Biol.* **43,** 123.
Hunt, R. T., Hunter, A. R., and Munro. A. J. (1969b). *Proc. Nutr. Soc.* (*Engl. Scot.*) **28,** 248.
Hurwitz, J. (1959). *J. Biol. Chem.* **234,** 2351.
Hütter, R., Poralla, K., Zachau, H. G., and Zähner, H. (1966). *Biochem. Z.* **344,** 190.
Hutzler, J., and Dancis, J. (1968). *Biochim. Biophys. Acta* **158,** 62.
Ichihara, A., and Koyama, E. (1966). *J. Biochem.* (*Tokyo*) **59,** 160.
Ikehara, M., Murao, K., Harada, F., and Nishimura, S. (1968). *Biochim. Biophys. Acta* **155,** 82.
Imamoto, F., and Ito, J. (1968). *Nature* **220,** 27.
Ingbar, S. H., and Woeber, K. A. (1968). *In* "Textbook of Endocrinology" (R. H. Williams, ed.), p. 105. Saunders, Philadelphia, Pennsylvania.
Ippen, K., Miller, J. H., Scaife, J. G., and Beckwith, J. R. (1968). *Nature* **217,** 825.
Ito, J., and Yanovsky, C. (1966). *J. Biol. Chem.* **241,** 4112.
Iwasaki, K., Sabol, S., Wahba, A. J., and Ochoa, S. (1968). *Arch. Biochem. Biophys.* **125,** 542.
Izawa, M., and Kawashima, K. (1969). *Biochim. Biophys. Acta* **190,** 139.
Jachymczyk, W. J., and Cherry, J. H. (1968). *Biochim. Biophys. Acta* **157,** 368.
Jacob, F., and Monod, J. (1961). *J. Mol. Biol.* **3,** 318.
Jacob, F., and Monod, J. (1965). *Biochem. Biophys. Res. Commun.* **18,** 693.
Jacob, S. T., Sajdel, E. M., and Munro, H. N. (1968a). *Biochim. Biophys. Acta* **157,** 421.
Jacob, S. T., Sajdel, E. M., and Munro, H. N. (1968b). *Biochem. Biophys. Res. Commun.* **32,** 831.
Jacob, S. T., Sajdel, E. M., and Munro, H. N. (1969a). Unpublished results.
Jacob, S. T., Sajdel, E. M., and Munro, H. N. (1969b). *European J. Biochem.* **7,** 449.
Jarett, L., and Hendler, R. W. (1967). *Biochemistry* **6,** 1693.
Jensen, E. V., Suzuki, T., Kawashima, T., Stumpf, W. E., Jungblut, P. W., and DeSombre, E. R. (1968). *Proc. Natl. Acad. Sci. U.S.* **59,** 632.
John, D. W., and Miller, L. L. (1966). *J. Biol. Chem.* **241,** 4817.
John, D. W., and Miller, L. L. (1969). *Biochem. Pharmacol.* **18,** 1135.
Johnston, I. R., Mathias, A. P., Pennington, F., and Ridge, D. (1968). *Biochem. J.* **109,** 127.
Joklik, W. K., and Becker, Y. (1965). *J. Mol. Biol.* **13,** 496 and 511.
Jondorf, W. R., Simon, D. C., and Avnilmelch, M. (1966). *Mol. Pharmacol.* **2,** 506.
Jorgensen, G. N., and Larson, B. L. (1968). *Biochim. Biophys. Acta* **165,** 121.
Jukes, T. H. (1967). *Biochem. Biophys. Res. Commun.* **27,** 573.
Julian, J. A., and Chytil, F. (1969). *Biochem. Biophys. Res. Commun.* **35,** 734.
Juva, K., Prockop, D. J., Cooper, G. W., and Lash, J. W. (1966). *Science* **152,** 92.
Kabat, D. (1968). *J. Biol. Chem.* **243,** 2597.
Kabat, D., and Rich, A. (1969). *Biochemistry* **8,** 3742.

Kadowaki, K., Yamaguchi, K., and Maruo, B. (1966). *J. Gen. Appl. Microbiol.* **12,** 157.

Kaempfer, R. O. R., Meselson, M., and Raskas, H. J. (1968). *J. Mol. Biol.* **31,** 277.

Kaji, H. (1967). *Biochim. Biophys. Acta* **134,** 134.

Kano-Sueoka, T., and Sueoka, N. (1966). *J. Mol. Biol.* **20,** 183.

Kaplan, N. O. (1965). *In* "Evolving Genes and Proteins" (V. Bryson and H. J. Vogel, eds.), p. 243. Academic Press, New York.

Karlson, P. (1968). *Humangenetik* **6,** 99.

Katz, J., Bonorris, G., Okuyama, S., and Sellers, A. L. (1967). *Am. J. Physiol.* **212,** 1255.

Kaufmann, E., Traub, A., and Teitz, Y. (1967). *Exptl. Cell Res.* **49,** 215.

Kay, E. R. M., Smellie, R. M. S., Humphrey, G. F., and Davidson, J. N. (1956). *Biochem. J.* **62,** 160.

Kazazian, H. H., and Freedman, M. L. (1968). *J. Biol. Chem.* **243,** 6446.

Kazazian, H. H., and Itano, H. A. (1968). *J. Biol. Chem.* **243,** 2048.

Kaziro, Y., and Kamiyama, M. (1967). *J. Biochem.* (*Tokyo*) **62,** 424.

Kedes, L. H., Koegel, R. J., and Kuff, E. L. (1966). *J. Mol. Biol.* **22,** 359.

Keller, P. J., Cohen, E., and Beeky, J. A. H. (1968). *J. Biol. Chem.* **243,** 1271.

Kendall, J. W. (1962). *Endocrinology* **71,** 452.

Kennell, D. (1964). *J. Mol. Biol.* **9,** 789.

Kerjean, P., Marchetti, J., and Szulmajster, J. (1967). *Bull. Soc. Chim. Biol.* **49,** 1139.

Kiho, W. (1968). *Japan. J. Microbiol.* **12,** 211.

King, H. W. S., and Fitschen, W. (1968). *Biochim. Biophys. Acta* **155,** 32.

Kinkade, J. M., and Cole, R. D. (1966). *J. Biol. Chem.* **241,** 5790 and 5798.

Kirsch, R., Frith, L., Black, E., and Hoffenberg, R. (1968). *Nature* **217,** 578.

Kleinfeld, R. G. (1966). *Natl. Cancer Inst. Monograph* **23,** 369.

Knox, W. E., and Piras, M. M. (1966). *J. Biol. Chem.* **241,** 765.

Knox, W. E., and Piras, M. M. (1967). *J. Biol. Chem.* **242,** 2959.

Köhler, K., and Arends, S. (1968a). *European J. Biochem.* **5,** 500.

Köhler, K., and Arends, S. (1968b). *Compt. Rend.* **266,** 170.

Kolakofsky, D., Dewey, K. F., Hershey, J. W. B., and Thach, R. E. (1968). *Proc. Natl. Acad. Sci. U.S.* **61,** 1066.

Koler, R. D., Vanbellinghen, P. J., Fellman, J. H., Jones, R. T., and Behrman, R. E. (1969). *Science* **163,** 1348.

Korenman, S. G., and Rao, B. R. (1968). *Proc. Natl. Acad. Sci. U.S.* **61,** 1028.

Korner, A. (1964). *Biochem. J.* **92,** 449.

Korner, A. (1969). *Biochim. Biophys. Acta* **174,** 351.

Korner, A., and Gumbley, J. M. (1966). *Nature* **209,** 505.

Korner, A., and Young, F. G. (1955). *J. Endocrinol.* **13,** 78.

Kretchmar, A. L., and Price, E. J. (1969). *Metab., Clin. Exptl.* **18,** 684.

Kretsinger, R. H., Manner, G., Gould, B. S., and Rich, A. (1964). *Nature* **202,** 438.

Kruh, J. (1968). *Biochim. Biophys. Acta* **169,** 511.

Kruh, J., and Blum, N. (1968). *Biochim. Biophys. Acta* **161,** 215.

Kume, F., and Chiga, M. (1967). *Lab. Invest.* **17,** 767.

Kume, F., and Chiga, M. (1968). *Gann* **59,** 151.

Küntzel, H., and Noll, H. (1967). *Nature* **215,** 1340.

Kurland, C. G., and Maaløe, O. (1962). *J. Mol. Biol.* **4,** 193.

Kvam, D. C., and Parks, R. E., Jr. (1960). *J. Biol. Chem.* **235,** 2893.
Kwan, S.-W., Webb, T. E., and Morris, H. P. (1968). *Biochem. J.* **109,** 617.
Lafarge, C., Frayssinet, C., and Simard, R. (1966). *Compt. Rend.* **D263,** 1011.
Laga, E. M., Baliga, B. S., and Munro, H. N. (1969). *Biochim. Biophys. Acta* (in press).
Lagerstedt, S. (1949). *Acta Anat.* Suppl. 9, 1.
Lane, R. S. (1967). *Nature* **215,** 161.
Lane, R. S. (1968). *Brit. J. Haematol.* **15,** 355.
Langan, T. A. (1967). *In* "Regulation of Nucleic Acid and Protein Biosynthesis" (V. V. Koningsberger and L. Bosch, eds.), p. 233. Elsevier, Amsterdam.
Langan, T. A. (1968a). *In* "Regulatory Mechanisms for Protein Synthesis in Mammalian Cells" (A. San Pietro, M. R. Lamborg, and F. T. Kenney, eds.), p. 101. Academic Press, New York.
Langan, T. A. (1968b). *Science* **162,** 579.
Langer, T., Strober, W., and Levy, R. I. (1970). *In* "Fifth International Symposium on Serum Proteins" (M. Rothschild, ed.). Academic Press, New York (in press).
Lavallé, R., and De Hauwer, G. (1968). *J. Mol. Biol.* **37,** 269.
La Via, M. F., Vatter, A. E., and Northrup, P. V. (1966). *Exptl. Mol. Pathol.* Suppl. 3, 124.
Lawford, G. R., and Schachter, H. (1966). *J. Biol. Chem.* **241,** 5408.
Lawford, G. R., Langford, P., and Schachter, H. (1966). *J. Biol. Chem.* **241,** 1835.
Lèbre, P., Bali, J. P., Heizman, P., Fontanges, R., and Louisot, P. (1969). *Compt. Rend.* **268,** 734.
Lee, J. C., and Ingram, V. M. (1967). *Science* **158,** 1330.
Lengyel, P., and Söll, D. (1969). *Bacteriol. Rev.* **33,** 264.
Leon, H. A., and Chackerian, M. J. (1968). *Endocrinology* **82,** 429.
Levere, R. D., and Granick, S. (1967). *J. Biol. Chem.* **242,** 1903.
Levinthal, C., Fan, D. P., Higa, A., and Zimmerman, R. (1963). *Cold Spring Harbor Symp. Quant. Biol.* **128,** 183.
Levitan, I. B., and Webb, T. E. (1968). *Biochim. Biophys. Acta* **155,** 632.
Li, Y. T., Li, S. C., and Shetlar, M. R. (1968). *J. Biol. Chem.* **243,** 656.
Liao, S., and Stumpf, W. E. (1968). *Endocrinology* **83,** 629.
Liau, M. C., and Perry, R. P. (1969). *J. Cell Biol.* **42,** 272.
Lim-Sylianco, C. Y., and Berg, C. P. (1969). *Arch. Biochem. Biophys.* **131,** 643.
Lin, C. Y., and Key, J. L. (1966). *J. Mol. Biol.* **26,** 237.
Lin, Y. C., Kinoshita, T., and Tanaka, N. (1968). *J. Antibiot.* **21,** 471.
Lindsay, D. T. (1966). *Arch. Biochem. Biophys.* **113,** 687.
Lipmann, F., and Levene, P. A. (1932). *J. Biol. Chem.* **98,** 109.
Lippe, B. M., and Szego, C. M. (1965). *Nature* **207,** 272.
Livingston, D. M., and Leder, P. (1969). *Biochemistry* **8,** 435.
Loeb, J. E. (1967). *Biochim. Biophys. Acta* **145,** 427.
Loening, U. E. (1968). *Ann. Rev. Plant Physiol.* **19,** 37.
Loftfield, R. B., and Eigner, E. A. (1968). *In* "Regulatory Mechanisms for Protein Synthesis in Mammalian Cells" (A. San Pietro, M. R. Lamborg, and F. T. Kenney, eds.), p. 231. Academic Press, New York.
Loftfield, R. B., and Eigner, E. A. (1969). *J. Biol. Chem.* **244,** 1746.
London, I. M., Tavill, A. S., Vanderhoff, G. A., Hunt, T., and Grayzel, A. I. (1967). *Develop. Biol.* Suppl. 1, 227.
Loud, A. V. (1968). *J. Cell Biol.* **37,** 27.
Low, R. B., and Wool, I. G. (1967). *Science* **155,** 330.
Luria, S. E. (1960). *In* "The Bacteria" (I. C. Gunsalus and R. Y. Stanier, eds.), Vol. 1, p. 15. Academic Press, New York.

Lyttleton, J. W. (1967). *Biochim. Biophys. Acta* **149,** 598.
McConkey, E. H., and Hopkins, J. W. (1969). *J. Mol. Biol.* **39,** 545.
McCormick, D. B., and Snell, E. S. (1961). *J. Biol. Chem.* **236,** 2085.
McElroy, W. D., DeLuca, M., and Travis, J. (1967). *Science* **157,** 150.
McFarlane, I. G., and von Holt, C. (1969). *Biochem. J.* **111,** 557 and 565.
McKeehan, W., Sepulveda, P., Lin, S. Y., and Hardesty, B. (1969). *Biochem. Biophys. Res. Commun.* **34,** 668.
Madras, B. K., and Sourkes, T. L. (1968). *Arch. Biochem. Biophys.* **125,** 829.
Maggio, R. (1966). *Natl. Cancer Int. Monograph* **23,** 213.
Maggio, R., Siekevitz, P., and Palade, G. E. (1963). *J. Cell Biol.* **18,** 267.
Maio, J. J., and Schildkraut, C. L. (1967). *J. Mol. Biol.* **24,** 29.
Maitra, U., and Hurwitz, J. (1965). *Proc. Natl. Acad. Sci. U.S.* **54,** 815.
Makman, M. H., Dvorkin, B., and White, A. (1968). *J. Biol. Chem.* **243,** 1485.
Malkin, L. I., and Rich, A. (1967). *J. Mol. Biol.* **26,** 329.
Malkin, M., and Lipmann, F. (1969). *Science* **164,** 71.
Malt, R. A., and LeMaitre, D. A. (1967). *Biochim. Biophys. Acta* **145,** 190.
Mandelstam, J. (1958). *Biochem. J.* **69,** 110.
Mandelstam, J. (1960). *Bacteriol. Rev.* **24,** 289.
Mangiarotti, G. (1969). *Nature* **222,** 947.
Marcus, A., Feeley, J., and Volcani, T. (1966). *Plant Physiol.* **41,** 1167.
Marcus, L., Bretthauer, R. K., Bock, R. M., and Halvorson, H. O. (1963). *Proc. Natl. Acad. Sci. U.S.* **50,** 782.
Marks, P. A., Burka, E. R., Conconi, F. M., Perl, W., and Rifkind, R. A. (1965). *Proc. Natl. Acad. Sci. U.S.* **56,** 701.
Martin, G. S., and Tocchini-Valentini, G. P. (1967). *In* "Regulation of Nucleic Acid and Protein Biosynthesis" (V. V. Koningsberger and L. Bosch, eds.), p. 91. Elsevier, Amsterdam.
Martin, R. G. (1963). *J. Biol. Chem.* **238,** 257.
Martin, S. J., and Brown, F. (1967). *Biochem. J.* **105,** 979.
Martin, T. E., and Wool, I. G. (1968). *Proc. Natl. Acad. Sci. U.S.* **60,** 569.
Marushige, K., and Bonner, J. (1966). *J. Mol. Biol.* **15,** 160.
Marushige, K., Brutlag, D., and Bonner, J. (1968). *Biochemistry* **7,** 3149.
Maruyama, H., and Mizuno, D. (1966). *Biochim. Biophys. Acta* **123,** 510.
Marver, H. S., Tschudy, D. P., Perlroth, M. G., and Collins, A. (1966a). *Science* **154,** 501.
Marver, H. S., Collins, A., and Tschudy, D. P. (1966b). *Biochem. J.* **99,** 31C.
Masukawa, H., Tanaka, N., and Umezawa, H. (1968). *J. Antibiotics (Tokyo)* **21,** 517.
Mathur, G. P., Ng, C. Y., and Henderson, L. M. (1964). *J. Biol. Chem.* **239,** 2184.
Matsushiro, A., Sato, K., Kida, S., and Imamoto, F. (1965). *J. Mol. Biol.* **11,** 54.
Maul, G. G., and Hamilton, T. H. (1967). *Proc. Natl. Acad. Sci. U.S.* **57,** 1371.
Maxwell, C. R., and Rabinovitz, M. (1969). *Biochem. Biophys. Res. Commun.* **35,** 79.
May, B. K., and Elliott, W. H. (1968). *Biochim. Biophys. Acta* **157,** 607.
Means, A. R., and Baker, C. A. (1969). *Biochim. Biophys. Acta* **182,** 461.
Mehta, S. L., Hadziyev, D., and Zalik, S. (1968). *Biochim. Biophys. Acta* **169,** 381.
Miller, A. O. A. (1968). *Biochem. Biophys. Res. Commun.* **30,** 267.
Miller, J. E., and Litwack, G. (1969). *Biochem. Biophys. Res. Commun.* **36,** 35.
Miller, O. L., Jr., and Beatty, B. R. (1969). *Science* **164,** 955.

Mimura, T., Yamada, C., and Swenseid, M. E. (1968). *J. Nutr.* **95,** 493.
Mittermayer, C., Braun, R., Chayka, T. G., and Rusch, H. P. (1966). *Nature* **210,** 1133.
Mizuno, D., and Anraku, Y. (1967). *Japan. J. Med. Sci. Biol.* **20,** 127.
Mizuno, S., Yamazaki, H., Nitta, K., and Umezawa, H. (1968). *Biochim. Biophys. Acta* **157,** 322.
Moldave, K., Ibuki, F., Rao, P. M., Schneir, M., Skogerson, L., and Sutter, R. P. (1968). *In* "Regulatory Mechanisms for Protein Synthesis in Mammalian Cells" (A. San Pietro, M. R. Lamborg, and F. T. Kenney, eds.), p. 194. Academic Press, New York.
Molnár, J. (1969). *Acta Biochem. Biophys. Acad. Sci. Hung.* **4,** 1.
Monneron, A., and Moulé, Y. (1968). *Exptl. Cell Res.* **51,** 531.
Monod, J., Changeux, J.-P., and Jacob, F. (1963). *J. Mol. Biol.* **6,** 306.
Monro, R. E., Celma, M. L., Battaner, E., Fernandez, R., and Vazquez, D. (1969a). *In* "Symposium on Polypeptides" (R. J. Beers, ed.), p. 3. Miles Lab. Publ., N.Y.
Monro, R. E., Celma, M. L., and Vazquez, D. (1969b). *Nature* **222,** 356.
Moore, P. B., Traut, R. R., Noller, H., Pearson, P., and Delius, H. (1968). *J. Mol. Biol.* **31,** 441.
Morris, A. J., and Liang, K. (1968). *Arch. Biochem. Biophys.* **125,** 468.
Morris, D. W., and DeMoss, J. A. (1966). *Proc. Natl. Acad. Sci. U.S.* **56,** 262.
Moscona, A. A., Moscona, M. H., and Saenz, N. (1968). *Proc. Natl. Acad. Sci. U.S.* **61,** 160.
Moses, V., and Sharp, P. B. (1966). *Biochim. Biophys. Acta* **119,** 200.
Mosier, D. E., and Coppleson, L. W. (1968). *Proc. Natl. Acad. Sci. U.S.* **61,** 542.
Moss, B. A., and Thompson, E. O. P. (1969). *Biochim. Biophys. Acta* **188,** 351.
Mosteller, R. D., Culp, W. J., and Hardesty, B. (1968). *J. Biol. Chem.* **243,** 6343.
Moulé, Y., and Frayssinet, C. (1968). *Nature* **218,** 93.
Mross, G. A., and Doolittle, R. F. (1967). *Arch. Biochem. Biophys.* **122,** 674.
Mueller, K., and Bremer, H. (1969). *J. Mol. Biol.* **43,** 89.
Muench, K. H., and Safille, P. A. (1968). *Biochemistry* **7,** 2799.
Mundell, R. D. (1967). *Biochem. Biophys. Res. Commun.* **28,** 117.
Munns, R., and Hallinan, T. H. (1968). *Arch. Biochem. Biophys.* **127,** 419.
Munro, H. N. (1968). *Federation Proc.* **27,** 1231.
Munro, H. N., and Clark, C. M. (1958). *Biochim. Biophys. Acta* **27,** 648.
Munro, H. N., and Downie, E. D. (1964). *Nature* **203,** 603.
Munro, H. N., and Drysdale, J. W. (1970). *Federation Proc.* (in press).
Munro, H. N., McLean, E. J. T., and Hird, H. J. (1964). *J. Nutr.* **83,** 186.
Munro, H. N., Waddington, S., and Begg, D. J. (1965). *J. Nutr.* **85,** 319.
Munro, H. N., Baliga, B. S., and Pronczuk, A. W. (1968). *Nature* **210,** 944.
Muramatsu, M., Smetana, K., and Busch, H. (1963). *Cancer Res.* **23,** 510.
Murty, C. N., and Hallinan, T. H. (1968). *Biochim. Biophys. Acta* **157,** 414.
Mushinski, J. F., and Potter, M. (1969). *Biochemistry* **8,** 1684.
Nadal, C., and Zajdela, F. (1967). *Exptl. Cell Res.* **48,** 518.
Nair, K. G., Rabinowitz, M., and Tu, M. H. C. (1967). *Biochemistry* **6,** 1898.
Nass, G., Poralla, K., and Zähner, H. (1969). *Biochem. Biophys. Res. Commun.* **34,** 84.
Nass, M. M. K., and Buck, C. A. (1969). *Proc. Natl. Acad. Sci. U.S.* **62,** 506.
Natori, S., and Mizuno, D. (1967). *Biochim. Biophys. Acta* **145,** 328.
Natori, S., Nozawa, R., and Mizuno, D. (1966). *Biochim. Biophys. Acta* **114,** 245.

Natori, S., Horiuchi, T., and Mizuno, D. (1967a). *Biochim. Biophys. Acta* **134,** 337.

Natori, S., Yogo, Y., and Mizuno, D. (1967b). *Biochim. Biophys. Acta* **145,** 621.

Neidhardt, F. C. (1964). *Progr. Nucleic Acid Res. Mol. Biol.* **3,** 145.

Neidhardt, F. C. (1966). *Bacteriol. Rev.* **30,** 701.

Nicholson, B., and Peacocke, A. (1966). *Biochem. J.* **100,** 50.

Nierlich, D. P. (1968). *Proc. Natl. Acad. Sci. U.S.* **60,** 1345.

Nierlich, D. P., and Vielmetter, W. (1968). *J. Mol. Biol.* **32,** 135.

Nishimura, S., Harada, F., and Ikehara, M. (1966). *Biochim. Biophys. Acta* **129,** 301.

Nitta, K., Mizuno, S., Yamazaki, H., and Umezawa, H. (1968). *J. Antibiotics (Tokyo)* **21,** 521.

Nomura, M., and Lowry, C. V. (1967). *Proc. Natl. Acad. Sci. U.S.* **58,** 946.

Noteboom, W. D., and Mueller, G. C. (1966). *Mol. Pharmacol.* **2,** 534.

Ogata, K., Terao, K., and Sugano, H. (1967). *Biochim. Biophys. Acta* **149,** 572.

Ohly, K. W., Mehta, N. G., Mourkides, G. A., and Alivisatos, S. G. A. (1967). *Arch. Biochem. Biophys.* **118,** 631.

Okamoto, T. (1967). *Biochim. Biophys. Acta* **138,** 198.

Okazaki, K., and Farber, E. (1969). Personal communication.

Okazaki, K., Shull, K. H., and Farber, E. (1968). *J. Biol. Chem.* **243,** 4661.

Oler, A., Farber, E., and Shull, K. H. (1969). *Biochim. Biophys. Acta* **190,** 161.

Ono, Y., Skoultchi, A., Klein, A., and Lengyel, P. (1968). *Nature* **220,** 1304.

Oppenheim, J., Scheinbuks, J., Biava, C., and Marcus, L. (1968). *Biochim. Biophys. Acta* **161,** 386.

Owens, L., and Bell, F. E. (1967). *Abstr. Papers, 154th Meeting Am. Chem. Soc.* p. 120.

Owens, L., and Blum, J. J. (1967). *J. Biol. Chem.* **242,** 2893.

Pardee, A. B. (1966). *J. Biol. Chem.* **241,** 5886.

Pardee, A. B., and Prestidge, L. (1956). *J. Bacteriol.* **71,** 677.

Parsons, J. T., and McCarty, K. S. (1968). *J. Biol. Chem.* **243,** 5377.

Partridge, S. M., Elsden, D. F., Thomas, J., Dorfman, A., Telser, A., and Ho, P.-L. (1964). *Biochem. J.* **93,** 31C.

Pastan, I., and Perlman, R. L. (1968). *Proc. Natl. Acad. Sci. U.S.* **61,** 1336.

Pastan, I., and Perlman. R. L. (1969). *J. Biol. Chem.* **244,** 2226.

Paul, J., and Gilmour, R. S. (1966). *Nature* **210,** 992.

Paul, J., and Gilmour, R. S. (1968). *J. Mol. Biol.* **34,** 305.

Paul, J., and Hunter, J. A. (1968). *Nature* **219,** 1362.

Paul, J. S., Reynolds, R. C., and Montgomery, P. O. (1967). *Nature* **215,** 749.

Pelc, S. R. (1968). *Nature* **219,** 162.

Pelling, C. (1966). *Proc. Roy. Soc.* **B164,** 279.

Penman, S. (1966). *J. Mol. Biol.* **17,** 117.

Penman, S. (1967). *New Engl. J. Med.* **276,** 502.

Penman, S., Vesco, C., and Penman, M. (1968). *J. Mol. Biol.* **34,** 49.

Penrose, W. R., Nichoalds, G. E., and Oxender, D. L. (1968). *Federation Proc.* **27,** 643.

Perlman, R. L., and Pastan, I. (1968). *J. Biol. Chem.* **243,** 5420.

Perry, R. P., and Kelley, D. E. (1968a). *J. Mol. Biol.* **35,** 37.

Perry, R. P., and Kelley, D. E. (1968b). *J. Cellular Physiol.* **72,** 235.

Perry, R. P., Srinivasan, P. R., and Kelley, D. E. (1964). *Science* **145,** 504.

Peters, J. H., and Alper, C. A. (1966). *J. Clin. Invest.* **45,** 314.

Peters, T. (1969). Personal communication.

Petrovic, S., Becarevic. A., and Petrovic, J. (1965). *Biochim. Biophys. Acta* **95,** 518.

Pettijohn, D., and Kamiyama, T. (1967). *J. Mol. Biol.* **29,** 275.

Pfuderer, P., and Swartzendruber, D. C. (1966). *J. Cell Biol.* **30,** 193.

Pfuderer, P., Cammarano, P., Holladay, D. R., and Novelli, G. D. (1965). *Biochim. Biophys. Acta* **109,** 595.

Piha, R. S., Cuenod, M., and Waelsch, H. (1966). *J. Biol. Chem.* **241,** 2397.

Piperno, J. R., and Oxender, D. L. (1966). *J. Biol. Chem.* **241,** 5732.

Piras, M. M., and Knox, W. E. (1967). *J. Biol. Chem.* **242,** 2952.

Pizer, L. I. (1963). *J. Biol. Chem.* **238,** 3934.

Pizer, L. I. (1964). *J. Biol. Chem.* **239,** 4219.

Plagemann, P. G. W. (1968a). *J. Biol. Chem.* **243,** 3029.

Plagemann, P. G. W. (1968b). *Biochim. Biophys. Acta* **155,** 202.

Pogo, A. O., Littau, V. C., Allfrey, V. G., and Mirsky, A. E. (1967). *Proc. Natl. Acad. Sci. U.S.* **57,** 743.

Pogo, B. G. T., Allfrey, V. G., and Mirsky, A. E. (1966). *Proc. Natl. Acad. Sci. U.S.* **55,** 805.

Pollock, M. R. (1963). *Biochim. Biophys. Acta* **76,** 80.

Pollock, M. R. (1967). *Bull. Soc. Chim. Biol.* **49,** 633.

Poralla, K., and Zähner, H. (1968). *Arch. Mikrobiol.* **61,** 143.

Prestayko, A. W., and Busch, H. (1968). *Biochim. Biophys. Acta* **169,** 327.

Prevec, L., and Graham, A. F. (1966). *Science* **154,** 522.

Prichard, P. M., Gilbert, J. M., Shafritz, D. A., and Anderson, W. F. (1970). *Nature* **226,** 511.

Pronczuk, A. W., Baliga, B. S., Triant, J. W., and Munro, H. N. (1968). *Biochim. Biophys. Acta* **157,** 204.

Pronczuk, A. W., Rogers, Q. R., and Munro, H. N. (1969). Unpublished results.

Ptashne, M. (1967). *Proc. Nat. Acad. Sci. U.S.* **57,** 306.

Radding, C. M., Bragdon, J. H., and Steinberg, D. (1958). *Biochim. Biophys. Acta* **30,** 443.

Raeburn, S., Goor, R. S., Schneider, J. A., and Maxwell, E. S. (1968.) *Proc. Natl. Acad. Sci. U.S.* **61,** 1428.

Raghupathy, E., Peterson, N. A., and McKean, C. M. (1968). *Biochim. Biophys. Acta* **161,** 575.

Ragnotti, G., Lawford, G. R., and Campbell, P. N. (1969). *Biochem. J.* **112,** 139.

Ramuz, M., Doly, J., Mandel, P., and Chambon, P. (1965). *Biochem. Biophys Res. Commun.* **19,** 114.

Reader, R. W., and Stanners, C. P. (1967). *J. Mol. Biol.* **28,** 211.

Reboud, J. P. (1969). *Compt. Rend.* **268,** 588.

Reddy, J., and Svoboda, D. (1968). *Lab. Invest.* **19,** 132.

Redgrave, T. G., and Zilversmit, D. B. (1969). *Am. J. Physiol.* **217,** 336.

Redman, C. M. (1968). *Biochem. Biophys. Res. Commun.* **31,** 845.

Reich, E., and Goldberg, I. H. (1964). *Progr. Nucleic Acid Res.* **3,** 183.

Reid, B. R. (1968). *Biochem. Biophys. Res. Commun.* **33,** 627.

Reid, B. R., Stellwagen, R. H., and Cole, R. D. (1968). *Biochim. Biophys. Acta* **155,** 593.

Reusser, F. (1969a). *Biochemistry* **8,** 3303.

Reusser, F. (1969b). *Biochem. Pharmacol.* **18,** 287.

Reynolds, R. C., and Montgomery, P. O. (1967). *Am. J. Pathol.* **51,** 323.

Richardson, J. P. (1966). *Proc. Natl. Acad. Sci. U.S.* **55,** 1616.

Richaud, F., and Cohen, G. (1968). *Biochem. Biophys. Res. Commun.* **30,** 45.
Richmond, M. H. (1962). *Bacteriol. Rev.* **26,** 398.
Richmond, M. H. (1963). *J. Mol. Biol.* **6,** 284.
Ris, H. (1967). *In* "Regulation of Nucleic Acid and Protein Biosynthesis" (V. V. Koningsberger and L. Bosch, eds.), p. 11. Elsevier, Amsterdam.
Ristow, H., and Köhler, K. (1966). *Biochim. Biophys. Acta* **123,** 265.
Rizzo, A. J., and Webb, T. E. (1968). Cited by Kwan *et al.* (1968).
Ro, T. S., Narayan, K. S., and Busch, H. (1966). *Cancer Res.* **26,** 780.
Robbins, E., and Borun, T. W. (1967). *Proc. Natl. Acad. Sci. U.S.* **57,** 409.
Rodionova, N. P., and Shapot, V. S. (1966). *Biochim. Biophys. Acta* **129,** 206.
Roedor, R. G., and Rutter, W. J. (1969). *Nature* **224,** 234.
Roels, H. (1966). *Intern. Rev. Cytol.* **19,** 1.
Roodyn, D. B. (1963). *Biochem. Soc. Symp. (Cambridge, Engl.)* **23,** 20.
Rothschild, M. A., Oratz, M., Evans, C. D., and Schreiber, S. S. (1966). *Am. J. Physiol.* **210,** 57.
Rothschild, M. A., Oratz, M., Mongelli, J., and Schreiber, S. S. (1968a). *J. Clin. Invest.* **47,** 2591.
Rothschild, M. A., Oratz, M., and Schreiber, S. S. (1968b). *J. Nucl. Biol. Med. (Turin)* **12,** 68.
Rothschild, M. A., Oratz, M., Mongelli, J., Fishman, L., and Schreiber, S. S. (1969a). *J. Nutr.* **98,** 395.
Rothschild, M. A., Oratz, M., Mongelli, J., and Schreiber, S. S. (1969b). *Am. J. Physiol.* **216,** 1127.
Rowley, P. T., and Morris, J. (1966). *Exptl. Cell Res.* **45,** 494.
Ruddon, R. W., and Johnson, J. M. (1968). *Mol. Pharmacol.* **4,** 258.
Ruderman, N. B., Richards, K. C., Valles de Bourges, V., and Jones, A. L. (1968). *J. Lipid Res.* **9,** 613.
Sabatini, D. D., Tashiro, Y., and Palade, G. E. (1966). *J. Mol. Biol.* **19,** 503.
Sachs, H. (1958). *J. Biol. Chem.* **233,** 650.
Samarina, O. P., Krichevskaya, A. A., and Georgiev, G. P. (1966). *Nature* **210,** 1319.
Samarina, O. P., Lukanidin, E. M., Molnár, J., and Georgiev, G. P. (1968). *J. Mol. Biol.* **33,** 251.
Samec, J., Jacob, M., and Mandel, P. (1968). *Biochim. Biophys. Acta* **161,** 377.
Sands, M. K., and Roberts, R. B. (1952). *J. Bacteriol.* **63,** 505.
Sarkar, S., and Thach, R. E. (1968). *Proc. Natl. Acad. Sci. U.S.* **60,** 1479.
Sarma, D. S. R., Reid, I. M., and Sidransky, H. (1969). *Biochem. Biophys. Res. Commun.* **36,** 582.
Scherrer, K., and Marcaud, L. (1965). *Bull. Soc. Chim. Biol.* **47,** 1697.
Scherrer, K., Marcaud, L., Zajdela, F., London, I. M., and Gros, F. (1966). *Proc. Natl. Acad. Sci. U.S.* **56,** 1571.
Schimke, R. T., Sweeney, E. W., and Berlin, C. M. (1965). *J. Biol. Chem.* **240,** 322 and 4609.
Schlesinger, S., and Schlesinger, M. J. (1967). *J. Biol. Chem.* **242,** 3369.
Schlessinger, D., and Ben-Hamida, F. (1966). *Biochim. Biophys. Acta* **119,** 171.
Schlessinger, D., Mangiarotti, G., and Apirion, D. (1967). *Proc. Natl. Acad. Sci. U.S.* **58,** 1782.
Schneider, W. C. (1948). *J. Biol. Chem.* **176,** 259.
Schneider, W. C., and Kuff, E. L. (1965). *Proc. Natl. Acad. Sci. U.S.* **54,** 1650.
Schneider, W. C., and Kuff, E. L. (1969). *J. Biol. Chem.* **244,** 4843.

Schoenmakers, J. G. G., Zweers, A., and Bloemendahl, H. (1967). *Biochim. Biophys. Acta* **145,** 120.

Schreml, W., and Burka, E. R. (1968). *Biochim. Biophys. Acta* **243,** 3573.

Schwartz, H. S., Sodergren, J. E., and Ambaye, R. Y. (1968a). *Cancer Res.* **28,** 192.

Schwartz, H. S., Sternberg, S. S., and Philips, F. S. (1968b). *In* "Actinomycin" (S. A. Waksman, ed.), p. 101. Wiley (Interscience), New York.

Scolnick, E., Tompkins, R., Caskey, T., and Nirenberg, M. (1968). *Proc. Natl. Acad. Sci. U.S.* **61,** 768.

Šebesta, K., and Horská, K. (1968). *Biochim. Biophys. Acta* **169,** 281.

Seed, R. W., and Goldberg, I. H. (1965). *J. Biol. Chem.* **240,** 764.

Segal, H. L., and Mologne, L. A. (1959). *J. Biol. Chem.* **234,** 909.

Seifart, K., and Sekeris, C. E. (1967). *Z. Physiol. Chem.* **348,** 1555.

Sekeris, C. E. (1967). *In* "Regulation of Nucleic Acid and Protein Synthesis" (V. V. Koningsberger and L. Bosch, eds.), p. 388. Elsevier, Amsterdam.

Sekeris, C. E., Schmid, W., Gallwitz, D., and Lukacs, I. (1966a). *Life Sci.* **5,** 969.

Sekeris, C. E., Schmid, W., Gallwitz, D., and Lukacs, I. (1966b). *Angew. Chem.* **78,** 601.

Sekeris, C. E., Sekeri, K. E., and Gallwitz, D. (1967). *Z. Physiol. Chem.* **348,** 1660.

Sekiguchi, M., and Cohen, S. (1963). *J. Biol. Chem.* **238,** 349.

Shaeffer, J. R. (1967). *Biochem. Biophys. Res. Commun.* **28,** 647.

Shapot, V. S., and Pitot, H. C. (1966). *Biochim. Biophys. Acta* **119,** 37.

Shea, J. R., Jr., and Leblond, C. P. (1966). *J. Morphol.* **119,** 425.

Shearer, R. W., and McCarthy, B. J. (1967). *Biochemistry* **6,** 283.

Shearn, A., and Horowitz, N. H. (1969). *Biochemistry* **8,** 295.

Shiflet, R. N., and Bucovaz, E. T. (1967). *Abstr. Papers, 154th Meeting Am. Chem. Soc., Chicago* Paper C119.

Shin, D. H., and Moldave, K. (1966). *J. Mol. Biol.* **21,** 231.

Shinozuka, H., Goldblatt, P. J., and Farber, E. (1968). *J. Cell Biol.* **36,** 313.

Shyamala, G., and Gorski, J. (1969). *J. Biol. Chem.* **244,** 1097.

Sidransky, H., Sarma, D. S. R., Bongiorno, M., and Verney, E. (1968). *J. Biol. Chem.* **243,** 1123.

Siev, M., Weinberg, R., and Penman, S. (1969). *J. Cell Biol.* **41,** 510.

Silber, R. H., and Porter, C. C. (1953). *Endocrinology* **52,** 518.

Silverstone, A. E., Magasanik, B., Reznikoff, W. S., Miller, J. H., and Beckwith, J. R. (1969). *Nature* **221,** 1012.

Simkin, J. L., and Jamieson, J. C. (1967). *Biochem. J.* **103,** 38P.

Simon, E. J., Van Praag, D., and Aronson, F. L. (1966). *Mol. Pharmacol.* **2,** 43.

Sinclair, J. H., Stevens, B. J., Gross, N., and Rabinowitz, M. (1967). *Biochim. Biophys. Acta* **145,** 528.

Sippel, A., and Hartmann, G. (1968). *Biochim. Biophys. Acta* **157,** 218.

Sirlin, J. L., and Loening, U. E. (1968). *Biochem. J.* **109,** 375.

Sluyser, M. (1966). *J. Mol. Biol.* **19,** 591.

Sluyser, M. (1969). *Biochim. Biophys. Acta* **182,** 235.

Smetana, K., Narayan, K. S., and Busch, H. (1966). *Cancer Res.* **26,** 786.

Smith, A E., and Marcker, K. A. (1970). *Nature* **226,** 607.

Smith, D. A., Martinez, A. M., Ratcliff, R. L., Williams, D. L., and Hayes, F. N. (1967). *Biochemistry* **6,** 3057.

So, A. G., Davie, E. W., Epstein, R. H., and Tissières, A. (1967). *Proc. Natl. Acad. Sci. U.S.* **58,** 1739.

Soeiro, R., Vaughan, M. H., Warner, J. R., and Darnell, J. E., Jr. (1968). *J. Cell Biol.* **39,** 112.

Sokawa, Y., Nakao, E., and Kaziro, Y. (1968). *Biochem. Biophys. Res. Commun.* **33,** 108.

Sokoloff, L., Roberts, P. A., Januska, M. M., and Kline, J. E. (1968). *Proc. Natl. Acad. Sci. U.S.* **60,** 652.

Soliman, A. G. M., and King, K. W. (1969). *J. Nutr.* **98,** 255.

Solomon, A. K. (1962). *Biophys. J.* Suppl. 2, 79.

Spahr, P. F. (1964). *J. Biol. Chem.* **239,** 3716.

Speakman, P. T. (1968). *Nature* **219,** 724.

Speigel, M. (1961). *Biol. Bull.* **121,** 547.

Speigel, M., and Speigel, E. S. (1964). *Biol. Bull.* **126,** 307.

Spirin, A. S. (1964). *Zh. Obshch. Biol.* **25,** 321.

Spirin, A. S. (1966). *Current Topics Develop. Biol.* **1,** 1.

Spiro, M. J., and Spiro, R. G. (1968). *J. Biol. Chem.* **243,** 6520 and 6529.

Spiro, R. G., and Spiro, M. J. (1966). *J. Biol. Chem.* **240,** 997.

Sporn, M. B., Dingman, C. W., Phelps, H. G., and Wogan, G. N. (1966). *Science* **151,** 1539.

Stadtman, E. R. (1968). *Ann. N.Y. Acad. Sci.* **151,** 516.

Staehelin, M. (1965). *Biochem. Z.* **342,** 459.

Staehelin, T., Wettstein, F. O., and Noll, H. (1963). *Science* **140,** 180.

Steele, W. J. (1968). *J. Biol. Chem.* **243,** 3333.

Steiner, D. F. (1969). *New Engl. J. Med.* **280,** 1106.

Stellwagen, R. H., and Cole, R. D. (1968a). *J. Biol. Chem.* **243,** 4452.

Stellwagen, R. H., and Cole, R. D. (1968b). *J. Biol. Chem.* **243,** 4456.

Stellwagen, R. H., and Cole, R. D. (1969). *Ann. Rev. Biochem.* **38,** 951.

Stenram, U. (1958a). *Acta Pathol. Microbiol. Scand.* **44,** 239.

Stenram, U. (1958b). *Exptl. Cell Res.* **15,** 174.

Stenram, U., and Willén, R. (1967). *Z. Zellforsch. Mikroskop. Anat.* **82,** 270.

Stent, G. S. (1966). *Proc. Roy. Soc.* **B164,** 181.

Stent, G. S., and Brenner, S. (1961). *Proc. Natl. Acad. Sci. U.S.* **47,** 2005.

Stevely, W. S., and Stocken, L. A. (1968). *Biochem. J.* **110,** 187.

Steward, D. L., Shaeffer, J. R., and Humphrey, R. M. (1968). *Science* **161,** 791.

Stewart, G. A., and Farber, E. (1967). *Science* **157,** 67.

Stewart, G. A., and Farber, E. (1968). *J. Biol. Chem.* **243,** 4479.

Stewart, J. A., and Papaconstantinou, J. (1967). *Proc. Natl. Acad. Sci. U.S.* **58,** 95.

Stirpe, F., and Fiume, L. (1967). *Biochem. J.* **105,** 779.

Stout, E. R., and Mans, R. J. (1967). *Biochim. Biophys. Acta* **134,** 327.

Stout, E. R., and Mans, R. J. (1968). *Plant Physiol.* **43,** 405.

Strehler, B. L., Hendley, D. D., and Hirsch, G. P. (1967). *Proc. Natl. Acad. Sci. U.S.* **57,** 1751.

Stroun, M., Charles, P., Anker, P., and Pelc, S. R. (1967). *Nature* **216,** 716.

Subak-Sharpe, H., Shepherd, W. M., and Hay, J. (1966). *Cold Spring Harbor Symp. Quant. Biol.* **31,** 538.

Subramanian, A. R., Ron, E. Z., and Davis, B. D. (1968). *Proc. Natl. Acad. Sci. U.S.* **61,** 761.

Summers, W. P., and Mueller, G. C. (1968). *Biochem. Biophys. Res. Commun.* **34,** 350.

Summers, W. P., Noteboom, W. D., and Mueller, G. C. (1966). *Biochem. Biophys. Res. Commun.* **22,** 399.

Sundharadas, G., Katze, J. R., Söll, D., Konigsberg, W., and Lengyel, P. (1968). *Proc. Natl. Acad. Sci. U.S.* **61,** 693.

Sykes, J. (1966). *J. Theoret. Biol.* **12,** 373.

Takagi, M., and Ogata, K. (1968). *Biochem. Biophys. Res. Commun.* **33,** 55.

Talal, N., and Exum, E. B. (1966). *Proc. Natl. Acad. Sci. U.S.* **55,** 1288.

Talal, N., and Kaltreider, H. B. (1968). *J. Biol. Chem.* **243,** 6504.

Tanaka, K., Teraoka, H., Nigara, T., and Tamaki, M. (1966). *Biochim. Biophys. Acta* **123,** 435.

Tanaka, N., Yamaguchi, H., and Umezawa, H. (1966). *J. Biochem. (Tokyo)* **60,** 429.

Tanaka, N., Kinoshita, T., and Masukawa, H. (1968). *Biochem. Biophys. Res. Commun.* **30,** 278.

Tanaka, N., Kinoshita, T., and Masukawa, H. (1969). *J. Biochem. (Tokyo)* **65,** 459.

Tanaka, T., and Knox, W. E. (1959). *J. Biol. Chem.* **234,** 1162.

Tata, J. R. (1967a). *Biochem. J.* **105,** 47P.

Tata, J. R. (1967b). *Nature* **213,** 566.

Tata, J. R. (1968). *Nature* **219,** 331.

Taubman, S. B., Jones, N. R., Young, F. E., and Corcoran, J. W. (1966). *Biochim. Biophys. Acta* **123,** 438.

Tavill, A. S., Grayzell, A. I., London, I. M., Williams, M. K., and Vanderhoff, G. A. (1968a). *J. Biol. Chem.* **243,** 4987.

Tavill, A. S., Craigie, A., and Rosenoer, V. M. (1968b). *Clin. Sci.* **34,** 1.

Taylor, M. W., Granger, G. A., Buck, C. A., and Holland, J. J. (1967). *Proc. Natl. Acad. Sci. U.S.* **57,** 1712.

Taylor, M. W., Buck, C. A., Granger, G. A., and Holland, J. J. (1968). *J. Mol. Biol.* **33,** 809.

Temmerman, J., and Lebleu, B. (1969). *Biochim. Biophys. Acta* **174,** 544.

Teng, C. S., and Hamilton, T. H. (1967). *Biochem. J.* **105,** 1091.

Terao, K., Sugano, H., and Ogata, K. (1968). *J. Biochem. (Tokyo)* **64,** 407.

Tester, C. F., and Dure, L. (1966). *Biochem. Biophys. Res. Commun.* **23,** 287.

Thiers, R. A., and Vallee, B. L. (1957). *J. Biol. Chem.* **226,** 911.

Tidwell, T., Allfrey, V. G., and Mirsky, A. E. (1968). *J. Biol. Chem.* **243,** 707.

Tissières, A., Bourgeois, S., and Gros, F. (1963). *J. Mol. Biol.* **7,** 100.

Tocchini-Valentini, G. P., Marino, P., and Colvill, A. J. (1968). *Nature* **220,** 275.

Toft, D., Shyamala, G., and Gorski, J. (1967). *Proc. Natl. Acad. Sci. U.S.* **57,** 1740.

Tolbert, B. M. (1963). "Respiratory Pattern Analysis," p. 1. Appl. Phys. Corp., Santa Monica, California.

Tomasi, T., Sledz, D., Wales, J. K., and Recant, L. (1967). *Proc. Soc. Exptl. Biol. Med.* **126,** 315.

Tomkins, G. M., and Ames, B. N. (1967). *Natl. Cancer Inst. Monograph* **27,** 221.

Tomkins, G. M., Yielding, K. L., Talal, N., and Curran, J. F. (1963). *Cold Spring Harbor Symp. Quant. Biol.* **28,** 461.

Tomkins, G. M., Yielding, K. L., Summers, M. R., and Bitensky, M. W. (1965). *J. Biol. Chem.* **240,** 3793.

Tóth, M., and Mányai, S. (1968). *Acta Biochim. Biophys. Acad. Sci. Hung.* **3,** 337.

Tourian, A., Goddard, J., and Puck, T. T. (1969). *J. Cell Physiol.* **73,** 159.

Tracht, M. E., Tallal, L., and Tracht, D. G. (1967). *Life Sci.* **6,** 2621.

Trakatellis, A. C., Axelrod, A. E., and Montjar, M. (1964a). *Nature* **203,** 1134.

Trakatellis, A. C., Axelrod, A. E., and Montjar, M. (1964b). *J. Biol. Chem.* **239,** 4237.

Trams, E. G., and Brown, E. A. (1966). *J. Theoret. Biol.* **12,** 311.

Traut, R. R., and Monro, R. E. (1964). *J. Mol. Biol.* **10,** 63.

Travers, A. A., and Burgess, R. R. (1969). *Nature* **222,** 537.

Trayer, I. P., Harris, C. I., and Perry, S. V. (1968). *Nature* **217,** 452.

Troll, W., Belman, S., Berkowitz, E., Chiemlewicz, Z. F., Ambrus, J. L., and Bardos, T. J. (1968). *Biochim. Biophys. Acta* **157,** 16.

Ueno, P., Hosoya, M., Morita, P., Ueno, I., and Tatsuno, T. (1968). *J. Biochem.* (*Tokyo*) **64,** 479.

Ul-Hassan, M., and Greenberg, D. M. (1952). *Arch. Biochem.* **39,** 129.

Umaña, R., and Dounce, A. L. (1964). *Exptl. Cell Res.* **35,** 277.

Umbarger, H. E. (1956). *Science* **123,** 848.

Umbarger, H. E. (1969). *Ann. Rev. Biochem.* **38,** 323.

Umbarger, H. E., and Brown, B. (1957). *J. Bacteriol.* **73,** 105.

Umbarger, H. E., and Brown, B. (1958). *J. Biol. Chem.* **233,** 415 and 1156.

Umezawa, H., Mizuno, S., Yamazaki, H., and Nitta, K. (1968). *J. Antibiot.* **21,** 234.

Unuma, T., Smetana, K., and Busch, H. (1967). *Exptl. Cell Res.* **48,** 665.

Unuma, T., Arendell, J. P., and Busch, H. (1968). *Exptl. Cell Res.* **52,** 429.

Valli, E. A., Sarma, D. S. R., and Sarma, P. S. (1968). *Indian J. Biochem.* **5,** 120.

Vendrely, R. (1955). *In* "The Nucleic Acids" (E. Chargaff and J. N. Davidson, eds.), Vol. 2, p. 155. Academic Press, New York.

Vesco, C., and Giuditta, A. (1968). *J. Neurochem.* **15,** 81.

Vesell, E. S. (1968). *Ann. N.Y. Acad. Sci.* **151,** 1.

Villa-Treviño, S., Shull, K. H., and Farber, E. (1963). *J. Biol. Chem.* **238,** 1757.

Villa-Treviño, S., Farber, E., Staehelin, T., Wettstein, F. O., and Noll, H. (1964). *J. Biol. Chem.* **239,** 3826.

Villa-Treviño, S., Shull, K. H., and Farber, E. (1966). *J. Biol. Chem.* **241,** 4670.

Vincent, W. S., and Miller, O. L., Jr. (1966). *Natl. Cancer Inst. Monograph* **23,** 1–610.

Vincent, W. S., Baltus, E., Løvlie, A., and Mundell, R. D. (1966). *Natl. Cancer Inst. Monograph* **23,** 235.

Vogel, H. J., and Vogel, R. H. (1967). *Ann. Rev. Biochem.* **36,** Part II, 519.

Vogel, H. J., Bacon, D. F., and Baich, A. (1963). *In* "Informational Macromolecules" (H. J. Vogel, V. Bryson, and J. O. Lampen, eds.), p. 293. Academic Press, New York.

von Euler, L. H., Rubin, R. J., and Handschumacher, R. E. (1963). *J. Biol. Chem.* **238,** 2464.

Wacker, W. E. C., and Vallee, B. L. (1964). *In* "Mineral Metabolism" (C. L. Comar and F. Bronner, eds.), Vol. 2, Part 2A, p. 483. Academic Press, New York.

Wade, W. E. (1961). *Biochem. J.* **78,** 457.

Wade, W. E., and Robinson, H. K. (1966). *Biochem. J.* **101,** 467.

Wagner, E. K., and Roizman, B. (1968). *Science* **162,** 569.

Waksman, A., and Rendon, A. (1968). *Arch. Biochem. Biophys.* **123,** 201.

Wang, J. H. C., and Matheson, A. T. (1966). *Biochem. Biophys. Res. Commun.* **23,** 740.
Wang, T. Y., and Johns, E. W. (1968). *Arch. Biochem. Biophys.* **124,** 176.
Wang, T. Y., and Patel, G. (1967). *Life Sci.* **6,** 413.
Ward, D. C., Reich, E., and Goldberg, I. H. (1964). *Science* **149,** 1259.
Waring, M. J. (1964). *Biochim. Biophys. Acta* **87,** 358.
Waring, M. J. (1965). *J. Mol. Biol.* **13,** 269.
Warner, J. R., and Soeiro, R. (1967). *Proc. Natl. Acad. Sci. U.S.* **58,** 1984.
Wasserman, R. H., and Taylor, A. N. (1968). *J. Biol. Chem.* **243,** 3987.
Wasserman, R. H., Corradino, R. A., and Taylor, A. N. (1968). *J. Biol. Chem.* **243,** 3978.
Watts, R. L., and Mathias, A. P. (1967). *Biochim. Biophys. Acta* **145,** 828.
Waxman, H. S., and Rabinovitz, M. (1966). *Biochim. Biophys. Acta* **129,** 369.
Webb, T. E. (1967). *Biochim. Biophys. Acta* **138,** 307.
Wehrli, W., Nüesch, J., Knüsel, F., and Staehelin, M. (1968a). *Biochim. Biophys. Acta* **157,** 215.
Wehrli, W., Knüsel, F., Schmid, K., and Staehelin, M. (1968b). *Proc. Natl. Acad. Sci. U.S.* **61,** 667.
Weinstein, I. B., and Finkelstein, I. H. (1967). *J. Biol. Chem.* **242,** 3757.
Weinstein, I. B., Chernoff, R., Finkelstein, I. H., and Hirschberg, E. (1965). *Mol. Pharmacol.* **1,** 297.
Weinstein, I. B., Carchman, R., Marner, E., and Hirschberg, E. (1967). *Biochim. Biophys. Acta* **142,** 440.
Weisberger, A. S., and Daniel, T. M. (1969). *Proc. Soc. Exptl. Biol. Med.* **131,** 570.
Weisblum, B., Gonano, F., von Ehrenstein, G., and Benzer, S. (1965). *Proc. Natl. Acad. Sci. U.S.* **53,** 328.
Weiss, P., and Grover, N. B. (1968). *Proc. Natl. Acad. Sci. U.S.* **46,** 1020.
Weiss, S. B. (1960). *Proc. Natl. Acad. Sci. U.S.* **46,** 1020.
Weiss, S. B., and Gladstone, L. (1959). *J. Am. Chem. Soc.* **81,** 4118.
Wérenne, J., Grosjean, H., and Chantrenne, H. (1966). *Biochim. Biophys. Acta* **129,** 585.
Wettstein, F. O., Staehelin, T., and Noll, H. (1963). *Nature* **197,** 430.
Whitehouse, H. K. L. (1967). *J. Cell Sci.* **2,** 9.
Whitson, G. L., Padilla, G. M., and Fisher, W. D. (1966). *Exptl. Cell Res.* **42,** 438.
Whittle, E. D., Bushness, D. E., and Potter, V. R. (1968). *Biochim. Biophys. Acta* **161,** 41.
Widnell, C. C., and Tata, J. R. (1964). *Biochem. J.* **92,** 313.
Windell, C. C., and Tata, J. R. (1966). *Biochim. Biophys. Acta* **123,** 478.
Wilkinson, J. M., Press, E. M., and Porter, R. R. (1966). *Biochem. J.* **100,** 303.
Williams, L. S., and Freundlich, M. (1969). *Biochim. Biophys. Acta* **179,** 515.
Williamson, J. R., Kreisberg, R. A., and Felts, P. W. (1966). *Proc. Natl. Acad. Sci. U.S.* **56,** 247.
Wilson, S. H., and Hoagland, M. B. (1967). *Biochem. J.* **103,** 556.
Wilson, S. H., Hill, H. Z., and Hoagland, M. B. (1967). *Biochem. J.* **103,** 567.
Windmueller, H. G. (1964). *J. Biol. Chem.* **239,** 530.
Windmueller, H. G., and Levy, R. I. (1967). *J. Biol. Chem.* **242,** 2246.
Windmueller, H. G., and Levy, R. I. (1968). *J. Biol. Chem.* **243,** 4878.
Winnick, T., Friedberg, F., and Greenberg, D. M. (1948). *J. Biol. Chem.* **173,** 189.

Wittmann, J. S., Lee, K. L., and Miller, O. N. (1969). *Biochim. Biophys. Acta* **174**, 536.

Woese, C., Naono, S., Soffer, R., and Gros, F. (1963). *Biochem. Biophys. Res. Commun.* **11**, 435.

Work, T. S., Coote, J. L., and Ashwell, M. (1968). *Federation Proc.* **27**, 1174.

Wunner, W. H., Bell, J., and Munro, H. N. (1966). *Biochem. J.* **101**, 417.

Wurtman, R. J., Axelrod, J., Vesell, E. S., and Ross, G. T. (1968). *Endocrinology* **82**, 584.

Wyatt, G. R., and Tata, J. R. (1968). *Biochem. J.* **109**, 253.

Wykes, J. R., and Smellie, R. M. S. (1966). *Biochem. J.* **99**, 347.

Yamaguchi, H., Yoshida, Y., and Tanaka, N. (1966). *J. Biochem.* (*Tokyo*) **60**, 246.

Yamamoto, S., and Hayaishi, O. (1967). *J. Biol. Chem.* **242**, 5260.

Yang, W. K., and Novelli, G. D. (1968). *Proc. Natl. Acad. Sci. U.S.* **59**, 208.

Yanovsky, C. (1967). *Ann. Rev. Genet.* **1**, 117.

Young, F. G. (1945). *Biochem. J.* **39**, 515.

Yu, C. T. (1966). *Cold Spring Harbor Symp. Quant. Biol.* **31**, 565.

Zamenhof, S., and Eichhorn, H. H. (1967). *Nature* **216**, 456.

Zelkowitz, L., Arimura, G. K., and Yunis, A. A. (1968). *J. Lab. Clin. Med.* **71**, 596.

Zimmerman, E. F. (1968). *Biochemistry* **7**, 3156.

Zimmerman, E. F., Hackney, J., Nelson, P., and Arias, I. M. (1969). *Biochemistry* **8**, 2636.

Zucker, W. V., and Schulman, H. M. (1967). *Biochim. Biophys. Acta* **138**, 400.

CHAPTER 31

Hormonal Regulation of Synthesis of Liver Enzymes[1]

Francis T. Kenney

Biology Division,
Oak Ridge National Laboratory,
Oak Ridge, Tennessee

I. Introduction

The old polemic on the relative significance of "genes" and "environment" as determinants in biology was long ago laid to rest in the realization that both of these play leading and interrelated roles in the drama of life. But it is only recently, i.e., within the memories of the present generation of biological scientists, that a systematic attempt has been launched to understand the molecular mechanisms involved in the interaction of genes and environment. The role of environmental influences in determining the efficacy of the genes can be considered, for the present

[1] Work from the author's laboratory discussed herein was supported jointly by the National Cancer Institute, National Institutes of Health and by the U. S. Atomic Energy Commission under contract with the Union Carbide Corporation.

purposes, to have two components, one of which involves influencing the *formation* of the gene product (protein), while the other embraces all the means by which the *function* of the gene product can be altered. The first of these encompasses the research area to be discussed here, and it reduces experimentally to the questions: What are the factors that govern the synthesis of specific proteins, and how do these factors work?

The major advances in regulation of protein synthesis have, of course, been made in studies on bacteria and bacterial viruses, where the investigator's ability to manipulate genes affords an enormous advantage. Still there is a growing number of researchers who recognize that the complexities of the mammalian organism cannot always be bypassed in favor of the simpler microbial systems; indeed, that it is partially by virtue of these complexities, such as the imposition of intercellular message systems (hormones) and intracellular membranes and compartments, that protein synthesis is controlled in mammalian cells. Due in large part to these and other complexities, however, much of the research activity in mammalian regulation has been descriptive rather than mechanistic, the researcher's efforts being confined to sorting out the switches rather than analyzing the mechanisms they activate. Only a few experimental systems have been studied in sufficient detail to allow meaningful discussion of the mechanisms involved, and in preparing this chapter I have concentrated on these few, making no attempt to review the literature or to cite all the contributions toward a particular subject. As befits a research area that is both vigorous and complex, there is much controversy here over the interpretation of experimental results. As an advocate of the view that the simplest interpretation is most probably correct, I have sought simplifying relationships and provided my interpretation of what others have reported—an interpretation often at odds with that of the original authors.

One of the areas of controversy is, unfortunately, the semantic one, and the unwary reader can be hopelessly confused by the various meanings given to terms such as "induction" and "repression." We deal here with enzyme proteins, and there are an almost infinite number of ways in which the *activity* of a given concentration of enzyme can be changed. Further, as discussed by Schimke in Chapter 32 of this volume, the *concentration* of an enzyme in mammalian tissues can be selectively altered by changes in the rate of its degradation, i.e., without altering its rate of synthesis. With the view that terms, to be useful, must represent clear concepts, I have used the terms "induction" and "repression" in a fashion that excludes both changes in activity and changes in concentration due to alterations in turnover, focusing attention instead on enzyme *synthesis* as the parameter involved in "induction" (a selective increase in the rate of enzyme synthesis) and "repression" (a selective decrease in this rate).

From these considerations it can be seen that, for a rigorous analysis, it is imperative that experimental means be taken to measure both the concentration of an enzyme protein and the rate of its synthesis. For this purpose several investigators have turned to immunochemical techniques, as the enzyme need be purified only once in sufficient quantity to serve as antigen for the preparation of a specific antiserum. Once such a serum is obtained, and its specificity established, it can be used as a specific reagent for isolation of the protein from crude extracts. This provides a direct measurement of concentration as well as, when combined with appropriate isotopic labeling, an estimate of the rate of synthesis of the protein. It is regrettable that this type of analysis has been made in only a few of the experimental systems examined, much experimentation relying instead on inhibitors of RNA and protein synthesis, which at best permit intelligent guesses but do not substitute for direct analyses.

Because of time limitations, I have found it necessary to omit discussion of several very interesting nonhepatic hormonal inductions and to limit the treatment of hepatic regulation to that initiated by a few hormones. But the hormones are the principal regulators of specific protein synthesis in mammalian systems; indeed, it remains to be seen whether molecules other than hormones (such as substrates and products) actually influence specific protein *synthesis* at all in mammals. Investigation of liver enzymes has been an overwhelmingly favorite choice for those interested in mammalian regulation; whether this is because the biochemist comes first upon the liver when opening the abdomen (as alleged by some) or because "that's where the action is" (as we contend), I shall leave for the reader to decide. For these reasons the hormonal regulation of synthesis of liver enzymes is not as restricted a category as it might seem, for it encompasses the bulk of the research that has been done in regulation of enzyme levels in mammals.

II. Glucocorticoids

The glucocorticoids are recognized by the physiologist as catabolic hormones, but in the first of many "paradoxical" effects to be discussed, these steroids promote marked increases in the synthesis of a few enzyme proteins in the liver, each of which catalyzes a significant reaction in the metabolism of amino acids.

A. Tryptophan Pyrrolase

This was the first liver enzyme found to be induced by glucocorticoids (Knox, 1951), and the mechanism of this induction, as well as of the comparable elevation of the enzyme by its substrate (Knox and Mehler, 1951) is still being vigorously pursued in several laboratories. While the reaction catalyzed by the enzyme is relatively simple (oxygenation of the pyrrole ring of tryptophan to yield formylkynurenine) the enzyme itself is far from simple, and there is still some controversy over the role of the substrate and other factors in determining the structure (and hence the stability) of the protein. As visualized by Knox and Piras (1966), tryptophan and hematin interact to activate the enzyme according to the scheme shown in Fig. 1.

Role of the Substrate in Vivo: Induction or Stabilization?

The mechanisms involved in elevating the enzyme *in vivo* by hydrocortisone or by tryptophan were shown to be different, as the two stimulators yielded additive effects on the pyrrolase level (Civen and Knox, 1959). In perceptive earlier studies, both Lee (1956) and Dubnoff and Dimick (1959) suggested that tryptophan acts by stabilizing the enzyme against continuous degradation, based upon considerations of the kinetics of the rise in enzyme level (Lee) and the observation that tryptophan protects the enzyme from inactivation in crude liver extracts (Dubnoff and Dimick). In later studies this was extended to show that the α-methyl analog of tryptophan (which is not a substrate) also stabilizes the enzyme in extracts, and significantly, elevates the enzyme when administered *in vivo* (Civen and Knox, 1960; Schimke *et al.*, 1965a).

With an immunochemical titration method, Feigelson and Greengard (1962) showed that the concentration of pyrrolase increases in proportion to the increase in activity of the enzyme, whether the increase is brought about by the substrate or by the hormone. This effort demanded purifica-

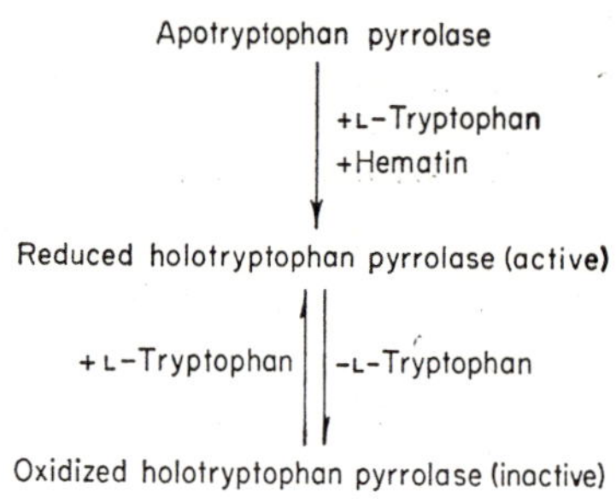

FIG. 1. Transformations of tryptophan pyrrolase. (From Knox and Piras, 1966.)

tion of the enzyme, and in the course of purification it was found that the purified pyrrolase is inactive without added hematin, the iron-porphyrin acting as a coenzyme in the reaction (Feigelson and Greengard, 1961; Greengard and Feigelson, 1962). This led to the important discovery that tryptophan administration *in vivo* is associated with conversion of the apoenzyme (inactive without added hematin) to the active holoenzyme; when the enzyme is elevated by hydrocortisone, the proportion present as apoenzyme remains unchanged (Greengard and Feigelson, 1961a,b; Knox *et al.,* 1966). A further significant distinction between the hormonal- and substrate-mediated elevations was made by Greengard *et al.* (1963a), who found that the hormonal effect was blocked by both puromycin and actinomycin, while the effect of the substrate was sensitive to the inhibitor of protein synthesis but not to actinomycin. From these studies the picture emerges that more newly synthesized enzyme is present in the liver after both the hormonal and the substrate treatment, but that synthesis of RNA is required only for the hormonal elevation while the substrate effect is associated with structural changes in the enzyme.

In a classic short paper Schimke *et al.,* (1964) elaborated upon the kinetic considerations introduced earlier by Lee (1956) and showed that since turnover processes are first order, an elevation of enzyme brought about by stopping turnover should follow a linear course in time, while increased synthesis without change in turnover rate should yield an exponential time course. Careful analysis of the nature of pyrrolase changes confirmed the theoretical considerations; hydrocortisone treatment gave an exponential response while tryptophan led to a linear increase in the enzyme level. If the level was elevated by hydrocortisone the usual subsequent decay to basal level was prevented by treatment with tryptophan, a result also in accordance with the view that the substrate protects the enzyme against inactivation *in vivo.*

Conclusive evidence that the hormone *induces* while the substrate *stabilizes* followed isolation of the enzyme in a form sufficiently pure to permit immunochemical–isotopic analyses of the rate of synthesis (Schimke *et al.,* 1965b). As shown by the data of Table I, hydrocortisone effected a five- to sevenfold increase in the rate of pyrrolase synthesis (estimated as the radioactivity incorporated into the enzyme during a 40-minute pulse), but tryptophan was virtually without effect on the incorporation of isotope. This result has recently been confirmed by Cho-Chung and Pitot (1968), who observed a twenty-twofold increase in synthesis with cortisone and a doubling of incorporation into the enzyme with tryptophan. In both experiments the labeling time (40 minutes) was a significant portion of the half-life of the enzyme (prob-

TABLE I
HYDROCORTISONE AND TRYPTOPHAN EFFECTS ON TRYPTOPHAN PYRROLASE SYNTHESIS[a]

Treatment[b]	Enzyme activity (units/gm liver)	Leucine-^{14}C incorporation	
		Tryptophan pyrrolase (total cpm)	Supernatant protein (cpm/mg)
None	4.2	1368	1190
Hydrocortisone	13.6	5640	1320
Tryptophan	8.2	1620	1564

[a] Adapted from Schimke *et al.* (1965b).

[b] Hydrocortisone and tryptophan were given 4 hours before measurements were made.

ably about 90 minutes) so that inhibition of turnover would be expected to cause a slight increase in the amount of labeled enzyme present when assays were made, as was observed. Schimke *et al.* (1965b) also made direct measurements of enzyme turnover under basal conditions, using an isotopic chase procedure, and confirmed that treatment with tryptophan brings about a virtual cessation of degradation of the enzyme. Thus in this, the classic example in mammalian regulation of the control by a substrate of an enzyme level, the mechanism of the substrate effect is stabilization rather than induction. This being so, the recurring suggestions of a translational mechanism of the tryptophan effect, all of which envision a mass-action type of regulation of pyrrolase *synthesis* (Greengard and Feigelson, 1961b; Greengard, 1963; Knox and Greengard, 1965; Knox, 1966) can be discounted. The data of Schimke *et al.* (1965b) and of Cho-Chung and Pitot (1968) show conclusively that the hormonal effect on tryptophan pyrrolase is, in fact, an induction. Addition of hydrocortisone to the circulating blood in experiments with isolated, perfused livers effected a marked increase in tryptophan pyrrolase (Goldstein *et al.*, 1962a), showing that the steroid hormone acts directly in hepatic cells to elevate pyrrolase synthesis. The mechanism of this hormonal induction will be discussed after a consideration of other enzymes that have been reported to be induced by adrenal steroids.

B. Tyrosine Transaminase

The second major object of research in regulation of liver enzyme synthesis has been tyrosine transaminase, an enzyme which is elevated quickly and markedly in response to glucocorticoids (Lin and Knox, 1957). This is a much simpler enzyme than tryptophan pyrrolase, and this simplicity, together with ease of assay and relative ease of purification, have made this enzyme a favorite experimental tool of several investigators, including the writer. Requiring only the coenzyme pyridoxal phosphate, the protein catalyzes transamination of tyrosine and α-ketoglutarate, and will also react to a limited extent with phenylalanine and tryptophan as amino donors (Jacoby and LaDu, 1964). The rat liver enzyme was purified to immunological homogeneity (Kenney, 1959, 1962a) and its crystallization has recently been reported (Hayashi *et al.*, 1967). The enzyme from induced livers is indistinguishable from the basal enzyme by immunological tests and by criteria such as kinetic characteristics and physical properties (Kenney, 1962a,b; unpublished observations, Hayashi *et al.*, 1967).

Massive amounts of the insoluble substrate, tyrosine, were reported to elevate this enzyme in intact animals but not after adrenalectomy; in the latter condition the response to hydrocortisone was doubled by simultaneous treatment with tyrosine (Lin and Knox, 1957). This potentiating effect was mimicked by a variety of insoluble materials such as Celite or bentonite and by tryptophan (Kenney and Flora, 1961), showing that the catalytic capacity of the transaminase is not related to the material causing potentiation. Thus there is no implication of "substrate induction" with this enzyme. The effect of stressing agents (tyrosine, Celite, bentonite) and of tryptophan can now be understood in terms of their capacity to bring about a second type of induction of tyrosine transaminase, probably by promoting secretion of glucagon, as will be discussed in Section III,B.

By using a specific antiserum against the purified transaminase, it was possible to demonstrate that glucocorticoid treatment caused an increase in enzyme concentration equivalent to the increased enzyme activity (Kenney, 1962b). With isotopic labeling *in vivo* and immunological isolation of the labeled enzyme, it was shown that the enzyme protein is synthesized at a measurable rate in both basal and hormone-treated conditions, and that the rate of synthesis is accelerated at least 4- to 5-fold by the hormone (Kenney, 1962c). The techniques used in these labeling experiments, the first to demonstrate that synthesis of a specific enzyme is regulated by a hormone, have been improved and can now pinpoint when the elevated synthesis begins and when it ceases. As shown in Fig. 2, administration of hydrocortisone is followed by a definite lag period during which transaminase synthesis may be slightly lower than normal. Increased synthesis is apparent about 1 hour after treatment, and the elevated rate is maintained for another 4 or 5 hours under these experimental conditions. The enzyme level rises shortly after the change in synthesis and continues to rise as long as synthesis is

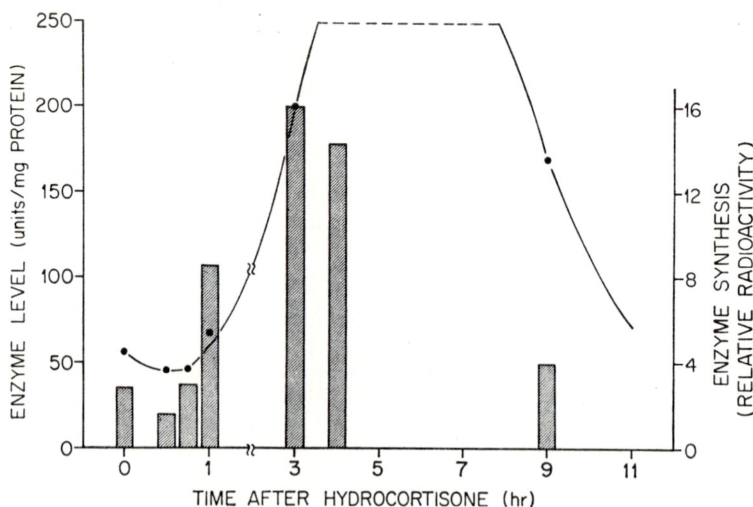

FIG 2. Immunochemical analyses of rate of synthesis of tyrosine transaminase after hydrocortisone treatment. The points represent enzyme activity plotted against the left ordinate. The bars represent relative enzyme radioactivity (counts per minute in isolated enzyme × 10^3/cpm per milligram of total soluble protein) and are plotted against the right ordinate. (From Kenney *et al.*, 1968b.)

elevated. After the elevated synthesis ceases, the enzyme level falls, returning to the basal level 10 or 12 hours after a single hormone treatment. During the phase in which the enzyme level falls, enzyme synthesis has returned to the basal rate, apparently owing to depletion of the steroid inducer. Changes in the transaminase level which follow glucocorticoid treatment can thus be attributed entirely to an elevation of synthesis of the enzyme by the steroid, an elevation which ceases when the hormone concentration falls below effective levels. It should be noted that these data show that there is no *necessity* to invoke repressors of enzyme synthesis (Garren *et al.*, 1964a) to explain the cessation of induced enzyme synthesis and the return of the enzyme to basal levels. Induction of tyrosine transaminase is due to a direct action of glucocorticoids in the liver, for these steroids will bring about a comparable elevation of the enzyme in experiments with isolated, perfused livers (Barnabei and Sereni, 1961; Goldstein *et al.*, 1962a; Hager and Kenney, 1968).

1. *Effects of Inhibitors*

Much confusion has been generated by several reports of paradoxical effects of inhibitors of RNA and protein synthesis on hepatic tyrosine transaminase. Induction by hydrocortisone of both tyrosine transaminase and tryptophan pyrrolase is essentially completely inhibited by large doses of puromycin (which releases incomplete peptides) or by moderate doses of actinomycin (an inhibitor of RNA synthesis), given shortly before or after the hormone (Nemeth and De la Haba, 1962; Greengard *et al.*, 1963a; Garren *et al.*, 1964a; Holten *et al.*, 1967). Cycloheximide (which blocks peptide synthesis) in amounts sufficient to block protein synthesis almost completely also blocks induction (Kenney, 1967a; Mavrides and Lane, 1967). This repeatedly confirmed inhibition of induction by these inhibitors has been challenged for the cycloheximide–transaminase effect (Fiala and Fiala, 1966) and for the actinomycin–pyrrolase interaction (Mishkin and Shore, 1967). However, these unusual findings can be discounted, as no evidence was presented in these reports that the materials administered (cycloheximide?, actinomycin?) did in fact inhibit the relevant synthetic processes.

Treatment with either puromycin or cycloheximide blocks transaminase induction but does not cause a marked change in the basal level of the enzyme, which led Grossman and Mavrides (1967) to conclude that tyrosine transaminase undergoes only a limited turnover in the basal condition. I made direct measurements of the basal rate of turnover, using an isotopic chase technique, and found that basal turnover is indeed rapid ($t_{1/2}$ about 1.5 hours); failure of protein synthesis-inhibitors to lower the enzyme level is due to inhibition of the turnover process (Kenney, 1967a). This finding—aside from its inherent interest in implicating synthetic processes in turnover—emphasizes the danger in attempting to interpret inhibitor experiments when just the enzyme level is measured. In this case, as in several others to be discussed, it is apparent that only by direct measurement of rates of synthesis and of turnover can a clear picture be obtained of the mechanisms involved in determining the amount of enzyme present in a given experimental situation.

With the realization that synthetic processes are required for the normal turnover of the transaminase, much of the confusion generated by inhibitor effects can be dissipated. For example, the slow elevation of the transaminase after long-term treatment with low doses of cycloheximide (Rosen and Milholland, 1966) or of actinomycin (Rosen *et al.*, 1964) could reflect differential inhibition of enzyme synthesis and turnover. If the turnover process is only slightly more sensitive to these inhibitors than is synthesis of the enzyme, passage of a period of time encompassing 8–50 times the normal half-life could easily result in marked elevation of the enzyme level. A similar interpretation can be made for the "paradoxical" effect of actinomycin in glucocorticoid-treated rats reported by Garren *et al.* (1964a) and later shown to occur in tissue cultures of hepatoma cells as well (Tomkins *et al.,* 1966); this "superinduction" will be discussed in the next section.

Grossman and Mavrides (1967), using a serial biopsy technique in which phenobarbital-anesthetized rats were given cortisol, confirmed the actinomycin effect of Garren *et al.* (1964a) but also added fresh fuel to the fire with their experiments showing that puromycin could elevate the transaminase if administered during the downturn phase of the induction response to the hormone. These authors consider the rapid reappearance of enzyme activity to result from either a reactivation of inactivated enzyme or from selective inhibition by puromycin of formation of a repressor of transaminase synthesis. The latter alternative now seems unlikely, for the argument that such a repressor exists is based upon a misinterpretation of the phenomenon called "superinduction" (see the next section). In the presence of puromycin the increased enzyme could hardly have resulted from enzyme synthesis, so the view that a reactivation process is responsible would appear to be correct. This finding, then, adds weight to the argument which holds that the process of enzyme turnover involves inactivation or denaturation of the enzyme (cf. Schimke, Chapter 32).

With regard to these "paradoxical" effects of antibiotics or other agents, it is significant that transaminase induction can be resumed if fresh steroid is given during the downturn phase; repeated administration of the hormone results in a plateau of high enzyme activity reflecting continuous elevation of enzyme synthesis (Fig. 3). Thus any experimental manipulation which resulted in a fresh flow of steroid to the liver could trigger a new induction phase and thereby a "paradoxical" effect of that manipulation. The existence of independent induction mechanisms for this enzyme, initiated by pancreatic hormones (Section II,B,C), further adds to the complexity of *in vivo* experiments; this complexity has led several investigators to seek simpler systems in which to study hormonal regulation of the synthesis of tyrosine transaminase.

2. Cell and Organ Culture

Experimentally induced liver tumors of the minimal deviation type were adapted to tissue culture by Pitot *et al.* (1964a), who quickly found that the tyrosine transaminase in cell cultures of the Reuber hepatoma (Reuber, 1961) could be elevated by supplementation of the medium with hydrocortisone. These cell cultures were made available to my laboratory by the Madison group, and the study of this response has continued both in Madison and at Oak Ridge (Potter *et al.,* 1967; Reel and Kenney, 1968a,b). Tomkins and his colleagues have developed a similar but not identical hepatoma cell line, which responds to gluco-

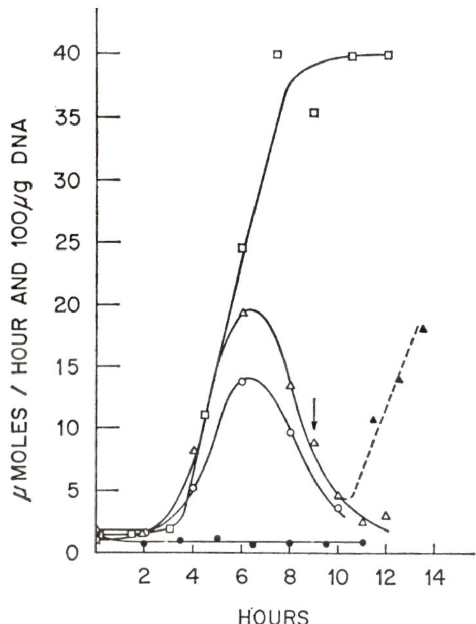

FIG. 3. Effects of various dosage schedules of cortisol on the kinetics of tyrosine transaminase induction in adrenalectomized rats. Cortisol was given in a single dose of 30 mg/kg at zero time, ○——○; in 4 doses of 3 mg/kg at 0, 1, 2, and 3 hours, △——△; in 12 hourly doses of 3 mg/kg, □——□; controls, ●——●. The arrow indicates the point at which an additional 30 mg/kg was administered to another group (▲——▲). (From Grossman and Mavrides, 1967.)

corticoids in a fashion nearly identical to that found with the Reuber cells (Tomkins *et al.*, 1966). A consistent finding in these experiments is the moderate concentration of steroid required (10^{-6} M is optimal) and the fact that induced enzyme synthesis continues until the hormone is removed by washing the cells and reincubating them in steroid-free medium (Fig. 4). This confirms the indication from *in vivo* experiments that enzyme synthesis will remain elevated as long as the steroid is present, and also suggests that the cultured tumor cells fail to degrade the hormone.

If RNA synthesis is blocked by actinomycin added at the same time or shortly after the steroid inducer, induced enzyme synthesis fails to occur and the enzyme level remains constant for several hours, as is the case *in vivo*. However, if cultured cells are preinduced to the plateau representing the full extent of hormonal induction, addition of actinomycin was found to effect a further elevation of the enzyme level (Thompson *et al.*, 1966). A similar "superinduction" effect can be observed in *in vivo* experiments (Garren *et al.*, 1964a), and it has been

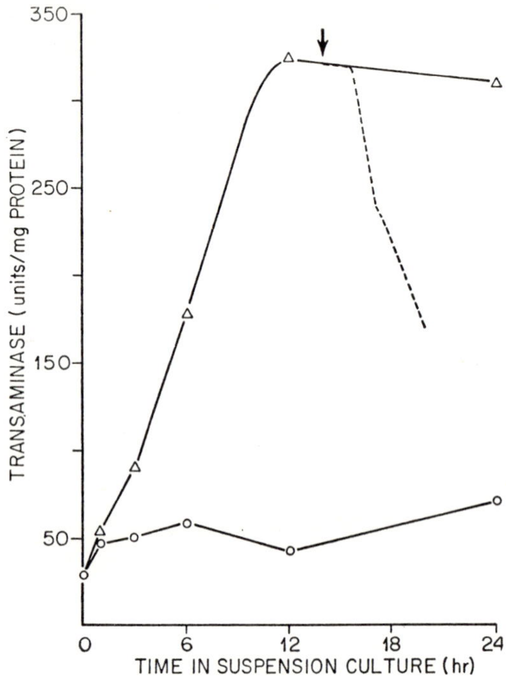

FIG. 4. Induction of tyrosine transaminase by hydrocortisone in suspension cultures of H35 hepatoma cells. Hydrocortisone (10^{-6} *M*) was added at zero time. The dashed line shows the drop in enzyme level which occurs when the steroid is removed. (From Kenney *et al.*, 1968b.)

attributed to inhibition by actinomycin of the synthesis of a labile repressor substance which is formed late in the response to steroids and which acts to limit enzyme synthesis by inhibiting translation. This important finding, implicating control mechanisms acting at the translational level, has excited much interest and has led to a spurt of reports of similar events occurring in other systems. The proposed repression mechanism demands that enzyme synthesis be stimulated after actinomycin treatment, and an initial report by Tomkins *et al.* (1966) presented isotopic-immunochemical evidence which appeared to confirm that synthesis is, indeed, elevated during "superinduction." However, the labeling time employed in this experiment (6.5 hours) was such that effects on enzyme turnover could not be excluded, since a blockade of degradation could also lead to increased labeling of the enzyme if the labeling interval is sufficiently long. Accordingly "superinduction" by actinomycin was reinvestigated, using brief "pulse" labeling times to measure synthesis and "chase" conditions to measure turnover (Reel and Kenney, 1968b). The results indicate clearly that enzyme synthesis is *not* stimulated after actinomycin treatment. Rather, enzyme turnover is quickly and effectively blocked, while enzyme synthesis declines more slowly at a rate probably indicative of the rate of degradation of the transaminase template (Fig. 5). Since turnover is inhibited quickly after actinomycin treatment, the enzyme level tends to rise despite the fact that the rate of synthesis is declining. These data show that the transaminase template in these cultured cells is quite labile after induction ($t_{1/2}$ about 3 hours, Fig. 5). Similar effects of actinomycin on transaminase synthesis and turnover *in vivo* have been observed in both induced and

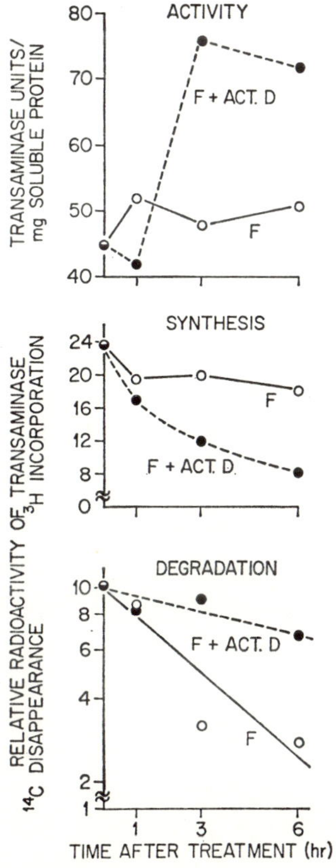

FIG. 5. Effect of actinomycin D on synthesis and degradation of tyrosine transaminase in preinduced HTC cultures in stationary phase. Hydrocortisone (F) (10^{-6} *M*) was added to monolayer cultures 24 hours before, and leucine-^{14}C 1 hour before, the experiments began. Just before zero time the ^{14}C-labeled medium was decanted and the cells were washed. At zero time fresh medium was added with actinomycin D at 5 μg/ml (filled circles). For rate of synthesis measurements leucine-^{3}H was added 15 minutes before cells were collected for analysis. (From Reel and Kenney, 1968b.)

basal conditions (Kenney, Kendrick, and Hager, unpublished experiments). Thus we have concluded that the "superinduction" of tyrosine transaminase reflects an actinomycin-mediated blockage of enzyme turnover and not the action of a cytoplasmic repressor.

Regulation of synthesis of tyrosine transaminase by glucocorticoids in fetal liver is of special interest, for fetuses respond very poorly or not at all to administered hydrocortisone (Sereni *et al.*, 1959). However, if fetal livers are placed in organ culture the hormone effects a marked induction after a refractory phase lasting about 12 hours (Wicks, 1968a). This has been interpreted as suggesting that the ineffectiveness of the hormone *in utero* may be due to the action of repressors which are dissipated in the organ culture environment. Induction by hydrocortisone in these cultures is like that in the hepatoma cell cultures in persisting until the hormone is removed from the medium.

C. Glutamic-Alanine Transaminase

Daily treatment of rats with cortisol causes a marked elevation of hepatic alanine transaminase (Rosen *et al.*, 1958). None of the substrates of the enzyme has any influence on the level of the enzyme *in vivo* (Rosen *et al.*, 1963). The time course of the response (Fig. 6) differs from that of either tyrosine transaminase or tryptophan pyrrolase, the enzyme level reaching a maximum about 5 times greater than the basal level only after several days of repeated steroid treatment (Rosen *et al.*, 1963; Segal and Kim, 1963). This difference in the time course of

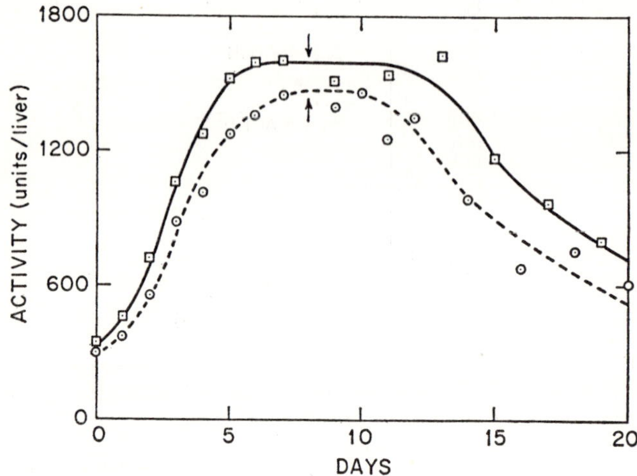

FIG. 6. Effect of administration of prednisolone and subsequent withdrawal on glutamic-alanine transaminase activity of rat liver. Upper curve, 4 mg prednisolone/day; lower curve, 2 mg/day. At the points marked by the arrows, hormone treatment was discontinued. (From Segal and Kim, 1963.)

response reflects the different turnover rates of these enzymes and is not the result of a different rate of response of enzyme synthesis (Segal and Kim, 1965; Berlin and Schimke, 1965).

This transaminase has been purified from rat liver by Segal *et al.* (1962a); there is no detectable difference in either physicochemical or immunochemical characteristics of the enzyme after induction (Gatehouse *et al.*, 1967). Immunochemical analyses demonstrated that increased activity reflects an increase in enzyme concentration (Segal *et al.*, 1962b), and pulse labeling analyses confirmed that increased synthesis of the enzyme follows hormone treatment (Segal and Kim, 1963). Elevated incorporation of leucine-^{14}C into the enzyme was apparent prior to any detectable change in the enzyme level, and was just sufficient to account fully for the ultimate increase in the enzyme level. Nevertheless the rate of degradation of the enzyme appeared to increase after hormone treatment; when treatment was stopped the enzyme level fell with a half-life of 3.5 days, while that calculated from the rate of increase was 1.2 days. Since the first of these measurements yields a true rate of degradation only if enzyme synthesis has dropped to a negligible rate, this apparent discrepancy probably reflects a residual elevation of synthesis that continued after treatment with prednisolone was stopped. Direct measurement of turnover rates will be necessary to resolve this question.

As yet there has been no report demonstrating that adrenal steroid hormones effect induction of this enzyme by direct action on liver cells; the possibility remains that the repeated steroid treatments caused the production of some other agent which is actually active in induction.

In the early reports describing the cortisol-mediated induction of glutamic-alanine transaminase there appeared to be promise in explaining the classical gluconeogenic effects of cortisol in terms of induced synthesis of this enzyme, which is presumably a key catalyst for the introduction of amino acid carbon into carbohydrate pathways. However, as Hubener (1960) and Segal and Gonzalez-Lopez (1963) have shown, increased glycogen deposition and the incorporation of isotopic carbon from alanine into glycogen is apparent within a few hours after cortisol while the transaminase level begins to increase much later. Thus it is clear that the *initiation* of gluconeogenesis cannot be dependent on induction of this enzyme, which may, however, contribute to glycogen deposition later on.

D. *Enzymes of Gluconeogenesis*

Conversion of pyruvate to glucose is limited by the "key enzymes of gluconeogenesis," glucose-6-phosphatase, fructose-1,6-diphosphatase, phosphopyruvate carboxykinase, and pyruvate carboxylase, which catalyze reactions allowing circumvention of the thermodynamic barriers to glucose formation from pyruvate (Krebs, 1954; Weber *et al.*, 1965). Much attention has thus been focused on the possibility that the gluconeogenic effect of glucocorticoids could result from hormonal induction of these enzymes. Daily treatment of intact rats with glucocorticoids elevates activity levels of all four of these enzymes, which were shown to increase at a rate and to an extent comparable to the glutamic-alanine transaminase response shown in Fig. 6 (Kvam and Parks, 1960; Mokrasch *et al.*, 1956; Shrago *et al.*, 1963; Henning *et al.*, 1963; Weber *et al.*, 1965). That the enzyme activities rise as a result of induced synthesis has been assumed from studies using inhibitors of protein and RNA synthesis, but no direct analyses have been made to establish this critical point. Treatment with these toxic compounds for long periods of time would appear to be an especially dangerous procedure, and it is clear that such experiments must be interpreted with a great deal of caution.

Indeed, Nordlie and his colleagues have taken issue with the conclusion that the hepatic concentration of glucose-6-phosphatase increases at all in response to glucocorticoid treatment (Nordlie *et al.*, 1965; Arion and Nordlie, 1967). In their experiments, daily glucocorticoid treatment for 3 days led to increased activity of the microsome-bound enzyme when assayed directly, but not when the microsomal membranes were dispersed by detergent treatment prior to assay. Further, when treatment with hormone plus actinomycin was compared to actinomycin alone, the apparent hormone effect on enzyme activity (assayed in the absence of detergent) was still present, since long-term actinomycin treatment lowered the enzyme activity as well as the total liver protein. These authors concluded that the reported effect of glucocorticoids on glucose-6-phosphatase involves an increased catalytic activity of existing enzyme molecules, an activation which may involve a reversal of inhibition by microsomal lipid components.

Short-term (3–6 hours) effects of adrenal steroid on hepatic phosphopyruvate carboxykinase have been analyzed by Foster *et al.* (1966), who concluded from a variety of observations that elevation of this enzyme by hydrocortisone is secondary to this hormone's effect on carbohydrate metabolism. Either glucose or insulin (which blocks glucagon effects in the liver) depress the enzyme level and reverse the effect of hydrocortisone. Administration of glucagon (or fasting) elevates the enzyme in a fashion very similar to that which follows hydrocortisone treatment (Shrago *et al.*, 1963), but the possibility that the reported effect of hydrocortisone is actually secondary to release of glucagon was apparently not considered, although this interpretation seems quite likely to me. Similar considerations can be made for the other gluconeogenic enzymes, for the elevation of each of these after hydrocortisone treatment is blocked by insulin (Weber *et al.*, 1965), a treatment which would be expected to stop the effect of glucagon in the liver. This relationship will be discussed in greater detail in Section III,B; for the present purposes, it is sufficient to note that there are serious, valid reservations to the conclusion that glucocorticoids *per se* are active in promoting synthesis of the gluconeogenic enzymes.

When rats are treated with sufficient actinomycin to block enzyme induction completely, glycogen deposition in the liver after glucocorticoid treatment is reduced by 50% or more (Greengard *et al.*, 1963b; Ray *et al.* 1964). Similar results of the antagonist, ethionine, were reported earlier (Kvam and Parks, 1960). Results of this kind have been cited as supporting the thesis that cortisol-enhanced gluconeogenesis depends upon enzyme induction (Greengard *et al.*, 1963b) and as firm evidence that enzyme induction is not required for gluconeogenesis (Lardy *et al.*, 1965). This important question has not yet been resolved. Glycogen deposition from glucose, catalyzed by glycogen synthetase, may be secondarily stimulated by glucocorticoids by an activation of the enzyme rather than induction of its synthesis (see Section III,A). If gluconeogenesis is defined as either (or both) conversion of amino acid precursors to carbohydrate or the flow of carbon from pyruvate to glucose, then the experiments cited were not adequate to determine the question, for these parameters were not assessed.

In short, there is as yet no evidence that glucocorticoids are effective *per se* in regulating synthesis of the enzymes involved in gluconeogenesis. Changes in activity of some enzymes that have been monitored after glucocorticoid treatment may be due to activation processes rather than to synthesis, and others may be secondary to steroid effects on other processes such as secretion of glucagon. And, the significance of enzyme induction in gluconeogenesis has not yet been determined.

E. Other Enzymes

A variety of other liver enzymes has been reported to increase in response to glucocorticoid treatment, but it is doubtful that these actually represent instances of steroid-mediated enzyme induction.

Picolinic carboxylase was elevated by cortisone treatment in the livers of diabetic-adrenalectomized, or hypophysectomized rats, but not in normal animals; in the latter tryptophan pyrrolase was increased 35-fold by the steroid during the same experiments (Mehler *et al.*, 1958a,b). This discrepancy seems to be a good indication that the steroid effect on picolinic carboxylase is indirect. Serine dehydrase is elevated by glucocorticoids *in vivo* (Freedland and Avery, 1964; Pitot and Peraino, 1964), but only in very high doses. When rats are fed a low protein diet serine dehydrase fails to respond to cortisol (Rosen and Nichol, 1964; Pitot *et al.,* 1965a) while the response of both tryptophan pyrrolase and tyrosine transaminase is essentially normal under these conditions (Pitot *et al.,* 1965a). A leucine-specific transaminase in liver is elevated by cortisol (Ichihara and Koyama, 1966), but again only at doses 5–10 times higher than required for induction of tyrosine transaminase (Ichihara, personal communication). These considerations, together with the documented role of glucagon and insulin in regulating most of these enzymes (Section III), lead to the conclusion that the reported effects of glucocorticoids on the enzymes discussed here (as well as the gluconeogenic enzymes) are probably secondary to alterations in circulating levels of the pancreatic hormones. Obviously studies in simplified systems in which these alternatives can be analyzed are required to clarify this situation.

F. Mechanism of Induction by Glucocorticoids

From the considerations discussed in Sections II,D and II,E, it would appear that induction of hepatic enzymes by glucocorticoids *per se* is probably much more limited than is suggested by the current literature. That both tryptophan pyrrolase and tyrosine transaminase are synthesized more rapidly in direct response to the steroid cannot be questioned. Direct evidence for increased synthesis after hydrocortisone treatment is available for glutamic-alanine transaminase, and inhibitor data are consistent with an increased synthesis of several other enzymes, but in none of these has the direct participation of the steroid in induction been established. As I have indicated, there is good reason to believe that the hepatic response is, in fact, limited to the induction of only a few enzyme proteins.

Administration of glucocorticoids to rats initiates a large increase in hepatic synthesis of RNA, as measured by incorporation of isotopically labeled precursors (Feigelson *et al.,* 1962). When pulse-labeling techniques were employed as a measure of synthetic rate the response could be seen to be limited to nuclear RNA, and to precede the first elevations of induced enzymes (Kenney and Kull, 1963). Further analysis demonstrated that labeling of all types of RNA—ribosomal, transfer, and "DNA-like"—was elevated by hydrocortisone (Wicks, *et al.,* 1965; Greenman *et al.,* 1965; Garren *et al.,* 1964b; Sekeris and Lang, 1964). The disparity between this general response of RNA synthesis and the specificity of enzyme induction (cf. Kenney *et al.,* 1965) led to much controversy, centered around the possibility that the hormone actually affects precursor pools and not the true synthesis of RNA (Feigelson and Feigelson, 1964; Kenney *et al.,* 1965). Our measurements showed that labeling (with ^{32}P) of the terminal adenylate of transfer RNA was unaffected; from this we argued that pool changes were not involved (Wicks *et al.,* 1965). Nevertheless, it must be admitted that it is virtually impossible to fashion experiments in which effects on pools can be rigorously excluded in *in vivo* labeling experiments. This is particularly difficult when the synthetic process takes place in a discrete cellular compartment which may contain separate precursor pools that cannot be measured directly.

The activity of crude RNA polymerase preparations (containing DNA and other nuclear components) is increased after glucocorticoid treatment (Lang and Sekeris, 1964). This is in accord with the conclusion that all the DNA-directed synthesis of RNA is elevated by the hormone. In a recent report Yu and Feigelson (1968) have indicated that much of the hormonally increased labeling of RNA reflects enhanced transport of labeled precursors; nevertheless there still appears to be accelerated labeling of RNA which cannot be attributed to precursor pools. And there the matter rests; at this writing, it is difficult to say whether the apparent burst in RNA synthesis which occurs *in vivo* following hydrocortisone administration is meaningful or not.

But it is clear that a massive change in RNA synthesis is not a necessary component of enzyme induction, for it has now been demonstrated incisively that induction occurs in cultured HTC cells without a detectable change in total RNA synthesis (Tomkins *et al.,* 1966). This result has been confirmed in cultures of the similar H35 cells (Reel and Kenney, manuscript in preparation) and in organ cultures of fetal liver (Wicks, 1968a). Gelehrter and Tomkins (1967) could find no change in labeling patterns of isolated total RNA during hormonal induction in HTC cells, even when a change was sought by a sensitive double-labeling technique. A very small but reproducible increase in labeling of the 45 S cytoplasmic ribonucleoprotein particle was observed in the hormone-treated cells (Gelehrter and Tomkins, 1967). This particle is labeled by RNA precursors faster than are the cytoplasmic polysomes (Sells and Takahashi, 1967), and this labeling was earlier shown to be stimulated by hydrocortisone *in vivo* (Henshaw *et al.,* 1965). There is evidence that some of the RNA of this particle is messenger-type RNA (Spirin and Nemer, 1965; Sells and Takahashi, 1967). Using competitive DNA-RNA hybridization techniques, Drews and Brawerman (1967a) found a change in the hybridizable fraction of nuclear RNA after hydrocortisone treatment. The validity of this result was strengthened by the demonstration that no change in this parameter was apparent after growth hormone treatment (Drews and Brawerman, 1967b); this hormone, like hydrocortisone, stimulates total RNA labeling *in vivo* (Korner, 1964; Sells and Takahashi, 1967). Thus the available evidence, although indirect and far from complete, points toward a role of the glucocorticoid-type steroids in controlling specific genetic transcriptions, resulting in increased cellular levels of the messenger RNA species coding for tyrosine transaminase, tryptophan pyrrolase, and perhaps a few other enzyme proteins. The kinetics and other properties of the induction response to hydrocortisone either *in vivo* or in cultured cells *in vitro* are consistent with this interpretation. For example, the pronounced lag in induction by hydrocortisone (Fig. 2) could be expected if new mRNA is formed but must undergo a limiting transition to the cytoplasm before elevated enzyme synthesis begins. Actinomycin experiments suggest a short-lived messenger for tyrosine transaminase in cultured hepatoma cells (Fig. 5) and in rat liver *in vivo* (Kenney, Kendrick and Hager, unpublished experiments). The induced synthesis of this enzyme ceases very quickly when the hormonal inducer is removed (Fig. 4), showing that the cellular content of some short-lived component is increased as long as the hormone is present in effective concentration, and the rational candidate for this component is, of course, the messenger.

Unfortunately direct chemical evidence to support this conclusion is lacking and probably cannot soon be expected. The steroid-mediated change in macromolecular synthesis, as it occurs in cultured cells, is virtually undetectable in terms of total RNA and protein or their rates of synthesis; indeed, it would not have been detected except for the catalytic capacity of the induced protein. Induction of tyrosine transaminase represents an increase in the enzyme from about 0.01% of the total cellular protein to about 0.1%. Given our current inability to isolate intact messenger RNA's, as well as to fractionate or assay these entities, it is apparent that resolution of a change of this magnitude at the RNA level presents a most formidable problem.

III. Insulin and Glucagon

These polypeptide hormones have been implicated in the regulation of a wide variety of liver enzymes, and their participation in control of hepatic enzyme synthesis is probably even more widespread than is indicated by the current literature. The two hormones are physiological antagonists, making it virtually impossible to manipulate the level of one without simultaneously changing the other. Surgical removal of the pancreas is rarely successful in laboratory rodents, making a joker, in this instance, of the endocrinologist's usual ace-in-the-hole. In view of these and other restrictions, and the fact that, as yet, very little effort has been made with these hormones in simplified experimental systems, it is not surprising that their role in regulating protein synthesis is poorly understood. The systems examined yield all possible combinations, for in addition to the large number of proteins which show no response to either hormone, there exist instances where insulin appears to act as inducer, but its action is blocked by glucagon, others where glucagon induces but is blocked by insulin, and a single case where each of these induces the same enzyme.

A. *Elevated by Insulin; Blocked by Glucagon*

1. *Glucokinase*

Control of glucose metabolism is accepted as the raison d'être of insulin, and for a time the hormone was thought to act by regulating the activity of hexokinase (Cori, 1945). In the current view it is agreed that regulation of the phosphorylation of glucose is indeed a significant aspect of the action of insulin, but that the hormone brings this about by regulating synthesis of a unique enzyme which is highly specific for glucose (thus, "glucokinase") but has a low affinity for this substrate.

Elevation of the glucose-phosphorylating hepatic enzymes after the feeding of carbohydrate to fasted rats was demonstrated by Niemeyer and his collaborators (1962, 1963). At about the same time it was found that liver hexokinases can be fractionated to distinguish the high K_m, glucose-specific isoenzyme that specifically responds to glucose feeding or to insulin (Viñuela *et al.*, 1963; Di Pietro *et al.*, 1962; Sharma *et al.*, 1963; Gonzalez *et al.*, 1964). These investigations showed that the glucokinase level falls after fasting or in diabetes, but can be restored by glucose (fasting) or by treatment with insulin (diabetes). As the glucose level is high in diabetes, these results were interpreted as indicating that both insulin and glucose are required to maintain the normal glucokinase level. Elevation of the enzyme in either experimental situation was blocked by inhibitors of protein synthesis and by actinomycin, but these are the only indications that a true induction is involved; as yet no direct measurements of rate of synthesis or of degradation have been reported. Actinomycin also blocks the fall in glucokinase which occurs upon fasting; indeed, there was even a "paradoxical" increase in the enzyme after actinomycin treatment (Sols *et al.*, 1965). That this may reflect an effect of the drug on enzyme turnover, analogous to that discussed in Section II,B,2, has not yet been investigated.

Glucagon (or epinephrine) does not affect the glucokinase level in fed animals (Salas *et al.*, 1963), so the fall in the enzyme which occurs

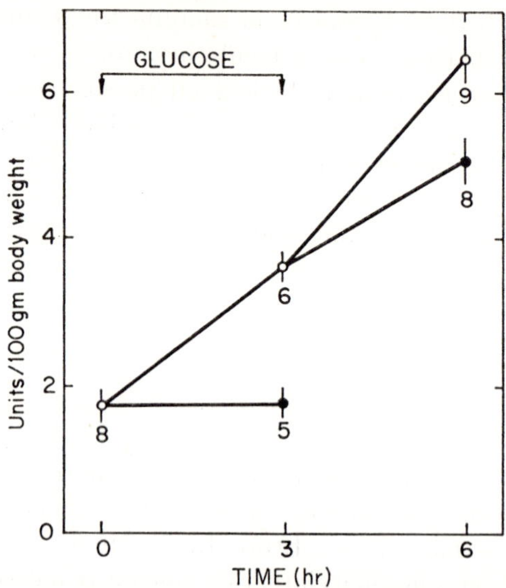

FIG. 7. Effects of glucose and glucagon on glucokinase activity in rat liver. All the rats were fed a carbohydrate-free diet for 6 days before treatment. Glucose (plus maltose-dextrin) given alone every 3 hours, ○——○; glucose (plus maltose-dextrin) given together with glucagon at 0 or at 3 hours, ●——●. The numbers are the numbers of experimental animals for each point. (From Niemeyer, *et al.*, 1966.)

upon fasting or in diabetes is not due to the high glucagon levels associated with these states. However, the increase in glucokinase which follows glucose administration to fasted rats is effectively blocked by either glucagon or epinephrine (Pitot *et al.*, 1964b). Inhibition by glucagon is identical to that by actinomycin, insofar as these inhibitors are fully effective (Fig. 7) only when given very early in the response (Niemeyer, *et al.*, 1966; Sols *et al.*, 1965). These data have been interpreted as indicating that both glucagon and actinomycin interfere with formation of template RNA involved in the induction, a template which must be relatively stable (Sols *et al.*, 1965; Niemeyer *et al.*, 1966).

Both glucose and insulin appear to be required to increase hepatic levels of glucokinase; assuming that an induction is involved, which of these is the inducer? Glucose will stabilize the enzyme in extracts and will prevent loss of the enzyme in liver slices in which protein synthesis is blocked (Sols *et al.*, 1965). This result, together with the observation that insulin has only a small effect in the fasted rat (where glucose levels would be low) led Sols and his colleagues to propose that the enzyme is induced by insulin, glucose being required to stabilize the otherwise labile enzyme synthesized in response to the hormone.

This view has been countered by Niemeyer *et al.* (1966), who point out that 2-deoxyglucose stabilizes the enzyme *in vitro* but does not affect the glucokinase level *in vivo*. After demonstrating that as much as 6 units of insulin was ineffective in rats fed a carbohydrate-free diet, and that insulin administered with glucose had no effect on the rate of response to the sugar, this group rejected the conclusion that insulin is the effective inducer and argued instead that glucose induces the enzyme, insulin being required in a "permissive" fashion, perhaps to allow glucose entry into the nucleus. A third alternative, touched upon by Niemeyer *et al.* (1966), is that glucagon is the active component in this system, acting to initiate repression of glucokinase synthesis, with insulin and/or glucose being required to block this effect in the liver and/or the release of glucagon from the pancreas. However, glucagon would be expected to lower the enzyme level in normal animals if this view were correct, and Salas *et al.* (1963) have shown that it does not. On the other hand, if insulin is the inducer, why is no response to insulin seen in normal animals in which insulin induces at least one other hepatic enzyme?

The role of the pancreatic hormones in regulating glucokinase synthesis thus remains obscure, and probably will remain so until a simplified experimental system is established in which the various regulators—glucose, insulin, and glucagon—can be dissected away from their intricate interrelationship *in vivo*. Also, the possibility remains that synthesis of glucokinase is not the parameter being altered in these experiments. Most of the data available would be equally compatible with hormonally mediated changes in synthesis of a polypeptide inhibitor, for example.

2. Glycogen Synthetase

Perhaps the complexity of mammalian regulatory phenomena is nowhere more apparent than in analyses of the multiple factors acting to determine the activity of the enzyme catalyzing glycogen deposition from glucose (UDPG-glycogen glucosyl transferase, or glycogen synthetase) and that catalyzing the reverse process of glucose formation from glycogen (phosphorylase). There is general agreement that the activity of glycogen synthetase is markedly stimulated by treatment with insulin, or with glucocorticoids, or after glucose is administered. But there the agreement ends, and there appear to be as many explanations for this increase as there are investigators examining it.

Steiner and King (1964) have described a four- to sixfold elevation of the enzyme which is maximal 6 hours after treatment of diabetic rats with insulin, an increase which they attribute to a selective effect of the hormone on synthesis of the enzyme, i.e., induction. The increase could be blocked by puromycin or by ethionine, but there has been no direct demonstration that the concentration of the enzyme is altered or that its rate of synthesis has been changed. Actinomycin at doses (80–150 μg/100 gm) sufficient to block the documented inductions by glucocorticoids did not effect a marked change in the response to insulin.

The analyses of Bishop and Larner (1967) and Mersman and Segal (1967) point instead to an insulin-dependent *activation* of a constant amount of enzyme protein which is usually in a form virtually inactive at physiological concentrations of UDPG and glucose 6-phosphate (b, or D form). Within a few minutes after insulin treatment a substantial portion of the b form is converted to the a or I form (Bishop and Larner, 1967), which functions at 50–100% of capacity under physiological conditions, thus providing a virtual on-off switch to begin the conversion of glucose to glycogen (Mersman and Segal, 1967). As only insulin will effect this conversion in diabetic animals, Kreutner and Goldberg (1967) concluded that both glucocorticoids and glucose are active only via insulin.

Recent experiments by DeWulf and Hers (1967a,b) have led to a new interpretation for the rapid elevation of glycogen synthetase in animals given a glucose load or treated with glucocorticoids. These investigators found that assays of crude liver extracts showed much greater activity when extracts were extensively diluted than when assays were made on concentrated homogenates at lower temperatures. A glucose load, or prednisolone treatment, caused a rapid increase in activity of the enzyme when assayed in concentrated form, but not when the enzyme was diluted. These findings suggest the action of an inhibitor of the enzyme which is diluted out and thus not apparent under ordinary assay conditions; treatment with glucose or prednisolone (and thereby, with insulin?) may bring about an inactivation of this inhibitor.

This brief sketch can only outline the fascinating work currently being carried out on the enzymology of glycogen synthetase and its regulation. Insulin appears to be the primary hormonal stimulus which increases activity of the enzyme; its effect can be blocked by glucagon, suggesting the involvement of changing levels of cyclic AMP (Bishop and Larner, 1967). However, the data show clearly that activation of existing enzyme is the mechanism for this response; that changes in synthesis of the enzyme also occur must be considered as unlikely.

3. Other Enzymes

A few other liver enzymes are decreased in the livers of diabetic animals and increased by insulin, including malic enzyme (Fitch and Chaikoff, 1962; Shrago, *et al.*, 1963) and the recently discovered form of pyruvic kinase found only in liver and erythrocytes (Tanaka *et al.*, 1967). Response to insulin occurs at a rate comparable to other hormonal inductions, but, like glucokinase, no response to insulin is apparent unless the normal enzyme levels are depleted by physiological manipulations such as prolonged fasting or experimental diabetes. Thus, there may be a group of liver proteins which require insulin in order to maintain their usual rate of synthesis, but which are not synthesized in excess in response to this hormone. This is in contrast to tyrosine transaminase, which is markedly elevated over basal levels owing to induction by insulin (Section III,C).

B. Elevated by Glucagon; Blocked by Insulin

A large group of liver enzymes has been found to increase after glucagon treatment or in diabetes. The glucagon effect is typically blocked by insulin, and the high enzyme level of the diabetic state is lowered by insulin treatment. Direct evidence for inclusion in this category exists for glucose-6-phosphatase (Langdon and Weakley, 1955), picolinic carboxylase (Mehler *et al.*, 1958a), ornithine transaminase (Pitot and Peraino, 1964; Peraino *et al.*, 1965), phosphopyruvate carboxykinase (Foster *et al.*, 1966), serine dehydrase (Freedland and Avery, 1964; Ishikawa *et al.*, 1965; Peraino *et al.*, 1966), and some others. These enzymes are characteristically elevated after prolonged fasting, a physiological condition associated with high glucagon levels. That changed rates of enzyme synthesis are responsible for the activity changes observed has been clearly demonstrated for serine dehydrase by Jost *et al.* (1968), and similar proof is available that glucagon alters tyrosine transaminase levels by increasing its rate of synthesis (Holten and Kenney, 1967). In both these cases it was demonstrated that marked increases in synthesis of the specific enzymes occurred while there was little or no effect on labeling of the total soluble proteins; thus the effects of glucagon are selective rather than general. A further indication of this important point comes from the fact that the physiologically labile tryptophan pyrrolase is not affected by glucagon; if synthesis of all the proteins were increased, this enzyme should also increase quickly (cf. Berlin and Schimke, 1965). Insulin-free glucagon induces tyrosine transaminase in perfusion studies with isolated livers (Hager and Kenney, 1968) and in hepatoma cell cultures (Lee and Kenney, unpublished). In organ cultures of fetal liver Wicks has demonstrated induction by glucagon of tyrosine transaminase (Wicks, 1968c) as well as phosphopyruvate carboxykinase (Wicks, unpublished). It is clear then, that glucagon acts directly on hepatic cells to induce synthesis of at least three enzymes, so there is little reason to doubt that the other enzymes which undergo *marked* increases after glucagon treatment (or in diabetes or fasting) do so by virtue of induced synthesis.

The changes that have been observed in the levels of these enzymes in diabetic or fasted animals point clearly to the conclusion that *any experimental manipulation which results in high glucagon or low insulin titers can be expected to increase these enzymes.* Conversely, levels of these enzymes can be expected to drop if experimental procedures are employed which elevate insulin or lower glucagon in the circulation. Further, evidence is accumulating that cyclic AMP is the effective intracellular inducing agent for these enzymes (see Section III,D), so it is clear that any manipulation which alters the intrahepatic concentration of this nucleotide—whether directly or via the hormones which affect it—will thereby change the level of these enzymes.

Recognition of these facts, i.e., of the hormonal and intracellular factors operating to regulate synthesis of these enzymes, provides a basis for understanding the bewildering variety of experimental procedures that have been reported to alter levels of these enzymes *in vivo*. For example, serine (threonine) dehydrase is elevated by glucocorticoids, but only after prolonged treatment with high doses of cortisol (Goldstein *et al.*, 1962b). The latter treatment is ineffective in rats on a 0% protein diet (Pitot *et al.*, 1965a). In contrast, tyrosine transaminase is inducible by low doses of steroid, even in animals fed the protein-deficient diet. Data of this kind provide a clear indication that the effect of excessive cortisol on serine dehydrase is indirect, i.e., secondary to other factors operating to regulate synthesis of this enzyme, for the alternative can only be that the same agent induces each enzyme by a separate mechanism, and this is unacceptable. That high doses of glucocorticoids can promote a secondary type of induction response in the liver can be seen from an early study (Kenney and Flora, 1961) of the response of tyrosine transaminase to varying doses of hydrocortisone (Fig. 8). This enzyme is now known to be induced both by hydrocortisone and by glucagon. Moderate steroid doses (0.5–2.5 mg per 100 gm) effect a typical tenfold elevation of the enzyme, comparable to that which occurs in tissue cultures supplemented with hydrocortisone. However, increasing the hydrocortisone dose to 10 mg per 100 gm nearly doubles the extent of enzyme induction—as does supplementation of a moderate dose of steroid with glucagon (Holten and Kenney, 1967). It is not unreasonable to conclude that the increased response to large doses of steroid is, in fact, due to promotion of glucagon release or to elevation of the hepatic cyclic AMP concentration via some other intermediate. And by the same argument, it is reasonable to conclude that serine dehydrase is not responsive to the steroid hormone *per se,* but rather

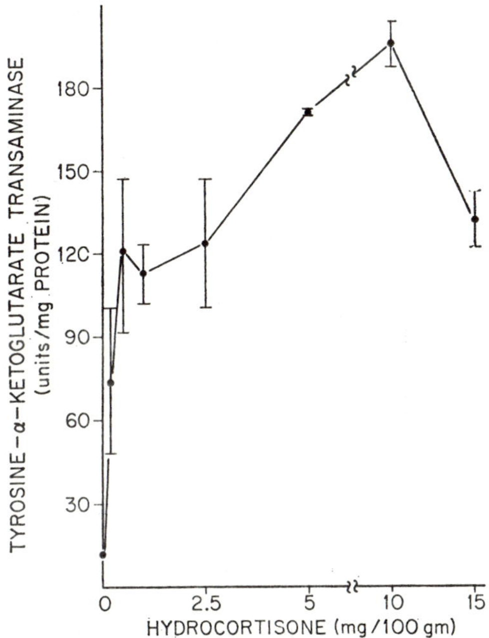

FIG. 8. Response of tyrosine transaminase to varying doses of hydrocortisone. (From Kenney *et al.*, 1968b.)

that the reported steroid effects on this enzyme (and on the others listed earlier in this section) are actually secondary to glucagon and/or cyclic AMP.

In animals fed a high protein diet, or given oral intubations of a casein hydrolyzate after several days of protein depletion, serine dehydrase, ornithine transaminase and tyrosine transaminase undergo marked increases in the liver (Pitot and Peraino, 1964); tryptophan is largely responsible for this "dietary induction" (Peraino *et al.*, 1965). The requirement for essential amino acids in the synthesis of any protein, together with the unique role of tryptophan in maintaining polysome structures (Munro, 1968) may complicate experiments wherein amino acids are withdrawn from the diet, for effects on total hepatic protein synthesis may be part of the responses observed. But the enzymes mentioned undergo very large increases in "dietary induction," comparable to those brought about by glucagon but far out of proportion to effects on general protein synthesis.

Understanding of the mechanisms operating in "dietary induction" may come from analyses of tyrosine transaminase, which is also elevated *in vivo* by casein hydrolyzates (Pitot and Peraino, 1964; Rosen, 1965) and by tryptophan (Kenney and Flora, 1961) and other indoles (Rosen and Nichol, 1964). The time course and extent of the response to these agents is virtually identical to those characterizing the response of this enzyme to glucagon, i.e., a maximum enzyme level 3–5 times above the basal level is reached within 2–4 hours of these treatments. Additive or synergistic effects are found when casein hydrolyzate treatment is combined with hydrocortisone (Rosen and Milholland, 1968), and when glucagon is combined with the steroid (Holten and Kenney, 1967; Csanyi *et al.*, 1967). Thus the "dietary induction" of this enzyme appears to be very similar, if not identical, to induction by glucagon. Since glucagon will induce in various simplified systems, e.g., organ cultures, or cell cultures richly supplemented with amino acids, it seems most reasonable to assume that glucagon, and not the materials administered in "dietary induction," is actually the effective agent. Alternatively, it might be supposed that casein hydrolyzates, tryptophan, etc., increase the hepatic cyclic AMP level by some means not involving glucagon (or catecholamines of adrenal origin, as adrenalectomized rats are used in all these experiments). Similar conclusions can be reached regarding the response of tyrosine transaminase to benzoate and related compounds (Singer and Mason, 1965) or to large doses of pyridoxine (Greengard and Gordon, 1963); in both cases the response observed is essentially identical to that which follows glucagon treatment. From these considerations it can be seen that both "dietary induction" and "cofactor induction" (i.e., by pyridoxine) can rationally be interpreted as a means of effecting a glucagon-mediated (or cyclic AMP-mediated) induction *in vivo*.

The term "glucose repression" has been employed to describe the fact that administration of glucose will block, and reverse, the amino acid-promoted inductions of serine dehydrase and ornithine transaminase (Pitot and Peraino, 1963; Peraino *et al.*, 1966). As insulin release has long been associated with a glucose load, and as insulin *per se* will lower the induced levels of these enzymes (Ishikawa *et al.*, 1965), the conclusion that "glucose repression" reflects a response to insulin is obvious. This is also suggested by the observation that "glucose repression" can be reversed by glucagon (Pitot and Peraino, 1964).

In summary, it appears that a good case can be made for the unifying conclusion that "dietary induction," "cofactor induction," "glucose repression," the reported inductions by indoles, benzoates, and excessive glucocorticoids, and the changes brought about by conditions such as prolonged fasting and diabetes, can all be considered as due to experimental manipulations which alter cyclic AMP levels in the liver, probably via changes in the circulating levels of the pancreatic hormones. The hormone active in inducing synthesis of these enzymes appears to be glucagon. Insulin blocks the response to glucagon; whether this effect is due to a shutdown of glucagon release or a direct effect of insulin in the liver remains to be determined. An impressive list of hepatic enzymes can be thought of as belonging in this category, including serine dehydrase, ornithine transaminase, phosphopyruvate carboxykinase, and probably pyruvate carboxykinase, fructose diphosphatase, and glucose-6-phosphatase as well as others not yet studied. That those which have been identified can all be considered as significant in gluconeogenesis is undoubtedly meaningful.

C. Elevated by Both Insulin and Glucagon

A single, and most puzzling, case has been reported of induction of an enzyme by both insulin and glucagon—this being the glucocorticoid-inducible tyrosine transaminase (Holten and Kenney, 1967; Csanyi *et al.*, 1967). Elevation of the transaminase level by either of these hormones is due entirely to a rapid acceleration of its synthesis, as shown for insulin in Fig. 9, and is clearly distinct from the induction by hydrocortisone. The kinetics and extent of response to either pancreatic hormone are virtually identical *in vivo* but differ in important respects from the response to hydrocortisone, particularly in displaying no significant lag period (cf. Fig. 9 and Fig. 2). Effects of insulin and glucagon on the rate of transaminase synthesis are not additive with one another but are additive with the effect of hydrocortisone (Holten and Kenney, 1967); in some experiments glucagon yields a clearly synergistic effect when given together with the steroid (Csanyi *et al.*, 1967).

Experimentation *in vivo* revealed that repeated treatment with either insulin or glucagon yielded induction kinetics that were essentially identical to the response to a single injection, i.e., the induction process ceased after 2.5 to 3 hours regardless of repeated hormone treatment (Holten and Kenney, 1967). The response to these hormones in experiments with isolated, perfused livers (Fig. 10) was like that seen *in vivo*, particularly in that induction ceased after 2–3 hours despite continuous infusion of the inducing hormones (Hager and Kenney, 1968). In these perfusion experiments it was possible to show that each of these hormones acted independently of the other, a conclusion confirmed in experimentation with organ cultures of fetal liver (Wicks, 1968c).

The identical responses obtained with insulin and with glucagon, together with their lack of additivity, led us to propose that there exist two induction mechanisms for tyrosine transaminase, one responsive to glucocorticoids and the other to either of the pancreatic hormones (Holten and Kenney, 1967; Hager and Kenney, 1968). The disparity between this proposal and the well-known antagonism of insulin and

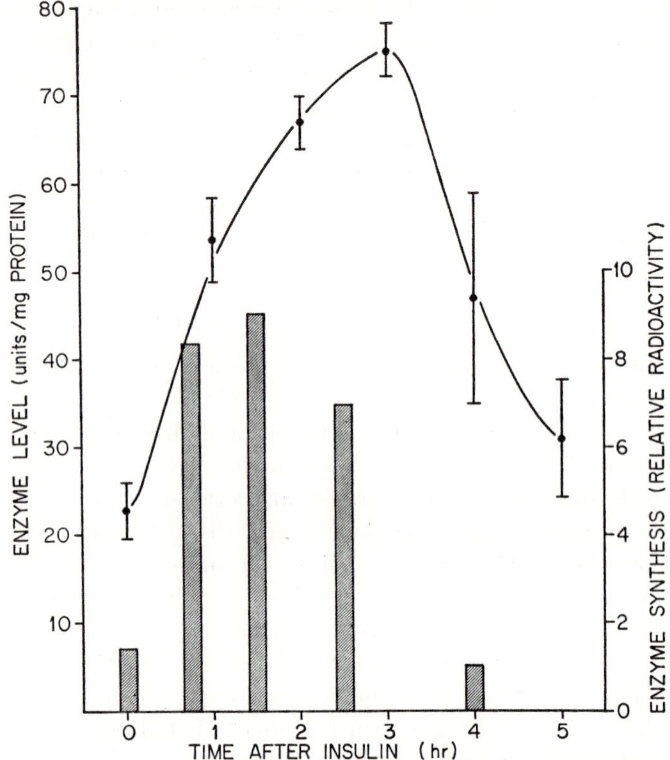

FIG. 9. Induction of tyrosine transaminase by insulin in adrenalectomized rats. The points are measurements of enzyme levels (± standard error for 3 rats), and the bars represent rate of synthesis measurements made at the times indicated. (From Kenney *et al.*, 1968b.)

glucagon in virtually all other experimental situations could be relieved by assuming that a common limiting factor was involved in inductions by these hormones, a factor which might be made available in quite different ways by insulin or by glucagon (Kenney *et al.*, 1968a). However, an alternative interpretation, more in keeping with the effect of insulin on other glucagon-mediated inductions, has become apparent in the realization that cyclic AMP is involved in induction by glucagon. As shown by Jefferson *et al.* (1968), insulin lowers the hepatic cyclic AMP concentration and blocks the usual elevation of this nucleotide by glucagon. This being so, one could not expect an additive response when insulin and glucagon are given together, for insulin would be expected to prevent glucagon from acting, and the response observed would be that due to insulin alone. Thus the data available at this time are fully consistent with the possibility that tyrosine transaminase is inducible by three mechanisms, initiated by hydrocortisone, glucagon, and insulin, respectively. That the properties of response to the latter two hormones are similar may be simply coincidental.

Evidence has been presented recently from several laboratories that the large diurnal variations in the level of tyrosine transaminase in the liver is independent of glucocorticoids but strongly dependent upon the feeding schedule of experimental animals (Potter *et al.*, 1966; Shambaugh *et al.*, 1967; Wurtman and Axelrod, 1967; Civen *et al.*, 1967). The circadian rhythm closely parallels changes in blood glucose, suggesting that either insulin or glucagon may be involved (Civen and Brown, 1968). On the other hand, Black and Axelrod (1968) have shown that drugs interfering with catecholamine metabolism can mimic the diurnal changes; these authors suggested that the nervous system may play

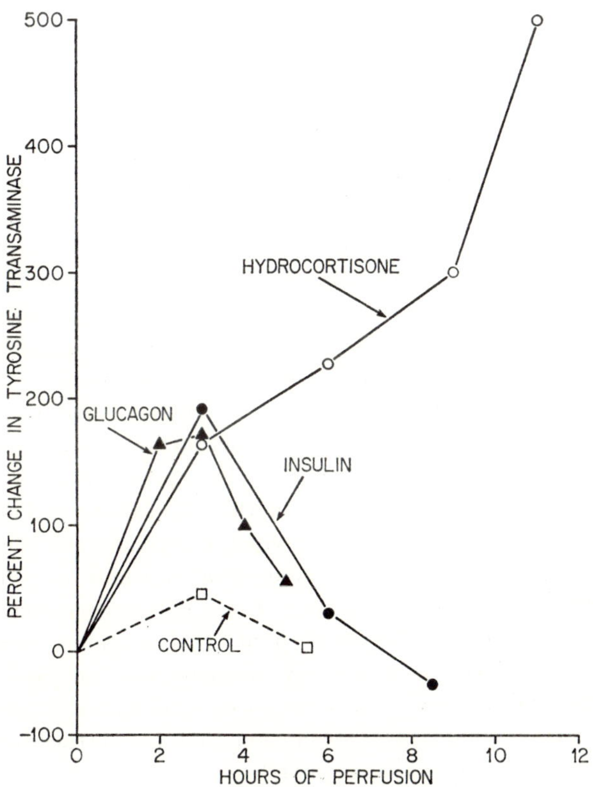

FIG. 10. Induction of tyrosine transaminase in isolated, perfused livers. (Adapted from Hager and Kenney, 1968.)

a role in this regulation of enzyme synthesis. Whether the regulating agents are the pancreatic hormones or catecholamines (which could originate at hepatic nerve endings) it seems possible that the cyclic AMP-mediated type of induction can be considered as the mechanism operating to govern the rate of enzyme synthesis. From this it can be concluded that the "basal" level of this enzyme, i.e., the level found in animals not treated with hormones or other agents, may in fact be a partially induced level, depending upon the physiological status of the factors governing the hepatic concentration of cyclic AMP. That the "basal" level varies as much as two- to threefold, even in animals assayed at the same time of day, is in agreement with this conclusion. Diurnal rhythms in enzymes associated with protein metabolism are discussed in detail in Chapter 36 by Wurtman.

D. Mechanisms of Inductions by Glucagon and Insulin

1. Glucagon

Glucagon stimulates a large increase in hepatic cyclic AMP levels within a few minutes of its administration (Rall and Sutherland, 1958), and a wide variety of metabolic effects of glucagon have proved to be mediated by cyclic AMP. That this nucleotide is involved in enzyme induction was first suggested by experiments of Peraino and Pitot (1964), who found that the effect of glucagon on serine dehydrase could be duplicated by epinephrine, another stimulator of hepatic adenylcyclase (for a review of hormonal-adenylcyclase interactions, see Sutherland *et al.*, 1965). Induction of tyrosine transaminase by cyclic AMP (or its dibutyryl derivative) was directly demonstrated by Wicks (1968b,c) in organ cultures of fetal liver, wherein the cyclic nucleotide *per se* was found to be as effective as glucagon. Induction by glucagon or by catecholamines was not additive with that by cyclic AMP, a finding to be expected if these hormones act via this intermediate. Theophylline, an inhibitor of the phosphodiesterase which inactivates the cyclic nucleotide, potentiated the response to suboptimal amounts of glucagon but had no effect on induction by insulin. These observations have recently been extended to include phosphopyruvate carboxykinase, which is also induced by cyclic AMP as well as by glucagon (but not by hydrocortisone) in organ cultures (Wicks, personal communication). There is thus ample evidence to support the conclusion that the intracellular effector in inductions by glucagon is cyclic AMP.

The task of formulating a reasonable concept of the mechanism of cyclic AMP-mediated inductions is considerably simplified by the studies made with tyrosine transaminase, for in this instance it is clear that *this type of induction is mechanistically different from that initiated by hydrocortisone.* As a good case can be made for selective gene activation, i.e., stimulation of the synthesis of specific mRNA's, as the mechanism of induction by the steroid, the inference can be made that glucagon (cyclic AMP) induces by some mechanism other than by promoting mRNA synthesis. Under some conditions induction by glucagon or cyclic AMP can be seen to be markedly synergistic with that by hydrocortisone (Csanyi *et al.*, 1967; Wicks, 1968c), suggesting that sequential modifications in the process of enzyme synthesis are involved in the two induction mechanisms. This result points to the conclusion that cyclic AMP acts to induce the synthesis of specific enzymes by an action at the translational level in protein synthesis, if we define this to include all the reactions subsequent to the readout of DNA-encoded information into mRNA.

In apparent opposition to this suggestion of a translational control mechanism is the virtually unanimous finding that inductions by glucagon (or cyclic AMP) are blocked when RNA synthesis is inhibited by actinomycin (Pitot and Peraino, 1964; Pitot, *et al.*, 1965b; Holten and Kenney, 1967; Csanyi *et al.*, 1967; Wicks, 1968c). This result may be misleading, from two points of view. First, it is widely recognized that actinomycin may inhibit enzyme induction through some means other than its effect on RNA synthesis; the drug is highly toxic and the amounts usually employed in these experiments are, in fact, ultimately fatal. The second objection is, however, probably the more pertinent one, since the toxicity question can usually be handled by various internal controls. This objection is that, if the templates for the enzymes induced are labile, a translational mechanism of induction will be effectively blocked when formation of new templates is stopped by actinomycin, for there shortly will be no template upon which a translational control can be effected.

Messenger RNA's are generally thought to be relatively stable in mammalian cells, but a careful analysis by Wilson and Hoagland (1967) showed there to be two populations in rat liver, one of which comprised two-thirds of the total and decayed with a half-life of 3–3.5 hours; the other third was very stable, decaying with a half-life of about 80 hours, and appears to function in the synthesis of "export" proteins, e.g., serum proteins (Wilson *et al.*, 1967). Now it is clear that the value of 3–3.5 hours represents an average half-time; within this population there undoubtedly exists messengers of considerably shorter or longer half-lives. But even accepting the value of 3 hours, it can be seen that in an induction experiment wherein actinomycin is given and analyses are made 6 hours later, the mRNA available for translational control will have been reduced to 25% of the original content by the end of the experiment, and the drug will thereby inhibit a translational mechanism of induction severely.

As yet there have been very few attempts at determining the lifetimes of mRNA's for individual proteins in liver. Pitot *et al.* (1965b) made measurements of actinomycin effects on the levels of various enzymes induced in liver and concluded that template lifetimes for serine dehydrase, ornithine transaminase, and tyrosine transaminase were 6–8 hours, 18–24 hours, and less than 3 hours, respectively. However, these measurements were made only in terms of the amounts of the enzymes present after actinomycin treatment; the very real possibility that turnover of the enzymes was blocked by actinomycin was not assessed, so these estimates may not be valid. As discussed earlier (Section II,F), actinomycin does block turnover of tyrosine transaminase in cultured hepatoma cells, leading to "superinduction" when cells are preinduced (Reel and Kenney, 1968b). Direct measurements of the rate of tyrosine transaminase synthesis after actinomycin treatment yielded a half-life of about 3 hours for the transaminase mRNA in the cultured hepatoma cells (Fig. 5). Comparable experiments to determine the lifetime of the transaminase mRNA *in vivo* have yielded variable results and we have not yet been able to determine this parameter with satisfactory precision. However, it can be said that the data indicate clearly that the transaminase mRNA decays very rapidly *in vivo,* with a half-life not much greater than 1 hour. An analysis by Tschudy *et al.* (1965) of the mRNA of δ-amino levulinic acid synthetase in rat liver yielded a half-life of about 40–70 minutes for this messenger. Thus some mRNA's in liver are quite labile; it may be that this is characteristic of the mRNA's coding for enzymes which are subject to control of synthesis. While our knowledge of the lifetimes of mRNA's for the various induced enzymes is incomplete, it is clear that very short lifetimes are a real possibility, and therefore actinomycin-inhibition studies cannot be considered an adequate means of distinguishing transcriptional from translational control mechanisms.

If the conclusion that cyclic AMP regulates translational events is accepted as best fitting the available data, how can we envision that this is effected at the molecular level? The term "envision" here is carefully chosen, for at this time there are almost no data upon which judgment can be based, and the search for a reasonable mechanism can be only conceptual. There are many possible sites of translational control, some of which are indicated in Fig. 11. The major problem with any translational control mechanism is the requirement for specificity; it

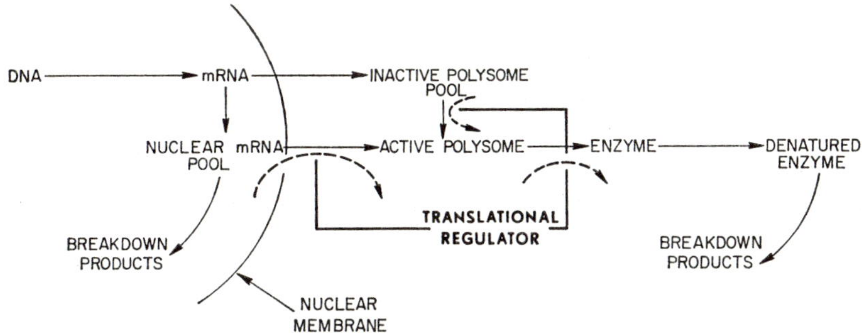

FIG. 11. Some possible sites of translational control mechanisms.

is very clear that cyclic AMP elevates the synthesis of a group of enzymes without exerting a marked effect on total protein synthesis. For this reason attention must be focused on the element conferring specificity in protein synthesis, which is (as nearly as we can tell) messenger RNA. Since mRNA synthesis may not be involved, it would seem that cyclic AMP either brings into play an otherwise inaccessible store of mRNA or that it accelerates the readout mechanism for some proteins and not for others.

One possibility for the first of these is based upon the realization that much of the messenger-type RNA in liver nuclei (i.e., DNA-like RNA) undergoes turnover within the nucleus and never reaches the cytoplasmic sites of protein synthesis (for a recent review, see Georgiev, 1968). If the dRNA is wholly or in part potential mRNA, a precise regulation of the process permitting passage of mRNA from nucleus to cytoplasm seems a virtual certainty. We have suggested that either or both of the pancreatic hormones may function in the regulation of this transmission process (Hager and Kenney, 1968), but experiments designed to test this have so far been uniformly negative. Pitot and his colleagues have suggested a role of the endoplasmic reticulum in determining availability of mRNA (Pitot *et al.*, 1965a); cyclic AMP may change the nature of the binding of some polysomal structures to these membranes. Support for the second alternative, that the ribosomal readout mechanisms is stimulated, comes from suggestive experiments reported by Khairallah and Pitot (1967), who found that cyclic AMP stimulated release of polypeptide chains from polysomes incubated *in vitro* with a purified "release enzyme." Further studies are being made to confirm this effect and in particular to assess its significance to regulation of the synthesis of specific proteins (Pitot, personal communication).

In systems wherein the role of cyclic AMP has been adequately studied, this nucleotide has often been found to regulate phosphorylation of proteins (Sutherland and Robison, 1966). In an exciting new development Langan (1968) has provided convincing evidence for the stimulation by cyclic AMP of the phosphorylation of certain histones from liver nuclei, thus providing a much-needed bridge between cyclic AMP effects on protein synthesis and its effects in other systems. Of course this effect is most consistent with a transcriptional control mechanism, since histones would be expected to bind DNA less readily after their phosphorylation and derepression of RNA synthesis should result. However, it must be admitted that our understanding of the role of histones in nuclear events is far from complete; it remains entirely possible that histones function in packaging or transport of mRNA's. This would bring these results into accord with the data indicating that cyclic AMP acts at some point subsequent to the synthesis of mRNA.

Another potential translation mechanism is suggested by the recent findings of Henshaw (1968), who made the discovery that the mRNA of rat liver polysomes (those not attached to membranes) exists in combination with some low density material, probably protein. A similar conclusion was reached by Perry and Kelley (1968), working with L cells. This result is in accord with experiments reported by Monroy *et al.* (1965), who found that the inactive polysomes of unfertilized sea urchin eggs could be activated by a brief trypsin treatment. Future work will surely test the possibility that the mRNA-protein interaction is sensitive to cyclic AMP, perhaps by regulation of phosphorylation of the binding protein.

Cyclic AMP is often characterized as "ubiquitous" and is found in certain bacteria (Makman and Sutherland, 1965). In a recent report Perlman and Pastan (1968) have presented evidence that this nucleotide stimulates synthesis of two inducible enzymes in *Escherichia coli*, tryptophanase and β-galactosidase. Optimal synthesis of these enzymes required both cyclic AMP and the substrate-type inducer, and certain aspects of the well known repression by glucose could be tentatively attributed to alterations in cyclic AMP levels. This promising development indicates that the cyclic AMP-sensitive reactions in liver have their counterparts in microbial systems; future studies of the latter may contribute enormously to understanding the mechanism of this type of induction in the liver.

It is apparent that recent research has opened several avenues which may lead to an understanding of the molecular mechanism by which glucagon induces certain enzymes. The availability of cell-free systems in which most aspects of translation can be examined, together with the recognition of cyclic AMP as the intracellular effector, holds the promise of a solution to this problem in the not-too-distant future.

2. Insulin

The chances of understanding, in molecular terms, the role of this hormone in regulating the synthesis of specific proteins are not so bright. At this time we can be certain that insulin accelerates synthesis of tyrosine transaminase; insulin may also regulate synthesis of glucokinase, one form of pyruvate kinase, malic enzyme, and perhaps other proteins as well. From the studies with tyrosine transaminase it can be said that insulin acts in some fashion different from the action of hydrocortisone, since the effects of these two hormones are additive in various experimental systems. It may be significant that, unlike glucagon, synergistic responses to the combination of insulin and hydrocortisone have not been observed in organ cultures of fetal liver (Wicks, 1968c). As mentioned above, the kinetics of the response of tyrosine transaminase to insulin are like that to glucagon, but the significance of this similarity is not yet clear.

Insulin clearly blocks inductions by glucagon of such enzymes as serine dehydrase and phosphopyruvate carboxykinase; in recent experiments Wicks has obtained evidence that insulin blocks the action of cyclic AMP and not only its formation (Wicks, personal communication). The studies with tyrosine transaminase are consistent with the conclusion that insulin blocks the glucagon (cyclic AMP)-mediated induction of this enzyme too, but in this instance insulin also induces the enzyme. This unique behavior of tyrosine transaminase is apparently not related to its inducibility by hydrocortisone, because the steroid-inducible tryptophan pyrrolase does not change in experiments where tyrosine transaminase is markedly elevated by insulin (Kenney, unpublished experiments). So we are faced with a situation where insulin simultaneously turns off one induction mechanism and turns on what appears to be another, with one enzyme being sensitive to both of these effects. It must be admitted that at this time we have no idea of how this comes about.

Insulin has been reported to influence protein synthesis in a variety of tissues (for a review see Wool, 1964). Recent experiments have implicated a change in muscle ribosomes which comes about very quickly after insulin administration (Wool, 1968) and which appears to account for the increased protein synthesis observed. This proposed translational action of insulin would be in accord with the data showing that insulin inductions are mechanistically different from those initiated by hydrocortisone, but it is very difficult to conceive of highly specific inductions occurring by this means. A satisfactory solution to the question of how insulin induces a few enzymes would also have to account for the role of this hormone in blocking inductions initiated by cyclic AMP.

IV. Somatotropic Hormone

Administration of the hypophyseal growth hormone increases the concentration of protein and RNA in the liver, especially if the test animals are hypophysectomized (Simpson, *et al.*, 1949; Geschwind *et al.*, 1950). Little specificity has been seen in the response of either RNA or protein synthesis, which appears to be completely general as befits a true growth response (Korner, 1965; Sells and Takahashi, 1967; Tata and Williams-Ashman, (1967). Korner has marshaled evidence that growth hormone affects protein synthesis at the translational level (Korner, 1965). However serious doubt remains that the increased synthesis of protein is actually due to a direct action of growth hormone in the liver; thus Sells and Jackson (1967) suggested that the hormone may act indirectly or may require a second component which acts synergistically to affect liver metabolism.

We have reported a repression of the synthesis of tyrosine transaminase in the livers of adrenalectomized rats treated either with agents causing severe stress (Kenney and Albritton, 1965) or with moderate amounts of growth hormone (Kenney, 1967b). Synthesis of the transaminase was found to drop to a rate that is impossible to distinguish from no synthesis at all, in the immunochemical analyses employed. The enzyme level fell, in most experiments, at approximately the same rate as the known rate of transaminase degradation, as expected if synthesis of the enzyme were blocked more or less completely. The enzyme level could not be reduced to zero, however, and ultimately it returned to the usual basal level even when growth hormone was given repeatedly (Kenney *et al.*, 1968a).

Tryptophan pyrrolase levels were unchanged in the livers of animals in which synthesis of tyrosine transaminase was repressed, showing that the repression response is not necessarily related to induction by glucocorticoids (Kenney and Albritton, 1965). However the steroid-inducible glutamic-alanine transaminase does fall in animals treated daily with growth hormone (Gaebler, 1955; Harding and Rosen, 1963). The half-life of this decay approximated that determined for the usual rate of degradation of this enzyme (Segal and Kim, 1963), in agreement with the conclusion that its synthesis, like that of tyrosine transaminase, was essentially stopped. Another result hinting vaguely at some relationship to steroid inductions is the finding that induction of tyrosine transaminase by hydrocortisone, but not by either glucagon or insulin, is

TABLE II
EFFECT OF GROWTH HORMONE ON INDUCTION OF TYROSINE TRANSAMINASE BY VARIOUS INDUCING HORMONES[a]

Hormone treatment[b]	Tyrosine transaminase (units/mg protein)[c]
None	23.7 ± 2.6
Hydrocortisone (F)	160.5 ± 14.1
F + growth hormone (STH)	69.0 ± 2.1
None	55.4 ± 3.3
Glucagon	157.4 ± 18.9
Glucagon + STH	143.3 ± 12.0
Insulin	125.3 ± 11.6
Insulin + STH	129.5 ± 6.8

[a] Kenney (unpublished data).
[b] Doses used were: hydrocortisone, 2.5 mg/100 gm; growth hormone, 200 μg/100 gm; glucagon, 150 μg/100 gm; insulin, 1 unit/100 gm. Assays were made 3 hours after treatment. The rats were adrenalectomized 2 days beforehand.
[c] Mean ± SE for 3 animals in the first series and 5 or 6 in the second.

markedly inhibited by simultaneous treatment with growth hormone (Table II). This effect was first detected some time ago by Harding and Rosen (1963), who also found that induction of glutamic-alanine transaminase by hydrocortisone was similarly inhibited, as is tryptophan pyrrolase (Csanyi and Greengard, 1968).

In our original interpretation of this repression (Kenney and Albritton, 1965) we argued that repression must occur at the translational level, as we assumed that the failure of the transaminase level to fall after actinomycin treatment indicated that the template must be stable. If templates are stable, repression is *ipso facto* a translational phenomenon. We now know this interpretation to have been based upon a false premise, since actinomycin treatment actually stops transaminase synthesis very quickly; the unexpected inhibition of enzyme turnover actually accounts for the stability of the enzyme level under these conditions. Thus at this time we have essentially no data upon which an intracellular mechanism of repression can be postulated. However, we do have clear indications that the repression response is not due to a direct action of growth hormone in hepatic cells. Growth hormone does not affect the tyrosine transaminase level in perfused livers (Hager and Kenney, 1968) or in hepatoma cell cultures (Lee and Kenney, unpublished) or in organ cultures of fetal liver (Wicks, personal communication). In the latter two experimental systems growth hormone was found to have no effect on induction by hydrocortisone. As each of the inducing hormones (insulin, glucagon, hydrocortisone) is effective in all three systems, the failure to respond to growth hormone would appear not to reflect loss of hepatic function in these artificial conditions, but rather that an extrahepatic component is involved in the result seen only *in vivo*.

It was mentioned earlier (Section II,C) that there are some indications that the "basal" level of tyrosine transaminase may actually be a partially induced level, maintained by a low level of operation of the cyclic AMP-mediated type of induction. If this is so, the capacity of growth hormone to lower the basal level markedly (but not completely) could be understood as reflecting a shutdown of the supply of an inducing hormone required to maintain the hepatic cyclic AMP concentration at levels high enough to support the basal rate of transaminase synthesis. If this interpretation is correct an important corollary is apparent in the inhibition by growth hormone of induction by hydrocortisone, this being that the transcriptional induction is not fully expressed unless the translational mechanism for accelerating synthesis is operating simultaneously.

V. Summary and Prospectives

From these studies we can now identify three types of hormones as the principal regulators of the synthesis of specific enzyme proteins in the liver. Glucocorticoids induce synthesis of a few enzymes; at present only two can be said to be definitely sensitive to induction by these steroids. While there may be others, it seems clear that this induction mechanism operates in regulating synthesis of a limited number of proteins. The available data favor the interpretation that glucocorticoids induce enzyme synthesis by promoting specific transcriptions, but this is not yet established conclusively. Glucagon, catecholamines, and other agents which increase hepatic cyclic AMP levels thereby accelerate synthesis of a rather large group of liver enzymes, most of which can be recognized as significant catalysts for production of glucose (or glycogen) from noncarbohydrate precursors. The principal agent acting to increase cyclic AMP levels appears to be glucagon. Cyclic AMP probably regulates enzyme synthesis at the translational level, but again further work is required to establish this point. Insulin blocks the cyclic AMP-mediated inductions and will lower the levels of the enzymes whose synthesis is stimulated by this nucleotide. Insulin also induces synthesis of at least one hepatic enzyme and probably a few others, but, like the glucocorticoids, does not appear to have widespread effects on hepatic enzyme synthesis. Little is known of the mechanism of induction by insulin, but there are indications that it may differ from either of the other two induction mechanisms. The hypophyseal growth hormone appears not to influence specific protein synthesis in the liver directly; effects seen only *in vivo* may be attributable to a secondary effect of growth hormone on circulating levels of the pancreatic hormones. Other hormones which participate in regulation of liver metabolism and may do so, in part, by regulating enzyme synthesis (e.g., thyroxine, androgens, estrogens) have not been considered in this chapter.

The frequency of evasive constructions ("appears to," "may," "probably," etc.) in the summarizing paragraph above indicates the magnitude of the sector that awaits solution in this research area. While this is obviously much larger than that which could be considered as solved, the prospects for major advances being made in the near future are rather bright. The introduction in the past few years of tissue culture and organ culture experimental systems in which hormonal regulation can be analyzed is a most promising development; in my view, this ranks together with the recognition of the heterogeneity and significance of enzyme turnover as the most important developments made since it was realized, nearly two decades ago, that hormones regulate enzyme concentrations. With simplified experimental systems it should be possible rather soon to distinguish the agents acting to regulate synthesis of each inducible enzyme; at present these are known with assurance for only a few and can be guessed at for some others. The culture systems can also be expected to yield particularly important dividends in understanding the mechanisms by which the various hormones influence synthesis of specific proteins, for only in such systems can the effect of any one regulator be analyzed in complete independence of the others.

The identification of cyclic AMP as the active intermediate in one of the major types of induction has opened the door to the study of that mechanism in molecular terms. It should be possible to identify the mode of action of this compound relatively easily, particularly if the conclusion is correct that it acts at the translational level, since many aspects of translational reactions are understood and can be analyzed satisfactorily in cell-free systems. On the other hand, a translational control mechanism poses the difficult problem of specificity—why synthesis of some proteins is increased and others are not. Since this implies that not all translations are equivalent, a solution may require a concerted cytological-biochemical approach to distinguish, for example, differing functional capacities of polysomes distributed in one or another part of the cell, or the possibility that movement of polysomes from one compartment to another may be responsible for altering rates of synthesis of specific proteins.

The extent to which hormonal regulation of transcription is comparable to the substrate-repressor systems operative in bacteria will undoubtedly be a major focus of future research in regulation by steroid hormones, if the deduction is correct that these hormones promote synthesis of specific mRNA's. This question is being actively investigated in several laboratories and an unequivocal answer can be expected soon. Perhaps it will become possible to adapt the key tool of the molecular biologists to mammalian systems, and to find mutations related to regulation of transcription. Such a development would be an enormous aid to understanding hormonal regulation of gene expression.

References

Arion, W. J., and Nordlie, R. C. (1967). *J. Biol. Chem.* **242,** 2207.
Barnabei, O., and Sereni, F. (1961). *Boll. Soc. Ital. Biol. Sper.* **36,** 1656.
Berlin, C. M., and Schimke, R. T. (1965). *Mol. Pharmacol.* **1,** 149.
Bishop, J. S., and Larner, J. (1967). *J. Biol. Chem.* **242,** 1355.
Black, I. B., and Axelrod, J. (1968). *Proc. Natl. Acad. Sci. U.S.* **59,** 1231.
Cho-Chung, Y. S., and Pitot, H. C. (1968). *European J. Biochem.* **3,** 401.
Civen, M., and Brown, C. B. (1968). *Federation Proc.* **27,** 394.
Civen, M., and Knox, W. E. (1959). *J. Biol. Chem.* **234,** 1787.
Civen, M., and Knox, W. E. (1960). *J. Biol. Chem.* **235,** 1716.
Civen, M., Ulrich, R., Trimmer, B. M., and Brown, C. B. (1967). *Science* **157,** 937.
Cori, C. F. (1945). *Harvey Lectures Ser.* **41,** 253.
Csanyi, V., and Greengard, O. (1968). *Arch. Biochem. Biophys.* **125,** 824.
Csanyi, V., Greengard, O., and Knox, W. E. (1967). *J. Biol. Chem.* **242,** 2688.
DeWulf, H., and Hers, H. G. (1967a). *European J. Biochem.* **2,** 50.
DeWulf, H., and Hers, H. G. (1967b). *European J. Biochem.* **2,** 57.
DiPietro, D. L., Sharma, C., and Weinhouse, S. (1962). *Biochemistry* **1,** 455.
Drews, J., and Brawerman, G. (1967a). *J. Biol. Chem.* **242,** 801.
Drews, J., and Brawerman, G. (1967b). *Science* **156,** 1385.
Dubnoff, J. W., and Dimick, M. (1959). *Biochim. Biophys. Acta* **31,** 541.
Feigelson, P., and Feigelson, M. (1964). *In* "Mechanisms of Hormone Action" (P. Karlson, ed.), p. 246. Academic Press, New York.
Feigelson, P., and Greengard, O. (1961). *Biochim. Biophys. Acta* **50,** 200.
Feigelson, P., and Greengard, O. (1962). *J. Biol. Chem.* **237,** 3714.
Feigelson, M., Gross, P. R., and Feigelson, P. (1962). *Biochim. Biophys. Acta* **55,** 495.
Fiala, S., and Fiala, E. (1966). *Nature* **210,** 530.
Fitch, W. M., and Chaikoff, I. F. (1962). *Biochim. Biophys. Acta* **57,** 588.
Foster, D. O., Ray, P. D., and Lardy, H. A. (1966). *Biochemistry* **5,** 555.
Freedland, R. A., and Avery, E. H. (1964). *J. Biol. Chem.* **239,** 3357.
Gaebler, O. H. (1955). *In* "Hypophyseal Growth Hormone, Nature and Actions" (R. W. Smith, O. H. Gaebler, and C. N. H. Long, eds.), p. 383. McGraw-Hill (Blakiston), New York.
Garren, L. D., Howell, R. R., Tomkins, G. M., and Crocco, R. M. (1964a). *Proc. Natl. Acad. Sci. U.S.* **52,** 1121.
Garren, L. D., Howell, R. R., and Tomkins, G. M. (1964b). *J. Mol. Biol.* **9,** 100.

Gatehouse, P. W., Hopper, S., Schatz, L., and Segal, H. L. (1967). *J. Biol. Chem.* **242,** 2319.

Gelehrter, T. D., and Tomkins, G. M. (1967). *J. Mol. Biol.* **29,** 59.

Georgiev, G. P. (1968). *In* "Regulatory Mechanisms for Protein Synthesis in Mammalian Cells" (A. San Pietro, M. R. Lamborg, and F. T. Kenney, eds.). Academic Press, New York, p. 25.

Geschwind, I., Li, C. H., and Evans, H. M. (1950). *Arch. Biochem.* **28,** 73.

Goldstein, L., Stella, E. J., and Knox, W. E. (1962a). *J. Biol. Chem.* **237,** 1723.

Goldstein, L., Knox, W. E., and Behrman, E. J. (1962b). *J. Biol. Chem.* **237,** 2855

Gonzalez, C., Ureta, R., Sanchez, R., and Niemeyer, H. (1964). *Biochem. Biophys. Res. Commun.* **16,** 347.

Greengard, O. (1963). *Advan. Enzyme Regul.* **1,** 61.

Greengard, O., and Feigelson, P. (1961a). *Nature* **190,** 446.

Greengard, O., and Feigelson, P. (1961b). *J. Biol. Chem.* **236,** 158.

Greengard, O., and Feigelson, P. (1962). *J. Biol. Chem.* **237,** 1903.

Greengard, O., and Gordon, M. (1963). *J. Biol. Chem.* **238,** 3708.

Greengard, O., Smith, M. A., and Acs, G. (1963a). *J. Biol. Chem.* **238,** 1548.

Greengard, O., Weber, G., and Singhal, M. (1963b). *Science* **141,** 160.

Greenman, D. L., Wicks, W. D., and Kenney, F. T. (1965). *J. Biol. Chem.* **240,** 4420.

Grossman, A., and Mavrides, C. (1967). *J. Biol. Chem.* **242,** 1398.

Hager, C. B., and Kenney, F. T. (1968). *J. Biol. Chem.* **243,** 3296.

Harding, H. R., and Rosen, F. (1963). *Federation Proc.* **22,** 409.

Hayashi, S., Granner, D. K., and Tomkins, G. M. (1967). *J. Biol. Chem.* **242,** 3998.

Henning, H. V., Seiffert, I., and Seubert, W. (1963). *Biochim. Biophys. Acta* **77,** 345.

Henshaw, E. C. (1968). *J. Mol. Biol.* **36,** 401.

Henshaw, E. C., Revel, M., and Hiatt, H. H. (1965). *J. Mol. Biol.* **14,** 241.

Holten, D., and Kenney, F. T. (1967). *J. Biol. Chem.* **242,** 4372.

Holten, D., Wicks, W. D., and Kenney, F. T. (1967). *J. Biol. Chem.* **242,** 1053.

Hubener, H. J. (1960). *Z. Physiol. Chem.* **322,** 135.

Ichihara, A., and Koyama, E. (1966). *J. Biochem.* (*Tokyo*) **59,** 160.

Ishikawa, E., Minagawa, T., and Suda, M. (1965). *J. Biochem.* (*Tokyo*) **57,** 506.

Jacoby, G. A., and LaDu, B. N. (1964). *J. Biol. Chem.* **239,** 419.

Jefferson, L. S., Exton, J. H., Butcher, R. W., Sutherland, E. W., and Park, C. R. (1968). *J. Biol. Chem.* **243,** 1031.

Jost, J. P., Khairallah, E. A., and Pitot, H. C. (1968). *J. Biol. Chem.* **243,** 3057

Kenney, F. T. (1959). *J. Biol. Chem.* **234,** 2707.

Kenney, F. T. (1962a). *J. Biol. Chem.* **237,** 1605.

Kenney, F. T. (1962b). *J. Biol. Chem.* **237,** 1610.

Kenney, F. T. (1962c). *J. Biol. Chem.* **237,** 3495.

Kenney, F. T. (1967a). *Science* **156,** 525.

Kenney, F. T. (1967b). *J. Biol. Chem.* **242,** 4367.

Kenney, F. T., and Albritton, W. L. (1965). *Proc. Natl. Acad. Sci. U.S.* **54,** 1693.

Kenney, F. T., and Flora, R. M. (1961). *J. Biol. Chem.* **236,** 2699.

Kenney, F. T., and Kull, F. J. (1963). *Proc. Natl. Acad. Sci. U.S.* **50,** 493.

Kenney, F. T., Wicks, W. D., and Greenman, D. L. (1965). *J. Cellular Comp. Physiol.* **66,** Suppl. 1, 125.

Kenney, F. T., Reel, J. R., Hager, C. B., and Wittliff, J. L. (1968a). *In* "Regulatory Mechanisms for Protein Synthesis in Mammalian Cells" (A. San Pietro, M. R. Lamborg, and F. T. Kenney, eds.). Academic Press, New York, p. 119.

Kenney, F. T., Holten, D., and Hager, C. B. (1968b). *Gunma Symp. Endocrinol.* **5,** 31.

Khairallah, E. A., and Pitot, H. C. (1967). *Biochem. Biophys. Res. Commun.* **29,** 269.
Knox, W. E. (1951). *Brit. J. Exptl. Pathol.* **32,** 462.
Knox, W. E. (1966). *Advan. Enzyme Regul.* **4,** 287.
Knox, W. E., and Greengard, O. (1965). *Advan. Enzyme Regul.* **3,** 247.
Knox, W. E., and Mehler, A. H. (1951). *Science* **113,** 237.
Knox, W. E., and Piras, M. M. (1966). *J. Biol. Chem.* **241,** 765.
Knox, W. E., Piras, M. M., and Tokuyama, K. (1966). *J. Biol. Chem.* **241,** 297.
Korner, A. (1964). *Biochem. J.* **92,** 449.
Korner, A. (1965). *J. Cellular Comp. Physiol.* **66,** Suppl. 1, 153.
Krebs, H. A. (1954). *Bull. Johns Hopkins Hosp.* **95,** 19.
Kreutner, W., and Goldberg, N. D. (1967). *Proc. Natl. Acad. Sci. U.S.* **58,** 1515.
Kvam, D. C., and Parks, R. E., Jr. (1960). *Am. J. Physiol.* **198,** 21.
Lang, N., and Sekeris, C. E. (1964). *Life Sci.* **3,** 391.
Langan, T. (1968). *Science* **162,** 579.
Langdon, R. G., and Weakley, D. R. (1955). *J. Biol. Chem.* **214,** 167.
Lardy, H. A., Foster, D. O., Young, J. W., Shrago, E., and Ray, P. D. (1965). *J. Cellular Comp. Physiol.* **66,** Suppl. 1, 39.
Lee, N. D. (1956). *J. Biol. Chem.* **219,** 211.
Lin, E. C. C., and Knox, W. E. (1957). *Biochim. Biophys. Acta* **31,** 541.
Makman, R. S., and Sutherland, E. W. (1965). *J. Biol. Chem.* **240,** 1309.
Mavrides, C., and Lane, E. A. (1967). *Science* **156,** 1376.
Mehler, A. H., McDaniel, E. G., and Hundley, J. M. (1958a). *J. Biol. Chem.* **232,** 323.
Mehler, A. H., McDaniel, E. G., and Hundley, J. M. (1958b). *J. Biol. Chem.* **232,** 331.
Mersman, E. G., and Segal, H. L. (1967). *Proc. Natl. Acad. Sci. U.S.* **58,** 1688.
Mishkin, E. P., and Shore, M. L. (1967). *Biochim. Biophys. Acta* **138,** 169.
Mokrasch, L. C., Davidson, W. D., and McGilvery, R. W. (1956). *J. Biol. Chem.* **222,** 179.
Monroy, A., Maggio, R., and Rinaldi, A. M. (1965). *Proc. Natl. Acad. Sci. U.S.* **54,** 107.
Munro, H. N. (1968). *Federation Proc.* **27,** 1231.
Nemeth, A. M., and De la Haba, G. (1962). *J. Biol. Chem.* **237,** 1190.
Niemeyer, H., Perez, N., Garcia, E., and Vergara, F. E. (1962). *Biochim. Biophys. Acta* **62,** 411.
Niemeyer, H., Clark-Turri, L., and Rabajille, E. (1963). *Nature* **198,** 1096.
Niemeyer, H., Perez, N., and Rabajille, E. (1966). *J. Biol. Chem.* **241,** 4055.
Nordlie, R. C., Arion, W. J., and Glende, E. A., Jr. (1965). *J. Biol. Chem.* **240,** 3479.
Peraino, C., and Pitot, H. C. (1964). *J. Biol. Chem.* **239,** 4308.
Peraino, C., Blake, R. L., and Pitot, H. C. (1965). *J. Biol. Chem.* **240,** 3039.
Peraino, C., Lamar, C., and Pitot, H. C. (1966). *Advan. Enzyme Regul.* **4,** 199.
Perlman, R. L., and Pastan, I. (1968). *Biochem. Biophys. Res. Commun.* **30,** 656.
Perry, R. P., and Kelley, D. E. (1968). *J. Mol. Biol.* **35,** 37.
Pitot, H. C., and Peraino, C. (1963). *J. Biol. Chem.* **238,** PC 1910.
Pitot, H. C., and Peraino, C. (1964). *J. Biol. Chem.* **239,** 1783.
Pitot, H. C., Peraino, C., Morse, P. A., Jr., and Potter, V. R. (1964a). *Natl. Cancer Inst. Monograph* **13,** 229.
Pitot, H. C., Peraino, C., Pries, N., and Kennan, A. L. (1964b). *Advan. Enzyme Regul.* **2,** 237.

Pitot, H. C., Cho-Chung, Y. S., Lamar, C., Jr., and Peraino, C. (1965a). J. Cellular *Comp. Physiol.* **66,** Suppl. 1, 163.

Pitot, H. C., Peraino, C., Lamar, C., and Kennan, A. L. (1965b). *Proc. Natl. Acad. Sci. U.S.* **54, 845.**

Potter, V. R., Gebert, R. A., and Pitot, H. C. (1966). *Advan. Enzyme Regul.* **4,** 247.

Potter, V. R., Watanabe, M., Becker, J. T., and Pitot, H. C. (1967). *Advan. Enzyme Regul.* **5,** 303.

Rall, T. W., and Sutherland, E. W. (1958). *J. Biol. Chem.* **232,** 1065.

Ray, P. D., Foster, D. O., and Lardy, H. A. (1964). *J. Biol. Chem.* **239,** 3396.

Reel, J. R., and Kenney, F. T. (1968a). *Federation Proc.* **27,** 641.

Reel, J. R., and Kenney, F. T. (1968b). *Proc. Natl. Acad. Sci. U.S.* **61,** 200.

Reuber, M. D. (1961). *J. Natl. Cancer Inst.* **26,** 891.

Rosen, F. (1965). *J. Cellular Comp. Physiol.* **66,** Suppl. 1, 146.

Rosen, F., and Milholland, R. J. (1966). *Federation Proc.* **25,** 285.

Rosen, F., and Milholland, R. J. (1968). *J. Biol. Chem.* **243,** 1900.

Rosen, F., and Nichol, C. A. (1964). *Advan. Enzyme Regul.* **2,** 115.

Rosen, F., Roberts, N. R., Rudnick, L. E., and Nichol, C. A. (1958). *Science* **127,** 287.

Rosen, F., Harding, H. R., Milholland, R. J., and Nichol, C. A. (1963). *J. Biol. Chem.* **238,** 3725.

Rosen, F., Raina, P. N., Milholland, R. J., and Nichol, C. A. (1964). *Science* **146,** 661.

Salas, M., Viñuela, E., and Sols, A. (1963). *J. Biol. Chem.* **238,** 3535.

Schimke, R. T., Sweeney, E. W., and Berlin, C. M. (1964). *Biochem. Biophys. Res. Commun.* **15,** 214.

Schimke, R. T., Sweeney, E. W., and Berlin, C. M. (1965a). *J. Biol. Chem.* **240,** 4609.

Schimke, R. T., Sweeney, E. W., and Berlin, C. M. (1965b). *J. Biol. Chem.* **240,** 322.

Segal, H. L., and Gonzales-Lopez, C. (1963). *Nature* **200,** 143.

Segal, H. L., and Kim, Y. S. (1963). *Proc. Natl. Acad. Sci. U.S.* **50,** 912.

Segal, H. L., and Kim, Y. S. (1965). *J. Cellular Comp. Physiol.* **66,** Suppl. 1, 11.

Segal, H. L., Beattie, D. S., and Hopper, S. (1962a). *J. Biol. Chem.* **237,** 1914.

Segal, H. L., Rosso, R. G., Hopper, S., and Weber, M. M. (1962b). *J. Biol. Chem.* **237,** PC3303.

Sekeris, C. E., and Lang, N. (1964). *Life Sci.* **3,** 625.

Sells, B. H., and Jackson, C. D. (1967). *Biochim. Biophys. Acta* **142,** 419.

Sells, B. H., and Takahashi, T. (1967). *Biochim. Biophys. Acta* **134,** 69.

Sereni, F., Kenney, F. T., and Kretchmer, N. (1959). *J. Biol. Chem.* **234,** 609.

Shambaugh, G. E., Warner, D. A., and Beisel, W. R. (1967). *Endocrinology* **81,** 811.

Sharma, C., Manjeshwar, R., and Weinhouse, S. (1963). *J. Biol. Chem.* **238,** 3840.

Shrago, E., Lardy, H. A., Nordlie, R. C., and Foster, D. O. (1963). *J. Biol. Chem.* **238,** 3188.

Simpson, M. E., Evans, H. M., and Li, C. H. (1949). *Growth* **13,** 151.

Singer, S., and Mason, M. (1965). *Biochim. Biophys. Acta* **110,** 370.

Sols, A., Sillero, A., and Salas, J. (1965). *J. Cellular Comp. Physiol.* **66,** Suppl. 1, 23.

Spirin, A. S., and Nemer, M. J. (1965). *Science* **150,** 214.

Steiner, D. F., and King, J. (1964). *J. Biol. Chem.* **239,** 1292.

Sutherland, E. W., and Robison, G. A. (1966). *Pharmacol. Rev.* **18,** 145.

Sutherland, E. W., Øye, I., and Butcher, R. W. (1965). *Rec. Progr. Hormone Res.* **21,** 623.

Tanaka, T., Harano, Y., Sue, F., and Morimura, H. (1967). *J. Biochem.* (*Tokyo*) **62,** 71.

Tata, J. R., and Williams-Ashman, H. G. (1967). *European J. Biochem.* **2,** 366.

Thompson, E. B., Tomkins, G. M., and Curran, J. F. (1966). *Proc. Natl. Acad. Sci. U.S.* **56,** 296.

Tomkins, G. M., Thompson, E. B., Hayashi, S., Gelehrter, T. D., Granner, D., and Peterkofsky, B. (1966). *Cold Spring Harbor Symp. Quant. Biol.* **31,** 349.

Tschudy, D. P., Marver, H. S., and Collins, A. (1965). *Biochem. Biophys. Res. Commun.* **21,** 480.

Viñuela, E., Salas, M., and Sols, A. (1963). *J. Biol. Chem.* **238,** PC1175.

Weber, G., Singhal, R. L., and Srivastava, S. K. (1965). *Proc. Natl. Acad. Sci. U.S.* **53,** 96.

Wicks, W. D. (1968a). *J. Biol. Chem.* **243,** 900.

Wicks, W. D. (1968b). *Science* **160,** 997.

Wicks, W. D. (1968c). *In* "Regulatory Mechanisms for Protein Synthesis in Mammalian Cells" (A. San Pietro, M. R. Lamborg, and F. T. Kenney, eds.), p. 143. Academic Press, New York.

Wicks, W. D., Greenman, D. L., and Kenney, F. T. (1965). *J. Biol. Chem.* **240,** 4414.

Wilson, S. H., and Hoagland, M. B. (1967). *Biochem. J.* **103,** 556.

Wilson, S. H., Hill, H. Z., and Hoagland, M. B. (1967). *Biochem. J.* **103,** 567.

Wool, I. G. (1964). *In* "Actions of Hormones on Molecular Processes" (G. Litwack and D. Kritchevsky, eds.), p. 422. Wiley, New York.

Wool, I. G. (1968). *In* "Regulatory Mechanisms for Protein Synthesis in Mammalian Cells" (A. San Pietro, M. R. Lamborg, and F. T. Kenney, eds.), p. 323. Academic Press, New York.

Wurtman, R. J., and Axelrod, J. (1967). *Proc. Natl. Acad. Sci. U.S.* **57,** 1594.

Yu, F. L., and Feigelson, P. (1968). *Abstr. 156th Ann. Meeting, Am. Chem. Soc. Div. Biol. Chem.* p. 138.

CHAPTER 32

Regulation of Protein Degradation in Mammalian Tissues[1,2]

ROBERT T. SCHIMKE

Departments of Pharmacology and Biology, Stanford University School of Medicine, Palo Alto, California

[1] Unpublished studies described in this review were supported by research grants from the USPHS (GM14931) and The American Cancer Society (P427).

[2] Considerable confusion exists in the use of "turnover" and "degradation." In the literature on bacteria, protein "turnover" is used to denote the degradation of protein (Mandelstam, 1960). In animal tissues, protein "turnover" has been used to denote the general phenomenon in which tissue constituents are continually synthesized and degraded, as well as used to denote the more limited process of degradation. In this review, we shall use "turnover" to denote the overall phenomenon of combined synthesis and degradation of tissue proteins. In this context, then, the concept of a "turnover rate" has little meaning, since turnover is made up of independent processes of synthesis and degradation. Only in a steady state is there justification for considering a "turnover rate." Since the steady state is a special and limited instance of the more general condition wherein levels of specific tissue proteins are continually changing (Knox *et al.*, 1956), it is suggested that the more specific and meaningful terms, "rates of synthesis and degradation" be used and that such rates be measured whenever possible.

I. Introduction

Knowledge of the events of protein synthesis and their control has increased immensely during the past decade (Introductory Chapter). In contrast, understanding of the mechanisms for breakdown of cell proteins is rudimentary, despite the fact that this process is equally important in the total economy of an organism. Awareness of the importance of protein degradation has developed in the past five years as a result of studies which demonstrate that control of degradation can be important in determining the levels of a number of specific proteins in intact animals and mammalian cells in culture. Such results have raised questions concerning the mechanism and control of protein degradation which have heretofore received little attention.

A. Purpose of Review

The purpose of this chapter, therefore, is to review selected recent studies concerning protein degradation with the purpose of focusing attention on the following questions: (1) What is the validity of various techniques used to measure degradation of specific cell proteins? (2) What is the significance of continual synthesis and degradation of proteins for the organism? (3) What is the mechanism(s) by which proteins are degraded within cells? (4) How is such degradation regulated? Although these questions will be posed largely from the standpoint of mammalian tissues, most specifically liver, investigations with microbial systems will be discussed when pertinent. Earlier studies on protein turnover in mammalian and microbial tissues are the subject of a number of excellent reviews, including those of Schoenheimer and Rittenberg (1940), Tarver (1954), Neuberger and Richards, Chapter 7 (see Vol. I), and Mandelstam (1960). The reader is referred to these sources for many studies which will not be mentioned here.

B. Development of the Concept of Protein Turnover

Early workers concerned with the metabolism of cell proteins made a distinction between those proteins which were metabolically inert and those which were derived from metabolism of food stuffs. Only the latter proteins were considered to be metabolically labile, or storage, proteins (see Voit, 1866; see also this treatise; Munro, Chapter 10; Neuberger and Richards, Chapter 7). A concept of breakdown of cellular proteins resulting from "wear and tear" was superimposed on this distinction by Folin in 1905. It was not until 1935 that Borsook and Keighley proposed that virtually all proteins in mammalian tissues were in a continual state of synthesis and degradation, a process they denoted as "continuing metabolism of nitrogen." More definitive demonstrations of continual protein turnover were not made until tracer techniques became available in the late 1930's. Thereafter an extensive series of studies was undertaken by many groups (Tarver, 1954; Schoenheimer and Rittenberg, 1940) which indicated an active incorporation and release of isotopic compounds from proteins of various tissues. Thus, the well-known phrase "dynamic state of body constituents" was coined by Schoenheimer (1942) to describe the concept of a continual replacement of cellular constitutents. Although such studies did, indeed, indicate replacement of protein, it was not clear to what extent this replacement was that of entire cells, synthesis and secretion of proteins from cells, or intracellular synthesis and degradation of proteins. The assumption that isotopic studies indicated a continual turnover of intracellular proteins in animal tissues was questioned by Hogness *et al.* (1955) following the demonstration of a lack of protein turnover in exponentially growing *Escherichia coli* by these workers as well as by Koch and Levy (1955), and Rotman and Spiegelman (1954). However, the early studies of Ussing with D_2O (1941), as well as the more recent investigations of Thompson and Ballou also with D_2O (1956), Swick using dietary ^{14}C-labeled carbonate (1958), Buchanan (1961b), who administered a diet consisting of uniformly ^{14}C-labeled algae, and Schimke (1964b), who administered an amino acid diet containing lysine-^{14}C have demonstrated convincingly that all liver proteins are being degraded, albeit at varying rates. These studies have employed the theoretically more sound method of continuous isotope administration (Buchanan, 1961a), which is not subject to the type of criticism of Hogness *et al.* (1955) of turnover experiments using single isotope administration and determining the decay in radioactivity of the isolated protein with time (see Section II). Buchanan (1961b) has estimated that approximately 70% of protein of rat liver is replaced every 4–5 days from a dietary source. Furthermore, since the life-span of hepatic cells is 160–400 days (Buchanan, 1961b; MacDonald, 1961; Swick *et al.,* 1956), it is clear that in liver most of the protein degradation takes place within cells rather than involving death and renewal of entire cells.

It is of interest that several years after the demonstration of a lack of protein degradation in exponentially growing *Escherichia coli,* Mandelstam (1958), and more recently, others (Pine, 1965; Willetts, 1967a,b) have shown that there is a rapid breakdown of protein in non-growing *E. coli* at a rate of 4–6% per hour. In addition, in some bacteria, such as *Bacillus cereus* (Urbá, 1959), there is a detectable rate of protein degradation in growing cells, a rate which increases markedly in the nongrowing state. Eagle *et al.* (1959) and more recently Pine (1967) have shown that protein degradation in cultured mammalian cells is largely intracellular and that, in contrast to *E. coli,* it occurs at a constant rate irrespective of cell growth rates. Thus if one were to compare bacterial and animal cells, it would appear that the animal cell is more correctly compared with a nongrowing bacterium with respect to protein degradation.

More recently the conclusion that virtually all proteins in an animal are in a continual flux has been extended to a number of specific cellular proteins. These studies will constitute, in large part, the material of Section III.

II. Methods for Measurements of Specific Protein Degradation

Obviously, information concerning the control of degradation of specific proteins and enzymes is only as good as the methods used. We shall therefore first review methods available for such measurements, and the assumptions and limitations their use entails. In this manner we can begin to understand a part of the degradation process and the multiple problems involved in making valid measurements.

A. Methods Based on Isotope Incorporation and Decay

Tracer techniques are used most commonly for estimating degradation rates. Although such techniques may seem quite direct, they are full of pitfalls which can result in ambiguous results. Theoretical formulations of the uses of isotope techniques for measurement of protein degradation have been made by Reiner (1953), Buchanan (1961a), and Koch (1962), and the reader is referred to these for a mathematical analysis of the problems and assumptions involved in these methods.

Isotope methods can be divided into two general techniques: (a) single isotope administration, and (b) continuous isotope administration.

1. Single Isotope Administration (Pulse) Method

This method involves the single intraperitoneal or intravenous administration of an isotopic form of a protein precursor, normally an amino acid. Theoretically the isotope should enter the pool from which the protein is synthesized rapidly (instantaneously), leave the pool rapidly (instantaneously), and should not return to that pool at any time following the initial isotope administration. The isotope administration, then, is a pulse, and the decay in specific radioactivity, *or total radioactivity if the protein is not in a steady state* (Koch, 1962), of the isolated protein is a measure of the rate of degradation. This is

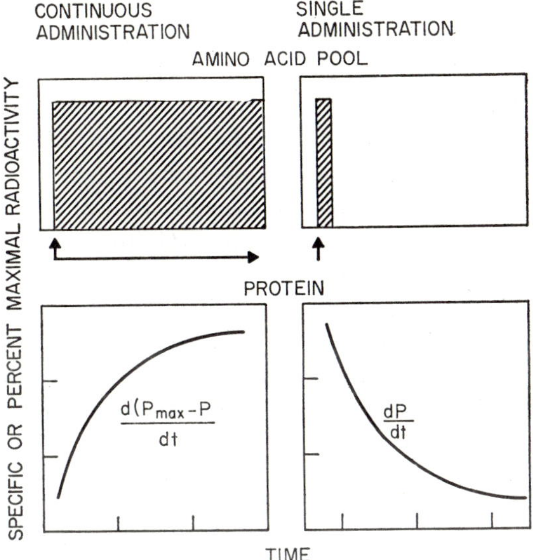

FIG. 1. Schematic representations of continuous administration and single administration methods for determining rates of protein degradation. P is used to designate specific activity or percent maximal activity.

depicted in Fig. 1. The specific protein may be isolated by standard purification procedures (Schimke, 1964b), or by means of a specific precipitating antibody (Schimke *et al.*, 1965a; Kenney, 1962; Jost *et al.*, 1968). Assumptions involved in the use of this method include the following: (1) The radioactivity isolated in the protein sample is representative of all the protein present in the isolated sample, rather than a minor contaminant. Thus it is imperative that great care is made to ensure absolute purity of the protein samples isolated. (2) There is no reutilization of isotope. The decay of isotope from protein is not only a function of the rate at which a labeled protein is degraded to constituent amino acids, but also a function of the rate at which such isotope reenters the amino acid pool from which the protein is synthesized (Reiner, 1953; Buchanan, 1961a; Koch, 1962). Isotope reutilization is a major limitation for obtaining accurate rates of protein degradation by the single administration method. Thus a lack of extensive uptake of a labeled amino acid into a specific protein or cell type, and its subsequent retention in such a protein may result from a slow rate of degradation. However, it may equally well indicate that, whatever the rate of intracellular protein degradation, the amino acid is not transported into the tissue readily, and once within a cell, it is not metabolized, or otherwise released from the cell. For example, there are any number of studies using intraperitoneal or intravenous administrations of isotope which indicate that proteins of the brain are slowly degraded, with estimated half-lives of 26–160 days (Beattie *et al.*, 1967; Kamin and Handler, 1951; Thompson and Ballou, 1956; Still, 1957). Yet from various other studies in which isotope has been administered by the intracisternal route and where no permeability barrier exists, it is evident that there is an active synthesis and degradation of brain proteins, with estimated half-lives as short as 6 days (Gaitonde and Richter, 1956; Maurer, 1957; Niklas *et al.*, 1958).

Isotope reutilization occurs in every tissue to varying degrees. Studies of Loftfield and Harris (1956) and Gan and Jeffay (1967) suggest that as much as 50% of the free amino acid pool of rat liver is derived from protein catabolism under normal dietary conditions. In muscle approximately 30% of the free amino acid pool is derived from tissue catabolism (Gan and Jeffay, 1967). Isotope reutilization can be minimized, theoretically, by the use of an isotope whose probability of reutilization is small. In liver of ureotelic animals, ^{14}C-labeled guanidino-L-arginine would appear to be such an isotope (Swick and Handa, 1956). This isotopic form of arginine has minimal reutilization because of the large amount of arginase activity in liver. Thus the administered isotope is rapidly hydrolyzed to labeled urea, which is excreted; likewise that labeled arginine which reenters the free amino acid pool will also be hydrolyzed rapidly, and hence not be reutilized. Guanidinoarginine-^{14}C has been used successfully in several experiments dealing with measurements of degradation rates (Hirsch and Hiatt, 1966; McFarlane, 1963; Schimke, 1964b).

The influence of isotope reutilization on the apparent measure of degradation can be seen in Table I, where the rates of degradation of cell fractions of rat liver are determined by decay of radioactivity of protein labeled with either guanidino-L-arginine-^{14}C or uniformly labeled L-arginine-^{14}C. Marked differences in mean degradation rates are evident in the different cell fractions studied. Furthermore the values obtained with guanidinoarginine-^{14}C are consistently lower than with uniformly labeled arginine. Several studies (Beattie *et al.*, 1967; Fletcher and Sanadi, 1961) have found mean half-lives of rat liver mitochondrial proteins of 9–11 days using isotopes subject to reutilization (methionine-^{35}S, leucine-4,5-^{3}H). Swick *et al.* (1968) have recently found that the half-life of mitochondrial protein is 5–6 days, a value close to that of Table I. Their technique involves the continuous administration of carbonate-^{14}C and excludes the problem of isotope reutilization (see below). Omura *et al.* (1967) found that the half-life of rat liver endoplasmic reticulum proteins was 3–4 days using L-leucine-1-^{14}C. The differences

TABLE I

MEAN HALF-LIFE IN DAYS OF PULSE-LABELED PROTEINS IN HOMOGENATES AND SUBCELLULAR FRACTIONS OF LIVER OF RATS WHICH RECEIVED A SINGLE INTRAPERITONEAL INJECTION OF EITHER ^{14}C-LABELED GUANIDINO-L-ARGININE (10 μC) OR ^{14}C-UNIFORMLY LABELED L-ARGININE (5 μC)[a]

Fractions	^{14}C-Guanidino-L arginine	^{14}C-Uniformly labeled arginine
Homogenate	3.3	4.5
Nuclear	5.1	6.5
Mitochondrial	6.8	9.0
Lysosomal	7.1	12.5
Microsomal	3.0	5.2
Supernatant	5.1	6.2
Smooth endoplasmic reticulum	2.1	5.2
Rough endoplasmic reticulum	2.0	5.8
Plasma membrane	1.8	3.1

[a] See Arias *et al.* (1969) for details.

between the results of Table I and those cited can be ascribed largely to differing degrees of isotope reutilization.

Results with pulse-labeling techniques are even more difficult to interpret when comparison of relative rates of degradation are made between two different proteins in the same tissue, the same protein in different tissues, or the same protein in a single tissue under different physiological or pharmacological conditions. In most instances valid comparisons of degradation rates cannot be made, since the differences in degrees of isotope reutilization cannot be assessed. One such example of this problem involves the mechanism of the increase in total microsomal protein and membrane-associated NADPH-cytochrome *c* reductase produced by repeated administrations of phenobarbital (Ernster and Orrhenius, 1965). Shuster and Jick (1966) found that phenobarbital increased the degree of isotope retention in microsomal protein and isolated NADPH-cytochrome *c* reductase (Jick and Shuster, 1966) following a pulse administration of L-leucine-4,5-^{3}H. They ascribed the effect of phenobarbital to a decreased degradation of the microsomes, i.e., a stabilization by phenobarbital. Schimke *et al.* (1968) have repeated this experiment and find that, when guanidino-L-arginine-^{14}C is used as the isotopic amino acid, phenobarbital has no effect on microsomal protein turnover in either mice or rats. Since phenobarbital is known to stimulate labeled amino acid incorporation into proteins of rodent liver (Kato *et al.*, 1966; Shuster and Jick, 1966), the results of Shuster and Jick can best be ascribed to an increased degree of reutilization of isotope associated with the increased rate of protein synthesis.

Despite the obvious advantage of guanidinoarginine-^{14}C for determining accurate rates of protein degradation, a major limitation is the small extent of incorporation following its administration. The specific activity of liver protein resulting from the administration of guanidinoarginine-^{14}C is $\frac{1}{15}$ to $\frac{1}{20}$th of that when an equal amount of radioactivity of a uniformly labeled arginine-^{14}C is administered. The difference, of course, is a crude measure of the difference in degree of reutilization, as well as the initial incorporation. In the author's hands, the use of guanidinoarginine has been limited to proteins, or cell fractions, which can be obtained in sufficiently large quantities to obtain accurate counting rates, i.e., about 1–2 mg.

Another approach to the problem of isotope reutilization has been employed with those proteins which contain a heme prosthetic group. Advantage is taken of the fact that δ-aminolevulinate is rapidly incorporated into the heme moeity of heme proteins and is subsequently degraded to bilirubin. Thus, following a single administration of the heme precursor, decay of radioactivity in isolated protein is determined as a measure of the rate of degradation. The major assumption of this method is that heme does not dissociate from protein until such time as the entire protein-heme is degraded. This assumption may not always be correct, since Bunn and Jandl (1966) have reported heme exchange in solutions of hemoglobin. However, Poole *et al.* (1969) and Druyan *et al.* (1970) have shown that rate of degradation of catalase and microsomal cytochrome b_5, respectively, to be the same by the use of the isotopic δ-aminolevulinate decay and other methods. This method has also been used to estimate the rates of degradation of mitochondrial cytochromes (Druyan *et al.*, 1970) and microsomal hemoprotein, presumably P450 (Levin and Kuntzman, 1969).

2. Continuous Isotope Administration Technique

In this technique the animal is continuously subjected to an isotope whose specific activity is constant (Fig. 1). That isotope may be D_2O, or a diet containing labeled protein, amino acid, or carbonate. The rate of degradation of a specific protein, cell fraction, or tissue is determined by the rate at which incorporation approaches a maximum value. Theoretically the isotope should saturate instantaneously the free amino acid pool from which the protein(s) is synthesized. This requirement is a major limitation of this method in determining accurate degradation rates, since the rate at which the labeled protein replaces the unlabeled protein will be a function, not only of the rate of degradation of the protein, but also of the rate at which the isotope saturates the appropriate free amino acid pool. In practice such saturation is not attained in less than 24 hours (Schimke, 1964b; Swick *et al.,* 1968). It follows that such a method will give more accurate turnover rates for proteins with longer half-lives than for proteins with short half-lives, since the rate of replacement of unlabeled with labeled protein cannot be more rapid than the replacement of the free amino acid pool. This method is also limited if two proteins are synthesized from different pools which becomes saturated with isotope at different rates. This may obviously be the case with different tissues, such as brain and liver, where permeability to amino acids is strikingly different (see Neuberger and Richards, Chapter 7). It may equally be the case, however, if different proteins in the same tissue are synthesized from different pools. Askonas and Humphrey (1958) have found that the kinetics of saturation of glycine pools in lung tissue during a constant infusion may suggest the presence of one pool that is rapidly equilibrated, and a second that is not. Green and Lowther (1959) found that granuloma slices incubated in a medium containing proline-^{14}C did not rapidly equilibrate between internal and external proline pools. Hendler (1965) has demonstrated that two pools exist for the synthesis of proteins in hen oviduct. The synthesis of ovalbumin does not occur from the internal pool of free amino acid but through a pool that is in more direct equilibrium with the incubating medium. Similar observations have been made in other systems (Kipnis *et al.,* 1961).

A representative example of the continuous dosage method is given in Fig. 2 (Schimke, 1964b). In this experiment a series of rats were fed an amino acid diet containing L-lysine-^{14}C of constant specific activity. Rats were sacrificed at varying intervals, and the incorporation of ^{14}C was determined in the trichloroacetic acid (TCA)-soluble fraction, total protein, and arginase of rat liver. Arginase was isolated by precipitation with a specific antibody. There was a maximal incorporation of counts per minute in the TCA-soluble fraction in 24 hours, a finding that is assumed to mean that the free pool from which proteins are synthesized was fully saturated from the dietary source of isotope in this time. The rate at which isotope is incorporated into total liver protein is initially rapid, and then begins to plateau after approximately 5–6 days. The rate at which isotope is incorporated into arginase is slower than the initial rate for total liver protein, indicating heterogeneity of degradation rates of liver proteins. A major advantage of the continuous dosage method is that it allows for an estimate of the percent of protein replaced in any given protein sample isolated. Thus when expressed as in Fig. 1, i.e., cpm/mg protein, one does not know whether the isotope

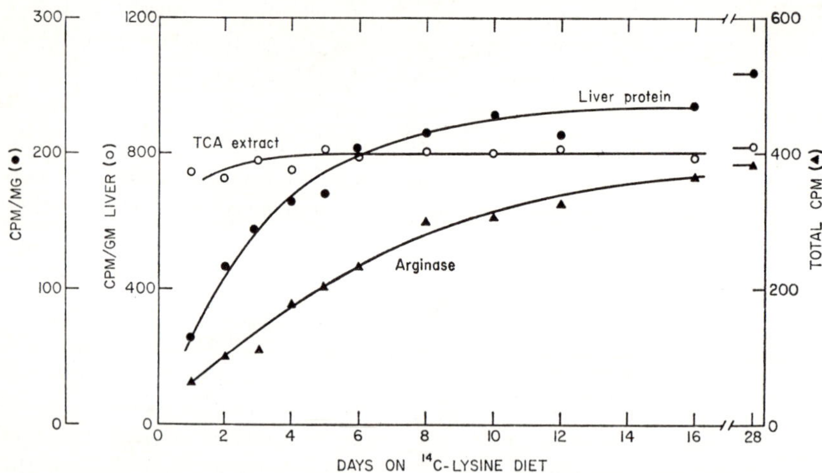

FIG. 2. Incorporation of continuously administered L-lysine-^{14}C into total protein, arginase, and trichloroacetic acid (TCA)-soluble extracts of rat liver. Details are given in the text and in Schimke (1964b).

incorporation represents 1% or 100% replacement of protein. Such information can be obtained by comparing the specific activity of lysine isolated from hydrolyzed protein with that of the initially administered diet. As shown in Table II, 75% of the lysine residues of total liver protein have been replaced from the diet in 20 days, and 95% of arginase. It should also be noted that essentially all of the radioactivity present in protein is in the form of lysine. If L-lysine-^{14}C were converted to other amino acids, and entered pools of these amino acids at different rates during the experiment, measurements of degradation rates would require determining the specific activity of protein lysine at each time point.

The method of determining degradation rates in rat liver devised by

TABLE II

SPECIFIC ACTIVITY OF L-LYSINE-^{14}C ISOLATED FROM TOTAL LIVER PROTEIN AND ARGINASE AFTER 20 DAYS OF CONTINUOUS ADMINISTRATION OF L-LYSINE-^{14}C DIET[a]

Source of lysine	Radioactivity recovered as lysine (%)	Lysine specific activity (cpm/μmole)	Percent replacement of lysine
Diet	98	1105	—
Total liver protein	95	829	75
Arginase	66	1106	91

[a] See text and Schimke (1964b).

Swick is particularly ingenious (Swick *et al.*, 1968). In this technique carbonate-^{14}C of known specific activity is fed continuously to rats. The isotope is rapidly incorporated into arginine as a result of the reactions of the urea cycle. At least 90% of radioactivity in arginine is present in the guanidino carbon. The specific activity of the intracellular pool of free arginine can be easily determined from the specific activity of excreted urea, since the guanidino carbon of arginine is the carbon of urea. Thus animals are killed at varying times after onset of continuous carbonate-^{14}C administration, liver protein samples are hydrolyzed, and the specific activity of the isolated arginine is determined, as well as that of urinary urea. The rate of degradation of the protein can be calculated from:

$$A = U(1 - e^{-kt}) \tag{1}$$

where A is the specific activity of the arginine, U is the specific activity of the urea, and k is the first-order rate constant of degradation.

3. Double Isotope Technique for Determining Heterogeneity of Degradation Rates

On many occasions it is desirable to know in the steady-state condition whether proteins are turning over at the same, or dissimilar rates. To answer this question a technique has been developed in the author's laboratory which involves the combined use of ^{3}H and ^{14}C forms of the same amino acid. The method (double-isotope method) is a modification of the standard pulse-labeling technique wherein the decay of specific radioactivity is determined (see above). The assumptions associated with the pulse-labeling technique are generally applicable to the double-isotope technique. Instead of establishing a number of time points along a decay curve, which involves isolation of the protein species at each time interval, the double-isotope technique allows for the identification of two time points on such a decay curve in the same protein in the same animal. Thus one isotopic form of an amino acid (^{14}C) is administered initially, and allowed to decay a specified length of time, usually 4–6 days. This is followed by a second administration of the same amino acid (^{3}H), and the animal is sacrificed a short time thereafter, usually 4–6 hours. The ^{3}H counts represent the initial time point and the ^{14}C counts represent the decay time point. Proteins with rapid degradation rates will have greater ^{3}H/^{14}C ratios. This is depicted graphically (Fig. 3) for two proteins which are present at the same concentration in a tissue, but one of which (protein A) is degraded at twice the rate of the other (protein B). The question of heterogeneity of degradation rate of proteins of a given organelle or cell fraction can be answered

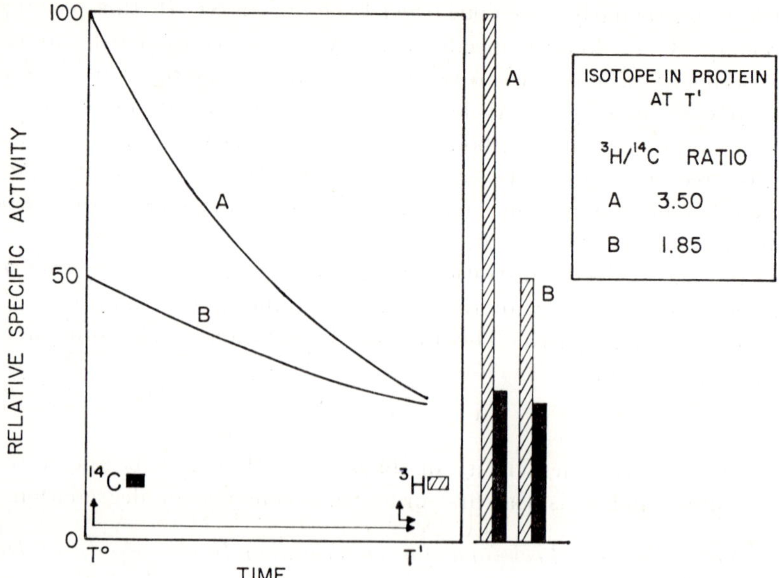

FIG. 3. Double-isotope technique for determining relative rates of protein degradation. Depicted here are the patterns of labeling with a ^{3}H and ^{14}C isotope of two proteins present at the same steady-state levels, of which one (A) is degraded at twice the rate of the other (B). From Arias *et al.* (1969).

without knowing the specific enzymic activity or nature of the proteins if they can be adequately separated, such as by gel electrophoresis or column chromatography. Thus if all proteins are degraded at the same rate, the $^{3}H/^{14}C$ ratios will be the same, whereas if they are degraded at different rates, those ratios will be dissimilar. It should be clear that this method is designed largely to determine the degree of heterogeneity of turnover rates of proteins. It does not allow for a measure of absolute rates of degradation.

Schimke *et al.* (1968) have used this method to demonstrate a marked heterogeneity of turnover of proteins of the endoplasmic reticulum. It has general applicability to any situation in which relative rates of degradation of proteins are to be measured. This method makes a number of specific assumptions: (1) the proteins will all follow exponential decay kinetics; (2) at the time of sacrifice all labeled proteins in the fraction under study are in a state of isotopic decay; (3) the rates of synthesis of the proteins are the same at the time of both isotope administrations; (4) the different proteins being compared are synthesized from the same amino acid pool; (5) the isotope used is not metabolized to other compounds which can enter other amino acid, carbohydrate, or lipid pools in a manner that allows them to be incorporated into the protein sample being counted.

4. Comments on the Interpretation of Isotope Decay or Accumulation Data

Perhaps the simplest pattern of isotope kinetics is that representative of a life-span. In this case, not depicted in Fig. 1, the continuous-administration technique would result in a linear increase in isotope accumulation with time to a maximum (see Buchanan, 1961b). The time required to obtain such a maximum would represent the life-span of the cell or protein. The single-administration technique would result in the presence of isotope at a given level in the tissue, with a relative precipitous decline to zero at the life-span. The kinetics of red cell labeling are classic examples of life-spans (see Schoenheimer and Rittenberg, 1940).

The kinetics of most cases of isotope incorporation or decay from proteins, tissues, or cell fractions follow more complex patterns that approximate exponential functions, as depicted in Fig. 1. The rates of degradation of such proteins, then, are generally described in terms of a half-life (Schoenheimer and Rittenberg, 1940). Exponential labeling patterns have been observed for mitochondria (Beattie *et al.*, 1967; Fletcher and Sanadi, 1961; Swick *et al.*, 1968), microsomes, and endoplasmic reticulum (Omura *et al.*, 1967; Schimke *et al.*, 1968; Shuster and Jick, 1966), and individual intracellular proteins, including arginase (Schimke, 1964b), tyrosine transaminase (Kenney, 1962, 1967), tryptophan pyrrolase (Schimke *et al.*, 1965a), and catalase (Poole *et al.*, 1969), and serine dehydratase (Jost *et al.*, 1968). The standard interpretation of exponential kinetics is that the newly synthesized (labeled) protein enters a pool of similar protein molecules, and that the subsequent degradation is random. Although this interpretation is probably correct in most cases, at least two other interpretations are possible. Thus exponential kinetics are also compatible with the synthesis of groups of molecules each with a different life-span, the summation of which gives the appearance of exponential kinetics. Another explanation of kinetics that may appear to be exponential may be especially applicable to proteins that are present in intracellular organelles, such as catalase in microbodies, or the protein constituents of mitochondria or membranes. Thus, we can picture such organelles as growing structures into which a particular protein is being continuously deposited as it is synthesized at a constant rate (enzyme/organelle/time). At some critical point in time or size, described as a "ripening time" by de Duve and Baudhuin (1966) the entire organelle is subject to degradation. Such a process has been proposed for catalase degradation (de Duve and Baudhuin, 1966), but can equally well represent the process of degradation of proteins of any organelle. This explanation would require that newly synthesized organelles be smaller. Furthermore, shortly after isotope incorporation, they should exhibit a higher specific radioactivity than larger (older) organelles.

B. Methods Based on Kinetics of Enzyme Activity Changes

A number of techniques have been devised during the past five years to estimate the rate of turnover of various enzymes on the basis of time courses of changes in enzyme levels. These methods have the advantage of not requiring the isolation of a specific enzyme or protein in a homogeneous state, or the manipulation of isotopes with its attendant assumptions. Such techniques, however, carry another set of assumptions and limitations. In order to more fully understand such techniques, it is necessary to present a general model for the regulation of enzyme levels which considers both enzyme synthesis and breakdown (Schimke *et al.*, 1964; Berlin and Schimke, 1965). Similar models have been proposed by Price *et al.* (1962) and Segal and Kim (1963). The majority of applications of this technique involve changes in enzyme levels in the rat liver. It has been a consistent finding in this organ that enzyme activities decay exponentially to basal levels after removal of some agent that has increased the level significantly over basal levels. Thus we may express the rate of enzyme degradation in terms of a first-order rate constant. Enzyme synthesis has conformed to zero-order kinetics (Schimke *et al.*, 1965a; Price *et al.*, 1962). Thus the most simplified model for a change in tissue content of an enzyme is:

$$dE/dt = S - kE \tag{2}$$

where E is the content of enzyme per unit weight, S the rate constant for synthesis,[3] expressed as enzyme units/time and k the first order rate constant for enzyme degradation, expressed as time^{-1}. In all experimental conditions where this model has been used, i.e., rat liver, no appreciable change in total liver mass occurs, and hence no expression is required to account for such a change.

At any time that a steady state for the enzyme level exists, i.e., $dE/dt = 0$, then

$$E = S/k \tag{3}$$

[3] It is obvious that the rate of synthesis will be determined by a large number of factors, including the number of ribosomes, amount of messenger RNA, levels of amino acids and of tRNA, content of initiation and soluble transfer factors (see Introductory Chapter, this volume). In this analysis the role of such variables has not been factored, and hence we include them all under a general notation of a rate of enzyme synthesis.

Thus under steady state conditions the level of E will be determined by the relative rates of synthesis and degradation. We may now theoretically determine the rate of degradation of the enzyme, E, in any one of four ways on the basis of changes in enzyme levels.

1. *Decay of Enzyme Activity after Elevation to High Levels by Administration of Appropriate Inducing Agents*

Thus as originally used by Feigelson *et al.* (1959), the levels of tryptophan pyrrolase were increased by the administration of hydrocortisone and/or tryptophan, and the rate of return to a basal level was determined. The rate constant of degradation (k) is given by the slope of the straight line obtained from a plot on semilogarithmic paper of $(E_t) - (E_{ss,o})$ versus time where E_t is the enzyme activity at any time, t, during the decay period, and $E_{ss,o}$ is that enzyme activity characteristic of the basal state (Segal and Kim, 1963). This technique has now been used with a number of enzymes (see Table III). The assumptions of this technique include: (a) a measure of the decay of enzyme activity is a true measure of the decay of enzyme protein, and not simply a reflection of the activity. Operationally, degradation and irreversible inactivation cannot be distinguished by this method—or by any other, for that matter. (b) The stimulus of the administered agent or altered physiological status ceases relatively abruptly, and the rate of synthesis of the protein immediately decreases to that characteristic of the basal state of enzyme synthesis. (c) The rate of decay of the elevated enzyme activity is similar to the rate of degradation of the enzyme in the basal state.

2. *Decay of Activity after Inhibition of Protein Synthesis*

Thus, the observed decay of enzyme activity following the inhibition of protein synthesis should reflect the inherent rate of enzyme degradation. This method carries the assumption that the inhibition of protein synthesis, such as produced by the administration of puromycin or cyclohexamide, does not itself alter the rate of enzyme degradation. Although inhibitors of protein synthesis do not appear to affect the rate of degradation of δ-aminolevulinic acid synthetase (Marver *et al.*, 1966) and tryptophan pyrrolase (Schimke *et al.*, 1965a) of rat liver, puromycin and cycloheximide do, indeed, diminish or abolish the breakdown (inactivation) of tyrosine transaminase in the same tissue (Kenney, 1967; Grossman and Mavrides, 1967), as well as carbamyl phosphate synthetase (Shambaugh *et al.*, 1969) and glutamate dehydrogenase (Balinsky *et al.*, 1970) of frog liver. In addition, the administration of actinomycin to rats inhibits the normally occurring decrease in the activity of mitochondrial α-glycerophosphate dehydrogenase that results from the cessation of thyroxine treatment in hypothyroid rats (Sellinger *et al.*, 1966). Therefore, the assumption that drugs will not alter degradation rates cannot be applied indiscriminately.

3. *Time Course of Approach of an Enzyme to a New Steady State When the Rate of Synthesis of the Enzyme Is Increased by Some Hormonal or Physiological Stimulus*

To understand this method, we must consider further the model for steady-state enzyme levels (Berlin and Schimke, 1965).

Consider the time course of the change of E with time where the rate of synthesis is changed from S to S' and the rate constant of degradation is changed from k to k'. Then:

$$dE/dt = S' - k'E$$
$$dE/(S' - k'E) = dt$$
$$\ln(S' - k'E) = -k't + c$$

At $t = 0$, $E = E_0$, c, the constant of integration is:

$$c = \ln(S' - k'E_0)$$

Substituting for c,

$$\ln(S' - k'E) - \ln(S' - k'E_0) = -k't$$
$$\frac{S' - k'E}{S' - k'E_0} = e^{-k't}$$

$$E/E_0 = \frac{S'}{k'E_0} - \left(\frac{S'}{k'E_0} - 1\right)^{e^{-k't}} \tag{4}$$

This equation represents a general solution of Eq. (1).

It should be emphasized that in Eq. (4) the only expression that determines the time course whereby an enzyme approaches a new steady state is the rate constant of degradation. The new steady-state level, on the other hand, is determined by the ratio of the rate constant of synthesis to the rate constant of degradation (S'/k').

It is possible to calculate, using Eq. (4), the time at which the enzyme has undergone one-half of its total change in content as follows:

$$\text{Steady state level at infinite time, } E' = S'/k'$$
$$\text{Initial enzyme level, } E_0 = 1$$
$$\text{One-half of the total change} = \tfrac{1}{2}(S'/k' - 1)$$

This expression is substituted for the left-hand side of Eq. (4) and solved for t, the time required to attain one-half the total change.

$$\tfrac{1}{2}(S'/k' - 1) = S'/k' - (S'/k' - 1)^{e^{-k't}}$$
$$\tfrac{1}{2} = e^{k't}$$
$$t = \ln 2/k'$$

The solution, i.e., ln $2/k'$, is of course, the familiar half-life definition. Thus the time taken to increase to one-half of the final increase at the steady state is equal to the half-life of the enzyme. In this method the rate constant of degradation (k) can also be obtained from the slope of the straight line obtained from a plot on semilogarithmic paper of $(E_{ss'}) - (E_t)$ where $E_{ss'}$ is the new (higher) steady state level, and E_t is the enzyme activity at any time, t, during the increase of activity resulting from treatment (Segal and Kim, 1963; Szepsi and Freedland, 1969). It is possible, then, to estimate experimentally the half-life of any enzyme by following its time course of increase to a new steady state level. This method assumes that the rate of synthesis of the given enzyme is increased rapidly to some constant rate, and is maintained subsequently at that level during the entire experimental period. It carries the further assumption that the increase in enzyme activity does, in fact, represent *de novo* enzyme synthesis.

Figure 4 illustrates the theoretical determination of the half-life from

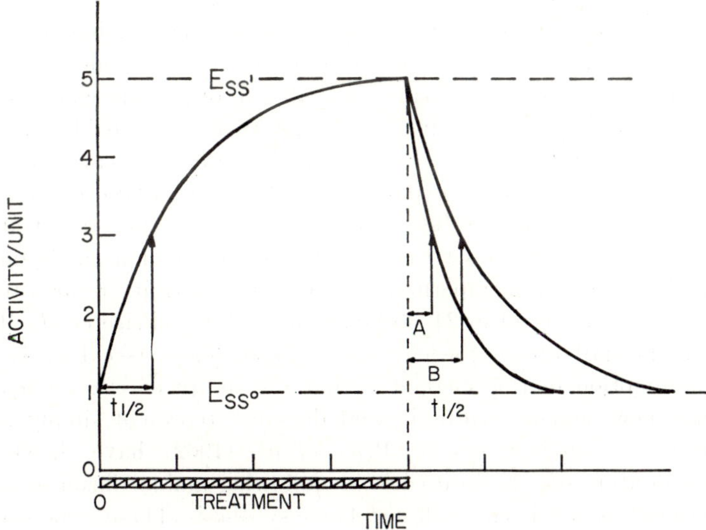

FIG. 4. Theoretical time course for changes in steady-state enzyme level during and after treatment.

the time course of change of enzyme level by both the decay and approach to steady-state techniques. This type of analysis may be of use in determining whether an observed change in enzyme level represents an effect largely due to increased enzyme synthesis, or a result of decreased enzyme degradation. Thus, if the half time of increase from one steady state, $E_{ss,0}$ to $E_{ss'}$ (Fig. 4), is the same under treatment as it is following the cessation of such treatment (representative of decay curve *B*) then the treatment has not altered the rate of degradation, and hence the treatment has increased the rate of synthesis only. If, on the other hand (case *A*), the half-life following cessation of treatment (presumed to be representative of the degradation rate in the basal state) is less than that during treatment, we can conclude that the treatment in part retarded the rate of enzyme breakdown. Such an analysis has been applied to cortisone induction of glutamate-alanine transaminase (Segal and Kim, 1963), phenobarbital induction of drug side-chain oxidation activity (Arias and De Leon, 1967), and dietary induction of alanine and ornithine transaminase (Swick *et al.*, 1968). In all cases the major effect was on enzyme synthesis.

4. *Time Course of Return of Enzyme Activity to the Steady State after Irreversible Inhibition*

The estimation of the rate of degradation (k) is similar to that described above for the time course of change from a lower to a higher steady state (Fig. 4). This method has considerable potential, but has been limited because of availability of suitable irreversible inhibitors of enzymes whose activities are not required for continued life. This method has been used to measure the half-life of plasma cholinesterase (Grob *et al.*, 1947; Neitlich, 1966) and brain monoamine oxidase activity (Barondes, 1966). It has also been applied to rat liver catalase (Price *et al.*, 1962) and has been exploited by Rechcigl (1968) in numerous studies on factors controlling levels of catalase in mice and rats. Catalase can be irreversibly inhibited by the administration of a single dose of 3-amino-1,2,4-triazole intraperitoneally. Catalase activity returns to half of the steady-state level in 24–30 hours in rat liver (Price *et al.*, 1962); therefore the half-life of the enzyme is 24–30 hours (see Fig. 4). The major assumption of this method is that the return of enzyme activity represents new enzyme synthesis, and does not represent simply return of activity of existing enzyme. Price *et al.* (1962) have shown that the incorporation of ^{59}Fe into catalase following aminotriazole administration to rats is consistent with new heme synthesis. These same workers have also shown that the repeated administration of allylisopropylacetamide to rats results in a progressive loss of catalase activity with a half-life of 24–30 hours, and have concluded that this drug inhibits the synthesis of catalase. Consistent with this result is the finding of Schmid *et al.* (1955) that the incorporation of glycine-^{14}C into the heme portion of catalase is inhibited by allylisopropylacetamide. These results can be criticized as relating only to the heme portion of catalase. More recently Ganschow and Schimke (1969) have obtained evidence with labeling of the protein portion of catalase that is consistent with this rapid rate of degradation.

No single method for determining turnover rates is completely free of assumptions and/or limitations, nor is any single method applicable to all proteins. The most reasonable course is to be aware of such assumptions and limitations and whenever possible to use more than one method for determining degradation rates. In a few cases different techniques have been applied to the measurement of degradation rates of the same protein. The rate of degradation of tyrosine transaminase of rat liver has been measured by decay of enzyme activity after prior induction with glucocorticoid (Lin and Knox, 1956; Berlin and Schimke, 1965; Grossman and Mavrides, 1967), by decay of enzyme activity under experimental conditions in which enzyme synthesis has been prevented by administration of growth hormone (Kenney, 1967), and by determining the decay of pulse-labeled protein precipitable by a specific antibody under conditions in which a basal, steady-state enzyme level exists (Kenney, 1962, 1967). By all methods, the half-life of the enzyme is 1.5–2 hours. The rate of degradation of tryptophan pyrrolase of rat liver has been estimated by the decay of induced enzyme activity (Feigelson *et al.*, 1959; Goldstein *et al.*, 1962; Nemeth, 1962; Schimke *et al.*, 1964) and by the decay of pulse labeled protein precipitated by a specific antibody under basal, steady-state conditions (Schimke *et al.*, 1965a). The half-life of this enzyme is about 2–4 hours. The rate of turnover of rat liver arginase has been measured by both continuous isotope and single-isotope administration methods, with resultant half-life measurements of 4–5 days (Schimke, 1964b).

III. Heterogeneity of Degradation of Specific Proteins

The question of the heterogeneity of rates of degradation of cells, tissues, cell organelles, and individual proteins, etc., is important in developing concepts of the mechanisms and regulation of specific protein degradation. Thus if all constituents of a cell or organelle were degraded at the same rate, one might consider quite different mechanisms for such degradation than, say, for a cell compartment in which the individual proteins were degraded at extremely varied rates. In the one case, we would most likely consider an all or none mechanism for degradation, whereas in the latter case, we would have to consider mechanisms which recognize different susceptibilities of individual proteins to degradation. The subject of heterogeneity of turnover in various tissues or organs of a given species has been discussed admirably by Neuberger and Richards (Chapter 7, see Vol. I). We shall, therefore, limit this discussion largely to intracellular proteins of liver and muscle.

A. Subcellular Organelles of Rat Liver

Table I presents data on estimated mean half-lives of cell fractions of rat liver as isolated by conventional techniques. As discussed previously, the data using guanidinoarginine-^{14}C should be considered to be more correct. The mitochondrial and lysosomal fractions have the greatest mean half-life, whereas the membrane fractions, including both the smooth and rough endoplasmic reticulum and the plasma membrane have very short half-lives of 1.8–2.3 days. In the case of each of these fractions, as well as that of the soluble proteins, heterogeneity of degradation rates exists. Thus Hirsch and Hiatt (1966) have shown that the ribosomal protein component of the rough endoplasmic reticulum has a mean half-life of approximately 5 days, whereas that of the proteins of the rough endoplasmic reticulum itself is 2–3 days. The proteins of the nucleus turn over at markedly different rates as studied in the nucleus of mouse liver. Thus Piha *et al.* (1966) have found that the half-life of a nuclear histone fraction was approximately 18 days, whereas NaCl-soluble and HCl-insoluble fractions were markedly heterogeneous, with mean half-lives varying from 1 to 15 days. Swick *et al.* (1968) and Druyan *et al.* (1970) have found heterogeneity in turnover rates of mitochondrial proteins. Of particular interest, because of its rapid rate of turnover, is the question of heterogeneity of turnover of proteins associated with the endoplasmic reticulum fraction of rat liver. Omura *et al.* (1967) have found that the half-life of cytochrome b_5 is 120 hours, whereas that of NADPH-cytochrome *c* reductase is 80 hours. These experiments were based on decay of single administrations of L-leucine-1-^{14}C. Schimke *et al.* (1968) using the double-isotope method described above have found a marked heterogeneity of degradation rates of the constituent proteins of the endoplasmic reticulum and have confirmed the results of Omura *et al.* (1967) showing differences in degradation rates of cytochrome b_5 and NADPH-cytochrome *c* reductase.

B. Specific Enzymes of Rat Liver

The best evidence for heterogeneity of turnover rates of individual proteins in a single tissue comes from studies on the turnover of identifiable enzymes of rat liver. A representative listing of the enzymes and estimated half-lives is given in Table III, together with their subcellular distribution and methods employed for their measurement. At one extreme, not indicated in Table III, is collagen, which is essentially inert in the adult animal (Neuberger and Slack, 1953). At the other extreme is δ-aminolevulinic acid synthetase, whose half-life has been estimated to be as short as 60 minutes (Marver *et al.*, 1966). Between these extremes is a remarkable heterogeneity of degradation rates. In addition there is no correlation between such rates and subcellular distribution. For instance δ-aminolevulinic acid synthetase ($t_{1/2}$ = 60–72 minutes) is associated with mitochondria ($t_{1/2}$ = 6–7 days), whereas the range of degradation rates of soluble enzymes is from 1.5 hours to 5 days, compared with a mean half-life of 3–3.5 days for total soluble proteins.

Another tissue in which the question of heterogeneity of degradation has been examined is muscle. Although the rate of degradation of total muscle may not be as rapid as that of liver, such degradation becomes important quantitatively, since the total muscle mass far exceeds that of the liver (45–50% *vs.* 4–5% of total body weight). Velick and his co-workers (Velick, 1956; Heimberg and Velick, 1954; Simpson and Velick, 1954) evaluated the turnover of a number of specific proteins in rabbit muscle by single administrations of various amino acids, including glycine, alanine, lysine, phenylalanine, and methionine. The specific proteins were isolated, and the decay of radioactivity was determined. In addition the specific activities of the various administered amino acids were compared in these proteins. These workers found that the relative specific activities of the amino acids in each fraction were constant, suggesting that the proteins were synthesized from the same amino acid pool. The estimated half-lives of the proteins were: glyceraldehyde-3-phosphate dehydrogenase, 100 days; actin, 67 days; aldolase and phosphorylase, 50 days; tropomyosin, 27 days; and L-meromyosin, 20 days. More recently Schapira *et al.* (1960) have studied the turnover of certain constituents in rat muscle. They found that the half-life of aldolase in rat muscle was 20 days, and suggested that the actual rate of degradation may be much less because of extensive reutilization of isotope. These same workers (Dreyfus *et al.*, 1960) found that the decay of glycine-^{14}C from the myofibrillar fraction of rat muscle followed life-span kinetics with a life-span of approximately 30 days when animals were maintained on a standard chow diet. However when the diet contained 24% protein, decay of isotope followed kinetics more consistent with a molecular turnover. Since isotopic glycine remained in the free amino acid pool for several weeks on the lower protein diet, it is not entirely clear to what extent the studies indicate life-span kinetics of myofibrils, or complex interactions involved in isotope reutilization. More recently Garlick (1969) has estimated the half-life of total

TABLE III
TURNOVER OF SPECIFIC CELL PROTEINS OF RAT LIVER

Protein	Intracellular site	Half-life	Methods of measurement	Reference
Acetyl CoA carboxylase	Soluble	48 hours	Isotope decay	Majerus and Kilburn (1970)
Alanine-aminotransferase	Mitochondrial	0.7–1.0 days	Decay of induced enzyme activity	Swick *et al.* (1968)
Arginase	Soluble	4–5 days	a. Continuous isotope incorporation	Schimke (1964b)
			b. Isotope (guanidino ^{14}C-arginine) decay	Schimke (1964b)
δ-Aminolevulinic acid synthetase	Mitochondrial	60–72 min	Decay following inhibition of protein synthesis	Marver *et al.* (1966)
Aspartate transcarbamylase	Soluble	2.5 days	Decay of induced enzyme activity	Bresnick *et al.* (1968)
Catalase	Peroxisomal	24–30 hours	a. Irreversible inhibitor-activity recovery	Price *et al.* (1962)
			b. Decay following inhibition of protein synthesis	Price *et al.* (1962)
Deoxythymidine kinase	Soluble	2.6–3.7 hours	Decay following inhibition of protein synthesis	Bresnick *et al.* (1967)
Cytochrome b_5	Microsomal	120 hours	Isotope (leucine-1-^{14}C) decay	Omura *et al.* (1967)
NADPH-cytochrome *c* reductase	Microsomal	80 hours	Isotope (leucine-1-^{14}C) decay	Omura *et al.* (1967)
Cytochrome P-450	Microsomal	50 hours	Decay of induced activity	Ernster and Orrhenius (1965)
Barbiturate side-chain oxidation	Microsomal	48 hours	Decay of induced enzyme activity	Arias and De Leon (1967)
Aminopyrine demethylase	Microsomal	48 hours	Decay of induced enzyme activity	Argyis and Magnus (1968)
Dihydroorotase	Soluble	12 hours	Decay of induced enzyme activity	Bresnick *et al.* (1968)
Glucokinase	Soluble	30 hours	Decay of induced enzyme activity	Niemeyer (1967)
Glucose 6-P dehydrogenase	Soluble	15 hours	Decay of induced enzyme activity	Freedland (1968)
Glutamic-alanine dehydrogenase	Soluble	84 hours	Decay of induced enzyme activity	Segal and Kim (1963)

TABLE III—(*Continued*)

Protein	Intracellular site	Half-life	Methods of measurement	Reference
α-Glycerophosphate dehydrogenase	Mitochondrial	4 days	Decay of induced enzyme activity	Tarentino *et al.* (1966)
Histidase	Soluble	2.5 days	Decay of induced enzyme activity	Schirmer and Harper (1970)
Malate dehydrogenase	Mixed	4 days	Decay of induced enzyme activity	Tarentino *et al.* (1966)
Ornithine aminotransferase	Mitochondrial	0.7–1.0 days	Time course of activity changes	Swick *et al.* (1968)
Serine dehydratase	Soluble	20 hours	Loss of prelabeled enzyme	Jost *et al.* (1968)
Thymidylate kinase	Soluble	18 hours	Decay of induced activity	Hiatt and Bojarski (1961)
Tryptophan oxygenase (pyrrolase)	Soluble	2–4 hours	a. Decay of induced enzyme activity	Feigelson *et al.* (1959) Schimke *et al.* (1964)
			b. Decay following inhibition of protein synthesis	Goldstein *et al.* (1962) Nemeth (1962)
			c. Isotope (U.L. leucine-^{14}C) decay in steady state	Schimke *et al.* (1965a)
Tyrosine transaminase	Soluble	2–3 hours	a. Decay of induced enzyme activity	Lin and Knox (1956)
			b. Decay following inhibition of protein synthesis	Kenney (1967)
			c. Isotope (U.L. leucine-^{14}C) decay	Kenney (1967)
Urocanase	Soluble	3.5 days	Decay of induced enzyme activity	Schirmer and Harper (1970)
Xanthine oxidase	Soluble	4 days	Decay of elevated enzyme activity	Rowe and Wyngaarden (1966)
D-3-Phosphoglycerate dehydrogenase	Soluble	14 hours	Decay following inhibition of protein synthesis	Fallon *et al.* (1966)
Lactate dehydrogenase isozyme-5	Soluble	16 days	Continuous isotope incorporation	Fritz *et al.* (1969)

muscle protein at 6 days, using a method involving continuous intravenous infusion of ^{14}C-glycine.

IV. Significance of Degradation in Control of Specific Protein (Enzyme) Levels

Two generalizations concerning degradation of proteins in mammalian tissues have been established thus far: (1) the level of any given protein represents a steady state, determined both by the rate of enzyme synthesis and the rate of enzyme degradation; (2) there is a marked heterogeneity of degradation rates of individual enzymes. In this section we shall examine certain aspects of these two principles which are important in understanding the dynamics of control of specific enzyme levels.

A. *Effects of Degradation Rates on the Time Course of Enzyme Induction*

The importance of degradation rates on the time course of enzyme induction is shown theoretically in Fig. 5, where the rate of synthesis

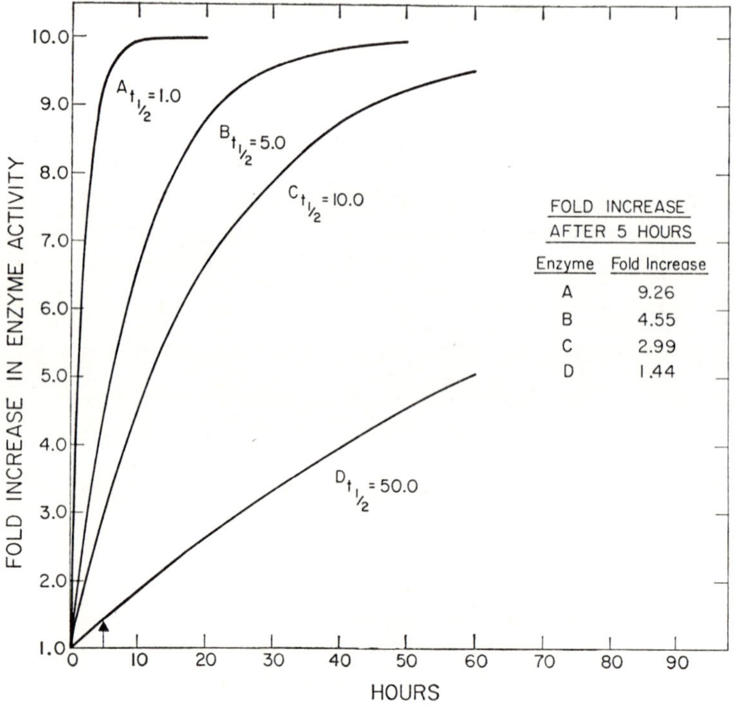

FIG. 5. Effect of differences in degradation rates on the response of an enzyme to a tenfold increase in the rate of synthesis. Degradation rates are expressed in terms of half-lives. From Berlin and Schimke (1965).

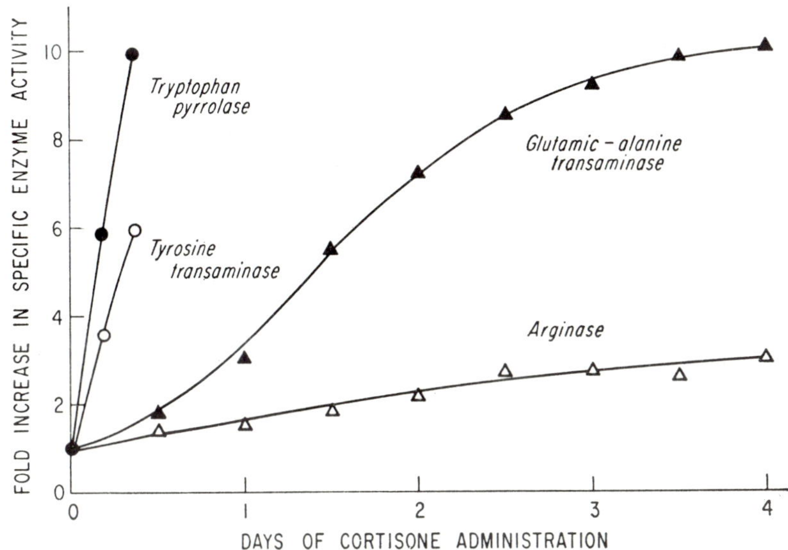

Fig. 6. Time course of the increase in tryptophan pyrrolase, tyrosine-glutamic transaminase, glutamic-alanine transaminase, and arginase with cortisone administration. Male, adult rats received 10 mg of cortisone acetate every 8 hours intramuscularly. Each value is the mean of 3 animals. Details are given in Berlin and Schimke (1965).

of four enzymes has been increased tenfold without altering their rates of degradation. The rate at which these enzymes approach a new steady state 10 times that of the original level depends on the rate of enzyme degradation. If enzyme assays were performed a short time after change in synthetic rate, the incorrect conclusion would be made that the effect was relatively specific for enzyme A (see inset of Fig. 5). This theoretical formulation has been applied to the effect of cortisone on the time course of reponse of tryptophan pyrrolase, tyrosine transaminase, glutamic-alanine transaminase, and arginase of rat liver (Fig. 6) (Berlin and Schimke, 1965). Cortisone specifically increases levels of tryptophan pyrrolase and tyrosine transaminase when one considers only the first 4–8 hours of administration. However when cortisone administrations are continued for longer periods, glutamic-alanine transaminase increased tenfold, and arginase levels increased threefold over control levels by day 4. When these time courses are considered in relation to the relative rates of turnover of the four enzymes, the results are consistent with the concept that cortisone has increased the rate of synthesis of all four enzymes approximately fourfold. Thus Table IV presents the calculated rates of synthesis of the four enzymes under basal conditions as

TABLE IV
COMPARISON OF RATES OF ENZYME SYNTHESIS UNDER BASAL CONDITIONS AND DURING CORTISONE TREATMENT[a]

Enzyme	Basal		Enzyme synthesized		
	Half-life (hours)	Enzyme activity (units)	Normal (units/hour)	Cortisone (units/hour)	Ratio cortisone: normal
Tryptophan pyrrolase	2.5	3.0	84	3.4	4.0
Tyrosine-glutamic transaminase	2.0	78	27	114	4.2
Glutamic-alanine transaminase	84	252	2.0	14.4	7.2
Arginase	96	13,800	138	534	3.9

[a] From Berlin and Schimke (1965).

estimated from Eq. (3) in which E, the steady-state level of enzyme, and k, the rate constant of degradation are known (k is derived from the half-life by the expression: $t_{1/2} = \ln 2/k$). The rate of synthesis of each enzyme during cortisone administration was estimated from the initial linear portion of the time course of increase in enzyme activity as obtained experimentally from Fig. 6. Such estimations are based on the assumptions that the initial increase in enzyme activity represents primarily the increased rate of synthesis rather than an effect produced by enzyme degradation. It can be seen that the ratios for the rate of enzyme synthesis during cortisone treatment to that under basal conditions were very similar for all four enzymes, varying from 4.0 for tryptophan pyrrolase to 7.0 for glutamic-alanine transaminase. Thus in spite of marked differences in the time course and apparent magnitude of response to cortisone, the extent of stimulation of enzyme synthesis is similar for all four enzymes. Furthermore the validity of the assumptions made in estimating the rates of enzyme synthesis from the time course is substantiated by the similarities of the calculated rates in Table IV to those obtained by isotope amino acid incorporation studies which indicate that glucocorticoids increase the rate of enzyme synthesis five- to sixfold for tryptophan pyrrolase (Schimke *et al.*, 1965a), four- to fivefold for tyrosine transaminase (Kenney, 1962), and fourfold for glutamic-alanine transaminase (Segal and Kim, 1963).

The effect of differing rates of turnover in controlling enzyme levels is portrayed graphically in Fig. 7, in which we have two enzymes, A and B, which are both synthesized at the same rate, i.e., 1 unit/hour per gram tissue. If the half-life of A is 0.693 hour, then at the steady state the level will be 1 unit. If the half-life of B is 69.3 hours, then at the steady state the level of B will be 100 units. Thus we can see that the level of an enzyme need not be a reflection of its rate of synthesis. Now if the rates of synthesis of both A and B are increased tenfold, then at the end of 1 hour A will have increased from 1 unit to 6.7 units, i.e., 6.7-fold. In contrast, the same increase in rate of synthesis will increase the level of B within the same hour only to 109 units. From this example and the experimental results of Fig. 6, it is evident that claims of specificity for a response to an increased rate of synthesis may be totally invalid unless the turnover rate of the specific enzyme in question is also considered.

On the basis of the above analysis, it might be suggested that in mammalian tissues the concepts of "constitutive" and "adaptive" enzymes, as applied to bacteria, may indicate primarily the relative rates of degradation of various enzymes. We might therefore suggest that enzymes whose levels are rate-limiting for a specific biochemical reaction *in vivo* will have a rapid rate of degradation. Such enzymes would re-

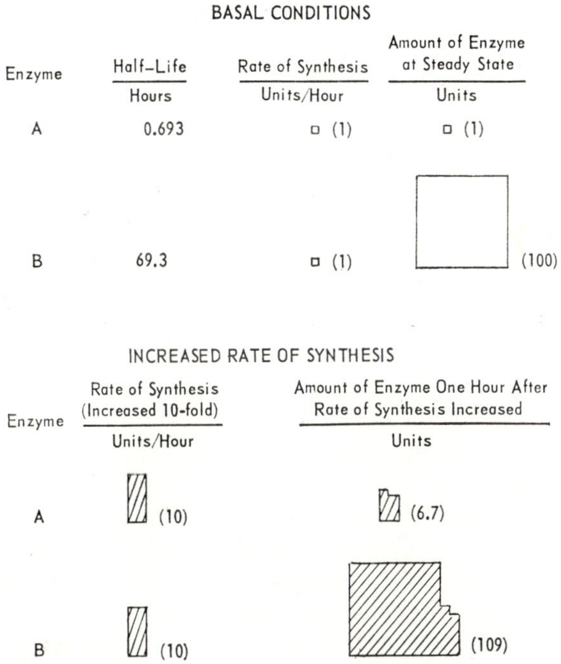

BASAL CONDITIONS

Enzyme	Half-Life (Hours)	Rate of Synthesis (Units/Hour)	Amount of Enzyme at Steady State (Units)
A	0.693	□ (1)	□ (1)
B	69.3	□ (1)	(100)

INCREASED RATE OF SYNTHESIS

Enzyme	Rate of Synthesis (Increased 10-fold) (Units/Hour)	Amount of Enzyme One Hour After Rate of Synthesis Increased (Units)
A	(10)	(6.7)
B	(10)	(109)

FIG. 7. Role of rate of degradation in determining the response of two hypothetical enzymes to the same increased rate of synthesis. From Berlin and Schimke (1965).

spond rapidly, both increasing and decreasing, in response to environmental alterations such as changes in diet, substrate, or hormonal levels. On the other hand, enzymes whose activity *in vivo* is controlled by feedback inhibition or by substrate availability, would not be required to have the ability to fluctuate rapidly, and hence would be those enzymes with slower degradation rates.

B. Effects of Altered Rates of Degradation on Specific Protein Levels

In this section we shall discuss certain examples in which altered rates of degradation or inactivation have been shown to result in a net accumulation of enzyme (induction) or a net decrease in enzyme (repression).

1. Tryptophan Oxygenase (Pyrrolase)

Perhaps the most studied example of an enzyme whose rate of degradation or inactivation can be altered is rat liver tryptophan pyrrolase. The levels of tryptophan pyrrolase can be increased by the administration of a glucocorticoid hormone or L-tryptophan or certain tryptophan analogs (Knox and Mehler, 1951; Knox and Auerbach, 1955). The increase in activity produced by both agents depends on *de novo* enzyme synthesis, since both increases are inhibited by inhibitors of protein synthesis (Greengard *et al.*, 1963). Nevertheless, mechanism (s) whereby the hormone and substrate result in accumulation of enzyme have been recognized as being different (Civen and Knox, 1960; Knox and Auerbach, 1955). Greengard and Feigelson (1961a,b) showed that tryptophan pyrrolase contains a readily dissociable prosthetic group, hematin, and that in the liver, the enzyme exists as varying proportions of apo- and holoenzyme. These workers showed that the administration of glucocorticoid increases both apo- and holoenzyme. L-Tryptophan administration, in contrast, resulted first in the conversion of apoenzyme to holoenzyme, followed by a net accumulation of total enzyme. The concept that L-tryptophan may alter, not the rate of synthesis, but the rate of enzyme degradation, was originally suggested by Lee and Williams (1952) on the basis of the time course of accumulation of tryptophan pyrrolase after administration of L-tryptophan. This concept has been supported subsequently by Schimke *et al.* (1964, 1965a,b). An analysis of the time course of increase in the enzyme following repeated administration of either hydrocortisone or L-tryptophan fit the theoretical model (Section III) in which hydrocortisone increased the rate of enzyme synthesis about four- to fivefold without altering the rate of enzyme degradation, whereas L-tryptophan administration did not affect enzyme synthesis, but decreased the rate of enzyme degradation (Schimke *et al.*, 1964).

The conclusions were further supported by studies utilizing combinations of isotopic and immunological techniques which indicated that, in control animals, tryptophan pyrrolase prelabeled with leucine-^{14}C in a single administration experiment decayed rapidly with a half-life of 3 hours whereas when L-tryptophan was administered to the animals, the prelabeled enzyme did not decay (Schimke *et al.*, 1965a). In other types of experiments L-tryptophan, as well as an entire series of tryptophan analogs including α-methyltryptophan, D-tryptophan, and *N*-acetyltryptophan, prevented the rapid decay of activity which normally occurs after tryptophan pyrrolase levels are increased to high levels by hydrocortisone (Schimke *et al.*, 1965b).

Schimke *et al.* (1965b) showed that substrate and substrate analogs protected the purified enzyme from denaturation by heat, organic solvents, urea, and proteolytic agents. They concluded that the holoenzyme was the most stable form. This is the form of the enzyme which accumulates *in vivo* after the administration of L-tryptophan (Greengard and Feigelson, 1961b). The stabilization of enzyme by substrate had been found previously by Dubnoff and Dimick (1959) and had been used by Feigelson and Greengard (1962) in their purification of this enzyme. More recently, Knox *et al.* (1966) as well as Greengard *et al.* (1966) have shown that previous studies with tryptophan pyrrolase were performed with an assay in which enzyme was not activated maximally. Knox and Piras (1967) have identified three forms of the enzyme, i.e., apoenzyme, holoenzyme, and activated holoenzyme. They have presented evidence that the reaction involving the conjugation of hematin with apoenzyme requires the presence of L-tryptophan, or certain tryptophan analogs such as α-methyltryptophan. They suggested that the tryptophan stabilization results from conversion of apo- to holoenzyme. These workers have confirmed the findings of Schimke *et al.* (1965a) that administration of L-tryptophan *in vivo* stabilizes the newly formed enzyme against subsequent degradation. They also suggested that L-tryptophan might have an additional effect in producing an increased accumulation of enzyme which results from an increased rate of enzyme synthesis.

2. Thymidylate Kinase

Thymidylate kinase activity is high in rapidly growing, and low in resting mammalian tissues. Hiatt and Bojarski (1961) and Bojarski and Hiatt (1960) found that when thymidine was administered to rats by intraperitoneal infusion, the content of thymidylate kinase increased seven- to eightfold along a linear time course for 100 hours. These workers showed also that the addition of thymidine to extracts of liver or kidney prevents the normally occurring inactivation of thymidylate kinase. These findings suggested to them that the thymidine infusion into the intact animal results in the accumulation of enzyme by retarding breakdown in the presence of continued enzyme synthesis. A similar phenomenon has been observed with cells in continuous culture. Kit *et al.* (1965), as well as Littlefield (1965), have observed that the decline of thymidine kinase activity which occurs when mouse fibroblast cells enter stationary phase can be abolished by the addition of the substrate, thymidine, to the medium. This suggests that the enzyme is continually synthesized, but in the absence of its substrate is rapidly inactivated.

3. Ferritin

Administration of iron to rats results in a marked increase in liver ferritin as a result of *de novo* synthesis (Loftfield and Harris, 1956). Drysdale and Munro (1966) have recently examined in more detail the mechanism of this accumulation. They found that the newly synthesized ferritin contains a low iron:protein ratio. Such new ferritin is unstable *in vivo* and has a half-life of approximately 18 hours. With repeated administrations of iron to animals, there was an increase in the iron:ferritin ratio, and a marked decrease in the degradation of prelabeled ferritin. The molecules of ferritin that are not degraded are those with a high iron:protein content, suggesting to these authors that the stabilized form of ferritin was that saturated with iron. They further suggested that the induction effect of iron may be solely one of stabilization of an extremely unstable precursor that is rapidly degraded in the absence of the stabilizing effect of iron.

4. Arginase

Schimke (1964b) has analyzed the roles of synthesis and degradation in controlling the steady-state levels of rat liver arginase under conditions where animals are maintained on diets containing 8% or 70% protein. He found that under steady-state conditions, where the levels of arginase were different by approximately two- to threefold, the rate of enzyme degradation was the same. Thus the factor that controls the steady-state level is the rate of enzyme synthesis as affected by the diet. However, when the dietary conditions of the animal were changed acutely, alterations in the rate of arginase degradation were observed. Thus, when animals were changed from an 8% protein diet to conditions of fasting, there was an approximately twofold increase in total enzyme in liver. By knowing the total amount of arginase, and the rate of decay of arginase labeled prior to onset of starvation with guanidinoarginine-^{14}C, he showed that the accumulation of enzyme was the result of continued enzyme synthesis in the absence of the normally occurring degradation. Conversely, when animals were changed from a diet containing 70% protein to one containing 8% protein, there was a decrease in total arginase to 40% of the level when animals were maintained on the 70% casein diet. This change resulted from both a decreased rate of enzyme synthesis and an increased rate of enzyme degradation. Thus, during steady-state conditions, the major factor determining the diet-induced differences in arginase levels is a difference in rate of enzyme synthesis. However, during acute changes in the metabolic milieu, changes in rates of degradation can occur, either increasing or decreasing, depending on the specific conditions involved.

5. Ribosomal Protein

The turnover of liver ribosomes can be modified by the dietary status of the rats, as shown by Hirsch and Hiatt (1966). These workers have shown that there is a marked decrease in the number of ribosomes during starvation. This effect results from both a decreased rate of ribosomal protein synthesis and an increased rate of degradation. They have also found that the ribosomal RNA follows the same pattern, and have concluded that the ribosome is degraded as a unit.

6. Muscle Protein

Goldberg (1969) has studied the effect of increased functional use and administration of growth hormone, which increase muscle mass, and of cortisone administration and denervation, which decrease muscle mass, on overall rates of synthesis and degradation of protein in rat soleus and gastrocnemius muscles. He finds that either rate can be controlled separately, the sum total determining muscle mass. In these experiments he determined the rate of loss of muscle prelabeled with ^{3}H-leucine under various experimental conditions. Work hypertrophy results from both an increased rate of muscle protein synthesis, and a decreased rate of degradation. The muscle growth resulting from growth hormone, in contrast, results only from increased synthesis, the rate of degradation being unchanged from control rats. Both denervation and cortisol administration result in decreased synthesis and increased degradation. These experiments show that the phenomenon of alterations in rates of degradation of proteins is not limited to liver.

7. Acetyl CoA Carboxylase

The levels of rat liver acetyl CoA carboxylase vary over a 25-fold range between animals maintained on a fat-free diet and those that are fasted for several days (Numa *et al.*, 1961). Majerus and Kilburn (1970) have recently examined this problem using combined immunologic and isotopic techniques. They found the rate of degradation of the enzyme on a fat-free diet had a half-life of 48 hours as measured by the loss of pulse-labeled isotope from immunologically isolated enzyme. When such animals were starved, degradation of the enzyme is accelerated such that the half-life was decreased to 18 hours. The rate of specific enzyme synthesis is approximately 5- to 10-fold greater in animals fed a fat-free diet than in fasted rats. However, there appears to be no relationship between the rate of synthesis and the rate of degradation, since in animals fed a diet containing 12% fat, the rate of enzyme synthesis is markedly diminished without affecting the rate of degradation ($t_{1/2}$ = 48 hours).

8. Lactate Dehydrogenase Isozymes (LDH-5)

One important question concerning heterogeneity of degradation rate constants of proteins is whether the same protein is degraded at the same rate in all tissues. This has been answered elegantly by Fritz *et al.* (1969) for the LDH-5 isozyme. These workers determined the rates of synthesis and degradation of the isozyme by the continuous dietary administration of ^{14}C amino acids and subsequent isolation of LDH-5 immunologically. They found that the half-lives for this isozyme in liver, heart muscle, and skeletal muscle, are 16, 1.6, and 31 days, respectively. Thus, the rate of degradation of a given enzyme is a function not only of the protein, but the tissue in which it resides.

9. Enzymes in Cell Culture Systems

Several examples of the potential importance of enzyme degradation or inactivation in determining steady-state enzyme levels can be found in cell culture systems. Cox and MacLeod (1964) found that fresh medium contained a substance that resulted in low steady-state levels of alkaline phosphatase activity in human cell lines. This substance, which was not present in medium of cells after several days of growth, was subsequently identified by them as cysteine. Its effect was ascribed to an alteration in enzyme synthesis, i.e., repression. However, De Mars (1964) examined the effect of cysteine under conditions in which no new enzyme synthesis could occur because several essential amino acids were omitted from the medium. Thus, if cysteine prevented the synthesis of new enzyme, but had no effect on enzyme inactivation, the amount of enzyme already existing would be unaffected by cysteine addition. De Mars found that cysteine increased the rate of inactivation of alkaline phosphatase. This finding suggests that the "repression" effect diminishes steady-state levels by increasing the rate of enzyme inactivation. Another similar example involves the effect of glutamine on glutamyl transferase activity of HeLa cells. De Mars (1958) first showed that substitution of glutamine for glutamic acid in the growth medium results in low, steady-state levels of glutamyl transferase in HeLa cells. He further examined the effect of glutamine on the kinetics of decay of glutamyl transferase following substitution of glutamine for glutamic acid. In an experiment similar to that outlined above for alkaline phosphatase, he showed that glutamine accelerated enzyme inactivation (De Mars, 1964). Paul and Fottrell (1963), using a somewhat different technique, also concluded that the effect of glutamine was not to affect enzyme synthesis, but rather, to accelerate enzyme inactivation. In these experiments it was not possible to determine whether the inactivation also involved actual degradation of the enzyme protein.

The addition of small molecules or ions to the culture medium can result in the accumulation of specific enzyme protein in which the mechanism seems to involve a stabilization phenomenon. Thus the addition of manganese ion to the culture medium increases the activity of arginase in HeLa (Schimke, 1964a) and Chang's liver (Eliasson and Strecker, 1966) cell lines, respectively, an effect ascribed to stabilization by Schimke (1964a). A stabilization phenomenon has also been proposed as one mechanism for the increase in folate reductase observed in cell culture (Hakala and Suolinna, 1966), and in human leukemic cells (Bertino *et al.*, 1965), following the administration of amethopterin, a folic acid analog that has a high affinity for the enzyme (Zakrzewski *et al.*, 1966).

10. Genetic Factors Influencing Protein Degradation

The availability and ready manipulation of mutants has been critical to the understanding and development of concepts of regulatory mechanisms in micoorganisms. We can expect that the study of mutational events may be important also to understanding the regulation of protein degradation. It is therefore of interest to review several mutational events in mice that affect protein degradation.

a. Hepatic Catalase of Inbred Mouse Strains. Rechcigl and Heston (1963, 1967) described two closely related strains of C57BL mice in which a mutational event resulted in an alteration in catalase degradation in one strain. Ganschow and Schimke (1969) studying this same mutational event have identified, in addition, a mutation affecting the catalytic activity of catalase. A summary of these results, and a proposed genetic model for the control of liver catalase activity is shown in Table V. The DBA/2 enzyme level is typical of most mouse strains studied

TABLE V
GENETIC CONTROL OF LIVER CATALASE ACTIVITY IN INBRED MICE[a]

Strain	Units per gram liver[b]	Specific activity[c]	Relative content of protein[d]	Relative rate of synthesis[e]	Relative rate of degradation[f]	Genetic model	
						Activity	Turnover
DBA-2	140 ± 3	4.3	A	X	Y	C_s+C_s+	CECE
C57BL/6	85 ± 2	2.7	A	X	Y	C_s57C_s57	CECE
C57BL/Ha	165 ± 3	2.7	2A	X	Y/2	C_s57C_s57	cece

[a] From Schimke *et al.* (1968).

[b] Micromoles of H_2O_2 per initial minute per gram wet weight of liver ± standard error.

[c] $M^{-1}sec^{-1} \times 10^{-7}$ enzyme concentration based on protein content or Soret adsorption of enzyme which is at least 99% homogeneous (Price *et al.*, 1962).

[d] Immunological titration of liver extracts.

[e] Uptake of ^{14}C-labeled amino acid after pulse of 4 hours.

[f] 3-Amino-1,2,4-triazole technique (Price *et al.*, 1962).

(Swiss-Webster, C_3H) and is considered to be representative of the wild type. The C57BL/6 strain has approximately 60% of the activity of DBA/2, and in the closely related C57RL/Ha strain, activity is slightly more than DBA/2. The specific activity of catalase purified from *both* C57BL strains is only 60% of that of the catalase purified from DBA/2 despite the fact that other parameters of the enzyme, including sedimentation constant, electrophoretic mobility, heat stability, and hematin content, are similar. As measured by immunological techniques, the livers of DBA/2 and the C57BL/6 strains have similar numbers of catalase molecules. Thus, the differences between these two strains can be accounted for by differences in catalytic activity. The C57BL/Ha strain has twice as many catalase molecules as the other two strains. Thus, although DBA/2 and C57BL/Ha have quite similar enzyme activities, the reasons for the apparent similarity are deceptive. By combinations of isotopic techniques, and the use of the aminotriazole technique for measuring catalase degradation, it was found that the rate of catalase turnover in C57BL/Ha is one-half that of catalase in the other two strains. To explain these results, as well as liver catalase activities of progeny from interstrain crosses, the following genetic model was proposed. Two codominant alleles (Cs^+ and Cs^{57}) of a genetic locus (*Cs*) determining the structure of catalase are responsible for the difference in specific activity of purified liver catalase between DBA/2 and C57BL animals. The model further states that C57BL/Ha carries a recessive allele (*ce*) at a locus (*CE*) affecting liver catalase degradation. The validity of the model is supported by the fact that a low frequency of individuals of the F_2 generation from the parental cross DBA/2 $\times$ C57BL/Ha have levels of liver catalase activity which are twice that of DBA/2. This phenotype corresponds to that predicted for the genotype Cs^+Cs^+cece. If the mutational event controlling catalase degradation were involved in the turnover of the entire peroxisome within which catalase is localized, then one would predict that other peroxisomal enzymes would show the same variation. However, Ganschow and Schimke found that liver activities of another peroxisomal enzyme, urate oxidase, are similar among the three inbred strains.

b. Hereditary Muscular Dystrophy of Mice. Simon *et al.* (1962) have shown that the rate of turnover of muscle protein of mice with genetically determined muscular dystrophy is more rapid than that of normal mice. Tappel *et al.* (1962) have studied the content of lysosomal enzymes in such animals and find that there is approximately a two-fold greater content in the dystrophic mice. The nature of this defect has not been elucidated, but may also give a clue to the signal for what determines the rate of turnover of muscle proteins.

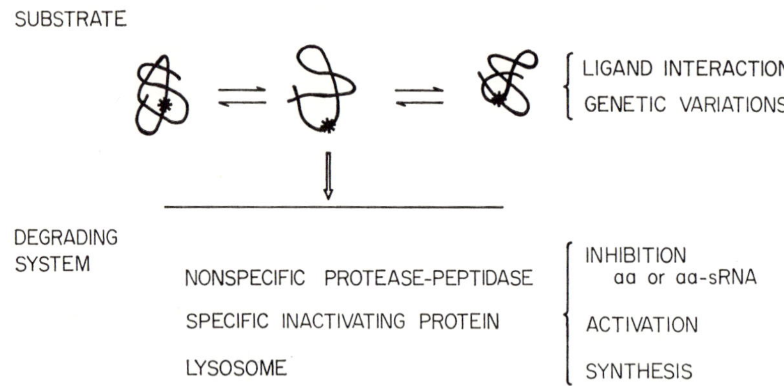

FIG. 8. Summary of possible means of regulating protein degradation.

V. Mechanisms for the Control of Protein Degradation[4]

Three general findings of importance for an understanding of control of protein degradation can be summarized from previous sections: (1) Protein degradation is extensive, involves essentially all proteins of a given tissue (liver), and is largely intracellular. (2) The rates at which individual proteins and cellular organelles are replaced vary markedly. (3) The rate of degradation of an individual protein can be altered.

It is clear from these studies that the degradation of specific proteins can be under various types of regulation.

We may ask the question of the regulation of degradation of a specific protein as follows: *What determines when a given protein molecule will undergo degradation?* Such a question can be answered in terms of whether this process is determined by the protein as a substrate for a degradative enzyme or whether the control resides in the degradative system(s). Figure 8 summarizes several possible mechanisms for control of protein degradation.

[4] It should be stated explicitly that a major problem exists in the operational use of the term "degradation." Thus, it is clear that total protein of liver is replaced from a dietary source, and hence proteins must be broken down, presumably to amino acids. However, when dealing with individual proteins, one cannot differentiate between degradation and inactivation. Thus, it is impossible to determine what has happened to a given protein species once it has lost enzymic activity or immunological reactivity. Such a process, which we operationally call "degradation," may in fact result only from a change in the conformation or aggregational state of the protein or its residence in a particular cell compartment or organelle, rather than a splitting of peptide bonds.

A. Alterations in Proteins as Substrates for Degradation

Perhaps the simplest concept that would explain when a given molecule will be subject to degradation would be that of an aging process, whereby because of use (or disuse) some change occurs in the structure of the molecule, and hence allows recognition for a degradative process. The red blood cell represents an example of "aging." Although it is not clear exactly why red blood cells have a specified life-span, various theories have suggested that the progressive decrease of certain enzyme activities, most notably glucose-6-phosphate dehydrogenase, may be critical (Beutler, 1966). Thus the loss of ability to produce glucose 6-phosphate, which in turn is necessary to maintain levels of reduced glutathione, may result in changes in internal proteins or surface sulfhydryl groups such that the reticuloendothelial cells can recognize an "aged" or otherwise abnormal cell (Beutler, 1966). In support of such a concept is the finding that genetic defects of glucose-6-phosphate dehydrogenase activity are associated with decreased red cell life-spans, particularly under conditions where the red cell requirements for reductive capacity have been increased by the administration of the antimalarial, primaquine (Beutler, 1966). Red cells which are subject to increased rates of degradation also result from genetic variations in hemoglobin structure, such as occurs with sickle-cell hemoglobinopathy. In this condition there is an abnormal hemoglobin which crystallizes at low oxygen tensions with the concomitant formation of deformed (sickled) cells which are more readily degraded (Lehman *et al.*, 1966).

The "aging" concept, however, does not appear applicable generally to the control of protein degradation. Thus virtually all studies using isotopic methods and changes in enzyme activities have found that kinetics of degradation are first order, both for individual enzymes and intracellular organelles (see Sections II and III). Although other interpretations are possible (see Section II), such kinetics most likely imply that once a molecule of a given protein species is synthesized, its chance of degradation is random. To explain this random process, then, we may conceive of a given protein molecule as existing as an equilibrium of a number of conformational states (see Fig. 8). Only in a certain state(s) or conformations would it be subject to degradation by a degrading system (proteases and peptidases). Such conformations could be altered by interactions with small molecules, similar or dissimilar peptides and subunits, or by associations with various intracellular organelles.

The concept that the rate of degradation of a protein is determined by its properties as a substrate for a nonspecific degradative system is supported by studies with various enzymes and cell organelles. These studies indicate that the rate of degradation (expressed as a first-order rate constant, or half-life) of various proteins is the same under conditions which affect rates of protein synthesis, and hence steady-state protein levels. Thus the rate of degradation of arginase ($t_{1/2}$ = 4–5 days) (Schimke, 1964b), xanthine oxidase ($t_{1/2}$ = 4 days) (Rowe and Wyngaarden, 1966), total mitochondrial protein ($t_{1/2}$ = 5–6 days) (Swick *et al.*, 1968), and total liver protein ($t_{1/2}$ = 3 days) (Schimke *et al.*, 1964) is the same in rats maintained on diets containing from 8 to 70–90% casein, conditions which markedly affect steady-state protein levels. Another example is that of catalase of rats (Price *et al.*, 1962). The rate of degradation of catalase is similar in both kidney and liver ($t_{1/2}$ = 24–30 hours), although the steady-state enzyme levels vary by at least threefold. Likewise under various dietary conditions, and in hepatomas with differing catalase levels, the rates of catalase degradation are similar (Rechcigl, 1968). The degradation of proteins of the endoplasmic reticulum provides another example. The half-life of such proteins is approximately 2 days under steady-state conditions (Schimke *et al.*, 1968; see also Table I). These proteins continue to be degraded at the same rate under conditions where the rate of synthesis has been increased as much as three- to fourfold (Arias and De Leon, 1967; Schimke *et al.*, 1968). These examples indicate that for certain proteins once that protein or organelle has been synthesized, its degradation proceeds at some fixed rate. However, note that the rate of ribosome degradation can vary with dietary conditions (Hirsch and Hiatt, 1966; Munro, 1968).

The general finding that rates of degradation are first order with respect to the concentration of protein (see Section III) would be consistent with the concept that the rate of degradation of a protein is a property of the protein *if* the rate-limiting step were the conversion of the protein to a form which was susceptible to degradation. The cleavage of an initial peptide bond, then, would be an irreversible step, following which the actions of proteases and peptidases could degrade the protein to free amino acids. The experimental and theoretical basis for this concept has been elaborated extensively by Linderstrøm-Lang and his colleagues (Linderstrøm-Lang, 1950) using specific proteins, such as ovalbumin and lactalbumin, and isolated proteolytic enzymes. The reader is referred to the excellent review of Green and Neurath (1954) for a discussion of these and related studies.

The exact nature of the states or conformations in which a protein could be subject to degradation is not clear, but would depend on the types of exposed peptide bonds and the specificity of the proteases involved. Multiple studies have indicated that denatured proteins are more readily degraded than are "native" proteins (Green and Neurath, 1954; Linderstrøm-Lang, 1950). Although such observations may serve as a model for considering why a protein is susceptible to degradation, there is no *a priori* reason for thinking that the initial step leading to degradation of a given enzyme *in vivo* first requires an *irreversible* heat denaturation. For this reason studies relating stabilities of purified enzymes to various inactivating agents *in vitro* may have little direct application to the process that occurs *in vivo.* For instance, arginase is extremely stable during purification (Schimke, 1964b), whereas tryptophan pyrrolase is extremely unstable in the absence of its substrate (Feigelson and Greengard, 1962), as is tyrosine transaminase in the absence of cofactor and substrate (Hayashi *et al.,* 1967). These findings may be interpreted to imply a correlation between stability *in vivo* and *in vitro.* However, this simple correlation does not fit when the effect is studied of trypsin on purified preparations of these three enzymes. Thus although tryptophan pyrrolase is unstable, and arginase stable, tyrosine transaminase activity is highly resistant to trypsin inactivation (Schimke *et al.,* 1965b), a finding which one would not predict from its instability *in vivo.* Clearly experiments under simulated conditions *in vivo,* including the complement of intracellular proteases and peptidases, will be required to study this process more definitively.

A more direct correlation between the stability of an enzyme *in vitro* and its rate of degradation *in vivo* is afforded by the example of Swiss acatalasia. Aebi (1967) and Matsubara *et al.* (1967) have presented evidence indicating that low levels of catalase in red blood cells of certain humans results from a mutation that affects the stability of catalase. These workers have demonstrated that new red blood cells contain considerable amounts of catalase activity, whereas older cells contain little catalase activity or catalase antigen. Studies with catalase purified from red blood cells of such patients reveals that the enzyme has a different electrophoretic mobility and is markedly less stable to heat than is catalase purified from humans with normal red blood cell levels of catalase. Aebi (1967) has proposed that the acatalasia results from a mutation in the catalase structure, and that the activity of catalase is low in red blood cells because it is degraded (inactivated) at a rapid rate. Thus catalase content of newly synthesized cells approaches that of normal humans. However the unstable catalase undergoes degradation during the life-span of the red blood cell, and since no protein synthesis occurs in mature erythrocytes, catalase activity is lost rapidly. It is quite likely that other examples exist in which genetically determined differences in levels of specific enzymes in various tissues may result from structural mutations affecting degradation rates. Perhaps the difference in catalase degradation in inbred mouse strains discussed above is such a mutation.

The rate of degradation of a protein can be altered by various agents and physiological conditions, as discussed in Section IV. Such alterations in degradation rates would be consistent with differences in protein–ligand interactions which shift equilibria of conformation states. There are numerous examples of substrates and products either stabilizing or labilizing specific enzymes against a host of inactivating agents (see Grisolia, 1964) which could be considered as models for such phenomena *in vivo*. Among specific examples pertinent to this review is the effect of tryptophan on the stability of tryptophan pyrrolase to trypsin inactivation. In these studies protection by tryptophan was most effective when the enzyme was completely conjugated with hematin (Schimke *et al.*, 1965b).

Consistent with a hypothesis that interactions of a protein with various ligands alters its stability is the finding of Mazur and Shorr (1950). They have shown that purified ferritin with a high iron content is less subject to inactivation by proteolysis than is ferritin with a low iron content. This finding would fit with the hypothesis of Drysdale and Munro (1966) that ferritin accumulation following administration of iron to the rat results from a stabilization of newly formed apoferritin molecules. Markus (1965) has made the interesting observation that the rate of digestion of serum albumin by various proteases can be inhibited markedly by binding of methyl orange to albumin. In this case no gross change in protein conformation could be detected by changes in sedimentation or optical rotatory dispersion properties. Thus the changes in protein–ligand interaction that alter susceptibility to degradation may be extremely subtle.

Further support of a concept that the rate of degradation of a protein is a function of the properties of the protein comes from certain studies in bacteria. One such line of evidence involves studies with thermophilic bacteria. Although such organisms contain thermostable proteins, such as an extracellular protease (Ohta, 1967; Ohta *et al.*, 1966), certain of the enzyme activities of extracts are thermolabile at temperatures at which such organisms normally grow (50–60°) (Allen, 1953). Recent studies by Bubela and Holdsworth (1966) indicate that there is a rapid turnover of proteins in growing *Bacillus steathothermophilis*. This suggests that thermolabile enzymes are labile *in vivo* and that they are replaced at a rapid rate. This system appears worthy of more study to exploit this interesting observation. Another example involves β-galactosidase of *E. coli*. In the original studies on protein degradation in *E. coli*, Mandelstam (1958) found that β-galactosidase was stable during a period of 3–4 hours during which total protein was degraded at a rate of 4–5% per hour. More recently, Willetts (1967a,b) has found that β-galactosidase is stable if magnesium is omitted from the medium, but is unstable in its presence. This is another example of a small molecule affecting the stability of a specific enzyme.

The model that emerges from these studies is one in which protein molecules (or organelles) are individually available to a degradative process, which is present in excess. Such molecules are in thermodynamic equilibrium among different states. In some state(s) they are subject to degradation, whereas in others, they are not. The effects of small molecules, then, is to shift the equilibrium to a form which is not (or is) subject to degradation. This model, then, can explain the heterogeneity of turnover of various proteins, and effects of small molecules on protein degradation. Shifting distributions and concentrations of substrates, cofactors, intermediates, products, etc., such as occur under various hormonal and physiological conditions, both in animals and microorganisms, may lead to a variety of effects on specific enzymes, either to stabilize or labilize, and thereby affect enzyme levels by altering rates of degradation. Such a concept has been expressed also by Grisolia (1964) and Pine (1966, 1967).

B. Alterations in Activity of Degradative Processes

Heretofore we have assumed that the activity of the degradative mechanism is not rate-limiting. We should also consider the situation in which the rate of degradation may be dependent on the activity of the degrading system. Thus degradation may be controlled by mechanisms such as activation and inhibition, translocation within a cell, or *de novo* synthesis of degrading enzymes.

Perhaps the clearest case in which protein degradation is controlled by the activity of a protease is during bacterial sporulation. When *Bacillus subtilis* cells are grown to the point of exhausting growth medium, spore formation commences. Mandelstam and colleagues have shown that sporulation is associated with a high rate of protein degradation (6–15% per hour), and the appearance of a specific exoprotease (1967, 1968). That this protease is involved in the degradation associated with sporulation was shown by selecting mutants which lacked the protease. Such mutants did not sporulate and did not have a rapid rate of protein degradation after exhaustion of the growth medium. Revertents which regained exoprotease, presumably as a result of a single mutation, also regained ability to sporulate and to initiate protein degradation in a manner similar to wild-type *B. subtilis*. An instructive comparison with the phenomenon described above for *B. subtilis* is the control of protein degradation in *E. coli.* Protein degradation has been found to be essentially nil during exponential growth of *E. coli* (Rotman and Spiegelman, 1954; Hogness *et al.,* 1955; Koch and Levy, 1955), whereas during nongrowing conditions, such as amino acid starvation, degradation occurs at a rate of 4–5% per hour (Mandelstam, 1958). In this case there is no evidence for an increase of proteolytic activity (Chaloupka and Liebster, 1959). Recently Pine (1965) and Willets (1967a,b) have found that even in growing *E. coli* there is a small fraction amounting to approximately 1% of total protein that is degraded at a rate of 20% per hour. During nongrowing conditions this fraction continues to be degraded at the same rate, but the number of proteins subject to degradation is increased. Pine (1966) ascribes this phenomenon to alterations of the proteins as substrates.

Another example in which a change in activity of degradative enzymes occurs is during amphibian metamorphosis. Weber (1967) has shown that the activity of acid proteases (cathepsins) of the tadpole tail increases severalfold during the time that tail resorption is taking place. The entire process of metamorphosis can be initiated by the administration of thyroxine. Tate (1966) has shown that the effects of thyroxine in increasing cathepsin activity can be prevented by the administration of antibiotics that inhibit protein or RNA synthesis in an isolated frog tail system. Such data have been interpreted as indicating that the increase in cathepsin activity requires both RNA and protein synthesis (Weber, 1967). The results, in fact, only indicate that some step between the administration of thyroxine, and the large number of physiological and biochemical events leading to tail resorption and increased cathepsin activity are sensitive to the action of these antibiotics. The question of whether *de novo* cathepsin synthesis is a *required* and *sufficient* event for the breakdown of tail constituents has not been answered. The increased breakdown of collagen following administration of various glucocorticoids to the rat is associated with increases in activities of collagenolytic and proteolytic activities in skin, increases that are inhibited by abministration of puromycin and actinomycin D (Houck *et al.,* 1968). Thus, in this case, the breakdown of proteins is associated with *de novo* synthesis of proteolytic enzymes. The appearance of collagenase during developmental sequences represents another example where a highly specialized proteolytic enzyme is elaborated for breakdown of a specific protein (Gross and Lapiere, 1962).

We have been assuming as a first approximation that the intracellular breakdown of protein is a simple matter of proteolysis, such as occurs with extracellular proteinases of microorganisms and animal tissues.

Available evidence, however, would suggest that the situation may not be as simple as this. Schimke *et al.* (1965b) have shown that the degradation of tryptophan pyrrolase as measured by the double criteria of loss of enzyme activity and immunological reactivity could be demonstrated only in structurally and metabolically intact tissues. Thus the inactivation of enzyme activity in crude homogenates was not accompanied by loss of immunological reactivity. The simplest system capable of loss of the immunological and catalytic activity involved use of liver slices. Inactivation of enzyme was prevented when liver slices were incubated under anaerobic conditions. These properties are strikingly similar to those found by Simpson (1953) and by Steinberg and Vaughn (1956), who studied the degradation of protein as measured by release of amino acids from prelabeled protein of liver slices. The degradation of intracellular protein was inhibited by agents that inhibited energy metabolism. In addition Steinberg and Vaughan (1956) found that several amino acid analogs inhibited degradation of protein in liver slices. Recently, Grossman and Mavrides (1967), Kenney (1967), and Schimke (1967) have observed that inhibition of protein synthesis by cycloheximide or puromycin prevents the normally occurring inactivation of tyrosine transaminase of rat liver. Grossman and Mavrides (1967) also observed that actinomycin D inhibited enzyme inactivation. Kenney (1967) has interpreted these findings to indicate that a specific protein that inactivates tyrosine transaminase turns over rapidly and in the absence of its synthesis tyrosine transminase is not inactivated. Other interpretations, such as an indirect effect of protein inhibition by increasing levels of peptides or free amino acids and thereby inhibiting proteolytic activity or a direct stabilization of tyrosine transaminase, have not been excluded.

A number of studies in microorganisms likewise implicate energy and protein metabolism in control of degradation. The inactivation of induced malate dehydrogenase of *Saccharomyces cerevisiae* is energy dependent and requires protein synthesis (Ferguson *et al.*, 1967). Likewise, the "deadaptation" of the galactozymase system in this same yeast is energy dependent (Robertson and Halvorson 1957). Furthermore the degradation of protein in *E. coli* is inhibited by agents that inhibit protein synthesis (Schlessinger and Ben-Hamida, 1966). Thus, certain characteristics of the degradation of protein that appear to be common to liver and microbial systems include (a) a dependence on energy metabolism, and (b) an inhibition of protein degradation by agents that inhibit protein synthesis. Several explanations for the apparent energy requirement for protein degradation might include: (1) a requirement for a necessary cofactor. Penn (1960, 1961) presented evidence for CoA as a necessary cofactor in the breakdown of protein in liver homogenate system; (2) a requirement for removal of degradation products that inhibit further protein degradation. Among such products may be amino acids or peptides; (3) a requirement for maintaining the integrity of specific structures, such as lysosomes, at least in animal tissues. An explanation for the effect of inhibitors of protein synthesis is more difficult. Mandelstam (1958) suggested that the inhibition of protein degradation may result from the accumulation of small molecular weight compounds that inhibit the degradative process. Schlessinger and Ben-Hamida (1966) have argued that the small molecules that accumulate may not be the amino acids themselves, but rather the aminoacyl-sRNA, or perhaps even peptide-sRNA species. Willetts (1967a,b) suggests that the hypothetical inhibitor of protein degradation may be an RNA species. Another explanation for the requirement for protein synthesis for protein degradation would be that a certain protein(s) that is turning over rapidly is required for protein degradation.

Throughout this review the assumption has been made that the products of protein breakdown are in all cases amino acids. There is no compelling evidence to assume that such is the case with intracellular protein degradation, especially since the various products of degradation have not been identified. An interesting hypothesis has been made by Walter (1960) to the effect that the acceptor of protein hydrolysis might not be water, but rather an sRNA, thus conserving the energy of the peptide bond.

Heretofore we have presented this discussion as if the degradative process constitutes a uniform system by means of which all protein degradation or inactivation within a cell or tissue takes place. In all likelihood this assumption is incorrect. There are increasing numbers of reports of specific proteins which inhibit or otherwise inactivate specific enzymes including 5′-nucleotidase in *E. coli* (Dvorak *et al.*, 1966), ornithine transcarbamylase (Bechet and Wiame, 1965; Messenguy and Wiame, 1969), and malate dehydrogenase (Ferguson *et al.*, 1967) of *S. cerevesiae,* glutamine synthetase of *Enterobacteriaceae* (Gancedo and Holzer, 1968), and rat liver ribonuclease (Blobel and Potter, 1966) and glucose-6-P dehydrogenase (Bonsignore *et al.*, 1968). At one extreme, then, we might propose that the degradation of each protein or enzyme requires a specific protein. This is logically impossible, since in liver there is a continual replacement of essentially all proteins. Thus, if a protein were necessary to degrade each specific protein, then it would follow that a protein would be necessary to degrade a protein to degrade a protein . . . , etc.

One clear candidate for a system capable of carrying out the breakdown of tissue constituents is the lysosome system, which occurs in essentially all cells. The significance of these intracellular digestive organelles has been reviewed recently by de Duve and Wattiaux (1966). Available evidence would indicate that lysosomes are involved in various types of gross intracellular autophagy, such as occurs with tail resorption during amphibian metamorphosis (Weber, 1967), the increased rate of turnover of muscle proteins in hereditary muscular dystrophy of mice (Tappel *et al.*, 1962), and in the increased rate of protein degradation that occurs following the administration of glucagon to intact animals (Ashford and Porter, 1962; Deter and de Duve, 1967) or when glucagon is added to the fluid perfusing isolated rat livers (Green and Miller, 1960; Miller, 1961), and during tissue involution following hormone withdrawal from sensitive tissues (Brandes *et al.*, 1962). Lysosomes are currently conceived as being involved in the autophagy of a discrete area of intracellular cytoplasm, which may include mitochondria, endoplasmic reticulum, etc., in a single autophagic vacuole. Such vacuoles contain the acid hydrolases (including cathepsins), and presumably function at a pH of approximately 5.0. It is difficult to envisage such autophagic vacuoles as being involved in the turnover of protein whose characteristics are those of heterogeneity and alteration by changes in the small molecule environment. Thus, we would have to design a mechanism whereby the recognition of whether a protein molecule will be degraded involves passage into a lysosome. Various proteolytic enzymes have been described in tissues, including brain (Marks and Lajtha, 1963, 1965), kidney and liver (Fruton, 1960; Green and Neurath, 1954; Smith, 1951), with properties, pH optima, and tissue distribution different from the cathepsins of lysosomes. Such findings might, then, suggest that more than one system for intracellular degradation of protein exists. It would seem most reasonable to propose that the system of lysosomes is important when cell involution or gross changes in rates of protein degradation occur, whereas the degradation that occurs in normal steady state conditions involves another system or systems which have not been defined well at present.

The extent to which activations or inhibitions of degradative systems may be instrumental in determining rates of protein degradation is not yet known. Various reports of tissue inhibitors of protease activity have been made (Finkenstaedt, 1957; Blackwood and Mandl, 1964), but at present their role in controlling protein degradation is not known. More interesting and potentially significant are those studies indicating an effect of certain steroid hormones on the lability of lysosomal membranes. Thus, glucocorticoids stabilize lysosomal membranes, whereas progesterone does the opposite (Weissman and Thomas, 1964). The extent to which the hormonal modulation of lysosome lability constitues a significant regulatory role in protein degradation remains to be determined.

VI. On the Significance of Protein Degradation

It should be evident to the reader that the area of the control of protein degradation requires considerably more study for a complete understanding. In this last section I should like to take a speculative bent in asking: *What is the significance of protein degradation to the organism?* We can look upon the continual degradation of protein and other cell constituents as part of the continuing adaptation of an organism to variations in its environment. Such adaptation may take several forms. Thus at one extreme is the case of total deprivation of ingested nutrients, in which body constituents must be degraded to provide energy or precursors for new cell constituents. The involution of tissues as part of a developmental sequence would be another such example. In addition to these gross forms of protein degradation are those more subtle and selective changes in individual enzymes and proteins associated with adaptation to a wide variety of nutritional and physiological conditions (Knox *et al.,* 1956). Such changes involve not only the synthesis of newly required proteins, but also their subsequent removal when no longer needed. In bacteria this removal process can be accomplished readily by rapid growth with subsequent dilution of the unneeded protein (enzyme). Cells in multicellular organisms, in contrast, often do not divide. Hence protein degradation can, in part, be considered as a "solution" to the problem of how to remove unneeded proteins in multicellular organisms. In this regard, it is of interest that protein degradation does become important in microorganisms under nongrowing conditions (Mandelstam, 1960).

If we accept protein degradation as a necessary part of adaptation to environmental stimuli, then we can ask: why must it be so continual and so extensive? At a first approximation, continual synthesis and degradation would appear to be extremely wasteful. To suggest an answer to this question, let us consider possible solutions to the problem of how to break down proteins, considering this from the standpoint of an evolutionary selection of potential mechanisms.

In the situation of starvation, where massive degradation of protein must occur, some mechanism would be required for relative indiscriminate degradation which could be turned on and off appropriately. Out of such a requirement might have developed the lysosomal system. The relative indiscriminate breakdown of cell constituents, resulting from lysosomal autophagy, however, would not be an appropriate solution for that protein degradation associated with selective and relatively specific removal of proteins or enzymes no longer required. Several "solutions" to the selective removal of proteins can be envisaged. Thus, specificity could have evolved through the elaboration of proteins specific for the removal (inactivation or degradation) of each protein in the cell requiring such removal. Such specific inactivating proteins may well exist for certain cases, as discussed in Section V. However, such a solution would appear to be very "expensive," in a genetic and evolutionary sense, since it would require the existence of a large number of specific inactivating proteins as well as the regulatory apparatus required to recognize when such proteins were to be synthesized.

An alternative solution is that specificity for degradation exists in the protein or enzyme itself. Only a few nonspecific degrading enzymes (proteases and peptidases) would be required. We can conceive, then, that during evolution, it became necessary to provide an ability for some proteins to fluctuate in content more rapidly than others. Thus, for those proteins whose levels need not change, or whose presence is required at all times, mutations in the peptide sequence leading to increased stability *in vivo* would be selected. Conversely, for those proteins whose levels must change, mutations increasing susceptibility to degradation would have been selected.

Thus, although the solution involving the continual synthesis and degradation of proteins is wasteful, this solution would appear to have been the predominant one to emerge in the mammal. However, it would appear that in some instances, those "solutions" involving highly specific inactivating proteins and nondiscriminating degradative systems, such as lysosomes, have also developed. Although such a hypothesis is highly speculative, many of the findings described in this review are consistent with it.

References

Aebi, H. (1967). *Proc. 3rd Intern. Congr. Human Genet.* p. 189. Johns Hopkins Press, Baltimore, Maryland.

Allen, M. B. (1953). *Bacteriol. Rev.* **17,** 125.

Argyris, T. S., and Magnus, D. R. (1968). *Develop. Biol.* **17,** 187.

Arias, I. M., and De Leon, A. (1967). *Mol. Pharmacol.* **3,** 216.

Arias, I. M., Doyle, D., and Schimke, R. T. (1969). *J. Biol. Chem.* **244,** 3303.

Ashford, T. P., and Porter, K. R. (1962). *J. Cell Biol.* **12,** 198.

Askonas, B. A., and Humphrey, J. H. (1958). *Biochem. J.* **68,** 252.

Balinsky, J. B., Shambaugh, G. E., and Cohen, P. P. (1970). *J. Biol. Chem.* **245,** in press.

Barondes, S. H. (1966). *J. Neurochem.* **13,** 721.

Beattie, D. S., Basford, R. E., and Koritz, S. B. (1967). *J. Biol. Chem.* **242,** 4584.

Bechet, J., and Wiame, J. B. (1965). *Biochem. Biophys. Res. Commun.* **21,** 226.

Berlin, C. M., and Schimke, R. T. (1965). *Mol. Pharmacol.* **1,** 149.
Bertino, J. R., Cashmore, A., Fink, M., Calabresi, P., and Lefkowitz, E. (1965). *Clin. Pharmacol. Therap.* **6,** 763.
Beutler, E. (1966). In "The Metabolic Basis of Inherited Disease" (J. B. Stanbury, J. B. Wyngaarden, and D. S. Fredrickson, eds.), 2nd ed., p. 1060. McGraw-Hill (Blakiston), New York.
Blackwood, C., and Mandl, I. (1964). *J. Cell Biol.* **23,** 11A.
Blobel, G., and Potter, V. R. (1966). *Proc. Natl. Acad. Sci. U.S.* **55,** 1283.
Bojarski, T. B., and Hiatt, H. H. (1960). *Nature* **188,** 112.
Bonsignore, A., De Flora, A., Mangiarotti, M. A., Lorenzoni, I., and Alema, S. (1968). *Biochem. J.* **106,** 147.
Borsook, H., and Keighley, G. L. (1935). *Proc. Roy. Soc.* **B118,** 488.
Brandes, D., Gyorkey, F., and Groth, D. P. (1962). *Lab. Invest.* **11,** 339.
Bresnick, E., Williams, S. S., and Mossé, H. (1967). *Cancer Res.* **27,** 469.
Bresnick, E., Mayfield, E. D., and Mossé, H. (1968). *Mol. Pharmacol.* **4,** 173.
Bubela, B., and Holdsworth, E. S. (1966). *Biochim. Biophys. Acta* **123,** 364.
Buchanan, D. L. (1961a). *Arch. Biochem. Biophys.* **94,** 489.
Buchanan, D. L. (1961b). *Arch. Biochem. Biophys.* **94,** 500.
Bunn, H. F., and Jandl, J. H. (1966). *Proc. Natl. Acad. Sci. U.S.* **56,** 974.
Chaloupka, J., and Liebster, J. (1959). *Folia Microbiol. (Prague)* **4,** 167.
Civen, M., and Knox, W. E. (1960). *J. Bio. Chem.* **235,** 1716.
Cox, R. P., and MacLeod, C. M. (1964). *Cold Spring Harbor Symp. Quant. Biol.* **29,** 233.
de Duve, C., and Baudhuin, P. (1966). *Physiol. Rev.* **46,** 323.
de Duve, C., and Wattiaux, R. (1966). *Ann. Rev. Physiol.* **28,** 435.
De Mars, R. (1958). *Biochim. Biophys. Acta* **27,** 435.
De Mars, R. (1964). *Natl. Cancer Inst. Monograph* **13,** 181.
Deter, R. L., and de Duve, C. (1967). *J. Cell Biol.* **33,** 437.
Dreyfus, J. C., Kruh, J., and Schapira, G. (1960). *Biochem. J.* **75,** 574.
Druyan, R., DeBarnard, B., and Rabinowitz, M. (1970). *J. Biol. Chem.* in press.
Drysdale, J. W., and Munro, H. N. (1966). *J. Biol. Chem.* **241,** 3630.
Dubnoff, J. W., and Dimick, M. (1959). *Biochim. Biophys. Acta* **31,** 541.
Dvorak, H. F., Anraku, Y., and Heppel, L. A. (1966). *Biochem. Biophys. Res. Commun.* **24,** 628.
Eagle, H., Piez, K. A., Fleischman, R., and Oyama, V. I. (1959). *J. Biol. Chem.* **234,** 592.
Eliasson, E. E., and Strecker, H. J. (1966). *J. Biol. Chem.* **241,** 5757.
Ernster, L., and Orrhenius, S. (1965). *Federation Proc.* **24,** 1190.
Fallon, H. J., Hackney, E. J., and Byrne, W. L. (1966). *J. Biol. Chem.* **241,** 4157.
Feigelson, P., and Greengard, O. (1962). *J. Biol. Chem.* **237,** 1908.
Feigelson, P., Dashman, T., and Margolis, F. (1959). *Arch. Biochem. Biophys.* **85,** 478.
Ferguson, J. J., Jr., Boll, M., and Holzer, H. (1967). *European J. Biochem.* **1,** 21.
Finkenstaedt, J. T. (1957). *Proc. Soc. Exptl. Biol. Med.* **95,** 302.
Fletcher, M. J., and Sanadi, D. R. (1961). *Biochim. Biophys. Acta* **51,** 356.
Folin, O. (1905). *Am. J. Physiol.* **13,** 117.
Freedland, R. A. (1968). *Life Sci.* **7,** 499.
Fritz, P. J., Vessell, E. S., White, E. L., and Pruitt, K. M. (1969). *Proc. Natl. Acad. Sci. U.S.* **62,** 558.
Fruton, J. S. (1960). *In* "The Enzymes" (P. D. Boyer, H. Lardy, and K. Myrbäck, eds.), Vol. 4, p. 233. Academic Press, New York.

Gaitonde, M. K., and Richter, D. (1956). *Proc. Roy. Soc.* **B145,** 83.

Gan, J. C., and Jeffay, H. (1967). *Biochim. Biophys. Acta* **148,** 448.

Gancedo, C., and Holzer, H. (1968). *European J. Biochem.* **4,** 190.

Ganschow, R., and Schimke, R. T. (1969). *J. Biol. Chem.* **244,** 4649.

Garlick, P. J. (1969). *Nature* **223,** 61.

Goldberg, A. L. (1969). *J. Biol. Chem.* **244,** 3217, 3223.

Goldstein, L., Stella, E. J., and Knox, W. E. (1962). *J. Biol. Chem.* **237,** 1723.

Green, M., and Miller, L. L. (1960). *J. Biol. Chem.* **235,** 3202.

Green, N. M., and Lowther, D. A. (1959). *Biochem. J.* **71,** 55.

Green, N. M., and Neurath, H. (1954). *In* "The Proteins" (H. Neurath and K. Bailey, eds.), Vol. 2, Part B, p. 1057. Academic Press, New York.

Greengard, O., and Feigelson, P. (1961a). *Nature* **190,** 446.

Greengard, O., and Feigelson, P. (1961b). *J. Biol. Chem.* **236,** 158.

Greengard, O., Smith, M. A., and Acs, G. (1963). *J. Biol. Chem.* **238,** 1548.

Greengard, O., Mendelsohn, N., and Acs, G. (1966). *J. Biol. Chem.* **241,** 304.

Grisolia, S. (1964). *Physiol. Rev.* **44,** 657.

Grob, D., Lillienthal, J. L., Jr., Harvey, A. M., and Jones, B. F. (1947). *Bull. Johns Hopkins Hosp.* **81,** 217.

Gross, J., and Lapiere, C. M. (1962). *Proc. Natl. Acad. Sci. U.S.* **48,** 1014.

Grossman, A., and Mavrides, C. (1967). *J. Biol. Chem.* **242,** 1398.

Hakala, M. T., and Suolinna, E. M. (1966). *Mol. Pharmacol.* **2,** 465.

Hayashi, S., Granner, D. K., and Tomkins, G. M. (1967). *J. Biol. Chem.* **242,** 3993.

Heimberg, M., and Velick, S. F. (1954). *J. Biol. Chem.* **208,** 725.

Hendler, R. W. (1965). *Proc. Natl. Acad. Sci. U.S.* **54,** 1233.

Hiatt, H. H., and Bojarski, T. B. (1961). *Cold Spring Harbor Symp. Quant. Biol.* **26,** 367.

Hirsch, C. A., and Hiatt, H. H. (1966). *J. Biol. Chem.* **241,** 5936.

Hogness, D. S., Cohn, M., and Monod, J. (1955). *Biochim. Biophys. Acta* **16,** 99.

Houck, J. C., Sharma, V. K., Patel, Y. M., and Gladner, J. A. (1968). *Biochem. Pharmacol.* **17,** 2081.

Jick, H., and Shuster, L. (1966). *J. Biol. Chem.* **241,** 5366.

Jost, J. P., Khairallah, E. A., and Pitot, H. C. (1968). *J. Biol. Chem.* **243,** 3057.

Kamin, H., and Handler, P. (1951). *J. Biol. Chem.* **188,** 193.

Kato, R., Jondorf, W. R., Loeb, L. A., Ben, T., and Gelboin, H. V. (1966). *Mol. Pharmacol.* **2,** 171.

Kenney, F. T. (1962). *J. Biol. Chem.* **237,** 3495.

Kenney, F. T. (1967). *Science* **156,** 525.

Kipnis, D. M., Reiss, E., and Helmreich, E. (1961). *Biochim. Biophys. Acta* **51,** 519.

Kit, S., Dubbs, D. R., and Frearson, P. M. (1965). *J. Biol. Chem.* **240,** 2565.

Knox, W. E., and Auerbach, V. H. (1955). *J. Biol. Chem.* **214,** 307.

Knox, W. E., and Mehler, A. (1951). *Science* **113,** 237.

Knox, W. E., and Piras, M. M. (1967). *J. Biol. Chem.* **242,** 2959.

Knox, W. E., Auerbach, V. H., and Lin, E. C. C. (1956), *Physiol. Rev.* **36,** 164.

Knox, W. E., Piras, M. M., and Tokuyama, K. (1966). *J. Biol. Chem.* **241,** 297.

Koch, A. L. (1962). *J. Theoret. Biol.* **3,** 283.

Koch, A. L., and Levy, H. R. (1955). *J. Biol. Chem.* **217,** 947.

Lee, N. D., and Williams, R. H. (1952). *Biochim. Biophys. Acta* **9,** 698.

Lehman, H., Huntsman, R. G., and Ager, J. A. M. (1966). *In* "The Metabolic Basis of Inherited Disease" (J. P. Stanbury, J. B. Wyngaarden, and D. S. Fredrickson, eds.), 2nd ed., p. 1100. McGraw-Hill (Blakiston), New York.

Levin, W., and Kuntzman, R. (1969). *J. Biol. Chem.* **244,** 3671.
Lin, E. C. C., and Knox, W. E. (1956). *Biochim. Biophys. Acta* **26,** 85.
Linderstrøm-Lang, K. (1950). *Cold Spring Harbor Symp. Quant. Biol.* **14,** 117.
Littlefield, J. W. (1965). *Biochim. Biophys. Acta* **95,** 14.
Loftfield, R. B., and Harris, A. (1956). *J. Biol. Chem.* **219,** 151.
MacDonald, R. A. (1961). *Arch. Internal Med.* **107,** 335.
McFarlane, A. S. (1963). *Biochem. J.* **89,** 277.
Majerus, P. W., and Kilburn, E. (1970). *J. Biol. Chem.* **245,** in press.
Mandelstam, J. (1958). *Biochem. J.* **69,** 110.
Mandelstam, J. (1960). *Bacteriol. Rev.* **24,** 289.
Mandelstam, J., Waites, W. W., and Warren, S. C. (1967). *Proc. 7th Intern. Cong. Biochem., Tokyo, Japan, 1967* p. 253.
Mandelstam, J., Waites, W. W., Warren, S. C., and Sterlini, J. M. (1968). *Biochem. J.* **106,** 28P.
Marks, N., and Lajtha, A. (1963). *Biochem. J.* **89,** 438.
Marks, N., and Lajtha, A. (1965). *Biochem. J.* **97,** 74.
Markus, G. (1965). *Proc. Natl. Acad. Sci. U.S.* **54,** 255.
Marver, H. S., Collins, A., Tschudy, D. P., and Rechcigl, M., Jr. (1966). *J. Biol. Chem.* **241,** 4323.
Matsubara, S., Suter, H., and Aebi, H. (1967). *Humangenetik* **4,** 29.
Maurer, W. (1957). *Wien. Z. Inn. Med. Grenzg.* **38,** 393.
Mazur, A., and Shorr, E. (1950). *J. Biol. Chem.* **182,** 607.
Messenguy, F., and Wiame, J.-M. (1969). *FEBS Letters* **3,** 47.
Miller, L. L. (1961). *Recent Progr. Hormone Res.* **17,** 539.
Munro, H. N. (1968). *Federation Proc.* **27,** 1231.
Neitlich, H. W. (1966). *J. Clin. Invest.* **45,** 380.
Nemeth, A. M. (1962). *J. Biol. Chem.* **237,** 3703.
Neuberger, A., and Slack, H. G. B. (1953). *Biochem. J.* **53,** 47.
Niemeyer, H. (1967). *Natl. Cancer Inst. Monograph* **27,** 29.
Niklas, A., Quincke, E., Maurer, W., and Neyen, H. (1958). *Biochem. Z.* **330,** 1.
Numa, S., Matsuhashi, M., and Lynen, F. (1961). *Biochem. Z.* **334,** 203.
Ohta, Y. (1967). *J. Biol. Chem.* **242,** 509.
Ohta, Y., Ogura, Y., and Wada, A. (1966). *J. Biol. Chem.* **241,** 5919.
Omura, T., Siekevitz, P., and Palade, G. E. (1967). *J. Biol. Chem.* **242,** 2389.
Paul, J., and Fottrell, P. F. (1963). *Biochim. Biophys. Acta* **67,** 334.
Penn, N. W. (1960). *Biochim. Biophys. Acta* **37,** 55.
Penn, N. W. (1961). *Biochim. Biophys. Acta* **53,** 490.
Piha, R. S., Cuénod, M., and Waelsch, H. (1966). *J. Biol. Chem.* **241,** 2397.
Pine, M. J. (1965). *Biochim. Biophys. Acta* **104,** 439.
Pine, M. J. (1966). *J. Bacteriol.* **92,** 847.
Pine, M. J. (1967). *Cancer Res.* **27,** 522.
Poole, B., Leighton, F., and de Duve, C. (1969). *J. Cell Biol.* **41,** 536.
Price, V. E., Sterling, W. R., Tarantola, V. A., Hartley, R. W., Jr., and Rechcigl, M., Jr. (1962). *J. Biol. Chem.* **237,** 3468.
Rechcigl, M., Jr. (1968). *Enzymologia* **34,** 23.
Rechcigl, M., Jr., and Heston, W. E. (1963). *J. Natl. Cancer Inst.* **30,** 855.
Rechcigl, M., Jr., and Heston, W. E. (1967). *Biochem. Biophys. Res. Commun.* **27,** 119.
Reiner, J. M. (1953). *Arch. Biochem.* **46,** 80.
Robertson, J. J., and Halvorson, H. O. (1957). *J. Bacteriol.* **73,** 186.

Rotman, B., and Spiegelman, S. (1954). *J. Bacteriol.* **68,** 419.
Rowe, P. B., and Wyngaarden, J. B. (1966). *J. Biol. Chem.* **241,** 5571.
Schapira, G., Kruh, J., Dreyfus, J. C., and Schapira, F. (1960). *J. Biol. Chem.* **235,** 1738.
Schimke, R. T. (1964a). *Natl. Cancer Inst. Monograph* **13,** 197.
Schimke, R. T. (1964b). *J. Biol. Chem.* **239,** 3808.
Schimke, R. T. (1967). *Natl. Cancer Inst. Monograph* **27,** 301.
Schimke, R. T., Sweeney, E. W., and Berlin, C. M. (1964). *Biochem. Biophys. Res. Commun.* **15,** 214.
Schimke, R. T., Sweeney, E. W., and Berlin, C. M. (1965a). *J. Biol. Chem.* **240,** 322.
Schimke, R. T., Sweeney, E. W., and Berlin, C. M. (1965b). *J. Biol. Chem.* **240,** 4609.
Schimke, R. T., Ganschow, R., Doyle, D., and Arias, I. M. (1968). *Federation Proc.* **27,** 1223.
Schirmer, M. D., and Harper, A. E. (1970). *J. Biol. Chem.* **245,** in press.
Schlessinger, D., and Ben-Hamida, F. (1966). *Biochim. Biophys. Acta* **119,** 171.
Schmid, R., Figen, J. F., and Schwartz, S. (1955). *J. Biol. Chem.* **217,** 263.
Schoenheimer, R. (1942). "The Dynamic State of Body Constituents." Harvard Univ Press, Cambridge, Massachusetts.
Schoenheimer, R., and Rittenberg, D. (1940). *Physiol. Rev.* **20,** 218.
Segal, H. L., and Kim, Y. S. (1963). *Proc. Natl. Acad. Sci. U.S.* **50,** 912.
Sellinger, O. Z., Lee, K. L., and Fesler, K. W. (1966). *Biochim. Biophys. Acta* **124,** 289.
Shambaugh, G. E., Balinsky, J. B., and Cohen, P. P. (1969). *J. Biol. Chem.* **244,** 5295.
Shuster, L., and Jick, H. (1966). *J. Biol. Chem.* **241,** 5361.
Simon, E. J., Gross, C. S., and Lessell, I. M. (1962). *Arch. Biochem. Biophys.* **96,** 41.
Simpson, M. V. (1953). *J. Biol. Chem.* **201,** 143.
Simpson, M. V., and Velick, S. F. (1954). *J. Biol. Chem.* **208,** 61.
Smith, E. L. (1951). *In* "The Enzymes" (J. B. Sumner and K. Myrbäck, eds.), Vol. 1, Part 2, p. 793. Academic Press New York.
Steinberg, D., and Vaughan, M. (1956). *Biochem Biophys. Acta* **19,** 584.
Still, J. W. (1957). *J. Heredit.* **48,** 202.
Swick, R. W. (1957). *J. Biol. Chem.* **231,** 751.
Swick, R. W., and Handa, D. T. (1956). *J. Biol. Chem.* **218,** 557.
Swick, R. W., Koch, A. L., and Handa, D. T. (1956). *Arch. Biochem.* **63,** 226.
Swick, R. W., Rexroth, A. K., and Stange, J. L. (1968). *J. Biol. Chem.* **243,** 3581.
Szepsi, B., and Freedland, R. A. (1969). *Arch. Biochem. Biophys.* **133,** 60.
Tappel, A. L., Zalkin, H., Caldwell, K. A., Desai, I. D., and Shibko, S. (1962). *Arch. Biochem. Biophys.* **96,** 340.
Tarentino, A. L., Richert, D. A., and Westerfeld, W. W. (1966). *Biochim. Biophys. Acta* **124,** 295.
Tarver, H. (1954). *In* "The Proteins" (H. Neurath and K. Bailey, eds.), Vol. II, Part B, p. 1199–1296. Academic Press, New York.
Tata, J. R. (1966). *Develop. Biol.* **13,** 77.
Thompson, R. C., and Ballou, J. E. (1956). *J. Biol. Chem.* **223,** 795.
Urbá, R. C. (1959). *Biochem. J.* **71,** 513.
Ussing, H. H. (1941). *Acta Physiol. Scand.* **2,** 209.
Velick, S. F. (1956). *Biochim. Biophys. Acta* **20,** 228.
Voit, C. (1866). *Z. Biol.* **2,** 307.

Walter, H. (1960). *Nature* **188,** 643.
Weber, R. (1967). *In* "The Biochemistry of Animal Development" (D. Weber, ed.), Vol. II, p. 227. Academic Press, New York.
Weissman, G., and Thomas, L. (1964). *Recent Progr. Hormone Res.* **20,** 215.
Willetts, N. S. (1967a). *Biochem. J.* **103,** 453.
Willetts, N. S. (1967b) *Biochem. J.* **103,** 462.
Zakrzewski, S. F., Hakala, M. T., and Nichol, C. A. (1966). *Mol. Pharmacol.* **2,** 423.

CHAPTER 33

Sites of Hormonal Regulation of Protein Metabolism

K. L. Manchester

Department of Biochemistry,
University College London,
London, England

I. Introduction

The concept of the sites of hormonal regulation can be interpreted either in a narrow sense or more broadly. Obviously every hormonal response results from a primary interaction of the hormone with some, presumably specific, tissue receptor. This interaction is strictly speaking the site of the hormone's action, though the effect may be manifest through the operation of a sequence of changes causally related to the first interaction. In a broader sense, therefore, the site of hormone action is the whole series of changes induced by the hormone and their metabolic consequences. It is usually through responses several stages removed from the primary action that the hormone's effects are going to be recognized; for example, the lowering of blood sugar by insulin results in part from increased uptake by responsive tissues; this in turn results from increased membrane permeability, which in turn must result from some change in functioning of the sugar transport apparatus, which in turn is somehow affected by the primary interaction of insulin with the tissue through an as yet unknown number of steps (Fig. 1).

The existence of several recognized effects of a hormone complicates the issue somewhat. We now have to consider the different possibilities as to whether the two responses represent two different steps in the same sequence—for example, lowering of blood sugar being a result of increased glucose consumption—or whether the different observed effects result from two different primary actions—as indeed the fall of plasma glucose may equally result from glucose retention by the liver. If the second, do we have to consider the possibility of two different initial sites of interaction of the hormone with presumably different receptor sites, or is it possible that a primary hormone receptor interaction initiates a sequence of changes which can ultimately lead to a number of different responses (Fig. 2)? As an example of the latter, again with insulin, we can ask whether effects on amino acid metabolism are secondary to effects on utilization of carbohydrate, or whether they arise separately, by a completely separate chain of events, or from a primary interaction sequence which divides to give different results. For instance, a hormonal action that increases the cell level of cyclic AMP ("a second messenger"), e.g., the action of ACTH, epinephrine, and glucagon in adipose tissue (Sutherland and Rall, 1960; Murad *et al.*, 1962; Sutherland *et al.*, 1965), could lead to several different changes—mobilization of lipid, activation of phosphorylase kinase and inhibition

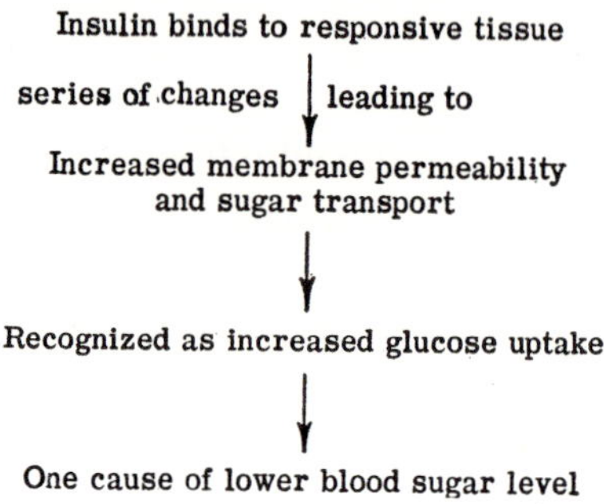

FIG. 1. Several effects of a hormone may be merely different facets of one essential action—illustrated by the influence of insulin on the blood sugar level.

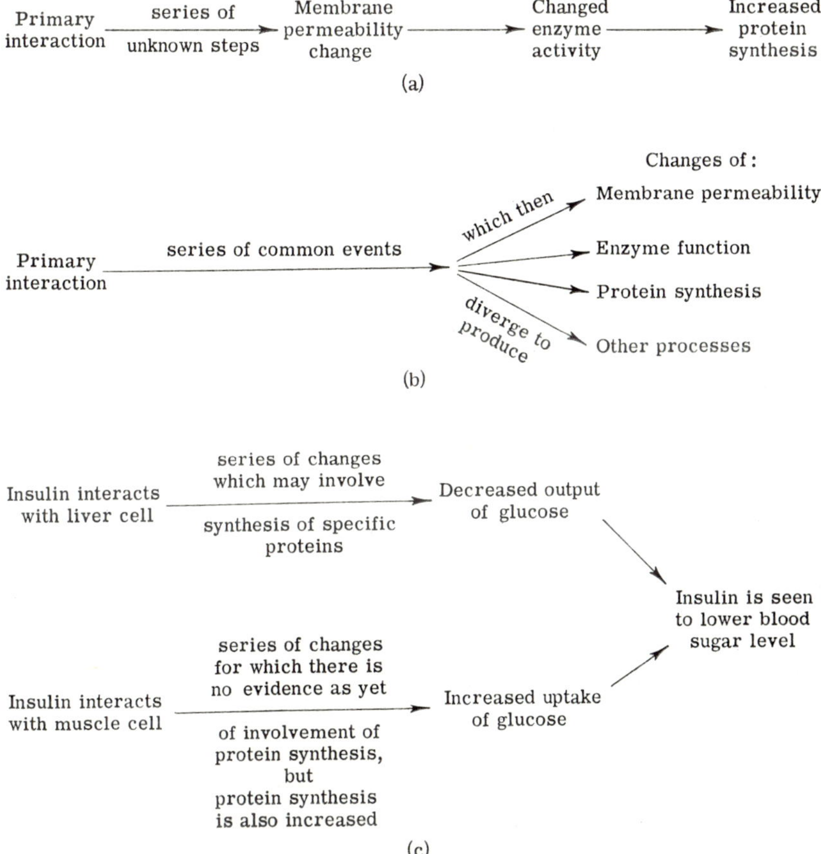

FIG. 2. Several effects of a hormone may be sequentially linked (a), or develop separately (b). Both possibilities arise in consideration of the action of insulin (c).

of glycogen synthetase kinase (Rizack, 1964; Huijing and Larner, 1966), each arising from the first primary action.

Obviously where the sequence is common, inhibition of any step will prevent all the consequences. If alternatively a step in one derivative sequence is inoperative, that particular action will cease, but other steps resulting from the primary interaction and not causally related through the inhibited step are equally operative. Again examples of both situations can be seen with insulin. Denervation, for instance, of diaphragm muscle in some unknown fashion desensitizes the tissue to all the usual responses to insulin (Buse and Buse, 1959, 1961; Buse *et al.*, 1965; Harris and Manchester, 1966). Presumably there is some fundamental change at an early stage in the hormone receptor interaction sequence. On the other hand, in studying the effect of the insulin on protein synthesis in muscle, we can selectively inhibit protein synthesis (though to be strictly accurate, not the hormonal sensitivity of this process) with specific inhibitors of protein synthesis, but we do not reduce hormonal stimulation of other parameters—uptake of glucose, etc. (Eboué-Bonis *et al.*, 1963; Burrow and Bondy, 1964; Carlin and Hechter, 1964; Wool and Moyer, 1964; Manchester, 1967a). In this case therefore, we are interfering with a step in the action-effect sequence not closely related to the primary interaction.

I have dealt with this point in some detail because delineation of action and effect is a major problem in endocrine research. A full biochemical understanding, moreover, of hormone action requires determination of each step in this sequence. For no hormone is it yet possible to give a complete account, but our knowledge continues to grow and as it grows tends to reveal increasing complexity in the numerous stages in the sequence. In general terms, more has been learned of the latter parts of the action sequences probably because this represents alteration in the rate of reactions whose mechanism has been of interest and subject to investigation for reasons unrelated to understanding of hormone action. Conversely, our knowledge of the primary sites of hormone interaction with tissues, or of the primary tissue receptors, is one of the least understood aspects of the subject—which is of course a pity since it is so fundamental to a complete understanding. Possibly the hormones most studied in this last respect are insulin, those stimulating cyclic AMP (cAMP) production, in neither instance from a viewpoint primarily concerned with protein metabolism, and more recently, this time very much concerned with growth effects, the steroid hormones.

In investigation of hormonal control of protein anabolism, attention has often focused around the possibility of hormone gene interactions and such regulation in various ways in concert with RNA metabolism. In this review we shall first look at the specific sites of hormone interaction with tissues, then examine the general question of hormonal control of RNA and protein metabolism. This will precede examination of the action of specific hormones.

II. Hormone Receptors

Since the binding sites of hormones to responsive tissues are not yet known in detail, it may be thought profitless to discuss the topic. So much consideration has been given to the subject, however, that it is useful to think about the types of mechanisms envisaged and their likely location.

A. *Insulin*

It is not of course immediately obvious whether the primary site of interaction of insulin with a tissue receptor is intracellular or on the external membrane, and it has not proved very easy from distribution studies (Stein and Gross, 1959; Izzo *et al.*, 1967) or electron microscopy to demonstrate whether insulin does indeed penetrate the cell membrane. One tends to think of the cell membrane as a simple shell to a cell, but in fact a much invaginated membranous structure with the added complexity, for example, of the tubular (T) system in muscle (Endo, 1964; Franzini-Armstrong and Porter, 1964; Huxley, 1964; Peachey, 1965) makes it less easy to define simply what is outside the cell and what inside. However, perhaps because of the evident binding of insulin to tissues (Stadie *et al.*, 1952, 1953; Bleehen and Fisher, 1954; Newerley and Berson, 1957; Garratt *et al.*, 1966a,b; Wohltman and Narahara, 1966), the analogy with catecholamine receptor sites (Ahlquist, 1948, 1967; Jenkinson and Morton, 1967a,b), and the role of insulin in regulating membrane permeability, it is generally assumed that there is a cell membrane receptor site for insulin. In one study (Edelman and Schwartz, 1966) 40% of labeled insulin after incubation with striated muscle was bound to the sarcolemma, and only a very small amount to microsomal and mitochondrial fractions. On the other hand, more recent studies (Maggi *et al.*, 1969) of insulin binding *in vivo* have shown very little localization of the hormone in muscle. What is there may be associated with the mitochondria. Distribution of labeled ACTH in adrenal tissue likewise shows little specific localization (Nishizawa *et al.*, 1965*[1]).

The structure of the insulin molecule does not suggest any groups obviously available for binding. There has been however considerable speculation as to whether, by analogy with vasopressin (Fong *et al.*, 1960; Rasmussen *et al.*, 1960; Schwartz *et al.*, 1960), binding might be through interaction of sulfhydryl groups in the membrane with the disulfide bridge groups of the insulin molecule, from which changes in permeability would arise, but the evidence for this is not convincing (Cadenas *et al.*, 1961; Carlin and Hechter, 1962; Edelman *et al.*, 1963; Hechter and Halkerston, 1964). Whereas a number of sulfhydryl blocking agents can interfere with the action of insulin these compounds only partly prevent the binding of the hormone to adipose tissue or muscle (Cadenas *et al.*, 1961; Mirsky and Perisutti, 1962), but this may be indicative only that there are several different groups by which the hormone can be bound, some more significant than others for exertion of physiological action (Garratt *et al.*, 1966a). Selective masking of

[1] An asterisk following a reference citation indicates that the reference is given in the list of supplementary references on pp. 297 and 298.

amino, carboxyl, and other groups and controlled degradation of the insulin molecule (Nicol, 1960; Carpenter, 1966) leads to a complex pattern of changed potency. Similar studies were made with ACTH (Dixon, 1964*).

What the insulin molecule actually does to the membrane with which it interacts is again a complex question which, given our inadequate understanding of membrane structure, it is difficult to answer. Rodbell (1966) and Rodbell and Jones (1966) have compared the action of insulin as being similar to that of phospholipase C, which breaks down phospholipids. They suggest that through alteration of membrane structure induced by the hormone there is increased glucose transport, fatty acid synthesis, and protein synthesis analogous to the effect of phospholipase C. A stimulation of protein synthesis by phospholipase is most striking since there is some evidence (see Section V,A) that control can be through membrane permeability. However it has to be pointed out that this parameter was that least investigated by Rodbell and my observations with muscle (Manchester and Wulwick, 1969) do not give any hint of a capacity of phospholipase A to stimulate amino acid accumulation. Phospholipase A, hydrolysing the fatty acyl esters (as opposed to phospholipase C which hydrolyses the phosphate esters), might be expected to have more significant effects on membrane structure than phospholipase C (Lenard and Singer, 1968). One word of caution perhaps needs to be made—a loss of membrane function might readily be seen to stimulate a process such as glucose consumption, but its effect on any anabolic activity, such as protein synthesis, more dependent on an integrated function of the cell can be disastrous (Krahl, 1966; Manchester, 1966). Similarly, we have to be cautious in interpreting changed metabolic behavior following alteration of the ionic composition of the supporting medium in *in vitro* incubations (Bhattacharya, 1961, 1964; Buse *et al.*, 1964; Beloff-Chain *et al.*, 1965) or of treating tissues with proteolytic enzymes (Pruitt *et al.*, 1966; Rieser, 1966; Kuo *et al.*, 1966, 1967) or catecholamine blocking agents (Bewsher *et al.*, 1966) unless it can be shown that the capacity for integrated behavior remains unimpaired.

B. Adenyl Cyclase Stimulators

Hormones whose actions appear to be mediated through elevation of the cellular level of cAMP stimulate the activity of the enzyme adenyl cyclase. Not much is known about the enzymology of cyclase. Its distribution is believed to be largely membranous, though the evidence for this is not as clear-cut as could be wished (Sutherland *et al.*, 1962; Davoren and Sutherland, 1963; Rodbell, 1967a; Chase and Aurbach, 1968), and it may be a lipoprotein. Thus of the numerous hormones that raise the cAMP level of adipose tissue (Robison *et al.*, 1967, 1968), we have no definite proof that their primary interaction is with a receptor in the membrane as opposed to at an intracellular location, and additionally whether the receptor is the same for each hormone or whether individual in each case—either situation raises its own problems. (It is also possible that *in vivo* this tissue is normally responsive to fewer hormones, but this is merely guessing.) Robison *et al.* (1967) have speculated on the way in which the adenyl cyclase system, particularly in its response to the catecholamines, may be related to the α and β adrenergic receptor sites. Insulin can suppress the catecholamine enhanced concentration of cAMP (Butcher *et al.*, 1966; Exton *et al.*, 1966) and it has been claimed to suppress the action of adenyl cyclase (Jungas, 1966), though this is questioned (Rodbell, 1967c; Williams *et al.*, 1968; Cryer *et al.*, 1969). Again we can ask whether it is necessary to envisage two separate sites of insulin interaction with the cell, one concerned with influencing cAMP concentrations as opposed to another governing membrane permeability, or whether it is simpler to conceive (in the absence of specific information) that a suppression of cyclase activity is yet one more manifestation of the hormone's primary action. One possibility would be that it is through suppressing cAMP synthesis that insulin influences protein synthesis. However Rodbell (1967b) concluded that although cAMP can lower glucose consumption, the nucleotide is not involved in the action of insulin on glucose and, by implication, amino acid transport. Grahame-Smith *et al.* (1967) have advanced the case for the primary role of cAMP as the mediator of ACTH action on the adrenal. A similar mechanism may be operative for several others of the anterior pituitary hormones, vasopressin and a variety of other agents (Handler *et al.*, 1961; Brown *et al.*, 1963; Pastan and Macchia, 1967; Robison *et al.*, 1967, 1968), but there is little evidence as yet for the involvement of cAMP in the action of hormones in protein synthesis.

C. Steroid Hormones

Jensen and Jacobson (1962) showed that when estradiol in very low quantities is administered to immature female rats it is preferentially bound by the uterus and vagina (Eisenfeld, 1967). The binding is less in the presence of other uterotropic steroids, but only biologically active estrogens inhibit—apparently competitively (Noteboom and Gorski, 1965). Thus we have evidence for the existence of specific binding sites in the uterus for estrogens, and calculation suggests of the order of 2500 receptors per uterine cell. Since the binding site is destroyed by extremes of pH, by proteolytic enzymes, and by —SH inactivating reagents (Terenius, 1967), but not nucleases, the receptor molecule has been identified as a protein (Noteboom and Gorski, 1965). They were unable to decide its location in the cell. Later work (Toft *et al.*, 1967) has suggested the existence of a specific binding protein of molecular weight of about 200,000 which is found in the supernatant fraction of uterine homogenates, but this obviously does not establish whether the cytoplasm is the normal location of the receptor. Autoradiography (Jensen *et al.*, 1966) shows that most of the estradiol in the tissue is localized in the nucleus and it appears to be bound to a protein component extractable with strong potassium chloride solution. This material has a sediment coefficient of about 5 S whereas the cytoplasmic factor has an S value of about 9.5. Whether the nuclear component is merely a subunit of the larger molecule disaggregated by the high salt concentration used in its extraction (Korenman and Rao, 1968) or is more complex (Jensen *et al.*, 1968; Shyamala and Gorski, 1969*) is not clear. Both components appear to have an association constant toward estradiol of around 10^9 M^{-1} (Toft *et al.*, 1967; Puca and Bresciani, 1968, 1969). A protein is likely to offer a high degree of specificity for hormone binding/interaction and it is interesting to note that evidence for the role of membrane proteins in the active transport of amino acids and ions and even for the facilitated diffusion of sugars has recently been put forward (Elsas and Rosenberg, 1967; Elsas *et al.*, 1967; Kahlenberg *et al.*, 1967; Langdon and Sloan, 1967), though the evidence of Langdon and Sloan (1967) for the last mentioned has been questioned by LeFevre (1967). Woolley and Gommi (1964) alternatively suggested a possible role of specific gangliosides as receptors. They also have the merit of being membrane constituents. Carroll and Sereda (1968) consider sialic acid in the uterine membrane as being a likely receptor for serotonin, and Formby (1967) suggested that in brain norepinephrine may bind to phosphatidyl serine. An amino acid binding protein has in fact been located in the membrane of *E. coli* cells (Nakane *et al.*, 1968*).

More direct evidence of interactions of steroid hormones with the nucleus has been obtained by Sekeris and colleagues. They have found that incorporation of precursor into RNA by isolated nuclei is stimulated on addition of low doses of cortisol to rat liver nuclei (Dukes and Sekeris, 1965) and also of ecdysone to isolated blowfly epidermis nuclei (Sekeris *et al.*, 1965). Begg and Munro (1965) found a similar effect on addition of the polypeptide thyroid-stimulating hormone to sheep thyroid nuclei. In each case the stimulation was seen within 15–20 minutes of the addition of the hormone. Dukes *et al.* (1966) and Schmid *et al.* (1967) also showed that nuclear RNA isolated after hormone treatment had increased template activity when added to ribosomes from rat liver. Cortisone was less active than cortisol, and progesterone, testosterone, and a number of other steroids had no effect. It would be of interest to know the effect of corticosterone since this is the predominant glucocorticoid in the rat. Nuclei from adrenalectomized rats were responsive to much smaller doses of cortisone than those from controls (Schmid *et al.,* 1967). An even shorter interval was found necessary to observe a stimulatory effect of steroids on RNA polymerase activity of isolated rat liver nuclei (Lukács and Sekeris, 1967). This time testosterone as well as cortisol was effective, and increases in enzyme activity were seen within 1 minute of application of the hormone—too short a period for appreciable *de novo* synthesis of enzyme to have taken place. The authors suggest that the hormone increases the template activity of the nuclear DNA by a derepression mechanism, but precisely how is not clear.

Autoradiographic evidence reveals the localization of administered testosterone and aldosterone within chromosomes (Edelman *et al.,* 1963; Loeb and Wilson, 1965). Siegel and Tobias (1966) found high concentrations of thyroxine in nuclei of cultured kidney cells. Goldberg and Atchley (1966) have studied the interaction of a variety of hormones with DNA from placental nuclei and showed changes in melting profiles. Whether hormones *in vivo* would actually interact with the nucleic acids of the nuclei or with proteins is less clear. Mangan *et al.* (1967) suggested that diethylstilbestrol taken up by prostatic nuclei binds to protein rather than DNA. Sluyser (1966a,b) found that cortisol and other steroids could be bound to purified histones. This observation opens up the possibility of hormonal control of nuclear activity through regulation of DNA–histone interactions (Dahmus and Bonner, 1965; Bonner *et al.,* 1968), an extension of which is the concept of derepression of DNA by acetylation of histones (Allfrey *et al.,* 1968). On the other hand, there is no change in the rate of histone turnover at rat liver during alternate enzyme induction and repression (Byvoet, 1967). Thyroxine treatment has been reported to increase template activity in chromatin from tadpole liver (Kim and Cohen, 1966) and Stackhouse *et al.* (1968) have shown that cortisol *in vitro* increases to a small extent the template efficiency of rat liver chromatin. It should be noted, however, that in *in vitro* studies Litwack *et al.* (1963, 1965) concluded that cortisol is bound in the liver almost exclusively by cytoplasmic proteins. Dingman and Sporn (1965) found very little binding of cortisol to nuclear structures by comparison with that achieved by actinomycin. Conversely, testosterone administered *in vivo* appears to be bound rapidly by prostatic nuclei to acidic nuclear protein and within the nuclei undergoes conversion to dihydrotestosterone through enzyme activity located in the chromatin (Bruchovsky and Wilson, 1968a,b). As with estradiol specific binding proteins are involved (Fang *et al.,* 1969*).

III. Tissue Specificity in Hormone Response

In addition to the complexity of responses evinced by responsive tissues, we can also ask how widespread are the influences of specific hormones and to what extent do demonstrable effects in infracellular systems tie up with grosser compositional changes. We must limit consideration to matters relevant to protein and nucleic acid metabolism. An obvious starting point, again with insulin (which it must be admitted indicates the author's personal interest in this hormone) is to consider the fact that, despite its well recognized stimulatory influence on amino acid incorporation into protein in muscle (Manchester, 1965), attempts to establish a growth-promoting action of insulin in normal animals, as judged by extra protein deposition, are unconvincing (for references see Manchester and Young, 1961). Of course it is true that diabetic animals lose nitrogen and that insulin therapy stems this loss, but in *in vitro* demonstrations normal cardiac or diaphragm muscle responds to the hormone and it is not necessary to start with diabetic material. Thus we have a possibly paradoxical situation where a well-documented *in vitro* effect does not necessarily have an *in vivo* counterpart, except in the negative instance of the effect of lack of the hormone. However, one of the consequences of the multiplicity of metabolic parameters influenced by a single hormone is that *in vivo* responses are bound to be the summation of extremely complex interactions which may show little final change because of the operation of homeostatic mechanisms. Likewise, turning to RNA metabolism, although the cellular RNA level as well as protein content shows a fall in diabetes, the influence of insulin on normal tissue is not marked (Manchester, 1967b; Wool *et al.*, 1968).

Another problem of some concern is the difference in time scale of responses. The time that injection of insulin takes to lower blood sugar is to be measured in minutes and correlates with the speed with which it influences metabolic processes in muscle and adipose tissue. Apart from the claim of Wool *et al.* (1968) of the existence of a rapidly turning over protein controlled by insulin, the hormone is not thought to influence greatly the *de novo* synthesis in muscle of any specific identifiable proteins. But when we consider the hormone's influence on the liver, we find first that many responses are much slower—a matter of hours for proteins—and that the metabolic changes can in part be ascribed to synthesis and repression of certain key enzymes (Weber *et al.*, 1966). Thus on the face of it we are dealing in different tissues with two very different types of response whose interrelationships are not at all clear. Again we have posed the question whether the primary action of the hormone is of necessity different in the two instances or whether the difference arises from an important alteration in the sequelae. Selective synthesis of specific proteins in the liver could be indicative of regulation of formation of specific messenger RNA molecules, but this seems unlikely to apply in muscle (Section V,A).

The number of tissues responsive to insulin are numerous (Field and Adams, 1965; Grossman and Manchester, 1966; Williams and Ensinck, 1966) though the extent of the response is variable. When we turn to other hormones, e.g., thyroxine and pituitary growth hormone, we encounter a different type of problem. Though from carcass analysis, etc., the deposition of protein under the influence of growth hormone (reviewed in Ketterer *et al.,* 1957) can be seen to be widespread in the musculature and visceral organs, technical considerations have necessitated that most of the *in vitro* tissue responses have been observed with muscle and adipose tissue, whereas work with subcellular systems has been mostly though not entirely with preparations from liver (Korner, 1965; Earl and Korner, 1965; Breuer and Florini, 1966; Florini and Breuer, 1966). There is no absolute guarantee that results so obtained interrelate. The possible role of growth hormone in control of cartilage synthesis again is likely to be of consequence for many tissues (Daughaday and Kipnis, 1966) and as such is important, for the proliferation of cartilage is required to support regular linear growth. A complex pattern of changes in composition of body constituents results from glucocorticoid treatment (Kochakian and Robertson, 1951; Roberts, 1953; Clark, 1953; Silber and Porter, 1953). Likewise there are interesting variations in the extent of synthesis of the M versus the H forms of lactic dehydrogenase following androgen and estrogen administration (Goodfriend and Kaplan, 1964). With the androgens there is also the problem of their well-known growth stimulating effects on tissues unrelated to their main target organs such as muscle, the extent of which vary considerably from one species to another, only the accessory sex organs responding similarly in each species (Frieden *et al.,* 1957; Kochakian *et al.,* 1957; Kochakian, 1965). Korner and Young (1955) made studies of the overall weight changes of many organs after androgen treatment and Breuer and Florini (1965) found castration reduced the synthetic capacity of muscle ribosomes. In the liver a "sex-associated" protein, which is present in the male but only to a very small extent in the female, has been identified (Bond, 1965; Barzilai and Pincus, 1965). The protein appears to decline in concentration following castration or treatment with estradiol, but its function is not known. Influence of androgen on kidney metabolism is also well established (Shaw and Koen, 1963; Kochakian, 1965). Probably we are wrong to think of hormone response as an all-or-nothing phenomenon but must consider it in graded terms. Thus both among species and among different tissues within one organism (and even within one tissue—e.g., effects of estrogen and progesterone on chick oviduct epithelium—P. O. Kohler *et al.*, 1968*), nature has produced a most complex pattern of differential response for which at the moment we have no adequate conceptual framework. Any possible explanation for this diversity seems likely ultimately to require an approach entirely different from that to most of the experimental work described in the following sections.

IV. General Considerations in Control of Protein Synthesis

A. Messenger-RNA

The general concept of a control of protein synthesis and of production of specific proteins by hormone interaction with genes is fairly obvious, given our present understanding of the process of genetic transcription to form messenger-RNA and the role of mRNA in polysome function and coding of amino acid sequences. (See introductory chapter to this volume for a description of this general area.)

The concept was first applied to a specific situation by Karlson (1963) as an explanation of the phenomena observable with ecdysone, a steroid secreted by the prothoracic glands of insect larvae that induces molting. Administration of ecdysone results in the appearance of characteristic puffs at two loci on the chromosomes of cells of the salivary gland (Clever and Karlson, 1960). The "puffs," which indicate the enlargement and unfolding of chromosomal material, appear to be sites of RNA synthesis, the RNA so produced being responsible for coding for the synthesis of the enzyme dopa carboxylase, which is required to bring about cuticle formation. Since pupation and enzyme and RNA synthesis are prevented by treatment with actinomycin (Sekeris, 1965; Clever and Romball, 1966), it is possible to conclude that the sequence of changes in the molting transformation is initiated by interaction of ecdysone with the nuclear material, from which the other changes follow. Rather unwisely in the event, the existence of similar mechanisms as the basis of the action of many other hormones was postulated by many workers. For some hormones such a mechanism was possible if unproven, in others clearly unlikely, but this seemed no deterrent. Before looking at individual cases, it is useful to ask the general question as to the extent to which initiation of messenger production could provide a suitable mechanism for control of protein metabolism in higher animals and what are the criteria by which such controls can be recognized.

Much of our knowledge of the mechanism of protein synthesis derives from work with microorganisms, though there is abundant evidence of similar functioning in eukaryotic cells. But though the mechanisms of protein formation may be similar in all forms of life, it does not necessarily follow that the control of these processes will be the same. In particular, the overall purpose of control in unicellular organisms is likely to be different from that of higher animals, who exhibit a greater emphasis on metabolic homeostasis which manifests itself in a much longer cell lifetime and a greater emphasis on protein turnover than on growth (Revel and Hiatt, 1964). Nonetheless the microorganism obviously has to adapt to different nutritional conditions and to be able to develop rapidly the necessary enzymes to cope with different nutritional supplies, as for example by the induction of the *lac* operon enzymes. The existence of the *lac* operon implies the simultaneous translation of a limited group of genes, possibly through the formation of a messenger-RNA molecule coding for several proteins. Several examples of polycistronic messengers are known for microorganisms (Ames and Martin, 1964), but the existence of similar material in animal cells has yet to be demonstrated. The concept of functional genic units in higher animals for control of glycolysis and gluconeogenesis (Weber *et al.*, 1966) might be regarded as indirect evidence for their existence (see also Granick and Kappas, 1967), but observations on the genetic origin of the chains of hemoglobin and lactic dehydrogenase do not support this concept (Huehns and Shooter, 1965; Wilkinson, 1965; Goodfriend *et al.*, 1966). However, whether polycistronic messenger exists in animal cells does not directly affect the question of whether hormones control mRNA production.

If hormones do control messenger-RNA production, and through this the rate of protein synthesis, it must mean that the rate of messenger production and its availability is a limiting step. It is obvious that supply of energy, nutrients, or amino acids for protein synthesis could themselves be limiting (e.g., Ove *et al.*, 1967), but this is less likely to be of consequence for animals than for unicellular organisms. However, imbalanced diets and certain amino acids do appear to exert a profound effect, which may or may not be exerted through stimulation of hormone secretion, on the protein synthetic capacity of liver polysomes (Wunner *et al.*, 1966; Sidransky *et al.*, 1967; Cammarano *et al.*, 1968; Freedman *et al.*, 1968; see also Chapter 34). Fallen *et al.* (1966), Earl *et al.* (1967), and Goldstein and Reddy (1967) have suggested that insulin controls protein synthesis in muscle by regulation of the rate of entry into the tissue of certain key amino acids. An additional point of regulation in muscle might arise from the extensive oxidation of branched chain amino acids normally present only in small amounts (Manchester, 1965*).

That messenger-RNA has a short life, or half-life, allows the potentiality for its availability readily to become limiting and moreover for its availability to be dependent on continual production. The half-life of messenger-RNA in animal tissues, measured by indirect indices, is much longer than for microorganisms and is to be measured in hours rather than minutes. Quite how many hours is debatable (see Manchester, 1968a, for a comparison of estimates), but it seems clear that the figures of 1–12 hours are substantially less than those of around 5 days for ribosomal RNA (Loeb *et al.,* 1965; Erdos and Bessada, 1966; Hadjiolov, 1966; Wilson and Hoagland, 1967). Of course it has to be remembered that estimates of messenger half-life must of necessity be an average figure combining both long and short and it has been suggested by Appel (1967) and Wilson and Hoagland (1967) that the figures for rat brain and liver fall into two groups of around 3 and 80 hours. (For *Escherichia coli* figures of 8 and 70 minutes have recently been put forward—Forchhammer and Kjeldgaard, 1967). Although there is no necessity for the stability of a messenger to be related to the life of the protein for which it codes, it appears that the half-lives of various proteins vary over a wide range from many days (according to the tissue and protein—Green and Miller, 1960; Richmond *et al.,* 1963; Schimke, 1964; Drysdale and Munro, 1966; Beattie *et al.,* 1967; see also Chapter 32) to as little as 8 minutes (Garren *et al.,* 1965). This latter figure is most likely to be less than that for the messenger coding for the protein and immediately suggests the possibility of a regulation of translation of the messenger by feedback. This introduces an alternative or additional possible means of control of protein synthesis which is considered later.

If availability of mRNA is a limiting step in protein synthesis, there must conversely be an excess of ribosomal-RNA and tRNA. Whether there are many free ribosomes available in the cell is questionable (Henshaw *et al.,* 1963; Mathias *et al.* 1964; Wilson and Hoagland, 1965; Drysdale and Munro, 1967) and is a difficult problem to investigate experimentally. Subcellular preparations often contain considerable quantities of monomers and dimers when analyzed on sucrose density gradients, but it is difficult to tell whether these result from breakup of polysomes during homogenization. That warming leads to loss of dimers to the monomer peak probably indicates that the dimers are temperature-dependent magnesium-linked complexes as opposed to being attached to pieces of messenger (Petermann and Pavlovec, 1967), but whether the monomers have appeared during preparation or not still remains uncertain. Some interesting calculations for the relative abundance of messenger strands and ribosomes in *E. coli* are provided by Goldstein *et al.* (1965). An estimate of the relative abundance of polysomes and free ribosomes in the average liver cell is shown in Fig. 6 of the introductory chapter of this volume. However, the whole matter becomes more complex if initiation of the translation of messenger uses ribosomal subunits rather than monomers (Girard *et al.,* 1965;

Joklik and Becker, 1965; Hogan and Korner, 1968). Indeed it is suggested that in muscle single ribosomes are not even equilibrating rapidly with the subunits (Kabat and Rich, 1969) and they may therefore not be functioning in protein synthesis.

If indeed there is no pool of free ribosomes in the cell waiting for messenger, a hormonal initiation of new messenger is unlikely to promote much extra protein synthesis, though it could of course alter the pattern of proteins being synthesized. For example, gluconeogenesis in the starving animal takes place at the expense of amino acids that have to come from protein or could be used for protein synthesis, but at the same time the concentration in the liver of enzymes associated with amino acid catabolism and gluconeogenesis rises (e.g., transaminases and phosphoenolpyruvate carboxykinase). The adaptive enzymes of liver are supposed to have relatively short half-lives—approximately 1.5–3 hours (Goldstein *et al.*, 1962; Kenney, 1962, 1967; see also Chapter 31)—and their appearance coupled with the decline of the key glycolytic enzymes (glucokinase, phosphofructokinase, pyruvate kinase) or *vice versa* could result from a change in the relative levels of the different messenger species. Otherwise, in situations where hormones are promoting increased growth and synthesis of protein, it has come to be recognized that the new RNA formed simultaneously is as much ribosomal as messenger.

B. Polymerase Activity

An index of rate of RNA synthesis is of course the rate of its radioactive labeling from suitable percursors. Another measure, even if indirect, is the activity of the enzyme responsible for joining together ribonucleotide units according to the DNA template, namely RNA polymerase. The elevation of its activity by various hormones is evidence of their influencing RNA synthesis as well as protein synthesis, but of course it does not directly follow from this that protein synthesis is increased *because* of elevation of RNA synthesis.

Polymerase is a complex enzyme system—as might be expected from its function. The effect of hormones on its activity in mammalian systems is usually measured in terms of the activity of isolated nuclei, though efforts at solubilization have been made (Ballard and Williams-Ashman, 1964, 1966; Ramuz *et al.*, 1965; Furth and Ho, 1965; Liao *et al.*, 1968). Seifert and Sekeris (1969) have commented on the low stability of their solubilized enzyme and the fact that it possesses only 1% of the specific activity of bacterial preparations. Cunningham and Steiner (1969) were able to purify in good yield a more stable form of polymerase from rat liver. Elevation of activity of hepatic nuclei has been shown following treatment with growth hormone, insulin, testosterone and estrogen, and cortisol (Hancock *et al.*, 1962; Gorski, 1964; Barnabei *et al.*, 1965, 1966; Pegg and Korner, 1965; Steiner and King, 1966; Widnell and Tata, 1966a) and with carcinogens (Bresnick, 1966). However, the interpretation of all these results is complicated by the fact that the type of RNA produced is dependent on the salt concentration and activating ion present. Thus the apparent activity of liver nuclei is greater in the presence of ammonium sulfate, and in its presence hormonal influences are sometimes no longer observed (Gorski, 1964; Pegg and Korner, 1965; Jacob *et al.*, 1969). [This is an effect reminiscent of that of fluoride on adenyl cyclase: addition of 10 m*M* fluoride enhances cyclase activity but obscures hormonal effects (Sutherland *et al.*, 1965)—and indeed fluoride can activate polymerase also (MacGregor and Mahler, 1967).] On the other hand, effects of various hormones on the ammonium sulfate-activated system have been seen in other studies, sometimes after longer periods of induction (Breuer and Florini, 1966; Widnell and Tata, 1966b; McGuire and O'Malley, 1967), so the picture is not clear-cut. Nor is the nature of the product formed or the reason for ammonium sulfate activation clear. Widnell and Tata (1964, 1966b) found that nuclei incubated in the presence of 4 m*M* Mn^{2+} and 10% saturated ammonium sulfate produced RNA with a DNA-like composition whereas with 5 m*M* Mg^{2+} alone ribosome-like material was formed. Nair *et al.* (1967) have made similar findings with heart nuclei. Mn^{2+} and ammonium sulfate are unlikely to be physiological stimulators; on the other hand, only a minute proportion of the genetic information of the nucleus codes for ribosomal and transfer RNA (Giacomini and Spiegelman, 1962; Goodman and Rich, 1962; Spiegelman and Hayashi, 1963; Morell *et al.*, 1967). The different forms of RNA have different origins in nucleate cells (Busch *et al.*, 1963; Brown and Gurdon, 1964; Perry *et al.*, 1964; Perry, 1967). Maul and Hamilton (1967) and Pogo *et al.* (1967) have put forward evidence that the Mg^{2+}-stimulated polymerase activity is primarily associated with the nucleolus whereas the Mn^{2+}-ammonium sulfate activity is chromosomal. Jacob *et al.* (1968, 1969), however, were able to show Mn^{2+}- as well as Mg^{2+}-dependent activity in the isolated nucleolus. Subsequent unpublished studies by this group show that RNA polymerase prepared in soluble (DNA-free) form from nucleoli differs in its divalent cation requirements from soluble polymerase prepared from whole nuclei which represents principally extranucleolar polymerase. The data are consistent with the presence of an Mg^{2+}-preferring enzyme in the nucleolus and an Mn^{2+}-activated enzyme in the chromosome fraction, although the differing action of these two divalent cations on the polymerases is not absolute. Cunningham and Steiner (1966) found that ammonium sulfate produced apparent stimulation of polymerase activity because of its capacity to inhibit a nuclear phosphodiesterase which can under *in vitro* conditions destroy a large portion of the newly synthesized *n*RNA, and similar conclusions have been drawn by Meisler and Tropp (1969). Certainly nucleotide incorporation proceeds longer and more linearly in media of high ionic strength. MacGregor and Mahler (1967) confirmed that substantial breakdown of RNA does take place during *in vitro* incubation of hepatic nuclei in media of low ionic composition, and Chambon *et al.* (1968a) found that the A-U rich material probably underwent degradation faster than G-C rich components. Such would explain why the Mg^{2+}-stimulated activity appears to produce "ribosomal-like" RNA, and, indeed, Johnson *et al.* (1969) noted a progressive change in the type of product as the concentration of ammonium sulfate was raised irrespective of whether Mg^{2+} or Mn^{2+} was the activating ion present. Incubation of nuclei with trypsin enhances polymerase activity (Pegg and Korner, 1967), but whether this results from removal of interfering histones or nucleases is not clear. Assay of polymerase activity is therefore not without its problems, and Chambon *et al.* (1968b) have raised doubts as to the validity of data derived from priming of polymerase with isolated chromatin because of the type of RNA produced which is abnormally rich in adenine and low in cytosine. Thus for the present it is not certain if there is more than one form of the enzyme in the nucleus (as claimed by Liao *et al.*, 1969 and Roeder and Rutter, 1969) the normal activating ions and the means by which ammonium sulfate, fluoride, and detergents can activate (Chambon *et al.*, 1968a,b). What is often overlooked is that very high concentrations of sulfate ions will complex with divalent cations, and high concentrations of ammonium ions will interfere with formation of divalent ion nucleoside triphosphate chelate complexes which are usually considered to be the preferred form of the nucleotides, so it is not clear what the real concentrations of free ions and complexes are likely to be in many of the media used. For present considerations it seems unlikely that rRNA requires simultaneous synthesis of mRNA or vice versa—formation of both would be independently controlled. Transfer RNA, by comparison with rRNA, is synthesized outside the nucleolus (Perry, 1962).

C. Ribosome Synthesis and Activity

Whether there is a close correlation between the rate of rRNA and mRNA synthesis is not clear, and the answer to this question depends partly on the way in which polysomes are formed. With nucleate cells there clearly has to be a migration of both mRNA and rRNA from the nucleus, where they are synthesized, to the cytoplasm, where they are functional. The way in which the RNA's cross the nuclear membrane is confused, and in particular whether messenger is in some way joined to ribosomal material has not been clearly resolved. In microorganisms without nuclei it is claimed that there exist in the cell free ribosomal subunits (30 S and 50 S particles) and that ribosomal movement along a messenger strand is initiated by the attachment of first the 30 S particle followed by attachment of the 50 S unit to the complex (Mangiarotti and Schlessinger, 1966, 1967). Such ribosomal particles as exist (70 S) are thought to be ribosomes that have detached from the messenger on completion of translation and may even possess changed subunits (Nakada and Kaji, 1967), though Dresden and Hoagland (1967a,b) presented evidence for their reusage. Joklik and Becker (1965) observed rapid attachment of free mRNA to 40 S subunits in HeLa cells, and Girard *et al.* (1965) of new rRNA entering polysomes before free monomeric ribosomes. Some ribosomal subunits are often found on sucrose gradients, and their presence is much enhanced by lack of Mg^{2+}. How many subunits are available *in vivo* will obviously depend partly on intracellular concentration of the Mg^{2+} ion (as opposed to the total amount of magnesium content). The implications of this are discussed again later (p. 278) but it is clear that the Mg^{2+} concentration is likely to be considerable less in bacteria than the conventional 5 mM of *in vitro* experimentation (Hurwitz and Rosano, 1967), and the same is true in animal cells (England *et al.*, 1967; Kerson *et al.*, 1967; Rose, 1968*). It is probably fair to say that the precise cytoplasmic content of ribosomal subunits is as yet not known, and with this it is difficult to say whether messenger is in excess or not. The role of subunits in initiation of translation of messenger in fact makes their availability of greater significance than that of monomers (R. E. Kohler *et al.*, 1968*).

It is apparent that in the nucleus there is a rapid turnover of RNA at a rate which is considerably in excess of the amount of RNA being synthesized and transferred to the cytoplasm (Bramwell and Harris, 1967; Georgiev, 1967; Shearer and McCarthy, 1967). There is evidence that ribosomal RNA for the two subunits is synthesized in nucleate cells in a single strand with a sedimentation coefficient of 45 S (Penman, 1966; Penman *et al.*, 1966) which in the nucleolus is cleaved following methylation first to 35 S then to 28 S and 18 S material. There has been confusion as to whether 18 S material is turning over more rapidly than the 28 S (Girard *et al.*, 1964; Henshaw *et al.*, 1965), whether there is more than one variety of 45 S material (Muramatsu *et al.*, 1966), and whether another rapidly labeled RNA fragment, associated with the 45 S material but not part of the 28 S or 18 S subunits, is messenger. An increased labeling of 45 S material under the influence of cortisol was interpreted by Finkel *et al.* (1966) as an increased synthesis of messenger, the transfer of which to the cytoplasm is brought about by the hormone, but they had difficulty in demonstrating distinct separation of label from the 18 S material. Weinberg *et al.* (1967) decided that the short-lived fragments of 45 S RNA not destined for 28 S and 18 S material were not connected with messenger. Warner *et al.* (1966b) recognized a heterogeneous nuclear fraction, distinct from 45 S RNA, which labeled rapidly, but it seems doubtful if this either is related to messenger RNA (Penman *et al.*, 1968). However, Parsons and McCarty (1968) believe that messenger RNA can be identified in rat liver nuclei as a rapidly labeling 45 S ribonucleoprotein complex distinct from ribosomal precursors. If indeed the nuclear sites of messenger and ribosomal RNA are separate, as they presumably must be (Prescott, 1964; Perry, 1967), then any association of messenger with 45 S material is likely to arise subsequent to transcription rather than to represent simultaneous synthesis. Peacock and Dingman (1967), King and Fitschen (1968), and McIndoe and Munro (1968*) have recently demonstrated by gel electrophoresis the existence of many new species of RNA in addition to those normally recognized by gradient centrifugation. Ultimate identification of the role of these may resolve some of the confusion of labeling experiments. A useful summary of recent literature on ribosomal genesis is given in Darnell (1968) and Maden (1968).

Another fairly obvious way in which to investigate the question of whether hormonal deprivation leads to a relative deficiency of messenger is to examine polysomal profiles and assess whether the proportion of polysomes to monomers changes. Thus for the hypophysectomized animal the proportion of polysomes to smaller forms appears to be lower than normal in both liver and muscle (Korner, 1965; Earl and Korner, 1966; Florini and Breuer, 1966). Likewise in diabetic muscle the proportion of polysomes is below normal but is raised by treatment with insulin (Stirewalt *et al.*, 1967). If ribosomal preparations from hormone-deficient animals are only short of messenger, addition of poly U should bring their amino acid incorporating activity up to that obtainable with preparations from normal animals. Indeed the apparent response to synthetic messenger should be greater, the greater the natural messenger deficiency. Good responses to poly U have been seen with ribosomes from both hypophysectomized and diabetic animals (Korner, 1965; Rampersad and Wool, 1965; Korner and Gumbley, 1966), but it appears that the maximal activities attainable with poly U never quite reach the values with preparations from normal animals, from which it is argued that ribosomes from hypophysectomized and diabetic animals suffer from some intrinsic defect unrelated to their content of natural messenger.

The response of subcellular systems to synthetic nucleotide sequences requires some consideration because observations so obtained (as above) provide one of the lines of evidence for the concept of control of protein synthesis at the translational as opposed to transcriptional level. To be sure that a difference in response to poly U resides in the ribosomes requires that poly U be present at saturating levels. Bearing in mind that in a polysome natural messenger will constitute at most about 1.5% of the total RNA present (90 nucleotide units/ribosome—Noll *et al.*, 1963; Staehelin *et al.*, 1964), it is curious to find that saturation with poly U occurs only when the poly U is present at a severalfold quantitative excess to the ribosomes (e.g., 250 μg of poly U to 30 μg of rRNA—Korner and Gumbley, 1966; 500 μg of poly U to 30 μg of rRNA—Wool and Cavicchi, 1967; 200 μg of poly U to 20–100 μg of rRNA—Tata and Williams-Ashman, 1967) despite the affinity of ribosomes for synthetic messenger (Okamoto and Takanami, 1963). Moreover in the presence of such excess one might expect the synthetic messenger on a mass action basis to displace the natural messenger, but the extent to which this happens, as judged by the lack of any interference of synthetic messenger to the incorporation of amino acids for which it does not code, is variable (Arnstein *et al.*, 1962; Weinstein *et al.*, 1963, 1966; Korner, 1964a; Wool *et al.*, 1968) and complicated by the presence of tRNA (Aaronson *et al.*, 1966). The optimal concentration of Mg^{2+} for poly-U directed protein synthesis (15 mM) is higher than for endogenous messenger (7.5 mM—Revel and Hiatt, 1965) and is near the range in which Szer and Ochoa (1964) and So and Davie (1965) observed ambiguous reading of the genetic code. Since a concentration of Mg^{2+} around 10 mM tends to promote dimer formation (Petermann, 1964), it may be that the high concentration of poly U required to saturate the system results because there is set up effectively a mass action equilibrium between poly U and Mg^{2+} for binding of the ribosomes. More poly U would tend to pull more ribosomes away from the dimer state, whereas excess Mg^{2+} leads to a fall away in the effectiveness of the poly U. A competition for ribosomes of synthetic messenger with Mg^{2+} rather than with natural messenger would explain why poly U does not interfere with the incorporation of other amino acids. It would also raise further complications of interpretation of observations using synthetic messenger. More recent papers on ambiguities in translation of artificial messenger are those of Gupta (1968), Friedman *et al.* (1968), and Deer and Scanlon (1969*).

D. Use of Inhibitors

Many studies of protein synthesis involve the use of specific inhibitors of protein and RNA formation. Since it is usually difficult, if not impossible but for immunological methods, to measure the actual quantity of protein present or being formed in the majority of systems, evidence that increased enzyme activity under hormonal treatment does indeed result from synthesis of new protein as opposed to other forms of enzyme activation can be most readily shown by failure of response in the presence of a specific inhibitor of protein synthesis (e.g., Steiner and King, 1964; Sols *et al.*, 1965). Details of the site and mode of action of inhibitors of RNA and protein synthesis are given in the introductory chapter of this volume.

Puromycin and cycloheximide are the two agents most commonly used with mammalian systems, but ethionine, *p*-fluorophenylalanine, proflavin, and a host of natural antibiotics have been tried on occasion (Manchester, 1961; Garren *et al.*, 1964; Clark and Chang, 1965; Sols *et al.*, 1965; Colombo *et al.*, 1966; Clark-Walker and Linnane, 1966; Weinstein and Finkelstein, 1967). Sensitivity to inhibition is also taken as evidence that incorporation by cell-free protein-synthesizing systems represents genuine ribosomal activity in forming peptide bonds, but most such systems show greater sensitivity to puromycin than to cycloheximide (Colombo *et al.*, 1965; Korner, 1966; Campagnoni and Mahler, 1967; Wool and Cavicchi, 1967; Means *et al.*, 1969). Demonstration of the dependence of enhanced protein synthesis on prior nucleic acid formation, for example messenger production, is often deduced from results obtainable with actinomycin, though in this instance it appears that actinomycin more effectively inhibits rRNA than mRNA synthesis (Perry, 1962, 1967; Georgiev *et al.*, 1963; Reich and Goldberg, 1964). Conversely failure of actinomycin to inhibit endocrine influences is taken as evidence that RNA formation is not involved (Eboué-Bonis *et al.*, 1963; Wool and Moyer, 1964; Korner, 1965; Martin and Young, 1965; Mayne and Barry, 1965; Dawson *et al.*, 1966; Kostyo, 1966; Wool and Cavicchi, 1966).

Use of inhibitors in this fashion is always open to the objection that the effects noted may not be specific or that there may be multiple points of effect with quite different end results. There is no reason at the present to suppose that the majority of findings based on the use of these materials are not justified. However a number of reservations should be borne in mind. As pointed out by Hechter and Halkerston (1965), the use of inhibitors *in vivo* may lead to various stress reactions. Lippe and Szego (1965) have suggested that much of the effect of actinomycin in preventing the action of estrogens on uterine metabolism may result from elevation of plasma glucocorticoids, which are known inhibitors of estrogen action (Planelles *et al.*, 1962). A similar explanation may lie behind the findings of Jondorf *et al.* (1966) of a stimulation of amino acid incorporation by isolated rat liver microsomes following injection of cycloheximide, though the ability of the inhibitor to promote polysome formation may also be of significance (Webb and Morris, 1969*). Another of the problems with the use of inhibitors is determination of a "suitable" dose. Actinomycin appears to be extremely toxic *in vivo* (Young, Robinson, and Sacktor, 1963) and to cells in culture (Soeiro and Amos, 1966), for reasons which are not clear; one therefore cannot be sure that its full proper effects are not most evident just at the time when its toxicity is in evidence. In addition to inhibiting RNA synthesis, actinomycin has been claimed both to shorten and lengthen the life-span of mRNA (Acs *et al.*, 1963; Kennell, 1964; Trakatellis *et al.*, 1964a,b; Chantrenne, 1965). It can also on occasion unexpectedly raise the levels of certain enzymes—for example, clearing-factor lipase (Eagle and Robinson, 1964; Wing and Robinson, 1968), duodenal alkaline phosphatase (Moog, 1964), transaminases, serine dehydrase, and tryptophan pyrrolase (Garren *et al.*, 1964; Rosen *et al.*, 1964)—as well as delaying the decay of induced glucokinase (Sols *et al.*, 1965). Some of these effects are also observable with puromycin or fluorophenylalanine (Moog, 1964; Garren *et al.*, 1964; Rosen, 1965; Sols *et al.*, 1965) which precludes the effects being peculiar to actinomycin. It is possible that there are rapidly turning over suppressors of protein and enzyme synthesis whose normal production is in these cases prevented by the inhibitors, but this postulate is quite unproven. It is also possible that inhibition of specific aspects of the general synthetic and control machinery leads to abnormal compensatory activities. For example inhibition of protein synthesis by cycloheximide can lead to abnormalities of RNA synthesis (Fiala and Davis, 1965; Fukuhara, 1965; Kay and Korner, 1966; Means and Hamilton, 1966a) and of ribosome formation (Ennis, 1966) though the latter is in conflict with the findings of Warner (1966) and Warner *et al.* (1966a) that only puromycin and actinomycin, but not cycloheximide, interfere with the joining together of ribosomal proteins and RNA.

The glycogenolytic action of puromycin is well established (Hofert *et al.*, 1962; Hofert and Boutwell, 1963), and it is possible that the antibiotic raises the intracellular cAMP through inhibition of phosphodiesterase (Appleman and Kemp, 1966). Such an action would obviously explain its capacity to prevent stimulation of glycogen synthesis by insulin (Sovik and Walaas, 1964; Sovik, 1967), which is at variance with its lack of influence on the stimulation by insulin of glucose and amino acid uptake (Carlin and Hechter, 1962; Fritz and Knobil, 1963). What also has to be recognized with puromycin, in contradiction to cycloheximide, is that though it prevents the formation of complete peptides it does not necessarily interfere with the movement of ribosomes along the messenger strand. A form of synthesis of peptides can to some extent still take place in which the puromycin is tacked onto the growing peptide chain and thus terminates its growth. Di- and tripeptides so formed (Smith *et al.*, 1965) will not be precipitated by the usual protein-precipitating agents and may therefore go unnoticed. This limited synthesis hardly leads to synthesis of recognizable proteins, but may be of consequence when the effect of puromycin on the capacity of insulin to stimulate amino acid uptake is considered (see p. 256). It has also been used by Wool and Kurihara (1967) to determine the proportion of ribosomes that are functional in normal versus diabetic preparations. Some histological changes following puromycin administration have been reported by Estensen *et al.* (1966) and Gambetti *et al.* (1968). Noteworthy was the observed swelling of mitochondria, as also found by Staehelin and Maier (1968). Injection of cycloheximide, by contrast, can enhance the glycogen content of the uterus, and other inhibitory agents seem to be even more effective in this respect (Bitman *et al.*, 1966*; Valadares *et al.*, 1968*).

The difference in mechanism by which puromycin as opposed to cycloheximide inhibits protein synthesis is reflected in the differential degrees of inhibition observable in various systems with the two inhibitors. Thus whereas puromycin is equally effective in tissues and subcellular systems, and in microorganisms as opposed to eukaryotic cells, the inhibitory effects of cycloheximide are not to be observed with, for example, microorganisms or mitochondria possessing 70 S as opposed to 78 S or 80 S ribosomes or with resistant strains of yeasts (Siegel and Sisler, 1965; Cooper *et al.*, 1967). Sensitivity to cycloheximide is in fact complementary to that to chloramphenicol, and it is possible that both inhibitors have a similar point of action, namely to interfere in some way with the growth of the nascent peptide chain on the ribosome (Das *et al.*, 1966; Lin *et al.*, 1966; Weber and DeMoss, 1966). It is not known at present whether this action is specifically concerned with inhibition of transfer of peptide-tRNA from one binding site to another (Das *et al.*, 1966), inhibition of the enzyme involved in peptide bond formation (Siegel and Sisler, 1964; Traut and Monro, 1964; Felicetti *et al.*, 1966; Lin *et al.*, 1966) or, since release of nascent peptides by puromycin is not prevented (Williamson and Schweet, 1965), inhibition of the movement of the ribosome along the messenger. This last action, proposed by Wettstein *et al.* (1964), would explain the "freezing" action of cycloheximide on polysomes (Wettstein *et al.*, 1964; Colombo *et al.*, 1965; Trakatellis *et al.*, 1965; Stanners, 1966), though it could of course equally arise from the other modes of inhibition. An inhibition of amino acid incorporation through specific inhibition of movement of the ribosome along the messenger would imply that a stuck ribosome cannot go on repetitively reading the same codon. Conversely it is arguable that a failure to respond to cycloheximide could suggest the occurrence of repetitive reading. The existence of separate binding sites on the ribosome for aminoacyl-tRNA and peptidyl-tRNA (Spyrides, 1964; Warner and Rich, 1964a; Cannon, 1967; Igarashi and Kaji, 1967; Leder and Nau, 1967; Seeds *et al.*, 1967) complicates the problem but in our present state of knowledge does not preclude the possibility. Such an occurrence might be the explanation of the varying sensitivity to cycloheximide of intact cells and subcellular systems. Thus Colombo *et al.* (1965) found the inhibitor to produce more complete inhibition of incorporation by intact reticulocytes than isolated polysomes derived therefrom, and similar findings were made with liver and L cells by Ennis and Lubin (1964) and cultured cells (Bennett *et al.*, 1964). The inhibitor brings about virtually complete cessation of protein synthesis by HeLa cells (Warner *et al.*, 1966) and isolated diaphragm muscle (Manchester, 1967a), ascites cells (Hogan, 1969), and also *in vivo* (Young *et al.*, 1963; Korner, 1966; Wool and Cavicchi, 1966). Conversely Siegel and Sisler (1963) found that cycloheximide *in vitro* did not inhibit all incorporation by yeast ribosomes, a finding confirmed by Korner (1966) for liver ribosomes, and Campagnoni and Mahler (1967) and Wool and Cavicchi (1967) found ribosomes from brain and muscle to be remarkable insensitive. The reasons for these discrepancies are not known, but it is not improbable that protein synthesis *in vitro* is more disorganized than *in vivo* (Nisman and Pelmont, 1964), and if this were the explanation sensitivity to cycloheximide might constitute an important criterion of performance. On the other hand, Munro *et al.* (1968) and Baliga *et al.* (1969) have evidence that cycloheximide interferes with the essential sulfhydryls of the ribosomal transferase II, and that sensitivity to inhibition is much reduced in the presence of protective agents such as glutathione. It is possible that cycloheximide has more than one site of action in inhibiting protein synthesis (Lin *et al.*, 1966).

E. Peptide Bond Formation in Vitro

In consideration of the action of ecdysone, Sekeris and Lang (1964) and Sekeris (1965) have offered as evidence that the hormone induces new messenger formation the fact that RNA synthesized under its influence can function as template to stimulate ribosomes from rat liver to synthesize dopa decarboxylase, an enzyme specifically induced by the hormone. This is indeed an important claim, for, short of specific isolation and characterization of messenger material, demonstration of its presence through its function is as good a title to proof as any and greatly superior to demonstration of production of rapidly labeling RNA, inhibition by actinomycin, etc., as so frequently shown. The question arises whether the claims can be substantiated, and this is discussed in some detail by Steiner (1966) and Korner (1967). Two other points could also be made. First, addition of RNA to ribosomal systems often has a stimulating effect on amino acid incorporation unrelated to its likely messenger content (Mathias *et al.*, 1964; Armentrout *et al.*, 1966*). Hunt and Wilkinson (1967) found that RNA from bone, liver, and reticulocytes all equally stimulated hemoglobin synthesis by a reticulocyte ribosomal system, and they suggested that the effect of the added RNA was to improve the use made of the endogenous messenger, possibly by limiting its destruction. Sekeris (1965) found that nRNA from control tissue also stimulated incorporation, though not as much as that extracted from ecdysone-treated tissue. In fact the degeneracy of the genetic code results in almost any sequence of bases, provided it does not contain the nonsense codons, coding for some sequence of amino acids. Thus whether *in vitro* a particular RNA species can function as a messenger probably depends primarily upon its physical configuration (Singer *et al.*, 1963; Haselkorn and Fried, 1964). Di Girolamo *et al.* (1967) could find no correlation between template activity and rapidity of labeling of RNA extracted from rat liver polysomes, ribosomes, and ribosomal subunits.

The second query is whether the rat liver ribosomal system, given a natural messenger from another species, is capable of carrying out the *full* synthesis of an enzyme protein [especially with the Mg^{2+} and ATP concentrations described by Sekeris and Lang (1964)]. As pointed out by Korner (1967) it has proved difficult to demonstrate the labeling, let alone the net synthesis, of albumin by liver ribosomes (Campbell, 1965). Serum albumin is of course a comparatively large molecule and labeling is more readily demonstrated with peptide hormones (Wagle, 1965; Adiga *et al.*, 1966, 1968*) as well as with hemoglobin (Dintzis, 1961; Warner and Rich, 1964b). For a larger molecule Morais and Goldberg (1967) have demonstrated labeling of thyroglobulin by a cell-free system, but labeling as such does not demonstrate complete *de novo* synthesis as opposed to completion of started chains. Moreover, figures of the actual amount of protein synthesis catalyzed by an isolated ribosome do not suggest that each ribosome synthesizes many peptide bonds, though this is bound to be an average value. Thus Lin *et al.* (1966) calculated that whereas intact reticulocytes synthesize about 110 peptide bonds per ribosome per minute, when the polysomes are isolated their incorporating capacity falls to about 1.3 amino acid residues per minute. A similar activity by liver ribosomes would preclude the possibility of their synthesizing a complete protein molecule in any reasonable incubation period. If dopa decarboxylase were to have subunits of molecular weight no greater than 35,000, this would still require a system capable of coupling 350 amino acid residues. Similar problems arise in consideration of claims for *de novo* synthesis of myosin by cell-free systems (Heywood and Nwagwu, 1968*, 1969*). Incorporating efficiencies similar to isolated reticulocyte polysomes have been claimed for a ribosomal system isolated from rat uterus (Teng and Hamilton, 1967a) and from muscle (Wool and Kurihara, 1967). It is interesting to note by contrast that from the rate of protein synthesis calculated by Manchester *et al.* (1959), Manchester (1961), and Hechter and Halkerston (1964), and assuming an rRNA content of about 1 mg per gram of tissue, ribosomes in isolated diaphragm muscle form at least 300 peptide bonds a minute. Maitra *et al.* (1967) have recently shown that RNA polymerase *in vitro* produces excessive amounts of small nucleotide sequences. It seems likely that here also *in vitro* activity is more disorganized than *in vivo*.

Gatica *et al.* (1966) have drawn attention to the fact that certain aminoacyl-tRNA species are rather unstable at a pH above 7 (Coles *et al.*, 1962; Wolfenden, 1963; Gatica *et al.*, 1966) and that their use in *in vitro* systems in place of free amino acids and pH 5 enzymes may introduce complications if the full complement of amino acids is necessary for true protein synthesis. On the other hand, Novelli (1967) has emphasized how the optimal conditions for loading tRNA with amino acids are very different for different amino acids. It is therefore possible that in *in vitro* systems the rate of protein synthesis is limited *inter alia* by deficiency of certain species of aminoacyl-tRNA.

V. Effects of Hormones on Protein and RNA Metabolism

A. Insulin

The effects of insulin on protein metabolism have been previously reviewed by the author (Manchester and Young, 1961; Manchester, 1965, 1968a), and references to earlier work can be found in these reviews. It is a long-recognized observation that diabetic animals waste and lose nitrogen, and that administration of insulin prevents this deterioration. However, it is not obviously clear from this sort of observation the extent to which insulin is acting directly on this area of metabolism, since protein depletion could equally arise from deranged carbohydrate metabolism. The diabetic animal represents a situation of effective glucose starvation, and under these conditions the rate of gluconeogenesis is markedly enhanced. (Since fatty acid mobilization and oxidation may also influence gluconeogenesis, this sort of situation demonstrates *par excellence* the intricate interaction between the different metabolic processes.) Just as insulin rapidly lowers the blood sugar level, so it lowers the plasma amino acid level, the corollary to this being that the hormone promotes the uptake and depresses the release of amino acids by tissues (Manchester, 1961; Scharff and Wool, 1965). Moreover, just as plasma glucose appears to be an effective trigger for pancreatic release of insulin (Hales, 1967), there is now evidence that certain amino acids can also promote insulin secretion (Floyd *et al.*, 1966a,b; Fajans *et al.*, 1967; Milner and Hales, 1967). Furthermore Cahill (1965) has provided evidence that in for example the toad fish the influence of insulin on amino acid metabolism may be more pronounced than on carbohydrate metabolism. Given a wider phylogenetic spectrum, there is evidence of an evolutionary role of insulin in controlling protein metabolism as great as for utilization of sugar.

This separate role can be best seen in isolated tissues and in particular in the capacity of the hormone to stimulate protein synthesis in isolated muscle and adipose tissue preparations (Manchester and Young, 1958; Wool and Krahl, 1959; Herrera and Renold, 1960; Mayne *et al.*, 1966). It should be emphasized that with these preparations we can see promotion of amino acid incorporation following almost immediately from addition of insulin to the suspending medium, and this situation should be distinguished from effects of the hormone on protein synthesis in the liver (discussed below), where a longer period is required for results to be seen. Consequently, insulin is usually administered *in vivo* before removal of the liver for examination. It is, however, important to distinguish between a stimulation of amino acid incorporation into protein and an increased synthesis of protein. Certainly the former is suggestive of the latter, but not unequivocal evidence, because any change in the specific activity of the free labeled amino acid being incorporated into protein will affect the amount of radioactivity that actually is incorporated. Thus in the case of incorporation of amino acids by diaphragm muscle, the exogenous labeled amino acid must first enter the intracellular contents of the tissue where it is likely to undergo dilution on mixing with the cell's amino acid pool before it is incorporated into protein. An increased rate of incorporation could result from (a) a real increase in the rate of protein synthesis; (b) an increased uptake of labeled amino acid, which means that specific activity in the tissue will rise more rapidly, hence more incorporation of label; (c) a shrinkage of the intracellular pool, which would mean that an unchanged rate of uptake of label would undergo less dilution and therefore retain a higher specific activity; (d) an expansion of the intracellular pool, which would mean the converse of (c) and thus indicate that an increased incorporation into protein represents, as in (a), an increased synthesis of protein. Differentiating between these possibilities has not proved easy (and it should be pointed out that this problem is not unique to the influence of insulin on muscle but arises wherever incorporation of label from precursor is studied in intact animals or in *in vitro* systems where pool sizes cannot be rigidly controlled).

Amino acids incorporated by muscle appear to enter a large number of different proteins, and there is no evidence of a selective stimulation by insulin (Goldstein and Reddy, 1967; Wool *et al.,* 1968). Although this incorporation is inhibited by the usual range of inhibitors of protein synthesis, such observations do not differentiate between the possibilities enumerated above. There is some evidence in diaphragm that insulin lowers the total amino acid resources of the tissue and medium (Manchester, 1961), which is presumptive evidence of increased synthesis of protein, but such measurements are complicated by the large amounts of taurine, etc., in muscle and similar observations were not found in the heart (Scharff and Wool, 1965). However, the more recent measurements of Pozefsky *et al.* (1968*) with an *in vivo* forearm preparation amply confirm the ability of insulin to suppress amino acid release by muscle.

That insulin increases accumulation of labeled amino acids by muscle (though apparently not by adipose tissue—Goodman, 1964, 1966) has been repeatedly demonstrated (Kipnis and Noall, 1958; Manchester and Young, 1960; Wool, 1964; Arvill and Ahrén, 1967a,b), but the significance of this observation is confused by the fact that whereas incorporation into protein of all normal amino acids is raised by insulin, enhanced accumulation is observable (in the absence of puromycin) for only about ten amino acids (summarized in Manchester, 1970a*,b*). Castles and Wool (1964) and Wool *et al.* (1965) found that in the presence of puromycin insulin promoted accumulation of all the amino acids. This is a very strange observation—doubly so in view of the inhibitory effects of puromycin and cycloheximide on amino acid accumulation when the tissue is incubated with the drugs for a longer period of time (Elsas *et al.,* 1967; Manchester, 1967a). It was explained by Wool on the basis that normally the increased entry of amino acids into protein under the influence of insulin depleted the cell pool of material so fast as to mask an effect of the hormone at this point. Although possible, this explanation does not seem to the present author very adequate because if (unlike the experiments of Wool *et al.,* 1965) the tissue is incubated with a relatively high concentration of amino acid, the proportion of the label entering the cell pool which goes into protein is small, yet stimulation of accumulation with puromycin remains visible. Though puromycin may stop protein synthesis it does not necessarily prevent peptide synthesis (Allen and Zamecnik, 1962; Rabinovitz and Fisher, 1962; Morris *et al.,* 1963; Nathans, 1964; Williamson and Schweet, 1964). It may be that, in the presence of puromycin, amino acids from the pool are getting incorporated into puromycin-linked peptides, not precipitable with TCA, but which are counted with the free amino acids. If this process, like protein synthesis, is still enhanced by insulin we get an apparent insulin effect on the accumulation of these amino acids due to the accumulation of peptides—which is indeed more apparent than real. However, addition of cycloheximide appears to show effects similar to puromycin on amino acid accumulation despite its different mechanism of inhibition of protein synthesis, so apparently other considerations must also be involved. Evidence that the interpretation of the significance of results obtainable with puromycin is less straightforward than originally suggested is discussed in more detail elsewhere (Manchester, 1970a*,b*).

Direct measurement of the free amino acid content of heart muscle shows no very consistent changes in the presence of insulin (Scharff and Wool, 1965). It is possible that the pool from which amino acids are taken for protein synthesis is not the same as that normally measured—evidence for and against this possibility is discussed by Kipnis *et al.* (1961) and Manchester and Wool (1963). As pointed out previously (Manchester, 1965), if protein synthesis *using isotopic amino acids* is more active near the cell membrane than in the depth of the tissue we may get an effective compartmentation of the amino acid pool which has no real physical basis. Examples of such compartmentation are discussed by Garfinkel and Heinmets in Chapter 27 of Volume III. In an attempt to get more precise information on the specific activity of the amino acid pool from which amino acids enter protein, Davey and Manchester (1969) measured the specific activity of amino acids contained in aminoacyl-tRNA. Although a certain number of assumptions have to be made in such calculations the results suggested that under the conditions employed in the perfused rat heart the specific activity of leucine and tyrosine bound to tRNA rose rapidly to levels similar to that of the perfusing amino acids. In view of the small pool of these amino acids in the tissue this result is not surprising. Unfortunately the instability of the aminoacyl-tRNA derivatives of glycine, proline, and methionine made their isolation difficult. There appeared to be some increase with insulin of the specific activity of the amino acid bound to tRNA.

Muscle is of course unique in possessing contractile elements, and these constitute a large bulk of its protein. Associated with the protein is RNA (Perry and Zydowo, 1959); this extensive association makes isolation of polysomal systems more difficult and can certainly reduce the yield of material available. Ribosomal systems have been isolated from muscle by Breuer *et al.* (1964), Earl and Korner (1965), Rampersad and Wool (1965), Earl and Morgan (1968*), and others. In order to extract more RNA, investigators recently have used media of high ionic strength (Heywood *et al.*, 1967; Chen and Young, 1968). Available evidence suggests that myofibrillar-bound RNA is essentially similar in kind to that found in the sarcoplasm (Zak *et al.*, 1967). Wool and colleagues have extensively studied the influence of insulin and diabetes on ribosomes isolated from cardiac and skeletal muscle. Ribosomes from diabetic tissue show roughly half the incorporating capacity of those from normal animals (Rampersad and Wool, 1965). Pretreatment of the diabetic animals with insulin for periods as short as 5–15 minutes brings about a dramatic recovery in the incorporating capacity (Stirewalt and Wool, 1966; Wool and Cavicchi, 1966). Variation in ability of ribosomes to synthesize protein correlates with change in sucrose density gradient profiles. Thus material of diabetic origin has a lower than normal quantity of polysomes and more monomers and dimers. A curious feature of the diabetic profile is the appearance of a peak sedimenting at 105 S, intermediate between monomers and dimers, which is made up of a monomer plus ribosomal subunit (Stirewalt *et al.,* 1967). Later work suggests that the extent of this peak is dependent on the concentration of potassium employed in preparation of the particles (Martin *et al.,* 1969). Pretreatment with insulin leads to disappearance of the 105 S peak and a return of the gradient profile to normal. This same abnormal pattern Stirewalt *et al.* (1967) found to be induced by treatment with puromycin or cycloheximide. This is in some ways unexpected since there is considerable evidence that cycloheximide, for example, "freezes" polysomal preparations and slows down the rate of their disaggregation (Wettstein *et al.,* 1964; Colombo *et al.,* 1965; Trakatellis *et al.,* 1965; Stanners, 1966). However, the present author has confirmed that administration of cycloheximide in quantities similar to those used by Stirewalt *et al.* (1967) does indeed lead after 90 minutes to a considerable shift of material from the heavy to the light end of gradient profiles of ribosomes from cardiac and skeletal muscle, liver, and brain. The mechanism of this change is uncertain, but could possibly be related to corticosteroid release in intact animals (Young *et al.,* 1968). Alternatively, it may indicate that cycloheximide has more than one point of action depending on concentration (Godchaux *et al.,* 1967) and indeed as shown by Jondorf *et al.* (1966) injection of small quantities of cycloheximide can improve subsequent incorporating capacity.

Wool and Cavicchi (1966) and Stirewalt *et al.* (1967) also found that if diabetic animals were pretreated with puromycin or cycloheximide (though not with actinomycin) then administration of insulin neither stimulated the capacity of the particles to incorporate nor changed the gradient profile. On the basis of these observations, they have suggested that one important aspect of the action of insulin on muscle is to promote the formation of a protein which is necessary for the proper functioning of the ribosomes both in production of polysomes and in capacity to incorporate amino acids. From various measurements they calculated the half-life of the postulated protein to be about 30 minutes. However, the author has found that injection of small doses of cycloheximide induces polysome formation even in muscle of diabetic rats, which suggests that the monomers in the diabetic tissue are not intrinsically incapable of forming polysomes.

These observations are extremely interesting in that they suggest the possibility of a regulation of protein synthesis at a translational level rather than transcription. However, the more important the claim the more acutely it requires scrutiny, and there are one or two other findings which suggest the need for caution in the interpretation of the above results. First, neither puromycin nor cycloheximide, though they markedly inhibit protein synthesis in isolated diaphragm muscle, interfere with the stimulation of this process by insulin until the rate of incorporation is down to unmeasurably low levels (Hechter and Halkerston, 1964; Wool *et al.*, 1965; Manchester, 1967a). Second, though ribosomes from normal animals treated with cycloheximide showed some diminution in incorporating capacity, the figure was nowhere as low as those for diabetic preparations, despite the fact that the conditions were such as to have converted the gradient profile to the diabetic picture (Wool and Cavicchi, 1966). Third, the amino acid-incorporating capacity of the ribosomal system used in these studies shows remarkable insensitivity to added cycloheximide (Wool and Cavicchi, 1967). Possible reasons for this were discussed earlier. Thus obviously at this stage it is not clear what can be concluded with certainty. The effects of puromycin and cycloheximide on diaphragm *in vitro* (which incidentally apply only to protein synthesis, not to any other parameters such as glucose consumption, etc.) suggest that, as the concentration of the inhibitor rises, an increasing proportion of the ribosomes are put out of action, but that those whose function can still be measured respond to insulin normally. It does not seem improbable that a system whose normal polysomal/ribosomal relationship is as disturbed as the gradients indicate in the inhibitor-treated cases should be unable to respond to a hormonal stimulus irrespective of whether the hormone itself induces synthesis of a particular protein (i.e., no response to hormone unless the cell is functional). However this possibility must imply that, despite similarity of gradient profiles, diabetic animals suffer from a different lesion than in the inhibitor-treated case because they can respond to insulin. This also is not improbable since the incorporating capacity of particles from diabetic rats is considerably less than those from normals treated with inhibitor (Wool and Cavicchi, 1966). Last, the way in which normal muscle responds to insulin is not obvious from the above observations.

Martin and Wool (1968) have subsequently shown that the defect in ribosomes of diabetic origin lies in the 60 S ribosomal subunits, and that it is possible to prepare hybrid ribosomes containing 40 S units from diabetic rats and 60 S from normal animals that possess normal incorporating capacity in the presence of poly U. The results obtained, however, are very much dependent on the magnesium concentration employed.

A decrease of polysomes to ribosomes can be indicative of a shortage of messenger. However, the effectiveness of insulin in the presence of actinomycin (Eboué-Bonis *et al.*, 1963; Kumar *et al.*, 1966; Wool and Cavicchi, 1966), coupled with the rapidity with which insulin acts, argues against a role of insulin operative in this manner. But there is evidence that diabetic muscle has a less than normal content of RNA (Wool *et al.*, 1968) and that insulin to some extent can promote RNA synthesis (Wool, 1963; Arvill and Ahrén, 1967b; Manchester, 1967b). It is not unlikely therefore that over longer periods of time insulin has a regulating effect on protein synthesis operating through increasing the cell's resources of RNA, but this is probably of all varieties, i.e. soluble, ribosomal, and messenger. There is as yet no evidence as to how this control could be exerted, and the author has had difficulty in showing any marked effect of the hormone on isotope incorporation *in vitro* (Manchester, 1967b).

The influence of insulin on protein synthesis in the liver has recently been reviewed by Steiner (1966). In this tissue, the effects of insulin on the control of sugar uptake and output have for long been enigmatic since the liver cell appears to be freely permeable to glucose (Cahill *et al.*, 1958; Steele, 1966), but at least a partial understanding may come from consideration of the hormone's control of protein synthesis in this tissue. An important contrast with muscle and adipose tissue is that, whereas the response to insulin for these tissues is almost immediate, the liver appears to respond more slowly. Moreover effects of addition of insulin to liver slices are seen only under rather restricted conditions of mild diabetes (Penhos and Krahl, 1962, 1963). Wittam *et al.* (1969) found that the hepatic polysome profile responded only slowly to diabetes although treatment with insulin led to its restoration to normal within an hour.

Liver microsomes from diabetic animals have impaired capacity to incorporate amino acids, and this deficiency can be corrected by treatment with insulin some hours previously (Korner, 1960; Robinson, 1961). Insulin treatment has been reported to stimulate the labeling of liver RNA (Kidson and Kirby, 1964) though there are recent reports of failure to confirm these experiments (Leader and Barry, 1967; Jackson and Sells, 1968). One virtue of the liver is that there is reason to believe that some of the proteins whose synthesis is raised by insulin can be specifically identified. Thus the diabetic animal has a very low level of liver glucokinase, and treatment with insulin leads after a few hours to its restoration (Salas *et al.*, 1963; Sharma *et al.*, 1963; Sols *et al.*, 1965; Niemeyer *et al.*, 1967). This action of insulin does not occur in the presence of actinomycin. Reduction in the level of glucokinase also occurs in fasting and is restored by feeding glucose, which under these conditions will promote secretion of insulin. Glucokinase is not the only liver enzyme to show adaptive changes in this way. The influence of glucocorticoids on gluconeogenic and protein catabolic enzymes are well recognized (discussed elsewhere in this volume), and the role of insulin as an inducer of glucose-using enzymes (glucokinase, glycogen synthetase, phosphofructokinase, and pyruvate kinase) and a suppresor of gluconeogenic enzymes (pyruvate carboxylase, phosphoenolpyruvate carboxykinase, fructose 1:6-diphosphatase, and glucose 6-phosphatase) (Fig. 3) has been elaborated by Weber *et al.* (1965, 1966) in the concept of

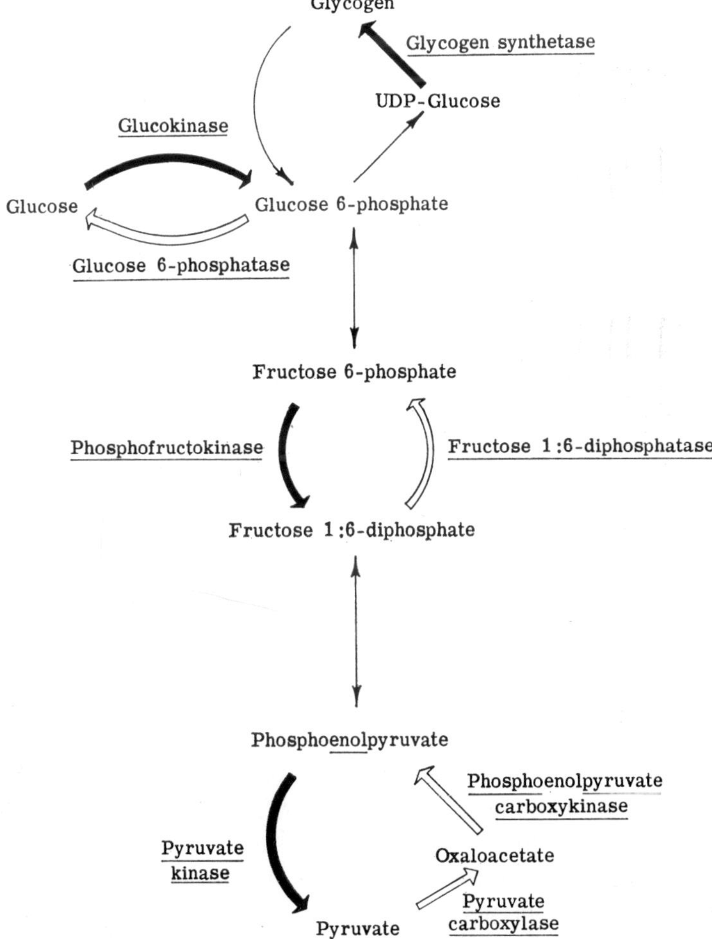

FIG. 3. The main reactions of hepatic glycolysis, glycogenesis, and gluconeogenesis. Enzymes whose activities are enhanced by the ready availability of insulin are named; they are indicated by the thick, solid arrows. Those whose levels rise in diabetes and are enhanced by glucocorticoid treatment are likewise named, they are indicated by open arrows. Other reactions are believed not to be greatly affected by these treatments.

functional genic units. Whether consideration in detail of the many studies of the time course and degree of change of level of different enzymes offer conclusive support for the concept of a single basic mechanism is uncertain; it is clear that the reciprocal interaction of insulin and glucocorticoids both raises and lowers the rate of synthesis, possibly even of decay, of important enzymes of the liver whose rate of turnover is clearly faster than average.

Consistent with these observations is the finding that treatment with insulin increases the polymerase activity of nuclei from the liver of alloxan-diabetic rats (Steiner and King, 1966). This action is prevented by the presence of puromycin, an observation suggesting that protein synthesis is involved. What type of RNA is being formed by the polymerase has not been determined. In the severely diabetic animal, administration of insulin leads after a day or so to marked liver cell hyperplasia and increase of synthesis of DNA (Younger *et al.*, 1966). How this relates to changes in levels of enzymes concerned with carbohydrate metabolism is uncertain. Lockwood *et al.* (1967) found that insulin elicited DNA polymerase activity in epithelial cells of mammary gland explants, thus initiating DNA synthesis. Effects of the hormone to maintain RNA levels in such cultures has also been seen (Mayne and Barry, 1967). It seems very likely that many hormonal manifestations will vary according to the particular metabolic/environmental circumstances of the responding cell.

The other secretion of the islets of Langerhans, glucagon, appears to stimulate catabolism of protein by the liver. The mechanism of this action, which involves specifically liver as opposed to plasma proteins, is not known, but it is prevented by insulin (Miller, 1965). The complicated relationship between insulin and glucagon in the induction of certain liver enzymes is discussed in Chapter 31 of this volume.

B. ACTH

The regulation of adrenal steroid production constitutes a fascinating problem (see Bransome, 1968, and Koritz, 1968, for useful accounts). The demonstration that ACTH raises the cyclic AMP level in the adrenal (Haynes, 1958) and the finding that cAMP stimulates steroidogenesis (Haynes *et al.*, 1959) suggested a basic mechanism of ACTH action similar to that for several other hormones. Grahame-Smith *et al.* (1967) found a correlation between the rate of steroidogenesis and the cyclic nucleotide concentration which also correlated with the level of ACTH employed. The hormone can also increase cAMP formation in homogenates.

The main steps in glucocorticoid production from cholesterol esters are indicated in Fig. 4. Hydroxylase reactions requiring NADPH are involved in no less than five or six steps in the sequence and in addition there appears to be considerable translocation of the steroids into and out of the mitochondrion during the transformations. The possibility of activation of adrenal phosphorylase by ACTH or cAMP at first suggested that stimulation of steroid production could result in increased availability of glucose 6-phosphate for NADPH formation (Haynes and Berthet, 1957). Although it is uncertain whether adrenal phosphorylase is influenced by cAMP (Cohen and Fagundes, 1966), Criss and McKerns (1968) have found a direct action of ACTH to lower the K_m of glucose 6-phosphate dehydrogenase toward NADP and glucose 6-phosphate. However, there is little correlation of steroidogenesis with NADPH levels, and Koritz (1966) has demonstrated that succinate can promote pregnenolone synthesis by isolated adrenal mitochondria—an observation which suggests that the NADPH required for the hydroxylase reactions does not necessarily come from extramitochondrial sources. Simpson and Estabrook (1969*) have suggested that the malic enzyme may play an important role in bringing NADPH into the mitochondrion. Stone and Hechter (1954) showed that ACTH acted principally at the stage between cholesterol and pregnenolone: ACTH does not stimulate further conversion of pregnenolone (Karaboyas and Koritz, 1965) despite the requirement for NADPH. Kowal (1969) has found with adrenal cells in culture a stimulation by ACTH or cAMP of 11β-hydroxylation but not of the activities of either 3β-hydroxysteroid dehydrogenase or 20α-hydroxysteroid dehydrogenase.

ACTH *in vivo* stimulates adrenal protein and nucleic acid synthesis (Bransome and Chargaff, 1962; Bransome and Reddy, 1963a, 1964; Farese and Reddy, 1963), but results on its addition to adrenal slices *in vitro* are more variable (Bransome and Reddy, 1963b; Halkerston *et al.*, 1964; Ferguson and Morita, 1964; Ferguson *et al.*, 1967). Since cortisol, progesterone and testosterone all inhibit amino acid incorporation (Morrow *et al.*, 1967), it is likely that steroidogenesis will *in vitro* interfere with ACTH stimulation of synthesis. Induction of steroidogenesis by ACTH is prevented by the presence of puromycin, cycloheximide, or chloramphenicol (Ferguson, 1963; Farese, 1964; Grahame-Smith *et al.*, 1967), but the inhibitors prevent neither the rise in cAMP formation (Grahame-Smith *et al.*, 1967), nor the conversion of cholesterol esters to cholesterol or pregnenolone to corticoids (Garren *et al.*, 1965; Davis and Garren, 1966, 1968). The inhibitory effect of chloramphenicol is particularly interesting since animal systems are on the whole rather insensitive to this antibiotic (though see Godchaux and Herbert, 1967, for effects on reticulocytes), but there is considerable evidence for its inhibition of protein synthesis by mitochondria (Mager, 1960; Kalf, 1963; Kroon, 1965). Garren *et al.* (1965) found that treatment of rat

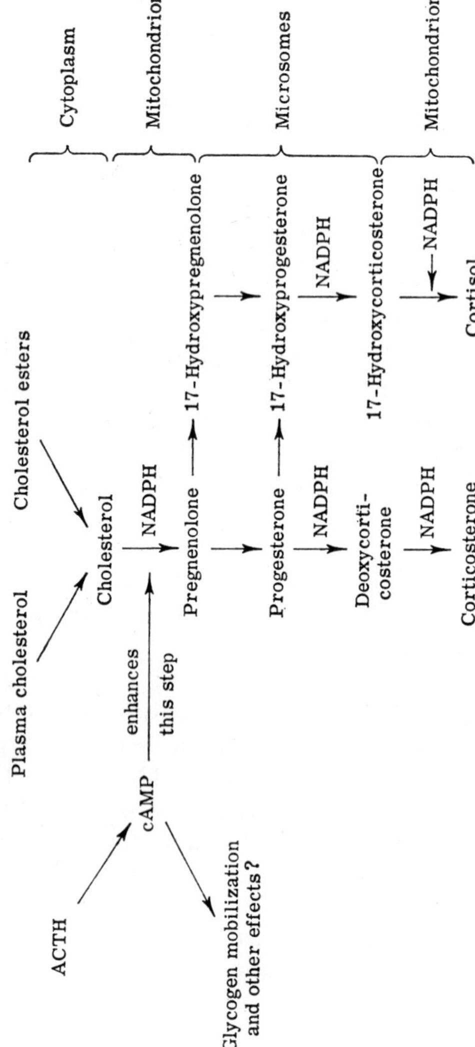

Fig. 4. The formation of glucocorticoids from cholesterol and its control by ACTH, indicating the intracellular location of the various reactions. The conversion of cholesterol to pregnenolone involves three steps, each of which requires the participation of NADPH and oxygen. Fuller details of the reactions are summarized in Samuels and Uchikawa (1967) and Bransome (1968).

adrenals *in vivo* with puromycin or cycloheximide not only prevented their response to ACTH, but led to a rapid decline in corticosterone production if the adrenals were already subject to ACTH stimulation. They suggested that ACTH induces the synthesis of a specific protein, of half-life calculated to be 7–10 minutes, that is a regulator of steroidogenesis. However an alternative explanation would be that steroidogenesis at one of its steps requires the presence of such a protein, but that if this protein disappears, as in the case of application of inhibitors, any stimulation of the pathway, irrespective of whether control is to be exercised at this particular point, will no longer be possible. It must be admitted, though, that the step in which the protein is involved is close to that regulated by ACTH/cAMP, namely the conversion of cholesterol to pregnenolone. In order for steroid formation to proceed, pregnenolone must leave the mitochondrion since its accumulation there inhibits further formation (Koritz and Hall, 1964; Urquhart *et al.*, 1968). If mitochondrial amino acid incorporation is mainly into membrane constituents (Roodyn, 1962; Roodyn *et al.*, 1962), it is conceivable that the rapidly turning over protein is perhaps involved in the passage of the steroids across the mitochondrial membrane, and the change of mitochondrial structure resulting from treatment with puromycin and cyloheximide is circumstantially consistent with this (Grief *et al.*, 1965; Staehelin and Maier, 1968). Experiments with disrupted mitochondria however do not clearly establish whether cAMP can regulate steroid translocation (Satoh *et al.*, 1966; Roberts *et al.*, 1967). Farese (1967) has put forward evidence for a cytoplasmic heat-labile factor which may be an intermediary in ACTH action and exert a control over mitochondrial activity. If proteinaceous, it will be interesting to find its relation to that postulated from the experiments with inhibitors.

In what way stimulation of polymerase activity results from the hormone's other actions (Farese and Schnure, 1967) is not clear, but may be connected with its important long-term tropic influence on the cortex. Treatment with ACTH stimulates the activity of adrenal aminoacyl transferase (Scriba and Reddy, 1965; Scriba and Fries, 1967) and of adrenal polysomes (Farese, 1966). Useful accounts of many aspects of ACTH action are to be found in McKerns (1968).

C. Growth Hormone

The role of a secretion of the anterior pituitary gland in the control of growth and directly or indirectly of many other metabolic processes is well established and well reviewed (Ketterer *et al.*, 1957; Engel and Kostyo, 1964; Knobil and Hotchkiss, 1964; Korner, 1967). The means by which growth hormone can control protein synthesis is less well established and is confused, first, by the fact that more than one mechanism may be involved and, second, by the existence of a number of *in vitro* actions of the hormone which do not clearly interdigitate with the consequences of administration of the hormone *in vivo.*

Not only does the protein content and capacity to synthesize protein of a hypophysectomized animal fall, but there is also a fall in content of RNA and its rate of synthesis which is reversed on treatment with growth hormone (Reid *et al.,* 1956; Di Stefano and Diermeier, 1959; Talwar *et al.,* 1962; Korner, 1965; Florini and Breuer, 1966). These observations suggest the possibility that growth hormone may regulate protein synthesis through some aspect of RNA metabolism; the possibility of synthesis of mRNA particularly comes to mind. There is evidence both in favor of and against this possibility (see Korner, 1967, for review) though it should be remembered that the evidence against means rather that there may be more than one defect in the hypophysectomized system repairable by growth hormone. Thus microsome and ribosome preparations from the livers of hypophysectomized rats have impaired capacity to incorporate amino acids into protein (Korner, 1959, 1961). However, when polysomes as opposed to ribosomes were prepared from liver of normal and hypophysectomized rats, little difference in specific incorporating capacity was found (Korner, 1964b). This observation suggested that the crucial difference between normal and hypophysectomized preparations lay in the proportion of polysomes to ribosomes, and indeed the gradients of homogenates of hypophysectomized rat liver revealed fewer polysomes and more monomeric ribosomes than controls (Korner, 1964b, 1968; Staehelin, 1965*). Garren *et al.* (1967), on the other hand, did not find markedly different polysome profiles following hypophysectomy, and Brewer *et al.* (1969) concluded that when polysome breakdown occurred it correlated with increased RNase activity. There seems to be general agreement however that the specific incorporating capacity of ribosomes from hypophysectomized animals is below normal.

Several attempts to show an increased synthesis under the influence of growth hormone of a specific RNA fraction equivalent to messenger have been unsuccessful (Korner, 1965; Jackson and Sells, 1967; Sells and Takahashi, 1967; Gupta and Talwar, 1968). Instead, administration of the hormone appears to stimulate the labeling of all types of RNA *in vivo.* In one sense this is to be expected because, if the total RNA content of a tissue is demonstrably raised by growth hormone, there is bound to be increase of RNA species other than messenger. On the other hand, given a more rapid turnover of messenger, a more pronounced effect on the synthesis of this species might have been expected with brief pulse labeling, but the results here again will be conditioned by any relationship between the rate at which messenger can be synthesized and be incorporated into functioning cytoplasmic particles, the factors controlling which are not known. Further evidence that growth hormone promotes RNA synthesis is the rise in RNA polymerase activity of isolated nuclei prepared from growth hormone-treated animals (Pegg and Korner, 1965; Breuer and Florini, 1966; Florini and Breuer, 1966), but as discussed earlier such stimulation is not evidence specifically of messenger-RNA formation.

Against these observations it should be noted that a stimulating effect of growth hormone can be seen despite concurrent administration of actinomycin (Korner, 1965). Also ribosomes from hypophysectomized liver, though responding to poly U and poly A, do not show as high an absolute capacity for nucleotide-stimulated incorporation as normal controls (Korner and Gumbley, 1966; Garren *et al.*, 1967; Korner, 1967), though Korner and Gumbley (1966) found that the deficiency could be made good by a large excess of cell sap. [This is an interesting contrast to the observations Scornik *et al.* (1967) of the presence in normal cell sap of an inhibitory fraction which disappears, for example, in regenerating liver.] These observations suggest the existence of cytoplasmic factors that can control translation and a changed response to these of ribosomes following hypophysectomy, but nothing is known of their nature. It should be emphasized that in the above studies growth hormone is always administered *in vivo*. Direct addition of the hormone to subcellular systems has no effect. When the hormone is added to a perfused liver system a promotion of incorporation of amino acid into protein can be seen under certain conditions within 30 minutes (Jefferson and Korner, 1967). Conversely, addition of growth hormone to liver slices does not affect RNA synthesis (Jackson and Sells, 1967).

In muscle, hypophysectomy leads to a reduced incorporation of amino acids into protein either in isolated tissues (Manchester *et al.*, 1959; Kostyo and Knobil, 1959a) or subcellular preparations (Earl and Korner, 1966; Florini and Breuer, 1966). Florini and Breuer (1966) found that treatment of the donor animals with growth hormone raised the proportion of polysomes in their preparations, as it did also after 12–18 hours of muscle RNA polymerase. Earl and Korner (1966) found that the proportion of polysomes to ribosomes did not change, but that the maximum activity attainable with poly U was less for ribosomes from hypophysectomized animals than from normal animals. However, the extraction of ribosomes from muscle with media of low ionic strength is poor, and different degrees of extraction may explain some of these observations.

Understanding of how growth hormone works is complicated by the so-called *in vitro* (insulin-like) actions. Thus, as far as amino acid metabolism is concerned. addition of the hormone to diaphragm muscle obtained from the hypophysectomized animal but not from the normal animal increases accumulation of amino acids and their incorporation into protein (Kostyo and Knobil, 1959b; Manchester and Young, 1959; Kostyo and Schmidt, 1962; Knobil and Hotchkiss, 1964; Hjalmarson and Ahrén, 1967). This action of the hormone is not prevented when nucleic acid synthesis is inhibited by actinomycin (Martin and Young, 1965), nor does the hormone stimulate nucleic acid synthesis under these conditions (Martin and Young, 1965; Dawson *et al.,* 1966; Kostyo, 1966). Very interestingly though, preincubation of the tissue with growth hormone prevents further response to the hormone, but this desensitizing effect of growth hormone is not seen if actinomycin is simultaneously present (Goodman, 1968). The significance of the *in vitro* effects is unclear at the moment, especially since the hormone can simultaneously promote glucose uptake and glycogen synthesis (Ketterer *et al.,* 1957). Mahler and Szabo (1969) have suggested that insulin-like effects of growth hormone *in vivo* may result from inhibition of insulin degradation. The effects of growth hormone however cease to be insulin-like when directed toward fat metabolism (Manchester, 1963*). Claims that growth hormone contains peptides responsible for the different *in vivo* activities (Bornstein and Hyde, 1960; Bornstein *et al.,* 1968a,b) lack independent confirmation, but may be related to other putative insulin antagonists of pituitary origin (Louis *et al.,* 1966*). Salaman *et al.* (1967) found that the fat mobilizing action of various growth hormone fractions *in vivo* correlated with their growth promoting action, but *in vitro* activity does not necessarily correlate with that *in vivo* (Manchester and Wallis, 1963*). Fain *et al.* (1965) have shown that promotion of mobilization of fatty acids from isolated adipose tissue cells is prevented by both puromycin and actinomycin. The effects of the hormone also normally take a few hours to become apparent. Involvement of specific protein and RNA metabolism seems likely.

D. *Thyroxine*

From a physiological standpoint, thyroxine has a variety of functions (see Tata, 1964a, for review), and it remains to be seen whether they are related. Like ecdysone, but unlike insulin and growth hormone, it has a function in metamorphosis, but is of course best known for its influence in higher animals on the basal metabolic rate (BMR). The importance of thyroxine in development and maturation is also well recognized, and thanks to the work largely of Tata and colleagues we can now see something of the relationship between the hormone's influence on protein synthesis, growth, and respiration.

Tata *et al.* (1963) found that low doses of thyroxine or 3,3,5-triiodothyronine administered to thyroidectomized rats raised the BMR of the animals but also led to a proliferation of mitochondria in the liver.

The effects of thyroid hormones, for reasons that are not known, are notoriously slow. Thus the effect of a single dose of triiodothyronine on oxygen uptake is visible some 2 days after its administration, but this effect is preceded, at about 30 hours, by an increased capacity of isolated liver microsomes and mitochondria to incorporate amino acids into protein (Tata *et al.*, 1963; Roodyn *et al.*, 1965). The slow pace of response by the thyroidectomized animal has made it a useful system in which to study sequential changes in RNA and protein metabolism. Thus the increased incorporation of amino acids into protein does not occur in the presence of actinomycin (Tata, 1964b), suggesting the dependence of the enhanced protein synthesis on elevated nucleic acid metabolism. In confirmation of this, Tata and colleagues have shown an increased synthesis of nuclear RNA, starting 3–4 hours after triiodothyronine injection and maximal at about 15 hours (see Tata, 1966, for a general summary). A little later there is a rise in Mg^{2+}-dependent RNA polymerase (which produces a ribosomal type of material). By 30 hours ribosomal synthesis is greatly elevated and is followed by an increase of protein synthesis. One unexpected feature is that a rise in Mn^{2+}/ammonium sulfate-dependent RNA polymerase (supposedly messenger producing) is neither marked nor rapid. Thus, though the observations clearly suggest an influence of thyroid hormones on nuclear RNA synthesis as a cause of subsequent developments, the results do not provide any support for the concept of hormones acting by stimulating specific messenger RNA production. On the other hand, the stimulation of nuclear metabolism leads to synthesis of specific proteins [e.g. mitochondrial α-glycerophosphate dehydrogenase (Sellinger and Lee, 1964; Lee and Miller, 1967)] which are different presumably from those produced by treatment with growth hormone or testosterone—in fact the effect of the three hormones on RNA polymerase is additive (Florini and Breuer, 1966; Widnell and Tata, 1966b). Presumably increased formation of ribosomes (if that is to be equated with increased synthesis of rRNA) must also require increased production of mRNA, on a similar proportional if not quantitative scale. Induction of carbamyl phosphate synthetase and formation of new RNA by tadpole liver following thyroxine treatment has been studied by Nakagawa and Cohen (1967) and Nakagawa *et al.* (1967), and the activity of cell-free systems from this source by Unsworth and Cohen (1968).

Direct effects of thyroxine added *in vitro* to stimulate amino acid incorporation in cell-free systems have been claimed (Roche *et al.*, 1962; Sokoloff *et al.*, 1963a,b; Brown, 1966). However the effects show several curious features; for example, triiodothyronine is not effective and only microsomes as opposed to ribosomes respond (Roche *et al.*, 1962), the stimulation is dependent on the presence of mitochondria as well as microsomes (Sokoloff *et al.*, 1963a,b). The significance of these observations is therefore uncertain.

E. Testosterone

The histological and grosser changes in metabolism of the seminal vesicles and prostate and its control by testosterone have been well summarized by Williams-Ashman (1965). Castration lowers the RNA content of the tissues and the activity of the ribosomes. Testosterone treatment conversely raises the quantity and activity of ribosomes and increases nuclear RNA polymerase activity (Williams-Ashman *et al.*, 1964). These authors also measured the priming or template activity of nuclear RNA added to *E. coli* ribosomes stripped of endogenous messenger and found it enhanced after testosterone treatment. The reason for this was not clear since the nuclear RNA had a similar base composition and sedimentation profile in the two cases. However, a small quantity of new mRNA would not necessarily be detectable by these means. Alternatively the priming capacity of the nRNA may depend upon its secondary structure (Singer *et al.*, 1963).

Later work by Liao and colleagues (Liao *et al.*, 1966a,b; Liao and Lin, 1967) suggests that the new RNA synthesized under the influence of testosterone may be primarily ribosomal. Thus testosterone stimulated formation by prostatic nuclei of material rich in guanine and cytosine, and its influence was inhibited by actinomycin. Actinomycin appeared to inhibit primarily the extra synthesis induced by the hormone, rather than the basal activity of the nuclei. Liao *et al.* (1966a,b) suggested that 70–80% of prostatic nuclear DNA is unavailable for transcription. The remaining 20–30% is functional, but this in turn can be divided into three regions—a small portion (about 1%), which is possibly nucleolar, is the region primarily responsive to testosterone. Liao *et al.* (1966) concluded that the hormone did not act by making increased amounts of DNA available for transcription. Increase in activity of RNA polymerase must therefore result from enhanced performance of the enzyme. This would be consistent with the inhibitory effects of puromycin on the androgen response if increased activity of polymerase resulted from increased synthesis. Autoradiography shows that testosterone treatment raises primarily nucleolar activity, not that of the rest of the nuclear chromatin, though this is quite active in RNA synthesis (Liao and Stumpf, 1968*). Testosterone, it must be remembered, also promotes amino acid incorporation in muscle of castrate animals and elevates polymerase activity, possibly by enhancement of the priming capacity of the muscle chromatin (Breuer and Florini, 1966, Florini and Breuer, 1966). The ability of anabolic steroids to counteract the inhibitory effects of glucocorticoids on protein synthesis has been explored by Bullock *et al.* (1968*).

F. Estrogen

Administration of estrogen to immature or ovariectomized mature female rats leads to a rapid growth of the uterus. A lot of the early increase in weight is due to uptake of water, but by between 6 and 12 hours after a single dose of estradiol the content of RNA, though not DNA, begins to increase, and by 12 hours there is increase in protein too (Mueller *et al.*, 1958). These changes are prevented by simultaneous administration of puromycin and actinomycin (Mueller *et al.*, 1961; Ui and Mueller, 1963). Noteboom and Gorski (1963) found that incorporation of labeled amino acids into protein is increased 2–4 hours after estradiol administration. Elevated incorporation of precursor into RNA and rise in activity of RNA polymerase occurs before 2 hours, and these increases are blocked by the presence of puromycin or cycloheximide (Noteboom and Gorski, 1963; Gorski, 1964; Gorski and Axman, 1964). Hamilton (1964) and Gorski *et al.* (1965) have suggested that, although a general increase in protein synthesis does not take place at first, in the early stages of estrogen action there is stimulated synthesis of certain specific proteins which are responsible for the increase in RNA synthesis. Response to estrogen is also inhibited by actinomycin (Ui and Mueller, 1963; Hamilton, 1964; Nicolette and Gorski, 1964a). The estrogen stimulation of RNA synthesis appears to apply to ribosomal and possibly to all types of RNA (Gorski and Nelson, 1965; Hamilton *et al.*, 1965), which, in view of the given hypertrophy of the tissue, is not unexpected—an increased content of ribosomes can be seen 4–24 hours later (Moore and Hamilton, 1964) and electron microscopy shows proliferation and enlargement of nucleoli (Laguens, 1964). Uptake of amino acids is also enhanced within an hour of estrogen injection (Noall, 1960; Noall and Allen, 1961; Roskoski and Steiner, 1967a,b), but this effect is prevented by prior injection of actinomycin or cycloheximide (Roskoski and Steiner, 1967a); this suggests the formation of more transport agent and is consistent with rise in the V_{max} rather than the K_M of the uptake process (Riggs *et al.*, 1968).

A useful summary of our knowledge of estrogen action is provided by Hamilton (1968). He suggests that the rapid binding of the steroid specifically to nuclear chromatin indicates that a likely primary site of action of the hormone is the cistrons that control synthesis of nuclear and ribosomal RNA. Within 2 minutes of injection, molecules of estradiol become bound to chromatin and after this period increased incorporation of precursor into RNA can be seen, although the effect is greatest by about 30 minutes (Means and Hamilton, 1966a,b). During this period, however, protein synthesis is decreased. Conversely the most potent stimulator of RNA synthesis appears to be cycloheximide, whereas actinomycin prevents response to estrogen. Administration of estradiol stimulates the priming activity of uterine, though not of liver or lung, chromatin and its activity also fluctuates in the menstrual cycle (Barker and Warren, 1966a). Thus the earliest effects of estradiol may be to stimulate transcription of selected regions of the genome, which results in very rapid synthesis of a few specific proteins (the early phase), which in turn increase RNA synthesis on a larger quantitative scale (identifiable for all forms of RNA, particularly rRNA—Hamilton *et al.*, 1968*; Billing *et al.*, 1969*), and ultimately general synthesis of protein is enhanced (the amplification phase)—for example increased synthesis of several enzymes has been shown (Barker and Warren, 1966b; Singhal *et al.*, 1967). Stimulation by estrogen and progesterone of synthesis of specific proteins in chicken oviduct can also be observed (O'Malley, 1967; O'Malley *et al.*, 1967), and in this tissue transfer RNA appears to be the form of RNA whose synthesis is most extensively enhanced (O'Malley *et al.*, 1968).

A curious feature of estrogen action is that it is seen only *in vivo*, and even when estrogen is administered *in vivo* and RNA synthesis is examined *in vitro* the response is lost, though enhancement of phospholipid synthesis could be observed (Gorski and Nicolette, 1963). This loss of response *in vitro* has led to suggestions that the *in vivo* effects of the hormone may be indirectly mediated (Hechter and Halkerston, 1965). That this could not be through glucocorticoid action has been demonstrated for protein and RNA synthesis by Nicolette and Gorski (1964b) and Nicolette and Mueller (1966), but the presence of glucocorticoid does inhibit water inhibition. The specific binding of estrogen by the uterus (discussed on p. 236) also argues against indirect mediation. It may be that other hormones or factors play a "permissive" role *in vivo*, and an alternative explanation exists in the capacity of estrogen to expand within 30–40 seconds the microcirculation of the uterus by local release of histamine (Lippe and Szego, 1965), which could be of considerable significance in allowing other plasma constituents greater access to the tissue. Since treatment of the isolated uterus with cyclic AMP has been claimed to stimulate RNA synthesis (Hechter *et al.*, 1967) and estrogen raises uterine cyclic AMP levels within 15 seconds (Szego and Davis, 1967) there is scope for invocation of the adenyl cyclase system, but again the rapid attachment of steroid to the binding protein and to nuclear chromatin limit the possibilities.

Greenman and Kenney (1964) found little change in amino acid-incorporating activity of uterine subcellular systems following ovariectomy, but the yield of ribosomes from the uterus dropped markedly. Treatment *in vivo* with estradiol raised both the quantity and the activity of particles. In longer operative periods Teng and Hamilton (1967b) found an increase in specific incorporating capacity of a ribosomal preparation after ovariectomy, though again there was a dramatic decrease in yield. Administration of estradiol first increased incorporating capacity, but if treatment continued incorporation fell by 72 hours to below the normal levels, the yield of material having meanwhile increased considerably. They also found that direct addition of high concentrations of estradiol to the ribosomal preparation increased incorporation if the ribosomes originated from normal animals but depressed incorporation by those from ovariectomized rats. Whether these last effects have any fundamental significance is not clear. Loecker *et al.* (1966) have reported stimulating effects on amino acid incorporation of estradiol-3,17 β-disulfate added to muscle microsomes, and Fahmy *et al.* (1967) have found activation of DNA polymerase from thymus nuclei by estradiol *in vitro*.

G. Glucocorticoids

The complexity of delineation of sites and mechanisms of hormone action is nowhere better demonstrated than with the glucorticoids (see Ashmore and Morgan, 1967, for one review). Such steroids are traditionally restrainers of protein formation and stimulators of protein catabolism (Long *et al.*, 1960). However, the pronounced enhancement of synthesis of key gluconeogenic and amino acid-degrading enzymes in the liver on glucocorticoid treatment, coupled with the stimulus of RNA synthesis in that organ, has attracted much attention. It is reviewed elsewhere in this volume (Chapter 31) and will not be considered here. The action of mineralocorticoids is reviewed by Edelman and Fanestil (1970*).

Elsewhere in the body the catabolic influence of glucocorticoids is more in evidence. Thus protein formation, as judged by isotope incorporation, is raised in muscle soon after adrenalectomy and reduced when the animals are treated with glucocorticoid (Manchester *et al.*, 1959; Wool and Weinshelbaum, 1959, 1960; Shimizu and Kaplan, 1964). Likewise steroid treatment reduces amino acid uptake by muscle (Wool, 1960), though *in vivo* amino acid levels rise (Kaplan and Shimizu, 1963) and corticoids raise the plasma amino acid level of eviscerated adrenalectomized animals (Bondy *et al.*, 1954). Incorporation of amino acids into protein is also depressed after steroid treatment in spleen, lymphocytes, reticulocytes, and isolated bone cells (Kit and Barron, 1963; Koritz and Dorfman, 1956; Blecher and White, 1958; Peck *et al.*, 1967) and in thymus tissue (Uete, 1966a; Makman *et al.*, 1967). Synthesis of collagen in chicken embryos is also impaired (Gould and Manner, 1967). Administration of cortisol to adrenalectomized rats raises release of amino acids and even of DNA from slices of thymus (Haynes and Sutherland, 1967; Sutherland and Haynes, 1967). These effects are seen following administration of the hormone *in vivo*, but similar observations have not been made on addition of the steroid *in vitro*. Munk (1962), however, has demonstrated an inhibitory effect of steroids *in vitro* on glucose uptake by adipose tissue,[2] and it appears that inhibition of amino acid uptake and incorporation can also be observed if diaphragm muscle is incubated for a sufficiently long time with the hormone (Kostyo and Redmond, 1966). In fact the inhibitory influence of the cortisone used was progressive with time and is similar to effects seen with puromycin. Although puromycin does not influence uptake of α-aminoisobutyrate studied in incubations of 1–2 hours (Fritz and Knobil, 1963; Burrow and Bondy, 1964; Carlin and Hechter, 1964), when the tissue is preincubated for several hours with puromycin before addition of amino acid, uptake of the latter is much reduced. Kostyo and Redmond (1966) suggested, first, that puromycin inhibits the formation of a rapidly turning over protein concerned with transport of amino acids [similar observations have been made in muscle by Adamson *et al.* (1966), Elsas *et al.* (1967), and Manchester (1967a)], and, second, that steroids could also inhibit the synthesis of this protein. A loss of amino acid from the cell could well promote a shift in the balance of protein turnover toward breakdown and hence explain some of the recognized protein catabolic effects. By way of contrast Kidson (1967) found with lymph node cells that, although cortisol always inhibited RNA synthesis, protein formation was first elevated then declined. He suggested the possibility of a general inhibition of protein synthesis, but also of removal of a repressor of protein synthesis, giving rise to a transient stimulation.

Precisely how steroids could interfere with protein synthesis remains to be determined. Ribosomes from thymus incorporate less amino acid when isolated after glucocorticoid injection, and the RNA polymerase activity of thymus nuclei is also depressed (Nakagawa and White, 1966; Peña *et al.*, 1966; Uete, 1966b). These changes are associated with decreased incorporation of labeled precursors into RNA and may result from inhibition of uptake of the precursors (Kit *et al.*, 1954; Gabourel and Fox, 1965; Brinck-Johnsen and Dougherty, 1965; Peck *et al.*, 1967); such effects can be seen on addition *in vitro* as well as *in vivo* and are prevented by the presence of puromycin and actinomycin (Kidson, 1965; Makman *et al.*, 1966; 1967). Thus we have a situation which parallels the effects of so many other hormones already discussed, yet is in reverse—with inhibition instead of stimulation. Paradoxically the liver can exhibit both facets—incorporation by ribosomes showing both stimulated or inhibited protein-synthesizing capacity depending on the size of dose (Korner, 1960; Leon *et al.*, 1962). Likewise the incorporating capacity by ribosomes from adrenalectomized liver is at first enhanced by adrenalectomy, but after longer periods following loss of the adrenal

[2] See also effects on mouse ear epidermis noted by Plager and Matsui (1966*).

incorporation it becomes subnormal (Korner, 1960). Incorporation by ribosomes from kidney appears to be enhanced at all times after adrenalectomy (De Venuto and Lange, 1967). As a final touch of contradiction, cortisol increases the uptake of the model amino acid aminosobutyrate by perfused liver (Riggs, 1964). This response of the perfused organ is not prevented by treatment with actinomycin, but is prevented by adrenergic blocking agents (Chambers *et al.*, 1965).

VI. Discussion

One of the purposes of the present presentation has been to examine the extent to which the mechanisms of hormone action, as we imperfectly understand them, fit the concept proposed by Karlson (1963) that hormones primarily operate through stimulation of specific genetic transcription with the consequent production of special or additional messenger-RNA. One of the supposed characteristics of mRNA is its rapid rate of turnover, and thus rapid uptake of labeled precursor. There is no doubt that an early effect of many hormones is the stimulation of incorporation of label into nuclear-RNA, but there seems very little evidence that the RNA so produced is specifically messenger precursor. Rather, studies of the nature of the destination of the nRNA, its base composition, the association of the Mg^{2+}-dependent polymerase activation with rRNA formation, and the fact that increase in bulk RNA can hardly be predominantly messenger, suggest that the hormonally induced stimulation of RNA synthesis is not specifically related to messenger production. This is readily comprehensible for hormones such as estrogen that stimulate the synthesis of large quantities of new material, but is more puzzling in instances such as the action of glucocorticoids on the liver where there is both marked specificity in the enzyme protein molecules produced and no marked change in the RNA content of the tissue (Peraino, 1967). Nevertheless the enhancement of labeling of RNA appears to apply equally to all forms of RNA (Greenman *et al.*, 1965; Wicks *et al.*, 1965). As pointed out by Kenney and Kull (1963), it is also strange that production of relatively few new enzymes in the liver (following cortisol) should require so great an upheaval in RNA metabolism, but if the enzymes whose synthesis is under endocrine control have much shorter than average half-lives, it could be that their synthesis employs a proportionately larger fraction of the synthetic capacity of the cell than their abundance would at first suggest (Berlin and Schimke, 1965).

That the nRNA whose production is promoted by hormones has enhanced template activity when added to ribosomal systems also seems clear, but the significance of these observations is less so. Given the limited capacity of subcellular systems available to date to incorporate amino acids, it is doubtful if we can expect to demonstrate complete *de novo* synthesis of any protein with a system of mammalian origin (*pace* Sekeris and Lang, 1964), although recently globin synthesis by *E. coli* ribosomes primed with RNA from reticulocytes has been demonstrated (Laycock and Hunt, 1969). Some of the difficulties of isolating fully characterized messenger molecules have been described by Chantrenne *et al.* (1967). In several studies (Mansour and Niu, 1965; Segal *et al.*, 1965; Fujii and Villee, 1967; Villee *et al.*, 1968) extracts of RNA from hormonally treated tissues have been shown to stimulate activity when applied to deficient tissues. The significance of this sort of experiment, though very interesting, is difficult to evaluate without better characterization of the RNA species present in the preparations. How fully formed RNA molecules would cross the cell membrane presents problems, but it could be argued that there is a precedent in nuclear secretion.

Thus we are left with uncertainties at both ends of the action sequence: granted that some hormones do act primarily at the nuclear level, what is the role of the nucleic acid first formed under their influence? and, second, again granted that they act at the nuclear level, do the hormones themselves interact with the nuclear receptors, or is their influence indirectly mediated? Thus Kroeger (1963, 1966) has questioned whether ecdysone really binds directly to the chromosomal apparatus in order to express its activity. Since various salts can also induce the puffs characteristic of ecdysone, as well as changing the permeability of the nuclear membrane, it could be that it is through this sort of change that ecdysone operates (Berendes *et al.*, 1965; Clever, 1965; Ito and Lowenstein, 1965). For mammalian hormones the evidence, though impressive in the case of estrogen and testosterone, is still short of full characterization of sites of binding and what happens immediately thereafter.

The discovery (Garren *et al.*, 1964) that administration of actinomycin or fluorouracil some hours after glucocorticoids can actually enhance the induction of tryptophan pyrrolase or tyrosine transaminase produced by the corticoid remains a puzzling phenomenon. It prompts the suggestion that there exist in animal cells rapidly turning over materials under genetic or metabolic control which can regulate the expression of activity of preformed messenger. Since actinomycin or fluorouracil injected with glucocorticoid prevent expression of its activation, synthesis of new nucleic acid is apparently necessary. Nevertheless, late administration of these inhibitors not only fails to restrain the decay, but actually leads to a further increase in production of the enzyme. Since it is improbable that actinomycin can be stimulating mRNA synthesis (since precursor incorporation is virtually completely inhibited), certain messengers must already exist, but in an inactive state. As stated by Tomkins *et al.* (1965), "there are therefore mechanisms in the cell which selectively inhibit the translation of certain m-RNA's without affecting general protein synthesis." Such a repressor, as well as being specifically directed, must be dependent on RNA synthesis (since it is inhibited by actinomycin), have a rapid rate of degradation, and involve a protein component. Similar consideration could equally apply to many of the other "abnormal" effects of actinomycin noted previously (p. 250). The problems of induction of hepatic enzymes are discussed more fully in Chapter 31 of this volume. The possibility of control at the translational level—that is, factors regulating the extent of use and translation of preexisting messengers—comes from several different sources. Much work (reviewed in Cohen, 1966; Tata, 1966) suggests that embryonic cells contain stable messenger molecules that are present as inactive polysomes in the egg cell prior to fertilization. How these polysomes are preserved and kept inactive until such time as they are required is not clear, but it is possible that their activation involves removal of a protein coat which interferes with normal functioning (Monroy *et al.*, 1965; Salb and Marcus, 1965). The experiments of Garren *et al.* (1964) showing unexpected stimulatory effects of actinomycin on tryptophan pyrrolase induction were explained on the basis of the existence of stable messenger but a rapidly turning over repressor molecule. Conversely Garren *et al.* (1965) found that puromycin and cycloheximide prevented the promotion of steroidogenesis by ACTH. The existence of a rapidly turning over protein (half-life approximately 10 minutes) was postulated which would in this case enhance messenger translation. The curious feature is that this putative protein appears to regulate only the ACTH-induced formation of steroids, not the basal activity.

Inability of ribosomes prepared from hormonally deficient tissues to respond to poly U to the same extent as their counterparts has also been put forward as evidence of impaired ribosomal capacity (Korner and Gumbley, 1966; Earl and Korner, 1966; Wool *et al.*, 1965; Garren *et al.*, 1967), though later results for insulin are less clear-cut (Wool and Cavicchi, 1967). Study of the isolated ribosomes in reconstituted systems presumably precludes simultaneous control by cytoplasmic protein factors, but the possibility of some ribosomal or microsomal proteins acting as regulators is suggested by the results of Hoagland *et al.* (1964) from studies in which they extracted from rat liver microsomes a heat-labile factor capable of inhibiting incorporation. There was much less of this material in microsomes from regenerating liver, where the amount of incorporation is much greater. Inhibitors of ribosome formation can apparently also be extracted from lysosomes and from endoplasmic reticulum of liver (Scornik *et al.*, 1967) and from reticulocyte cytoplasm (Beard and Armentrout, 1967).

In general terms, rapidly acting hormones, e.g., insulin, ACTH, might be expected to act through modulation of existing machinery rather than becoming involved in promoting *de novo* synthetic processes. ACTH in particular regulates secretory activity rather than growth in its minute-by-minute control. Moreover, as we realize increasingly the integration and interaction of metabolic activities in determining metabolic homeostasis, we have to ask whether effects of hormones on other aspects of the metabolic activity of their target cells do not also bring about changes which affect protein synthesis. The implicit assumption behind most of the work described here is that many of the metabolic consequences of hormone action result from prior production of specific proteins which in turn result from prior stimulation of RNA synthesis. For several hormones, this sequence, apart from a few important gaps, seems realistic. However, with the faster acting hormones, particularly insulin acting on muscle, where many aspects of metabolism are simultaneously affected, there may well be complex interactions between, say, effects on carbohydrate metabolism and protein synthesis. For example protein synthesis by isolated ribosomes, which is many times less efficient than *in vivo,* is demonstrably dependent on the concentration of magnesium present, and it is open to question whether in either bacteria or animal cells the concentration of free Mg^{2+}, because of chelation by nucleotides and carboxylic acids and binding by protein and nucleic acids, is at the optimal level. Any other metabolic factors affecting the availability of Mg^{2+} may be expected to affect protein synthesis. This consideration applies even more forcefully to the initial activation of amino acids to aminoacyl-tRNA esters (Novelli, 1967). It has always to be borne in mind that optimal magnesium concentrations noted for ribosomal systems are the total magnesium concentration present—due to complex formation of one sort and another the concentration of free Mg^{2+} will usually be very much less. Hurwitz and Rosano (1967) Salas *et al.* (1967), and Tanner (1967) have provided evidence that spermidine and polyamines may substitute for Mg^{2+} *in vivo* and lower the ionic requirement. It is also possible that binding of ribosomes to membranes may alter the ion requirements. The fact that detergents are not absolutely necessary for the preparation of polysomes from muscle, uterus, or reticulocytes does not necessarily imply that *in vivo* the particles have no association with structural elements, though the nature of the association is presumably different from those degraded by detergents. In fact it must surely be highly probable that spatial orientation and configuration of the polysome is one of the factors responsible for its greater activity *in vivo*. Moreover even in reticulocytes a substantial proportion of the cell's RNA appears to be associated with the membrane (Burka *et al.,* 1967), and the connection of RNA with membranes in many cells (Attardi and Attardi, 1967; Blobel and Potter, 1967a,b; Teng and Hamilton, 1967a) suggests this may be of physiological importance both for function and control.

One ribosome can bind up to 2800 atoms of magnesium, i.e., one Mg^{2+} per two phosphate residues (Goldberg, 1966), but the amount of magnesium bound depends on the concentration of added Mg^{2+} and of K^+. The recently proposed model for the ribosome in which helical regions of the RNA project from the surface of the particle (Cotter *et al.*, 1967) exposes the magnesium binding regions and makes it very likely that changes in the intracellular mileu could have considerable influence on the conformation of the particles. This would determine *inter alia* the extent of binding to messenger and attachment of tRNA (Cannon *et al.*, 1963; Takanami and Okamoto, 1963; Moore, 1966). Moreover Millar and Steiner (1966) have shown that the conformation of tRNA also varies with the magnesium concentration. In regulation of many enzymes of intermediary metabolism the relative levels of the adenine nucleotides and inorganic phosphate are of great importance (Atkinson, 1966). In recent years improved methods of estimation have given lower likely values for free ions, not only for magnesium, but also for inorganic phosphate (Seraydarian *et al.*, 1961). That protein synthesis in tissues is inhibited by agents that interfere with ATP production is well known. That protein synthesis is *in vivo* affected by the more normal changes in relative nucleotide levels has not been demonstrated, but *in vitro* comparatively high ratios of Mg^{2+}/ATP seem optimal (Sachs, 1957; Florini, 1964; Pronczuk *et al.*, 1968). It is not at all improbable that the same sensitivity to nucleotide levels and other metabolic pacemaking regulators may apply to the enzymes involved in amino acid activation and peptide synthesis. Novelli (1967) has pointed out how optimal conditions of pH, ionic strength, and especially Mg/ATP is markedly different for each different activating enzyme both within a species and between species. Most studies have used enzymes from bacteria or yeast, so how far the same findings apply in higher organisms is not yet known. Nirenberg and Leder (1964) and Kaji and Kaji (1965) have found that uncharged tRNA can interfere with peptide synthesis in cell-free systems, though GTP appears to reverse this inhibition (Seeds and Conway, 1966). Ceccarini *et al.* (1967) have provided evidence of regulation of protein synthesis in embryos by change in activity of aminoacyl synthetases. In microorganisms, withholding amino acids can suppress RNA synthesis. Stent and Brenner (1961) and Kurland and Maaløe (1962) proposed that uncharged tRNA could be a specific repressor of RNA synthesis, and there is much evidence for a role of tRNA in enzyme repression (Vogel and Vogel, 1967). Inhibition of polymerase has also been observed (Bremer *et al.,* 1966). A similar possibility in higher organisms could relate amino acid transport and protein and nucleic acid synthesis and provide another possible area through which hormonal control could be exercised.

It seems to the author that the role of transfer RNA in the possible control of protein synthesis deserved more study. For most cells about 80% of RNA is ribosomal. In molar terms this means that the ratio of tRNA to rRNA is about 10–20:1, depending on the proportion of the remainder that is tRNA. Given that at least 20 forms of tRNA are required for protein synthesis and possibly 60, the quantity of specific tRNA's becomes quite small. Somehow the cell must be able to ensure that all the appropriate species are charged with their amino acids and available to the ribosomes at the right time.

In vitro incorporation of amino acids not only flourishes best with relatively high Mg^{2+}/ATP ratios, but also at a pH (often about 7.6) that is appreciably higher than pertains *in vivo*. A pH more alkaline than *in vivo* also appears optimal for several enzymes of carbohydrate metabolism that are normally under complex metabolic control (Krebs *et al.,* 1964; Uyeda and Racker, 1965; Lowry and Passonneau, 1966). A possible reason for this is that the nearer physiological pH is to the p*K* of ATP and the lower the free Mg^{2+} concentration, the more fluctuation in relative concentration there is likely to be in the cell between the different ATP (and GTP) ions and their Mg chelates. Differing activities and competitive inhibition between the various forms offers considerable scope for enzyme control, which will additionally be related to intracellular pH (Manchester, 1968b). Recent electrode measurements of intracellular pH have been as low as 6.0 (Carter *et al.,* 1967), though the validity of these has been questioned by Butler *et al.* (1967) on the basis of measurements made by chemical methods.

Krahl (1957, 1961) and Hechter and Lester (1960), adopting the concept of a cellular "cytoskeleton" (Peters, 1956), have suggested a fundamental hormone action to be alteration of the organization of the cytoskeleton. The problem with this concept has always been lack of definitive evidence. However, it is not difficult to envisage in general terms how a primary interaction of a hormone with its receptor site could change cellular orientation in such a way as to alter, for example, local Mg/adenine nucleotide ratios, from which a host of metabolic sequalae, including changed rates of RNA and protein synthesis, could follow. However, the need to postulate compartmentation in the cell, the existence of which is hardly to be doubted, unfortunately prejudices the possibility of readily investigating such hypotheses since it is impossible to measure directly local concentration changes. Moreover, an early consequence of changed intracellular membrane permeability, e.g., of mitochondria and nuclei, will obviously be of consequence both in the regulation in higher organisms of any possible interaction of, say, tRNA with the sites of RNA synthesis and in the regulation of RNA transfer to the cytoplasm, as well as influencing the distribution of ions, ATP, and other metabolites (including possibly H^+) within the cell. Thus we are as yet a long way from fully understanding how hormonal control could be exercised *in vivo,* let alone knowing the details in any specific case.

References

Aaronson, S. A., Korner, A., and Munro, A. J. (1966). *Biochem. J.* **101,** 448.

Acs, G., Reich, E., and Valanju, S. (1963). *Biochim. Biophys. Acta* **76,** 68.

Adamson, L. F., Langeluttig, S. G., and Anast, C. S. (1966). *Biochim. Biophys. Acta* **115,** 355.

Adiga, P. R., Rao, P. M., Hussa, R. O., and Winnick, T. (1966). *Biochemistry* **5,** 3850.

Ahlquist, R. P. (1948). *Am. J. Physiol.* **153,** 586.

Ahlquist, R. P. (1967). *Ann. N.Y. Acad. Sci.* **139,** 549.

Allen, D. W., and Zamecnik, P. C. (1962). *Biochim. Biophys. Acta* **55,** 865.

Allfrey, V. G., Pogo, B. G. T., Littau, V. C., Gershey, E. L., and Mirsky, A. E. (1968). *Science* **159,** 314.

Ames, B. N., and Martin, R. G. (1964). *Ann. Rev. Biochem.* **33,** 235.

Appel, S. H. (1967). *Nature* **213,** 1253.

Appleman, M. M., and Kemp, R. G. (1966). *Biochem. Biophys. Res. Commun.* **24,** 564.

Arnstein, H. R. V., Cox, R. A., and Hunt, J. A. (1962). *Nature* **194,** 1042.

Arvill, A., and Ahrén, K. (1967a). *Acta Endocrinol.* **56,** 279.

Arvill, A., and Ahrén, K. (1967b). *Acta Endocrinol.* **56,** 295.

Ashmore, J., and Morgan, D. (1967). *In* "The Adrenal Cortex" (A. B. Eisenstein, ed.), p. 249. Churchill, London.

Atkinson, D. E. (1966). *Ann. Rev. Biochem.* **35,** 85.

Attardi, B., and Attardi, G. (1967). *Proc. Natl. Acad. Sci. U.S.* **58,** 1051.

Baliga, B. S., Pronczuk, A. W., and Munro, H. N. (1969). *J. Biol. Chem.* **244,** 4480.

Ballard, P. L., and Williams-Ashman, H. G. (1964). *Nature* **203,** 150.

Ballard, P. L., and Williams-Ashman, H. G. (1966). *J. Biol. Chem.* **241,** 1602.

Barker, K. L., and Warren, J. C. (1966a). *Proc. Natl. Acad. Sci. U.S.* **56,** 1298.

Barker, K. L., and Warren, J. C. (1966b). *Endocrinology* **78,** 1205.

Barnabei, O., Romano, B., and Di Bitonto, G. (1965). *Arch. Biochem. Biophys.* **109,** 266.

Barnabei, O., Romana, B., Di Bitonto, G., and Tomasi, V. (1966). *Arch. Biochem. Biophys.* **113,** 478.

Barzilai, D., and Pincus, G. (1965). *Proc. Soc. Exptl. Biol. Med.* **118,** 57.

Beard, N. S., and Armentrout, S. A. (1967). *Proc. Natl. Acad. Sci. U.S.* **58,** 750.

Beattie, D. S., Basford, R. E., and Koritz, S. B. (1967). *J. Biol. Chem.* **242,** 4584.

Begg, D. J., and Munro, H. N. (1965). *Nature* **207,** 483.

Beloff-Chain, A., Betto, P., Bleszynski, W., Catanzaro, R., Chain, E. B., Dimitrovskii, A. A., Longinotti. L., and Pocchiari, F. (1965). *Biochem. J.* **97,** 565.

Bennett, L. L., Smithers, D., and Ward, C. T. (1964). *Biochim. Biophys. Acta* **87,** 60.

Berendes, H. D., van Breugel, F. M. A., and Holt, Th. K. H. (1965). *Chromosoma* **16,** 35.

Berlin, C. M., and Schimke, R. T. (1965). *Mol. Pharmacol.* **1,** 149.

Bewsher, P. D., Hillman, C. C., and Ashmore, J. (1966). *Mol. Pharmacol.* **2,** 227.

Bhattacharya, G. (1961). *Biochem. J.* **79,** 369.

Bhattacharya, G. (1964). *Biochim. Biophys. Acta* **93,** 644.

Blecher, M., and White, A. (1958). *J. Biol. Chem.* **233,** 1161.

Bleehen, N. M., and Fisher, R. B. (1954). *J. Physiol. (London)* **123,** 260.

Blobel, G., and Potter, V. R. (1967a). *J. Mol. Biol.* **26,** 279.

Blobel, G., and Potter, V. R. (1967b). *J. Mol. Biol.* **26,** 293.

Bond, H. E. (1965). *J. Cellular Comp. Physiol.* **66,** Suppl. 1, 120.

Bondy, P. K., Ingle, D. J., and Meeks, R. C. (1954). *Endocrinology* **55,** 354.

Bonner, J., Dahmus, M. E., Fambrough, D., Huang, R. C., Marushige, K., and Tuan, D. Y. H. (1968). *Science* **159,** 47.

Bornstein, J., and Hyde, D. (1960). *Nature* **187,** 125.

Bornstein, J., Krahl, M. E., Marshall, L. B., Gould, M. K., and Armstrong, J. McD. (1968a). *Biochim. Biophys. Acta* **156,** 31.

Bornstein, J., Armstrong, J. McD., and Jones, M. D. (1968b). *Biochim. Biophys. Acta* **156,** 38.

Bramwell, M. E., and Harris, H. (1967). *Biochem. J.* **103,** 816.

Bransome, E. D. (1968). *Ann. Rev. Physiol.* **30,** 171.

Bransome, E. D., and Chargaff, E. (1964). *Biochim. Biophys. Acta* **91,** 180.

Bransome, E. D., and Reddy, W. J. (1963a). *Endocrinology* **73,** 540.

Bransome, E. D., and Reddy, W. J. (1963b). *Arch. Biochem. Biophys.* **101,** 21.

Bransome, E. D., and Reddy, W. J. (1964). *Endocrinology* **74,** 495.

Bremer, H., Yegian, C., and Konrad, M. (1966). *J. Mol. Biol.* **16,** 94.

Bresnick, E. (1966). *Mol. Pharmacol.* **2,** 406.

Breuer, C. B., and Florini, J. R. (1965). *Biochemistry* **4,** 1544.

Breuer, C. B., and Florini, J. R. (1966). *Biochemistry* **5,** 3857.

Breuer, C. B., Davis, M. C., and Florini, J. R. (1964). *Biochemistry* **3,** 1713.

Brewer, E. N., Foster, L. B., and Sells, B. H. (1969). *J. Biol. Chem.* **244,** 1389.

Brinck-Johnsen, T., and Dougherty, T. F. (1965). *Acta Endocrinol.* **49,** 471.

Brown, D. M. (1966). *Endocrinology* **78,** 1252.

Brown, D. D., and Gurdon, J. B. (1964). *Proc. Natl. Acad. Sci. U.S.* **51,** 139.

Brown, E., Clarke, D. L., Roux, V., and Sherman, G. H. (1963). *J. Biol. Chem.* **238,** PC 852.

Bruchovsky, N., and Wilson, J. D. (1968a). *J. Biol. Chem.* **243,** 2012.

Bruchovsky, N., and Wilson, J. D. (1968b). *J. Biol. Chem.* **243,** 5953.

Burka, E., Schreml, W., and Kick, C. (1967). *Biochem. Biophys. Res. Commun.* **26,** 334.

Burrow, G. N., and Bondy, P. K. (1964). *Endocrinology* **75,** 455.

Busch, H., Byvoet, P., and Smetana, K. (1963). *Cancer Res.* **23,** 313.

Buse, M. G., and Buse, J. (1959). *Diabetes* **8,** 218.

Buse, M. G., and Buse, J. (1961). *Diabetes* **10,** 134.

Buse, M. G., Buse, J., McMaster, J., and Krech, L. H. (1964). *Metab. Clin. Exptl.* **13,** 339.

Buse, M. G., McMaster, J., and Buse, J. (1965). *Metab. Clin. Exptl.* **14,** 1220.

Butcher, R. W., Ho, R. J., Meng, H. C., and Sutherland, E. W. (1965). *J. Biol. Chem.* **240,** 4515.

Butcher, R. W., Sneyd, J. G. T., Park, C. R., and Sutherland, E. W. (1966). *J. Biol. Chem.* **241,** 1651.

Butler, T. C., Waddell, W. J., and Pool, D. T. (1967). *Federation Proc.* **26,** 1327.

Byvoet, P. (1967). *Mol. Pharmacol.* **3,** 303.

Cadenas, E., Kaji, H., Park, C. R., and Rasmussen, H. (1961). *J. Biol. Chem.* **236,** PC 63.

Cahill, G. F. (1965). *Proc. 5th Intern. Diabetes Federation.* Excerpta Med. Intern. Congr. Ser. No. 84, p. 116. Amsterdam.

Cahill, G. F., Ashmore, J., Earle, A. S., and Zottu, S. (1958). *Am. J. Physiol.* **192,** 491.

Cammarano, P., Chinali, G., Gaetani, S., and Spadoni, M. A. (1968). *Biochim. Biophys. Acta* **155,** 302.

Campagnoni, A. T., and Mahler, H. R. (1967). *Biochemistry* **6,** 956.

Campbell, P. N. (1965). *Progr. Biophys. Mol. Biol.* **15,** 1.

Cannon, M. (1967). *Biochem. J.* **104,** 934.

Cannon, M., Krug, R., and Gilbert, W. (1963). *J. Mol. Biol.* **7,** 360.

Carlin, H., and Hechter, O. (1962). *J. Biol. Chem.* **237,** PC 1371.

Carlin, H., and Hechter, O. (1964). *Proc. Soc. Exptl. Biol. Med.* **115,** 127.

Carpenter, F. H. (1966). *Am. J. Med.* **40,** 750.

Carroll, P. M., and Sereda, D. D. (1968). *Nature* **217,** 667.

Carter, N. W., Rector, F. C., Jr., Campion, D. S., and Seldin, D. W. (1967). *Federation Proc.* **26,** 1322.

Castles, J. J., and Wool, I. G. (1964). *Biochem. J.* **91,** 11C.

Ceccarini, C., Maggio, R., and Barbata, G. (1967). *Proc. Natl. Acad. Sci. U.S.* **58,** 2235.

Chambers, J. W., Georg, R. H., and Bass, A. D. (1965). *Mol. Pharmacol.* **1,** 66.

Chambon, P., Ramuz, M., Mandel, P., and Doly, J. (1968a). *Biochim. Biophys. Acta* **157,** 504.

Chambon, P., Karon, H., Ramuz, M., and Mandel, P. (1968b). *Biochim. Biophys. Acta* **157,** 520.

Chantrenne, H. (1965). *Biochim. Biophys. Acta* **95,** 351.

Chantrenne, H., Burny, A., and Marbaix, G. (1967). *Progr. Nucleic Acid Res.* **7,** 173.

Chase, L. R., and Aurbach, G. D. (1968). *Science* **159,** 545.

Chen, S. C., and Young, V. R. (1968). *Biochem. J.* **106,** 61.

Clark, I. (1953). *J. Biol. Chem.* **200,** 69.

Clark, J. M., and Chang, A. Y. (1965). *J. Biol. Chem.* **240,** 4734.

Clark-Walker, G. D., and Linnane, A. W. (1966). *Biochem. Biophys. Res. Commun.* **25,** 8.

Clever, U. (1965). *Chromosoma* **17,** 309.

Clever, U., and Karlson, P. (1960). *Exptl. Cell Res.* **20,** 625.

Clever, U., and Romball, C. G. (1966). *Proc. Natl. Acad. Sci. U.S.* **56,** 1470.

Cohen, N. R. (1966). *Biol. Rev. Cambridge Phil. Soc.* **41,** 503.

Cohen, R. B., and Fagundes, L. A. (1966). *Endocrinology* **78,** 220.

Coles, N., Bukenberger, M. W., and Meister, A. (1962). *Biochemistry* **1,** 317.

Colombo, B., Felicetti, L., and Baglioni, C. (1965). *Biochem. Biophys. Res. Commun.* **18,** 389.

Colombo, B., Felicetti, L., and Baglioni, C. (1966). *Biochim. Biophys. Acta* **119,** 109.

Cooper, B., Banthorpe, D. V., and Wilkie, D. (1967). *J. Mol. Biol.* **26,** 347.

Cotter, R. I., McPhie, P., and Gratzer, W. B. (1967). *Nature* **216,** 864.

Criss, W. E., and McKerns, K. W. (1968). *Biochemistry* **7,** 2364.

Cryer, P. E., Jaratt, L., and Kipnis, D. M. (1969). *Biochim. Biophys. Acta* **177,** 586.

Cunningham, D., and Steiner, D. F. (1966). *Federation Proc.* **25,** 788.

Cunningham, D., and Steiner, D. F. (1969). *Biochim. Biophys. Acta* **171,** 67.

Dahmus, M. E., and Bonner, J. (1965). *Proc. Natl. Acad. Sci. U.S.* **54,** 1370.

Darnell, J. E. (1968). *Bacteriol. Rev.* **32,** 262.

Das, H. K., Goldstein, A., and Kanner, L. C. (1966). *Mol. Pharmacol.* **2,** 158.

Daughaday, W. H., and Kipnis, D. M. (1966). *Recent Progr. Hormone Res.* **22,** 49.

Davey, P. J., and Manchester, K. L. (1969). *Biochim. Biophys. Acta* **182,** 85.

Davis, W. W., and Garren, L. D. (1966). *Biochem. Biophys. Res. Commun.* **24,** 805.

Davis, W. W., and Garren, L. D. (1968). *J. Biol. Chem.* **243,** 5153.

Davoren, P. R., and Sutherland, E. W. (1963). *J. Biol. Chem.* **238,** 3016.

Dawson, K. G., Patey, P., Rubenstein, D., and Beck, J. C. (1966). *Mol. Pharmacol.* **2,** 269.

DeVenuto, F., and Lange, R. J. G. (1967). *Biochim. Biophys. Acta* **134,** 443.

Di Girolamo, A., Di Girolamo, M., Gaetani, S., and Spadoni, M. A. (1967). *European J. Biochem.* **1,** 164.

Dingman, C. W., and Sporn, M. B. (1965). *Science* **149,** 1251.

Dintzis, H. (1961). *Proc. Natl. Acad. Sci. U.S.* **47,** 247.

Di Stefano, H. S., and Diermeier, H. F. (1959). *Endocrinology* **64,** 448.

Dresden, M. H., and Hoagland, M. B. (1967a). *J. Biol. Chem.* **242,** 1065.

Dresden, M. H., and Hoagland, M. B. (1967b). *J. Biol. Chem.* **242,** 1069.

Drysdale, J. W., and Munro, H. N. (1966). *J. Biol. Chem.* **241,** 3630.

Drysdale, J. W., and Munro, H. N. (1967). *Biochim. Biophys. Acta* **138,** 616.

Dukes, P. P., and Sekeris, C. E. (1965). *Z. Physiol. Chem.* **341,** 149.

Dukes, P. P., Sekeris, C. E, and Schmid, W. (1966). *Biochim. Biophys. Acta* **123,** 126.

Eagle, G. R., and Robinson, D. S. (1964). *Biochem. J.* **93,** 10C.

Earl, D. C. N., and Korner, A. (1965). *Biochem. J.* **94,** 721.

Earl, D. C. N., and Korner, A. (1966). *Arch. Biochem. Biophys.* **115,** 445.

Earl, D. C. N., Broadus, A. E., and Morgan, H. E. (1967). *Abstr. 5th Federation European Biochem. Soc. Meeting, Oslo.*

Eboué-Bonis, D., Chambaut, A. M., Volfin, P., and Clauser, H. (1963). *Nature* **199,** 1183.

Edelman, I. S., Bogoroch, R., and Porter, G. A. (1963). *Proc. Natl. Acad. Sci. U.S.* **50,** 1169.

Edelman, P. M., and Schwartz, I. L. (1966). *Am. J. Med.* **40,** 695.

Edelman, P. M., Rosenthal, S. L., and Schwartz, I. L. (1963). *Nature* **197,** 878.

Eisenfeld, A. J. (1967). *Biochim. Biophys. Acta* **136,** 498.

Elsas, L. J., and Rosenberg, L. E. (1967). *Proc. Natl. Acad. Sci. U.S.* **57,** 371.

Elsas, L. J., Albrecht, I., Koehne, W., and Rosenberg, L. E. (1967). *Nature* **214,** 916.

Endo, M. (1964). *Nature* **202,** 1115.

Engel, F. L., and Kostyo, J. L. (1964). *Hormones* **5,** 69.

England, P. J., Denton, R. M., and Randle, P. J. (1967). *Biochem. J.* **105,** 32C.

Ennis, H. L. (1966). *Mol. Pharmacol.* **2,** 543.

Ennis, H. L., and Lubin, M. (1964). *Science* **146,** 1474.

Erdos, T., and Bessada, R. (1966). *Biochim. Biophys. Acta* **129,** 628.

Estensen, R. D., and Baserga, R. (1966). *J. Cell Biol.* **30,** 13.

Exton, J. H., Jefferson, L. S., Butcher, R. W., and Park, C. R. (1966). *Am. J. Med.* **40,** 709.

Fahmy, A. R., Griffiths, K., Mahler, R., and Williams, A. R. (1967). *Biochem. J.* **105,** 6C.
Fain, J. N., Kovacev, V. P., and Scow, R. O. (1965). *J. Biol. Chem.* **240,** 3522.
Fajans, S. A., Floyd, J. C., Knopf, R. F., and Conn, J. W. (1967). *Recent Progr. Hormone Res.* **23,** 617.
Fallen, E., Tremblay, G. M., and Gorlin, R. (1966). *Federation Proc.* **25,** 441.
Farese, R. V. (1964). *Biochim. Biophys. Acta* **87,** 699.
Farese, R. V. (1966). *Endocrinology* **78,** 125.
Farese, R. V. (1967). *Biochemistry* **6,** 2052.
Farese, R. V., and Reddy, W. J. (1963). *Endocrinology* **73,** 294.
Farese, R. V., and Schnure, J. J. (1967). *Endocrinology* **80,** 872.
Felicetti, L., Colombo, B., and Baglioni, C. (1966). *Biochim. Biophys. Acta* **119,** 120.
Ferguson, J. J. (1963). *J. Biol. Chem.* **238,** 2754.
Ferguson, J. J., and Morita, Y. (1964). *Biochim. Biophys. Acta* **87,** 348.
Ferguson, J. J., Morita, Y., and Mendelsohn, L. (1967). *Endocrinology* **80,** 521.
Fiala, E. S., and Davis, F. F. (1965). *Biochem. Biophys. Res. Commun.* **18,** 115.
Field, R. A., and Adams, L. C. (1965). *Biochim. Biophys. Acta* **106,** 474.
Finkel, R. M., Henshaw, E. C., and Hiatt, H. H. (1966). *Mol. Pharmacol.* **2,** 221.
Florini, J. R. (1964). *Biochemistry* **3,** 209.
Florini, J. R., and Breuer, C. B. (1966). *Biochemistry* **5,** 1870.
Floyd, J. C., Fajans, S. S., Conn, J. W., Knopf, R. F., and Rull, J. A. (1966a). *J. Clin. Invest.* **45,** 1479.
Floyd, J. C., Fajans, S. S., Conn, J. W., Knopf, R. E., and Rull, J. A. (1966b). *J. Clin. Invest.* **45,** 1487.
Fong, C. T. O., Silver, L., Christman, D., and Schwartz, I. L. (1960). *Proc. Natl. Acad. Sci. U.S.* **46,** 1273.
Forchhammer, J., and Kjeldgaard, N. O. (1967). *J. Mol. Biol.* **24,** 459.
Formby, B. (1967). *Mol. Pharmacol.* **3,** 284.
Franzini-Armstrong, C., and Porter, K. R. (1964). *J. Cell Biol.* **22,** 675.
Freedman, M. L., Fisher, J. M., and Rabinovitz, M. (1968). *J. Mol. Biol.* **33,** 315.
Frieden, E. H., Laby, M. R., Bates, F., and Layman, N. W. (1957). *Endocrinology* **60,** 290.
Friedman, S. M., Berezney, R., and Weinstein, I. B. (1968). *J. Biol. Chem.* **243,** 5044.
Fritz, G. R., and Knobil, E. (1963). *Nature* **200,** 682.
Fujii, T., and Villee, C. (1967). *Proc. Natl. Acad. Sci. U.S.* **57,** 1468.
Fukuhara, H. (1965). *Biochem. Biophys. Res. Commun.* **18,** 297.
Furth, J. J., and Ho, P. (1965). *J. Biol. Chem.* **240,** 2602.
Gabourel, J. D., and Fox, K. E. (1965). *Biochem. Biophys. Res. Commun.* **18,** 81.
Gambetti, P., Gonatas, N. K., and Flexner, L. B. (1968). *J. Cell Biol.* **36,** 379.
Garratt, C. J., Cameron, J. S., and Menzinger, G. (1966a). *Biochim. Biophys. Acta* **115,** 179.
Garratt, C. J., Jarrett, R. J., and Keen, H. (1966b). *Biochim. Biophys. Acta* **121,** 143.
Garren, L. D., Howell, R. R., Tomkins, G. M., and Crocco, R. M. (1964). *Proc. Natl. Acad. Sci. U.S.* **52,** 1121.
Garren, L. D., Ney, R. L., and Davis, W. W. (1965). *Proc. Natl. Acad. Sci. U.S.* **53,** 1443.
Garren, L. D., Richardson, A. P., Jr., and Crocco, R. M. (1967). *J. Biol. Chem.* **242,** 650.

Gatica, M., Allende, C. C., Mora, G., Allende, J. E., and Medina, J. (1966). *Biochim. Biophys. Acta* **129,** 201.

Georgiev, G. P. (1967). *Progr. Nucleic Acid Res.* **6,** 259.

Georgiev, G. P., Samarina, O. P., Lerman, M. I., Smirnov, M. N., and Severtzov, A. N. (1963). *Nature* **200,** 1291.

Giacomoni, D., and Spiegelman, S. (1962). *Science* **138,** 1328.

Girard, M., Penman, S., and Darnell, J. E. (1964). *Proc. Natl. Acad. Sci. U.S.* **51,** 205.

Girard, M., Latham, H., Penman, S., and Darnell, J. E. (1965). *J. Mol. Biol.* **11,** 187.

Godchaux, W., and Herbert, E. (1966). *J. Mol. Biol.* **21,** 537.

Godchaux, W., Adamson, S. D., and Herbert, E. (1967). *J. Mol. Biol.* **27,** 57.

Goldberg, A. (1966). *J. Mol. Biol.* **15,** 663.

Goldberg, M. L., and Atchley, W. A. (1966). *Proc. Natl. Acad. Sci. U.S.* **55,** 989.

Goldstein, A., Kirschbaum, J. B., and Roman, A. (1965). *Proc. Natl. Acad. Sci. U.S.* **54,** 1669.

Goldstein, L., Stella, E. J., and Knox, W. E. (1962). *J. Biol. Chem.* **237,** 1723.

Goldstein, S., and Reddy, W. J. (1967). *Biochim. Biophys. Acta* **141,** 310.

Goodfriend, T. L., and Kaplan, N. O. (1964). *J. Biol. Chem.* **239,** 130.

Goodfriend, T. L., Sokol, D. M., and Kaplan, N. O. (1966). *J. Mol. Biol.* **15,** 18.

Goodman, H. M. (1964). *Am. J. Physiol.* **206,** 129.

Goodman, H. M. (1966). *Am. J. Physiol.* **211,** 815.

Goodman, M. H. (1968). *In* "Growth Hormone" (A. Pecile and E. E. Müller, eds.) *Excerpta Med. Intern. Congr. Ser.* **158,** 153.

Goodman, H. M., and Rich, A. (1962). *Proc. Natl. Acad. Sci. U.S.* **48,** 2101.

Gorski, J. (1964). *J. Biol. Chem.* **239,** 889.

Gorski, J., and Axman, M. C. (1964). *Arch. Biochem. Biophys.* **105,** 517.

Gorski, J., and Nelson, N. J. (1965). *Arch. Biochem. Biophys.* **110,** 284.

Gorski, J., and Nicolette, J. A. (1963). *Arch. Biochem. Biophys.* **103,** 418.

Gorski, J., Noteboom, W. D., and Nicolette, J. A. (1965). *J. Cellular Comp. Physiol.* **66,** Suppl. 1, 91.

Gould, B. S., and Manner, G. (1967). *Biochim. Biophys. Acta* **138,** 189.

Grahame-Smith, D. G., Butcher, R. W., Ney, R. L., and Sutherland, E. W. (1967). *J. Biol. Chem.* **242,** 5535.

Granick, S., and Kappas, A. (1967). *Proc. Natl. Acad. Sci. U.S.* **57,** 1463.

Green, M., and Miller, L. L. (1960). *J. Biol. Chem.* **235,** 3202.

Greenman, D. L., and Kenney, F. T. (1964). *Arch. Biochem. Biophys.* **107,** 1.

Greenman, D. L., Wicks, W. D., and Kenney, F. T. (1965). *J. Biol. Chem.* **240,** 4420.

Grief, R. L., Song, C. S., and Chipkin, D. (1965). *Endocrinology* **77,** 223.

Grossman, S. H., and Manchester, K. L. (1966). *Nature* **211,** 1300.

Gupta, N. K. (1968). *J. Biol. Chem.* **243,** 4959.

Gupta, N. K., and Talwar, G. P. (1968). *Biochem. J.* **110,** 401.

Hadjiolov, A. A. (1966). *Biochim. Biophys. Acta* **119,** 547.

Hales, C. N. (1967). *In* "Essays in Biochemistry" (P. N. Campbell and G. D. Greville, eds.), Vol. 3, p. 73. Academic Press, New York.

Halkerston, I. D. K., Feinstein, M., and Hechter, O. (1964). *Endocrinology* **74,** 649.

Hamilton, T. H. (1968). *Science* **161,** 649.

Hamilton, T. H. (1964). *Proc. Natl. Acad. Sci. U.S.* **51,** 83.

Hamilton, T. H., Widnell, C. C., and Tata, J. R. (1965). *Biochim. Biophys. Acta* **108,** 168.

Hancock, R. L., Zelis, R. F., Shaw, M., and Williams-Ashman, H. G. (1962). *Biochim. Biophys. Acta* **55,** 257.

Handler, J. S., Butcher, R. W., Sutherland, E. W., and Orloff, J. (1961). *J. Biol. Chem.* **240,** 4524.

Harris, E. J., and Manchester, K. L. (1966). *Biochem. J.* **101,** 135.

Haselkorn, R., and Fried, V. A. (1964). *Proc. Natl. Acad. Sci. U.S.* **51,** 1001.

Haynes, R. C. (1958). *J. Biol. Chem.* **233,** 1220.

Haynes, R. C., and Berthet, L. (1957). *J. Biol. Chem.* **225,** 115.

Haynes, R. C., and Sutherland, E. W. (1967). *Endocrinology* **80,** 297.

Haynes, R. C., Peron, F. G., and Koritz, S. B. (1959). *J. Biol. Chem.* **234,** 1421.

Hechter, O., and Halkerston, I. D. K. (1964). *Hormones* **5,** 697.

Hechter, O., and Halkerston, I. D. K. (1965). *Ann. Rev. Physiol.* **27,** 133.

Hechter, O., and Lester, G. (1960). *Recent Progr. Hormones Res.* **16,** 139.

Hechter, O., Yoshinago, K., Halkerston, I. D. K., and Birchall, K. (1967). *Arch. Biochem. Biophys.* **122,** 449.

Henshaw, E. C., Bojarski, T. B., and Hiatt, H. H. (1963). *J. Mol. Biol.* **7,** 122.

Henshaw, E. C., Revel, M., and Hiatt, H. H. (1965). *J. Mol. Biol.* **14,** 241.

Herrera, M. G., and Renold, A. E. (1960). *Biochim. Biophys. Acta* **44,** 165.

Heywood, S. M., Dowben, R. M., and Rich, A. (1967). *Proc. Natl. Acad Sci. U.S.* **57,** 1002.

Hjalmarson, Å., and Ahrén, K. (1967). *Acta Endocrinol.* **56,** 347.

Hoagland, M. B., Scornik, O. A., and Pfefferkorn, L. C. (1964). *Proc. Natl. Acad. Sci. U.S.* **51,** 1184.

Hofert, J., and Boutwell, R. K. (1963). *Arch. Biochem. Biophys.* **103,** 338.

Hofert, J., Gorski, J., Mueller, G. C., and Boutwell, R. K. (1962). *Arch. Biochem. Biophys.* **97,** 134.

Hogan, B. L. M. (1969). *Biochim. Biophys. Acta* **182,** 85.

Hogan, B. L. M., and Korner, A. (1968). *Biochim. Biophys. Acta* **169,** 139.

Huehns, E. R., and Shooter, E. M. (1965). *J. Med. Genet.* **2,** 1.

Huijing, Fr., and Larner, J. (1966). *Proc. Natl. Acad. Sci. U.S.* **56,** 647.

Hunt, J. A., and Wilkinson, B. R. (1967). *Biochemistry* **6,** 1688.

Hurwitz, C., and Rosano, C. L. (1967). *J. Biol. Chem.* **242,** 3719.

Huxley, H. E. (1964). *Nature* **202,** 1067.

Igarashi, K., and Kaji, A. (1967). *Proc. Natl. Acad. Sci. U.S.* **58,** 1971.

Ito, S., and Loewenstein, W. R. (1965). *Science* **150,** 909.

Izzo, J. L., Bartlett, J. W., Roncone, A., Izzo, M. J., and Bale, W. F. (1967). *J. Biol. Chem.* **242,** 2343.

Jackson, C. D., and Sells, B. H. (1967). *Biochim. Biophys. Acta* **142,** 419.

Jackson, C. D., and Sells, B. H. (1968). *Biochim. Biophys. Acta* **155,** 417.

Jacob, S. T., Sajdel, E. M., and Munro, H. N. (1968). *Biochim. Biophys. Acta* **157,** 421.

Jacob, S. T., Sajdel, E. M., and Munro, H. N. (1969). *European J. Biochem.* **7,** 449.

Jefferson, L. S., and Korner, A. (1967). *Biochem. J.* **104,** 826.

Jenkinson, D. H., and Morton, I. K. M. (1967a). *J. Physiol. (London)* **188,** 387.

Jenkinson, D. H., and Morton, I. K. M. (1967b). *Ann. N.Y. Acad. Sci.* **139,** 762.

Jensen, E. V., and Jacobson, H. I. (1962). *Recent Progr. Hormone Res.* **18,** 387.

Jensen, E. V., Jacobson, H. I., Flesher, J. W., Saha, N. N., Gupta, G. N., Smith, S., Colucci, V., Shiplicoff, D., Neumann, H. G., DeSombre, E. R., and Jungblut, P. W.

(1966). *In* "Steroid Dynamics" (G. Pincus, T. Nakao, and J. F. Tait, eds.). p. 133. Academic Press, New York.

Jensen, E. V., Suzuki, T., Kawashima, T., Stumpf, W. E., Jungblut, P. W., and DeSombre, E. R. (1968). *Proc. Natl. Acad. Sci. U.S.* **59,** 632.

Johnson, J. D., Jant, B. A., Sokoloff, L., and Kaufman, S. (1969). *Biochim. Biophys. Acta* **179,** 526.

Joklik, W. K., and Becker, Y. (1965). *J. Mol. Biol.* **13,** 511.

Jondorf, W. R., Simon, D. C., and Avnilmelch, M. (1966). *Mol. Pharmacol.* **2,** 506.

Jungas, R. L. (1966). *Proc. Natl. Acad. Sci. U.S.* **56,** 757.

Kabat, D., and Rich, A. (1969). *Biochemistry* **8,** 3742.

Kahlenberg, A., Galsworthy, P. R., and Hokin, L. E. (1967). *Science* **157,** 434.

Kaji, H., and Kaji, A. (1965). *Proc. Natl. Acad. Sci. U.S.* **54,** 213.

Kalf, G. F. (1963). *Arch. Biochem. Biophys.* **101,** 350.

Kaplan, S. A., and Shimizu, C. S. N. (1963). *Endocrinology* **72,** 267.

Karaboyas, G. C., and Koritz, S. B. (1965). *Biochemistry* **4,** 462.

Karlson, P. (1963). *Perspectives Biol. Med.* **6,** 212.

Kay, J. E., and Korner, A. (1966). *Biochem. J.* **100,** 815.

Kennell, D. (1964). *J. Mol. Biol.* **9,** 789.

Kenney, F. T. (1962). *J. Biol. Chem.* **237,** 1610 and 3495.

Kenney, F. T. (1967). *Science* **156,** 525.

Kenney, F. T., and Kull, F. J. (1963). *Proc. Natl. Acad. Sci. U.S.* **50,** 493.

Ketterer, B., Randle, P. J., and Young, F. G. (1957). *Ergeb. Physiol. Biol. Chem. Physiol.* **66,** Suppl. 1, 125.

Kerson, L. A., Garfinkel, D., and Mildvan, A. S. (1967). *J. Biol. Chem.* **242,** 2124.

Ketterer, B., Randle, P. J., and Young, F. G. (1967). *Ergeb. Physiol. Biol. Chem. Exptl. Pharmakol.* **49,** 127.

Kidson, C. (1965). *Biochem. Biophys. Res. Commun.* **21,** 283.

Kidson, C. (1967). *Nature* **213,** 779.

Kidson, C., and Kirby, K. S. (1964). *Nature* **203,** 599.

Kim, K., and Cohen, P. P. (1966). *Proc. Natl. Acad. Sci. U.S.* **55,** 1251.

King, H. W. S., and Fitschen, W. (1968). *Biochim. Biophys. Acta* **155,** 32.

Kipnis, D. M., and Noall, M. W. (1958). *Biochim. Biophys. Acta* **28,** 226.

Kipnis, D. M., Reiss, E., and Helmreich, E. (1961). *Biochim. Biophys. Acta* **51,** 519.

Kit, S., and Barron, E. S. G. (1953). *Endocrinology* **52,** 1.

Kit, S., Bacila, M., and Barron, E. S. G. (1954). *Biochim. Biophys. Acta* **13,** 516.

Knobil, E., and Hotchkiss, J. (1964). *Ann. Rev. Physiol.* **26,** 47.

Kochakian, C. D. (1965). *In* "Mechanisms of Hormone Action" (P. Karlson, ed.), p. 192. Academic Press, New York.

Kochakian, C. D., and Robertson, E. (1951). *J. Biol. Chem.* **190,** 495.

Kochakian, C. D., Tillotson, C., and Austin, J. (1957). *Endocrinology* **60,** 144.

Korenman, S. G., and Rao, B. R. (1968). *Proc. Natl. Acad. Sci. U.S.* **61,** 1028.

Koritz, S. B. (1966). *Biochem. Biophys. Res. Commun.* **24,** 805.

Koritz, S. B. (1968). *In* "Protein and Polypeptide Hormones" (M. Margoulies, ed.), Part 1, Excerpta Med. Intern. Conf. Ser. No. 161, p. 171. Amsterdam.

Koritz, S. B., and Dorfman, R. I. (1956). *Arch. Biochem. Biophys.* **65,** 491.

Koritz, S. B., and Hall, P. F. (1964). *Biochemistry* **3,** 1298.

Korner, A. (1959). *Biochem. J.* **73,** 61.

Korner, A. (1960). *J. Endocrinol.* **20,** 256.

Korner, A. (1961). *Biochem. J.* **81,** 292.

Korner, A. (1964a) *Nature* **201,** 501.

Korner, A. (1964b). *Biochem. J.* **92,** 449.

Korner, A. (1965). *Recent Progr. Hormone Res.* **21,** 205.

Korner, A. (1966). *Biochem. J.* **101,** 627.

Korner, A. (1967). *Progr. Biophys. Mol. Biol.* **17,** 63.

Korner, A. (1968). *Ann. N.Y. Acad. Sci.* **148,** 408.

Korner, A., and Gumbley, J. M. (1966). *Nature* **209,** 505.

Korner, A., and Young, F. G. (1955). *J. Endocrinol.* **13,** 78.

Kostyo, J. L. (1966). *Biochim. Biophys. Acta* **129,** 294.

Kostyo, J. L., and Knobil, E. (1959a). *Endocrinology* **65,** 395.

Kostyo, J. L., and Knobil, E. (1959b). *Endocrinology* **65,** 525.

Kostyo, J. L., and Redmond, A. F. (1966). *Endrocrinology* **79,** 531.

Kostyo, J. L., and Schmidt, J. E. (1962). *Endocrinology* **70,** 381.

Kowal, J. (1969). *Biochemistry* **8,** 1821.

Krahl, M. E. (1957). *Perspectives Biol. Med.* **1,** 69.

Krahl, M. E. (1961). "The Action of Insulin on Cells." Academic Press, New York.

Krahl, M. E. (1966). *Federation Proc.* **25,** 832.

Krebs, E. G., Love, D. S., Bratvold, G. E., Trayser, K. A., Meyer, W. L., and Fischer, E. H. (1964). *Biochemistry* **3,** 1022.

Kroeger, H. (1963). *Nature* **200,** 1234.

Kroeger, H. (1966). *Exptl. Cell Res.* **41,** 64.

Kroon, A. M. (1965). *Biochim. Biophys. Acta* **108,** 275.

Kuo, J. F., Holmlund, C. E., Dill, I. K., and Bohonos, N. (1966). *Arch. Biochem. Biophys.* **117,** 269.

Kuo, J. F., Dill, I. K., and Holmlund, C. E. (1967). *J. Biol. Chem.* **242,** 3659.

Kurland, C. G., and Maaløe, O. (1962). *J. Mol. Biol.* **4,** 193.

Kumar, M., Singh, V. N., and Chaikoff, I. L. (1966). *Biochim. Biophys. Acta* **129,** 640.

Laguens, R. (1964). *J. Ultrastruct. Res.* **10,** 578.

Langdon, R. G., and Sloan, H. R. (1967). *Proc. Natl. Acad. Sci. U.S.* **57,** 401.

Latham, H., and Darnell, J. E. (1965). *J. Mol. Biol.* **14,** 1.

Laycock, D. G., and Hunt, J. A. (1969). *Nature* **221,** 1118.

Leader, D. P., and Barry, J. M. (1967). *Nature* **215,** 1374.

Leder, P., and Nau, M. N. (1967). *Proc. Natl. Acad. Sci.* **58,** 774.

Lee, K.-L., and Miller, O. N. (1967). *Mol. Pharmacol.* **3,** 44.

LeFevre, P. G. (1967). *Science* **158,** 274.

Lenard, J., and Singer, S. J. (1968). *Science* **159,** 738.

Leon, H. A., Arrhenius, E., and Hultin, D. (1962). *Biochim. Biophys. Acta* **63,** 423.

Liao, S., and Lin, A. H. (1967). *Proc. Natl. Acad. Sci. U.S.* **57,** 379.

Liao, S., Barton, R. W., and Lin, A. H. (1966a). *Proc. Natl. Acad. Sci. U.S.* **55,** 1593.

Liao, S., Lin, A. H., and Barton, R. W. (1966b). *J. Biol. Chem.* **241,** 3869.

Liao, S., Sagher, D., and Fang, S.-M. (1968). *Nature* **220,** 1336.

Liao, S., Sagher, D., Lin, A. H., and Fang, S. (1969). *Nature* **223,** 297.

Lin, S-Y., Mosteller, R. D., and Hardesty, B. (1966). *J. Mol. Biol.* **21,** 51.

Lippe, B. M., and Szego, C. M. (1965). *Nature* **207,** 272.

Litwack, G., Sears, M. L., and Diamonstone, T. I. (1963). *J. Biol. Chem.* **238,** 302.

Litwack, G., Fiala, E. S., and Filosa, R. J. (1965). *Biochim. Biophys. Acta* **111,** 569.

Lockwood, D. H., Voytovich, A. E., Stockdale, F. E., and Topper, Y. T. (1967). *Proc. Natl. Acad. Sci. U.S.* **58,** 658.

Loeb, P. M., and Wilson, J. D. (1965). *Clin. Res.* **13,** 45.

Loeb, J. N., Howells, R. R, and Tomkins, G. M. (1965). *Science* **149,** 1093.
Loecker, W. C. de, Brooks, S. C., and De Wever, F. (1966). *Biochim. Biophys. Acta* **119,** 655.
Long, C. N. H., Smith, O. K., and Fry, E. G. (1960). *In* "Metabolic effects of Adrenal Hormones" (G. E. W. Wolstenholme and M. O'Connor, eds.), p. 4. Churchill, London.
Lowry, O. H., and Passonneau, J. V. (1966). *J. Biol. Chem.* **241,** 2268.
Lukács, I., and Sekeris, C. E. (1967). *Biochim. Biophys. Acta* **134,** 85.
MacGregor, R. R., and Mahler, H. R. (1967). *Arch. Biochem. Biophys.* **120,** 136.
McGuire, W. L., and O'Malley, B. W. (1968). *Biochim. Biophys. Acta* **157,** 187.
McKerns, K. W., ed. (1968). "Functions of the Adrenal Cortex." Appleton, New York.
Maden, B. E. H. (1968). *Nature* **219,** 685.
Mager, J. (1960). *Biochim. Biophys. Acta* **38,** 150.
Maggi, V., Franks, L. M., Wilson, P. D., and Carbonell, A. W. (1969). *Diabetologia* **5,** 67.
Mahler, R. J., and Szabo, O. (1969). *Diabetes* **18,** 550.
Maitra, U., Nakata, Y., and Hurwitz, J. (1967). *J. Biol. Chem.* **242,** 4908.
Makman, M. H., Dvorkin, B., and White, A. (1966). *J. Biol. Chem.* **241,** 1646.
Makman, M. H., Nakagawa, S., and White, A. (1967). *Rec. Progr. Hormone Res.* **23,** 195.
Manchester, K. L. (1961). *Biochem. J.* **81,** 135.
Manchester, K. L. (1965). *Proc. 5th Intern. Diabetes Federation,* Excerpta Med. Intern. Congr. Ser. No. 84, p. 101. Amsterdam.
Manchester, K. L. (1966). *Biochem. J.* **98,** 711.
Manchester, K. L. (1967a). *Nature* **216,** 394.
Manchester, K. L. (1967b). *Biochem. J.* **105,** 13C.
Manchester, K. L. (1968a). *In* "The Biological Basis of Medicine" (E. E. Bittar and N. Bittar, eds.), Vol. 2, p. 221. Academic Press, New York.
Manchester, K. L. (1968b). *Proc. Roy. Soc. Med.* **61,** 812.
Manchester, K. L., and Wool, I. G. (1963). *Biochem. J.* **89,** 202.
Manchester, K. L., and Wulwick, C. (1969). *In* "Protein and Polypeptide Hormones" (M. Margoulies, ed.), Excerpta Med. Intern. Congr. Ser. 161, p. 828. Amsterdam.
Manchester, K. L., and Young, F. G. (1959). *J. Endocrinol.* **18,** 381.
Manchester, K. L., and Young, F. G. (1960). *Biochem. J.* **75,** 487.
Manchester, K. L., and Young, F. G. (1961). *Vitamins Hormones* **19,** 95.
Manchester, K. L., and Young, F. G. (1968). *Biochem. J.* **70,** 353.
Manchester, K. L., Randle, P. J., and Young, F. G. (1959). *J. Endocrinol.* **18,** 395.
Mangan, F. R., Neal, G. E., and Williams, D. C. (1967). *Biochem. J.* **104,** 1075.
Mangiarotti, G., and Schlessinger, D (1966). *J. Mol. Biol.* **20,** 123.
Mangiarotti, G., and Schlessinger, D. (1967). *J. Mol. Biol.* **29,** 395.
Mansour, A. M., and Niu, M. C. (1965). *Proc. Natl. Acad. Sci. U.S.* **53,** 764.
Martin, T. E., and Wool, I. G. (1968). *Proc. Natl. Acad. Sci. U.S.* **60,** 659.
Martin, . E., and Poung, F. G. (1965). *Nature* **208,** 684.
Martin, T. E., Rolleston, F. S., Low, R. B., and Wool, I. G., (1969). *J. Mol. Biol.* **43,** 135.
Mathias, A. P., Williamson, R., Huxley, H. E., and Page, S. (1964). *J. Mol. Biol.* **9,** 154.
Maul, G. G., and Hamilton, T. H. (1967). *Proc. Natl. Acad. Sci. U.S.* **57,** 1371.

Mayne, R., and Barry, J. M. (1965). *Biochim. Biophys. Acta* **107,** 160.

Mayne, R., and Barry, J. M. (1967). *Biochim. Biophys. Acta* **138,** 195.

Mayne, R., Barry, J. M., and Rivera, E. M. (1966). *Biochem. J.* **99,** 688.

Means, A. R., and Hamilton, T. H. (1966a). *Proc. Natl. Acad. Sci. U.S.* **56,** 686.

Means, A R., and Hamilton, T. H. (1966b). *Proc. Natl. Acad. Sci. U.S.* **56,** 1594.

Means, A. R., Hall, P. F., Nicol, L. W., Sawyer, W. H., and Baker, C. A. (1969). *Biochemistry* **8,** 1488.

Meisler, A. I., and Tropp, B. E. (1969). *Biochim. Biophys. Acta* **174,** 476.

Millar, D. B., and Steiner, R. F. (1966). *Biochemistry* **5,** 2289.

Miller, L. L. (1965). *Federation Proc.* **24,** 737.

Milner, R. D. G., and Hales, C. N. (1967). *Diabetologia* **3,** 47.

Mirsky, I. A., and Perisutti, G. (1962). *Biochim. Biophys. Acta* **62,** 490.

Monroy, A., Maggio, R., and Rinaldi, A. M. (1965). *Proc. Natl. Acad. Sci. U.S.* **54,** 107.

Moog, F. (1964). *Science* **144,** 414.

Moore, P. B. (1966). *J. Mol. Biol.* **18,** 8.

Moore, R. J., and Hamilton, T. H. (1964). *Proc. Natl. Acad. Sci. U.S.* **52,** 439.

Morais, R., and Goldberg, I. H. (1967). *Biochemistry* **6,** 2538.

Morell, P., Smith, I., Dubnan, D., and Marmur, J. (1967). *Biochemistry* **6,** 258.

Morris, A. J., Arlinghaus, R., Flavelukes, S., and Schweet, R. (1963). *Biochemistry* **2,** 1084.

Morrow, L. B., Burrow, G. N., and Mulrow, P. J. (1967). *Endocrinology* **80,** 883.

Mueller, G. C., Herranen, A. M., and Jervell, K. F. (1958). *Rec. Progr. Hormone Res.* **14,** 95.

Mueller, G. C., Gorski, J., and Aizawa, Y. (1961). *Proc. Natl. Acad. Sci. U.S.* **47,** 164.

Munk, A. (1962). *Biochim. Biophys. Acta* **57,** 318.

Munro, H. N., Baliga, B. S., and Pronczuk, A. W. (1968). *Nature* **219,** 944.

Murad, F., Chi, Y.-M., Rall, T. W., and Sutherland, E. W. (1962). *J. Biol. Chem.* **237,** 1233.

Muramatsu, M., Hodnett, J. L., Steele, W. J., and Busch, H. (1966). *Biochim. Biophys. Acta* **123,** 116.

Nair, K. G., Rabinowitz, M., and Chen Tu, M. (1967). *Biochemistry* **6,** 1898.

Nakada, D., and Kaji, A. (1967). *Proc. Natl. Acad. Sci. U.S.* **57,** 128.

Nakagawa, H., and Cohen, P. P. (1967). *J. Biol. Chem.* **242,** 642.

Nakagawa, S., and White, A. (1966). *Proc. Natl. Acad. Sci. U.S.* **55,** 900.

Nakagawa, H., Kim, K., and Cohen, P. P. (1967). *J. Biol. Chem.* **242,** 635.

Nathans, D. (1964). *Proc. Natl. Acad. Sci. U.S.* **51,** 585.

Newerly, K., and Berson, S. A. (1957). *Proc. Soc. Exptl. Biol. Med.* **94,** 751.

Nicol, D. S. H. W. (1960). *In* "The Mechanism of Action of Insulin" (F. G. Young, W. A. Broom, and F. W. Wolff, eds.), p. 3. Blackwell, Oxford.

Nicolette, J. A., and Gorski, J. (1964a). *Arch. Biochim. Biophys.* **107,** 279.

Nicolette, J. A., and Gorski, J. (1964b). *Endocrinology* **74,** 955.

Nicolette. J. A., and Mueller, G. C. (1966). *Endocrinology* **79,** 1162.

Niemeyer, H., Pérez, N., and Codoceo, R. (1967). *J. Biol. Chem.* **242,** 860.

Nirenberg, M., and Leder, P. (1964). *Science* **145,** 1399.

Nisman, B., and Pelmont, J. (1964). *Progr. Nucleic Acids Mol. Biol.* **3,** 236.

Noall, M. W. (1960). *Biochim. Biophys. Acta* **40,** 180.

Noall, M. W., and Allen, W. M. (1961). *J. Biol. Chem.* **236,** 2987.

Noll, H., Staehelin, T., and Wettstein, F. O. (1963). *Nature* **198,** 632.

Noteboom, W. D., and Gorski, J. (1963). *Proc. Natl. Acad. Sci. U.S.* **50,** 250.
Noteboom, W. D., and Gorski, J. (1965). *Arch. Biochem. Biophys.* **111,** 559.
Novelli, G. D. (1967). *Ann. Rev. Biochem.* **36,** 449.
Okamato, T., and Takanami, M. (1963). *Biochim. Biophys. Acta* **76,** 266.
O'Malley, B. W. (1967). *Biochemistry* **6,** 2546.
O'Malley, B. W., McGuire, W. L., and Korenman, S. G. (1967). *Biochim. Biophys. Acta* **145,** 204.
O'Malley, B. W., Aronow, A., Peacock, A. C., and Dingman, C. W. (1968). *Science* **162,** 567.
Ove, P., Takai, S., Umeda, T., and Lieberman, I. (1967). *J. Biol. Chem.* **242,** 4963.
Parson, J. T., and McCarty, K. S. (1968). *J. Biol. Chem.* **243,** 5377.
Pastan, I., and Macchia, V. (1967). *J. Biol. Chem.* **242,** 5757.
Peachey, L. D. (1965). *J. Cell Biol.* **25,** 209.
Peacock, A. C., and Dingman, C. W. (1967). *Biochemistry* **6,** 1818.
Peck, W. A., Brandt, J., and Miller, I. (1967). *Proc. Natl. Acad. Sci. U.S.* **57,** 1599.
Pegg, A. E., and Korner, A. (1965). *Nature* **205,** 904.
Pegg, A. E., and Korner, A. (1967). *Arch. Biochim. Biophys.* **118,** 362.
Peña, A., Dvorkin, B., and White, A. (1966). *J. Biol. Chem.* **241,** 2144.
Penhos, J. C., and Krahl, M. E. (1962). *Am. J. Physiol.* **202,** 349.
Penhos, J. C., and Krahl, M. E. (1963). *Am. J. Physiol.* **204,** 140.
Penman, S. (1966). *J. Mol. Biol.* **17,** 117.
Penman, S., Smith, I., and Holtzman, E. (1966). *Science* **154,** 786.
Penman, S., Vesco, C., and Penman, M. (1968). *J. Mol. Biol.* **34,** 49.
Peraino, C. (1967). *J. Biol. Chem.* **242,** 3860.
Perry, R. P. (1962). *Proc. Natl. Acad. Sci. U.S.* **48,** 2179.
Perry, R. P. (1967). *Progr. Nucleic Acid Res. Mol. Biol.* **6,** 219.
Perry, R. P., Srinivasan, P. R., and Kelley, D. E. (1964). *Science* **145,** 504.
Perry, S. V., and Zydowo, M. (1959). *Biochem. J.* **72,** 682.
Petermann, M. L. (1964). "The Physical Properties of Ribosomes." Elsevier, Amsterdam.
Petermann, M. L., and Pavlovec, A. (1967). *Biochemistry* **6,** 2950.
Peters, R. A. (1956). *Nature* **177,** 426.
Planelles, J., Ozeretskovsky, N., and Djeksenbaev, O. (1962). *Nature* **195,** 713.
Pogo, A. O., Littau, V. C., Allfrey, V. G., and Mirsky, A. E. (1967). *Proc. Natl. Acad. Sci. U.S.* **57,** 743.
Prescott, D. M. (1964). *Progr. Nucleic Acid Res. Mol. Biol.* **3,** 33.
Pronczuk, A. W., Baliga, B. S., and Munro, H. N. (1968). *Biochem. J.* **110,** 783.
Pruitt, K. M., Boshell, B. R., and Kreisberg, R. A. (1966). *Diabetes* **15,** 342.
Puca, G. A., and Bresciani, F. (1968). *Nature* **218,** 967.
Puca, G. A., and Bresciani, F. (1969). *Nature* **223,** 745.
Rabinovitz, M., and Fisher, J. M. (1962). *J. Biol. Chem.* **237,** 477.
Rampersad, O. R., and Wool, I. G. (1965). *Science* **149,** 1102.
Ramuz, M., Doly, J., Mandel, P., and Chambon, P. (1965). *Biochem. Biophys. Res. Commun.* **19,** 114.
Rasmussen, H., Schwartz, I. L., Schoessler, M. A., and Hochster, G. (1960). *Proc. Natl. Acad. Sci. U.S.* **46,** 1278.
Ray, P. D., Foster, D. O., and Lardy, H. A. (1964). *J. Biol. Chem.* **239,** 3396.
Reich, E., and Goldberg, I. H. (1964). *Progr. Nucleic Acid Res. Mol. Biol.* **3,** 184.
Reid, E. M., O'Neal, M. A., Stevens, B. M., and Burnop, V. C. E. (1956). *Biochem. J.* **64,** 33.

Revel, M., and Hiatt, H. H. (1964). *Proc. Natl. Acad. Sci. U.S.* **51,** 810.

Revel, M., and Hiatt, H. H. (1965). *J. Mol. Biol.* **11,** 467.

Richmond, J. E., Shoemaker, W. C., and Elwyn, D. H. (1963). *Am. J. Physiol.* **205,** 848.

Rieser, P. (1966). *Am. J. Med.* **40,** 759.

Riggs, T. R. (1964). *In* "Actions of Hormones on Molecular Processes" (G. Litwack and D. Kritchevsky, eds.), p. 1. Wiley, New York.

Riggs, T. R., Pan, M. W., and Feng, H. W. (1968). *Biochim. Biophys. Acta* **150,** 92.

Rizack, M. A. (1964). *J. Biol. Chem.* **239,** 392.

Roberts, S. (1953). *J. Biol. Chem.* **200,** 77.

Roberts, S., McCune, R. W., Creange, J. E., and Young, P. L. (1967). *Science* **158,** 372.

Robinson, W. S. (1961). *Proc. Soc. Exptl. Biol. Med.* **106,** 115.

Robison, G. A., Butcher, R. W., and Sutherland, E. W. (1967). *Ann. N.Y. Acad. Sci.* **139,** 703.

Robison, G. A., Butcher, R. W., and Sutherland, E. W. (1968). *Ann. Rev. Biochem.* **37,** 149.

Roche, J., Michel, R., and Kamei, T. (1962). *Biochim. Biophys. Acta* **61,** 647.

Rodbell, M. (1966). *J. Biol. Chem.* **241,** 130.

Rodbell, M. (1967a). *J. Biol. Chem.* **242,** 5744.

Rodbell, M. (1967b). *J. Biol. Chem.* **242,** 5751.

Rodbell, M. (1967c). *Biochem. J.* **105,** 2P.

Rodbell, M., and Jones, A. B. (1966). *J. Biol. Chem.* **241,** 140.

Roeder, R. G., and Rutter, W. J. (1969). *Nature* **224,** 234.

Roodyn, D. B. (1962). *Biochem. J.* **85,** 177.

Roodyn, D. B., Suttie, J. W., and Work, T. S. (1962). *Biochem. J.* **83,** 29.

Roodyn, D. B., Freeman, K. B., and Tata, J. R. (1965). *Biochem. J.* **94,** 628.

Rosen, B. (1965). *J. Mol. Biol.* **11,** 845.

Rosen, F., Raina, P. N., Milholland, R. J., and Nichol, C. A. (1964). *Science* **146,** 661.

Roskoski, R., and Steiner, D. F. (1967a). *Biochim. Biophys. Acta* **135,** 347.

Roskoski, R., and Steiner, D. F. (1967b). *Biochim. Biophys. Acta* **135,** 727.

Sachs, H. (1957). *J. Biol Chem.* **228,** 23.

Salaman, M. R., Manchester, K. L., and Wallis, M. (1967). Excerpta Med. Intern. Congr. Ser. 142, 37. Amsterdam.

Salas, M., Viñuela, E., and Sols, A. (1963). *J. Biol. Chem.* **238,** 3535.

Salas, M., Hille, M. B., Last, J. A., Wahba, A. J., and Ochoa, S. (1967). *Proc. Natl. Acad. Sci. U.S.* **57,** 387.

Salb, J. M., and Marcus, P. I. (1965). *Proc. Natl. Acad. Sci. U.S.* **54,** 1353.

Samuels, L. T., and Uchikawa, T. (1967). *In* "The Adrenal Cortex" (A. B. Eisenstein, ed.) Churchill, London.

Satoh, P., Constantopoulos, G., and Tchen, T. T. (1966). *Biochemistry* **5,** 1646.

Scharff, R., and Wool, I. G. (1965). *Biochem. J.* **97,** 257.

Schimke, R. T. (1964). *J. Biol. Chem.* **239,** 3808.

Schmid, W., Gallwitz, D., and Sekeris, C. E. (1967). *Biochim. Biophys. Acta* **134,** 80.

Schwartz, I. L., Rasmussen, H., Schoessler, M. A., Silver, L., and Fong, C. T. O. (1960). *Proc. Natl. Acad. Sci. U.S.* **46,** 1288.

Scornik, O. A., Hoagland, M. B., Pfefferkorn, L. C., and Bishop, E. A. (1967). *J. Biol. Chem.* **242,** 131.

Scriba, P. C., and Fries, M. (1967). *Nature* **214,** 91.

Scriba, P. C., and Reddy, W. J. (1965). *Endocrinology* **76,** 745.

Seeds, N. W., and Conway, T. W. (1966). *Biochem. Biophys. Res. Commun.* **23,** 111.

Seeds, N. W., Retsema, J. A., and Conway, T. W. (1967). *J. Mol. Biol.* **27,** 421.

Segal, S. J., Davidson, O. W., and Wada, K. (1965). *Proc. Natl. Acad. Sci. U.S.* **54,** 782.

Seifert, K. H., and Sekeris, C. E. (1969). *European J. Biochem.* **7,** 408.

Sekeris, C. E. (1965). *In* "Mechanism of Hormone Action" (P. Karlson, ed.), p. 149. Academic Press, New York.

Sekeris, C. E., and Lang, N. (1964). *Life Sci.* **3,** 625.

Sekeris, C. E. Dukes, P. P., and Schmid, W. (1965). *Z. Physiol. Chem.* **341,** 152.

Sellinger, O. Z., and Lee, K.-L. (1964). *Biochim. Biophys. Acta* **91,** 183.

Sells, B. H., and Takahashi, T. (1967). *Biochim. Biophys. Acta* **134,** 69.

Seraydarian, K., Mommaerts, W. F. H. M., Wallner, A., and Guillory, R. J. (1961). *J. Biol. Chem.* **236,** 2071.

Sharma, C., Manjeshwar, R., and Weinhouse, S. (1963). *J. Biol. Chem.* **238,** 3840.

Shaw, C. R., and Koen, A. L. (1963). *Science* **140,** 70.

Shearer, R. W., and McCarthy, B. J. (1967). *Biochemistry* **6,** 283.

Shimizu, C. S. N., and Kaplan, S. A. (1964). *Endocrinology* **74,** 709.

Sidransky, H., Bongiorno, M., Sarma, D. S. R., and Verney, E. (1967). *Biochem. Biophys. Res. Commun.* **27,** 242.

Siegel, E., and Tobias, C. A. (1966). *Science* **153,** 763.

Siegel, M. R., and Sisler, H. D. (1963). *Nature* **200,** 675.

Siegel, M. R., and Sisler, H. D. (1964). *Biochim. Biophys. Acta* **87,** 83.

Siegel, M. R., and Sisler, H. D. (1965). *Biochim. Biophys. Acta* **103,** 558.

Silber, R. H., and Porter, C. C. (1953). *Endocrinology* **52,** 518.

Singer, M. F., Jones, O. W., and Nirenberg, N. W. (1963). *Proc. Natl. Acad. Sci. U.S.* **49,** 392.

Singhal, R. L., Valadares, J. R. E., and Ling, G. M. (1967). *J. Biol. Chem.* **242,** 2593.

Sluyser, M. (1966a). *J. Mol. Biol.* **19,** 591.

Sluyser, M. (1966b). *J. Mol. Biol.* **22,** 411.

Smith, J. D., Traut, R. R., Blackburn, G. M., and Monro, R. E. (1965). *J. Mol. Biol.* **13,** 617.

So, A. G., and Davie, E. W. (1965). *Biochemistry* **4,** 1973.

Soeiro, R., and Amos, H. (1966). *Biochim. Biophys. Acta* **129,** 406.

Sokoloff, L., Campbell, P. L., Francis, C. M., and Klee, C. B. (1963a). *Biochim. Biophys. Acta* **76,** 329.

Sokoloff, L., Kaufman, S., Campbell, P. L., Francis, C. M., and Gelborn, H. V. (1963b). *J. Biol. Chem.* **238,** 1432.

Sols, A., Sillero, A., and Salas, J. (1965). *J. Cellular Comp. Physiol.* **66,** Suppl. 1, 23.

Søvik, O. (1967). *Biochim. Biophys. Acta* **141,** 190.

Søvik, O., and Walaas, O. (1964). *Nature* **202,** 396.

Spiegelman, S., and Hayashi, M. (1963). *Cold Spring Harbor Symp. Quant. Biol.* **28,** 161.

Spyrides, G. J. (1964). *Proc. Natl. Acad. Sci. U.S.* **51,** 1220.

Stackhouse, H. L., Chetsanga, C. J., and Tan, C. H. (1968). *Biochim. Biophys. Acta* **155,** 159.

Stadie, W. E., Haugaard, N., and Vaughan, M. (1952). *J. Biol. Chem.* **199,** 729.

Stadie, W. E., Haugaard, N., and Vaughan, M. (1953). *J. Biol. Chem.* **200,** 745.

Staehelin, M., and Maier, R. (1968). *In* "Protein and Polypeptide Hormones" (M. Margoulies, ed.), Part 1, p. 193. Excerpta Med. Intern. Conf. Ser. No. 161, Amsterdam.

Staehelin, T., Wettstein, F. O., Oura, H., and Noll, H. (1964). *Nature* **201,** 264.

Stanners, C. P. (1966). *Biochim. Biophys. Res. Commun.* **24,** 758.

Stavey, L., and Gross, P. R. (1967). *Proc. Natl. Acad. Sci. U.S.* **57,** 735.

Steele, R. (1966). *Ergeb. Physiol. Biol. Chem. Exptl. Pharmakol.* **57,** 91.

Stein, O., and Gross, J. (1959). *Endocrinology* **65,** 707.

Steiner, D. F. (1966). *Vitamins Hormones* **24,** 1.

Steiner, D. F., and King, J. (1964). *J. Biol. Chem.* **239,** 1292.

Steiner, D. F., and King, J. (1966). *Biochim. Biophys. Acta* **119,** 510.

Stent, G. S., and Brenner, S. (1961). *Proc. Natl. Sci. U.S.* **47,** 2005.

Stirewalt, W. S., and Wool, I. G. (1966). *Science* **154,** 284.

Stirewalt, W. S., Wool, I. G., and Cavicchi, P. (1967). *Proc. Natl. Acad. Sci. U.S.* **57,** 1885.

Stone, D., and Hechter, O. (1954). *Arch. Biochem. Biophys.* **51,** 457.

Sutherland, E. W., Øye, I., and Butcher, R. W. (1965). *Recent Progr. Hormone Res.* **21,** 623.

Sutherland, E. W., and Haynes, R. C. (1967). *Endocrinology* **80,** 288.

Sutherland, E. W., and Rall, T. W. (1960). *Pharmacol. Rev.* **12,** 265.

Sutherland, E. W., Rall, T. W., and Menon, T. (1962). *J. Biol. Chem.* **237,** 1220.

Szego, C. M., and Davis, J. S. (1967). *Proc. Natl. Acad. Sci. U.S.* **58,** 1711.

Szer, W., and Ochoa, S. (1964). *J. Mol. Biol.* **8,** 823.

Takanami, M., and Okamoto, T. (1963). *J. Mol. Biol.* **7,** 323.

Talwar, G. P., Panda, N. L., Sarin, G. S., and Tola, A. J. (1962). *Biochem. J.* **82,** 173.

Tanner, M. J. A. (1967). *Biochemistry* **6,** 2686.

Tata, J. R. (1964a). *In* "Actions of Hormones on Molecular Processes" (G. Litwack and D. Kritchevsky, eds.), p. 58. Wiley, New York.

Tata, J. R. (1964b). *Biochim. Biophys. Acta* **87,** 528.

Tata, J. R. (1966). *Progr. Nucelic Acid Res. Mol. Biol.* **5,** 191.

Tata, J. R., and Williams-Ashman, H. G. (1967). *European J. Biochem.* **2,** 366.

Tata, J. R., Ernster, L., Linberg, O., Arrhenius, E., Pedersen, S., and Hedman, R. (1963). *Biochem. J.* **86,** 408.

Teng, C.-S., and Hamilton, T. H. (1967a). *Biochem. J.* **105,** 1091.

Teng, C.-S., and Hamilton, T. H. (1967b). *Biochem. J.* **105,** 1101.

Terenius, L. (1967). *Mol. Pharmacol.* **3,** 423.

Toft, D., and Gorski, J. (1966). *Proc. Natl. Acad. Sci. U.S.* **55,** 1574.

Toft, D., Shyamala, G., and Gorski, J. (1967). *Proc. Natl. Acad. Sci. U.S.* **57,** 1740.

Tomkins, G. M., Garren, L. D., Howell, R. R., and Peterkofsky, B. (1965). *J. Cellular Comp. Physiol.* **66,** Suppl. 1, 137.

Trakatellis, A. C., Axelrod, A. E., and Montjar, M. (1964a). *Nature* **203,** 1134.

Trakatellis, A. C., Axelrod, A. E., and Montjar, M. (1964b). *J. Biol. Chem.* **239,** 4237.

Trakatellis, A. C., Montjar, M., and Axelrod, A. E. (1965). *Biochemistry* **4,** 2065.

Traut, R. R., and Monro, R. E. (1964). *J. Mol. Biol.* **10,** 63.

Ui, H., and Mueller, G. C. (1963). *Proc. Natl. Acad. Sci. U.S.* **50,** 256.
Uete, T. (1966a). *Biochim. Biophys. Acta* **121,** 386.
Uete, T. (1966b). *Biochim. Biophys. Acta* **121,** 395.
Unsworth, B. R., and Cohen, P. P. (1968). *Biochemistry* **7,** 2581.
Urquhart, J., Krall, R. L., and Li, C. C. (1968). *Endocrinology* **83,** 390.
Uyeda, K., and Racker, E. (1965). *J. Biol. Chem.* **240,** 4682.
Villee, D. B., Rettig, J., and Greenough, L. (1968). *Science* **159,** 1365.
Vogel, H. J., and Vogel, R. H. (1967). *Ann. Rev. Biochem.* **36,** 519.
Wagle, S. R. (1965). *Biochim. Biophys. Acta* **107,** 524.
Warner, J. R. (1966). *J. Mol. Biol.* **19,** 383.
Warner, J. R., and Rich, A. (1964a). *Proc. Natl. Acad. Sci. U.S.* **51,** 1134.
Warner, J. R., and Rich, A. (1964b). *J. Mol. Biol.* **10,** 202.
Warner, J. R., Girard, M., Latham, H., and Darnell, J. E. (1966a). *J. Mol. Biol.* **19,** 373.
Warner, J. R., Soeiro, R., Birnboim, H. C., Girard, M., and Darnell, J. E. (1966b). *J. Mol. Biol.* **19,** 349.
Warren, J. C., and Barker, K. L. (1967). *Biochim. Biophys. Acta* **138,** 421.
Weber, M. J., and DeMoss, J. A. (1966). *Proc. Natl. Acad. Sci. U.S.* **55,** 1224.
Weber, G., Singhal, R. L., and Srivastava, S. K. (1965). *Proc. Natl. Acad. Sci. U.S.* **53,** 96.
Weber, G., Singhal, R. L., Stamm, N. B., Lea, M. A., and Fisher, E. A. (1966). *Advan. Enzyme Regulation* **4,** 59.
Weinberg, R. A., Loening, U., Willems, M., and Penman, S. (1967). *Proc. Natl. Acad. Sci. U.S.* **58,** 1088.
Weinstein, I. B., and Finkelstein, I. H. (1967). *J. Biol. Chem.* **242,** 3757.
Weinstein, I. B., Schechter, A. N., Burka, E. R., and Marks, P. A. (1963). *Science* **140,** 314.
Weinstein, I. B., Friedman, S. M., and Ochoa, M. (1966). *Cold Spring Harbor Symp. Quant. Biol.* **31,** 671.
Wettstein, F. O., Noll, H., and Penman, S. (1964). *Biochim. Biophys. Acta* **87,** 525.
Wicks, W. D., Greenman, D. L., and Kenney, F. T. (1965). *J. Biol. Chem.* **240,** 4414.
Widnell, C. C., and Tata, J. R. (1964). *Biochim. Biophys. Acta* **87,** 531.
Widnell, C. C., and Tata, J. R. (1966a). *Biochem. J.* **98,** 621.
Widnell, C. C., and Tata, J. R. (1966b). *Biochim. Biophys. Acta* **123,** 478.
Wilkinson, J. H. (1965). "Isoenzymes." Spon, London.
Williams, R. H., and Ensinck, J. W. (1966). *Diabetes* **15,** 623.
Williams, R. H., Walsh, S. A., Hepp, D. K., and Ensinck, J. W. (1968). *Metab., Clin. Endocrinol.* **17,** 653.
Williams-Ashman, H. G. (1965). *J. Cellular Comp. Physiol.* **66,** Suppl. 1, 111.
Williams-Ashman, H. G., Liao, S., Hancock, R. L., Jurkowitz, L., and Silverman, D. A. (1964). *Recent Progr. Hormone Res.* **20,** 247.
Williamson, A. R., and Schweet, R. (1964). *Nature* **202,** 435.
Williamson, A. R., and Schweet, R. (1965). *J. Mol. Biol.* **11,** 358.
Wilson, S. H., and Hoagland, M. B. (1965). *Proc. Natl. Acad. Sci. U.S.* **54,** 600.
Wilson, S. H., and Hoagland, M. B. (1967). *Biochem. J.* **103,** 556.
Wing, D. R., and Robinson, D. S. (1968). *Biochem. J.* **106,** 667.
Wittam, J. S., Lee, K.-L., and Miller, O. N. (1969). *Biochim. Biophys. Acta* **174,** 536.
Wohltmann, H. J., and Narahara, H. T. (1966). *J. Biol. Chem.* **241,** 4931.

Wolfenden, R. (1963). *Biochemistry* **2,** 1090.
Wool, I. G. (1960). *Am. J. Physiol.* **199,** 715.
Wool, I. G. (1963). *Biochim. Biophys. Acta* **68,** 28.
Wool, I. G. (1964). *Nature* **202,** 196.
Wool, I. G., and Cavicchi, P. (1966). *Proc. Natl. Acad. Sci. U.S.* **56,** 991.
Wool, I. G., and Cavicchi, P. (1967). *Biochemistry* **6,** 1231.
Wool, I. G., and Krahl, M. E. (1959). *Am. J. Physiol.* **196,** 961.
Wool, I. G., and Kurihara, K. (1967). *Proc. Natl. Acad. Sci. U.S.* **58,** 2401.
Wool, I. G., and Moyer, A. N. (1964). *Biochim. Biophys. Acta* **91,** 248.
Wool, I. G., and Weinshelbaum, E. I. (1959). *Am. J. Physiol.* **197,** 1089.
Wool, I. G., and Weinshelbaum, E. I. (1960). *Am. J. Physiol.* **198,** 360.
Wool, I. G., Castles, J. J., and Moyer, A. N. (1965). *Biochim. Biophys. Acta* **107,** 333.
Wool, I. G., Stirewalt, W. S., Kurihara, K., Low, R. B., Bailey, P., and Oyer, D. (1968). *Recent Progr. Hormone Res.* **24,** 139.
Woolley, D. W., and Gommi, B. W. (1964). *Nature* **202,** 1074.
Wunner, W. H., Bell, J., and Munro, H. N. (1966). *Biochem. J.* **101,** 417.
Young, C. W., Robinson, P. F., and Sacktor, B. (1963). *Biochem. Pharmacol.* **12,** 855.
Young, V. R., Chen, S. C., and MacDonald, J. (1968). *Biochem. J.* **106,** 913.
Younger, L. R., King, J., and Steiner, D. F. (1966). *Cancer Res.* **26,** 1408.
Zak, R., Rabinowitz, M., and Platt, C. (1967). *Biochemistry* **6,** 2493.

References Added in Proof

Adiga, P. R., Hussa, R. O., Robertson, M. C., Hohl, H. R., and Winnick, T. (1968). *Proc. Natl. Acad. Sci. U.S.* **60,** 606.
Armentrout, S. A., Mills, W. A., and Simmons, L. R. (1966). *Proc. Natl. Acad. Sci. U.S.* **60,** 606.
Bitman, J., Trezise, L. A., and Cecil, H. C. (1966). *Arch. Biochem. Biophys.* **113,** 414.
Billing, R. J., Barbiroli, B., and Smellie, R. M. S. (1969). *Biochem. J.* **112,** 563.
Bullock, G., White, A. M., and Worthington, J. (1968). *Biochem. J.* **108,** 417.
Deer, R. F., and Scanlon, K. S. (1969). *Biochim. Biophys. Acta* **182,** 253.
Dixon, H. B. F. (1964). *In* "The Hormones" (G. Pincus, ed.), Vol. 5. Academic Press, New York.
Earl, D. C. N., and Morgan, H. E. (1968). *Arch. Biochem. Biophys.* **128,** 460.
Edelman, I. S., and Fanestil, D. D. (1970). *In* "Biochemical Actions of Hormones" (G. Litwack, ed.), Vol. 1, p. 321. Academic Press, New York.
Fang, S., Anderson, K. M., and Liao, S. (1969). *J. Biol. Chem.* **244,** 6584.
Hamilton, T. H., Widnell, C. C., and Tata, J. R. (1968). *J. Biol. Chem.* **243,** 408.
Heywood, S. M., and Nwagwu, M. (1968). *Proc. Natl. Acad. Sci. U.S.* **60,** 229.
Heywood, S. M., and Nwagwu, M. (1969). *Biochemistry* **8,** 3839.
Kohler, P. O., Grimley, P. M., and O'Malley, B. W. (1968). *Science* **160,** 86.
Kohler, R. E., Ron, E. Z., and Davis, B. D. (1968). *J. Mol. Biol.* **36,** 71.
Liao, S., and Stumpf, W. E. (1968). *Endocrinology* **83,** 629.
Louis, L. H., Conn, J. W., and Minick, M. C. (1966). *Metab. Clin. Exptl.* **15,** 309.
McIndoe, W., and Munro, H. N. (1967). *Biochim. Biophys. Acta* **134,** 461.
Manchester, K. L. (1963). *Biochim. Biophys. Acta* **70,** 531.

Manchester, K. L. (1965). *Biochim. Biophys. Acta* **100,** 295.

Manchester, K. L. (1970a). *Biochem. J.* **117,** 457.

Manchester, K. L. (1970b). *In* "Biochemical Actions of Hormones" (G. Litwack, ed.), Vol. 1, p. 267. Academic Press, New York.

Manchester, K. L., and Wallis, M. (1963). *Nature* **200,** 888.

Nakane, P. K., Nichoalds, G. E., and Oxender, D. L. (1968). *Science* **161,** 182.

Nishizawa, E. E., Billiar, R. B., Karr, J., and Eik-Nes, K. B. (1965). *Can. J. Biochem.* **43,** 1489.

Plager, J. E., and Matsui, N. (1966). *Endocrinology* **78,** 1154.

Pozefsky, T., Felig, P., Soeldner, J. S., and Cahill, G. F. (1968). *Trans. Assoc. Am. Physicians* **81,** 258.

Rose, I. A. (1968). *Proc. Natl. Acad. Sci. U.S.* **61,** 1079.

Shyamala, G., and Gorski, J. (1969). *J. Biol. Chem.* **244,** 1097.

Simpson, E., and Estabrook, R. W. (1969). *Advan. Enzyme Regulation* **7,** 259.

Staehelin, M. (1965). *Biochem. Z.* **342,** 459.

Valadares, J. R. E., Singhal, R. L., and Parulekar, M. R. (1968). *Arch. Biochem. Biophys.* **123,** 417.

Webb, T. E., and Morris, H. P. (1969). *Biochem. J.* **115,** 575.

CHAPTER 34

Free Amino Acid Pools and Their Role in Regulation

H. N. MUNRO

Physiological Chemistry Laboratories,
Department of Nutrition and Food Science,
Massachusetts Institute of Technology,
Cambridge, Massachusetts

I. Introduction

The first volume of this treatise, dealing with the basic biochemistry of protein metabolism, was permeated by the assumption that free amino acids are the currency of protein metabolism. In particular, it was concluded in Chapter 2 that the final products of digestion of dietary protein passing into the portal vein are free amino acids, and that peptides do not play a significant part in amino acid transport, except that, as pointed out in Chapter 38, plasma proteins may represent a vehicle of limited capacity for amino acid transport between the liver and other tissues. It follows that the supply of free amino acids to the tissues must assume a central position in the course of protein metabolism. Consequently, one way in which protein metabolism may be regulated is through variations in the supply of free amino acids to the tissues. This could take the form of a general effect or a local effect. An example of a general change in amino acid supply occurs when the level of protein in the diet is altered, thus varying the influx of amino acids into the body. A localized change in amino acid supply limited to one or a few tissues is achieved by mechanisms such as hormonal stimulation of a target tissue. It is easy to demonstrate from the published literature that both diet and hormones can affect uptake of free amino acids by various tissues, and it is well known that both these agents also cause changes in tissue protein metabolism (see Chapters 10 and 33). The problem that we must examine in this chapter is (a) whether these metabolic changes result from the increased or diminished flow of free amino acids into the tissue, (b) whether instead the alterations in free amino acid levels in the tissue are the consequence of the changes in cell metabolism, or (c) whether both events are related but independent aspects of the action of diet and hormones on the tissues. For example, insulin is known to increase the concentration of certain free amino acids in muscle and also to stimulate protein synthesis in the same tissue. What are the relationships between these events, and why does insulin affect only some free amino acid levels in muscle?

II. The Free Amino Acid Pools of the Body

In order to evaluate the metabolic significance of changes in tissue free amino acid levels, it is first necessary to summarize what is known about the concentrations of free amino acids in the tissues and fluids of the body and to discuss the rate of renewal of amino acids in different tissues. In general, this description will apply to the rat, for which much of the data on amino acid pools have been gathered. The question of free amino acid pools in other mammalian species has been considered in Chapter 25, in which comparative aspects of mammalian protein metabolism are examined.

A. Concentrations of Free Amino Acids in Body Fluids and Tissues

1. General Features of the Free Amino Acid Pool

Free amino acids are found in all fluids and tissues of the body, which contain a number of ninhydrin-reacting compounds additional to the 20 amino acids found in tissue proteins. Thus, in human plasma Stein and Moore (1954) identified 28 ninhydrin-reacting components which accounted for 90–100% of the total free amino N of plasma. In the nonprotein fraction of mammalian tissues, Tallan *et al.* (1954) found 40 discrete compounds that reacted with ninhydrin. Some of these compounds may be small peptides; in addition to glutathione, nucleotide-peptides (Steinberg *et al.*, 1960; Wilken and Hansen, 1961; Ondarza, 1965) and lipid-bound peptides (Barnabei and Ferrari, 1961; Tria and

Barnabei, 1963; Schwartzman *et al.*, 1966a) have been isolated from tissues. The identification of ninhydrin-reacting substances present in the plasma and urine of the dog is discussed by Wright and Nicholson (1965). Details of other ninhydrin-positive compounds in tissues will be found in the treatise "Amino Acid Pools" by Holden (1962). In general, the present chapter will deal only with the 20 amino acids commonly found in proteins.

When compared with the quantities of amino acids present in the tissues in the form of protein, the concentrations of free amino acids found in the body are small. The carcass of the rat contains about 20% protein; that is to say, protein-bound amino acids are present in the tissues at a concentration of some 2 *M*. In contrast to this, the average tissue concentration of free amino acids in the fasting rat is only 0.01 *M* (Herbert *et al.*, 1966), so that free amino acids represent about 0.5% of the total amount of amino acids in the animal's body. Consequently, quite a small change in the protein content of the body caused by increased protein synthesis or breakdown could have a considerable effect on free amino acid levels.

Individual free amino acids differ widely in the size of their pools. Table I displays the concentrations of each free amino acid in micromoles/100 gm body weight in the whole body of the fasting rat. The levels of the essential amino acids are all low, whereas four nonessential amino acids (alanine, glutamic acid, glutamine and glycine) account for 80% of the total free amino acid nitrogen of the body. It follows that measurements of tissue changes in total free amino-N are not very meaningful. In Table I the percentage of the total body free amino acid pool present in several major tissues has been calculated from the data of Herbert *et al.* (1966). For each amino acid, skeletal muscle is seen to be the largest reservoir in the body, often accounting for more than half and sometimes as much as 80% of the total amount in the whole animal. To some extent, this reflects the large proportion (nearly half) of body weight accounted for as muscle. In contrast, the plasma contains a very small proportion of the total free amino acid pool, varying from 0.2 to 6% for individual amino acids. Because of its accessibility, plasma is frequently sampled for free amino acid determination, but clearly may not be representative of the changes occurring in the free amino acid pool as a whole. For example, Pawlak and Pion (1968b) have shown that increasing intakes of lysine in the diet can increase the level of free lysine in plasma 7-fold, whereas in muscle the levels change 28-fold. This also confirms the unusually large affinity of muscle for storing free lysine (compare Table I).

TABLE I

AMOUNTS OF FREE AMINO ACID IN THE WHOLE BODY OF THE RAT, THE PROPORTION OF EACH AMINO ACID POOL IN THE MAJOR TISSUES, AND A COMPARISON OF THE TOTAL POOL SIZE OF EACH FREE AMINO ACID WITH THE POOL OF PROTEIN-BOUND AMINO ACID, THE AMOUNT OF AMINO ACID REQUIRED IN THE DIET, AND THE AMOUNT SUPPLIED BY 1 GM OF WHOLE EGG PROTEIN

Amino acid	Total body content of amino acids (μmoles/100 gm rat)		Daily amino acid requirement (μmoles/100 gm rat)[c]	Amino acid content of 1 gm whole egg protein (μmoles/gm)[d]	Percentage of free amino acid pool in individual tissues[e]					
	Free[a]	Protein bound[b]			Plasma (2% BW)	Kidney (1% BW)	Liver (4% BW)	Gut (5% BW)	Muscle (45% BW)	Brain (1% BW)
Essential										
Arginine	7	8400	—	340	6	1	<1	27	55	<1
Histidine	24	3600	140	140	2	1	11	16	66	1
Isoleucine	10	8400	400	500	3	1	9	40	35	<1
Leucine	14	16500	500	690	2	2	8	43	39	1
Lysine	15	8900	600	460	2	1	4	11	78	<1
Methionine	6	4050	350	220	3	1	6	34	44	1
Phenylalanine	9	5800	450	380	2	1	7	49	35	<1
Threonine	20	7550	400	470	2	1	6	12	71	2
Tryptophan	2	980	55	60	—	—	—	—	—	—
Tyrosine	8	3550	—	210	2	1	5	39	45	<1
Valine	12	9400	500	560	3	4	9	29	48	1
Nonessential										
Alanine	100	13500	—	730	0.8	1	8	10	75	1
Aspartic acid	19	11300	—	760	0.2	4	11	20	39	12
Glutamic acid	132	17700	—	880	0.4	4	9	17	50	8
Glutamine	223	—	—	—	1	<1	11	6	75	2
Glycine	323	24700	—	480	0.4	1	6	5	81	<1
Serine	20	12400	—	770	2	2	5	19	65	1

[a] Values for analysis of whole carcass of fasting rat weighing 50 gm (Herbert *et al.*, 1966).

[b] Calculated from data given by Block and Bolling (1951) for mixed proteins of whole rat. For amino acids not analyzed in whole animal, the mean amino acid concentrations for liver and muscle have been used. The rat has been assumed to have 20% protein in its body.

[c] Calculated from Rama Rao *et al.* (1959), assuming that a 100-gm rat consumes 10 gm of diet per day. The value for phenylalanine includes tyrosine; the value for methionine includes cystine.

[d] Calculated from FAO (1963) compilation of values published in the literature.

[e] Calculated from concentrations of amino acids in individual tissues of the fasting rat (Herbert *et al.*, 1966) and the weight of each tissue as a percentage of total body weight (BW) (Donaldson, 1924), which is shown in parentheses below each tissue. In each case, the animals used were male rats weighing 200 gm.

2. Differences in Amino Acid Concentrations in Tissues

The tissues tend to show higher levels of free amino acids than occur in plasma. This was first demonstrated by measurements of total amino-N concentrations; as shown in Table II for the rat and the dog, there are much higher concentrations of free amino-N in all the tissues sampled than in the plasma. This difference is largely due to the high tissue levels of five amino acids, namely glycine, glutamic acid, glutamine, alanine, and aspartic acid, as shown in Fig. 1 by data collected on fasting rats and alligators (Herbert *et al.*, 1966). This diagram gives the tissue levels of various amino acids as the mean concentrations in the combined tissues of the body. These tissue concentrations have been corrected for amino acids retained in extracellular fluid and for the presence of solids in the tissue, so that the figures provide true levels in intracellular fluid. For comparison, Fig. 1 also shows the concentrations of the same amino acids in the plasma of the rat and the alligator. In the case of the rat, the intracellular concentrations of alanine, aspartic acid, glutamic acid, glutamine, and glycine vary from 58 to 10 times greater than their levels in the plasma. The tissue content of 17 other free amino acids, including all the essential amino acids, were combined; they averaged 1.6 times the level found in the plasma of the fasting rat (Fig. 1). Figure 1 also demonstrates that the tissues and plasma of the alligator present an essentially similar pattern.

When the individual tissues of the fasting rat were examined, Herbert *et al.* (1966) found essentially the same pattern for each tissue. In most tissues, the four nonessential amino acids named earlier (alanine, glutamic acid, glutamine, and glycine) constituted the bulk of the intracel-

TABLE II

TOTAL FREE AMINO NITROGEN IN THE PLASMA AND TISSUES OF THE RAT AND THE DOG[a]

Body constituent	Rat[b]	Dog[c]
Plasma	6.4	4.7
Liver	32.1	23.0
Kidney	44.0	24.0
Spleen	36.0	31.3
Skeletal muscle	19.2	24.8
Brain	43.1	23.1

[a] Data given in milligrams of NH_2-N per 100 ml or per 100 gm.

[b] Data of Friedberg and Greenberg (1947) for groups of rats.

[c] Data of Hamilton (1945) on tissues from 2–6 dogs.

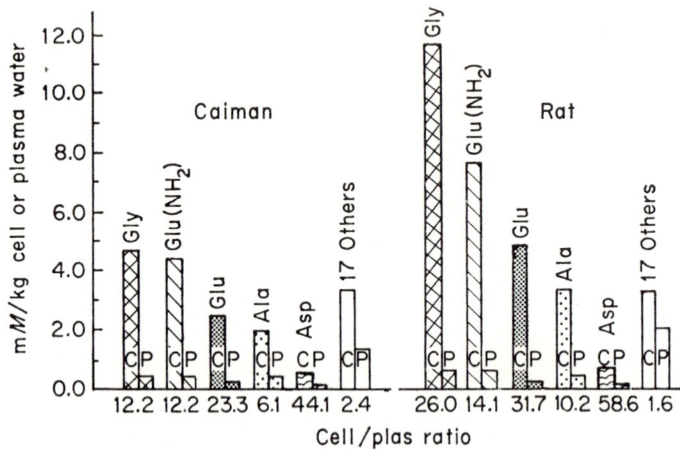

FIG. 1. The relationship between certain intracellular (C) and extracellular (P) amino acids in the entire carcasses of one 50-gm rat and one 50-gm alligator (caiman). The intracellular concentrations have been corrected for cell solid content and extracellular fluid. (From Herbert *et al.,* 1966.)

lular amino-N, and with aspartic acid were the main amino acids showing high concentration ratios relative to plasma. Among the essential amino acids, concentrations larger than twice the plasma levels were observed only for kidney, spleen, and gut. It is possible that continued digestion of endogenous protein in these fasting animals may have contributed to the high levels of free amino acids recovered from the gut. Tissue to plasma concentration ratios of some amino acids in other mammals are cited earlier in this treatise by Christensen (Chapter 4), who was the first to draw attention to the high concentration ratios of glycine and glutamine in the tissues of the mammal (Christensen *et al.,* 1948a).

What is the reason for this unequal distribution of free amino acids? Herbert *et al.* (1966) considered that the three or four nonessential amino acids present in high concentrations throughout the body represent amino-N that accumulated in the course of degradation of other amino acids. They support this view by an experiment in which alligators were fed on protein; after 24 hours they showed an increase in free amino-N confined mainly to alanine, glutamine, and glycine. Even if we accept this proposition as an explanation for the large amount of free nonessential amino acids, do the essential amino acids show any meaningful concentration pattern? Solomon *et al.* (1951) compared the relative abundance of individual free and total essential amino acids in several tissues of the rat and found no relationship between the pattern of amino acids in each group. These studies were performed by microbiological assay, but measurement by column chromatography of the free amino acid pool of the whole rat (Table I) also shows that the relative concentrations of different amino acids bear no resemblance to the pattern of amino acids in the mixed proteins of the body of the rat. Table I displays in addition the quantities of amino acids provided by 1 gm of whole egg protein, which has an ideal amino acid pattern for the nutrition of the rat (see Chapter 12), and also the daily requirements of the growing rat for essential amino acids. Once more, the free amino acid concentrations in the carcass of the fasting rat do not follow the pattern of intake or of requirements, except that tryptophan is present in by far the lowest concentrations in the tissues and the dietary sources.

The evidence regarding free amino acid pools in the fasting rat can be summarized as showing (a) that amino acids in the free form account for only a small fraction of the total amino acid content of the body; (b) that most of the free amino-N consists of four nonessential amino acids; (c) that half or more of the free amino acid pool of each amino acid is present in skeletal muscle; (d) that cells tend to contain higher concentrations of amino acids than are found in the plasma, mainly in the case of nonessential amino acids; (e) that the relative amounts of free amino acids found in the body bear little relationship to either the average composition of tissue proteins, to the amino acid pattern of dietary protein, or to the known requirements of the rat for essential amino acids. This last feature is of considerable interest. If the rate of protein synthesis is stimulated, amino acids will be withdrawn from the free amino acid pool in the proportions needed for synthesis of new protein molecules. In consequence, the concentrations of free amino acids could be disproportionately reduced, since the pattern in the free amino acid pool is not the pattern for protein synthesis. Under some circumstances, this could lead to depletion of one free amino acid to the point at which it became a limiting factor in the rate of protein synthesis. Accordingly, one question in the regulation of protein synthesis must be whether this occurs, and if so, whether it is limited to specific tissues or is general to all. (f) An exception to the lack of relationship between tissue free amino acid patterns and those of the dietary and tissue proteins is tryptophan, which is uniquely low in all three locations. The significance of this in the regulation of protein synthesis will be considered later.

B. Amino Acid Transport

Amino acids pass into and out of cells; in addition, free amino acids cross membranous barriers within the cells. Consequently, such transfers from one compartment to another provide potential sites for the regulation of amino acid metabolism. We must therefore consider briefly the features of amino acid transport. First, what are the characteristics of mechanisms used to transport amino acids into and out of mammalian cells? Secondly, what is the evidence regarding the occurrence of discrete subcellular pools of free amino acids? Finally, what are the rates of renewal of amino acids in different body compartments?

1. Mechanisms for Transporting Free Amino Acids

Present concepts of amino acids transport across cell barriers have been largely developed by Christensen, who described the basic principles in Chapter 4 and elsewhere (e.g., Christensen, 1966, 1968). The study of amino acid transport has been greatly aided by the use of model amino acids which are subject to intracellular transport and concentration but are not incorporated into protein or further metabolized. A commonly used model amino acid is α-aminoisobutyric acid (AIBA), which can be employed in a ^{14}C-labeled form to study transport, and which resembles glycine in its transport properties. Other model amino acids are constantly being described (e.g. Christensen and Handlogten, 1968; Christensen *et al.*, 1969).

These studies indicate that there are several mechanisms for transporting amino acids into cells, and that consequently, the amino acids can be arranged into transport groups. Presumably each transport group represents a separate carrier protein in the cell membrane that has an affinity for a certain series of amino acids of designated charge and structure (e.g., acidity). In the case of Ehrlich ascites tumor cells, there appear to be six transport families of amino acids (Christensen, 1968) as follows: (1) an alanine-preferring (A) system carries alanine, glycine, and other neutral amino acids with small or polar side chains; this system is linked to the sodium pump and is energy-dependent, (2) a second (L) system transporting neutral amino acids with branched chains (e.g., leucine) or with aromatic rings (phenylalanine). This system is not linked to the Na^+ pump or to energy needs and exchanges amino acids within the cell for those outside, (3) a system that tolerates basic amino acids, (4) a system that is exclusively reserved for basic amino acids, (5) a transport system for dicarboxylic amino acids, and (6) a concentrative system for taurine and β-alanine. Under special circumstances, it has even been possible to recognize four instead of two transport systems for neutral amino acids (Eavenson and Christensen, 1967). As shown in Table III, Blasberg and Lajtha (1965) also recognize six families of amino acids involved in amino acid accumulation by mouse brain cells, two each for neutral and for basic amino acids, one for acidic amino acids, and one for penetration of γ-amino-

TABLE III
SPECIFICITY OF CEREBRAL AMINO ACID ACCUMULATION AT STEADY STATE BY MOUSE BRAIN[a]

Amino acid class	Amino acid class inhibitor					
	Neutral		Basic			
	Small	Large	Small	Large	Acidic	Amide
Small neutral	++	+++	+	0	+	+
Large neutral	+	++++	−	0	0	0
Large basic	+	+++	−	++	0	++
Acidic (+ GABA)	+	0	+++	0	++++	++

[a] From Blasberg and Lajtha (1965). + = inhibition; 0 = no effect; − = stimulation.

butyric acid. Table III illustrates how amino acids compete for transport into mouse brain. It will be noted that amino acids from the same transport class often inhibit uptake of others, whereas in other instances, amino acids from another transport group can stimulate uptake of an amino acid from a different transport group because it favors exchange (class 2 above). Consequently, changing amino acid levels in the extracellular fluid can have complex effects on intracellular accumulations. This is notably demonstrated by Christensen and Cullen (1968), who observed extensive reductions in the levels of most neutral free amino acids 2 hours after giving AIBA to rats; AIBA appears to act by competing with these neutral amino acids for transport into the liver. Amino acid exit from cells is also regulated by active processes (Oxender and Christensen, 1963), a substantial portion of the amino acids leaving through reversal of the processes for amino acid uptake (Christensen and Handlogten, 1968). However, the factors affecting amino acid exit are not necessarily identical to those influencing uptake. For example, passage of amino acids out of kidney cells seems to be specifically inhibited by phlorhizin (Segal *et al.*, 1963) and cysteine exit from the same cells is inhibited by basic amino acids (Schwartzman *et al.*, 1966b). The steady state levels of amino acids in tissues thus become a balance between influx and efflux. Amino acids with a high influx and low efflux can accumulate to higher intracellular levels than amino acids with a low influx and high efflux (Lajtha *et al.*, 1968).

A further complication is that the same amino acid can differ in its transport characteristics from cell to cell. For example, the transport of glycine into Ehrlich ascites tumor cells is inhibited by methionine,

ethionine, or proline whereas the transport of glycine into rat brain is not subject to the same inhibitory action of these amino acids (Abadom and Scholefield, 1962). Similarly, the lysine transport mechanism in rat kidney and intestinal mucosa differ in several respects (Segal *et al.*, 1968). An extensive comparison of L-histidine entry into a variety of tissues (intestinal mucosa, testis, spleen, kidney cortex, and brain) has shown that inhibition due to the presence of other amino acids can vary from one tissue to another (Neame, 1966). This means that an excessive level in the extracellular fluid of one amino acid may affect entry of other amino acids into some tissues but not necessarily into all. Even in different areas of the brain, concentrations of free amino acids vary and this has been traced to regional differences in the intensity of the transport mechanisms (Kandera *et al.*, 1968). The response of transport mechanisms to hormones also appears to vary from tissue to tissue. A striking example is the uptake of AIBA into the parathyroid glands. The activity of this process is inversely related to the calcium level in the plasma, whereas no such relationship has been found between calcium level and the rate of uptake of AIBA into liver or muscle (Raisz and O'Brien, 1963). Blood calcium concentration regulates the amount of parathormone secreted by the parathyroid glands, and the finding with AIBA suggests that the rate of free amino acid uptake into the parathyroid glands is linked to rate of hormone production.

From these observations, it can be concluded that amino acid transport into and out of cells provides one potent means through which regulation of protein metabolism could be obtained. First, the influx of amino acids into the body after meals could alter intracellular amino acid concentrations in a complicated fashion due to acceleration or inhibition of transport. Second, high levels of individual amino acids in the diet will distort the transport mechanisms. Third, hormones and other factors could affect metabolism in selected target organs through a change in rate of amino acid uptake that is specific for these cells and is not a widespread action throughout the body. Finally, metabolic changes in the tissue itself may change transport mechanisms.

2. Subcellular Amino Acid Pools

The presence within the mammalian cell of subcellular structures bounded by membranes (nuclei, mitochondria, endoplasmic reticulum vesicles) makes it likely that the concentrations of metabolites and other small molecules may not be uniform throughout the cell. In the case of mitochondria, there is plenty of evidence for selective permeability of the mitochondrial membrane so that some small molecules are excluded while others can penetrate and thus act as substrates for the enzymic reactions within this organelle. Evidence of this kind shows that metabolites can accumulate in subcellular compartments within cells, and we have elsewhere referred to these as "micropools" (Munro and Portugal, 1970). Intracellular micropools of metabolites have been described or postulated for free fatty acids in adipose tissue (Rubinstein *et al.*, 1965), pyridine nucleotides in mitochondria and cell sap of ascites cells (Kohen, 1964), iodide in the thyroid gland (Nagataki and Ingbar, 1963), and potassium in erythrocytes (Solomon and Gold, 1955) among other examples.

Is there similar compartmentation of free amino acids within mammalian cells? Three lines of evidence indicate that this may be so. First, there is reason to suspect that organelles within the cell offer compartmentation. It is known that isolated nuclei can accumulate free amino acids by an energy-dependent process (Allfrey *et al.*, 1961) which is specifically inhibited by adenine nucleotides (Karjalainen, 1966). Mitochondria prepared from brain cells accumulate glycine by an active process which differs in certain respects from glycine transport across the cell membranes of the neuron (Nukada, 1965). Consequently, there is every reason to expect that nuclei and mitochondria may have pools of free amino acids that are not part of the general amino acid pool of the cell. The second line of evidence favoring the occurrence of micropools of free amino acids comes from direct examination of the specific activities of free amino acids in these subcellular fractions after the administration of a labeled amino acid. This can arise not only from differences in rate of penetration of the amino acid into the subcellular compartment, but also because it is possible that the dilution of the labeled amino acid with endogenously released amino acids differs in the various compartments. Intracellular dilution can be demonstrated by continuous intravenous infusion of labeled amino acid in order to achieve a steady state. Loftfield and Harris (1956) first used this technique to examine the specific activity of liver free leucine after repeated injections of large amounts of leucine-^{14}C over an 80-minute period; they demonstrated that, due to dilution with endogenous amino acids, the liver free leucine never rose above 40% of the specific activity of the injected leucine. Gan and Jeffay (1967) confirmed that endogenously liberated amino acids dilute incoming amino acids 50% in the liver of fed rats but only 30% in muscle. A similar approach has been used by Portugal and Jeffay (1966) to determine whether the dilution with endogenous amino acids may differ in different subcellular compartments. Portugal and Jeffay (1966) infused rats with lysine-^{14}C over a 3-hour period and obtained a constant specific activity of free lysine in the plasma. Nevertheless, they found that the free lysine of the nuclear, mitochondrial, microsomal, and cell sap fractions of the liver differed in specific activity. This argues in favor of different degrees of dilution with unlabeled endogenous amino acids in the various compartments.

The final source of evidence of compartmentation comes from studies in which proteins and other products synthesized from radioactive amino acids have been found to have specific activities that are not compatible with their origin from the general free amino acid pool of the cell. For example, Green and Lowther (1959) observed that proline-^{14}C incorporated into collagen often showed a higher specific activity than its proline precursor in the cells. Similarly, hippuric acid has been synthesized *in vivo* from glycine-^{14}C with a specific activity in the product which was higher than that of free glycine within the cell where the hippuric acid is made (Garfinkel and Lajtha, 1963). Since hippurate synthesis from glycine is known to occur in the mitochondria, this might be thought to represent compartmentation of the free labeled glycine by the mitochondria; however, it seems more likely that the hippurate is rich in glycine-^{14}C because the mitochondria are congregated at the edge of the cell where the glycine enters (see Chapter 27).

Again, a number of studies have been carried out on the metabolism of labeled free amino acids in the brain; Garfinkel (1966) has recently analyzed some of the published results in order to quantitate the integrated metabolism of glutamate, aspartate, glutamine, glutathione, γ-aminobutyric acid, and related compounds. By using a digital computer to analyze these metabolic data, Garfinkel has shown that published results of amino acid metabolism in the brain can best be explained if he postulates one large and one small intracellular compartment, each of which contains separate pools of free glutamate, aspartate, and γ-aminobutyric acid. The large compartment is considered to consist of the bodies of neurons, whereas the small compartment is thought to be made up of nerve endings and glial cells. The author points out the functional uses of such compartmentation in brain cell metabolism. The application of computers to the analysis of pathways and compartments in protein metabolism is decribed in detail by Garfinkel and Heinmets in Chapter 27. Finally, Ito *et al.* (1964) have examined the serum proteins of dogs injected with labeled phenylalanine or tyrosine and have found that the time course of tyrosine labeling in these proteins differs according to whether it was derived directly from the injected tyrosine or from the phenylalanine. Their data suggest complexity in the pools of tyrosine available for serum protein biosynthesis within the liver. Compartmentation of the cytoplasmic free amino acid pool is claimed for a variety of tissues, the inference being that protein is made from a small pool separate from the bulk of the intracellular free amino acid. Early evidence on mammalian cells comes from experiments with labeled amino acids in which there was a lack of concordance between the counts in the free amino acid pool and their uptake by protein under various circumstances (e.g. Kipnis *et al.*, 1961; Rosenberg *et al.*, 1963). Since these authors did not measure the specific activity of the free amino acid under consideration, their evidence of compartmentation is not conclusive. However, Roscoe *et al.* (1968) incubated ascites tumor cells with leucine-^{14}C and found that the specific activity of free leucine in the cell could be severely depressed by adding D-phenylalanine to the medium, without causing any change in incorporation of the label into tissue protein. In these studies, the control curve for uptake of leucine-^{14}C into the intracellular pool was unorthodox and complicates interpretation. Recently, Hider *et al.* (1969) incubated rat muscle, first with ^{14}C-labeled leucine, followed by leucine-^{3}H. On changing from one form of labeled leucine to the other, incorporation into muscle protein switched abruptly from ^{14}C to ^{3}H, as if the protein were being made from amino acids obtained directly from the incubation medium without passing through the intracellular pool, which still retained much of its leucine-^{14}C. They therefore suggest that activation of amino acids occurs at the muscle cell membrane and that amino acids destined for protein synthesis are thus segregated into a separate metabolic compartment at the time of penetration of the cell membrane.

Subcellular compartmentation of the free amino acids raises some obvious problems in the interpretation of experiments in which labeled amino acids are used to trace intracellular reactions. Proteins are said to be synthesized by cell nuclei (Allfrey, 1963) and by mitochondria (Roodyn, 1965) as well as by the microsomes. The occurrence of micropools of free amino acids within these organelles could complicate the study of nuclear and mitochondrial protein synthesis from labeled free amino acids injected into the whole animal.

3. Rate of Free Amino Acid Renewal

It is obvious from Table I that the free amino acids of the plasma must be rapidly renewed, since the daily intake of amino acids is large in comparison with the total amounts of each in the plasma (e.g., the isoleucine content of 1 gm of egg protein is 500 μmoles, whereas the total amount of isoleucine in the plasma of a 100-gm rat is only 0.3 μmoles, that is, 3% of the 10 μmoles in the animal's body). In addition, amino acids are recycled between tissues.

When labeled amino acids are injected into the bloodstream of animals, they are rapidly removed, indicating a high rate of free amino acid transfer between body tissues. This implies rapid responses to changes in amino acid supply. One of the earliest and most extensive studies with labeled amino acids was made by Borsook and his colleagues. Borsook *et al.* (1950) injected ^{14}C-labeled L-leucine, L-histidine, or glycine into the tail veins of mice. Less than 3% of the dose of radioactivity remained in the bloodstream at 10 minutes after injection. It was recovered, however, as nonprotein radioactivity in the tissue of the carcass. At 30 minutes after injection, this carcass activity had fallen considerably and had now been replaced by activity in the proteins of the visceral organs and by $^{14}CO_2$. These findings suggest that some fraction of the carcass, probably skeletal muscle, concentrates amino acids rapidly but does not metabolize them further. The free amino acids therefore pass out again into the blood and are captured by other tissues, such as the viscera, which form proteins rapidly and also dispose of the free amino acids in other ways.

This evidence of rapid equilibration of the free amino acid pool between plasma and other body compartments is supported by other studies. Barton (1951) injected glycine-^{14}C into mice and found that activity in the plasma fell rapidly. Peters (1962) injected leucine-^{14}C into the tail vein of the rat and found 5 minutes later that the blood contained only 2% of the injected radioactivity. In the course of studying the labeling of pancreatic proteins after intravenous injection of DL-4,5 leucine-^{3}H into guinea pigs, Caro and Palade (1964) observed that the blood level of radioactivity had fallen to 20% of its peak value within 5 minutes. They also computed the half-life of free leucine in the pancreas to be about 15 minutes. In the rabbit, Maurer *et al.* (1954) observed an initial half-life of about 20 minutes for plasma methionine activity after intravenous injection of methionine-^{35}S. Elwyn (1966) has published curves for the disappearance of glycine-^{14}C from dog blood after intravenous injection which show a half-life of glycine in systemic arterial blood of about 10 minutes. From evidence presented in Chapter 25, it is likely that amino acid turnover will be progressively slower with increasing body size of the species; the available data are insufficient to test this prediction.

These studies provide rough indices of the considerable rapidity of the rates of transfer of the free amino acids between the various compartments of the body of the mammal. Presumably rates of amino acid transfer will vary somewhat with the status of the animal, e.g., fasting or fed. A more detailed study was made by Henriques *et al.* (1955), who injected glycine-1-^{14}C intravenously into rabbits and isolated the free glycine from the plasma, liver, and skeletal muscle at various times up to 18 hours thereafter. This overcomes the objection to many of the earlier studies that radioactivity present in cold acid extracts of tissues is likely to include degradation products as well as the amino acid that was injected. The results obtained by Henriques *et al.* for free glycine in plasma, liver, and muscle are plotted in Fig. 2. At 5 minutes after injection the specific activity of the free glycine in the plasma of their rabbits was very high; it decreased sharply within the next 40 minutes, and thereafter declined more slowly. These changes cannot be described by a single exponential equation from which a half-life can be deduced. This difficulty occurs because the glycine of the plasma is mixing at different rates with many separate compartments in the body. Henriques *et al.* were able to obtain a reasonable fit between their findings and an equation which assumed three such body compartments, each exchanging at widely different rates. Complex curves were also found for the disappearance of phenylalanine-^{14}C from the plasma of human subjects (Grümer *et al.*, 1961).

The results obtained by Henriques *et al.* also throw light on the rate of exchange of glycine between the plasma and tissues. In the liver, the activity of the free glycine rose to a maximum level at 15 minutes after injection; by 4 hours it had fallen to 5% of the maximum activity. In the case of skeletal muscle, the highest levels of radioactivity reached were less than in the case of the liver, and there appeared to be two maxima, one at 30 minutes and the second at 3 hours after injection. Henriques *et al.* considered whether the initial peak might be due to trapping by muscle of highly radioactive blood; however, perfusion of muscle failed to remove this initial high activity and the authors concluded that skeletal muscle may have more than one amino acid pool. It will be remembered that Borsook *et al.* (1950) found that various labeled amino acids disappeared from the plasma of mice within 10 minutes after injection and were found as nonprotein radioactivity in the carcass, which then quickly lost this radioactivity again. This rapid deposition and removal of amino acids may correspond to the early and transient peak of labeling observed by Henriques *et al.* in muscle, a tissue which provides a major part of the carcass. Henriques *et al.* also attempted a mathematical treatment of the curves shown in Fig. 2 to provide a description of the rates of exchange of free glycine between plasma and liver or muscle. In the case of liver, the renewal time of free glycine was estimated to be 17 minutes. In the case of muscle, they neglected the initial peak of activity and computed a value for the remaining observations, which fitted a renewal time of 1400 minutes. The significance of the initial uptake into muscle thus remains unresolved. It is obvious that further studies with other amino acids should be undertaken before accepting the observations of Henriques *et al.*

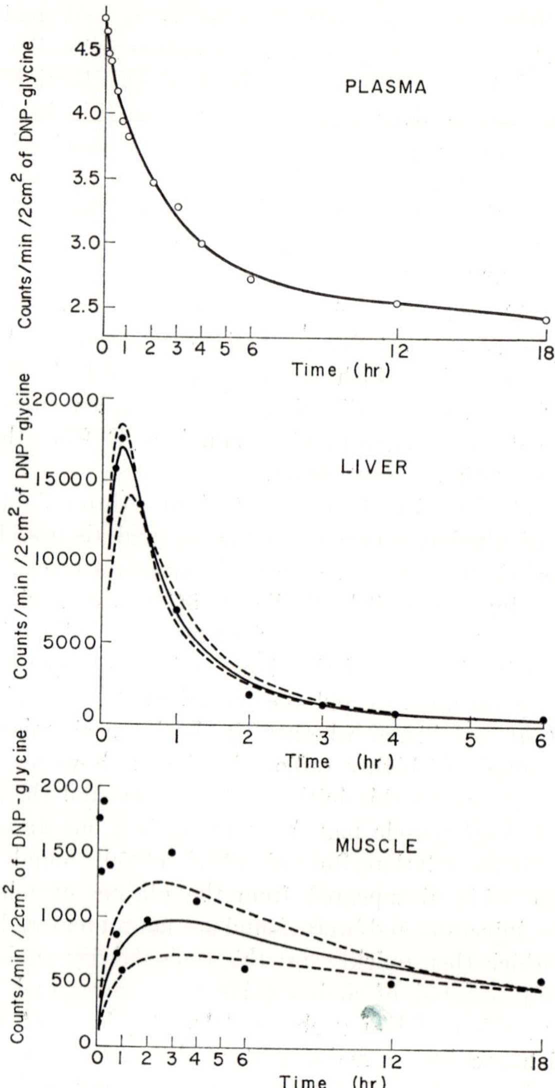

FIG. 2. Changes with time of the specific radioactivity of free glycine in plasma, liver, and muscle after intravenous administration of glycine-^{14}C to rabbits. (From Henriques *et al.*, 1955.) Note that the top curve (plasma) is on a logarithmic scale. The curves on the middle panel (liver) represent the best fit for a renewal time of 17 minutes (solid line); the upper and lower lines are calculated for renewal times of 13.5 and 29 minutes respectively. The curves on the lower panel (muscle) represent the best fit assuming a renewal time of 1400 minutes; the upper and lower lines are calculated for renewal times of 1000 and 2000 minutes respectively.

as representative of all amino acids. These problems are discussed further by Neuberger and Richards in Chapter 7. The evidence of micro-pools of amino acids in muscle, discussed in Section II,B,2 above, is also relevant.

4. *Identification of Changes in Free Amino Acid Pools*

The criteria for identifying changes in amino acid transport are discussed in Volume I, Chapter 4. Before proceeding to consider factors that affect the free amino acid pools, it is as well to remind ourselves of the types of evidence which can be used to indicate responses in the metabolism of free amino acids: (a) The absolute concentrations of free amino acids in the body fluids and tissues. (b) The concentration gradients between tissues and plasma. This distinguishes between changes in tissue concentration due to alterations in plasma levels, and changes due to differences in transport into and out of cells. For example, corticosteroids are known to increase the transport of amino acids into the liver since the concentration gradient changes (Chapter 4). (c) The tissue distribution of AIBA or other nonmetabolizable amino acids. These have the advantage that metabolic removal from or addition to the pool is not involved. For example, removal of amino acids for protein synthesis can perhaps mask the increased passage of amino acids into muscle treated with insulin (see Chapter 33). (d) Rates of exchange of labeled amino acids between body compartments. It must be confessed that most of the information to follow will be confined to the first of these criteria.

III. Factors Affecting Free Amino Acid Concentrations

The free amino acids represent intermediates in protein metabolism. In consequence, the size of the pool of each amino acid is the resultant of a balance between input and removal. As suggested in the introductory chapter to this volume, the pool of free amino acids in the body may be considered to have three metabolic outlets: (a) protein synthesis, (b) synthesis of a number of compounds of low molecular weight, such as creatine and including the nonessential amino acids, and (c) degradation through the pathways of amino acid catabolism. It is likely that these competing metabolic uses of amino acids are in equilibrium, so that a change in one route of disposal is compensated by reciprocal changes in others. Thus the newborn animal shows intense protein synthesis but limited degradation of circulating amino acids (see Chapter 16, Volume II). Similarly, the administration of hydrazine to animals impedes amino acid degradation by interfering with transamination, an effect that can be reversed by providing pyridoxal phosphate (McCormick and Snell, 1961). In animals treated with hydrazine, the greatly reduced catabolism results in elevated plasma and tissue levels of free amino acids (Korty and Coe, 1968; Cornish and Wilson, 1968), and liver protein synthesis is stimulated (Amenta and Johnston, 1963), a clear case of cause and effect which by itself is strong evidence that enhanced free amino acid levels can stimulate protein synthesis within the cell.

This general division into three major routes of amino acid disposal is, of course, a simplification. In the case of most amino acids, pathways of entry into, and exit from, the pool are numerous. Figure 3 shows the metabolic pathways involved in serine metabolism in rat liver (see Chapter 5; also Walsh and Sallach, 1966) and also shows the relative magnitudes assigned to some of them by Nemer *et al.* (1960). The free serine pool of the liver is the resultant of these reactions and changes in pool size can clearly occur from a variety of causes:

1. Alterations in the rate of transfer of the amino acid between the extracellular and intracellular compartments.

2. Changes in rate of release of serine into the pool from degradation of proteins in the liver.

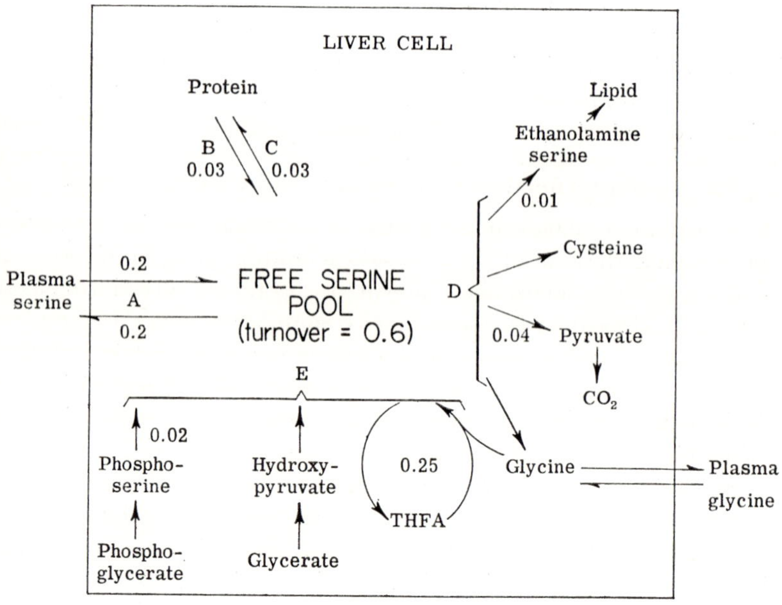

FIG. 3. Pathways of serine metabolism in rat liver assembled from the data of Nemer *et al.* (1960) and of Walsh and Sallach (1966). Rates of reactions are given in micromoles per minute per gram of liver.

3. Changes in rate of utilization of serine for protein biosynthesis.
4. Variations in other reactions consuming serine (lipid synthesis, cysteine formation, glycine formation, degradation to pyruvate and CO_2).
5. Changes in rate of serine biosynthesis from glycine and tetrahydrofolic acid, and from glyceric acid either through hydroxypyruvate or phosphohydroxypyruvate (Walsh and Sallach, 1966).

Not only is the level of free serine in the cell determined by the balance between these pathways. Some at least of these pathways are known to be under regulatory control. Thus, the formation of serine from phosphoserine is inhibited by high levels of free serine (Nemer *et al.*, 1960), whereas the formation of serine from glucose by cells, in tissue culture at least, is not affected by the amount of free serine (Eagle *et al.*, 1965). It is also known that the enzyme regulating serine removal to carbohydrate, serine dehydrase, is readily induced under a variety of circumstances (see Chapter 35). Finally, rate of removal of serine for protein synthesis will vary according to conditions which promote or reduce removal of free amino acids in general from the pool. Thus, an alteration in level of free serine in the liver cell indicates that a metabolic pathway leading to or from serine has changed and that a series of regulatory processes caused by this change has set the free serine concentration at a new value. In order to make a complete analysis of this situation, it would be necessary to provide much more information than is displayed in Fig. 3, which gives only a rough approximation to the magnitude of these pathways under one set of steady-state conditions. Eventually, by amassing data under a variety of circumstances, it may be possible to set up a series of equations which by computer analysis (Chapter 27) will yield a comprehensive quantitative picture for the metabolism of serine. Even when this is achieved, it is not the end of the story. Pathways of amino acid biosynthesis and catabolism are not identical in all tissues or in all mammals; Walsh and Sallach (1966) have shown that the relative magnitudes of the two pathways of serine biosynthesis from carbohydrate vary from tissue to tissue and from species to species (see Chapter 25). In order to provide a complete picture for one amino acid, it will be necessary to obtain such metabolic equations for each tissue in the body.

By comparison with this objective, our knowledge of factors affecting free amino acid pools and the role of these pools in regulating protein metabolism is mostly fragmentary. In the remainder of this chapter, we shall consider published information about the response of the free amino acid pools to dietary conditions, to hormonal action and to other factors known to affect protein metabolism. This will be assembled with three questions in mind. First, how do these factors affect the metabolism of individual free amino acids? To answer this, we shall survey changes in free amino acid concentration in the body fluids and tissues, and where known, any alterations in concentration gradients, in uptake of model amino acids, and in rate of turnover of free amino acids. Second, can we unmask the mechanisms by which diet, hormones, etc., cause these changes in amino acid pools? Third, can we identify any metabolic sequelae to the changes in the metabolism of the free amino acid pools? It must at once be admitted that in no case are the data conclusive on all these points, and in many instances we must be content with reporting that free amino acid concentrations in the plasma or tissues have undergone alterations. Nevertheless, evidence of such changes is a challenge to future investigators, and for that reason alone deserves attention here.

A. Diet and Free Amino Acid Pools

The concentrations of free amino acids in the tissues and body fluids are influenced by many dietary constituents. This is not surprising, since dietary factors will affect not only the supply of amino acids entering the free amino acid pool, but also the rate at which these amino acids are subsequently used in the body. Consequently, the concentrations of free amino acids in the body change in response to the amount of protein fed, and also in relation to intake of carbohydrate, fat, vitamins, and minerals, which affect the fate of the absorbed amino acids. Reported changes in free amino acid pools caused by each of these dietary factors will now be summarized, and the relationship of these changes to metabolic events within the cell will be considered. In Table IV some of the published effects of dietary factors on the levels of free amino acids in the plasma have been recalculated as percentage changes from values obtained in control animals or human subjects.

1. Response to a Meal of Protein

The levels of amino acids in the plasma have frequently been reported to rise after a meal containing protein has been given to a fasting animal or human subject. An increase in total plasma amino nitrogen was first observed by Van Slyke and Meyer (1912). In recent years, studies have been made of individual amino acid concentrations, both in portal blood and systemic blood, after feeding various diets. Denton *et al.* (1953) and Denton and Elvehjem (1954) measured the concentrations of essential amino acids in the portal and systemic blood of dogs after a meal containing protein. When casein or beef was used as the protein source, the portal blood concentrations rose rapidly; the systemic blood levels also increased, though to a lesser extent, and attained their maximum

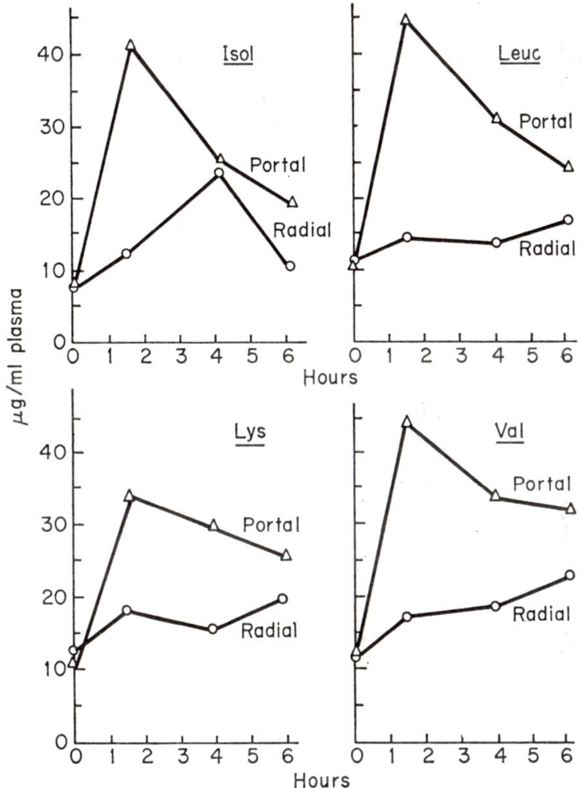

FIG. 4. Concentrations of isoleucine, leucine, lysine, and valine in the portal and radial veins of dogs after feeding a meal containing 38 gm of casein at zero time. (From Denton *et al.*, 1953.)

some hours later (Fig. 4). Similar experiments by other authors have demonstrated that administration of protein increases amino acid concentration more extensively in portal plasma than in systemic plasma (Ganapathy and Nasset, 1962; Porter and Williams, 1963; Peraino and Harper, 1963; Elwyn, 1966). This picture is documented in detail in Chapter 38. The effect of feeding protein on free amino acid levels in the tissues is less well documented. Sheffner and Bergeim (1952) fed peptone to fasting rats and observed that lysine, which is abundant in peptone, rose considerably in concentration in muscle and liver, whereas other amino acids did not increase; in the spleen and in the kidney the levels of all amino acids tended to rise. These authors used microbiological assays for amino acid estimation, and moreover filtration through paper of the protein free filtrate may have added free amino acids. A study by ion exchange chromatography of the tissue concentra-

TABLE IV

Percentage Changes in the Levels of Free Amino Acids in Plasma Caused by Various Diets

Dietary state	Species	Duration (days)	Essential amino acid change (%)												Nonessential amino acid change (%)			
			Arg	His	Ile	Leu	Lys	Met	Phe	Thr	Trp	Val	Tyr	Cys	Ala	Asp	Glu	Gly
Specific amino acid deficiency	Mouse	14	−20	−22	−33	−18	−30	+5	−15	−4	−65	−24	−25	—	−7	—	−11	−17
		14	±0	±0	+16	−9	−27	−46	−15	+47	−22	−11	−13	—	+21	—	−19	−6
	Rat	23	−18	—	+28	+28	−91	−44	—	+29	—	−50	−53	+36	−16	—	−11	−23
	Rat	28	+35	+75	+25	+35	±0	−38	+68	+400	−14	+55	+25	−79	—	—	—	—
		28	−35	+7	−36	−28	−86	−39	−33	+410	−84	−24	−52	−40	—	—	—	—
	Rat	3	−21	+37	+415	−53	+2	+78	+122	+22	—	+565	+87	+87	—	—	—	—
		3	−17	+5	−21	−9	−21	+27	+47	−34	—	+11	+48	+145	—	—	—	—
		3	−3	+68	+25	+1	+9	+45	+62	+10	—	−48	+96	+61	—	—	—	—
		3	+28	+46	+8	−10	+23	+45	+81	−74	—	+6	−14	+32	—	—	—	—
	Rat	23	—	—	—	−50	−82	—	—	−61	—	—	—	—	—	—	+28	—
	Dog	322	−16	+72	−38	−33	−16	−14	+27	−5	−30	−40	−24	+29	—	—	—	—
	Pig	21	−57	+9	—	—	—	—	—	+20	—	—	—	—	—	—	—	—
			+20	−23	—	—	—	—	—	+160	—	—	—	—	—	—	—	—
			−27	+28	—	—	—	—	—	−9	—	—	—	—	—	—	—	—
			−7	+80	—	—	—	—	—	−80	—	—	—	—	—	—	—	—
	Man	5	−7	+5	−23	−17	−9	+13	±0	+60	—	−52	−3	−14	−7	—	+51	+26
	Man	28	−3	−3	−25	−21	−20	+16	−13	−19	—	−23	−18	−28	+11	—	−38	−6
Amino acid imbalance and antagonism	Rat	8	+66	−90	+9	+12	+140	+290	+106	+99	—	+24	+106	—	+31	—	+94	+7
	Rat	12	—	—	−25	+120	−14	±0	—	+58	—	−39	—	—	—	—	—	—
Amino acid toxicity	Man	12	+3	+9	−4	+12	+30	+66	—	+1	—	+29	—	—	—	—	—	—
	Rat	28	+25	—	—	—	−4	+490	—	—	−32	−22	—	—	—	—	—	—
	Rat	28	+155	+150	+67	−8	+83	+2390	+4	+7	—	−24	+11	—	+38	+51	−4	−37
	Rat	49	—	—	—	—	—	+1700	—	−40	—	—	—	—	—	—	—	—
	Monkey	58–170	−22	+15	+8	+46	+21	+19	+1900	+35	—	+31	+540	—	±0	+35	±0	+17
	Rat	28	—	−20	—	—	—	—	—	—	±0	—	—	—	+570	—	—	−4
			+460	−20	—	—	—	—	—	—	−60	—	—	—	—	—	—	−14
			−80	−10	—	—	—	—	—	—	−60	—	—	—	—	+560	—	+18

			—	+20	—	—	—	—	—	—	−20	—	—	+870	—	—	—	−5
			+33	+10	—	—	—	—	—	—	−20	—	—	—	—	—	+38	−27
			−40	−10	—	—	—	—	—	—	±0	—	—	—	—	—	—	+210
			+20	+1240	—	—	—	—	—	—	+20	—	—	—	—	—	—	+13
			−33	+10	+230	—	—	—	—	—	−20	—	—	—	—	—	—	−5
			+27	−10	—	+125	—	—	—	—	±0	—	—	—	—	—	—	−5
			+53	−5	—	—	+640	—	—	—	−60	—	—	—	—	—	—	−18
			+80	−30	—	—	—	+1490	—	—	±0	—	—	—	—	—	—	−32
			—	−10	—	—	—	—	+420	—	+20	—	—	—	—	—	—	−4
			+13	+10	—	—	—	—	—	+1320	±0	—	—	—	—	—	—	+91
			−53	−35	—	—	—	—	—	—	+1100	—	—	—	—	—	—	±0
			−20	−30	—	—	—	—	—	—	—	—	+6000	—	—	—	—	−23
	Rat	28	−54	−50	−29	−33	−35	−22	−44	−74	−59	−36	−64	−60	—	—	—	—
Protein deficiency	Rat	3	−23	−1	−73	−51	−7	−63	−47	−41	−32	−32	−44	−6	+155	+24	+40	−16
	Rat	7	+3	±0	−44	−10	−2	−33	−50	−55	−25	−18	−55	—	—	—	—	—
	Rat	28	+25	+78	−33	−35	−9	−54	−20	−41	−65	−35	−41	−89	—	—	—	—
	Dog	130	−51	+100	−49	−37	+10	−60	−13	−27	−61	−47	−63	−72	—	—	—	—
	Pig	8	−23	+8	−59	−32	−1	−36	−19	−60	—	−53	−64	—	+2	−23	−39	+10
	Man	60	—	—	−35	−35	±0	−25	−44	−44	—	−47	−25	−25	—	—	—	—
	Man	7–10	−5	−11	−19	−16	−13	—	−8	−7	—	−30	−13	−10	+73	+44	+33	+30
	Man	14	+15	+11	−8	+1	−3	+29	+9	+8	—	−12	−6	−39	+64	+17	+38	+45
Restricted food intake	Guinea pig	28	+71	+24	+17	+33	+130	+62	+4	+87	—	+27	+35	—	+6	−11	−40	−45
Carbohydrate administered	Rat	2 hr	—	—	−58	−59	—	—	—	−52	−18	—	—	—	—	—	—	—
	Man	2 hr	−29	−18	−57	−52	−16	−45	−31	−25	—	−36	−38	−29	−7	−52	−49	−14
Vitamin deficiencies	Guinea pig	28	−2	−13	+7	+3	−16	−8	+36	−22	—	+2	+25	—	−3	+11	−13	−24
	Rat	28	−46	−2	−26	+1	−37	−17	−27	−27	−56	−26	−34	−70	−32	+74	+6	+195
	Rat	42	+3	−2	+1	+1	−2	−4	+78	−38	—	±0	+72	±0	−2	+1	+2	−3
Fasting	Rat	7	+15	+20	+50	+20	±0	+10	+10	−30	+10	+12	−35	—	—	—	—	
	Rat	9	—	—	+150	+150	+47	+25	—	−12	—	+80	+22	+72	−24	—	−22	−67
	Rabbit	3.5	−37	−8	+45	+15	−7	−17	−21	−18	—	+28	−43	+46	−20	−45	−53	−16
	Man	2	−18	—	—	+55	−14	−5	—	−36	−29	+31	—	—	—	—	—	—
	Man	14	—	—	+30	+30	—	—	—	—	—	+30	—	—	—	—	—	—
	Man	7	−7	−11	+85	+64	−12	+20	−5	+7	—	+29	−10	−14	−41	−8	−18	+22

(*continued*)

TABLE IV—*Continued*

Percentage Changes in the Levels of Free Amino Acids in Plasma Caused by Various Diets

Dietary state	Species	Duration (days)	Nonessential amino acid change (%) Pro	Ser	$AspNH_2$	$GluNH_2$	Experimental diet fed	Animals used as controls	Authors
Specific amino acid deficiency	Mouse	14	−14	—	—	—	Low tryptophan	On complete diet	Steele *et al.* (1950)
		14	+10	—	—	—	Low methionine	On complete diet	
	Rat	23	—	−5	—	−22	Lysine-tryptophan-free (zein)	On casein diet	Wu (1954)
	Rat	28	—	—	—	—	Low methionine (casein)	On same diet with methionine	Swendseid *et al.* (1963b)
		28	—	—	—	—	Low lysine (gluten)	On same diet with lysine	
	Rat	3	—	—	—	—	Leucine-free (amino acids)	On complete amino acids	Clark *et al.* (1966)
		3	—	—	—	—	Isoleucine-free (amino acids)	On complete amino acids	
		3	—	—	—	—	Valine-free (amino acids)	On complete amino acids	
		3	—	—	—	—	Threonine-free (amino acids)	On complete amino acids	
	Rat	23	—	—	—	+15	Low lysine (gluten)	On casein	Vandermeers-Piret *et al.* (1966)
	Dog	322	—	—	—	—	Low lysine (gluten)	Same dog at start	Longenecker (1961)
	Pig	21	—	—	—	—	Arginine-free (amino acids)	On complete amino acids	Baker *et al.* (1966)
			—	—	—	—	Histidine-free (amino acids)	On complete amino acids	
			—	—	—	—	Lysine-free (amino acids)	On complete amino acids	
			—	—	—	—	Threonine-free (amino acids)	On complete amino acids	
	Man	5	−4	+23	+16	+16	Valine-free (amino acids)	On same diet with valine	Swendseid *et al.* (1966)
	Man	28	+16	+2	+2	+2	Low lysine (gluten)	Same men at start	Zimmermann-Telschow and Jekat (1965)
Amino acid imbalance and antagonism	Rat	8	+44	−4	—	—	Histidine imbalance (basal diet + amino acids less histidine)	On basal diet	Sanahuja and Harper (1963)
	Rat	12	—	+30	—	—	Casein + 5% leucine	On casein diet alone	Tannous *et al.* (1966)
Amino acid toxicity	Man	12	—	—	—	—	Excess methionine	On basal diet	Kirsner *et al.* (1949)
	Rat	28	—	—	—	—	Excess methionine	On basal diet + riboflavin	Harrill and Gifford (1962)
	Rat	28	+61	−18	−49	−49	Excess methionine	On basal diet (pair-fed)	Klavins (1965)
	Rat	49	—	—	—	—	Excess methionine	On basal diet (absorbing)	Girard *et al.* (1968)
	Monkey	58–170	+13	+59	—	±0	Excess phenylalanine	On basal milk formula	Kerr *et al.* (1968)
	Rat	28	—	—	—	—	5% alanine	On basal diet (6% casein)	Sauberlich (1961)
			—	—	—	—	5% arginine	On basal diet (6% casein)	
			—	—	—	—	5% aspartic acid	On basal diet (6% casein)	

			—	—	—	—	5% cystine	On basal diet (6% casein)	
			—	—	—	—	5% glutamic acid	On basal diet (6% casein)	
			—	—	—	—	5% glycine	On basal diet (6% casein)	
			—	—	—	—	5% histidine	On basal diet (6% casein)	
			—	—	—	—	5% isoleucine	On basal diet (6% casein)	
			—	—	—	—	5% leucine	On basal diet (6% casein)	
			—	—	—	—	5% lysine	On basal diet (6% casein)	
			—	—	—	—	2% methionine	On basal diet (6% casein)	
			—	—	—	—	5% phenylalanine	On basal diet (6% casein)	
			—	—	—	—	5% threonine	On basal diet (6% casein)	
			—	—	—	—	5% tryptophan	On basal diet (6% casein)	
			—	—	—	—	5% tyrosine	On basal diet (6% casein)	
	Rat	28	—	—	—	—	7.5% glycine	On basal diet (8% casein)	Swendseid *et al.* (1963b)
Protein deficiency	Rat	3	+54	+56	—	—	Protein-free	Rats fasting 3 days	Wiss (1948, 1949a)
	Rat	7	−25	—	—	—	Protein-free	Same rats at start	Henderson *et al.* (1949)
	Rat	28	—	—	—	—	8% casein	On 18% casein diet	Swendseid *et al.* (1963b)
	Dog	130	—	—	—	—	Protein-free	Same dog at start	Longenecker (1961)
	Pig	8	−2	−7	—	—	Protein-free	On 30% protein diet	Richardson *et al.* (1965)
	Man	60	—	—	—	—	Low protein	Same man at start	Tuttle *et al.* (1962)
	Man	7–10	—	+5	—	+5	Protein-free	Subjects at start	Young and Scrimshaw (1968)
	Man	14	+2	+40	—	—	Protein-free	Subjects at start	Adibi (1968)
Restricted food intake	Guinea pig	28	−36	−28	—	+7	Reduced food intake	On same diet ad libitum	Schønheyder and Lyngbye (1962)
Carbohydrate administered	Rat	2 hr	—	—	—	—	Dose of glucose	Fasting group	Munro and Thomson (1953)
	Man	2 hr	−19	−25	−29	−29	Dose of glucose	Same subjects fasting	Zinneman *et al.* (1966)
Vitamin deficiencies	Guinea pig	28	+34	+11	—	−23	Scorbutic diet	Pair fed on a diet with ascorbic acid	Schønheyder and Lyngbye (1962)
	Rat	28	−29	−38	−7	−7	Vitamin B_6 deficient	On diet with vitamin B_6	Swendseid *et al.* (1964a)
	Rat	42	—	+5	—	—	Vitamin A deficient	Pair fed on diet with vitamin A	Malathi *et al.* (1961)
Fasting	Rat	7	−50	—	—	—	Fasting	Same rats at start	Henderson *et al.* (1949)
	Rat	9	—	−43	—	−32	Fasting	Rats on casein diet	Wu (1954)
	Rabbit	3.5	−45	−50	−30	−30	Fasting	Same rabbits at start	Block and Hubbard (1962)
	Man	2	—	—	—	—	Fasting	Same subjects on full diet	Charkey *et al.* (1955)
	Man	14	—	—	—	—	Fasting	Same subject at start	Swendseid *et al.* (1961)
	Man	7	+5	−5	—	—	Fasting	Same subject at start	Adibi (1968)

tions of all amino acids throughout the period of absorption after a protein meal would thus fill in a gap in our knowledge. No comprehensive data of this kind are available, but in the course of studying amino acid imbalance, Ellison and King (1968) studied rats fed *ad libitum* on a balanced (corrected) diet containing large amounts of amino acids. Immediately after giving such a meal to fasting animals, there was an immediate and extensive rise in the free histidine content of liver, plasma, and muscle and this was sustained for as long as the diet was fed. On the other hand, the levels of free leucine in these three locations in the body showed little change with feeding. In the absence of definitive information, it must be assumed that free amino acids are removed sufficiently rapidly to prevent *large* accumulations in the tissues during absorption. What we do not know is the general magnitude of the increase in tissue levels after a meal, a critical factor in evaluating the stimulus to tissue protein synthesis after a meal.

Several authors have explored whether there is any demonstrable relationship between the amino acid composition of the protein fed and the pattern of free amino acids appearing in the plasma during absorption. A priori, one would expect increases in the plasma levels of free amino acids that would reflect the relative amounts of these amino acids in the protein undergoing digestion. On this basis, amino acids present in most abundance in the diet should cause the largest rises. This relationship has been confirmed by several authors, most of whom compared the levels of amino acids in portal blood during absorption with the amino acid content of the diet undergoing digestion (Denton and Elvehjem, 1954; Goldberg and Guggenheim, 1962; McLaughlan *et al.*, 1963; Pion *et al.*, 1964; Buraczewski *et al.*, 1967). However, Nasset and his colleagues (Ganapathy and Nasset, 1962; Nasset *et al.*, 1963) and Levenson *et al.* (1959) could see no relationship between the amino acid composition of the diet absorbed and changes in the free amino acid levels in the portal blood of the dog; as pointed out in Chapter 38, Nasset's data may in fact be compatible with some relationship of portal blood levels to dietary amino acid content during the early absorptive period, which was not measured. There has also been a failure by some investigators to establish a relationship between the relative abundance of dietary amino acids and the rise in level of these amino acids in *peripheral* blood during the absorptive period (Schreier and Remsperger, 1951; Frame, 1958; Yearick and Nadeau, 1967). This is perhaps not surprising since the magnitude of the changes in peripheral blood levels are much smaller than those in the portal vein after a meal is consumed (Fig. 4), and furthermore the liver has a modifying action on amino acid pattern due to its inability to remove branched-chain as compared with other essential amino acids (Chapter 38); this accounts for the excessively large increases in branched-chain amino acids levels after feeding a meal of protein even to ruminants (Hogan *et al.*, 1968). In addition, if carbohydrate is given along with the protein, the carbohydrate tends to lower the blood amino acid levels (see Section III,A,5). This may account for the failure of Anderson and Linkswiler (1969) to observe any rise in plasma essential amino acid levels during the first 2 hours after consuming a mixed meal providing a rather small amount of protein along with carbohydrate and fat. Indeed, many amino acids showed a reduction in plasma level in their experiment.

Such direct relationships of diet to blood amino acid levels are also obscured by differences in the rates of removal of amino acids from the plasma to meet the requirements of the tissues. This has been elegantly demonstrated by the studies of Longenecker (Longenecker and Hause, 1959, 1961; Longenecker, 1963). He has pointed out that the rise in plasma free amino acid levels during absorption will be counterbalanced by removal of amino acids, largely to provide for tissue protein synthesis. The pattern of requirements of amino acids for this latter purpose does not change with diet but is a constant spectrum. Thus if one essential amino acid is present in relative low concentrations in the dietary protein, it will tend to be removed by the tissues as rapidly as it is absorbed from the intestine in order to allow maximal protein synthesis to proceed. Consequently, the most limiting amino acid of the food protein will be indicated by showing the smallest rise in plasma concentration, when this is expressed as a ratio relative to the amino acid requirements of the animal. Conversely, the amino acid present in largest excess in the food protein will show the greatest increase in this ratio. In order to make these comparisons, it is necessary to quantitate the amino acid requirements of the experimental animal; the rises in plasma essential amino acid concentrations after a meal are then expressed in relation to these requirements as a series of ratios (plasma amino acid ratios). For example, when wheat gluten was fed to dogs, the lysine concentration in the systemic plasma actually fell, giving a negative plasma amino acid ratio and thereby indicating that lysine is the most limiting amino acid in wheat gluten (Table V). On the other hand, the plasma amino acid ratios are seen to be positive for other amino acids, according to their relative abundance in gluten. Thus, methionine is considered to be the second limiting amino acid because it shows the smallest rise relative to requirement (Table V). This method of evaluating dietary protein has been confirmed on man by Yearick and Nadeau (1967), who found no relationship between the dietary pat-

TABLE V

CALCULATION OF THE PLASMA AMINO ACID (PAA) RATIOS FOR WHEAT GLUTEN FED TO DOGS[a]

Amino acid	Plasma amino acid levels mg./100 ml. Fasting (A)	During absorption (B)	Differences (C)	Amino acid requirements of dogs (D)	PAA ratio (C/D × 100) (E)	Order of limiting amino acids (F)
Lysine	2.02	1.25	−0.77	7.1	−10.8	1
Methionine	0.66	0.81	+0.15	5.1	2.9	2
Arginine	1.50	1.68	+0.18	5.7	3.2	3
Threonine	2.56	3.11	+0.55	4.0	13.8	4
Valine	1.77	2.79	+1.02	6.4	15.9	5
Isoleucine	0.87	1.94	+1.07	6.2	17.3	6
Phenylalanine	0.94	2.59	+1.65	9.2	17.9	7
Leucine	1.05	3.07	+2.02	8.5	23.8	8
Tryptophan	0.95	1.35	+0.40	1.1	36.4	9
Histidine	0.76	1.61	+0.85	2.0	42.5	10

[a] From Longenecker and Hause (1959).

tern of amino acids and the rise in plasma concentrations of individual free amino acids, but were able to show such a relationship when the plasma amino acid ratios were calculated from their data. The plasma amino acid ratio has also been modified by feeding the test protein for several days instead of in a single meal (Chapter 29; also McLaughlan, 1963, 1964; McLaughlan and Venkat Rao, 1967; McLaughlan *et al.*, 1967; Smith and Scott, 1965*a,b*; Smith, 1966). The concept of a smaller rise or even a fall during absorption in the plasma level of the most limiting amino acid in the diet is supported by the extensive evidence obtained on animals consuming amino acid imbalanced diets (Chapter 13). For example, rats receiving an imbalanced amino acid diet deficient in histidine showed an extensive fall in blood histidine within 1 hour of starting to eat the diet (Chapter 13, Table XV). Such changes are not confined to the systemic blood, but are also seen in the portal blood (Pion and Rérat, 1967).

These findings raise some interesting basic questions concerning the relationship of free amino acid supply to amino acid utilization. Since the level of the least abundant amino acid can actually fall during the absorptive phase, this suggests that amino acid utilization is accelerated during absorption, particularly protein synthesis. This increased utilization after meals is compatible with much other data. Thus Table I demonstrates that the daily intake of amino acids by the rat is many times larger than the pools of free amino acids in its body, particularly the pools of essential amino acids. Consequently, each meal of protein must be followed by a rapid metabolic adjustment in order to prevent this influx from causing a large accumulation of free amino acids. This is strikingly shown after the intravenous infusion of amino acids; despite the large influx, the plasma levels of amino acids show only transient increments (Knauff *et al.,* 1966). As demonstrated from the data given in Chapter 38, much of this adjustment following the influx of amino acids takes place in the liver, where a considerable proportion of a large incoming amino acid load can be disposed of through degradative reactions before it can reach the systemic circulation. There is good reason to believe that amino acids which escape this destruction are rapidly removed from circulation through an increase in the rate of liver protein synthesis during the absorptive period. This is implied in the picture presented in Chapter 38, which shows that more absorbed amino acids are removed from circulation by the liver than can be accounted for by urea formation. It is also supported by direct evidence. Thus C. M. Clark *et al.* (1957) injected glycine-^{14}C into rats that were either (a) fasting after a period of feeding on a protein-free diet, (b) fasting after a normal diet, or (c) absorbing amino acid after a meal of casein. At 3, 6, and 9 hours after injection of the labeled amino acid, animals from each group were killed and the uptake of activity into liver protein was measured and corrected for the level of radioactivity in the free glycine pool and for the amount of protein in the liver; Clark *et al* (1957) considered that computations based on such data would provide a measure of the number of new liver protein molecules per cell made under these nutritional conditions. Figure 5 shows that the two fasting groups of rats had essentially similar rates of liver protein formation, whereas the group absorbing amino acids showed about twice the rate of uptake of free glycine into liver protein. This implies that the main impact of dietary amino acids on liver protein synthesis is limited to the absorptive period after a meal. Such a conclusion is, of course, compatible with the extensive evidence already reviewed (Chapter 10) of the lability of liver protein. The protein content of the liver changes rapidly when the dietary protein level is increased; some of this labile protein undoubtedly consists of enzymes involved in amino acid metabolism, such as tryptophan pyrrolase, whose rapid induction and turnover are discussed in Chapters 31 and 32. Indeed, the whole concept of labile body protein is probably a reflection of changes in tissue enzyme levels needed to cope with nutrients.

Is there any evidence about the mechanism through which an increase in the free amino acid content of the portal blood after a meal causes

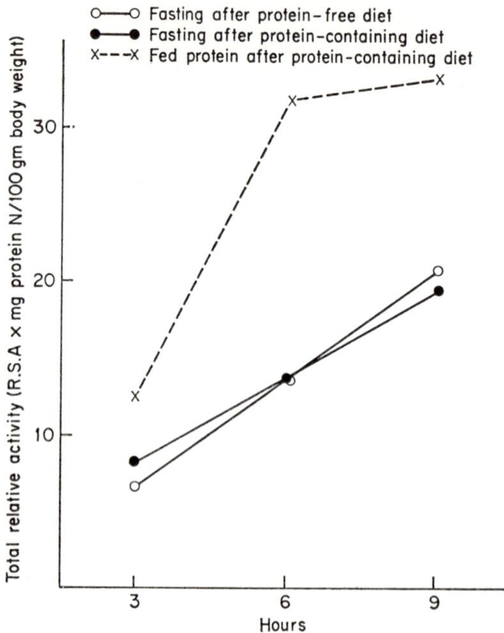

FIG. 5. Incorporation of glycine-^{14}C into liver protein under various nutritional circumstances. Rats were fasted after either a protein-free diet or a protein-containing diet and others were fed a meal of protein. All were injected with glycine, and the liver was excised at different times thereafter. The uptake into the liver protein is given as the specific activity of glycine in the protein expressed as a percentage of the specific activity in the free glycine pool of the liver and multiplied by the total amount of protein in the liver per 100 gm of body weight. This is termed total relative activity and is a measure of the total number of new protein molecules made in the time-interval, since it corrects uptake for precursor activity and for dilution by the amount of protein already in the liver. (From C. M. Clark *et al.*, 1957.)

an acceleration of protein synthesis in the liver cell? It might be anticipated that the response of the liver would depend on a complete set of essential amino acids being absorbed from the meal, and indeed this is so. In experiments performed several years ago (Munro and Clark, 1959), we fed rats by stomach-tube with either a complete or a tryptophan-deficient mixture of amino acids and observed that there was an immediate difference in liver RNA metabolism during the period of amino acid absorption. This was taken as evidence that the mixture of amino acids supplied to the liver influences the RNA metabolism of that organ because of some aspect of RNA function in protein synthesis. It was subsequently demonstrated (Fleck *et al.*, 1965) that microsomes prepared from the livers of rats absorbing the tryptophan-deficient amino acid mixture were less effective for *in vitro* protein synthesis than were microsomes prepared from animals absorbing the complete mixture. This difference in the efficiency of the microsomes was traced to changes in the polysome population of the liver cell (Fleck *et al.*, 1965; Wunner *et al.*, 1966; Drysdale and Munro, 1967). Figure 6 shows polysome patterns obtained 1 hour after feeding such amino acid mixtures to rats. Absence of tryptophan from an otherwise complete amino acid mixture resulted in a loss of heavy polysomes and an accumulation of monosomes and other oligosomes; there was also an increase in ribosome subunits, not shown on the gradient of Fig. 6 (Wunner *et al.*, 1966). Such changes due to tryptophan deficiency are evident within an hour of feeding the amino acid mixtures, and persist throughout the period of amino acid absorption, but are readily reversible by giving tryptophan to the animals (Wunner *et al.*, 1966). Several other authors (Webb *et al.*, 1966; Sox and Hoagland, 1966; Staehelin *et al.*, 1967) have confirmed that polysomes increase in abundance soon after a meal of protein is fed to a fasting animal. This response is rapid, unlike the polysome reaggregation following glucose administration to fasting rats, which does not occur for several hours and is presumably due to an indirect action on the liver (Wittman *et al.*, 1969). On the basis of these findings, it has been concluded (Munro, 1966; 1968) that the supply of free amino acids entering the liver cell regulates the proportion of ribosomes in the form of polysomes, oligosomes, and monosomes, and as subunits. An ample and complete supply of amino acids results in extensive polysome formation, with few oligosomes and subunits; on the other hand, an influx of amino acids deficient in tryptophan is not compatible with increased protein synthesis and therefore the polysomes do not aggregate and in consequence oligosomes, monosomes, and subunits accumulate. The relationship of polysome aggregation to amino

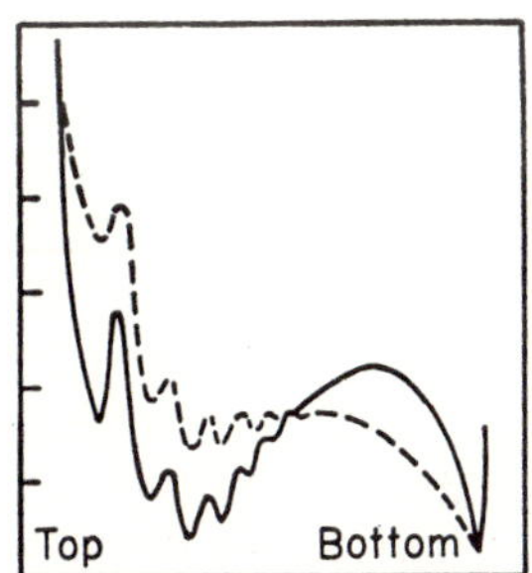

FIG. 6. Polysome profiles obtained from the livers of rats fed amino acid mixtures, either nutritionally complete (——) or deficient in tryptophan (------). (From Drysdale and Munro, 1967.)

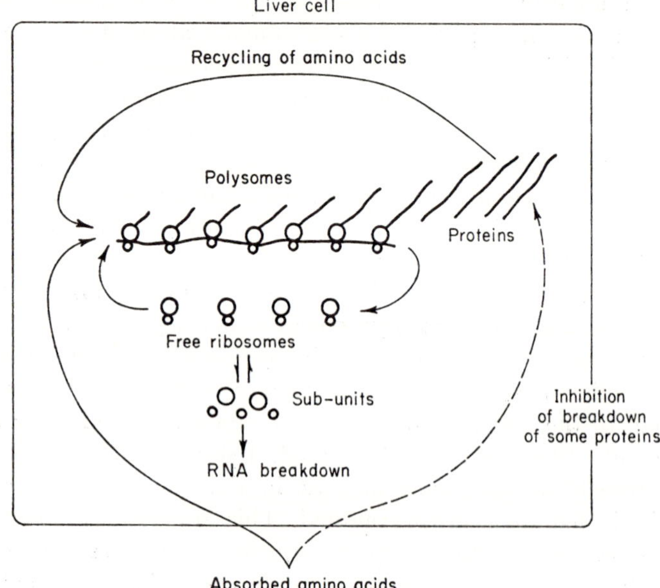

FIG. 7. Scheme showing relationship of amino acid supply after a protein meal to protein synthesis and breakdown and RNA breakdown in the liver cell. Recent evidence suggests that the subunits may recycle directly to the polysomes.

acid supply is displayed diagrammatically in Fig. 7. It was also concluded from these studies that amino acid supply affects liver RNA metabolism because latent ribonuclease becomes activated when the ribosome subunits break apart; consequently lack of tryptophan leads to an increase in RNA catabolism because a larger proportion of the ribosome population is held in subunit form. The immediate acceleration in RNA catabolism on withdrawing protein from the diet has been confirmed by Enwonwu and Munro (1970).

The site of this regulatory response to amino acid supply was considered to be exclusively cytoplasmic for two reasons. First, it was found that the changes in polysome profile caused by feeding the tryptophan-deficient mixture could also be induced in rats pretreated with actinomycin D at a level sufficient to prevent synthesis of new messenger RNA (Fleck *et al.*, 1965; Staehelin *et al.*, 1967). More compelling evidence, however, came from the study of polysome disaggregation and reaggregation in a cell-free protein-synthesizing system (Baliga *et al.*, 1968). Liver polysomes were incubated with amino acid activating and transferring enzymes and tRNA (depleted of free amino acids) in a medium containing guanosine triphosphate and adenosine triphosphate. Because of almost complete elimination of free amino acids, the system was capable of very limited protein synthesis, and after 20 minutes of incubation the polysomes had undergone extensive nonspecific disaggregation (Fig. 8). When a complete mixture of amino acids was now added to the medium, amino acid incorporation into peptides restarted and extensive polysome reaggregation occurred. Omission of tryptophan from the amino acid mixture prevented this response from taking place. The pictures *in vivo* (Fig. 6) and *in vitro* (Fig. 8) are almost identical. Finally, a mechanism involving adrenocortical stimulation can be excluded, since adrenalectomized mice still show a response of liver polysomes to tryptophan administration (Sidransky *et al.*, 1968).

These experiments suggest that the protein-synthesizing machinery in the liver cell cytoplasm responds directly to changes in amino acid supply through a mechanism that can distinguish the influx of a nutri-

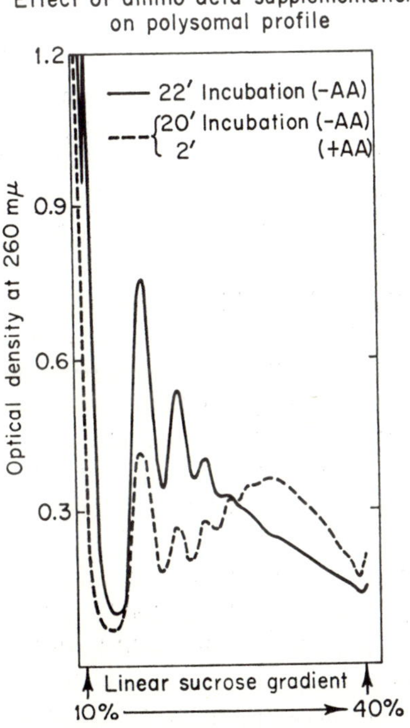

FIG. 8. Effect of delayed supplementation with 20 amino acids on the polysome profile incubated *in vitro* in a system for amino acid incorporation. The system was prepared from liver polysomes and cell-sap factors freed of nonprotein amino acids. It was incubated for 20 minutes without amino acids and showed extensive breakdown of polysomes. A complete amino acid mixture was then added for 2 minutes, and the polysome profiles were compared. (From Baliga *et al.*, 1968.)

tionally complete amino acid mixture from one lacking tryptophan. When this essential amino acid is omitted from the mixture, synthesis of protein on liver polysomes presumably becomes retarded and there is an accumulation of free ribosomes awaiting attachment to messenger RNA. It would thus be logical to conclude that deficiency of any other essential amino acid would similarly slow the rate of translation and consequently cause disaggregation of liver polysomes. However, studies with rats tube-fed on amino acid mixtures deficient in one essential amino acid show that only lack of tryptophan gives rise to a reduction in polysome aggregates and accumulation of oligosomes (Pronczuk *et al.*, 1968). The unique action of tryptophan has been also confirmed with mice by Sidransky *et al.* (1968) who observed that 5-hydroxytryptophan, 5-hydroxytryptamine, indole, and 3-hydroxyanthranilic acid are also somewhat active. On the other hand, omission of any one of a large number of amino acids from the mixture added to a cell-free system of liver polysomes results in failure of the amino acid mixture to support reaggregation (Baliga *et al.*, 1968). Our studies thus demonstrate that the unique effect of tryptophan in causing changes in polysome pattern in the whole animal cannot be attributed to some special role of tryptophan in the mechanism of protein synthesis. Consequently we must look for some factor in the whole animal through which tryptophan deficiency alone exerts an effect. There is no reason to believe that free tryptophan concentration in the liver is especially sensitive to lack of tryptophan in the diet. When rats were fed amino acid mixtures deficient in one essential amino acid and the free amino acid pools in the liver were examined an hour later, absence of any one essential amino acid from the mixture fed usually caused a depression in the level of that amino acid in the intracellular pool; in several instances (e.g., valine), the percentage reduction resulting from absence of that amino acid from the meal was greater than the percentage fall in tryptophan level induced by feeding a tryptophan-deficient mixture to rats (Munro and Zak, 1968).

The unique sensitivity of the liver polysome pattern in the whole animal to tryptophan deficiency is probably related to the fact that tryptophan is the least abundant amino acid both in the liver free amino acid pool and in the proteins formed from the pool (Table I). This could have two consequences. First, it is possible that the low level of tryptophanyl-tRNA in the cell could determine the rate of protein synthesis, simply because insertion of tryptophan at the correct sites in the peptide chain will involve the greatest delay in translation of the message. In the liver of the normally nourished animal it does not seem possible to reduce the intracellular concentration of other amino acids to the point at which they become more rate-limiting in translation than tryptophan, because the extensive recycling of amino acids in the liver maintains a fairly constant pattern of free amino acids in that tissue. However, we have recently restricted this recycling by depleting rats of liver protein; when such animals are fed meals containing imbalanced amino acid mixtures, changes in polysome profile can now be achieved with amino acids other than tryptophan (Pronczuk *et al.*, 1969). Finally, Eliasson *et al.* (1967) have shown that Chang liver cells in tissue culture undergo a loss of polysomes and a gain in monosomes when arginine or glutamine is withdrawn from the medium. This is strong evidence that tryptophan is normally the amino acid whose concentration determines polysome aggregation *in the liver of the intact animal.* It might be supposed that perfusion of rat liver with various amino acid mixtures might offer a situation in which the effects of deficiency of amino acids other than tryptophan could be made more severe and thus more clear-cut. The evidence is, however, contradictory. Jefferson and Korner (1969) found that addition to the perfusate of a complete mixture of amino acids at 10 times the normal plasma concentration promoted polysome aggregation and protein synthesis, and that omission of any one of 11 amino acids from this mixture prevented this response. On the other hand, Rothschild *et al.* (1969) found that albumin synthesis by the perfused livers of fasting rats was especially sensitive to availability of tryptophan in the perfusate. The difficulty of reducing the intracellular supply of amino acids other than tryptophan is illuminated by a recent paper by Allen *et al.* (1969) which demonstrates that the charging of tRNA with tryptophan is especially sensitive to dietary deficiencies. Using amino acid mixtures similar to those of Wunner *et al.* (1966), they tube-fed rats on mixtures from which one amino acid at a time was deleted. The charging of tRNA in the liver was not significantly reduced by deletion of phenylalanine, lysine, leucine, or arginine from the mixture, was moderately reduced by deletion of valine, isoleucine, and methionine, but was severely reduced by deletion of tryptophan. Thus, absence of tryptophan from the diet will more severely limit charging of its tRNA species than will shortage of any other amino acid.

Hori *et al.* (1967) suggested an additional explanation for the apparent unique regulatory action of tryptophan, based on their studies on protein synthesis in reticulocytes. They pointed out that the infrequency of tryptophan in most proteins means that considerable stretches of peptide chain will not include tryptophan, and in consequence lack of tryptophanyl-tRNA will not restrict the rate of translation in these areas. For example, if the mixed proteins of the liver are assumed to have an average molecular weight of 70,000, corresponding to about 600 amino acid residues, then the average protein with 1% tryptophan will have 6 such residues and, again on average, the terminal 100 amino acid residues of such a protein will not contain any tryptophan. Consequently, ribosomes on the terminal part of the messenger for the protein will roll off unimpeded by tryptophan deficiency, but will be unable to become reattached to messenger RNA because movement along the messenger is retarded by lack of tryptophan. Consequently, there is an accumulation of free ribosomes along with slowing down of recycling. Other amino acids occur much more frequently in proteins, and only short stretches of chain will be without a given amino acid; in the average protein, such stretches will be too short to allow release of ribosomes from the end of the messenger if any one amino acid is not available for protein synthesis. Those who wish to pursue further the effects of amino acid supply, and particularly of tRNA supply, on protein synthesis and on polysome function will find a detailed account in the introductory Chapter of this volume, Section III,A,3. It leaves no doubt that tRNA availability is a potent regulator of protein synthesis.

In evaluating the unique effect of tryptophan deficiency on liver polysome pattern in the whole animal, several related observations should be kept in mind. As discussed in Chapter 35, the levels of a number of liver enzymes can be greatly increased by administration of tryptophan alone in rats; in addition to tryptophan pyrrolase, these enzymes include tyrosine transaminase, serine dehydrase, and others. Second, there are many studies in which a protein or amino acid mixture lacking one essential amino acid has been administered at one time of day and the missing amino acid given at a different time (Berg and Rose, 1929; Elman, 1939; Henry and Kon, 1946; Cannon *et al.*, 1947; Geiger, 1947, 1948, 1950; Henderson and Harris, 1949; Yang *et al.*, 1961, 1963; Howe and Dooley, 1963; Mitchell and Morrison, 1965; Lapeze *et al.*, 1965; Tilton *et al.*, 1965; Deinken *et al.*, 1966; Yang *et al.*, 1968). From these investigations only tryptophan consistently emerges as an amino acid that must be consumed in the same meal as the others in order to ensure their utilization. There is some evidence that methionine may also fall into this category; it is significant that methionine is the second least abundant essential amino acid in the free amino acid pool (Table I). Recently, Rothschild *et al.* (1969) have shown with perfused rabbit liver that serum albumin synthesis is stimulated by addition of tryptophan or isoleucine to the perfusate, provided the donor was fasting at the time the liver was excised. These various observations suggest that availability of tryptophan may play a special part in regulating protein synthesis in the liver of the intact animal not only because this amino acid is the least abundant in the free amino acid pool, but also because increased activity in the enzyme system for tryptophan catabolism is readily brought about by excess substrate. Intracellular free tryptophan thus becomes the amino acid limiting rate of protein synthesis. On the other hand, it must be admitted that amino acid deficiencies other than tryptophan may cause immediate effects on liver protein synthesis. Benevenga *et al.* (1968) found that addition of histidine to a histidine-deficient (imbalanced) meal caused an increase in *total* histidine-^{14}C uptake into liver protein within a few hours; a similar picture was obtained by Hartman and King (1967). Although both of these groups examined the amount of isotopic label in the total free amino acid pool of liver, they did not measure activity per micromole of free histidine; consequently, it is not possible to exclude completely a change in the specific activity of the precursor free amino acid to account for their findings. There is also the rather curious finding of Hanking and Roberts (1964, 1965) that addition of excess of single amino acids to liver slices can stimulate incorporation of leucine into liver protein and also increase the incorporating activity of ribosomes isolated from such tissue. Some free amino acids can stimulate transport of others into ascites tumor cells (Schafer and Jacquez, 1967).

From the information presented above and in other chapters of this treatise, we can assemble a partial picture of the effect of a meal of protein on liver protein metabolism (Fig. 7). In response to a change in amino acid supply, the liver gains or loses protein rapidly, and in consequence it is a major site of labile body protein deposition (Chapter 10). Some of the mechanisms involved in this adjustment to amino acid supply are understood. Digestion of dietary protein is so efficient (Chapter 2) that, within an hour of consuming a protein meal, the levels of free amino acids in the portal blood are elevated far above their fasting concentrations, which gives the cells of the liver a stimulus not shared by other tissues in the body with the exception of the intestinal mucosa. As shown in Chapter 38, much of this influx of amino acids does not pass into the general systemic circulation. When Elwyn (Chapter 38) fed dogs on large amounts of beef, part retained by the liver was used for the synthesis of plasma proteins and of liver proteins, but more than half of the amino acid load passing up the portal vein was rapidly degraded through the action of liver enzymes, such as tryptophan pyrrolase and the enzymes of urea synthesis. In his studies, very large meals of protein were used and it seems likely that amino acid catabolism after smaller meals of protein might be less dramatic. Nevertheless, his data illustrate the dominant position of the liver in protecting the systemic circulation against violent fluctuations after each meal containing protein.

The amount of protein in the liver increases during the absorption period and this probably occurs for two reasons. First, there is adequate documentation (e.g., Fig. 5) that the influx of amino acids after a meal increases the rate of liver protein synthesis. The role of individual essential amino acids in modifying this response is less clear, but there is evidence summarized earlier to suggest that availability of tryptophan normally plays a special role in the process through being the least abundant amino acid. Second, the breakdown of liver protein is probably reduced during this influx of amino acids; since protein synthesis still continues, protein will therefore accumulate in the liver. There are several reasons for drawing this conclusion. As discussed in Chapters 32 and 35, the concentrations of many degradative enzymes in the liver increase with the influx of their substrates, and in several cases this has been shown to occur through reduction in enzyme protein breakdown resulting from substrate stabilization. Thus, Fig. 9 illustrates the response of rat liver tryptophan pyrrolase activity to a dose of tryptophan; enzyme level rises after the concentration of tryptophan in the free amino acid pools has risen. This action of tryptophan is due to stabilization of the enzyme (Chapter 32). If this type of regulation is an extensive phenomenon, it means that individual amino acids can influence the amount of protein in the liver cell without requiring the simultaneous participation of all amino acids demanded for protein synthesis, since each amino acid will independently affect the turnover of enzymes in its own pathway of degradation. This is not true of all such enzymes, since a number of amino acid degrading enzymes (e.g., serine dehydratase) appear to respond to an increase in free tryptophan level and not to an increase in substrate concentration (see Chapter 35). As pointed out in an earlier section of this chapter, recycling of amino acids within the liver cell is an important contributor to the free amino acid pool of the cell. Reduction of the turnover of liver protein after a protein meal would thus reduce this endogenous component of amino acid supply and in consequence make it easier for the liver to accommodate the influx of amino acids from the diet. When absorption of amino acids has ceased, the consequent instability and breakdown of liver enzyme protein will now flood the cell with amino acids and the intracellular recycling process will once more become a major source of amino acids for the liver amino acid pool. This has in fact been demonstrated experimentally by Gan and Jeffay (1967), who injected rats continuously with lysine-^{14}C to provide a constant level of activity in the plasma and compared the specific activities of free lysine in liver and in plasma

(Fig. 10). In the case of fed rats, they calculated that endogenous amino acids from the catabolism of liver protein provided 50% of the intracellular lysine pool, whereas in the early stages of fasting 90% of the intracellular amino acid pool came from this source. As fasting progressed, the endogenous amino acid supply decreased again, as one would expect from exhaustion of the labile liver protein accumulated during the absorptive period. The changes during fasting are particularly convincing evidence of variations in the rate of breakdown of liver proteins. It will be noted from Fig. 10 that muscle does not show the same fluctuations as liver does during starvation.

It has already been pointed out that the impact of a protein meal on the peripheral tissues is much less evident than on the liver

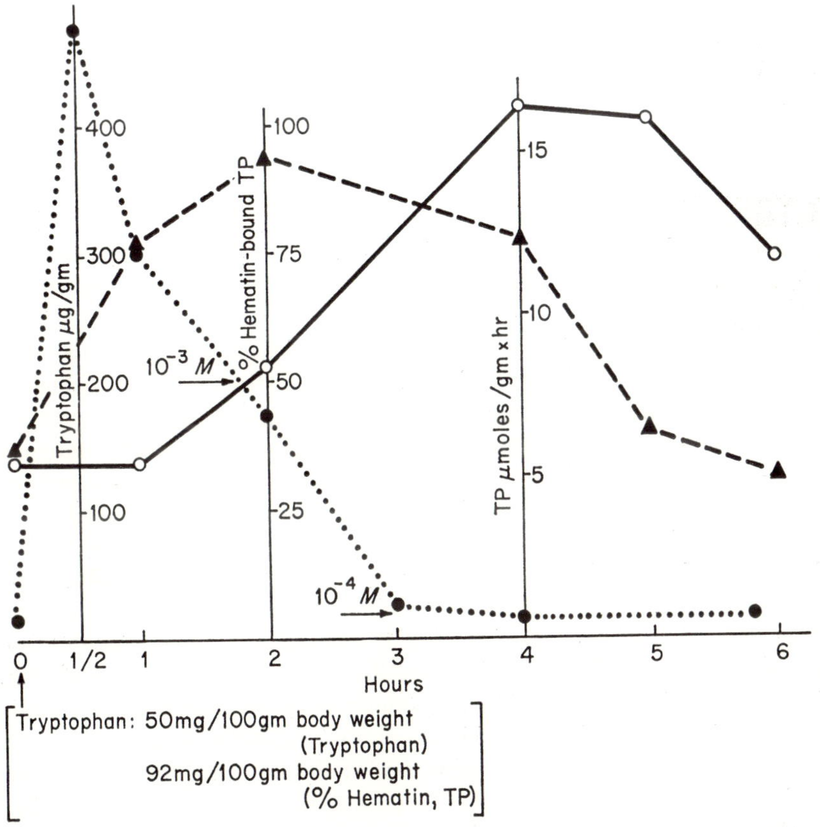

FIG. 9. Effect of a dose of tryptophan on liver free tryptophan (....), tryptophan pyrrolase (TP) activity (0——0), and percent conjugation of hematin with enzyme (▲ — — ▲). Reprinted with permission from Knox and Greengard, *Advan. Enzyme Regulation* **3,** 247 (1965), Pergamon Press Ltd.

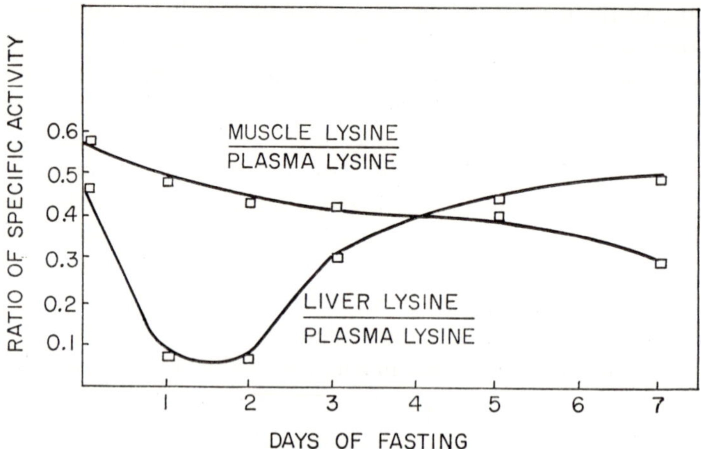

FIG. 10. Effect of fasting on the sources of intracellular lysine. Animals were infused each with lysine-^{14}C every hour for 10 hours, and the specific activities in liver, muscle, and plasma were measured. The diagram shows changes in the specific activities of lysine in the two tissues relative to the specific activity in plasma. (From Gan and Jeffay, 1967).

(Fig. 4). It should be admitted that we know very little about the effect of lack of specific amino acids in the diet on synthesis of protein in tissues other than the liver. The enormous increase in free amino acid concentrations in the portal vein during absorption gives the liver a stimulus not shared by other areas of the body, and the response of the liver may well be quantitatively unique. However, a severe fall in systemic blood concentration of the most limiting amino acid in the diet, such as Table V shows for lysine, must have some consequences for the rate of protein synthesis in the peripheral tissues of the body. Thus appetite is influenced within a few hours of feeding a meal low in one essential amino acid (see Chapter 13); since this effect is abolished by destruction of the hypothalamic region of the brain (Nasset *et al.*, 1967), an action of plasma free amino acid level on cell function in that area of the brain can be assumed. Leung and Rogers (1969) have recently demonstrated that the appetite-depressing action of a diet low in one essential amino acid can be alleviated by infusing the missing amino acid directly into the blood supply to the brain, thus supporting the concept of an action of amino acid deficiency through systemic free amino acid concentrations.

Direct evidence of regulation of rate of protein synthesis through changes in amino acid supply has been provided by experiments on mammalian cells in tissue culture. Eagle *et al.* (1961) observed with a variety of human cell strains in culture that the critical intracellular concentrations of valine, lysine, and threonine to permit protein synthesis were 0.01–0.04 m*M* and that optimal rates of synthesis were achieved at 0.03–0.15 m*M* levels. Due to the high concentrative power of the cells, such concentrations could be obtained with one-tenth these levels in the extracellular fluid. Riggs and Walker (1963) confirmed with Ehrlich ascites tumor cells that protein synthesis is sensitive to the rate of passage of amino acids into the cells from the medium. Several factors were found to act differently on amino acid transport and protein synthesis; for example, transport was much less sensitive to energy supply. Recently, Schingoethe *et al.* (1967) have cultured mammary tissue *in vitro* under conditions that favor continued formation of milk proteins over a period of several hours. They have demonstrated that each of the essential amino acids is required for synthesis of β-lactoglobulin and casein by rat mammary cells and that an increment in amino acid content of the medium accelerates output of these proteins. These various studies demonstrate regulation of the rate of tissue protein synthesis by amino acid supply. In order to relate such findings to the whole animal, it would be necessary to know the diurnal amplitude of variations in free amino acid supply to the mammary gland and the other tissues studied *in vitro*. Diurnal variations in plasma amino acid levels are discussed in Chapter 36. Finally, there is a series of studies on reticulocyte hemoglobin synthesis which shows that inadequate availability of tRNA can retard the rate (Hunt *et al.*, 1969) and that certain tRNA species are particularly critical for effective translation in cell-free lysates of reticulocytes (Anderson and Gilbert, 1969).

In summary, there is reason to believe that each influx of amino acids into the body causes changes in the rates of protein synthesis and turnover in various tissues. The impact on the liver is extensive and readily demonstrable. It is also likely that some at least of the peripheral tissues must share in the stimulus of the increase amino acid supply after each meal. A *cytoplasmic* regulation mechanism sensitive to amino acid supply has been proposed (Fig. 9). This is, however, an oversimplification of the impact of the absorbed amino acids on the tissues of the body, since we know that a meal of protein also causes rapid changes in RNA metabolism within the nucleus of the liver cell (Stenram, 1958; Munro *et al.*, 1964). Nuclear regulation of RNA metabolism in relation to amino acid supply is also suggested by the loss of different messenger species within the liver cell when no dietary protein is available. Xanthine oxidase activity disappears from the liver cell within a few days of withdrawing protein from the diet (Litwack *et al.*, 1950), whereas the templates for plasma albumin (Wilson *et al.*, 1967) and for ferritin (Drysdale *et al.*, 1968) continue to be available in adequate amounts. While this may be due to different stabilities of the messenger for these three proteins, it could also arise from different effects of amino acid supply on the rates of synthesis of the various messenger RNA species in the nucleus. Finally, rate of cell division also appears to be responsive to amino acid supply. In the intestinal mucosa, where cells replicate rapidly, the rate of cell division appears to be influenced by the protein content of the diet (Munro and Goldberg, 1964). This also seems to be the case for the immunocompetent cells of the spleen, which replicate less rapidly when a protein-depleted animal is challenged with an antigen (Kenney *et al.*, 1968; Cooper and Munro, 1969).

2. Changes Induced by Long-Term Feeding of Various Amino Acid Sources

The experiments described above represent the immediate responses of the free amino acid pools occurring at the time of absorption of a single meal of protein. There is, however, abundant evidence that diet-induced changes in free amino acid patterns can persist when proteins low in one or more essential amino acids are fed for longer periods (Table IV). In assessing the significance of such data, it should be remembered that the food intake of animals on deficient diets generally declines. This means that a fall in total protein and total energy intake may be responsible for some of the reported changes in free amino acid levels. It is difficult to provide appropriate control groups to overcome this objection. Animals on the adequate control diet can be restricted to the same total food intake as that of the deficient animals by pair-feeding, but such restricted controls tend to consume the limited amount of food quickly and thus starve for most of the day. In consequence, one may be comparing the levels of free amino acids in the plasma and tissues of a fasting group with the levels in a group still absorbing diet. Force-feeding of amino acid-deficient diets overcomes this problem, but as discussed later, metabolic changes in such animals have proved to be complex and difficult to interpret. Human subjects receiving experimental diets deficient in amino acids can usually be maintained voluntarily on a constant intake of food and provide us with the least complicated evidence of changes induced in plasma amino acid levels by the deficiency (Table IV).

Numerous studies on mice and rats (Steele *et al.*, 1950; Wu, 1954; Morrison *et al.*, 1961b; Swendseid *et al.*, 1963b; McLaughlan, 1964; A. J. Clark *et al.*, 1966; Vandermeers-Piret *et al.*, 1966; McLaughlan and Illman 1967) demonstrate that the giving of a diet low in one essential amino acid results in a reduction in the plasma concentration of the deficient amino acid without changes of corresponding magnitude in the plasma levels of other amino acids. Other species have been less extensively studied. Pigs fed on different dietary proteins (Puchal *et al.*, 1962) or on amino acid mixtures lacking one amino acid (Baker *et al.*, 1966) undergo a depression in the plasma level of the amino acid least abundant in the diet. The same observation has been made with chicks (Zimmerman and Scott, 1965). Birds were fed a diet containing a complete mixture of amino acids and the concentrations of lysine, arginine, or valine in this diet were varied. When any one of these dietary amino acids was added in increasing amounts to an otherwise adequate diet, concentration of the limiting amino acid remained low until the dietary need was satisfied, at which point (e.g., 0.9% lysine in Table VI), the plasma level suddenly increased markedly. These experiments demonstrate a close correlation between ability to grow and the level of the limiting essential amino acid in the plasma. Also illustrated in Table VI is the finding that other amino acids can undergo wide variation in plasma concentration without apparently influencing growth rate (e.g., threonine levels behave in the opposite way to lysine). This implies that rate of tissue protein synthesis is regulated by the level of the most limiting amino acid in the pool. The approach to amino acid requirements illustrated in Table VI has been adapted to pig studies (Mitchell *et al.*, 1968). Young pigs were fed diets in which the levels of lysine, isoleucine, leucine, or histidine were varied one at a time. Addition of any one of these amino acids to the diet did not raise the plasma level of that amino acid until the dietary need was satisfied, indicated by a sharp increase in plasma level. Requirements evaluated in this way corresponded closely to needs for these amino acids as measured by the nitrogen balance methods. Similar studies have also been performed on rats (Pawlak and Pion, 1968a,b). Indeed, these authors (Pawlak and Pion, 1968b) found that the free lysine content of muscle was an even more sensitive index of adequacy than plasma lysine level. When rats were fed increasing intakes of lysine, the lysine level in muscle varied 28-fold, with a sharp rise at 0.8% lysine in the diet, which supports maximal growth.

Studies have been made of the long-term action of diet on the plasma amino acids of man whose food intake can more easily be kept constant on a voluntary basis. When wheat gluten was fed as the sole dietary protein to two men, plasma lysine underwent a modest reduction which was further accentuated by increasing the protein demands through heavy exercise (Zimmermann-Telschow and Jekat, 1965). In similar experiments in which the diet was low in sulfur amino acids, Zimmermann-Telschow (1965) found a reduction in plasma methionine. It will

TABLE VI

EFFECT OF FEEDING GRADED AMOUNTS OF LYSINE ON PLASMA AMINO ACIDS AND WEIGHT GAIN OF CHICKS[a]

Dietary lysine[b] (%)	Weight gain (gm/day)	Plasma amino acid concentrations (mg/ml)													
		Lys	Arg	His	Ile	Leu	Met	Phe	Thr	Tyr	Val	Cys	Glu	Gly	Pro
0.6	8.2	6.8	66.0	21.2	17.1	21.3	17.2	15.5	141.4	23.9	31.9	21.2	44.4	108.0	25.2
0.7	11.1	9.7	69.7	15.9	14.6	21.3	17.4	14.4	192.1	17.3	31.0	—	58.4	97.1	24.4
0.8	12.2	7.2	59.5	10.4	13.5	17.2	15.5	12.4	108.8	13.9	29.3	32.7	44.3	73.8	22.0
0.9	13.6	15.5	61.8	5.0	9.2	14.6	14.2	9.1	82.8	9.7	20.4	25.6	66.2	84.1	22.4
1.0	12.8	23.1	45.2	4.1	8.8	12.6	13.5	4.9	47.6	5.4	16.9	—	73.5	125.1	25.8
1.2	12.9	65.7	25.0	1.7	8.2	10.7	10.6	—	37.0	—	15.2	30.4	32.3	60.4	22.2
1.4	13.2	106.7	25.3	3.4	9.8	13.1	12.9	4.7	44.8	4.7	15.1	26.4	42.8	78.1	22.0
1.8	10.7	233.7	6.1	10.2	12.0	14.6	12.4	8.4	126.1	11.2	22.7	33.2	27.7	64.8	19.5

[a] From Zimmerman and Scott (1965).

[b] The birds received a synthetic amino acid mixture providing various levels of lysine.

be noted (Table IV) that dogs receiving wheat gluten for long periods also showed only a 20% fall in level of plasma lysine (Longenecker, 1961). A more clear-cut effect was obtained by feeding two human subjects on an amino acid mixture lacking valine (Swendseid *et al.*, 1966). After 3 days on this diet, the plasma valine levels were only half the control values. The blood samples used in this experiment were taken in the morning in the fasting state and in consequence it can be concluded that the low plasma levels of the limiting dietary amino acid observed in this and other experiments described in this paragraph are not due to a temporary effect during absorption but represent a permanent reduction throughout the absorptive and postabsorptive periods.

Free amino acid concentrations in the *tissues* are affected by prolonged administration of diets deficient in one or more essential amino acids. One might anticipate that these would be less severe in some tissues than in the plasma, because of the contribution to the intracellular amino acid pool from recycling of tissue proteins. For example, recycling of amino acids is much more extensive in liver than in muscle (see Section III,A,1). This predicted difference has been confirmed experimentally. Vandermeers-Piret *et al.* (1966) fed rats for about 3 weeks on a diet containing 7% wheat gluten, which is especially deficient in lysine and threonine. The levels of lysine and threonine in the plasma fell by 82 and 61% respectively, whereas the concentration of free leucine did not fall below 50% of the control level. In the liver, the levels of free lysine, threonine, and leucine were each reduced only about 40%, and the levels in the pancreas were less affected than those in the plasma. The studies of Pawlak and Pion (1967) complete the picture by demonstrating that rats fed on wheat gluten show as severe a reduction in muscle free lysine as in plasma lysine concentration. Not all investigators agree with this picture. In a study involving force-feeding rats for 3 days on diets containing amino acid mixtures lacking either leucine, isoleucine, valine, or threonine it was found by A. J. Clark *et al.* (1966) that the free amino acids of the plasma and the liver underwent extensive reductions corresponding to the free amino acid deleted from the diet, but skeletal muscle was not affected. Denton *et al.* (1950) found that force-feeding of a methionine-deficient diet to rats caused an extensive reduction in liver methionine and arginine levels, whereas in the brain arginine and tryptophan levels fell but methionine remained unaffected. However, the concept of less effect of dietary amino acid pattern on liver than on the peripheral tissues receives strong support from the imbalance and antagonism studies described below.

The effects of force-feeding amino acid-deficient diets on body protein synthesis have been examined in depth by Sidransky and his colleagues.

They observed that force-feeding a diet containing a threonine-deficient amino acid mixture to young rats for several days resulted in an increase in liver size, uptake of amino acids into protein and RNA content and a shift of polysomes towards heavier aggregates (Sidransky and Farber, 1958; Sidransky and Verney, 1964; Sidransky *et al.*, 1964; Staehelin *et al.*, 1967). These phenomena are compatible with increased liver protein synthesis in animals receiving the threonine-deficient diet; this was confirmed by demonstrating more extensive incorporation of labeled amino acids into microsomes prepared from the livers of these animals (Sidransky *et al.*, 1964). At the same time, administration of the threonine-deficient amino acid mixture caused a loss of protein from muscle and diminished capacity of muscle ribosomes to incorporate amino acids into protein (Sidransky and Verney, 1967). This pattern of increased liver protein synthesis and diminished muscle protein content is characteristic of adrenocorticoid action (see Chapter 10, Section IV). Nevertheless, such a mechanism can be discounted since Wagle and Sidransky (1965) have shown that adrenalectomized rats given maintenance doses of cortisone still exhibit increased hepatic protein synthesis when fed the threonine-deficient diet. However, Leon *et al.* (1965) have demonstrated that liver microsomes show an increased capacity for protein synthesis when physical stress is applied to adrenalectomized rats, and it is still possible that the force-feeding of excessive amounts of a threonine-deficient diet may constitute a stress not imposed by allowing the animals free access to a deficient diet. Under *normal* nutritional conditions, the prolonged administration of diets low in one essential amino acid does *not* result in maintenance of the total protein content of the liver (e.g., the action of wheat gluten on liver protein content shown in Fig. 3, Chapter 12).

Finally, we must consider changes in the levels of free amino acid in plasma and the tissues induced by a *general* reduction in protein intake below requirements. Experimental studies on rats (Wiss, 1948, 1949a; Henderson *et al.*, 1949), dogs (Longenecker, 1961), pigs (Richardson *et al.*, 1965), and man (Tuttle *et al.*, 1962; Swendseid *et al.*, 1966, 1968; Adibi, 1968) show that prolonged feeding of a protein-deficient diet usually causes a fall in the *plasma* levels of most essential amino acids except histidine and lysine, which may even rise in concentration. The nonessential amino acid levels tend to remain constant or increase (Wiss, 1949a; Swendseid *et al.*, 1963a, 1966, 1968; Young and Scrimshaw, 1968), and in consequence the ratio of essential to nonessential amino acids (E/N ratio) declines during protein depletion. This picture agrees reasonably well with the aminogram of protein malnutrition as it occurs in poor communities. In an extensive survey of children receiving low protein intakes in various countries, Holt *et al.* (1968) concluded that (1) there is a decline in the concentration of the plasma essential amino acids, most pronounced for the branch chain amino acids and least for phenylalanine and lysine, (2) among nonessential amino acids tyrosine, arginine, citrulline, and α-amino-n-butyric acid decline, and (3) other nonessential amino acids (glycine, serine, proline, histidine, aspartic acid) do not decline to the same extent, or may even rise above the values found in healthy subjects. Because of these changes, the ratio of essential to nonessential amino acids (E/N ratio) is reduced in protein-deficient children. The significance of the E/N ratio in diagnosing human protein malnutrition is discussed in Chapter 28.

The *intracellular* concentrations of free amino acids have been examined in rats subjected to depletion on a protein-free diet. Thompson *et al.* (1950) found that most of the essential amino acids in the liver, muscle, spleen, and brain of the rat underwent a reduction in level. In all tissues, however, histidine and arginine increased and in liver and muscle threonine also rose in concentration. Studies on the livers of protein-depleted rats by Krueger and Wiss (1949), Wiss and Krueger (1949), and Wiss (1949b) have confirmed the general fall in essential amino acid levels and the increase in histidine level. In the case of the brain of the adult rat, Mark and Mandel (1964) also noted that protein deficiency reduced the levels of some essential amino acids, but increased that of histidine. These extensive changes in cerebral amino acid levels can be contrasted with the resistance of the brain in adult animals to protein depletion (see Chapters 10 and 12), and emphasize that we know little about the relationship of free amino acid concentrations to the regulation of protein content in an organ.

The response of free amino acid levels in ruminants to dietary protein supply (Shimbayashi *et al.*, 1965; Shimbayashi and Yonemura, 1965; Purser *et al.*, 1966; Oltjen *et al.*, 1967; Schelling *et al.*, 1967) presents a more complicated picture, probably due in part to changes in amino acid synthesis occurring in the rumen (Oltjen and Putnam, 1966). Nevertheless, lambs fed on different dietary proteins show significantly different plasma amino acid patterns, indicating that the rumen does not entirely obliterate amino acid patterns of food proteins (Theurer *et al.*, 1968).

3. *Free Amino Acid Patterns During Amino Acid Imbalances, Antagonisms and Toxicities.*

Further complexity is introduced when the diet fed results in imbalances and antagonisms between amino acids. The concept of amino acid imbalance is discussed earlier in this treatise by Harper (Chapter 13) and the effects of dietary amino acid imbalance on plasma amino acid levels have been reviewed by Harper and Rogers (1965). An imbalance is created by changing the proportions of amino acids in the diet so as to cause a depression in growth which can then be alleviated by adding the essential amino acid present in least amount in relation to requirements. For example, a histidine-deficient imbalanced diet can be produced by adding an amino acid mixture minus histidine to a diet containing 6% fibrin as its protein component (Kumta and Harper, 1960). On the unsupplemented fibrin diet, rats grow slowly; the addition of the amino acid mixture depresses growth still further, even though all the amino acids of the fibrin are still present in the diet. Growth can be restored by adding the limiting amino acid (histidine) to the imbalanced diet. When rats were fed the diet supplement with the histidine-deficient amino acid mixture, the concentration of free histidine in the plasma fell within a few hours to a very low level and remained depressed so long as the imbalanced diet continued to be fed (Kumta and Harper, 1962; Sanahuja and Harper, 1963). This suggests that the growth failure of the rats on the imbalanced diet arises from lack of adequate amounts of free histidine in the tissues. Presumably the abundance of other amino acids in the histidine-imbalanced diet stimulates utilization for protein synthesis of almost all the available free histidine and thus causes a severe reduction in the free histidine pools. Since animals on an imbalanced diet quickly reduce their food intake, prolonged feeding on the imbalanced diet is likely to produce further changes in free amino acid levels resulting from lower food consumption.

Harper and Rogers (1965; Rogers and Harper, 1968) report a study of the immediate effects of amino acid imbalance associated with threonine deficiency, in which they examined the levels of free amino acids in tissues as well as plasma (Fig. 11). This study shows that the response in free amino acid pattern varies from tissue to tissue. Two groups of rats were killed at 5 hours after eating either a meal containing 6% casein or a similar meal supplemented with an amino acid mixture lacking threonine. By comparison with the animals receiving only 6% casein, the rats fed the meal supplemented with amino acids showed a lower level of free threonine in plasma, whereas the concentrations of most other amino acids were elevated. The skeletal muscle of the imbalanced group showed a severe depression in the levels of most free amino acids, especially threonine. In contrast, the levels of all free essential amino acids in the intestinal wall and liver where *higher* than the concentrations found for the control group; even threonine was elevated, though less so than the other amino acids. In the same series of experiments, threonine-^{14}C was injected into some of the animals, and they

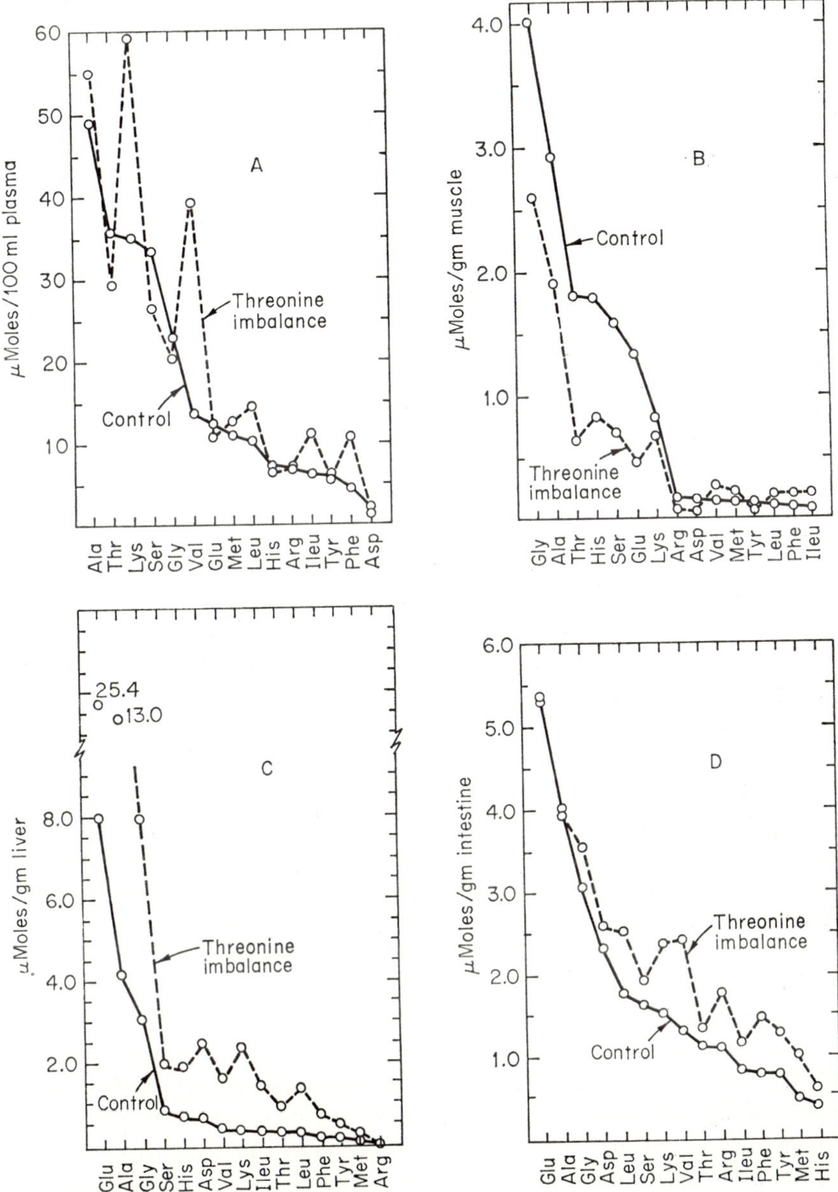

FIG. 11. Patterns of free amino acids in plasma (A), muscle (B), liver (C), and intestine (D) of rats following administration of meals containing either 6% casein (control group) or casein with an amino acid mixture lacking threonine (imbalanced group). (From Rogers and Harper, 1968.)

were killed several hours later. Rats on the imbalanced diet showed an elevated incorporation into liver protein, whereas uptake by muscle protein was not increased. These observations suggest that the general increase in the free amino acid pool of the liver caused by threonine imbalance stimulates protein synthesis in that organ. The problem thus becomes one of trying to identify the cause of the increase flow of amino acids to the liver. It is possible that a reduced uptake of amino acids by muscle as a result of reduced protein synthesis in this bulky tissue could cause deviation of the amino acids to the viscera and notably the liver.

Two other adverse effects of dietary amino acid patterns, namely antagonisms and toxicities, are also recognized (Chapter 13). Extensive changes in free amino acid pools have been reported and may provide further conditions under which to test the relationship of amino acid concentrations to the regulation of protein metabolism. Antagonism involves structurally related amino acids, notably leucine, isoleucine, and valine. When rats are given a diet containing a large excess of leucine, there is a depression of growth which can be relieved by adding isoleucine and valine to the diet (Chapter 13). Rogers *et al.* (1962) found that feeding a single low protein meal containing excess leucine raised the plasma leucine level in rats, and depressed the isoleucine and valine levels; this latter could be corrected by administration of isoleucine and valine. A. J. Clark *et al.* (1968) have made the interesting observation that administration to rats of a single dose of leucine causes a reduction in the plasma levels of isoleucine and valine only when there is an intact insulin secreting apparatus. As discussed later, leucine can cause release of insulin, which diverts plasma amino acids into muscle; the antagonism between leucine, isoleucine and valine may thus be complicated by stimulation of hormonal secretion, a theme dealt with later in the chapter. Tannous *et al.* (1966) studied the plasma and tissue free amino acids over longer periods of excess leucine administration on a low protein diet. The levels of free leucine in all tissues continued to be considerably elevated in the case of the leucine-fed animals. In the plasma and skeletal muscle, isoleucine and valine concentrations were depressed by the toxic level of leucine in the diet, but were partly restored towards normal by addition of isoleucine and valine to the diet. On the other hand, excessive intake of leucine did not depress the levels of isoleucine or valine in the liver or intestinal wall, although the free leucine concentration in both tissues was elevated. Thus, as in the case of threonine imbalance cited earlier, the tissues of the carcass and the viscera show different responses. Again, the more active recycling of amino acids in the liver may make it less susceptible to dietary extremes.

An antagonism of some interest is that between arginine and lysine. Excess lysine in the diet increases the arginine requirement of the growing chick (Jones, 1964; Lewis, 1967). As shown in Table VI and confirmed by Squibb (1968), the plasma levels of arginine and lysine undergo reciprocal changes as the lysine intake of the chick is increased. The reduction in arginine level due to raising lysine intake may occur because a high level of lysine in the plasma increases the activity of kidney arginase (Jones *et al.*, 1967). Certain strains of chick with a high requirement for dietary arginine have been found to have excessive amounts of lysine in their blood; this may cause the need for extra dietary arginine by inducing high levels of kidney arginase, although the evidence is contradictory (Chapter 25, Section III,B).

Amino acid toxicities can be induced by adding a large dose of one amino acid to the diet. In general, a *single* large dose of an amino acid causes a rapid increase in the level of that amino acid in the plasma (Steele and LeBovit, 1951; Block *et al.*, 1969), often accompanied by depression in the levels of other amino acids (Hier, 1947; Schreier and Plückthun, 1950; Swendseid *et al.*, 1965; Holton, 1968; McKean *et al.*, 1968; Snyderman *et al.*, 1968a). On the other hand, while chronic administration of large amounts of individual amino acids generally causes enormous increases in the plasma level of the administered amino acid (Table IV) toxicities can sometime be produced without gross changes in plasma concentration (Sauberlich, 1961), presumably because of adaptive enzyme increases sufficient to deal with the excess. The efficiency of such adaptation is evidenced by the return to normal plasma levels of amino acids fed in excess to infants, who show a rise in the plasma level of the amino acid administered only on the first day (Snyderman *et al.*, 1968a). As in the case of single doses, repeated administration of toxic levels of an amino acid generally leads to changes in the levels of other amino acids in the plasma (Table IV). In the case of the chick, (Richardson *et al.*, 1953b) or the rat (Girard *et al.*, 1968), quite small excesses of lysine or methionine were sufficient to affect the levels of other free amino acids in the plasma. The cause of toxicity from amino acid excess is uncertain and probably varies for each amino acid given in excess (Chapter 13). In the case of toxicity produced in rats by adding tyrosine to a diet low in threonine, the toxic effects can be partly reversed by increasing the intake of threonine (Alam *et al.*, 1966a). The data show (Alam *et al.*, 1966b) that the level of free tyrosine in plasma and muscle is greatly elevated by feeding excess tyrosine and partly reduced by simultaneous administration of threonine. Once more, the liver showed much smaller changes in free amino acid content when excess tyrosine was fed. Godin (1967) has shown with tyrosine-^{14}C that growing rats given a large intake of tyrosine exhibit an adaptive increase in the rate of tyrosine breakdown to CO_2. Addition of threonine to the diet both lowered the blood tyrosine level and diverted some of the excess tyrosine to protein synthesis. However, it is impossible to decide which is cause and which is effect. It should also be noted that endocrine function is sensitive to free amino acid levels. Large doses of leucine or methionine have been found to increase adrenocortical secretion with consequent changes in liver protein synthesis (Munro and Mukerji, 1962; Munro *et al.*, 1963, 1965). It is therefore of interest to note that Leung *et al.* (1968) were able to partially correct the plasma amino acid changes caused by amino acid imbalance when cortisol was injected into their animals. Other endocrine glands are also sensitive to variations in the levels of the circulating amino acids; as discussed in Section III,B, secretion of insulin, glucagon, and growth hormone can be elicited by infusing amino acids. Consequently, changes in endocrine secretory rates should be considered among the adaptations involved in adjusting to an influx of amino acids. Although studies to demonstrate such effects of amino acid load on endocrine function have usually employed large doses, it does not necessarily mean that this mechanism responds only to toxic levels of amino acids. In the case of the action of methionine and of leucine involving adrenocortical secretion, Munro and Mukerji (1962) were able to obtain a linear dose–response curve, indicating that these amino acids affect adrenocortical function at all levels of administration including the physiological range of intakes (see Fig. 26, Chapter 10, Volume I for illustrative data). This also emerges in studies of the effect of methionine intake on mammary gland function in the lactating rat; lactation is reduced linearly in proportion to methionine intake over a range from 1.5% to 20% of the diet of the rat (Daniel *et al.*, 1968).

Finally, it is convenient in this context to consider the adaptation of free amino acid levels to excessive intakes of protein. Holt and his colleagues (Holt *et al.*, 1968; Snyderman *et al.*, 1968b) fed infants for several weeks on various levels of milk protein from 1 gm/kg through an adequate level of 2 gm/kg up to 9 gm/kg. Although the highest level of protein intake was quite excessive, growth was normal but the concentrations of amino acids in the plasma were mostly elevated, particularly the branched-chain amino acids and methionine. Unlike the majority of the essential amino acids, which are mainly or exclusively degraded in the liver, the transaminase for the branched-chain amino acids is most active in muscle (Mimura *et al.*, 1968), and methionine is transformed to cysteine in a wide variety of tissues. Consequently, it is not surprising that the levels of these amino acids in the systemic plasma should rise to high levels since they have to be transported to the peripheral tissues; the levels of other amino acids are more effectively restrained from rising in the systemic circulation by the action of liver enzymes on amino acids coming to that organ in the portal blood. This selective action of the liver is elegantly documented by the direct studies of Elwyn on amino acid flow to and from the liver following a meal of protein (Chapter 38). The response of plasma amino acids to increasing levels of dietary protein intake has also been examined with rats. Using diets varying from 5 to 75% casein content, Anderson *et al.* (1968) found that plasma amino acid levels were greatly elevated on the first day of feeding the highest protein levels, but thereafter decreased considerably except in the case of the branched-chain amino acids which remained raised. This decrease in the concentrations of the majority of amino acids coincided with an elevation of the level of liver serine dehydratase, a typical enzyme of amino acid degradation. This inverse correlation between liver enzyme induction and free amino acid levels is also seen when starved rats are treated with phlorhizin; this drug causes catabolism of peripheral proteins and in consequence, the levels of serine and threonine in the liver increase for a few hours until liver serine dehydratase activity rises, after which the concentrations of these free amino acids decline again (Bojanowska and Williamson, 1968).

4. The Influence of Energy-Yielding Nutrients

The influence of the energy content of the diet on free amino acid levels in the blood and tissues has not received much attention although, as discussed earlier, transport of amino acids across membranes can involve expenditure of energy. Munro *et al.* (1962) examined the effect of the energy content of the preceding diet on the uptake of glycine-^{14}C by rat liver slices. They came to the conclusion that the rate of transfer of glycine into the liver cell is influenced by the energy intake of the animal.

The first question to be answered is the effect on free amino acid pools of a period of fasting or starvation, which, of course, involves withdrawal of all nutrients. One would anticipate that the free amino acids of the body would be severely depleted by fasting and thereby the rate of protein synthesis would be retarded. In fact, gross reductions in the concentrations of free essential amino acids have not been recorded. In the rat, the plasma levels of most essential amino acids actually increase during the first 9–12 hours of fasting and then decline again to about their initial level (Henderson *et al.*, 1949; Harker *et al.*, 1968). This could represent breakdown of labile tissue proteins deposited during the absorptive period after the preceding meal. A 24-hour fast has also been found to cause only minor fluctuations in the plasma amino acid levels of the pig (Richardson *et al.,* 1965) though several plasma amino acids in the fasting chick rise more quickly (Shao and Hill, 1967; Zimmerman and Scott, 1967). Boomgaardt and McDonald (1969) confirm this species difference. As shown in Table IV, the changes produced by prolonged fasting (up to 14 days) on plasma concentrations of free amino acids have been recorded for rats (Henderson *et al.,* 1949; Wu, 1954), rabbits (Block and Hubbard, 1962), steers (Brown *et al.,* 1961), lambs (Leibholz and Cook, 1967), and man (Charkey *et al.,* 1955; Swendseid *et al.,* 1961). Except for the lambs, all species showed similar patterns for the essential amino acids. The plasma levels of leucine, isoleucine, valine, and cystine increased during the fast, and that of threonine generally decreased, whereas the levels of the remaining essential amino acids were not consistently influenced (Table IV). It may be observed that increased levels of the branched-chain amino acids have also been noted with human subjects receiving diets of high fat content (Swendseid *et al.,* 1967b), thus suggesting that raised levels of branched-chain amino acids in starvation may be associated with the increased fat metabolism of inanition. With longer periods of fasting, the levels of the branched-chain amino acids decrease again (Tuttle *et al.,* 1962; Adibi, 1968; Swendseid *et al.,* 1969; Felig *et al.,* 1969a).

Among the nonessential series, the scanty data (Table IV) indicate a general reduction in concentration during fasting. Consequently, the E/N ratio increases. This tendency for the E/N ratio to rise during chronic starvation has been confirmed on human subjects (Drenick *et al.,* 1964). Thus, the E/N ratio is affected by starvation in the opposite direction to that characteristic of protein deficiency without caloric insufficiency. This difference is probably related to the observation that gluconeogenesis and the enzymes involved in amino acid catabolism are considerably stimulated by starvation, whereas a protein-free (high carbohydrate) diet has the opposite effect. The use of the E/N ratio to distinguish clinical kwashiorkor (protein deficiency) from marasmus (energy deficiency) is discussed in Chapter 28. As discussed later (Section III,B), lack of insulin secretion may contribute to the increased levels of the branched-chain amino acids during starvation. Since prolonged starvation causes a much more extensive fall in alanine than in the levels of other plasma nonessential amino acids, Felig *et al.* (1969a) suggest that this amino acid is the principal carrier of amino N to the liver, a conclusion in agreement with the large amount of alanine constantly coming from muscle (London *et al.,* 1965) and from the intestine after a meal of protein (Chapter 38).

Much less work has been done on the reaction of plasma-free amino acids to experimental caloric undernutrition as opposed to total starvation. So far as is known, the trends are similar to those of total food withdrawal. A study of the effects of restricted food intake on the plasma amino acids of the guinea pig (Schønheyder and Lyngbye, 1962) showed that all essential amino acids increased in concentration, whereas the majority of the nonessential series diminished. Interestingly, the effects of increments in energy intake *above* maintenance caused by infusion of propionic acid have been studied in the lactating cow (Halfpenny *et al.,* 1969). This caused a fall in the plasma essential amino acid levels, probably because milk secretion was stimulated by the increased energy intake, and in consequence more amino acids were removed from the circulation.

Do plasma concentrations during fasting reflect similar changes in the tissues? Wu (1954) found in rats starved for several days that the levels of free essential and nonessential amino acids in liver and muscle followed the general plasma patterns, namely an increase in branched-chain essential amino acids, methionine and cystine, with a reduction in the levels of most others. Thompson *et al.,* (1950) used microbiological assays to examine the levels of all free essential amino acids and proline in the liver, muscle, spleen, and brain of rats fasted for 7 days. Proline, the only true nonessential amino acid measured, underwent some reduction in all tissues. Lysine levels fell in muscle, spleen, and brain; histidine in muscle and brain, phenylalanine and valine in brain; and threonine in spleen. The levels of other essential amino acids remained unchanged or even increased. The liver showed no significant reduction in the concentrations of any essential amino acid. Although these two sets of studies on the tissue levels during fasting are not in complete agreement, and, furthermore, the methods of assay used are not precise, they nevertheless do show that no single free amino acid undergoes such an extensive reduction during fasting that it might lead to severe restraint on tissue protein synthesis. As was seen in the previous section, lack of one essential amino acid in the diet can cause very extensive reductions in the plasma levels of that amino acid, and lack of all amino acids (protein-free diet) induces reductions in the levels of most essential amino acids (Table IV). No changes of similar magnitude have been noted in the free essential amino acid pools of fasting animals, although the fasting animal loses more body protein than the animal on a protein-free diet (see Chapters 10 and 11). The much more severe reduction in plasma and tissue levels of essential amino acids on a protein-free diet than during starvation is strikingly illustrated by the studies of Henderson *et al.* (1949) and Thompson *et al.* (1950). Consequently, starvation must affect protein metabolism through mechanisms additional to lack of dietary protein. It is likely that the absence of all nutrients during fasting affects protein metabolism through a complex of factors which include lack of energy-yielding nutrients and changes in hormonal secretory rates, such as the well-known hyperactivity of the adrenal cortex during fasting, and reduced insulin secretion.

5. Specific Action of Dietary Carbohydrate

For a few hours after a meal consisting of carbohydrate has been eaten, there is a reduction in the levels of free amino acids in the plasma (see Chapter 10, Section III,B for a review of the literature). This action is specific to carbohydrates, since the giving of fat does not have this effect. This phenomenon has been observed to occur not only in monogastric mammals, but also in ruminants (Purser *et al.*, 1966). In extending these ruminant experiments, Potter *et al.* (1968) showed that administration of butyrate or acetate caused only minor changes in the plasma-free amino acid levels of sheep, whereas there was a substantial reduction after infusing propionate, presumably because the latter is a carbohydrate precursor. Reverting to the nonruminant, Munro and Thomson (1953) found that all the essential amino acids measured by them underwent reductions in level after glucose administration (Table IV). Recent studies by ion-exchange chromatography (Zinneman *et al.*, 1966) have extended the list to cover 22 free amino acids which can be separated from human plasma; as shown in Table IV, most of these underwent a reduction in concentration after glucose administration, though not all the nonessential amino acids show this fall (Swendseid *et al.*, 1967a). Munro and Thomson (1953) examined the pattern of removal of several essential amino acids from the plasma and concluded that the amounts removed are in the same proportions as they occur in mixed muscle protein. Evidence was later obtained using labeled amino acids that skeletal muscle is the major tissue in which the amino acids are deposited after a dose of glucose is administered to rats (Munro *et al.*, 1959). The mechanism responsible for removal of the amino acids appears to involve insulin secretion (see Chapter 10; Section III,B). Knipfel *et al.* (1969) confirm preferential deposition of labeled amino acids in muscle after giving glucose to rats.

This phenomenon is of some nutritional interest, since it is usual to include carbohydrate in most meals, and the action of the carbohydrate will influence utilization of the amino acids simultaneously absorbed from the protein of the same meal. Indeed, there is evidence that the N balance of rats and man is transiently improved when the carbohydrate of the diet is consumed along with dietary protein, as compared with

N balance during separate consumption of these nutrients (Chapter 10, Section III,B). This presumably occurs because a proportion of the incoming amino acids after each mixed meal are deposited in muscle as a result of the stimulus provided by the carbohydrate absorbed from the same meal. Rabinowitz *et al.* (1966) showed by direct assay that insulin secretion is much more enhanced by a meal containing carbohydrate and protein than by giving carbohydrate or protein singly.

6. Effects of Vitamin and Mineral Deficiencies on Amino Acid Pools

Plasma and tissue free amino acid concentrations have been recorded in a number of vitamin deficiency states (Table IV). Malathi *et al.* (1961) found by paper chromatography that the free phenylalanine and tyrosine concentrations were increased and that threonine concentration was decreased in the plasma, liver, and kidneys of vitamin A-deficient rats as compared with pair-fed animals. Also using paper chromatography, Rao (1961) observed that administration of vitamin D to rachitic rats rapidly decreased the concentrations of some free amino acids in tibial cartilage, whereas the levels of others rose. The relationship of these changes to bone protein formation is not known. Vitamin E deficiency in rats had no effect on the plasma levels of arginine, methionine, or tyrosine (Harrill and Gifford, 1966), but vitamin E-deficient rabbits showed increased transport of a model amino acid (AIBA) and of glycine into the dystrophic muscle, in which the rate of protein synthesis was also accelerated (Diehl, 1963, 1965, 1966). The relationship between increased transport and increased protein synthesis is not clear; the appearance of new cell types in the dystrophic muscle may be responsible for both phenomena (Nichoalds *et al.*, 1969).

Several authors have examined the free amino acid concentrations in the blood, muscle, and liver of scorbutic guinea pigs (Christensen and Lynch, 1948; Ginter, 1957; Rangneker and Dugal, 1958; Schønheyder and Lyngbye, 1962). The most extensive studies are those of Schønheyder and Lyngbye, who gave their animals either a scorbutic diet, the same diet fed ad libitum with ascorbic acid added, or the diet supplemented with ascorbic acid but fed in restricted amounts to give the same weight loss as for the scorbutic group. Lack of vitamin C caused significant increases in the concentrations of phenylalanine in the plasma, and of phenylalanine, tyrosine, leucine, and aspartic acid in muscle when compared with the partially starved group receiving adequate amounts of ascorbic acid.

Some years ago, Richardson *et al.* (1953a) reported changes in plasma free amino acid concentrations in chicks deficient in the B complex vitamins. More recently, Swendseid *et al.* (1964a) examined the free amino acid levels in the plasma, liver, and muscle of rats deficient in vitamin B_6. In all tissues, the levels of essential amino acids declined, whereas the concentrations of nonessential amino acids either remained unchanged or increased. Free glycine level rose, whereas serine fell; the glycine level was further increased by adding glycine to the diet of the deficient group, presumably because of reduced transamination through lack of pyrodoxal phosphate. Patients with pernicious anemia undergo a rise in plasma methionine and a fall in valine when treatment with vitamin B_{12} or folic acid is instituted (Parry, 1969). These are known cofactors in methionine methyl transfer and valine degradation via methylmalonate.

A few observations link the intake of potassium with the tissue levels of the basic amino acids. The levels of free lysine, arginine and histidine are higher in the skeletal muscles of rats made deficient in potassium (Iacobellis *et al.*, 1956; Brandt *et al.*, 1960; Leibholz *et al.*, 1966). Under similar conditions, potassium-deficient dogs do not show this response (Iacobellis *et al.*, 1957). Interactions between lysine, arginine, and potassium have been noted in the nutrition of the growing chick (O'Dell and Savage, 1966). Riley *et al.* (1969) record the changes in plasma amino acid concentrations when the intake of zinc, protein, and vitamin D are varied.

B. Hormones and Free Amino Acid Pools

The molecular sites of action of hormones on protein metabolism have been considered in Chapter 33. Here we must attempt to view the actions of hormones as factors changing the amounts and flow of free amino acids. Hormonal action could influence free amino acids through (a) rate of transfer between the extracellular and intracellular pools, (b) rate of utilization for protein synthesis, transamination and other means of disposal, (c) rate of degradation of protein with release of amino acids into the intracellular pool, and (d) rate of synthesis of the amino acid from precursors, if it is a nonessential one. The balance between these reactions will decide the action of the hormone on the levels of free amino acids in tissues susceptible to its action.

The interaction of factors is well illustrated by the effect of insulin on the free amino acid levels in skeletal and cardiac muscle. As discussed in detail in Chapter 10, Section III,B, administration of insulin causes a rapid reduction in the plasma levels of all amino acids, and there is much evidence to suggest that the major site at which they are deposited is muscle. However, study of free amino acids within the muscle cell shows that insulin increases the accumulation of only 6, whereas the levels of 13 others remain unchanged (Wool, 1964). On the basis of further investigation (Castles and Wool, 1964; Scharff and Wool, 1965; Wool and Scharff, 1968), it has been suggested that this latter group fail to accumulate because protein synthesis removes them too rapidly. Wool's evidence for this was that, when sufficient puromycin was administered to suppress protein synthesis, the free intramuscular level of most amino acids now increased after insulin treatment. The validity of the evidence obtained with the use of puromycin is criticized in Chapter 33, Section V,A. Even if we accept that the passage of amino acids in general into muscle may be facilitated by insulin, there is evidence to show that the stimulus of insulin on muscle protein synthesis is not caused by the increased availability of the amino acids in the tissue (Manchester and Krahl, 1959; Wool and Krahl, 1959, 1964; Stirewalt and Wool, 1966; Hay and Waterlow, 1967; Wool and Scharff, 1968) although Goldstein and Reddy (1968) disagree. We shall later see that thyroxine causes stimulation of a selected group of amino acids into cartilage, without any action on the rate of protein synthesis. The general nature of an effect on amino acid transport may thus be important in determining whether it is an adequate stimulus to protein synthesis in the target organ.

The extent and mechanism of action of insulin on amino acid metabolism in the liver remains uncertain and at best is more sluggish than muscle. In experiments on perfused rat liver, Mondon and Mortimore (1967) observed that addition of insulin reduced the rate at which amino acids accumulated in the perfusate. This was not due to increased urea formation or augmented synthesis of liver protein; the action of insulin was therefore attributed to inhibition of protein degradation in the liver, so that fewer amino acids were released. As pointed out in Chapter 31, Section V,A, insulin induces some liver enzymes but suppresses others, so that the overall picture is more complex than a simple statement of decreased protein degradation would imply. In the whole animal, Sanders and Riggs (1967) have demonstrated increased penetration of model amino acids into the liver following insulin administration; however, this seems to be secondary to release of other hormones consequent on the hypoglycemic action of insulin. Even if the liver cell can accumulate amino acids better in the presence of insulin, the small contribution of the liver to the total free amino acid pool of the body (Table I) makes it quantitatively much less important than muscle as a site of amino acid deposition. Indeed, there is evidence (Chapter 10, Section III,B, Fig. 24) to suggest that the reduction in plasma amino acid levels caused by insulin-directed deposition in muscle may impede the supply of amino acids to other tissues such as liver.

The action of adrenocortical hormones on free amino acid levels in the tissues is inherently likely to be more complex, since the known actions of these hormones on protein metabolism vary from tissue to tissue. Thus there is a loss of protein from the carcass, whereas the liver and other viscera gain protein (for literature, see Chapter 10, Section IV). It was therefore suggested in Chapter 10 that there may be a flow of amino acids from the carcass to the viscera following administration of these steroid hormones. There is general agreement that administration of cortisol or cortisone results in an increase in the levels of most free amino acids in muscle (Kaplan and Shimizu, 1963; Ryan and Carver, 1963; Betheil *et al.*, 1965). This could occur because of an increased uptake of amino acids into muscle from the extracellular fluid, a decrease in the rate of muscle protein synthesis, or an increase in the rate of muscle protein breakdown. It has been demonstrated both by *in vivo* studies (Wool, 1960; Eichhorn *et al.*, 1961) and by *in vitro* experiments (Kostyo and Schmidt, 1963; Kostyo, 1965; Kostyo and Redmond, 1966) that adrenal steroid hormones decrease the capacity of muscle to concentrate free amino acids, including nonmetabolizable amino acids. The accumulation of free amino acids in corticosteroid-treated muscle appears to be due to inhibition of muscle protein synthesis (Wool and Weinshelbaum, 1960), which probably is an action independent of the reduced amino acid penetration (Kostyo and Redmond, 1966). It may be noted that the diminished muscle protein synthesis occurs in spite of the higher intracellular free amino acid levels, indicating that factors other than amino acid supply regulate the rate of protein formation in muscle.

The action of corticosteroid hormones on liver free amino acid levels is complex. As suggested above, one might anticipate that the increased flow of amino acids from the carcass would act like a meal of protein and flood the liver with amino acids. Indeed, Noall *et al.* (1957) found that transport of the model amino acid AIBA into liver was markedly increased a few hours after hydrocortisone administration, and Weber *et al.* (1965) found that injection of corticosteroid hormones into fed rats caused a rapid increase in the level of free amino nitrogen in the liver. On the other hand, Betheil *et al.* (1965) and Enwonwu and Munro (1969) injected hydrocortisone into fasting rats and observed elevations in the liver concentrations of only three amino acids, namely glutamic acid, aspartic acid, and alanine. The difference in response appears to be related to the nutritional state of the animal; Kaplan and Shimizu (1962) found that cortisol caused a general reduction in liver free amino acid levels in fasting adrenalectomized mice, but observed an increase in free amino acid concentration when cortisol was given to fed mice. It is thus important to specify the nutritional conditions under which free amino acid concentrations are studied. Goodlad and Munro (1959) noted that the increase in liver protein content caused by cortisone administration is dependent on an adequate energy intake from the diet, thus reinforcing this argument.

The action of thyroxine on protein metabolism bears some resemblance to that of corticosteroids. As discussed in Chapter 10, Section IV,B, production of hyperthyroidism in rats causes a loss of carcass protein which is often accompanied for a period by an increase in the protein content of the liver. This could be the result of a flow of amino acids from carcass to viscera, and indeed Foley *et al.* (1966) report an increased arteriovenous difference in the amino acids coming from the forearm of human subjects undergoing thyroid therapy. Unlike the action of cortisone on muscle, there is neither diminished transport of amino acids into muscle (Tonoue and Yamamoto, 1967) nor decreased muscle protein synthesis (Brown, 1966) to account for the catabolic action of thyroid hormone. Presumably tissue protein breakdown is accelerated. Wellers and Leblanc (1966) have measured the pool sizes of free essential amino acids in several tissues of rats receiving large doses of thyroxine. In the liver, the levels of all amino acids were considerably increased, whereas in the carcass only the concentrations of histidine, lysine, and tryptophan were significantly greater than in the control animals. The plasma showed elevated levels of these three amino acids and also of isoleucine, leucine, valine, and cystine. The picture is thus the converse of that observed after corticosteroid administration, which increases mainly the free amino acid pools of the carcass rather than the viscera. In thyroidectomized rats, the pools of all free amino acids are diminished in the body as a whole, but in the case of the plasma there are significant elevations in the levels of histidine, tryptophan, and valine (Wellers and Leblanc, 1968). However, in chicks treated with thiouracil, the levels of several plasma amino acids were found to fall (Shao and Hill, 1968, 1969).

The difference in action of the two catabolic hormones cortisone and thyroxine on free amino pools is partly explicable because of their different actions on enzymes that degrade amino acids. The importance of rate of amino acid removal is illustrated by the differing responses of free tyrosine levels to corticosteroids and to thyroxine administration. Administration of corticosteroids reduces plasma tyrosine level, probably owing to the considerable increase in tyrosine transaminase activity in the liver caused by these hormones (Betheil *et al.*, 1965). On the other hand, plasma tyrosine levels are raised after thyroxine administration because tyrosine removal by liver transaminase is not accelerated by this hormone whereas the outflow of tyrosine from the carcass *is* increased (Rivlin, 1963; Rivlin and Levine, 1963; Rivlin and Melmon, 1965). These studies on free amino acid pool sizes after administration of insulin, corticosteroids, and thyroid hormones demonstrate that the level of a given free amino acid is the resultant of a balance between input and outgo, either or both of which can change as a result of hormonal action.

The action of thyroid hormones on amino acid transport and on protein synthesis can be dissociated. The action of thyroid hormones on amino acid uptake by embryonic chick cartilage (Adamson and Ingbar, 1967) is limited to amino acids accumulated in the intracellular pool by the L-transport mechanism, and does not stimulate protein synthesis. In contrast, serum added to cartilage stimulates uptake of *all* amino acids and also protein synthesis. Yang and Sanadi (1969) note that the injection into thyroidectomized rats of thyroxine changed the pattern of loading of isoaccepting tRNA species within 30 minutes.

There is a good deal of evidence that other hormones also change the metabolism of free amino acids in their target organs. Noall *et al.* (1957) found that growth hormone increased penetration of AIBA into all tissues except cardiac muscle. This does not necessarily mean that the stimulant action of this hormone on body protein synthesis is due to the greater uptake of free amino acids by the tissues; in fact, the converse may be true, uptake being increased due to enhanced protein synthesis (Kostyo, 1968). Kostyo (1964) found that incorporation of glycine by diaphragm muscle protein is still enhanced by growth hormone, even when the action of this hormone on glycine transport into the muscle is inhibited by replacing Na^+ in the incubation by choline. Noall *et al.* (1957) observed that injection of estradiol specifically increases AIBA uptake into the uterus after a 24-hour time interval. Later, Noall and Allen (1961) and Riggs *et al.* (1968) obtained an increased amino acid uptake as early as 30 to 60 minutes after estrogen injection. However, at similar time intervals (30–40 minutes) after estradiol administration, Means and Hamilton (1966a) noted a depression of methionine uptake into the uterus although RNA synthesis had already increased long before this time (Means and Hamilton, 1966b); this implies that increased penetration of amino acids into the uterus is not an early event and is not necessary for initiation of hormonal action. The rate of uptake of AIBA and other amino acids into the parathyroid gland varies inversely with the calcium level in the blood (Raisz and O'Brien, 1963; Raisz, 1967). This links blood calcium level to amino acid metabolism in a gland which synthesizes parathormone, a polypeptide hormone regulating calcium metabolism. However, we have no evidence indicating that secretion of the hormone from the parathyroid gland is determined by the rate of passage of free amino acids into the gland, or whether they are independently affected by blood calcium concentration. Finally, it has been claimed (Dubons and Pitman, 1962) that thyrotropin increases the rate of passage of AIBA into slices of thyroid gland, but Raghupathy *et al.* (1963) failed to confirm this observation. Similarly Bradley and Wissig (1966) failed to show a greater uptake of leucine-^{3}H into thyroid tissue as a result of thyrotropine administration, although a stimulus to protein synthesis in the gland was evident.

We can now consider the relationship between these reported actions on free amino acid concentrations and concomitant hormonal effects of these hormones on tissue protein metabolism. The accumulation of free amino acids in muscle under the action of insulin does not appear to be directly related to the stimulant action of insulin on muscle protein synthesis. Thyroid hormones stimulate uptake of only some amino acids into chick cartilage, and fail to promote protein synthesis in that target organ. Although growth hormone is credited with increasing passage of free to protein synthesis in these tissues. In the case of estrogens and of amino acids into many organs, we do not know how this action is related thyrotropin, it seems that the mechanism of protein synthesis in the target organs can undergo stimulation prior to or even without an increase in uptake of free amino acids. Accordingly, there is no evidence to support the suggestion (Noall *et al.*, 1957) that increased passage of free amino acids is the trigger mechanism through which these hormones act on their target tissues. At present, it would seem that any increase in flow of amino acids is likely to be a concomitant but not a causal action. This conclusion still leaves us with the status of the liver following administrations of corticosteroids, thyroid hormones, and other catabolic agents that cause a loss of protein from the carcass and a gain of visceral protein. Is the latter entirely due to direct action of these hormones on liver protein synthesis, or does the flow of amino acids from the carcass play even a small part? In the case of the perfused liver, addition of cortisol to the perfusate is known to elevate uptake of the model amino acid AIBA (Chambers *et al.*, 1965) and to stimulate synthesis of tyrosine transaminase (Barnabei and Sereni, 1964). These direct actions on the isolated surviving liver eliminate flow of amino acids from the peripheral tissues as the major mechanism of action of the steroid hormones on hepatic enzyme synthesis. This is compatible with the extensive evidence (e.g., Greengard and Acs, 1962) that corticosteroids increase enzyme levels in the liver by stimulation of liver RNA synthesis, which can be blocked by actinomycin D. Nevertheless, Enwonwu and Munro (1968) have found that administration of hydrocortisone to rats receiving large doses of actinomycin D can still cause small but reproducible changes in the polysome pattern within the liver cell. This suggests that corticosteroids have an action on the protein-synthesizing apparatus of the liver other than through increased formation of RNA. Since free tryptophan levels have not been assayed in any of these studies of corticosteroid action, it is not possible to exclude a change in the concentration of this amino acid as a contributory cause of liver polysome aggregation, just as tryptophan appears to be the dominant factor in causing polysome aggregation in the liver following a meal of protein. Recently, Clemens and Korner (1969) have observed that a stimulant action of growth hormone on protein synthesis by liver slices could only be obtained with high levels of amino acids in the medium, thus suggesting a synergistic action. Such high levels of amino acids occur in the portal vein after meals, so that Clemens and Korner's observation may have a physiological basis deserving further exploration.

Finally, it should not be forgotten that amino acid levels in the plasma can influence secretion of hormones by endocrine glands. In susceptible subjects, an oral dose of leucine causes hypoglycemia due to release of insulin (Cochrane *et al.*, 1956), and in normal individuals this response can be evoked by intravenous administration of any one of several amino acids (Floyd *et al.*, 1966). The last-named authors consider that the secretion of insulin after a meal of protein is due to the rise in blood amino acid level. In premature infants, infusion of glucose fails to cause insulin release, whereas infused amino acids show this releasing effect (Grasso *et al.*, 1968). Studies on the isolated pancreas *in vitro* (Edgar *et al.*, 1969; Milner, 1969) show that glucose, a primary inducer of *in vitro* insulin secretion, must also be present in the medium in order to obtain insulin release on adding amino acids, of which arginine, leucine, lysine and perhaps histidine are the only effective ones. It is suggested that these amino acids cause insulin release by first stimulating glucagon secretion (Milner, 1969); glucagon is known to lower plasma amino acid levels (Landau and Lugibihl, 1969). It is also suggested that leucine can act directly on the β-cells of the pancreas without requiring a preliminary stimulus from glucose (Milner, 1969). Intravenous administration of large amounts of arginine is a potent means of increasing the plasma levels of growth hormone (Knopf *et al.*, 1965; Parker *et al.*, 1968) and insulin (Cremer *et al.*, 1968) although Best *et al.* (1968) believe that spontaneous variations in blood levels of growth hormone have been confused with amino acid stimulation. Oral administration of 1-gram quantities of leucine or methionine to rats has been found to stimulate secretion of corticosterone by the adrenal cortex (Munro and Mukerji, 1962; Munro *et al.*, 1965). Although the amounts of amino acid used to provide endocrine stimulation in these various studies were mostly excessive, it is possible that similar although less extreme responses of the glands may occur from intakes of these amino acids that fall more nearly within the physiological range (see Chapter 10; also A. J. Clark *et al.*, 1968).

These studies suggest that the levels of certain amino acids in the plasma may act as regulators for the secretion of insulin, glucagon, growth hormone, and corticosteroids. This area thus provides interesting possibilities for revealing hitherto unsuspected feedback control mechanisms in which amino acid levels are involved. For example, Felig *et al.* (1969b) have examined the interrelationships between plasma insulin content and blood amino acid levels in a series of obese and nonobese subjects. The two groups of subjects did not differ in mean fasting blood sugar concentration, but the fasting obese cases had raised plasma levels of insulin, and also significant elevations in the plasma concentrations of the branched-chain amino acids, tyrosine, and phenylalanine. After giving glucose to the obese subjects, there was an excessive rise in plasma insulin level but this failed to reduce the amino acid concentrations to the levels found in nonobese subjects given glucose. These investigators conclude that insulin resistance associated with obesity results in a tendency for both glucose and certain amino acids to accumulate in the blood because they are not taken up by muscle. This will apply particularly to the branched-chain amino acids, since they are mainly degraded in muscle. This hyperaminoacidemia stimulates secretion of insulin which is sufficient in nondiabetic obese subjects to ensure a normal fasting blood sugar level but is inadequate to normalize the concentration of blood amino acids. Consequently, the continuing high levels of amino acids may provide the feedback signal to the islet tissue cells that insulin is not effective and thus could be responsible for the persistant secretion of high levels of insulin in these obese patients. There is some circumstantial evidence to support this interpretation. As discussed elsewhere in this chapter, subjects who are starving or are consuming diets rich in fat show elevated plasma levels of these amino acids and not of others; this may occur because of insufficient secretion of insulin in the absence of dietary carbohydrate to ensure the passage of amino acids into muscle. After a meal containing protein, the three branched-chain amino acids often show the largest increments in the peripheral blood, which may have relevance for the increase in blood insulin levels found after meals of protein. Conversely, the blood levels of leucine, isoleucine, and valine are especially depressed in cases of protein malnutrition, a condition which is said to be associated with impaired secretion of insulin and thus of glucose tolerance

(Chapter 21, Volume II). In normal subjects, free amino acids probably regulate hormone secretion in response to changes in the flow of amino acids into the plasma. One can hypothesize that an influx of amino acids, such as a meal provides, causes a rise in blood amino acid level sufficient to produce a sharp rise in insulin secretion, especially if the meal also raises blood glucose level. In consequence of this additional secretion of insulin, the excess of amino acids is rapidly transferred to muscle and the signal ceases. The hyperaminoacidemia after meals probably also results in increased activity of adrenal cortex which after a lapse of time causes enhanced activity of the liver enzymes degrading amino acids and thus provides for the eventual disposal of those amino acids that fail to be used for synthetic purposes.

C. Miscellaneous Factors Affecting Free Amino Acid Pools

1. Physiological Factors

a. Sex. Two groups of authors (Wheeler and Morgan, 1958; Morrison *et al.*, 1961a) found higher levels of several free amino acids in the blood of female than of male rats, whereas glycine has been shown to be lower (Christensen and Streicher, 1949). Among adult women, a lower level of amino acids generally was observed by Lacy and Crofford (1964) but not by Westall (1962); no difference was found between the sexes at birth (Berry and Leonard, 1966). Nyhan *et al.* (1968) find more free histidine in the plasma and muscle of girls than of boys. Soupart (1960) has recorded a significant lowering of plasma concentrations of lysine, threonine, alanine, serine, and proline at the change from the follicular to the luteal phase of the menstrual cycle of women. The plasma amino acid levels in pregnant women tend to be lower (Lindblad and Baldesten, 1967), whereas the red cell levels rise (Björnesjö, 1968). Churchill *et al.* (1969) claim that pregnant mothers with a low plasma amino N have children that are less well developed, notably in cranial capacity. Certain amino acids show a reduced plasma concentration with the onset of lactation in cows (Halfpenny and Rook, 1968), presumably due to uptake by the mammary gland. Serial studies on the plasma of several subjects over long periods suggest that there are no genetically determined differences between individuals (Zimmermann-Telschow and Jekat, 1965).

b. Age. Changes in plasma amino acid levels have been recorded as age advances. The pattern in early postnatal life is reviewed in Chapter 26 (Volume III). Comparison in several species of fetal blood with maternal blood has shown that the amino acid concentrations are higher in the fetus (Christensen and Streicher, 1948; Foley *et al.*, 1967). It is thus not surprising to find that the plasma levels of a number of free amino acids are higher in the human infant at birth than the adult human subject (Schreier and Plückthun, 1950; Crumpler *et al.*, 1950; Schreier, 1962; Dickinson *et al.*, 1965; Lindblad and Baldesten, 1967; Björnesjö, 1968) though the published evidence does not agree in detail for all amino acids (see Kerr, 1968 for literature survey). These higher levels reflect the high fetal concentrations and do not persist beyond the third postnatal month (Schreier, 1962). The plasma levels of most amino acids are generally slightly lower in the child than in the adult (Schreier, 1962; Scriver and Davies, 1965). At puberty in boys, Zachmann *et al.* (1966) noted an increase in the levels of isoleucine, leucine, valine, methionine and phenylalanine which could be related to androgen output. As regards old age, Wehr and Lewis (1966) consider that 12 of 18 plasma amino acids tended to be elevated in the elderly, but other investigators of geriatric human subjects found a significant lowering of most amino acids (Ackerman and Kheim, 1964a) or limited to the branched-chain amino acids with an increase in the nonessential amino acids (Theimer, 1964). The former group (Ackermann and Kheim, 1964b) were able to increase the plasma levels of the branched-chain amino acids and phenylalanine by giving testosterone to geriatric patients, thus confirming the observations of Zachmann noted above. It would be necessary to demonstrate that these variations in amino acid levels with age were not due to dietary habits, and especially protein intakes, of the groups studied (see Introduction to Part II, Volume II, showing a fall in the protein intake of older subjects). Lack of control of such factors makes evidence on man subject to doubts about its value and probably explains its contradictory nature.

As pointed out earlier in this chapter, the plasma, although easily accessible, contains only a small proportion of the total free amino acid pool of the body. It is known from paper chromatography that important changes in tissue amino acid concentration occur in many tissues during the early stages of postnatal development (Roberts and Simonsen, 1962). Quantitative studies confirm this. Thus, the fetal liver is unable to concentrate nonmetabolizable model amino acids, but within 24 hours of birth this capacity makes its appearance (Feldman and Christensen, 1962). This change coincides with the appearance of several adaptive enzymes in the liver, but we are left with the question unanswered: Does the increased inflow of free amino acids into the liver cell cause the development of these enzymes, or result from their synthesis, or are the two events unrelated? In skeletal muscle on the other hand, concentrative capacity for free amino acids is greater in the fetus than in the mother (Christensen and Streicher, 1948). In the course of growth and development, the patterns of amino acid concentrations change in various tissues. The relationship of the changes to protein synthesis is described in Chapter 26. In the case of brain, several groups of investigators (Berl and Purpura, 1963; Agrawal *et al.,* 1966, 1967, 1968; Oja and Piha, 1966) are in fair agreement in showing a marked increase in the concentrations of glutamic and aspartic acids, glutamine and γ-aminobutyric acid up to 20–30 days of age, accompanied by a reduction in the levels of proline, valine, isoleucine, leucine, tyrosine, and phenylalanine. The free amino acids of human plasma and muscle have also been studied at various ages by Zachmann *et al.* (1966). These authors found that at puberty, males show significant increases in the plasma levels of isoleucine, leucine, methionine, phenylalanine, and valine; the magnitude of the increase could be correlated with their urinary output of androgens. However, muscle failed to show corresponding changes in the concentrations of these free amino acids except in the case of phenylalanine. In view of the fact that muscle is a major target for androgens, this lack of correlation is unexpected, unless the free amino acids are as rapidly removed as they penetrate the muscle cell. In old age, the capacity of rat muscle to take up AIBA is reduced, whereas the liver concentrates it more extensively (Kipnis *et al.,* 1959).

c. Cold Stress and Exercise. Williams *et al.* (1950) have reported the acute effects of a few hours of chilling or of exercise on the free amino acid levels in rat plasma, liver, muscle, and brain. The animals were fasted throughout the experiment. The effects of both stressing agents were in the same direction, but exposure to cold caused less pronounced changes. Both agents decreased the plasma levels of some amino acids and raised the levels of others. Most free amino acids in the liver tended to increase, but the levels of only a few muscle amino acids were elevated. This is not simply the result of stress activation of adrenocortical secretion, for it is known that administration of corticosteroids to the fasting rat raises the levels of most amino acids in muscle but lowers the levels in the liver (Kaplan and Shimizu, 1963; Betheil *et al.,* 1965). As pointed out earlier, stress has actions on protein metabolism other than through the adrenal cortex (Leon *et al.,* 1965). It has also been found that chilling causes increased excretion in the urine of the nonmetabolizable model amino acid AIBA (Bavetta and Nimni, 1964).

The effects of prolonged exercise (training) on plasma amino acid levels has been investigated with human subjects (Zimmermann-Telschow, 1965; Zimmermann-Telschow and Jekat, 1965). When the subjects were receiving a diet in which sulfur amino acids were limiting, training caused a fall in the plasma levels of methionine and cystine. When lysine was the limiting amino acid in the diet, this amino acid underwent reduction relative to the levels of other plasma amino acids during the period of muscle work. These studies appear to show that the anabolic stimulus of muscle training has an effect similar to feeding a meal of protein: withdrawal of free amino acids for muscle protein synthesis causes a depression in the plasma level of the essential amino acid present in least abundance in the diet (compare Table V).

Finally, the stress of high altitude causes a decrease in the levels of several essential amino acids in the plasma, but an increase in glutamic acid concentration (Whitten, *et al.*, 1968).

d. Biological Rhythms. The occurrence of cyclical changes in free amino acid levels in plasma has been noted by several investigators. With rats Rapoport *et al.* (1966) observed diurnal fluctuations in the plasma levels of free tryptophan and corticosterone and of liver tryptophan pyrrolase activity. The objective of these studies was to determine whether an increase in plasma corticosterone content might have been responsible for mobilizing tryptophan from the tissues and thus inducing more of the catabolic enzyme in the liver. However, the cyclical changes observed did not substantiate a simple relationship of this kind. In growing chicks, Squibb (1966) found diurnal rhythms in the free amino acids not only of the plasma, but also in liver and muscle. His diagrams show a lack of correlation between the cycles in plasma and tissues, particularly the liver. This finding demonstrates that correlations between the rhythms in plasma free amino acid levels and the activities of tissue degradative enzymes are unlikely to emerge. Diurnal variations in free amino acid levels in the chick are apparently not due to food intake, since they are still seen in birds with severe inanition due to virus infection (Squibb, 1966). This conclusion is supported by studies on sheep (Purser *et al.*, 1966) in which the diurnal rhythm in plasma amino acid levels continued during a 24-hour fast. From these observations, it is possible to conclude (a) that there are regular rhythms in plasma and tissue free amino acid concentrations, (b) that these are not determined solely by food intake, and (c) that the relationship of these variations in amino acid level to adrenocortical function and to tissue enzymes degrading amino acids requires further elucidation. Wurtman and his colleagues have recently carried out extensive investigations of the diurnal rhythms of plasma amino acids in man and have found that all amino acids share in the change, but the greatest amplitudes are shown by tyrosine, phenylalanine, tryptophan, methionine, cystine, leucine, and isoleucine (Wurtman *et al.*, 1968a). The rhythm persists in subjects receiving a diet low in protein and is not related to physical exercise (Wurtman *et al.*, 1967); it can, however, be rephased by administration of cortisone in large doses (Wurtman *et al.*, 1968b). On the other hand, although diurnal rhythms in the tyrosine transaminase content of the liver are also extensive, they appear to be generated by the dietary intake of protein (Wurtman *et al.*, 1968c). In this mechanism, tryptophan seems to play the role of chief mediator, since there is an absence of response of transaminase levels to tryptophan-deficient diets, and tryptophan given alone can induce a response in the liver level of the transaminase. From these observations, it would appear that the rhythms of liver enzymes involved in amino acid degradation are determined chiefly by diet, a conclusion that conforms closely to earlier discussion in this chapter of the considerable role of the liver in degrading amino acids during absorption of each meal. In contrast, the rhythm in plasma free amino acid level is probably mainly determined by a complex series of controls which include the influence of carbohydrate in each meal which causes intermittent deposition of amino acids in muscle, the pattern of amino acids passing through the liver after the meal, and changes in hormonal secretory rate. The question of diurnal rhythms in mammalian protein metabolism is dealt with in detail by Wurtman in Chapter 36.

In the case of hibernating hedgehogs and golden hamsters, Kristoffersson and Broberg (1968a,b) have shown that lysine, ornithine, and the branched-chain amino acids tend to be elevated, whereas other amino acids tend to fall. These changes are, of course, those observed in animals fasting generally. During hibernation of the dormouse, the brain levels of free tyrosine, threonine, aspartic, and glutamic acids decreased, whereas the levels of γ-aminobutyric acid and glutamine increased (Mandel *et al.*, 1966). Arousal from hibernation causes a sharp rise in plasma amino acids to levels above those of the normal animals (Klain and Whitten, 1968).

2. *Pathological Factors*

The free amino acid concentrations in only a few diseases causing general changes in metabolism can be considered here. Disease of the liver might be expected to have predictable effects on the metabolism of free amino acids, and this is borne out by reported changes in free amino acid levels in patients with hepatic cirrhosis. According to Miller (1962) the liver appears to be the almost exclusive site of degradation of arginine, histidine, leucine, threonine, phenylalanine, and tryptophan, whereas leucine, isoleucine, and valine and the nonessential amino acids in general are also metabolized to CO_2 by other tissues. It is thus not surprising that the blood of the cirrhotic patient shows elevated levels of some amino acids, but a tendency for the levels of others, notably the branched-chain series, not to rise or even to fall. For example, Knauff *et al.* (1964, 1966) report increased plasma levels for methionine, phenylalanine, tyrosine, alanine, and proline, the increase being proportional to the severity of the disease, and Schreier (1962) lists elevations in children with liver disease of arginine, lysine, methionine, phenylalanine, tyrosine, and tryptophan among the amino acids he measured. Levine and Conn (1967) report abnormally high plasma tyrosine levels in cirrhotic subjects after a test dose of tyrosine. The rate of disappearance (clearance) of amino acids from the plasma of cirrhotic subjects after a meal of protein has also revealed differences in metabolism (Iob *et al.*, 1966). Clearing of tryptophan was retarded, whereas removal of leucine, isoleucine, valine, and threonine was accelerated. This last was correlated with reduced concentrations of branched-chain amino acid in the muscle of cirrhotics (Iob *et al.*, 1967). The selective interference of cirrhosis with the catabolism of some amino acids and acceleration of others is compatible with the exclusive role of the liver for the degradation of only certain amino acids. However, the cirrhotic subject can show depression as well as increases in free amino acid levels, varying from case to case (Cachin *et al.*, 1952; Iber *et al.*, 1957; Richmond and Girdwood, 1962; Wu *et al.*, 1955; Ning *et al.*, 1967; Krawitt and Clifton, 1968; Zinneman *et al.*, 1969). The picture is thus not consistent from case to case, but in general the level of tyrosine is often raised, and those of leucine, isoleucine, and valine are reduced below the levels observed normally. There are distinctive changes in the plasma amino acid levels of chronic alcoholics, whether sober or after imbibing ethanol (Siegel *et al.*, 1963, 1964).

Other types of liver injury have also been explored. Liver damage induced by carbon tetrachloride is accompanied by a rise in the plasma levels of most free amino acids, but there is a marked reduction in plasma tryptophan level; this latter effect is proportional to the dose of carbon tetrachloride given (Truhaut *et al.*, 1966). It is therefore rather interesting to note that disintegration of liver polysomes also results from treatment with this toxin (Smuckler and Benditt, 1965). Further study would be needed to establish a relationship to the low free tryptophan level. Following partial hepatectomy in the rat, there is an increase in free amino acid concentration in the liver during the period of maximum regeneration (Christensen *et al.*, 1948b). This appears to represent mainly an increase in glutamic and aspartic acids and lysine (Ferrari and Harkness, 1954). Roberts and Simonsen (1962) confirm that there are surprisingly few changes in free amino acid levels in regenerating liver. Lieberman and his colleagues have examined the factors leading to the increase in free lysine content after partial hepatectomy. The increase in lysine level in the remaining portion of the liver occurs soon after partial hepatectomy (Fujioka *et al.*, 1963), and is corrected some time later when the enzymes for lysine degradation increase in activity (Higashino and Lieberman, 1965), indicating that accelerated destruction now accompanies increased uptake. In an interesting inherited anomaly, familial protein intolerance, the subject shows an abnormal increase in plasma free amino acids and ammonia after each meal containing protein (Kekomäki *et al.,* 1967a,b). The failure to metabolize free amino acids is due to insufficiency of urea synthesis, possibly because of inadequate liver ornithine; in turn this seems to be linked to excessive urinary loss of basic amino acids, including arginine.

Changes in tissue amino acid metabolism are to be expected in many pathological states. Injury induces variations in plasma free amino acid levels (see Chapter 19). Thus, Schreier and Karsh (1954) observed that operation caused a reduction a few hours later in the plasma levels of a number of amino acids, except that the value for leucine and isoleucine tended to be raised. The many hormonal changes occurring shortly after injury (Chapter 19) make it difficult to assign a cause to this phenonenon, except to note that once more the branched-chain amino acids behave differently from the others. Following injection of tetanus vaccine (Stöckl *et al.,* 1967) or during the prodromal phase of infection with *Pasteurella tularensis* (Feigin and Dangerfield, 1967), there is a fall in the plasma free amino acids of human subjects. In the latter series of cases, onset of fever was frequently accompanied by a rise in free amino acid levels. However, children with fever show lower plasma amino acids (Holton *et al.,* 1968). Depression during the incubation period with Newcastle virus disease of chicks (Squibb, 1968) and a similar fall with onset of florid disease has been noted in chicks infected with tuberculosis (Squibb *et al.,* 1968); reduced food intake may have contributed to the picture. Infection of dogs with distemper virus causes changes in liver free amino acids that depend on the level of diet previously fed (Newberne *et al.,* 1969). This may be due to loss of appetite of infected animals at the higher level of intake. In experimental diabetes mellitus, insulin deficiency causes a complex series of changes in the concentrations of free amino acids in plasma and muscle; these have been partly traced (Scharff and Wool, 1966; Wool and Scharff, 1968) to the various hormonal and metabolic changes resulting from insulin deficiency. In diabetic subjects (Carlsten *et al.,* 1967) and diabetic dogs (Ivy *et al.,* 1951), and in fasting, grossly obese subjects (Swendseid *et al.,* 1964b), the plasma levels of the branched-chain amino acids are moderately raised. Again, these conditions are associated with deviations in fat metabolism. Nyhan *et al.* (1968) record changes in free amino acid concentrations in the plasma and muscle of children suffering from various endocrine diseases.

There are a number of diseases in which the cells show an enhanced uptake or concentration of free amino acids. It is well known that many malignant cells take up free amino acids avidly (see Chapters 4 and 20), but the relationship of this process to rapid tumor growth is not known. The pattern of free amino acids in tumor cells generally differs from that of the tissue of origin (Roberts and Simonsen, 1962). As a result of the avidity of the ascites tumor for amino acids, the blood of the host can be severely depleted of amino acids (Drewes and McKee, 1967). On the other hand, the Walker tumor seems to elevate the plasma amino acids of the host (Wannemacher and Yatvin, 1965).

Finally, some observations on miscellaneous diseases have been recorded. In essential thrombocytosis, a disease in which the blood platelets are increased in size and number, there appears to be an enhanced content of free amino acids in the platelets (Nachman *et al.*, 1966). A study by Taguchi (1965) shows that the production of cardiac hypertrophy in dogs results in an increase in the glutamine content of the myocardium, but a decrease in its free glutamic acid content. Reid and Berjak (1967) present evidence that low plasma tryptophan levels may be connected with the frequency of cardiomyopathy in a maize-eating African population. In the case of skeletal muscle, atrophy due to denervation is accompanied by increased uptake of AIBA into the affected muscles (Diehl and Jones, 1966; Bombara and Bergamini, 1968). In the Huntingdon's chorea, Perry *et al.* (1969) have found a significant depression of plasma tyrosine, proline, alanine, and the branched-chain amino acids, apparently not caused by an inadequate protein intake. It has been suggested that very low methionine and cystine levels in the plasma of subjects with nutritional ataxic neuropathy may be due to cyanide released from their cassava diet (Osuntokun *et al.*, 1968).

IV. General Conclusions about the Role of Free Amino Acids in the Regulation of Protein Metabolism

It has been assumed in this chapter that, for all practical purposes, free amino acids are the currency through which protein metabolism operates, and furthermore that no significant storage of free amino acids takes place. As shown for the rat in Table I, the daily intake of each amino acid as dietary protein is many times greater than the pool of the corresponding free amino acid in the body, especially in the case of the essential amino acids; in addition, the tissue proteins release each day considerable amounts of amino acids that are reutilized in the same tissue and elsewhere in the body. In view of this extensive traffic passing through the free amino acid pools, sensitive control mechanisms must exist if the magnitude of each pool is to be maintained within tolerable limits. Although mechanisms regulating amino acid metabolism will be discussed in detail in Chapter 35, it can be pointed out here that both input into the pool and disposal from the pool are under control. The complexity of the metabolic pathways involved in such control mechanisms is illustrated for serine in Fig. 3. These mechanisms demand both feedback control of serine biosynthesis and induction of enzymes regulating removal of serine. Regulatory mechanisms such as these must be capable of coping with the rapid influx of large amounts of amino acids from the diet, and also with an increased flow of amino acids from any tissue that is discharging amino acids at an excessive rate (e.g., skeletal muscle following corticosteroid administration).

The role of free amino acids in the regulation of protein metabolism has not emerged from this literature survey with any finality, but a number of features can be gathered together here.

1. The liver appears to act as the first line of defense against the influx of amino acids after each meal, and in this way prevents excessive fluctuations in the amino acid levels in the systemic circulation. As discussed in Chapter 38, when a large meal of protein is given to a dog, a considerable part of the amino acids passing to the liver are immediately metabolized and do not appear in the general circulation. Presumably hepatic disposal of incoming amino acids is less intense when a smaller meal of protein is given, but Elwyn's technique of continuously monitoring blood amino acid flow through the liver has not yet been extended to these conditions. The proportion of the absorbed amino acids used for liver protein synthesis appears to be determined by the pattern of amino acids coming to the liver. The absorbed amino acids mix in the liver cell with amino acids derived from the breakdown of liver protein. In the normally nourished animal, tryptophan is probably the least abundant amino acid among those available for reutilization from endogenous turnover of liver protein, and thus becomes the amino acid limiting the rate of translation of messenger RNA. Consequently, the tryptophan content of the amino acid mixture absorbed from the intestine determines the intensity of protein synthesis in the liver during the absorptive period. In addition, liver protein content increases after each meal of protein through a reduction in the rate of breakdown of some of the enzymes involved in amino acid catabolism. In this way, recycling of amino acids within the liver cell becomes less intense during the absorptive period.

2. The rates of protein synthesis in tissues served by the systemic circulation are undoubtedly affected by the quantity and quality of the amino acid supply, but there is no information about their immediate

responses to alterations in free amino acid pattern. In the case of many of these tissues (e.g., muscle), hormones are known to cause an increase in protein synthesis that is accompanied by an augmented flow of amino acids into the tissue, but in no case has it been possible to demonstrate that the amino acid influx triggers the increased tissue protein formation, and in several instances the evidence is against this interpretation. Consequently, they may be concomitant events not causally related.

3. The major sites of degradation vary from one amino acid to another. In the case of the essential amino acids, most are degraded in the liver; exceptions are the branched-chain amino acids which are chiefly transaminated in muscle, and methionine which is transformed to cysteine in several tissues. Consequently, after a meal of protein, especially a large amount of protein, the branched-chain amino acids and methionine pass through the liver into the general circulation and cause a much greater increment in systemic blood levels than in the case of the other essential amino acids, which are removed efficiently by the liver. This no doubt also accounts for the changes in levels of branched-chain amino acids in the systemic circulation which tend to occur independently of variations in the levels of other amino acids (e.g., in starvation, diabetes, obesity). Alanine may be a major carrier for returning amino N to the liver from other tissues after degradation of nonessential and branched-chain amino acids.

4. Endocrine function is sensitive to amino acid intake. It is especially noteworthy that methionine and leucine, two amino acids that pass readily through the liver into the general circulation, are particularly potent in causing increased adrenocortical activity a few hours after their administration. This effect probably occurs following ingestion of amounts of these amino acids within the physiological range of intakes, and indeed it has been shown that a protein-rich diet increases plasma albumin turnover in normal subjects but not in patients without adrenal glands (Iber *et al.,* 1958). Secretion of insulin, glucagon, and growth hormone may also occur as a result of increased levels of circulating amino acids, but there is insufficient evidence to indicate that normal amplitudes of blood amino acid levels can cause these hormones to be released. This area of control deserves further consideration, especially the interesting thesis that the levels of some blood amino acids, notably the branched-chain amino acids, act as feedback regulators of secretion of insulin, a hormone that stimulates uptake of amino acids into muscle and thus removes them from the blood. Indeed, it may well turn out that methionine and the branched-chain amino acids act as monitors signaling an influx of amino acids, usually from a meal, and causing

transfer of the amino acid load immediately into muscle (insulin) and later removal by a slower increase in catabolic enzymes in the liver (corticosteroid action).

5. Hormones can change the flow of amino acids from one tissue to another and also their subsequent disposal. For example, corticosteroids not only induce a release of tyrosine from muscle, but also induce an increase in liver tyrosine transaminase activity so that the tyrosine is rapidly removed and does not accumulate in the plasma. Thyroxine also causes release of tyrosine from peripheral tissues but it does *not* increase the activity of liver tyrosine transaminase, so that the plasma level of tyrosine rises. Thus, the action of hormones on the magnitude of free amino acid pools is determined by differentials between their effects on amino acid release from the tissues and on amino acid disposal.

6. Each meal containing carbohydrate causes release of insulin which results in deposition of amino acids in muscle and lowering of their levels in the plasma. If the meal also contains protein, the release of insulin is augmented and the amino acid deposition is presumably accentuated. Consequently, the diurnal rhythms observed in plasma free amino acids represent mainly a rather complex response to meals, the changes probably being due to (a) rapid secretion of insulin caused by the carbohydrate in each meal, with deposition of amino acids in muscle, (b) an augmentation in insulin level from the rise in blood amino acid concentrations after the meal, (c) a somewhat later stimulation of adrenocortical activity from increased levels of branched-chain amino acids and methionine in the general circulation, and (d) a pattern of individual essential amino acid levels that to some extent reflects the spectrum found in the dietary protein modified in relation to the tissue needs for these amino acids.

7. Skeletal muscle, the largest tissue in the body, is a major site of free amino acid deposition and also a target for many of the metabolic hormones. Consequently, it is not surprising that changes in the plasma amino acid levels should often reflect those found in the free amino acid pool of muscle. It is also likely, from what has been said in the preceding section, that the diurnal rhythm in plasma amino acids arises from changes in the rate of release from, or deposition of, amino acids in skeletal muscle. On the other hand, the free amino acids of the liver often do not follow the changes taking place in the plasma. This occurs for two reasons. First, the liver proteins undergo extensive turnover and the endogenous amino acids so released are reutilized to provide a major part of the free amino pool of that organ. Second, amino acids from the diet are the major exogenous source of liver amino acids during the absorptive phase after each meal. Consequently, diurnal rhythms

in liver free amino acid levels and in the activities of liver degradative enzymes such as tyrosine transaminase are related to dietary protein supply rather than to rhythmic changes in systemic plasma amino acid levels.

Acknowledgements

The work of the author described in this chapter was supported by grant CA-08893 from the National Institutes of Health.

References

Abadom, P. N., and Scholefield, P. G. (1962). *Can J. Biochem. Physiol.* **40,** 1591.
Ackermann, P. G., and Kheim, T. (1964a). *Clin. Chem.* **10,** 32.
Ackermann, P. G., and Kheim, T. (1964b). *J. Gerontol.* **19,** 207.
Adamson, L. F., and Ingbar, S. H. (1967). *Endocrinology* **81,** 1362 and 1372.
Adibi, S. A. (1968). *J. Appl. Physiol.* **25,** 52.
Agrawal, H. C., Davis, J. M., and Himwich, W. A. (1966). *J. Neurochem.* **13,** 607.
Agrawal, H. C., Davis, J. M., and Himwich, W. A. (1967). *Brain Res.* **3,** 374.
Agrawal, H. C., Davis, J. M., and Himwich, W. A. (1968). *J. Neurochem.* **15,** 917.
Alam, S. Q., Becker, R. V., Stucki, W. P., Rogers, Q. R., and Harper, A. E. (1966a). *J. Nutr.* **89,** 91.
Alam, S. Q., Rogers, Q. R., and Harper, A. E. (1966b). *J. Nutr.* **89,** 97.
Allen, R. E., Raines, P. L. and Regen, D. M. (1969). *Biochim. Biophys. Acta* **190,** 323.
Allfrey, V. G. (1963). *Proc. 5th Intern. Congr. Biochem., Moscow* **2,** 127.
Allfrey, V. G., Meudt, R., Hopkins, J. W., and Mirsky, A. E. (1961). *Proc. Natl. Acad. Sci. U.S.* **47,** 907.
Amenta, J. S., and Johnston, E. H. (1963). *Lab. Invest.* **12,** 921.
Anderson, H. L., and Linkswiler, H. (1969). *J. Nutr.* **99,** 91.
Anderson, H. L., Benevenga, N. J., and Harper, A. E. (1968). *Am. J. Physiol.* **214,** 1008.
Anderson, W. F., and Gilbert, J. M. (1969). *Biochem. Biophys. Res. Commun.* **36,** 456.
Baker, D. H., Becker, D. E., Norton, H. W., Jensen, A. H., and Harmon, B. G. (1966). *J. Nutr.* **88,** 382.
Baliga, B. S., Pronczuk, A. W., and Munro, H. N. (1968). *J. Mol. Biol.* **34,** 199.
Barnabei, O., and Ferrari, R. (1961). *Arch. Biochem. Biophys.* **94,** 79.
Barnabei, O., and Sereni, F. (1964). *Biochim. Biophys. Acta* **91,** 239.
Barton, A. D. (1951). *Proc. Soc. Exptl. Biol. Med.* **77,** 481.
Bavetta, L. A., and Nimni, M. E. (1964). *J. Nutr.* **82,** 379.
Benevenga, N. J., Harper, A. E., and Rogers, Q. R. (1968). *J. Nutr.* **95,** 434.
Berg, C. P., and Rose, W. C. (1929). *J. Biol. Chem.* **82,** 479.
Berl, S., and Purpura, D. P. (1963). *J. Neurochem.* **10,** 237.
Berry, H. K., and Leonard, C. S. (1966). *Am. J. Clin. Nutr.* **19,** 99.
Best, J., Catt, K. J., and Burger, H. G. (1968). *Lancet* **ii,** 124.
Betheil, J. J., Feigelson, M., and Feigelson, P. (1965). *Biochim. Biophys. Acta* **104,** 92.
Björnesjö, K. B. (1968). *Clin. Chim. Acta* **20,** 11.
Blasberg, R., and Lajtha, A. (1965). *Arch. Biochem. Biophys.* **112,** 361.
Block, R. J., and Bolling, D. (1951). "Amino Acid Composition of Proteins and Foods," 2nd ed. Thomas, Springfield, Illinois.

Block, W. D., and Hubbard, R. W. (1962). *Arch Biochem. Biophys.* **96,** 557.
Block, W. D., Markovs, M. F., and Steele, B. F. (1969). *Am. J. Clin. Nutr.* **22,** 33.
Bojanowska, K., and Williamson, D. H. (1968). *Biochim. Biophys. Acta* **159,** 560.
Bombara, G., and Bergamini, E. (1968). *Biochim. Biophys. Acta* **150,** 226.
Boomgaardt, J., and McDonald, B. E. (1969). *Can. J. Physiol. Pharmacol.* **47,** 392.
Borsook, H., Deasy, C. L., Hagen-Smit, A. J., Keighley, C., and Lowy, P. H. (1950). *J. Biol. Chem.* **187,** 839.
Bradley, A. S., and Wissig, S. L. (1966). *J. Cell Biol.* **30,** 433.
Brandt, I. K., Matalka, V. A., and Combs, J. T. (1960). *Am. J. Physiol.* **119,** 39.
Brown, D. M. (1966). *Endocrinology* **78,** 1252.
Brown, H. E., Kunkel, H. O., and Prescott, J. M. (1961). *J. Animal Sci.* **20,** 967.
Buraczewski, S., Porter, J. W. G., Westgarth, D. R., and Williams, A. P. (1967). *Proc. Nutr. Soc. (Eng. Scot.)* **26,** viii.
Cachin, M., Durlach, J., and Blass, J. (1952). *Semaine Hop. Paris* **28,** 3231.
Cannon, P. R., Steffee, C. H., Frazier, L. E., Rawley, D. A., and Stepto, R. C. (1947). *Federation Proc.* **6,** 390.
Carlsten, A., Hallgren, B., Jagenburg, R., Svanborg, A., and Werkö, L. (1967). *Acta Med. Scand.* **181,** 195 and 199.
Caro, L. G., and Palade, G. E. (1964). *J. Cell Biol.* **20,** 473.
Castles, J. J., and Wool, I. G. (1964). *Biochem. J.* **91,** 11C.
Chambers, J. W., Georg, R. H., and Bass, A. D. (1965). *Mol Pharmacol.* **1,** 66
Charkey, L. W., Kano, A. K., and Hougham, D. F. (1955). *J. Nutr.* **55,** 469.
Christensen, H. N. (1966). *Federation Proc.* **25,** 850.
Christensen, H. N. (1968). *In* "Protein Nutrition and Free Amino Acid Patterns" (J. H. Leathem, ed.), p. 40. Rutgers Univ. Press, New Brunswick, New Jersey.
Christensen, H. N., and Cullen, A. M. (1968). *Biochim. Biophys. Acta* **150,** 237.
Christensen, H. N., and Handlogten, M. E. (1968). *J. Biol. Chem.* **243,** 5428.
Christensen, H. N., and Lynch, E. L. (1948). *J. Biol. Chem.* **172,** 107.
Christensen, H. N., and Streicher, J. A. (1948). *J. Biol. Chem.* **175,** 95.
Christensen, H. N., and Streicher, J. A. (1949). *Arch. Biochem.* **23,** 96.
Christensen, H. N., Streicher, J. A., and Elbinger, R. L. (1948a). *J. Biol. Chem.* **172,** 515.
Christensen, H. N., Rothwell, J. T., Sears, R. A., and Streicher, J. A. (1948b). *J. Biol. Chem.* **175,** 101.
Christensen, H. N., Handlogten, M. E., Lam, I., Trager, H. S., and Zand, R. (1969). *J. Biol. Chem.* **244,** 1510.
Churchill, J. A., Moghissi, K. S., Evans, T. N., and Frohman, C. (1969). *Obstet. Gynecol.* **33,** 492.
Clark, A. J., Yamada, C., and Swendseid, M. E. (1968). *Am. J. Physiol.* **215,** 1324.
Clark, A. J., Peng, Y., and Swendseid, M. E. (1966). *J. Nutr.* **90,** 228.
Clark, C. M., Naismith, D. J., and Munro, H. N. (1957). *Biochim. Biophys. Acta* **23,** 587.
Clemens, M. J., and Korner, A. (1969). *Biochem. J.* **113,** 10P.
Cochrane, W. A., Payne, W. W., Simpkiss, M. J., and Woolf, L. I. (1956). *J. Clin. Invest.* **35,** 411.
Cooper, W. C., and Munro, H. N. (1969). Unpublished results.
Cornish, H. H., and Wilson, C. E. (1968). *Toxicol. Appl. Pharmacol.* **12,** 265.
Cremer, G. M., Bilstad, J. M., Faiman, C., and Moxness, K. E. (1968). *Mayo Clin. Proc.* **43,** 776.
Crumpler, H. R., Dent, C. E., and Lindan, O. (1950). *Biochem. J.* **47,** 223.

Daniel, P. M., Moorhouse, S. R., and Pratt, O. E. (1968). *J. Physiol.* (*London*) **196,** 106P.

Deinken, E. A., Yang, S. P., and Tilton, K. S. (1966). *Federation Proc.* **25,** 239.

Denton, A. E., and Elvehjem, C. A. (1954). *J. Biol. Chem.* **206,** 449.

Denton, A. E., Williams, J. N., and Elvehjem, C. A. (1950). *J. Biol. Chem.* **186,** 377.

Denton, A. E., Gershoff, S. N., and Elvehjem, C. A. (1953). *J. Biol. Chem.* **204,** 731.

Dickinson, J. C., Rosenblum, H., and Hamilton, P. B. (1965). *Pediatrics* **36,** 2.

Diehl, J. F. (1963). *Biochem. Z.* **337,** 333.

Diehl, J. F. (1965). *Biochim. Biophys. Acta* **115,** 239.

Diehl, J. F. (1966). *Nature* **209,** 75.

Diehl, J. F., and Jones, R. R. (1966). *Am. J. Physiol.* **210,** 1080.

Donaldson, H. H. (1924). "The Rat," 2nd ed. Mems. Wistar Inst., Philadelphia, Pennsylvania.

Drenick, E. J., Swendseid, M. E., Blahd, W. H., and Tuttle, S. G. (1964). *J. Am. Med. Assoc.* **187,** 100.

Drewes, P. A., and McKee, R. W. (1967). *Nature* **213,** 411.

Drysdale, J. W., and Munro, H. N. (1967). *Biochim. Biophys. Acta* **138,** 616.

Drysdale, J. W., Olafsdottir, E., and Munro, H. N. (1968). *J. Biol. Chem.* **243,** 552.

Dubons, A. F., and Pitman, J. A. (1962). *Endocrinology* **70,** 937.

Eagle, H., Piez, K. A., and Levy, M. (1961). *J. Biol. Chem.* **236,** 2039.

Eagle, H., Washington, C. L., and Levy, M. (1965). *J. Biol. Chem.* **240,** 3944.

Eavenson, E., and Christensen, H. N. (1967). *J. Biol. Chem.* **242,** 5386.

Edgar, P., Rabinowitz, D., and Merimee, T. J. (1969). *Endocrinology* **84,** 835.

Eichhorn, J., Scully, E., Halkerson, I. D., and Hechter, O. (1961). *Proc. Soc. Exptl. Biol. Med.* **106,** 153.

Eliasson, E., Bauer, G. E., and Hultin, T. (1967). *J. Cell Biol.* **32,** 287.

Ellison, J. W., and King, K. W. (1968). *J. Nutr.* **94,** 543.

Elman, R. (1939). *Proc. Soc. Exptl. Biol. Med.* **40,** 484.

Elwyn, D. H. (1966). *Federation Proc.* **25,** 584.

Enwonwu, C. O., and Munro, H. N. (1968). *Federation Proc.* **27,** 416.

Enwonwu, C. O., and Munro, H. N. (1970). *Arch. Biochem. Biophys.* (in press).

FAO (1963). Tables of Amino Acid Composition of Foods. Unpublished.

Feigin, R. D., and Dangerfield, H. G. (1967). *J. Infect. Diseases* **117,** 346.

Feldman, B., and Christensen, H. N. (1962). *Proc. Soc. Exptl. Biol. Med.* **109,** 700.

Felig, P., Owen, O. E., Wahren, J., and Cahill, G. F. (1969a). *J. Clin. Invest.* **48,** 584.

Felig, P., Marliss, E., and Cahill, G. F. (1969b). *New Engl. J. Med.* **281,** 811.

Ferrari, V., and Harkness, R. D. (1954). *J. Physiol.* (*London*) **124,** 443.

Fleck, A., Shepherd, J. and Munro, H. N. (1965). *Science* **150,** 628.

Floyd, J. C., Fajans, S. S., Conn, J. W., Knopf, R. F., and Rull, J. A. (1966). *J. Clin. Invest.* **45,** 1479 and 1487.

Foley, T. H., London, D. R., and Prenton, M. A. (1966). *J. Clin. Endocrinol. Metab.* **26,** 781.

Foley, T. H., Holm, L. W., London, D. R., and Young, M. (1967). *Am. J. Obstet. Gynecol.* **99,** 1106.

Frame, E. G. (1958). *J. Clin. Invest.* **37,** 1710.

Friedberg, F., and Greenberg, D. M. (1947). *J. Biol. Chem.* **168,** 405.

Fujioka, M., Koga, M., and Lieberman, I. (1963). *J. Biol. Chem.* **283,** 3401.

Gan, J. C., and Jeffay, H. (1967). *Biochim. Biophys. Acta* **148,** 448.

Ganapathy, S. N., and Nasset, E. S. (1962). *J. Nutr.* **78,** 241.
Garfinkel, D. (1966). *J. Biol. Chem.* **241,** 3918.
Garfinkel, D., and Lajtha, A. (1963). *J. Biol. Chem.* **238,** 2429.
Geiger, E. (1947). *J. Nutr.* **34,** 97.
Geiger, E. (1948). *J. Nutr.* **36,** 99 and 813.
Geiger, E. (1950). *Science* **111,** 594.
Ginter, E. (1957). *Csl. Gastroent. Vyz.* **11,** 329.
Girard, A., Robin, P., and Jacquot, R. (1968). *Compt. Rend. Acad. Sci. Paris* **266,** 2160.
Godin, C. (1967). *J. Nutr.* **92,** 503.
Goldberg, A., and Guggenheim, K. (1962). *Biochem. J.* **83,** 129.
Goldstein, S., and Reddy, W. J. (1968). *Biochim. Biophys. Acta* **150,** 733.
Goodlad, G. A. J., and Munro, H. N. (1959). *Biochem. J.* **73,** 343.
Grasso, S., Saporito, N., Messina, A., and Reitano, G. (1968). *Lancet* **ii,** 755.
Green, N. M., and Lowther, D. A. (1959). *Biochem. J.* **71,** 55.
Greengard, O., and Acs, G. (1962). *Biochim. Biophys. Acta* **61,** 652.
Grümer, H. D., Koblet, H., and Woodard, C. (1961). *J. Clin. Invest.* **40,** 1758.
Halfpenny, A. F., and Rook, J. A. F. (1968). *Proc. Nutr. Soc. (Eng. Scot.)* **27,** 19A.
Halfpenny, A. F., Smith, G. H., and Rook, J. A. F. (1969). *Proc. Nutr. Soc. (Engl. Scot.)* **28,** 29A.
Hamilton, P. B. (1945). *J. Biol. Chem.* **158,** 397.
Hanking, B. M. and Roberts, S. (1964). *Nature* **204,** 1194.
Hanking, B. M. and Roberts, S. (1965). *Nature* **207,** 862.
Harker, C. S., Allen, P. E., and Clark, H. E. (1968). *J. Nutr.* **94,** 495.
Harper, A. E. and Rogers, Q. R. (1965). *Proc. Nutr. Soc. (Eng. Scot.)* **24,** 173.
Harrill, I., and Gifford, E. D. (1962). *J. Nutr.* **78,** 320.
Harrill, I., and Gifford, E. D. (1966). *J. Nutr.* **89,** 247.
Hartman, D. R., and King, K. W. (1967). *J. Nutr.* **92,** 455.
Hay, A. M., and Waterlow, J. C. (1967). *J. Physiol. (London)* **191,** 111P.
Henderson, L. M., Schurr, P. E., and Elvehjem, C. A. (1949). *J. Biol. Chem.* **177,** 815.
Henderson, R., and Harris, R. S. (1949). *Federation Proc.* **8,** 382.
Henriques, O. B., Henriques, S. B., and Neuberger, A. (1955). *Biochem. J.* **60,** 409.
Henry, K. M., and Kon, S. K. (1946). *J. Dairy Res.* **14,** 330.
Herbert, J. D., Coulson, R. A., and Hernandez, T. (1966) *Comp. Biochem. Physiol.* **17,** 583.
Hider, R. C., Fern, E. B., and London, D. R. (1969). *Biochem. J.* **114,** 171.
Hier, S. W. (1947). *J. Biol. Chem.* **171,** 813.
Higashino, K., and Lieberman, I. (1965). *Biochim. Biophys. Acta* **111,** 346.
Hogan, J. P., Weston, R. H., and Lindsay, J. R. (1968). *Australian J. Biol. Sci.* **21,** 1263.
Holden, J. T., ed. (1962). "Amino Acid Pools." Elsevier, Amsterdam.
Holt, L. E., Jr., Snyderman, S. E., Norton, P. M., and Roitman, E. (1968). *In* "Protein Nutrition and Free Amino Acid Patterns." (J. H. Leathem, ed.), p. 32. Rutgers Univ. Press, New Brunswick, New Jersey.
Holton, J. B. (1968). *Clin. Chim. Acta* **21,** 241.
Holton, J. B., Jones, S., and Small, N. A. (1968). *Clin. Sci.* **35,** 425.
Hori, M., Fisher, J. M., and Rabinowitz, M. (1967). *Science* **155,** 83.
Howe, E. E., and Dooley, C. L. (1963). *J. Nutr.* **81,** 379.

Hunt, R. T., Hunter, A. R., and Munro, A. J. (1969). *Proc. Nutr. Soc.* (*Engl. Scot.*) **28,** 248.
Iacobellis, M., Muntwyler, E., and Dodgen, C. L. (1956). *Am. J. Physiol.* **185,** 275.
Iacobellis, M., Griffin, G. E., and Muntwyler, E. (1957). *Proc. Soc. Exptl. Biol. Med.* **96,** 64.
Iber, F. L., Rosen, H., Levenson, S. M., and Chalmers, T. C. (1957). *J. Lab. Clin. Med.* **50,** 417.
Iber, F. L., Nassau, K., Plough, I. C., Berger, F. M., Meroney, W. H., and Fremont-Smith, K. (1958). *J. Clin. Invest.* **37,** 1442.
Iob, V., Coon, W. W., and Sloan, M. (1966). *J. Surg. Res.* **6,** 233.
Iob, V., Coon, W. W., and Sloan, M. (1967). *J. Surg. Res.* **7,** 41.
Ito, T., Guroff, G. and Udenfriend, S. (1964). *J. Biol. Chem.* **239,** 3385.
Ivy, J. H., Svec, M., and Freeman, S. (1951). *Am. J. Physiol.* **167,** 182.
Jefferson, L. S., and Korner, A. (1969). *Biochem. J.* **111,** 703.
Jones, J. D. (1964). *J. Nutr.* **84,** 313.
Jones, J. D., Petersburg, S. J., and Burnett, P. C. (1967). *J. Nutr.* **93,** 103.
Kandera, J., Levi, G., and Lajtha, A. (1968). *Arch. Biochem. Biophys.* **126,** 249.
Kaplan, S. A., and Shimizu, C. S. N. (1962). *Am. J. Physiol.* **202,** 695.
Kaplan, S. A., and Shimizu, C. S. N. (1963). *Endocrinology* **72,** 267.
Karjalainen, E. (1966). *Acta Chem. Scand.* **20,** 586.
Kekomäki, M. P., Visakorpi, J. K., Perheentupa, J., and Saxén, L. (1967a). *Acta Paediat. Scand.* **56,** 617.
Kekomäki, M. P., Räihä, N. C. R., and Perheentupa, J. (1967b). *Acta Paediat. Scand.* **56,** 631.
Kenney, M. A., Roderuck, C. E., Arnrich, L. and Piedad, F. (1968). *J. Nutr.* **95,** 173.
Kerr, G. R. (1968). *Pediat. Res.* **2,** 187.
Kerr, G. R., Chamove, A. S., Harlow, H. F., and Waisman, H. A., (1968). *Pediatrics* **42,** 27.
Kipnis, D. M., Galvao, P. A. A., Greene, G., and Daughaday, W. H. (1959). *J. Lab. Clin. Med.* **54,** 914.
Kipnis, D. M., Reiss, E., and Helmreich, E. (1961). *Biochim. Biophys. Acta* **51,** 519.
Kirsner, J. B., Sheffner, A. L., and Palmer, W. L. (1949). *J. Clin. Invest.* **28,** 716.
Klain, G. J., and Whitten, B. K. (1968). *Comp. Biochem. Physiol.* **27,** 617.
Klavins, J. V. (1965). *Biochim. Biophys. Acta* **104,** 554.
Knauff, H. G., Seybold, D., and Miller, B. (1964). *Klin. Wochschr.* **42,** 326.
Knauff, H. G., Hamelmann, H., Seybold, D., and Kanters, A. (1966). *Klin. Wochschr.* **44,** 147.
Knauff, H. G., Mayer, G., and Drücke, F. (1966). *Klin. Wochschr.* **44,** 929.
Knipfel, J. E., Botting, H. G., Noel, F. J., and McLaughlan, J. M. (1969). *Can. J. Biochem.* **47,** 323.
Knopf, R. F., Conn, J. W., Fajans, S. S., Floyd, E. M., Guntsche, E. M., and Rull, J. A. (1965). *J. Clin. Endocrinol.* **25,** 1140.
Knox, W. E., and Greengard, O. (1965). *In* "Advances in Enzyme Regulation" (G. Weber, ed.), Vol. 3, p. 247. Pergamon, Oxford.
Kohen, E. (1964). *Exptl. Cell Res.* **35,** 303.
Korty, P., and Coe, F. L. (1968). *J. Pharmacol. Exptl. Therap.* **160,** 212.
Kostyo, J. L. (1964). *Federation Proc.* **23,** 512.
Kostyo, J. L. (1965). *Endocrinology* **76,** 604.
Kostyo, J. L. (1968). *Ann. N.Y. Acad. Sci.* **148,** 389.
Kostyo, J. L., and Redmond, A. F. (1966). *Endocrinology* **79,** 531.

Kostyo, J. L., and Schmidt, J. E. (1963). *Am. J. Physiol.* **204**, 1031.
Krawitt, E. L., and Clifton, J. A. (1968). *Gastroenterology* **54**, 866.
Kristoffersson, R., and Broberg, S. (1968a). *Experientia* **24**, 148.
Kristoffersson, R., and Broberg, S. (1968b). *Ann. Acad. Sci. Fennicae, Ser. A. IV. Biol.* p. 130.
Krueger, R., and Wiss, O. (1949). *Helv. Chim. Acta* **32**, 1341.
Kumta, U. S., and Harper, A. E. (1960). *J. Nutr.* **70**, 141.
Kumta, U. S., and Harper, A. E. (1962). *Proc. Soc. Exptl. Biol. Med.* **110**, 512.
Lacy, W. W., and Crofford, O. B. (1964). *J. Lab. Clin. Med.* **64**, 828.
Landau, R. L., and Lugibihl, K. (1969). *Metabolism Clin. Exptl.* **18**, 265.
Lajtha, A., Blasberg, R., and Levi, G. (1968). *In* "Protein Nutrition and Free Amino Acid Patterns" (J. H. Leathem, ed.), p. 187. Rutgers Univ. Press, New Brunswick, New Jersey.
Lapeze, B. L., Yang, S. P., Tilton, K. S., and Deinken, E. A. (1965). *Federation Proc.* **24**, 168.
Leibholz, J. M., and Cook, C. F. (1967). *J. Nutr.* **93**, 561.
Leibholz, J. M., McCall, J. T., Hays, V. W., and Speer, V. C. (1966). *J. Animal Sci.* **25**, 37.
Leon, H. A., Feller, D. D., Neville, E. D., and Daligcon, B. (1965). *Life Sci.* **4**, 737.
Leung, P. M. B., and Rogers, Q. R. (1969). *Life Sci.* **8**, 1.
Leung, P. M. B., Rogers, Q. R., and Harper, A. E. (1968). *J. Nutr.* **96**, 139.
Levenson, S. M., Rosen, H., and Upjohn, H. L. (1959). *Proc. Soc. Exptl. Biol. Med.* **101**, 178.
Levine, R. J., and Conn, H. O. (1967). *J. Clin. Invest.* **46**, 2012.
Lewis, D. (1967). *In* "Protein Utilization by Poultry" (R. A. Morton and E. C. Amoroso, eds.), p. 29. Oliver & Boyd, Edinburgh and London.
Lindblad, B. S., and Baldesten, A. (1967). *Acta Paediat. Scand.* **56**, 37.
Litwack, G., Williams, J. N., Feigelson, P., and Elvehjem, C. A. (1950). *J. Biol. Chem.* **187**, 605.
London, D. R., Foley, T. H., and Webb, C. G. (1965). *Nature* **208**, 588.
Loftfield, R. B., and Harris, A. (1956). *J. Biol. Chem.* **219**, 151.
Longenecker, J. B. (1961). *In* "Meeting Protein Needs of Infants and Children," Natl. Acad. Sci. Publ. No. 843, p. 469. Washington, D.C.
Longenecker, J. B. (1963). *In* "Newer Methods of Nutritional Biochemistry" (A. A. Albanese, ed.), Vol. 1, p. 113. Academic Press, New York.
Longenecker, J. B., and Hause, N. L. (1959). *Arch. Biochem. Biophys.* **84**, 46.
Longenecker, J. B., and Hause, N. L. (1961). *Am. J. Clin. Nutr.* **9**, 356.
McCormick, D. B., and Snell, E. S. (1961). *J. Biol. Chem.* **236**, 2085.
McKean, C. M., Boggs, D. E., and Peterson, N. A. (1968). *J. Neurochem.* **15**, 235.
McLaughlan, J. M. (1963). *Federation Proc.* **22**, 1122.
McLaughlan, J. M. (1964). *Can. J. Biochem.* **42**, 1353.
McLaughlan, J. M., and Illman, W. I. (1967). *J. Nutr.* **93**, 21.
McLaughlan, J. M., and Venkat Rao, S. (1967). *Proc. 7th Intern. Congr. Nutr., Hamburg.*
McLaughlan, J. M., Noel, F. J., Morrison, A. B., and Campbell, J. A. (1963). *Can. J. Biochem. Physiol.* **41**, 191.
McLaughlan, J. M., Venkat Rao, S., Noel, F. J. and Morrison, A. B. (1967). *Can. J. Biochem.* **45**, 31.
Malathi, P., Sastry, P. S., and Ganguly, J. (1961). *Nature* **189**, 660.

Manchester, K. L., and Krahl, M. E. (1959). *J. Biol. Chem.* **234**, 2938.
Mandel, P., Godin, Y., Mark, J., and Kayser, C. (1966). *J. Neurochem.* **13**, 533.
Mark, J., and Mandel, P. (1964). *Compt. Rend. Soc. Biol.* **158**, 2478.
Maurer, W., Niklas, A., and Lehnert, G. (1954). *Biochem. Z.* **326**, 28.
Means, A. R., and Hamilton, T. H. (1966a). *Biochim. Biophys. Acta* **129**, 432.
Means, A. R., and Hamilton, T. H. (1966b). *Proc. Natl. Acad. Sci. U.S.* **56**, 686 and 1594.
Miller, L. L. (1962). *In* "Amino Acid Pools" (J. T. Holden, ed.). p. 708. Elsevier, Amsterdam.
Milner, R. D. G. (1969). *Lancet* **i**, 1075.
Mimura, T., Yamada, C., and Swendseid, M. E. (1968). *J. Nutr.* **95**, 493.
Mitchell, E. M., and Morrison, M. A. (1965). *Federation Proc.* **24**, 499.
Mitchell, J. R., Becker, D. E., Jensen, A. H., Harmon. B. G., and Norton, H. W. (1968). *J. Animal Sci.* **27**, 1327.
Mondon, C. E., and Mortimore, G. E. (1967). *Am. J. Physiol.* **212**, 173.
Morrison, A. B., Middleton, E. J., and McLaughlan, J. M. (1961a). *Can. J. Biochem. Physiol.* **39**, 1675.
Morrison, A. B., McLaughlan, J. M., Noel, F. J., and Campbell, J. A. (1961b). *Can. J. Biochem. Physiol.* **39**, 1681.
Munro, H. N. (1966). *Nutr. Dieta* **8**, 179.
Munro, H. N. (1968). *Federation Proc.* **27**, 1231.
Munro, H. N., and Clark, C. M. (1959). *Biochem. Biophys. Acta* **13**, 551.
Munro, H. N., and Goldberg, D. M. (1964). *In* "The Role of the Gastrointestinal Tract in Protein Metabolism" (H. N. Munro, ed.), p. 189. Blackwell, Oxford.
Munro, H. N., and Mukerji, D. (1962). *Biochem. J.* **82**, 520.
Munro, H. N., and Portugal, F. H. (1970). *In* "International Encyclopedia of Food and Nutrition" (E. J. Bigwood, ed.), Vol. 11. Pergamon, Oxford. In press.
Munro, H. N., and Thomson, W. S. T. (1953). *Metab., Clin. Exptl.* **2**, 354.
Munro. H. N., and Zak, M. (1968). Unpublished results.
Munro, H. N., Black, J. G., and Thomson, W. S. T. (1959). *Brit. J. Nutr.* **13**, 475.
Munro, H. N., Chisholm, J., and Naismith, D. J. (1962). *Brit. J. Nutr.* **16**, 245.
Munro, H. N., Steele, M. H., and Hutchison, W. C. (1963). *Nature* **199**, 1182.
Munro, H. N., McLean, E. J. T., and Hird, H. J. (1964). *J. Nutr.* **83**, 186.
Munro, H. N., Steele, M. H., and Hutchison, W. C. (1965). *Brit. J. Nutr.* **19**, 137.
Nachman, R. L., Horowitz, H. I., and Silver, R. T. (1966). *Blood* **27**, 715.
Nagataki, S., and Ingbar, S. H. (1963). *Endocrinology* **73**, 479.
Nasset, E. S., Ganapathy, S. N., and Goldsmith, D. P. J. (1963). *J. Nutr.* **81**, 343.
Nasset, E. S., Ridley, P. T., and Schenk, E. A. (1967). *Am. J. Physiol.* **213**, 645.
Neame, K. D. (1966). *J. Physiol.* (*London*) **185**, 627.
Nemer, M. J., Wise, E. M., Jr., Washington, F. M., and Elwyn, D. H. (1960). *J. Biol. Chem.* **235**, 2063.
Newberne, P. M., Young, V. R., and Gravlee, J. F. (1969). *Brit. J. Exptl. Pathol.* **50**, 172.
Nichoalds, G. E., Jones, R. R., Diehl, J. F., and Fitch, C. D. (1969). *J. Nutr.* **99**, 27.
Ning, M., Lowenstein, L. M., and Davidson, C. S. (1967). *J. Lab. Clin. Med.* **70**, 554.
Noall, M. W., and Allen, W. M. (1961). *J. Biol. Chem.* **236**, 2987.
Noall, M. W., Riggs, T. R., Walker, L. M., and Christensen, H. N. (1957). *Science* **126**, 1002.
Nukada, T. (1965). *Can. J. Biochem.* **43**, 1119.

Nyhan, W. L., Yujnovsky, A. O., and Wehr, R. F. (1968). In "Human Growth" (D. B. Cheek, ed.), p. 396. Lea & Febiger. Philadelphia, Pennsylvania.
O'Dell, B. L., and Savage, J. E. (1966). *J. Nutr.* **90,** 364.
Oja, S. S., and Piha, R. S. (1966). *Life Sci.* **5,** 865.
Oltjen, R. R., and Putnam, P. A. (1966). *J. Nutr.* **89,** 385.
Oltjen, R. R., Kozak, A. S., Putnam, P. A., and Lehmann, R. P. (1967). *J. Animal Sci.* **26,** 1415.
Ondarza, R. N. (1965). *Biochem. Biophys. Acta* **107,** 112.
Osuntokun, B. O., Durowoju, J. E., McFarlane, H., and Wilson, J. (1968). *Brit. Med. J.* **iii,** 647.
Oxender, D. L., and Christensen, H. N. (1963). *J. Biol. Chem.* **238,** 3686.
Parker, M. L., Hammond, J. M., and Daughaday, W. H. (1967). *J. Clin. Endocrinol.* **27,** 1129.
Parry, T. E. (1969). *Brit. J. Haematol.* **16,** 221.
Pawlak, M., and Pion, R. (1967). *Compt. Rend. Acad. Sci.* **264,** 380.
Pawlak, M., and Pion, R. (1968a). *Compt. Rend. Acad. Sci.* **266,** 1993.
Pawlak, M., and Pion, R. (1968b). *Ann. Biol. Animale, Biochim. Biophys.* **8,** 517.
Peraino, C., and Harper, A. E. (1963). *J. Nutr.* **80,** 270.
Perry, T. L., Hansen, S., Diamond, S., and Stedman, D. (1969). *Lancet* **i,** 806.
Peters, T. (1962). *J. Biol. Chem.* **237,** 1186.
Pion, R., and Rérat, A. (1967). *Compt. Rend. Acad. Sci.* **264,** 632.
Pion, R., Fauconneau, G., and Rérat, A. (1964). *Ann. Biol. Animale Biochim. Biophys.* **4,** 383.
Porter, J. W. G., and Williams, A. P. (1963). *Biochem. J.* **87,** 7P.
Portugal, F. H., and Jeffay, H. (1966). *Federation Proc.* **25,** 709.
Potter, E. L., Purser, D. B., and Cline, J. H. (1968). *J. Nutr.* **95,** 655.
Pronczuk, A W., Baliga, B. S., Triant, J. W., and Munro, H. N. (1968). *Biochim. Biophys. Acta* **157,** 204.
Pronczuk, A. W., Rogers, Q. R., and Munro, H. N. (1969). Unpublished results.
Puchal, F., Hays, V. W., Speer, V. C., Jones, J. D., and Catron, D. V. (1962). *J. Nutr.* **76,** 11.
Purser, D. B., Klopfenstein, T. J., and Cline, J. H. (1966). *J. Nutr.* **89,** 226.
Rabinowitz, D., Merimee, T. J., Mafezzoli, R., and Burgess, J. A. (1966). *Lancet* **ii,** 454.
Raghupathy, E., Tong, W., and Chaikoff, I. L. (1963). *Endocrinology* **72,** 620.
Raisz, L. G. (1967). *Biochim. Biophys. Acta* **148,** 460.
Raisz, L. G., and O'Brien, J. (1963). *Am. J. Physiol.* **205,** 816.
Rama Rao, P. B., Metta, V. C., and Johnson, B. C. (1959). *J. Nutr.* **69,** 387.
Rangneker, P. V., and Dugal, L. P. (1958). *Can. J. Biochem. Physiol.* **36,** 25.
Rao, G. V. G. K. (1961). *Nature* **192,** 269.
Rapoport, M. I., Feigen, R. D., Bruton, J., and Beisel. W. R. (1966). *Science* **153,** 1642.
Reid, J. V. O., and Berjak, P. (1967). *Am. Heart J.* **74,** 337.
Richardson, L. R., Blaylock, L. G., and Lyman, C. M. (1953a). *J. Nutr.* **49,** 21.
Richardson, L. R., Blaylock, L. G., and Lyman, C. M. (1953b). *J. Nutr.* **51,** 515.
Richardson, L. R., Hale, F., and Ritchey, S. J. (1965). *J. Animal Sci.* **24,** 368.
Richmond, J., and Girdwood, R. H. (1962). *Clin. Sci.* **22,** 301.
Riggs, T. R., and Walker, L. M. (1963). *J. Biol. Chem.* **238,** 2663.
Riggs, T. R., Pan, M. W., and Feng. H. W. (1968). *Biochim Biophys. Acta* **150,** 92.
Riley, D. R., Harrill, I., and Gifford, E. D. (1969). *J. Nutr.* **98,** 351.

Rivlin, R. S. (1963). *J. Biol. Chem.* **238,** 3341.

Rivlin, R. S., and Levine, R. J. (1963). *Endocrinology* **73,** 103.

Rivlin, R. S., and Melmon, K. L. (1965). *J. Clin. Invest.* **44,** 1690.

Roberts, E., and Simonsen, D. G. (1962). *In* "Amino Acid Pools" (J. T. Holden, ed.), p. 289. Elsevier, Amsterdam.

Rogers, Q. R., and Harper, A. E. (1968). *In* "Protein Nutrition and Free Amino Acid Patterns" (J. H. Leatham, ed.), p 107. Rutgers Univ. Press, New Brunswick, New Jersey.

Rogers, Q. R., Spolter, P. D., and Harper, A. E. (1962). *Arch. Biochem. Biophys.* **97,** 497.

Roodyn, D. B. (1965). *Biochem. J.* **97,** 782.

Roscoe, J. P., Eaton, M. D., and Chin-Choy, G. (1968). *Biochem. J.* **109,** 507.

Rosenberg, L. E., Berman, M., and Segal, S. (1963). *Biochim. Biophys. Acta* **71,** 664.

Rothschild, M. A., Oratz, M., Mongelli, J., Fishman, L., and Schreiber, S. S. (1969). *J. Nutr.* **98,** 395.

Rubinstein, D., Daniel, A. M., Chiu, S., and Beck, J. C. (1965). *Can. J. Biochem.* **43,** 271.

Ryan, W. L., and Carver, M. J. (1963). *Proc. Soc. Exptl. Biol. Med.* **114,** 816.

Sanahuja, J. C., and Harper, A. E. (1963). *Am. J. Physiol.* **204,** 686.

Sanders, R. B., and Riggs, T. R. (1967). *Endocrinology* **80,** 29.

Sauberlich, H. (1961). *J. Nutr.* **75,** 61.

Schafer, J. H., and Jacquez, J. A. (1967). *Biochim. Biophys. Acta* **135,** 741.

Scharff, R., and Wool, I. G. (1965). *Biochem. J.* **97,** 257.

Scharff, R., and Wool, I. G. (1966). *Biochem. J.* **99,** 173.

Schelling, G. T., Hinds, F. C., and Hatfield, E. E. (1967). *J. Nutr.* **92,** 339.

Schingoethe, D. J., Hageman, E. C., and Larson, B. L. (1967). *Biochim. Biophys. Acta* **148,** 469.

Schønheyder, F., and Lyngbye, J. (1962). *Brit. J. Nutr.* **16,** 75.

Schreier, K. (1962). *In* "Amino Acid Pools" (J. T. Holden, ed.), p. 263. Elsevier, Amsterdam.

Schreier, K., and Karch, H. L. (1954). *Arch. Klin. Chir.* **280,** 516.

Schreier, K., and Plückthun, H. (1950). *Biochem. Z.* **320,** 447.

Schreier, K., and Remsperger, H. (1951). *Biochem. Z.* **322,** 298.

Schwartzman, L., Crawhall, J., and Segal, S. (1966a). *Biochim. Biophys. Acta* **124,** 62.

Schwartzman, L., Blair, A., and Segal, S. (1966b). *Biochem. Biophys. Res. Commun.* **23,** 220.

Scriver, C. R., and Davies, E. (1965). *Pediatrics* **36,** 592.

Segal, S., Blair, A., and Rosenberg, L. E. (1963). *Biochim. Biophys. Acta* **71,** 676.

Segal, S., Lowenstein, L. M., and Wallace, A. (1968). *Gastroenterology* **55,** 386.

Shao, T. C., and Hill, D. C. (1967). *Can. J. Physiol. Pharmacol.* **45,** 225.

Shao, T. C., and Hill, D. C. (1968). *Poultry Sci.* **47,** 1806.

Shao, T. C., and Hill, D. C. (1969). *Poultry Sci.* **48,** 697.

Sheffner, A. L., and Bergeim, O. (1952). *J. Nutr.* **48,** 139.

Shimbayashi, K., and Yonemura, T. (1965). *Nat. Inst. Animal Health Quart.* **4,** 202 and 213.

Shimbayashi, K., Ide, Y., and Yonemura, T. (1965). *Agr. Biol. Chem.* **29,** 774.

Sidransky, H., and Farber, E. (1958). *Arch. Pathol.* **66,** 119 and 135.

Sidransky, H., and Verney, E. (1964). *Arch. Pathol.* **78,** 134.

Sidransky, H., and Verney, E. (1967). *Biochim. Biophys. Acta* **138,** 426.

Sidransky, H., Staehelin, T., and Verney, E. (1964). *Science* **146,** 766.

Sidransky, H., Sarma, D. S. R., Bongiorno, M., and Verney, E. (1968). *J. Biol. Chem.* **243,** 1123.

Siegel. F. L., Roach, M. K., and Deville, W. B. (1963). *Federation Proc.* **22,** 680.

Siegel, F. L., Roach, M. K., and Pomeroy, L. R. (1964). *Proc. Natl. Acad. Sci. U.S.* **51,** 605.

Smith, R. E. (1966). *J. Nutr.* **89,** 271.

Smith, R. E., and Scott, H. M. (1965a). *J. Nutr.* **86,** 37.

Smith, R. E., and Scott, H. M. (1965b). *J. Nutr.* **86,** 45.

Smuckler, E. A., and Benditt, E. P. (1965). *Biochemisty* **4,** 671.

Snyderman, S. E., Holt, L. E., Jr., Norton, P. M., and Roitman, E. (1968a). *In* "Protein Nutrition and Free Amino Acid Patterns" (J. H. Leathem, ed.), p. 19. Rutgers Univ. Press, New Brunswick, New Jersey.

Snyderman, S. E., Holt, L. E., Jr., Norton, P. M., Roitman, E., and Phansalkar, S. V. (1968b). *Pediat. Res.* **2,** 131.

Solomon, A. K., and Gold, G. L. (1955). *J. Gen. Physiol.* **38,** 371.

Solomon, J. D., Johnson, C. A., Scheffner, A. L., and Bergeim, O. (1951). *J. Biol. Chem.* **189,** 629.

Soupart, P. (1960). *Clin. Chim. Acta* **5,** 235.

Sox, H. C., and Hoagland, M. B. (1966). *J. Mol. Biol.* **20,** 113.

Squibb, R. L. (1966). *J. Nutr.* **90,** 71.

Squibb, R. L. (1968). *Poultry Sci.* **47,** 199.

Squibb, R. L., Siegel, H., Solotorovsky, M., and Lyons, M. M. (1968). *Poultry Sci.* **47,** 519.

Staehelin, T., Verney, E., and Sidransky, H. (1967). *Biochim. Biophys. Acta* **145,** 105.

Steele, B. F., and LeBovit, C. B. (1951). *J. Nutr.* **45,** 325.

Steele, B. F., Reynolds, M. S., and Baumann, C. A. (1950). *Arch. Biochem.* **25,** 124.

Stein, W. H., and Moore, S. (1954). *J. Biol. Chem.* **211,** 915.

Steinberg, D., Vaughan, M., Sherman, F. G., and O'Dell, B. L. (1960). *Biochim. Biophys. Acta* **40,** 225.

Stenram, U. (1958). *Acta Pathol. Microbiol. Scand.* **44,** 239.

Stirewalt, W. S., and Wool, I. G. (1966). *Science* **154,** 284.

Stöckl, W., Stacher, A., Weiser, M., Zacherl, M. K., and Müller, H., (1967). *Z. Immunitaetsforsch.* **134,** 276.

Swendseid, M. E., Friedrich, B. W., and Tuttle, S. G. (1961). *Federation Proc.* **20,** 8.

Swendseid, M. E., Griffith, W. H., and Tuttle, S. G. (1963a). *Metab. Clin. Exptl.* **12,** 96.

Swendseid, M. E., Villalobos, J., and Friedrich, B. (1963b). *J. Nutr.* **80,** 99.

Swendseid, M. E., Villalobos, J., and Friedrich, B. (1964a). *J. Nutr.* **82,** 206.

Swendseid, M. E., Villalobos, J., and Drenick, E. J. (1964b). *Federation Proc.* **23,** 448.

Swendseid, M. E., Villalobos, J., Figueroa, W. G., and Drenick, E. J. (1965). *Am. J. Clin. Nutr.* **17,** 317.

Swendseid, M. E., Tuttle, S. G., Figueroa, W. G., Mulcare, D., Clark, A. J., and Massey, F. J. (1966). *J. Nutr.* **88,** 239.

Swendseid, M. E., Tuttle, S. G., Drenick, E. J., Joven, C. B., and Massey, F. J. (1967a). *Am. J. Clin. Nutr.* **20,** 243.

Swendseid, M. E., Yamada, C., Vinyard, E., Figueroa, W. G., and Drenick, E. J. (1967b). *Am. J. Clin. Nutr.* **20,** 52.
Swendseid, M. E., Yamada, C., Vinyard, E., and Figueroa, W. G. (1968). *Am. J. Clin. Nutr.* **21,** 1381.
Swendseid, M. E., Umezawa, C. Y., and Drenick, E. J. (1969). *Am. J. Clin. Nutr.* **22,** 740.
Taguchi, K. (1965). *Japan Circ. J.* **29,** 746.
Tallan, H. H., Moore, S., and Stein, W. H. (1954). *J. Biol. Chem.* **211,** 927.
Tannous, R. I., Rogers, Q. R., and Harper, A. E. (1966). *Arch. Biochem. Biophys.* **113,** 357.
Theimer, V. W. (1964). *Naturwissenchaften* **51,** 465.
Theurer, B., Woods, W., and Poley, G. E. (1968). *J. Animal Sci.* **27,** 1059.
Thompson, H. T., Schurr, P. E., Henderson, L. M., and Elvehjem, C. A. (1950). *J. Biol. Chem.* **182,** 47.
Tilton, K. S., Yang, S. P., and Deinken, E. A. (1965). *Federation Proc.* **24,** 168.
Tonoue, T., and Yamamoto, K. (1967). *Endocrinology* **81,** 101.
Tria, E., and Barnabei, O. (1963). *Nature* **197,** 598.
Truhaut, R., Delarue, J. C., and Bohuon, C. (1966). *Ann. Biol. Clin.* (*Paris*) **24,** 727.
Tuttle, S. G., Swendseid, M. E., Friedrich, B., and Griffith, W. H. (1962). *Federation Proc.* **21,** 395.
Vandermeers-Piret, M. C., Pokorni, E., Wodon, C., and Christophe, J. (1966). *Bull. Soc. Chim. Biol.* **48,** 525.
Van Slyke, D. D., and Meyer, G. M. (1912). *J. Biol. Chem.* **12,** 399.
Wagle, D. S., and Sidransky, H. (1965). *Metab. Clin. Exptl.* **14,** 932.
Walsh, D. A., and Sallach, H. J. (1966). *J. Biol. Chem.* **241,** 4068.
Wannemacher, R. W., and Yatvin, M. B. (1965). *J. Nutr.* **85,** 393.
Webb, T. E., Blobel, G., and Potter, V. R. (1966). *Cancer Res.* **26,** 253.
Weber, G., Srivastava, S. K., and Singhal, R. L. (1965). *J. Biol. Chem.* **240,** 750.
Wehr, R. F., and Lewis, G. T. (1966). *Proc. Soc. Exptl. Biol. Med.* **121,** 349.
Wellers, G., and Leblanc, M. (1966). *Compt. Rend. Soc. Biol.* **160,** 1785.
Wellers, G., and Leblanc, M. (1968). *Compt. Rend. Soc. Biol.* **162,** 39.
Westall, R. G. (1962). *In* "Amino Acid Pools" (J. T. Holden, ed.), p. 195. Elsevier, Amsterdam.
Wheeler, P., and Morgan, A. F. (1958). *J. Nutr.* **64,** 137.
Whitten, B. K., Hannon, J. P., Klain, G. J., and Chinn, K. S. K. (1968). *Metab. Clin. Exp.* **17,** 360.
Wilken, D. R., and Hansen, R. G. (1961). *J. Biol. Chem.* **236,** 1051.
Williams, J. N., Schurr, P. E., and Elvehjem, C. A. (1950). *J. Biol. Chem.* **182,** 55.
Wilson, S. H., Hill, H. Z., and Hoagland, M. B. (1967). *Biochem. J.* **103,** 567.
Wiss, O. (1948). *Helv. Chim. Acta* **31,** 2148.
Wiss, O. (1949a). *Helv. Chim. Acta* **32,** 153.
Wiss, O. (1949b). *Helv. Chim. Acta* **32,** 1344.
Wiss, O., and Krueger, R. (1949). *Helv. Chim. Acta* **32,** 527.
Wittman, J. S., Lee, K. L., and Miller, O. N. (1969). *Biochim. Biophys. Acta* **174,** 536.
Wool, I. G. (1960). *Am. J. Physiol.* **199,** 715.
Wool, I. G. (1964). *Nature* **202,** 196.
Wool, I. G., and Krahl, M. E. (1959). *Nature* **183,** 1399.

Wool, I. G., and Krahl, M. E. (1964). *Biochim. Biophys. Acta* **82,** 606.
Wool, I. G., and Scharff, R. (1968). *In* "Protein Nutrition and Free Amino Acid Patterns" (J. H. Leathem, ed.), p. 157. Rutgers Univ. Press, New Brunswick, New Jersey.
Wool, I. G., and Weinshelbaum, E. I. (1960). *Am. J. Physiol.* **198,** 1111.
Wright, L. A., and Nicholson, T. F. (1965). *Can. J. Physiol. Pharmacol.* **43,** 961.
Wu, C. (1954). *J. Biol. Chem.* **207,** 775.
Wu, C., Bollman, J. L., and Butt, H. R. (1955). *J. Clin. Invest.* **34,** 845.
Wunner, W. H., Bell, J., and Munro, H. N. (1966). *Biochem. J.* **101,** 417.
Wurtman, R. J., and Axelrod, J. (1967). *Proc. Natl. Acad. Sci. U.S.* **57,** 1594.
Wurtman, R. J., Chou, C., and Rose, C. M. (1967). *Science* **158,** 660.
Wurtman, R. J., Rose, C. M., Chou, C., and Larin, F. (1968a). *New Engl. J. Med.* **279,** 171.
Wurtman, R. J., Rose, C. M., Rose, L., Williams, G., and Lauler, D. (1968b). *Clin. Res.* **16,** 355.
Wurtman, R. J., Shoemaker, W. J., and Larin, F. (1968c). *Proc. Natl. Acad. Sci. U.S.* **59,** 800.
Yang, S. P., Clark, H. E., and Vail, G. E. (1961). *J. Nutr.* **75,** 241.
Yang, S. P., Steinhauer, J. E., and Masterson, J. E. (1963). *J. Nutr.* **79,** 257.
Yang, S. P., Tilton, K. S., and Ryland, L. L. (1968). *J. Nutr.* **94,** 178.
Yang, S. S., and Sanadi, D. R. (1969). *J. Biol. Chem.* **244,** 5081.
Yearick, E. S., and Nadeau, R. G. (1967). *Am. J. Clin. Nutr.* **20,** 338.
Young, V. R., and Scrimshaw, N. S. (1968). *Brit. J. Nutr.* **22,** 9.
Zachmann, M., Cleveland, W. W., Sandberg, D. H., and Nyhan, W. L. (1966). *Am. J. Diseases Child.* **112,** 283.
Zimmerman, R. A., and Scott, H. M. (1965). *J. Nutr.* **87,** 13.
Zimmerman, R. A., and Scott, H. M. (1967). *J. Nutr.* **91,** 507.
Zimmerman-Telschow, H. (1965). *Nutr. Dieta* **7,** 37.
Zimmerman-Telschow, H., and Jekat, F. (1965). *Nutr. Dieta* **7,** 283.
Zinneman, H. H., Nuttall, F. Q., and Goetz, F. C. (1966). *Diabetes* **15,** 5.
Zinneman, H. H., Seal, U. S., and Doe, R. P. (1969). *Am. J. Digest. Diseases,* [n.s.] **14,** 118.

CHAPTER 35

The Regulation of Intermediary Amino Acid Metabolism in Animal Tissues[1]

Joel H. Kaplan[2] and Henry C. Pitot[3]

Departments of Oncology and Pathology,
McArdle Memorial Laboratory,
The Medical School,
University of Wisconsin,
Madison, Wisconsin

[1] The work emanating from the authors' laboratory described in this review was supported in part by grants from the National Cancer Institute (CA-07175) and the American Cancer Society (P-314).

[2] Postdoctoral Fellow of the National Cancer Institute (CA-21666). Present address: General Electric Research and Development Center, Schenectady, New York.

[3] Career Development Awardee of the National Cancer Institute (CA-29,405).

I. Introduction

A consideration of the regulation of intermediary amino acid metabolism in a multicellular organism is quite different from a discussion of the same subject in microorganisms. By its very nature the microorganism must be designed for survival in an environment wherein it is wholly dependent on relatively simple molecular species to maintain its biochemical homeostasis. Thus, in the case of many wild-type microorganisms "essential" metabolites and amino acids are nonexistent since the bacteria are capable of producing these molecules from elements of ammonia, carbon dioxide, and the like. Some microorganisms, such as the streptococcus, have quite fastidious requirements approaching those seen in mammalian tissues. It is of interest that a large number of such fastidious organisms belong to the class that we call pathogenic. Thus, by existing as parasites, they have evolved a dependence of their nutrition upon the environment available to them within the host that they parasitize.

On the other hand, the multicellular mammalian organism has evolved to the point where its nutritional needs are met by mixtures of amino acids and other essential nutrients which, during evolution, it has apparently lost the ability to synthesize (see Chapter 25, Volume III). Therefore, unlike the wild-type microorganism, the mammal is entirely dependent upon its environment for the majority of the amino acids necessary to build its proteins since it lacks the biosynthetic mechanisms necessary to produce such "essential" amino acids. It has been suggested that, because of this fact, the regulation of intermediary amino acid metabolism in mammalian tissues may well be geared to a multivalent type of control rather than a univalent type (Peraino *et al.*, 1965). The regulation of the synthesis of the so-called nonessential amino acids in the mammal is relatively primitive compared to the fine mechanisms developed in microorganisms. However, the mammal has at its disposal a regulatory mechanism not available to the unicellular organism; that is, the hormonal control of metabolism.

It has become apparent from many studies that the liver and kidney of the mammal are the major sites of intermediary amino acid metabolism and thus of its regulation. In the liver, a number of enzymes involved in intermediary amino acid metabolism are found in the cell sap. Many of these enzymes may be altered by changes in diet or hormonal status of the animal. With the exception of the transaminases, tissues other than these have relatively low levels of enzymes involved in intermediary amino acid metabolism, particularly gluconeogenesis. In liver as well as kidney, a few enzymes, such as ornithine-δ-transaminase, are present in the mitochondrial fraction and recent studies on the microbody have suggested that this organelle may play a role in gluconeogenesis (de Duve and Baudhuin, 1966). Undoubtedly, in the mammalian organism the location of certain enzymes in itself acts as a regulatory mechanism. Some enzymes, such as glutamic-pyruvic transaminase, occur both in the soluble form as well as the mitochondrial form. Recent studies by Swick *et al.* (1968) have shown that the mitochondrial form of the enzyme is also subject to environmental regulation.

As with bacteria, there are certain examples in mammalian tissues of direct effect of substrates and regulators on enzyme activity by so-called allosteric effects. The enzyme tryptophan pyrrolase, which has been studied by numerous authors, has been shown to be allosterically inhibited by NADPH (Cho-Chung and Pitot, 1967). Several other examples are also now known, although it would appear that the extensive allosteric regulation demonstrable in microorganisms has not yet been demonstrated in mammalian systems.

This review deals with the regulation of intermediary amino acid metabolism in mammalian organisms. As with most reviews, this presentation is not meant to be comprehensive, but rather to cover the field in areas not discussed in other chapters of these volumes and to serve as an introduction to the reader for further intensive research on the subject.

II. Gluconeogenesis and Amino Acid Catabolism

Gluconeogenesis is the synthesis of carbohydrate, especially glucose, from noncarbohydrate compounds. The precursors include amino acids as well as the intermediates of the tricarboxylic acid cycle, glycerol, and lactate. A modified scheme depicting the entry of some of these compounds into the gluconeogenic pathway is shown in Fig. 1. The two main entry points into the pathway are either at the level of conversion of precursor to pyruvate or to oxaloacetate. Five of the amino acids which directly or indirectly undergo such conversion (serine, threonine, alanine, aspartic acid, and ornithine) are catalyzed by enzymes which have been extensively studied in regard to their regulation by various

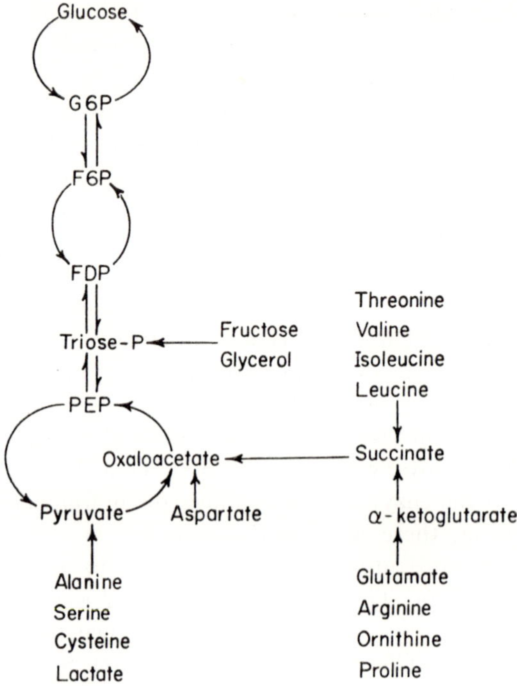

FIG. 1. Outline of pathway of gluconeogenesis from various precursors.

dietary and hormonal factors. Some of these studies will be reviewed in this section, with emphasis on the dietary induction of enzyme activity. Where applicable, the influence of these enzymes and their substrates on the gluconeogenic process itself also will be discussed. For a comprehensive discussion of enzyme regulation by hormones, the reader is referred to the review by Kenney in this volume (Chapter 31). Eisenstein (1967), Scrutton and Utter (1968), and Pontremoli and Grazi (1968) have recently reviewed the subject of gluconeogenesis, covering the role of enzymes not involved in amino acid metabolism.

A. Threonine-Serine Dehydratase

It is accepted that the activity of gluconeogenesis from amino acids in the liver is enhanced when the utilization of glucose is limited under various hormonal and dietary conditions such as diabetes, starvation, or administration of a high-protein diet free of carbohydrate. There is much evidence at present (Henning *et al.,* 1963; A. D. Freedman and Kohn, 1964; Lardy *et al.,* 1965b) to show that the sequence of gluconeogenic reactions is initiated by carboxylation of pyruvate to oxaloacetate by the enzyme pyruvate carboxylase (Utter and Keech, 1963;

Nordlie and Lardy, 1963). Pyruvate is therefore considered to be of great importance as a carbon source in gluconeogenesis (Fig. 1). From this point of view, an important physiological role is suggested for the enzyme serine dehydratase. Changes in serine dehydratase activity in rat liver under various conditions are parallel to those found for such key gluconeogenic enzymes as glucose-6-phosphatase, fructose-1,6-diphosphatase, phosphoenolpyruvate carboxykinase, and pyruvate carboxylase. Furthermore, these changes are in an intimate reciprocal relation to those of glucokinase, which catalyzes the first step of glucose utilization.

Serine dehydratase is a soluble enzyme which catalyzes the nonoxidative deamination of serine to pyruvic acid and ammonia (Meister, 1965): L-serine $\rightarrow$ pyruvate + NH_3. In rats, appreciable serine dehydratase activity is found only in liver (Freedland and Avery, 1964). The enzyme has recently been purified to homogeneity and shown to have a molecular weight of 63,500 (Nakagawa *et al.*, 1967). However, other investigators have found the molecular weight to be smaller (Nagabhushanam and Greenberg, 1965). This in part may be due to differences in purification procedures. The enzyme has a cofactor requirement for pyridoxal phosphate, but, unlike its counterpart in bacteria (Whitely and Tahara, 1966; Phillips and Wood, 1964; Hayaishi *et al.*, 1963), its activity is not affected by the presence of various purine and pyrimidine nucleotides (Nakagawa *et al.*, 1967; J. H. Kaplan and Pitot, 1969). Since the pure enzyme also catalyzes the conversion of threonine to α-ketobutyrate, it also can be referred to as threonine dehydratase. In addition, dietary and hormonal factors affecting the activity of serine dehydratase cause a proportional change in threonine dehydratase (Freedland and Avery, 1964). This is interpreted as being indicative that both these activities are due to a single protein.[4]

Serine dehydratase is an enzyme that can be considerably altered in amount by changing the dietary conditions of the animal (Pitot *et al.*, 1961; Bottomley *et al.*, 1963). Feeding *ad libitum* of a 90% casein diet for 7 days increased serine dehydratase activity by 80-fold as compared to chow-fed rats, presumably on a normal protein intake (Pitot and Peraino, 1964a). If rats that were fed a protein-depleted diet for 5 days were repeatedly tube-fed with 1-gm aliquots of enzymically hydrolyzed casein in water, a marked increase in serine dehydratase (300-fold) occurred over a 24-hour period. The induction of the enzyme was

[4] In a recent report, Veneziale *et al.* (1969) have presented results which suggest that the role of serine dehydratase in serine metabolism requires reevaluation. Therefore, the major function of the enzyme would be to catalyze the conversion of threonine to α-ketobutyrate.

dependent on the amount of casein hydrolyzate given between 0.1 and 1.0 gm per dose. A single dose of 1.0 gm of hydrolyzate resulted in induction up to 12 hours with an almost complete return to the control levels at 18 hours. Since these animals were protein-depleted, it was assumed that a complete mixture of amino acids must be fed before hepatic enzyme synthesis could occur. Threonine administered alone to protein-depleted or to chow-fed rats (Goldstein *et al.*, 1962) produced little or no threonine dehydratase. Freedland and Avery (1964) also showed that this enzyme was increased in activity by giving a high-protein diet, but not by feeding excess dietary threonine.

Further studies were carried out to determine whether or not a degree of specificity existed in the dietary amino acid requirements for induction of threonine dehydratase as well as for ornithine-δ-transaminase. The latter enzyme had been shown previously to increase 20-fold in activity over a 24-hour period by multiple intubations of casein hydrolyzate. In these experiments (Peraino *et al.*, 1965), threonine dehydratase and ornithine-δ-transaminase activities were measured as a function of the qualitative and quantitative composition of a mixture of free amino acids administered to rats by stomach tube. These investigators found that a mixture consisting of the 10 essential amino acids in equimolar proportion produced an induction which was 2- and 4-fold greater than the induction produced by a mixture consisting of these amino acids in proportions similar to their concentration in mammalian liver. Feeding the equimolar mixture lacking lysine and arginine caused good induction of threonine dehydratase but poor induction of ornithine-δ-transaminase, whereas the equimolar mixture lacking threonine and phenylalanine had the opposite effect. Feeding the equimolar mixture lacking tryptophan produced little or no induction of threonine dehydratase or ornithine-δ-transaminase in rats which had not been deprived of food previously. Prior fasting lessened the tryptophan requirement for threonine dehydratase induction and abolished the tryptophan requirement in the case of ornithine-δ-transaminase induction. The feeding of tryptophan alone produced a significant but submaximal induction of both enzymes. None of the other amino acids produced induction of either enzyme when administered singly. From these results it became apparent that the factors that control the levels of certain amino acid catabolizing enzymes are independent of the general nitrogen balance of the animal.

The use of inhibitors of RNA synthesis (actinomycin D and fluoroorotic acid) and of protein synthesis (puromycin) have provided indirect evidence that these increases in enzyme activity are the result of protein synthesis *de novo* (Pitot and Peraino, 1964a). Puromycin administration at zero time prevented enzyme induction of threonine dehydratase activity. When given 12 hours after the first intubation of casein hydrolyzate (in animals repeatedly intubated every 6 hours), puromycin inhibited further induction. Actinomycin D and 5-fluoroorotic acid given at zero time also prevent induction; however, these compounds had little or no effect when given from the 12-hour point on, as shown in Fig. 2. These results were interpreted to mean that, at the initial point of induction, the synthesis of new messenger RNA was required for induction, but that at the 12-hour point the messenger RNA was stable and thus the antibiotic had no effect on the increasing enzyme level at this time. In more recent experiments, J. H. Kaplan *et al.* (1969) have measured the net rate of enzyme synthesis in casein-induced rats by pulse labeling *in vivo* with valine-^{3}H and quantitative precipitation of the partially purified serine dehydratase antigen with pure anti-serine dehydratase antibody. It was found that actinomycin D inhibited enzyme induction when given at the beginning of the induction period, but had little or no effect when administered between 1.5 and 9 hours after a single dose of casein was intubated. Thus, one may define the template lifetime of serine dehydratase as being about 6–8 hours. This use of immunochemical methods to demonstrate insensitivity of enzyme induction to delayed actinomycin D administration takes on further significance in light of recent findings (Tomkins *et al.*, 1966; Reel and Kenney, 1968; Kenney, this volume, Chapter 31) of a "paradoxical" effect by this antibiotic; that is, under certain conditions enzyme *activity* may increase although the synthesis of new messenger RNA is apparently blocked.

Utilizing the technique of pulse labeling of the enzyme, serine dehydratase, *in vivo* with valine-^{14}C, Jost *et al.* (1968) showed that administration of mixtures of amino acids dramatically increase the rate of synthesis of the enzyme *in vivo* without appreciably enhancing the synthesis of soluble proteins, thus indicating that a specific induction of synthesis of this enzyme occurred. As seen in Table I, the amino acid L-tryptophan, as well as the hormone glucagon, effect an increased rate of synthesis, the latter being the most efficient inducer of the enzyme among those compounds studied (see Chapter 31).

B. Glutamic-Alanine Transaminase

Glutamic-alanine transaminase catalyzes the transamination of alanine to pyruvate (Segal *et al.*, 1962)

$$\text{L-alanine} + \alpha\text{-ketoglutarate} \rightleftharpoons \text{pyruvate} + \text{L-glutamate}$$

and therefore would seem to occupy a key position in the metabolic interelationships involving gluconeogenesis and intermediary protein and

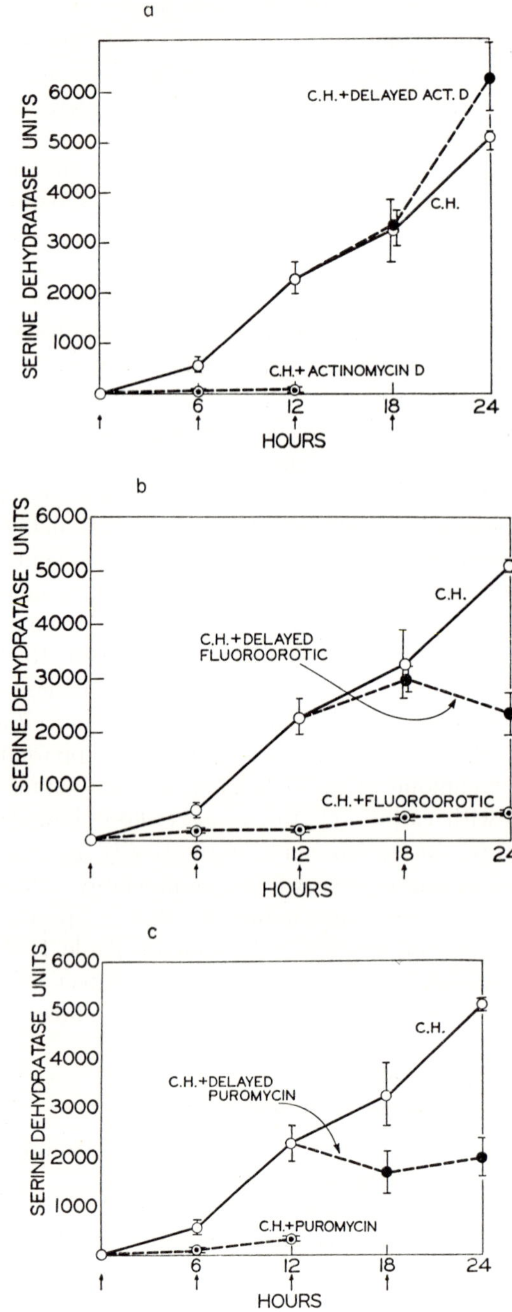

FIG. 2. The effect of actinomycin D (a), 5-fluoroorotic acid (b), and puromycin (c) administration on the dietary induction of threonine dehydratase. See Pitot and Peraino (1964a) for experimental details.

TABLE I

INDUCTION AND REPRESSION OF SERINE DEHYDRATASE IN LIVERS OF INTACT RATS[a]

			Valine-^{14}C-incorporation	
Treatment	Number of animals	Serine dehydratase activity (units/gm liver)	Serine dehydratase per gram liver (dpm)	Soluble Proteins per gram liver (dpm)
Protein control, 0%	5	4.4 ± 0.4	66.0 ± 23.0	320,000
Amino acids				
6 hours	6	60.0 ± 14.3	836.0 ± 137.0	
10 hours	7	125.6 ± 27.1	1816.0 ± 396.0	325,000
18 hours	3	260.0 ± 20.8	1850.0 ± 427.0	
Amino acids + glucose[b] (10 hours)	5	5.6 ± 0.5	40.5 ± 24.3	324,000
Amino acids (10 hours) + delayed glucose[b] (6 hours only)	4	43.6 ± 13.4	150.1 ± 45.9	489,000
L-Tryptophan (10 hours)	3	71.9 ± 6.1	1365.0 ± 60.0	273,000
Glucagon (10 hours)	3	509.7 ± 57.0	5858.0 ± 706.0	465,000

[a] The experimental details may be found in Jost *et al.* (1968). An incomplete mixture of essential amino acids, lacking valine and methionine, was administered at the times indicated. L-Tryptophan, 0.37 mmole was administered at 0 time and 6 hours to the group receiving this single amino acid. Glucagon was administered intraperitoneally at a dose of 0.2 mg/100 gm body weight at 0 time and again at 6 hours. The hours in parentheses denote the time of death in relation to 0 time, with the exception of the "6 hours only" designation for delayed glucose, which signifies the only time of glucose administration.

[b] See text in Section III, A for details. The data in this chart are taken from the data of Jost *et al.* (1968).

carbohydrate metabolism. Rosen *et al.* (1959) observed an increase in activity of this enzyme in livers of rats subjected to several conditions (hydrocortisone, diabetes, fasting, high-protein diet) associated with enhanced gluconeogenesis. Further studies (Waldorf *et al.,* 1963) on the influence of diet on glutamic-alanine transaminase revealed that increases in enzyme activity depended on increased casein intake, not on decreased carbohydrate intake; feeding low-carbohydrate, high-fat diets caused no observable change in enzyme activity. This observation was also made for glutamic-aspartate transaminase (Section II,C).

Segal and Kim (1963) showed that the corticosteroid induction of hepatic glutamic-alanine transaminase results from an increased rate of enzyme synthesis. Adrenal cortical hormones when administered *in vivo* may also cause an increased conversion of alanine to glucose by stimulating an increase in the intrahepatic concentration of this gluconeogenic amino acid (Noall *et al.,* 1957; Betheil *et al.,* 1965). Herrera *et al.* (1966) have reported that increasing the concentration of alanine in the medium in a perfused rat liver preparation increases net hepatic glucose production. These workers proposed that the physiological mobilization of endogenous amino acids may regulate hepatic gluconeogenesis by determining the amount of substrate immediately available and by increasing or diminishing the activity of rate-limiting gluconeogenic enzymes by a protracted increase or decrease in substrate concentration.

Garcia *et al.* (1966) observed that glucagon stimulated incorporation of labeled alanine, lactate, and pyruvate into glucose by the isolated perfused liver. The concomitant rise of urea formation with increased glucose production from alanine during glucagon administration suggested that the hormone may act by stimulating glutamic-alanine transaminase activity (Curry and Beaton, 1958).

Freedland and co-workers have reported (1968) on the influence of high-protein diet and glucocorticoids on several enzymes associated with amino acid catabolism (glutamic-alanine transaminase, glutamic-asparate transaminase, and serine dehydratase). Data are presented showing that the percentage increases in the two transaminases and serine dehydratase in response to a high-protein diet were similar in both adrenalectomized and intact rats. This is indicative that glucocorticoids are either not essential or play no role in the increased activities of these enzymes caused by high-protein feeding. These investigators made the important observation that the effects of hormones on enzyme activity cannot be properly evaluated without considering the nutritional condition of the animal, in particular the dietary regimen.

C. Glutamic-Aspartate Transaminase

The role that glutamic-aspartate transaminase plays in gluconenogenesis is to catalyze the conversion of aspartate to oxaloacetate (Velick and Vavra, 1962)

$$\text{L-aspartate} + \alpha\text{-ketoglutarate} \leftrightharpoons \text{oxaloacetate} + \text{L-glutamate}$$

Oxaloacetate in turn may be converted to phosphoenolpyruvate in the extramitochondrial compartment of the cell by the enzyme phosphoenolpyruvate carboxykinase (PEPCK) (Seubert and Huth, 1965; Utter and

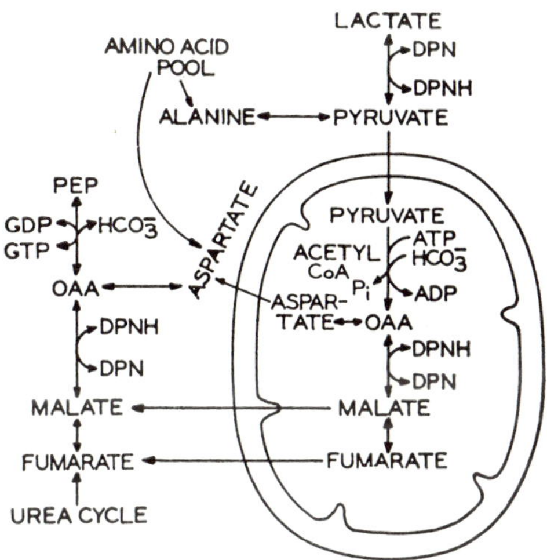

FIG. 3. Proposed scheme for phosphoenolpyruvate formation in rat liver from dicarboxylic and amino acids. (From Shrago and Lardy, 1966; reproduced by courtesy of the *Journal of Biological Chemistry*.)

Kurahashi, 1954). Amino acids that yield pyruvate follow a somewhat different pathway. Pyruvate carboxylase (PC) is located primarily in liver mitochondria while PEPCK occurs in the soluble fraction of the cell (Shrago and Lardy, 1966) in rat, mouse, and hamster liver. In addition, oxaloacetate does not diffuse out of the mitochondria. Therefore, the formation of this metabolite from pyruvate would seem to provide an obstacle to the synthesis of phosphoenolpyruvate. Instead, oxaloacetate is reduced to malate or transaminated with glutamate to form asparate (Lardy *et al.*, 1965b). Malate, aspartate, and α-ketoglutarate readily diffuse from the mitochondria to the soluble part of the cell, where they are reconverted to oxaloacetate (Fig. 3). The latter compound is transformed to phosphoenolpyruvate by PEPCK. Shrago and Lardy (1966) demonstrated that the enzymes necessary for reconversion of malate (malic dehydrogenase) and aspartate (glutamic-asparate transaminase) to oxaloacetate are present in the soluble component of the cell and that the activities of these enzymes rise when gluconeogenesis occurs at an increased rate.

D. Amino Acid Catabolism and the Regulation of Gluconeogenesis during Development

The fetal rat liver does not synthesize glucose or glycogen until birth, at which time the gluconeogenic pathway begins to function (Ballard and Oliver, 1963, 1965). Ballard and Hanson (1967) presented evidence suggesting that hepatic gluconeogenesis is initiated in the newborn rat by a sharp increase in the activity of phosphoenolpyruvate carboxykinase (PEPCK). Yeung and Oliver (1967) employing enzymic and isotopic techniques, studied the utilization of amino acids for carbohydrate synthesis in neonatal rat liver. After a short-term incubation of neonatal rat liver homogenates with α-ketoglutarate and a ^{14}C-labeled amino acid mixture, oxaloacetate, pyruvate and α-ketoglutarate together were found to account for 94% of the radioactive keto acids that were formed. The total radioactivity was distributed as follows: 67% in oxaloacetate, 15% in pyruvate, 12% in α-ketoglutarate, and 6% unidentified. It thus seemed likely that the major contribution to the observed gluconeogenesis was made by asparate, alanine, and glutamate. In agreement with this conclusion was the fact that a marked postnatal increase in soluble glutamic-aspartate transaminase activity was observed, whereas glutamic-alanine transaminase, although weakly active in neonatal rat liver, increased significantly only in late postnatal life.

The coordinated increase in activity between the key gluconeogenic enzymes, PEPCK and pyruvate carboxylase (Ballard and Hanson, 1967) and the catabolizing enzymes of the gluconeogenic amino acids in neonatal liver (Yeung and Oliver, 1967) makes it tempting to propose that all these enzymes have a common site of regulation. Such a mechanism of regulation might be analogous to the action of repressor and inducer molecules on the operon sites of microorganisms (Jacob and Monod, 1961; Ames and Hartman, 1963), or to the turning on and off of blocks of cistrons by nucleohistones (Bonner and Ts'o, 1964). On the other hand, regulation could occur at the translational level of protein synthesis through the stabilization of specific messenger RNA molecules (Pitot *et al.*, 1965b; Pitot, 1969a) at various times during fetal and postnatal development.

III. The Glucose Effect and Amino Acid Regulation

Repression of enzyme synthesis by catabolites or glucose is well known in microorganisms (Magasanik, 1961; Adhya and Echols, 1966; Moses and Prevost, 1966). Several examples of glucose or carbohydrate-induced inhibition of enzyme levels *in vivo* have also been reported in liver. Tschudy *et al.* (1964) have shown that carbohydrate administration decreases the level of the mitochondrial enzyme δ-aminolevulinic acid synthetase in liver. Dietary carbohydrate affects hepatic dimethylaminoazobenzene reductase in a similar manner (Jervell *et al.*, 1965), and elevated levels of hepatic phosphoenolypyruvate carboxykinase may be suppressed by administration of glucose or glycerol (Lardy *et al.*, 1965a;

Shrago *et al.*, 1967). Studies in microorganisms have indicated that glucose may be exerting its effect of repressing enzyme synthesis by an action either at the RNA template level (Hauge *et al.*, 1961) or at the level of genetic transcription (Nakada and Magasanik, 1964). Magasanik (1961) and Loomis and Magasanik (1966) have suggested that the concentration of certain catabolites in repressed cells determines the degree of repression of a specific enzyme.

The purpose of this section is to review some studies where an effect of glucose has been observed in mammalian liver on the rate of enzyme increase stimulated by amino acids. Emphasis will be placed on several amino acid-catabolizing enzymes.

A. Glucose Repression of Threonine-Serine Dehydratase Induction

Serine dehydratase appears to play a key role in the metabolic interrelationships involving gluconeogenesis and intermediary protein and carbohydrate metabolism (Section, II, A). Therefore, the ability of glucose or a metabolite thereof to regulate the synthesis of this enzyme would be of great importance to an organism where a homeostatic balance must be maintained between oxidation and assimilation of carbon compounds.

Ten years ago, early studies by Pitot (1959) demonstrated that the level of serine-threonine dehydratase in rat liver is an almost direct function of the protein content of the diet beyond a protein level of approximately 12%. Since that time several authors have confirmed these studies (Fallon 1968; Harper, 1968). Later Pitot and Peraino (1963) reported evidence for a carbohydrate effect on enzyme induction in the mammal which resembles the glucose effect on enzyme induction as seen in microorganisms (Magasanik, 1961). They found that, if glucose is given along with dietary casein, an almost complete suppression of threonine dehydratase induction occurs. This appeared to be specific for carbohydrate since fat had no effect. Later results (Peraino and Pitot, 1964) indicated that the effect was probably not due to altered absorption of dietary amino acids from the gastrointestinal tract, nor did it arise from a generalized depression of protein synthesis in the liver or from an ATP deficiency. The administration of fructose could effect a repression of this enzyme as efficiently as glucose.

Using specific immunochemical techniques and the quantitative precipitation of the enzyme by its specific antibody, Jost *et al.* (1968) were able to demonstrate that administration of glucose resulted in a complete cessation of serine dehydratase synthesis whether it was administered from zero time or at a time interval thereafter at which synthesis of the enzyme is completely resistant to the action of actinomycin D (Table I). Studies of the composition of amino acids pools in the intact animal showed that the different rates of synthesis seen in rats fed amino acids with respect to those fed amino acids plus glucose were not the result of differential labeling of the labeled precursor pool. By prelabeling the enzyme and administering glucose with and without amino acids, it was shown that glucose appeared to effect primarily a cessation of enzyme synthesis as well as an increase of degradation of the enzyme. The effects of glucose administration on protein synthesis did not appear to be general since serum protein synthesis was not affected by this treatment. Earlier work by Pitot *et al.* (1964) and the data of Tepperman and Tepperman (1958) indicated that certain enzymes are induced by the administration of glucose and thus must be regulated by different mechanisms.

Since the administration of large amounts of glucose to animals stimulates insulin release from the pancreas (see Chapter 31), it is reasonable to suggest that this hormone plays a major role in the repression by glucose of serine dehydratase synthesis. This is especially true since Suda and his associates (Ishikawa *et al.,* 1965) have demonstrated that serine dehydratase levels in liver of alloxan-diabetic rats is markedly higher than that seen in normal animals. Administration of low-protein, high-carbohydrate diet does not appear to alter the level of this enzyme in the liver of the alloxan-diabetic animals. Furthermore, glucagon has been shown to relieve the glucose effect (Peraino and Pitot, 1964). Therefore, glucose itself does not appear to be the mediator of the glucose effect. Rather, both glucose and insulin appear to be required. In more recent studies Söling *et al.* (1968) have shown that when alloxan-diabetic rats are fed a high-protein diet there is a dramatic increase in the level of serine dehydratase activity. This high level of activity is prevented by simultaneous administration of glucose but is not affected by a 48-hour administration of insulin. Glucose and insulin administered together do not have an additive effect. This would suggest that the effect of glucose in repressing the synthesis of serine dehydratase is *not* primarily mediated by insulin, but is probably an extremely complex mechanism which involves several hormones, such as glucagon, hydrocortisone, and possibly others.

A metabolite that may play an important role in the regulation of enzyme levels is cyclic-3′,5′-AMP. (Robison *et al.,* 1968; Perlman and Pastan, 1968; Kenney, this volume, Chapter 31). Khairallah and Pitot (1967) have presented evidence which suggests that this metabolite may regulate the rate of enzyme release from the polysome, and thus the rate of enzyme synthesis at the template level. Pitot and Jost (1968) using specific immunochemical techniques, have found that, when glucose is administered to animals during a time of enzyme induction, there is an increase in polysome-bound serine dehydratase in liver. These data suggest that glucose, or some metabolite thereof, may affect the release of completed polypeptide chains from the translating unit. An action of a metabolite at the translational level is supported by the fact that the dietary induction of serine dehydratase is prevented when glucose is administered at a time when synthesis of the enzyme is insensitive to actinomycin D (Jost *et al.*, 1968; Pitot *et al.*, 1965a).

What part cyclic AMP plays in the induction and repression of enzyme synthesis has now become an important question for researchers concerned with the problems of metabolic regulation. Kenney in this volume (Chapter 31) has presented a detailed discussion of the possible role that cyclic AMP may play in hormonal regulation of enzyme synthesis.

B. Glucose Repression of Ornithine-δ-Transaminase Induction

Much of what has been said about the glucose effect on threonine-serine dehydratase induction can be applied to ornithine transaminase. Although serine dehydratase is much more sensitive to the repressive effects of glucose than ornithine-δ-transaminase, both enzymes respond qualitatively in the same way to carbohydrate repression (Peraino and Pitot, 1964). Thus the dietary induction of this transaminase is prevented when glucose is administered either from zero time or at a time interval thereafter at which the apparent synthesis of the enzyme is resistant to actinomycin D. Administration of glucagon at the 0.2-mg level could restore the induction of the enzyme in the presence of 2.0-gm doses of glucose to approximately 75% of that observed with glucagon alone. When alloxan-diabetic rats were fed a high-protein diet (Söling *et al.*, 1968) there was an increase in the level of ornithine transaminase activity which could be prevented by simultaneous feeding of glucose. By contrast, the administration of insulin for a 2-day period to these animals produced no significant change in enzyme level.

Experiments utilizing the techniques of pulse labeling ornithine-δ-transaminase *in vivo* with ^{14}C-labeled amino acids and quantitative precipitation of the enzyme by its specific antibody now have been performed (Park, 1969). Administration of mixtures of amino acids increases the rate of synthesis of the enzyme in a manner similar to that found for serine dehydratase. Preliminary data indicate that delayed administration of glucose results in inhibition of enzyme synthesis.

C. A Possible Glucose Effect on Glutamic-Alanine Transaminase

Ruderman and Herrera (1968) have reported that, in perfused rat liver, both amino acid utilization and urea production were diminished by high glucose concentrations. This has been suggested also by the association between rates of gluconeogenesis and urea production (Miller, 1961; Mortimore, 1963).

The site(s) at which glucose suppresses gluconeogenesis from alanine is still not known. Exton and Park (1967) reported that the step limiting the maximum rate of gluconeogenesis from lactate appeared to be located in the sequence of reactions converting pyruvate to phosphoenolpyruvate (Fig. 1). They also found that high concentrations of glucose did not suppress gluconeogenesis from saturating concentrations of lactate. These results taken in conjunction with the findings of Ruderman and Herrera (1968) might suggest that glucose was inhibiting the activity of glutamic-alanine transaminase and thus suppressing gluconeogenesis from the point of conversion of alanine to pyruvate. Of course other possibilities cannot be ruled out.

It thus appears that the enzymes which catalyze the catabolism of the gluconeogenic amino acids as well as at least one key gluconeogenic enzyme in the common pathway, namely phosphoenolpyruvate carboxykinase (Lardy *et al.*, 1965a), can be repressed by glucose administration. These same enzymes also can be induced in their activity and synthesis by the hormone glucagon (Peraino and Pitot, 1964; Shrago *et al.*, 1963), which has been found to relieve the glucose effect. Numerous studies have shown that one of the metabolic effects of this hormone on liver is to increase the intercellular level of cyclic AMP (Kenney, this volume, Chapter 31). In mammalian systems it is well known that cyclic AMP catalyzes reactions that make energy available to the cell. Muscle phosphorylase (Krebs, 1966) is an excellent example of such an enzyme. Thus there exists an antagonism between cyclic AMP and glucose, the former stimulating enzymic reactions which supply glucose to the cell and the latter inhibiting such reactions. The action of these two metabolites at the molecular level is now becoming the subject of much exciting speculation and experimentation (Jost *et al.*, 1969; Mallette *et al.*, 1969).

IV. Regulation of the Intermediary Metabolism of Specific Amino Acids and Groups of Amino Acids in Mammalian Tissues

A. Arginine, Ornithine, Citrulline, and the Urea Cycle

In all ureotelic animals a metabolic cycle, known as the urea cycle, functions in the production of urea to be excreted as a waste product of nitrogen metabolism. The enzymes involved in the cycle are five: carbamyl phosphate synthetase, ornithine transcarbamylase, argininosuccinate synthetase, argininosuccinase, and arginase (Meister, 1965). The metabolic pathways involved are shown in Fig. 4.

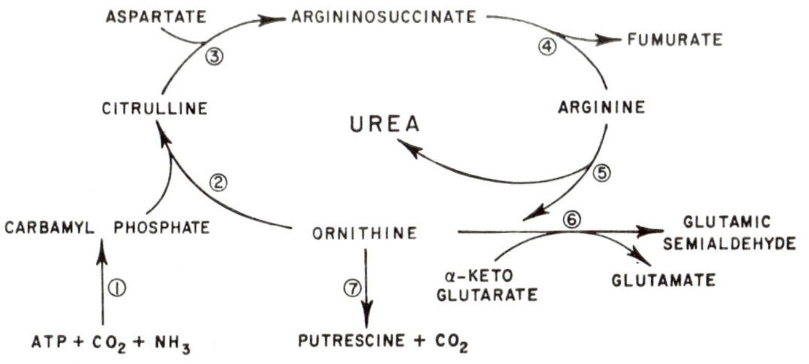

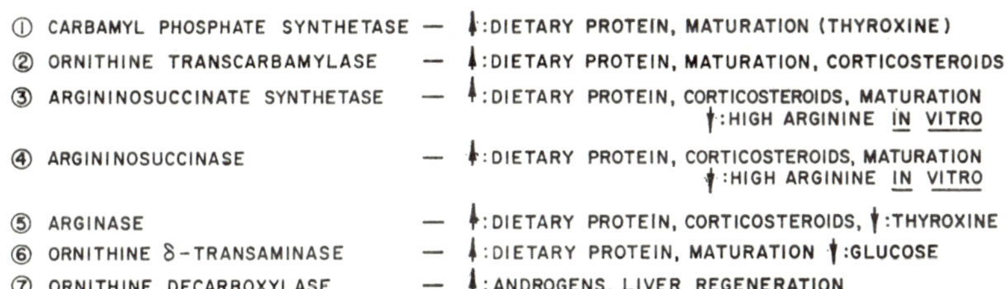

FIG. 4. Regulation of enzymes and pathways of arginine, ornithine, and citrulline metabolism.

It is readily apparent, however, that the regulation of these various enzymes in mammalian tissues is not linked as is that of the enzymes of arginine biosynthesis in microbial tissues. In liver the ratio of arginase, ornithine transcarbamylase, argininosuccinase, and a fourth enzyme not directly involved in the urea cycle but quite active in ornithine metabolism, ornithine-δ-transaminase are found in ratio of their activities 3000:414:10:1, respectively. In kidney the ratio of arginase to ornithine-δ-transaminase to arginine-glycine-transamidinase is of the order of 24:2.7:1.0 (Civen *et al.*, 1967). Of these various enzymes Civen and his associates were able to demonstrate that the enzyme most sensitive to dietary amino acids was ornithine-δ-transaminase, which had the lowest specific activity of the enzymes studied in liver. On the other hand, Schimke (1964b) demonstrated that the change in levels of arginase in liver brought about by changes in dietary protein intake was probably the result of alteration in the rates of enzyme degradation with concomitantly no change or decrease in the rates of synthesis. It thus may be that the variation in activities of these various enzymes seen in the organs is primarily a result of the rate of turnover of each of the enzymes. To substantiate this type of thinking, the work of Berlin and Schimke (1965) demonstrated that, while cortisone administration dramatically increased the activity of tryptophan pyrrolase and tyrosine transaminase in a very short period, it only slowly increased the activity of glutamic-alanine transaminase and arginase. However, when the rate of enzyme synthesis was determined for each of the enzymes, it was found that cortisone stimulated an increase in this rate to almost equal amounts. The reason for the discrepancy between the effect on specific activity and on enzyme synthesis in response to cortisone administration is fully explained in Chapter 32 by Schimke, who points out that enzymes with a short half-life will inevitably show a large effect of a given chain in protein synthesis, whereas those with a long half life will not show a comparable increase when the same number of new enzyme molecules is added to the pool.

The dietary regulation of ornithine-δ-transaminase has been studied rather extensively by Waldorf and Harper (1963) as well as Peraino and Pitot (1964). These studies demonstrated that the force-feeding of casein hydrolyzate or a mixture of certain amino acids including arginine and lysine to protein-depleted rats resulted in rather dramatic increases in the level of this enzyme over a period of 6–24 hours. If only a single dose of amino acids or casein hydrolyzate was administered, the enzyme increased over the 24-hour period but at a very slow rate, leveling off toward the end of this time period. Thus, there was no evidence of the usual sigmoid-shaped curve as seen with enzymes such as tryptophan pyrrolase and tyrosine transaminase after the administration of either hormone or amino acid. The reason for this phenomenon was argued to be that the enzyme, which occurs in the mitochondrion, once synthesized assumes the decay period of the mitochondrial protein itself, i.e., 9.6 days (Wilson and Dove, 1965), although recent data presented by Swick, *et al.* (1968) suggest a half-life of 1 day compared with 4–6 days for total mitochondrial protein. The fact that ornithine-δ-transaminase is probably involved primarily in degradative reactions has been suggested by Strecker (1965) and also by our own work. However, in a recent publication, A. D. Smith *et al.* (1967) demonstrated that the enzyme was probably involved in the conversion of proline to ornithine and thus could also be considered as a synthetic pathway enzyme. Recent studies from several laboratories (Katunuma *et al.,* 1964) as well as from our own laboratory (Park, 1969) have demonstrated that certain amino acids, notably valine and isoleucine, inhibit the enzyme competitively with respect to ornithine. The function of such a regulation in the overall organismal economy is not understood at the present time and it would appear that this mechanism does not fit into the so-called allosteric effects that have been described. Recently Herzfeld and Knox (1968) have shown that, during the first 3 months after birth in rat liver and kidney, ornithine transaminase increased in level in the female to twice that in the male. Ovariectomy prevented the sex difference whereas estrogen administration increased the enzyme in both sexes. Rat liver arginase, as indicated earlier, has been shown to vary with protein content of the diet as well as with urinary urea excretion (Ashida and Harper, 1961). In a recent review (Knox and Greengard, 1965), the effects of substrate protein and manganese on liver arginase activity have been described. Specifically, arginase activity was markedly decreased in manganese-deficient rats. However, as the reviewers point out, there appears to be some discrepancy in the various results reported from different laboratories. An interesting note by Cittadini *et al.* (1964) suggested that administration of L-lysine *in vivo* inhibits arginase activity in liver measured *in vitro*. The inhibition was maximal (56%) at 1 hour after lysine administration. In chicks, lysine increases kidney arginase activity (Nesheim, 1968).

Several studies on the effects of hormones on the urea cycle enzymes have also been reported. Adrenalectomy causes significant decreases in arginase and argininosuccinate synthetase activity (Freedland, 1964) in rats on a 65% dextrin–25% casein diet fed *ad libitum*. Arginase activity is markedly increased in the thyrotoxic rat, whereas carbamyl phosphate synthetase is unchanged (Grillo, 1964). It is of interest that this result is somewhat in contrast to that seen in frogs where the latter enzyme increases in the developing tadpole after injection of thyroxine (Paik and Cohen, 1960). In this system, the product of the enzyme, carbamyl phosphate, is used primarily for pyrimidine biosynthesis. Recently McLean and Gurney (1963) extended the data of Freedland (1964) in demonstrating that all enzymes of the urea cycle are decreased in adrenalectomized animals. Administration of cortisone to normal or adrenalectomized animals increases the level of argininosuccinate synthetase, argininosuccinase and arginase. If animals are treated with growth hormone, there appears to be a reduction in the activity of the argininosuccinate synthetase activity. Peraino (1968) has demonstrated the interesting effect that cortisone administration to animals on a high-protein diet actually inhibited the usual dietary stimulation of ornithine-δ-transaminase. This is in distinct contrast to a number of the soluble enzymes which are increased by the administration of cortisone, thus suggesting that this mitochondrial enzyme is regulated in a different manner when this steroid is administered.

In the developing mammal, Jones *et al.* (1961) had demonstrated that both carbamyl phosphate synthetase and ornithine transcarbamylase only appeared late in fetal life in the rat and did not reach adult levels until after birth. However, arginase occurs throughout the life span of the fetus. In the human fetus argininosuccinase, carbamyl phosphate synthetase, and ornithine transcarbamylase reach adult values during the course of fetal development (Volume III, Chapter 26). However, the levels of arginase apparently do not reach their adult levels until after birth. In the rat, as mentioned, a similar phenomenon occurs such that at the time of birth the levels of arginase and ornithine transcarbamylase in liver are essentially identical (Illnerova, 1968a). The level of argininosuccinate synthetase, however, is considerably lower than this, approximately 0.3% of these values. About 14 days after birth the activity of argininosuccinate synthetase in liver begins to rise and reaches adult levels shortly after weaning. Ornithine-δ-transaminase is at extremely low levels before birth in the rat. On the day of birth there is a transient increase in the level of this enzyme to about 20% of the adult level, which then decreases again to very low levels and does not begin to show significant increases then until after the tenth postnatal day. Administration of the corticosteroid analog triamcinolone prior to the natural increase in the enzyme after birth causes a marked elevation in the enzyme within 24 hours. It is of interest that, in fetal and adult animals, triamcinolone does not produce this increase (Raiha and Kekomaki, 1968). As has been shown in other systems, the level of the enzyme aspartate transcarbamylase is reciprocal with that of ornithine transcarbamylase. The activity of this enzyme is high during fetal life and falls rapidly after birth (Illnerova, 1968b). In the chick embryo, arginase activity increases more or less with total protein synthesis through the third day of development. However, injection of arginine during this period results in a rise in the specific activity of arginase in the embryo. This effect is only transient (Eliasson, 1963).

A recent interesting pathway of metabolism of ornithine described in mammalian tissues is that of the decarboxylation of ornithine to form putrescine. Russell and Snyder (1968) have demonstrated that, shortly after the removal of 70% of the rat liver, there is a dramatic increase in the level of ornithine decarboxylase. This increase is 25-fold by 16 hours after the operation. Williams-Ashman (1968) has also demonstrated this activity to occur at highest levels in the prostate and seminal vesicles of rats. Castration of the animal decreases the level of the enzyme, and administration of androgens to the castrated animal results in dramatic increases within a relatively few hours, although maximally only after a couple of days. This regulatory mechanism may be related to the increases in RNA polymerase seen in the same tissue. The polymerase is stimulated by polyamines formed from decarboxylated amino acids.

Some studies have been carried out on enzymes of the urea cycle in cultured mammalian cells. Studies by Schimke (1963a) demonstrated that addition of arginine or manganese to the media of cultured cells resulted in continued increases in the level of arginase. However, he suggested again that this increase was due primarily to stabilization of the enzyme as it was synthesized in the cultured cells. In contrast, this same author (Schimke, 1964a) demonstrated that addition of arginine to the media of several cultured cell lines, including the HeLa cell and mouse fibroblasts, caused a decrease in the activity of the enzymes argininosuccinate synthetase and argininosuccinase. It would appear from these studies that in fact these enzymes for arginine biosynthesis are actually repressed in this situation much as the enzymes of the arginine biosynthetic pathway are repressed in bacterial cells grown in the presence of arginine. The enzyme arginase, which increases in activity, is probably primarily degradative in its functions whereas the other enzymes of the cycle may be considered as synthetic with respect to arginine itself.

B. Phenylalanine and Tyrosine Metabolism

A consideration of the regulation of phenylalanine and tyrosine metabolism in mammalian tissues will necessarily involve a divergence of metabolic pathways (Fig. 5). However, since, in another chapter of this volume, the enzyme reaction, tyrosine transaminase, and its regulation by hormones and other effectors is considered in detail (Kenney, Chapter 31) we will only mention it in this discussion. Rather, we will be more concerned with the regulation of the conversion of phenylalanine to tyrosine and of metabolic steps beyond tyrosine transaminase.

The hydroxylation of phenylalanine to tyrosine in mammalian tissues

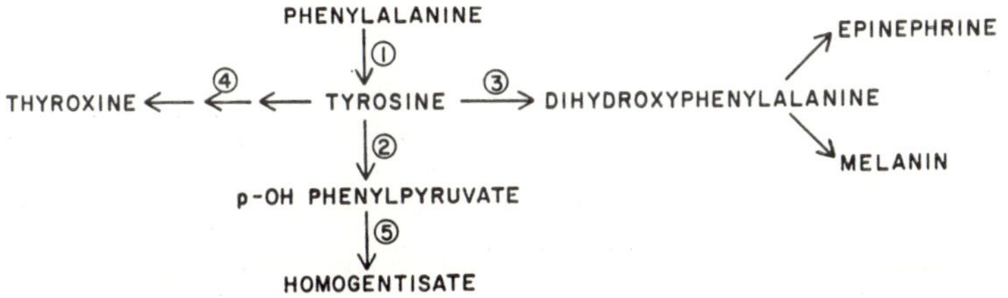

PHENYLALANINE
↓ ①
THYROXINE ← ④ ← TYROSINE —③→ DIHYDROXYPHENYLALANINE → EPINEPHRINE, MELANIN
↓ ②
p-OH PHENYLPYRUVATE
↓ ⑤
HOMOGENTISATE

① PHENYLALANINE HYDROXYLASE — ↑:MATURATION ↓:DIETARY PHENYLALANINE
② TYROSINE TRANSAMINASE — ↑:CORTICOSTEROIDS, GLUCAGON ↓:GROWTH HORMONE
③ TYROSINE HYDROXYLASE — ↓:SCURVY, NOREPINEPHRINE (?)
④ TYROSINE IODINATING AND COUPLING ENZYMES — ↑:TSH (?)
⑤ p-HYDROXYPHENYLPYRUVATE OXIDASE — ↓:THYROXINE

FIG. 5. Regulation of enzymes and pathways of aromatic amino acid metabolism.

has now been extensively studied by several authors, primarily Kaufman (1964, 1966). Kaufman and his associates showed earlier that a cofactor other than NADP was required for this conversion. It is now apparent that this cofactor is a reduced pteridine, specifically dihydrobiopterin (Kaufman, 1963). It is of interest that analogs of pteridines, specifically amethopterin, interfere with phenylalanine metabolism in man (Goodfriend and Kaufman, 1961). Freedland *et al.* (1961) as well as Renson *et al.* (1962) have demonstrated that the enzyme which hydroxylates phenylalanine is also capable of hydroxylating tryptophan. It is apparent, however, that there is a specific tryptophan hydroxylase producing 5-hydroxytryptophan. Evidence for this may be seen from the fact that patients having the metabolic error phenylketonuria, although lacking phenylalanine hydroxylase, still produce normal amounts of serotonin. Furthermore, Chari-Bitron (1963) demonstrated that adrenalectomy of rats significantly lowered the level of the tryptophan-hydroxylating enzyme in liver, but did not affect phenylalanine hydroxylase. Similarly, injection of cortisone increased tryptophan hydroxylation while not significantly affecting tyrosine hydroxylation. Freedland *et al.* (1964) demonstrated that excess phenylalanine in the diet actually caused a marked decrease in phenylalanine hydroxylase. Tryptophan or tyrosine added to the diet, however, increased the enzyme activity in rat liver.

In contrast to the hydroxylation of phenylalanine, the further hydroxylation of tyrosine to norepinephrine or melanin requires ascorbic acid rather than a pteridine cofactor. This difference is exemplified in ascorbic acid-deficient guinea pigs who are given tyrosine. These animals excrete large amounts of *p*-hydroxyphenylpyruvic acid, apparently because the further hydroxylation of this material is blocked due to the fact that the cofactor of the enzyme involved becomes limiting (Goswami and Knox, 1963).

As indicated above, the hormonal regulation of aromatic amino acid metabolism has been discussed in Chapter 31. However, it is of interest to note that Litwack *et al.* (1964) demonstrated that the administration of thyroid hormone to rats caused a dramatic decrease in the activity of *p*-hydroxyphenylpyruvate oxidase in liver. Withdrawal of the hormone resulted in a rapid increase to above normal levels seen in this enzyme. The conversion of tyrosine to norepinephrine also appears to be under the feedback regulation of the intracellular level of norepinephrine (Spector *et al.*, 1967).

The role of tyrosine in the production of another amino acid hormone, thyroxine, has been reviewed (DeGroot, 1965). The regulation of the formation of iodinated aromatic amino acids appears to be primarily related to the availability of iodine in the organism. There is also evidence for a role of the thyroid-stimulating hormone (Field, 1968), but the exact mechanism of the action of this latter hormone is not understood. The recent studies of McKenzie *et al.* (1968) have demonstrated that inhibitors of RNA and protein synthesis, when administered *in vivo,* inhibit the release of thyroid radioiodine stimulated after thyrotropin injection. However, a time lapse of 8 or more hours was required before the effect was seen. This is probably a result of the fact that thyroglobulin synthesis occurs independently of RNA synthesis (Seed and Goldberg, 1963), as well as the fact that thyroglobulin is synthesized from a precursor molecule, possibly analogous to the trypsinogen–trypsin conversion, this synthesis being unaffected by puromycin. Apparently for the thyroid-stimulating hormone to have an effect on the thyroid, the external phospholipid configuration of the thyroid cell must remain intact, as shown by the recent work of Macchia and Pastan (1967).

Early studies by Kenney *et al.* (1958) demonstrated that phenylalanine hydroxylase appeared in mammalian liver only after birth. The enzymic ability to reduce the pteridine cofactor, however, was present during fetal life. Recent studies by Ryan and Orr (1966) have demonstrated that there appears to be a low level of phenylalanine hydroxylase in human fetuses, as demonstrated by radioactive means. Tyrosine transaminase, as indicated earlier, is also extremely low during fetal life, but increases dramatically at birth. In the liver of the newborn, the remaining enzymes of the tyrosine-oxidizing system appear to be at levels approaching those of adult liver (Kretchmer and McNamara, 1956). Tyrosine transaminase in several animals appears to act quite similarly in that the enzyme is low in the fetus but increases after birth. Furthermore, administration of cortisone during fetal life causes no change whatsoever in the level of the enzyme (Litwack and Nemeth, 1965), although recent studies by Greengard and Dewey (1967) have indicated that administration of glucagon to fetal animals does cause an increase in the level of this enzyme.

C. Regulation of Serine, Glycine, and Threonine Metabolism in Mammals

The interconversion of glycine and serine and the relationship of their metabolism has been known for some years (Stokstad and Koch, 1967). Recently the relationship between this enzymic conversion and the threonine–glycine interconversion has become apparent (Schirch and Gross, 1968). It thus seems appropriate to consider these three amino acids as an interrelated group (Fig. 6).

Since serine is a so-called "nonessential" amino acid for the mammalian organism *in vivo* (although this is not always so in tissue cultures

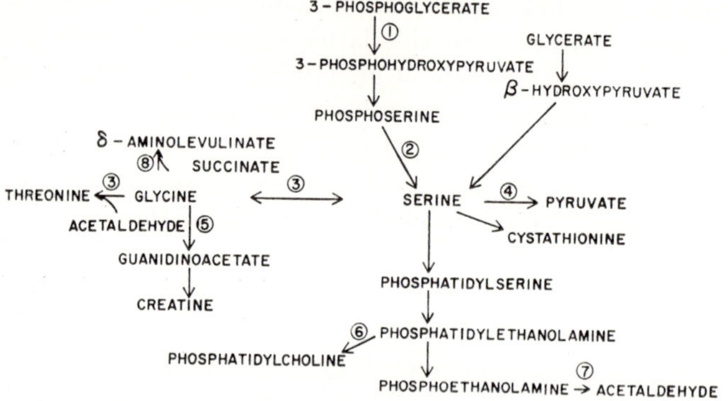

① 3-PHOSPHOGLYCERATE DEHYDROGENASE — ↑:LOW SERINE IN DIET ↓:HIGH SERINE *IN VITRO*
② PHOSPHOSERINE PHOSPHATASE — ↑:LOW SERINE IN DIET ↓:HIGH SERINE *IN VITRO* ↑:METHIONINE *IN VIVO*
③ SERINE TRANSHYDROXYMETHYLASE (THREONINE ALDOLASE) — (?)
④ SERINE DEHYDRATASE — ↑:DIETARY PROTEIN, GLUCAGON ↓:GLUCOSE
⑤ ARGININE-GLYCINE TRANSAMIDINASE — ↑:DIETARY PROTEIN ↓:DIETARY CREATINE
⑥ PHOSPHATIDYLETHANOLAMINE TRANSMETHYLASE — ↑:ESTROGENS (?), CHOLINE DEFICIENCY
⑦ PHOSPHOETHANOLAMINE DEAMINASE — ↑:DIETARY PROTEIN, AGING, CORTICOSTEROIDS ↓:B_6 DEFICIENCY (?)
⑧ δ-AMINOLEVULINATE SYNTHETASE — ↑:DRUGS, ESTROGENS ↓:GLUCOSE, HEMIN

FIG. 6. Regulation of enzymes and pathways of serine, glycine, and threonine metabolism.

in vitro), pathways for the biosynthesis of this amino acid have not been lost during evolution. Ichihara and Greenberg (1957) initially described a pathway of serine synthesis which involved phosphorylated intermediates, while more recently Willis and Sallach (1962) have described enzyme activities which interconvert nonphosphorylated compounds including glycerate and serine. Using radioactive phosphoglycerate and glycerate, Pizer (1966) has demonstrated that, in mammalian cells both *in vitro* and *in vivo,* the so-called "phosphorylated pathway" appeared to be the major pathway for the synthesis of serine from carbohydrate. Furthermore, the extremely low or nondetectable levels of enzymes in the phosphorylated pathway seen in rat liver may be increased to readily measurable levels by feeding animals a synthetic diet lacking in serine and glycine (Tanaka, 1965). Such an apparent repressive effect has also been described in cells grown *in vitro* (Pizer, 1964). Cells grown in millimolar L-serine have lower levels of 3-phosphoglycerate dehydrogenase and serine phosphate phosphatase activities compared to cells grown in normal media. Furthermore, the activity of the latter enzyme is inhibited by L-serine. This is true of the enzyme in cultured cells as well as in liver and brain (Subrahmanyam, 1963). The work of

Tanaka (1965) has also reasonably been confirmed by Fallon and Byrne (1966) in rats fed liquid synthetic diets. These authors clearly indicated that it is only the enzymes of the phosphorylated pathway that are responsive to dietary changes. More recent studies (Fallon, 1968) have shown that methionine supplementation prevents the rise in 3-phosphoglycerate dehydrogenase seen in protein depletion.

Early studies by Stetten (1942) demonstrated that the carbon and nitrogen of serine was converted to the same elements of the ethanolamine molecule *in vivo* in the rat. More recently Wilson *et al.* (1960) and Nemer and Elwyn (1960) have further established this fact, and the former group has demonstrated that a probable intermediate may be phosphatidylserine. Ansell and Spanner (1967) have shown that labeled ethanolamine given intracerebrally is incorporated directly into the base moiety of the ethanolamine lipids with no conversion to lipid-bound choline or serine. This is quite different from liver, whereas a significant proportion of the labeled ethanolamine is incorporated into lipid-bound choline. Thus, this would argue that, in the brain, ethanolamine incorporation into lipid is primarily through the cytidine pathway whereas liver has other pathways available to it. It is of interest that the conversion of ethanolamine phosphatides to choline phosphatides by direct methylation, a membrane-bound enzymic activity, is dependent on added methylated aminoethanol phosphatides for lecithin formation. Thus, in a nutritional situation in which methyl groups are limiting, not only will the substrate be unavailable, but also the activators for the actual formation of phosphatidylcholine will be scarce (Rehbinder and Greenberg, 1965).

Ethanolamine phosphate is readily synthesized using ATP in several tissues. However, choline is a potent inhibitor of ethanolamine phosphorylation, but the converse is not true (Sung and Johnstone, 1967). An interesting mechanism regulating phosphorylethanolamine and choline formation appears in the skin wherein these two compounds are equally labeled with ^{32}P whether skin cells are under optimal nutritional conditions or not (Carney *et al.*, 1966). The deamination of phosphoryl ethanolamine may occur by the action of alkaline or acid phosphatase (Smith and Rossi, 1958) or by direct deamination or transamination to give phosphoryl glycolaldehyde (Horie and Shimazono, 1961).

The degradation of ethanolamine was reported to occur by deamination and eventual formation of glyoxylate (Weissbach and Sprinson, 1953). However, the amount of radioactivity incorporated into glycine via this pathway appeared to be very small. In microorganisms, ethanolamine is converted to acetaldehyde by a reaction requiring the cobamide coenzyme (B. H. Kaplan and Stadtman, 1968). In the authors' laboratory, Fleshood (1969) has demonstrated the existence of an enzyme-converting phosphoryl ethanolamine to acetaldehyde, inorganic phosphate and ammonia in equimolar amounts. The activity of this enzyme varies dramatically in various species, is absent before birth although increasing to quite high levels within 1 or 2 days after birth, thereafter decreasing to low adult levels rather markedly. The level may be increased by the administration of steroid hormones or the feeding of a high protein diet.

The degradation of serine has largely been considered in an earlier section and will not be further considered here. However, a possible degradative pathway of serine is its conversion to glycine through the enzyme serine transhydroxymethylase. The recent demonstration by Schirch and Gross (1968) of the identity of serine transhydroxymethylase with threonine aldolase presents an interesting area for further study in the regulation of amino acid metabolism. Early studies by Karasek and Greenberg (1957) as well as Gilbert (1957) demonstrated the existence of threonine and allothreonine aldolases in rat liver. The former transforms threonine to glycine and acetaldehyde (Chapter 5). Preliminary attempts by the former authors indicated, however, that the dietary requirement for threonine could not be replaced by glycine and ethanol. The enzyme required pyridoxal phosphate and also, apparently for the interconversion of serine and glycine, tetrahydropteroylglutamic acid (Blakley, 1957). The studies by Schirch and Gross (1968) appear to clarify the situation and to bring closer together the relationship between glycine, serine, and threonine. Their studies indicate that serine is the preferred substrate, having a considerably greater affinity for enzyme than threonine. The fact that dietary glycine and a source of acetaldehyde cannot replace the essential requirement of threonine in the diet indicates the enzyme is probably primarily degradative in nature. Unfortunately, at present little is known of the regulation of the synthesis or activity of this enzyme. Another interrelationship between threonine and glycine may be seen through the enzyme threonine dehydrogenase, which is mitochondrial and apparently leads to the formation of aminoacetone. Aminoacetone may also be formed by the condensation of glycine and acetyl-CoA (Hartshorne and Greenberg, 1964; Elliott, 1959).

The regulation of glycine by synthesis is obviously related to a number of pathways. The most important pathway is probably that of the serine transhydroxymethylase. However, in addition, glycine has many other roles including the formation of creatine, porphobilinogen, and various aspects of "active formaldehyde" metabolism via folic acid. The regulation of the formation of creatine in mammalian avian tissues has been the subject of considerable investigation. The pathways for the formation of creatine requires not only the liver, but also the kidney and pancreas (Koszalka, 1967). The enzyme apparently involved in the primary regulation of the formation of creatine is arginine-glycine-transamidinase. This enzyme is decreased in activity in the kidney of the rat by feeding creatine for 1 week (J. B. Walker, 1959). However, the actual role of creatine in the regulation of the level of the transamidinase was disputed by Van Pilsum and Canfield (1962), who demonstrated that addition of high levels of arginine together with creatine seems to prevent the decrease in the transamidinase. Furthermore, the level of transamidinase varies with the protein content of the diet. A more clear-cut demonstration of the metabolic control of creatine biosynthesis by dietary creatine was demonstrated in the chick by J. B. Walker (1960, 1961). The feeding of dietary creatine to chicks resulted in a dramatic decrease in liver transamidinase activity. This activity could be restored by removing creatine from the diet and the restoration was inhibited by the administration of ethionine. The inhibition by ethionine could be reversed by adding methionine to the diet. These workers (M. S. Walker and Walker, 1962) also demonstrated that administration of creatine to developing chick embryos dramatically decreased the level of transamidinase in very early development when transamidinase levels were normally high.

Another pathway of glycine metabolism is in the synthesis of porphobilinogen. Specifically, the enzyme δ-aminolevulinate synthetase in liver mitochondria has been studied rather extensively by several authors. Marver *et al.* (1966) have demonstrated that administration of allylisopropylacetamide causes a marked increase in the synthesis of δ-aminolevulinate synthetase in liver as evidenced by an inhibition by actinomycin D or 5-fluorouracil administration. Puromycin administration results in a dramatic decrease in the level of the enzyme which has a half-life of approximately 70 minutes. Studies with actinomycin or fluorouracil suggest that the messenger RNA for the enzyme turns over rather rapidly. Glucose administration markedly inhibits the induction of hepatic δ-aminolevulinate synthetase. The activity of δ-aminolevulinate synthetase may also be increased by the administration of ethanol (Shanley *et al.*, 1968). When animals were given hydrocortisone or triiodothyronine along with allylisopropylacetamide, δ-aminolevulinate synthetase was stimulated to a much greater degree than administration of the latter chemical alone. Administration of the hormone alone, however, had little, if any, effect on the level of the enzyme (Matsuoka *et al.*, 1968). Tschudy *et al.* (1967) demonstrated that administration of estrogen to rats resulted in marked oscillations of hepatic δ-amino levulinate synthetase. These authors felt that the hormone interfered with some sort of a closed negative feedback loop regulating the synthesis of the enzyme and thus resulted in the oscillations in the activity of the enzyme. Wada *et al.* (1967) demonstrated that δ-aminolevulinate synthetase in spleen reacted to various stimuli such as hypoxia, and compounds elevating erythropoiesis stimulated increase in the level of the enzyme in the spleen but not in the liver, whereas the chemical 3,5-dicarbethoxy-1,4-dihydrocollidine caused a marked increase in the enzyme in the liver but no change in the spleen enzyme. Administration of hemin or bilirubin to animals given allylisopropylacetamide suppressed the induction of δ-aminolevulinate synthesis (Hayashi *et al.*, 1968). Calissano *et al.* (1966) found that heme and related compounds inhibited δ-aminolevulinate dehydratase isolated from blood cells. Russell and Coleman (1963) demonstrated that in mice, δ-aminolevulinate dehydratase activity is high in fetal liver and falls to a low level shortly before birth, then rises again to the adult level at 3 to 6 weeks. Two strains of mice were obtained having different levels of the enzyme, one being more readily induced than the other. Granick and Kappas (1967) have demonstrated the stimulation of porphyrin biosynthesis in chick embryo liver cells, growing in primary culture, by steroid hormones, especially the sex hormones and their derivatives. Earlier studies by Granick (1963) had established an apparent induction of δ-aminolevulinate synthetase by allylisopropylacetamide and related compounds in cultured liver cells. In a recent provocative report by Hickman *et al.* (1968), these authors suggested that RNA isolated from animals fed allylisopropylacetamide stimulates the level of δ-aminolevulinate synthetase in chick embryonic liver cells when added *in vitro*. The importance of the regulation of δ-aminolevulinate synthetase by the mechanisms indicated in this section in human disease has been reviewed by Levere (1965).

The role of glycine in folic acid metabolism has been reviewed by others (Stokstad and Koch, 1967) and will not be considered here. The role of glycine in purine synthesis will also not be taken up in this particular review since there is a considerable amount of other literature on this subject.

D. *Regulation of Methionine, Cysteine, and Choline Metabolism*

In a consideration of the regulation of sulfur amino acid metabolism in the mammal the major pathway (Fig. 7) is the interconversion of methionine to cysteine. Earlier studies in nutrition demonstrated that administration of high levels of cysteine could "spare" methionine in the diet. Studies by Kato *et al.* (1964) first suggested that in mammalian

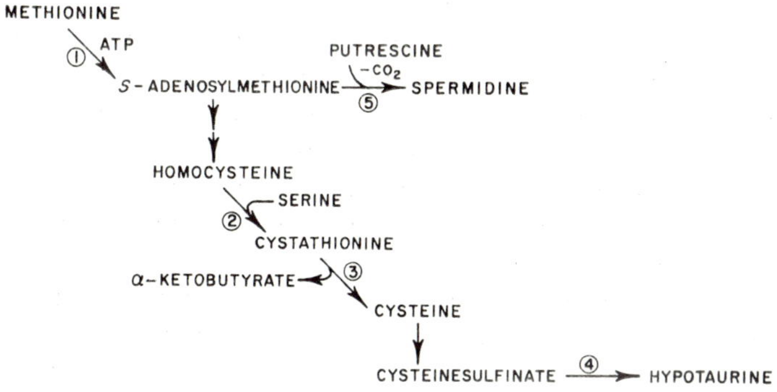

METHIONINE
① ATP
S-ADENOSYLMETHIONINE → ⑤ ($-CO_2$) → SPERMIDINE
PUTRESCINE
HOMOCYSTEINE
② SERINE
CYSTATHIONINE
α-KETOBUTYRATE ← ③
CYSTEINE
CYSTEINESULFINATE → ④ → HYPOTAURINE

① *S*-ADENOSYLMETHIONINE SYNTHETASE — ↑:FASTING, STEROIDS ↓:DIETARY CYSTINE
② CYSTATHIONINE SYNTHETASE — ↑:DIETARY PROTEIN, MATURATION, GLUCAGON ↓:DIETARY CYSTINE
③ CYSTATHIONASE — ↑:DIETARY PROTEIN, MATURATION, GLUCAGON, ETHIONINE
④ CYSTEINE SULFINATE DECARBOXYLASE — ↑:CORTICOSTEROIDS ↓:PYRIDOXINE DEFICIENCY, ESTROGENS
⑤ *S*-ADENOSYLMETHIONINE DECARBOXYLASE — ↑:ANDROGENS

FIG. 7. Regulation of enzymes and pathways in sulfur amino acid metabolism.

liver the activity of the enzyme cystathionine synthetase, which normally varied rather considerably with the time of day in animals fed a relatively normal diet, decreased to very low levels when a relatively large amount of cysteine was added to the diet. Later studies by this group (Kato *et al.*, 1966a,b) showed that the apparent inhibitor was elemental sulfur produced by the final enzyme of the pathway, cystathionase. More recent studies by Finkelstein and Mudd (1967) demonstrated a similar effect on both cystathionine synthetase and the methionine-activating enzyme. They also showed that both of these enzymes were inhibited *in vitro* by L-cysteine. However, they felt that the most likely cause of their results was an actual effect on enzyme synthesis. Earlier, Finkelstein (1962) also had investigated the effect of age, diet, and hormones on the enzymes of the methionine to cysteine pathway. During other earlier studies (Nagabhushanam and Greenberg, 1965), the enzymes serine dehydratase and cystathionine synthetase were thought to be identical. However, more recent studies (Brown *et al.*, 1966; Nakagawa and Kimura, 1968) have demonstrated that rat liver serine dehydratase and cystathionine synthetase are separable activities, each apparently due to a different protein species. In further substantiation of this latter finding are the data of Cihak *et al.* (1966), who demonstrated the inhibition of rat liver cystathionine synthetase by the chemical, 3,5-dimethylthio-1,2,4-triazine. In further support of this separation of the two activities are the data of Mudd *et al.* (1965), who demonstrated that in human patients suffering from the inborn error homocystinuria, which is most probably due to a deficiency of cystathionine synthetase, these people still retain demonstrable threonine dehydratase activities. In contrast to these effects, Trautmann and Chatagner (1964) demonstrated that cystathionase, which also has the ability to deaminate homoserine and desulfurate cysteine, is increased in activity in animals fed a high protein diet. An odd effect in the regulation of cystathionase is seen by the fact that a single injection of L-ethionine, the ethyl analog of methionine, stimulates a rapid increase in cystathionase activity within a few hours (Durieu-Trautmann and Chatagner, 1966). This increase occurs in both male and female rats except when they are adrenalectomized, and then only male rats respond to ethionine administration. Hydrocortisone injection has little effect except possibly in male rats. The increase in the enzyme is apparently not explained on the hypothesis that ethionine is incorporated into the enzyme in significant amounts (Loiselet and Chatagner, 1966). The further metabolism of cysteine ultimately to sulfate is an extremely complex area of biochemistry (Meister, 1965). A recent review by Jacobsen and Smith (1968) has exemplified a number of these areas including certain regulatory phenomena. The enzyme cysteine sulfinate decarboxylase, a pyridoxal phosphate-requiring enzyme, has been shown to be subject to several regulatory phenomena. Preparations of the enzyme from livers of pyridoxine-deficient rats are unstable in cell-free systems (Hope, 1955), but stability is not restored by addition of pyridoxal phosphate *in vitro*. However, stable activity is restored to normal levels within 6 hours after administration of large doses of pyridoxine to the deficient animal, and it would appear that such stabilization is accompanied by enzyme synthesis (Greengard and Gordon, 1963). As indicated in the review by Jacobsen and Smith (1968), administration of estrogens and thyroid hormone causes decrease in the level of cysteine sulfinate decarboxylase. On the other hand, administration of hydrocortisone actually increases the activity of this enzyme.

The regulation of methionine metabolism in mammalian tissue is extensively involved with the formation of *S*-adenosylmethionine. The conversion of this compound to cysteine has already been discussed. The involvement of this intermediate in the formation of spermidine is established in microbial systems, and recently it has been described as a pathway in rat ventral prostate (Williams-Ashman, 1969). These authors found that, in the presence of excess putrescine, *S*-adenosylmethionine decarboxylase decreased to approximately 5% of normal activity 7 days after castration. Injection of testosterone significantly increased the level of this enzyme within 3 hours and causes a 9-fold increase 24 hours after administration of the hormone. Although the amino acid analog ethionine is not a normal metabolite in the animal, its effects *in vivo* are quite remarkable. Administration of L-ethionine *in vivo* results in a dramatic decrease in the intracellular level of ATP in mammalian liver within 6 hours after injection (Villa-Treviño *et al.*, 1963). Administration of the analog inhibits protein synthesis in male rats and also markedly decreases RNA synthesis. Polysome patterns are lost but may be restored, as well as the effects on protein and RNA synthesis reversed, by the administration of adenine with or without methionine (Gordon and Farber, 1965). Farber (1963) has reviewed the effects of this analog in mammals.

S-Adenosylmethionine also acts as a major methyl donor in numerous biochemical reactions including methylation of nucleic acids, of drugs, and of various biochemical amino groups including the formation of choline. The biosynthesis of choline-containing phosphatides occurs either through the reaction of CDP-choline with diglyceride or the successive methylation of phosphatidylethanolamine by *S*-adenosylmethionine (Bremer and Greenberg, 1961). The latter pathway is increased in activity in rats receiving a choline-deficient diet (Wells, 1963). Cysteine (Kwong and Barnes, 1967) and ethanolamine (Mulford *et al.*, 1959) aggravate a choline deficiency under certain dietary conditions. Wells and Remy (1965) concluded that choline deficiency affects primarily the free choline present in liver and kidney since the turnover of lipid-bound choline was unchanged in deficient rats, whereas the concentration of free choline decreased during the deficiency. Such data would also be supported by that of Corredor *et al.* (1967), who demonstrated that the decrease in fatty acid oxidation seen in choline-deficient animals is probably the result of a depletion in intracellular carnitine. Thus, it is apparent that the basic mechanisms of the effects produced by choline deficiency have still to be elucidated, although it would appear from available evidence that they must involve a number of different metabolic control mechanisms.

E. Regulation of the Metabolism of Histidine and Lysine

Although histidine is not synthesized to a significant extent from precursors in mammalian tissues, it does undergo several rather important reactions (Fig. 8). Histidine-pyruvate transaminase is present in rat liver in two forms, one associated with mitochondria and the other a soluble form (Spolter and Baldridge, 1964). The activity of these combined enzymes is about three times greater than that of histidase, the other major pathway of histidine degradation in mammalian liver. In the face of dietary protein alterations, however, only the activities of the enzymes histidase and urocanase were altered in a direct manner by

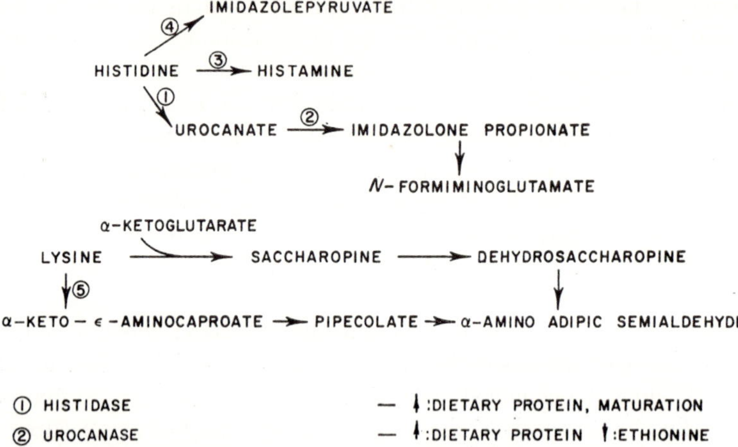

FIG. 8. Regulation of enzymes and pathways in histidine and lysine metabolism.

changing the protein intake, whereas the activity of histidine-pyruvate transaminase was unaffected by these environmental changes (Rao *et al.*, 1965). When ethionine is given to animals on a low-protein diet, there is a suppression of liver urocanase activity with no effect on the activity of histidase (Kolenbrander and Berg, 1967). It is of interest to speculate, therefore, that the histidase degradative pathway is primarily under dietary control whereas the transaminase pathway may be primarily under corticosteroid control.

A further pathway of histidine degradation is that of histidine decarboxylase. The product of this reaction, histamine, has considerable pharmacological effects *in vivo*. Schayer *et al.* (1963) demonstrated that *Salmonella* endotoxin markedly suppressed the activity of rabbit kidney histidine decarboxylase when given *in vivo*. Histidine decarboxylase activity in the stomach varies quite dramatically, being increased 3- to 4-fold a few hours after feeding fasting animals (Castellucci, 1965). Snyder and Epps (1968) demonstrated that the administration of gastrin markedly increased gastric histidine decarboxylase within 2 hours after administration of the hormone. Puromycin and cycloheximide completely prevented this increase, whereas actinomycin D actually stimulated it.

The synthesis of lysine is essentially nonexistent in mammalian tissues and our knowledge of its degradative pathways is at present not very great. Lysine transamination in mammalian tissues occurs only to a very small extent or not at all. Its oxidative deamination is also a very slow reaction by the general amino acid oxidases (Meister, 1965). There is evidence, however, that lysine is degraded in animal tissues via an intermediate, saccharopine (Hutzler and Dancis, 1968). In a recent study utilizing doubly labeled lysine (Gupta and Spenser, 1969), a relative unequivocal conversion of lysine to pipecolic acid via the α-keto-ϵ-aminocaproic acid pathway, both in mammals and plants was shown. Lysine is a relatively toxic amino acid and because of its metabolic inertness, dietary changes have not been readily demonstrable. Recent studies in the author's laboratory (Weiss and Pitot, 1969) have shown the existence of a pyridine nucleotide-linked lysine α-amino dehydrogenase. This enzyme only occurs in liver, is absent in the fetus, and increases slightly when the animal is given large amounts of protein in the diet.

During fetal development histidase activity is barely detectable and remains low for the first 2 to 3 weeks of life. This is not true of histidine-pyruvate transaminase which increases dramatically during the latter third of gestation and by 24 hours after birth is found at levels identical to that in adult rats (Makoff and Baldridge, 1964). Although fetal tissues contain high histidine decarboxylase activity, Hakanson (1963) presented data suggesting that this enzymic activity in fetal tissues is different from the enzyme present in adult rabbit kidney.

F. Regulation of the Metabolism of Leucine, Isoleucine, and Valine

Again, as with lysine, the branched-chain amino acids are not synthesized by mammalian tissues. Their degradation involves primarily three specific steps (Fig. 9), the first of which is a transaminase which appears to be one enzyme for all three branched-chain amino acids (Ichihara and Koyama, 1966). This transaminase decreases in activity in rat liver from fetal to adult life, but increases transiently after partial hepatectomy (Ichihara and Takahashi, 1968). Recently Mimura *et al.* (1968) have shown that increases in dietary protein from an initial level of zero percent results in decreased activity of branched-chain amino acid transaminase, both in liver and muscle. They confirm the high specific activity of the enzyme in kidney and muscle. Other authors have also reported this transaminase in peripheral tissues (Dancis *et al.*, 1967). In contrast, the decarboxylation of the branch-chain α-keto acids may occur by two different enzymic pathways (Connelly *et al.*, 1968). Since, in the human, defects involving only one or two of the branched amino acids have been demonstrated, such a divergence of enzymic steps is entirely possible (Budd *et al.*, 1967). A similar report by Dancis *et al.* (1967) might suggest even that the transaminases are different en-

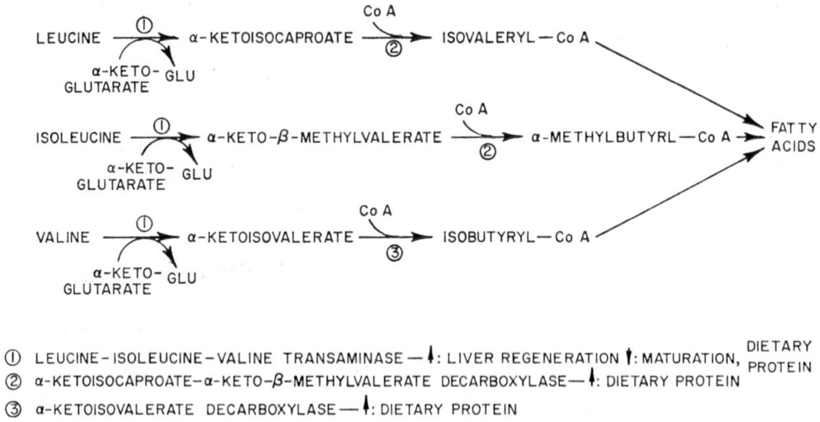

LEUCINE →(1)→ α-KETOISOCAPROATE →(2, Co A)→ ISOVALERYL—Co A → FATTY ACIDS
(α-KETO-GLUTARATE → GLU)

ISOLEUCINE →(1)→ α-KETO-β-METHYLVALERATE →(2, Co A)→ α-METHYLBUTYRL—Co A → FATTY ACIDS
(α-KETO-GLUTARATE → GLU)

VALINE →(1)→ α-KETOISOVALERATE →(3, Co A)→ ISOBUTYRYL—Co A → FATTY ACIDS
(α-KETO-GLUTARATE → GLU)

(1) LEUCINE-ISOLEUCINE-VALINE TRANSAMINASE—↓: LIVER REGENERATION ↑: MATURATION, DIETARY PROTEIN
(2) α-KETOISOCAPROATE-α-KETO-β-METHYLVALERATE DECARBOXYLASE—↓: DIETARY PROTEIN
(3) α-KETOISOVALERATE DECARBOXYLASE—↓: DIETARY PROTEIN

FIG. 9. Regulation of enzymes and pathways of branched-chain amino acid metabolism.

zymes. That the level of one of the decarboxylases may be regulated by diet was described briefly by McFarlane and von Holtz (1966).

G. Regulation of the Metabolism of Aspartic and Glutamic Acids and Proline

The dicarboxylic acids play major roles in amino acid metabolism and interconversions (Fig. 10). These acids also are important in the formation of asparagine and glutamine, the two major amino acid amides found in protein. Aspartic acid, in addition to being involved in transamination reactions, is also a precursor of nucleic acid pyrimidines through the transcarbamylation reaction. As indicated earlier, aspartic transcarbamylase undergoes different regulatory mechanisms than ornithine transcarbamylase. In fetal livers the level of aspartic transcarbamylase is high in early fetal life and decreases during development to low levels in the liver of the adult animal (Kim and Cohen, 1965). After partial hepatectomy aspartic transcarbamylase rises 2-fold or more 24 hours after the operation. As might be expected, tissues exhibiting the highest activity of aspartic transcarbamylase in the rat are those undergoing the greatest degree of cellular turnover, such as testis, intestine, and bone marrow (J. E. Young *et al.*, 1967).

Until the recent popular use of asparaginase in the treatment of lymphomas (Broome, 1968), the synthesis of asparagine was assumed to occur identically to that of glutamine. However, the studies by Patterson and Orr (1968) have indicated that glutamine is a better amide donor than is ammonia in the reaction. Furthermore, L-asparagine inhibits its own synthesis. It would appear that those lymphomas having substantial intracellular asparagine synthetase activity are not affected by the administration of asparaginase to the host. However, those tumors dependent on host asparagine are sensitive to administration of the enzyme (Horowitz *et al.*, 1968).

Glutamic acid metabolism plays perhaps a more central role than does aspartic acid in nitrogen metabolism. Its role in transamination in many reactions is an obvious one and transaminated ammonia from other amino acids in most instances will end up in glutamic acid which then, through the glutamic dehydrogenase reaction, may allow the ammonia to ultimately enter the urea cycle via carbamyl phosphate synthetase. The key reaction in these conversions is, of course, glutamic dehydrogenase. Wergedal and Harper (1964), in an extensive study of the effect of protein intake on glutamic dehydrogenase activity, showed that feeding an 80% casein diet increased the level of this enzyme in liver and kidney, but only from 11% to 40% over control levels. Since glu-

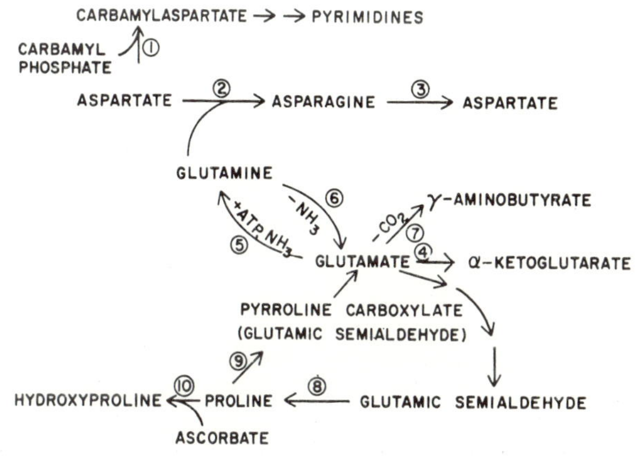

① ASPARTIC TRANSCARBAMYLASE	—↑:LIVER REGENERATION	↓:MATURTION
② ASPARAGINE SYNTHETASE	—	↓ HIGH ASPARAGINE (?)
③ ASPARAGINASE	— (?)	
④ GLUTAMATE DEHYDROGENASE	—↑:DIETARY PROTEIN	
⑤ GLUTAMINE SYNTHETASE	—↑:MATURATION	↓:LIVER REGENERATION, METHIONINE SULFOXIMINE
⑥ GLUTAMINASE	—↑:ACIDOSIS	
⑦ GLUTAMATE DECARBOXYLASE	—↑:MATURATION	
⑧ PYRROLINE CARBOXYLATE REDUCTASE	— (?)	
⑨ PROLINE OXIDASE	—↑:CORTICOSTEROIDS	
⑩ PROLINE HYDROXYLASE	—↓:SCURVY	

FIG. 10. Regulation of enzymes and pathways of aspartate, glutamate, and proline metabolism.

tamic dehydrogenase is a mitochondrial enzyme, it might be considered to be somewhat less reactive to dietary protein although, as we have seen earlier, ornithine-δ-transaminase and δ-aminolevulinate synthetase, both mitochondrial enzymes, are quite responsive to environmental changes.

The other major metabolic pathway of glutamic acid is the formation and degradation of its amide, glutamine. Glutamine synthetase is low in the nervous system of the mammal and chick, although increasing after birth. In chick retina Rudnick and Waelsch (1955) demonstrated the enzyme's appearance and rapid increase after day 17 of gestation. Reif and Amos (1966) demonstrated that this increase is probably regulated by a dialyzable inducer appearing in the embryo. Kirk (1965) demonstrated that, after the appearance of the enzyme, its further synthesis becomes independent of concomitant RNA synthesis. In the brain of the kitten, glutamine synthetase does not reach adult levels until 6 weeks after birth (Berl, 1966). In adults, the level of hepatic glutamine synthetase is decreased after partial hepatectomy (Wu *et al.*, 1965). Lamar and Sellinger (1965) demonstrated that the administration of methionine sulfoximine *in vivo* caused a dramatic decrease in the level of the enzyme, both in liver and in brain. The decrease in the level of the enzyme in brain coincided with the appearance of convulsions in the animals. This inhibition of enzyme activity appears to be an irreversible reaction of methionine sulfoximine or a metabolite thereof with the enzyme. Ronzio and Meister (1968) demonstrated that the analog could be phosphorylated by glutamine synthetase.

The degradation of glutamine by the enzyme glutaminase has been relatively well studied (Knox and Greengard, 1965). A major interesting control mechanism is that of the increase in renal glutaminase seen in acidosis. The increase in the enzyme produced by making animals acidotic is inhibited by the administration of actinomycin D (Goldstein, 1965). Glutaminase occurs as isozymes with at least two forms which are found in kidney. Liver has only one of these forms in the adult, but during embryonic development fetal liver also possesses the other kidney type not found in adult liver (Katsunuma *et al.*, 1968). Work from this same laboratory (Katsunuma *et al.*, 1968) has demonstrated that the glutaminase isozyme not requiring phosphate is apparently an allosteric enzyme. The role of glutaminase in renal ammonia formation is discussed further in Chapter 39.

Another important pathway of glutamate metabolism is its decarboxylation to γ-aminobutyric acid. It would appear that this reaction is primarily limited to neurons and that γ-aminobutyric acid (GABA) is important in the regulation of neuronal transmissions. However, Van den Berg *et al.* (1965) demonstrated that glutamic decarboxylase is low in activity at birth in the brain of the rat but increases to adult level within 30 days. Recent studies have also demonstrated (Fonnum, 1968) that glutamic decarboxylase is localized to the synaptosome fraction of the brain.

Proline metabolism in mammals consists both of synthesis via the reduction of cyclized glutamic semialdehyde, a soluble enzyme, and the oxidation of proline via the proline oxidase complex of the mitochondrial fraction. The former reaction may be inhibited *in vitro* by proline and also by adenine nucleotides (Peisach and Strecker, 1962). In unpublished studies from this laboratory (Peraino and Pitot, 1969), feeding animals a high-protein diet had little effect on the synthesis of proline from glutamic semialdehyde. However, administration of cortisone stimulated the activity of mitochondrial proline oxidase in the kidney. The degradation of hydroxyproline appears to follow the same pathway as that of proline and probably also pipecolic acid oxidation is catalyzed by the enzyme system known as the proline oxidase complex. However, it is now apparent that the synthesis of hydroxyproline occurs during the biosynthesis of collagen (Peterkofsky and Udenfriend, 1963). This hydroxylation involves ascorbic acid (Stone and Meister, 1962), thus accounting for the inability of scorbutic animals to properly synthesize collagen.

H. Conclusions

From this brief survey of the regulation of amino acid metabolism in mammalian tissues, it is obvious that the pathways open to animal tissues for the metabolism and regulation of metabolism of amino acids is not nearly so great as those open to microorganisms and more primitive forms of life. The reasons for this were outlined in the introduction to this review. Furthermore, it appears to be of rather significant interest that the affinity of amino acids for a number of enzymes involved in their degradation appears to be relatively low. Table II lists the more important rate-limiting steps in the degradation of several amino acids in the mammal. It will be noted that the K_m for most of these is in the range of 1–10 mM. In one instance the K_m ranges above 100 mM. It would thus appear that the animal is geared rather to a conservation of amino acids than to their degradation, and only when the amount of amino acid available to the system becomes rather excessive does the mammalian organism begin to degrade these amino acids for gluconeogenesis. Such an economy is undoubtedly not unique to the mammal, but again is left over from the situation seen in more primitive forms.

TABLE II

MICHAELIS CONSTANTS OF SOME ENZYMES OF AMINO ACID DEGRADATION

Enzyme	K_m	Reference
Glutamate dehydrogenase	2×10^{-3}	Frieden (1963)
Glutaminase	5×10^{-3}	Roberts (1960)
Histidase	3×10^{-3}	Spolter and Baldridge (1963)
Histidine-pyruvate transaminase	1.2×10^{-2} (pyruvate)	Spolter and Baldridge (1963)
	7×10^{-3} (histidine)	Spolter and Baldridge (1963)
Lysine dehydrogenase	2×10^{-1}	Weiss and Pitot (1969)
Arginase	1.1×10^{-2}	Greenberg (1960)
Ornithine δ-transaminase	7.2×10^{-3} (ornithine)	Peraino and Pitot (1963)
Serine dehydratase	5.9×10^{-2}	Nakagawa and Kimura (1968)
Threonine aldolase	1.5×10^{-2}	Schirch and Gross (1968)
Cystathionase	3×10^{-3}	Greenberg (1962)

V. The Role of Tryptophan in the Regulation of Intermediary Amino Acid Metabolism

A. Tryptophan and Enzyme Levels in Liver

As early as 1950, Williams and Elvehjem (1950) demonstrated that inducing a tryptophan deficiency in the rat results in a markedly decreased liver xanthine oxidase activity as well as a lowering of the endogenous respiration. These authors also demonstrated that the activity of succinate oxidase was decreased whereas "fatty acid oxidase" and liver protein remained unchanged during the deficiency. In 1965 Peraino *et al.* demonstrated that of all the amino acids essential to the nutrition of the rat only tryptophan caused a significant increase in the activities of the enzymes threonine dehydratase and ornithine-δ-transaminase (Peraino *et al.,* 1965). These authors studied a number of mixtures of amino acids in attempts to determine which amino acids were required for the induction of these two enzymes. It was found that for threonine dehydratase the amino acids threonine, phenylalanine, and tryptophan were required to give optimal enzyme induction whereas, in the case of ornithine-δ-transaminase, lysine, arginine, and tryptophan were required. Omissions of tryptophan from either mixture resulted in virtually no increases in the level of the enzyme. Although the administration of tryptophan resulted in dramatic increases in both enzymes, the kinetics of the induction are somewhat different from that seen when a complete mixture of amino acids was given (Fig. 11). It can be noticed from this figure that there is a lag period of about 6 hours before any increase is seen in the level of the enzyme. A maximal change occurs between 6 and 12 hours, whereas after 12 hours even continued administration of tryptophan does not result in any further increase in the level of either enzyme. In order to determine whether or not this increase seen in the activity of these enzymes was the result of changes in the rate of synthesis, Jost *et al.* (1968), utilizing specific antibodies for serine dehydratase, determined whether or not the administration of tryptophan resulted in an increased rate of synthesis. As seen earlier in Table I, tryptophan administration after 10 hours does significantly increase the rate of synthesis of serine dehydratase. However, the increased rate is not nearly so great as that seen with a complete amino acid mixture or with the hormone glucagon. In unpublished studies from this laboratory (Pitot, 1969b), it was shown that the administration of tryptophan inhibits the induction of the enzyme glucokinase by the administration of glucose. Such an inhibition is similar to that seen in the case of the glucagon inhibition of glucokinase induction (Pitot *et al.*, 1964).

Later studies by Foster *et al.* (1966) demonstrated that when L-tryptophan was administered to either intact or adrenalectomized rats, a marked increase in the activity of hepatic phosphoenolpyruvate carboxykinase activity resulted within a relatively short period of time after the initial administration of the amino acid. The *in vitro* addition of L-tryptophan had no effect on enzyme activity. This increase in enzyme activity was inhibited both by puromycin and actinomycin D. However, the effect of these inhibitors could only be demonstrated after the first

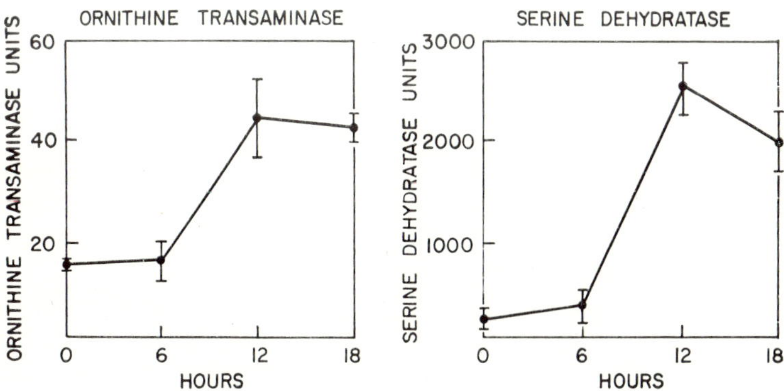

FIG. 11. Induction of ornithine-δ-transaminase and threonine dehydratase by the administration of tryptophan alone. See Peraino *et al.* (1965) for details.

hour following administration of tryptophan. During this first hour there was a marked increase, usually a doubling, of activity which was insensitive to the effects of these inhibitors and thus, presumably was due to some sort of enzyme activation. In a related study (Ray *et al.,* 1966), when tryptophan was administered to intact or adrenalectomized animals, a rapid increase in the hepatic contents of lactate, pyruvate, citrate, aspartate, malate, and oxaloacetate occurred. However, the concentration of phosphoenolpyruvate and all succeeding intermediates on the path to glycogen synthesis was decreased to one-half of normal concentrations or less. These authors therefore suggested that, after tryptophan administration, gluconeogenesis is blocked because phosphoenolpyruvate carboxykinase becomes nonfunctional. At first glance this is rather difficult to rationalize with the earlier studies demonstrating an increase in the level of the enzyme.

These experiments were repeated in the isolated perfused liver (Veneziale *et al.,* 1967), and essentially the same results were found. In this instance, however, certain metabolites of tryptophan, including L-kynurenine and 3-hydroxyanthranilic acid, as well as quinolinic acid metabolites, were found to duplicate the effects of tryptophan on gluconeogenesis in the perfused liver. These data were later explained as due to direct inhibition of phosphoenolpyruvate carboxykinase by an effect of chelation by tryptophan metabolites on metals, such as manganese ions required for the activity of the enzyme (Foster *et al.,* 1967).

In this laboratory it has also been found that the administration of tryptophan will induce the enzyme histidase and in all probability a number of other gluconeogenic enzymes. It is entirely possible that the induction of these enzymes leads to an increase in the metabolites of the Krebs cycle such as Lardy and his associates showed, but that the inhibition of phosphoenolpyruvate carboxykinase by tryptophan metabolites then causes the blockages seen. The reason for the induction of phosphoenolpyruvate carboxykinase is not clear, and it would be of considerable interest to determine whether or not another key intermediate enzyme beyond this, such as fructose-1,6-diphosphatase would also increase after the administration of tryptophan. If such a phenomenon would occur, this might substantiate the suggestions by Weber *et al.* (1964) that the key enzymes in gluconeogenesis are all regulated as if they belonged to a single regulating unit. Exactly how tryptophan is producing its effects, however, still remains a mystery. Tryptophan administration also causes accumulation of fatty acids in the liver within $2\frac{1}{2}$ hours (Hirata *et al.,* 1967). This appears to be due to a reduction in liver ATP level.

B. Tryptophan and Polyribosome Content in Cells

Although this review is primarily involved in the regulation of specific enzymes in amino acid metabolism, recent studies demonstrating peculiar effects of tryptophan *in vivo* on protein-synthesizing machinery will briefly be discussed in this section. One of the earliest demonstrations that tryptophan was related to polysome patterns seen in liver was described by Fleck *et al.* (1965). These workers demonstrated that omission of tryptophan from an amino acid mixture given to rats resulted in a marked loss of heavy C-ribosomes in liver within 1 hour after administration of the mixture. Extension of these studies by Wunner *et al.* (1966) confirmed these earlier results and demonstrated that in animals fed a tryptophan-deficient amino acid mixture, a large increase in dimer population of the polysomes resulted. Furthermore, preparations of polysomes from animals fed the tryptophan-free mixture showed a decreased amino acid incorporating activity as compared with animals fed the complete mixture. Lack of tryptophan also resulted in a decrease in orotic acid labeling of cytoplasmic RNA. Extension of these studies by Sidransky *et al.* (1967) demonstrated that administration of tryptophan alone or together with incomplete mixtures of essential amino acids resulted in an enhancement of hepatic protein synthesis as well as a buildup of polyribosomes. In a most recent publication, Munro (1968) has been able to demonstrate that polysome profiles from animals fed amino acid mixtures deficient in amino acids other than tryptophan did not show the marked shift to monosomes and disomes. Only when tryptophan was omitted from the mixture was this decrease seen. Furthermore, in a completely cell-free system supplemented with amino acids, omission of tryptophan resulted in a marked inhibition of leucine-^{14}C incorporation into protein. If tryptophan were added to the mixture after 20 minutes of incubation, a marked enhancement of further incorporation occurred.

Such *in vitro* studies are comparable to those reported by M. L. Freedman *et al.* (1968) in which tryptophan deficiency causes a disaggregation of reticulocyte polyribosomes which may be reversed by the administration of low doses of puromycin. Earlier studies had suggested that the decrease in polyribosome patterns seen in tryptophan-deficient reticulocytes was due to the fact that tryptophan occurred near the amino-terminal ends of the hemoglobin molecule. Whether or not such arguments can be used for the tryptophan effect on polyribosome patterns in liver remains to be seen, although it is difficult to imagine that every protein in liver has tryptophan near its amino-terminal end. In recent unpublished studies (Bushnell and Potter, 1969), omission of any essential amino acid resulted in a marked decrease in the polyribosome pattern in cultured cells *in vitro*. In this situation it would appear that the inhibition of protein synthesis by decreased tryptophan levels is primarily a result of a deletion of a single amino acid, since any other essential amino acid will cause the same effect. Munro (1968) has also suggested a similar phenomenon to occur in liver wherein the tryptophan pool is, or may be, rate limiting and thus a deletion of this specific amino acid from the diet may inhibit the synthesis of new protein in liver. This may be particularly true if the mobilization of tryptophan from peripheral tissues, especially muscle, is more difficult either because there is less of it or for some other reason. Furthermore, the significant increases seen in tryptophan pyrrolase after slight increases in tryptophan administration may serve as a careful regulator for the already low levels of tryptophan present within the organism. These speculations are further elaborated in Chapter 34.

VI. Regulation of Amino Acid Metabolism in Neoplasia

Neoplasia is associated with an alteration in the control of cellular metabolism and function in multicellular organisms. In Table III are

TABLE III

CONTROL SYSTEMS IN LIVER AND HEPATOCELLULAR CARCINOMAS[a]

Type	Example	Stimulus	Comparison of liver and tumor
Substrate induction	Tryptophan pyrrolase	Tryptophan	Absent in majority of tumors and in all adrenalectomized hosts
Hormonal induction	Tyrosine transaminase	Cortisone	Increased "sensitivity" to hormone in tumors
Dietary (multiple substrate) induction	Serine dehydratase	Amino acids	Absent in most tumors. Differential actinomycin D sensitivity in some tumors
Product repression	Hydroxymethylglutaric-CoA reductase	Cholesterol	Absent in all tumors studied
Glucose repression	Serine dehydratase, ornithine-δ-transaminase	Glucose	Absent in all tumors studied

[a] Reproduced from Pitot (1968a).

cited specific examples of both normal mechanisms seen in the control of rat liver enzyme synthesis and their abnormal counterparts (Pitot, 1966; 1968a) as seen in several recently developed highly differentiated hepatocellular carcinomas (Morris, 1963; Reuber, 1966). The defects appear to be independent of the growth rate of the tumor since changes seen in the very early transplants in any particular neoplasm appear to be identical to those found in much later, more rapidly growing transplants. Numerous studies (Pitot, 1966; 1968a) have led to the generalization that in these hepatomas most, if not all, of the mechanisms regulating enzyme synthesis are abnormal when compared with normal liver and that each specific neoplasm appears to have its own set of altered control mechanisms. In this section we will review some of those studies which have demonstrated differences in control mechanisms among experimental neoplasms as reflected in their varying responses to hormonal and dietary environmental stimuli.

At the experimental level, numerous examples of alterations in control mechanisms of amino acid catabolizing enzymes have been found in the highly differentiated neoplasms. It is the opinion of these authors that these derangements in the regulatory systems of multicellular organisms are some of the biochemical essentials which determine the critical difference between a normal and neoplastic cell (Pitot, 1963b).

A. Regulation of Threonine-Serine Dehydratase in Hepatomas

A study of the dietary induction of specific hepatic enzymes in hepatomas was first undertaken with the Morris hepatoma 5123 (Pitot *et al.*, 1961). In this study the effect of dietary protein on the levels of threonine and serine dehydratase in host liver and hepatoma was determined. Threonine dehydratase in the 5123 hepatoma growing in intact female hosts was found to be at very high levels. Shifting of the dietary regimen had little effect on the enzyme in the neoplasm; however, control livers showed dramatic effects, the enzyme level approaching zero on a 2% protein diet while values up to 500 enzyme units or more were obtained with high-protein feeding. In contrast, the threonine dehydratase of the host liver was at negligible values in animals on a laboratory chow diet. Feeding 91% protein for 7 days raised the level of the enzyme to that seen in control animals on chow diet. If the tumor-bearing animals were adrenalectomized, the threonine dehydratase in the neoplasm dropped to almost negligible values and was still unresponsive when the animal was placed on a high-protein diet. Later work (Bottomley *et al.*, 1963) indicated that the unresponsiveness to diet of the dehydratase was almost ubiquitous in the hepatomas under study. However, one tumor, the Morris 7800, did show a significant rise in threonine dehydratase after 7 days feeding of a high-protein diet. In addition to being an apparent exception to the rule, the fact that one neoplasm was inducible by feeding a high-protein diet strongly suggested that blood supply differences were not the reason for a failure of induction with the other unresponsive neoplasms. With adrenalectomy of the tumor-bearing host, the 7800 hepatoma lost its capacity for dietary induction. Thus, any demonstrable induction of threonine dehydratase in this tumor is completely dependent on the presence of the adrenal steroids.

In view of the high levels of threonine dehydratase in the 5123 and the marked tumor-host effect on liver dehydratase levels, one might suggest that the 5123 is acting as a threonine trap, effectively robbing the host of considerable quantities of threonine by its high degradative capacity. Experiments have shown (Pitot *et al.*, 1963) that in animals bearing the 5123 hepatoma, plasma threonine levels are decreased by 50% from the normal. Thus, the 5123 may be acting as an amino acid "trap," in complete analogy to the tumor "nitrogen trap" concept advanced by Mider (1951). In the 5123 hepatoma the trapping mechanism appears to reflect the insensitivity of this neoplasm to specific metabolic controls. These data suggest that the relatively uncontrolled metabolism of the neoplasm is dramatically affecting the relatively well-regulated control mechanisms in the normal organ from which the tumor arose.

That one is actually dealing with altered regulations of enzyme synthesis, not just with activity, was demonstrated by measuring the actual rate of serine dehydratase synthesis in hepatoma 5123 through the use of specific immunochemical techniques (Jost *et al.*, 1968; Pitot, 1968a; Jost and Pitot, 1969). These measurements showed that the tumor is devoting 4–5% of its total soluble protein synthesis to the synthesis of this enzyme. In normal liver, even in the fully induced state, the radioactivity incorporated into serine dehydratase per gram is less than 1% of that incorporated into total soluble protein per gram liver.

B. Regulation of Tryptophan Pyrrolase in Hepatomas

Studies with the Morris 5123 hepatoma (Pitot and Morris, 1963) demonstrated that no substrate or hormonal induction of tryptophan pyrrolase (TP) occurred in this hepatoma. In addition, unlike threonine dehydratase, the levels of this enzyme in the hepatoma were very low. The difference in blood supplies of liver and tumor was ruled out as a factor in the failure of the tumor to respond, by inducing the enzyme with tryptophan-^{14}C and examining the incorporation of the inducer into protein and the free amino acid pool. Later studies (Pitot, 1963a; Cho *et al.*, 1964) substantiated the interpretation that the loss of control of TP levels in the 5123 hepatoma was a defect inherent in the tumor. Other tumors, the Morris 7316 and 7800, were also shown to be incapable of a response, but the Reuber 35 and Morris 7793 hepatomas were responsive to tryptophan administration *in vivo* by a substantial, though low, increase in tryptophan pyrrolase activity. Dyer *et al.* (1964) extended these results to show that the Morris 7794B and 7795 hepatomas were also responsive to tryptophan. However, Cho *et al.* (1964) were able to show that, with the H-35 and 7793, adrenalectomy of the host abolished the response in the tumor, but not in the host liver. In the H-35, the tumor response was restored by a maintenance dose of cortisone given to the adrenalectomized tumor-bearing host. This finding would suggest a permissive effect of cortisone on the substrate induction in the neoplasm which is similar to that reported previously with this enzyme (Lee and Baltz, 1962; Lin and Knox, 1957).

Pitot and Morris (1963) demonstrated that, in the case of tryptophan pyrrolase, the Morris hepatoma 5123 showed no change in the levels of this enzyme after cortisone administration. Later work (Pitot, 1963a; Cho *et al.,* 1964), however, indicated that this neoplasm growing in adrenalectomized hosts was capable of a 2- to 5-fold increase in tryptophan pyrrolase after cortisone administration. However, both the 7316 and 7793 hepatoma gave significant, although low levels of induction after cortisone administration in intact hosts. In adrenalectomized hosts only the 5123 and 7793 hepatomas showed significant increases in tryptophan pyrrolase after cortisone administration. The 7793 was almost unique in that the induction of tryptophan pyrrolase in the tumor was as good as that seen in the liver. Dyer *et al.* (1964) also showed a high degree of cortisone induction of tryptophan pyrrolase in the 7793 hepatoma and less in the Morris 7795. Thus, it would appear that in contrast to the virtual absence of the substrate induction of tryptophan pyrrolase in tumors of adrenalectomized hosts, as well as its rarity in tumors of intact hosts, the cortisone induction of this enzyme in hepatomas is not altogether defective in all cases.

It should be noted that studies by the Pitot group (Pitot, 1968b) on purified tryptophan pyrrolase from both the Reuber hepatoma H-35 and the Morris hepatoma 7793 have not disclosed any differences in kinetic properties of the tumor enzyme, as related to substrate or substrate analogs of this enzyme, and to its behavior on gel electrophoresis or in immunodiffusion. The tumor enzymes react to derivatives of nicotinic acid in a manner identical to that seen with the enzymes from normal liver. Thus, there does not appear to be any alteration in the structure of the enzyme itself, indicating that the defects in control may not be secondary to structural mutations.

C. Regulation of Tyrosine α-Ketoglutarate Transaminase in Hepatomas

In contrast to the low response of tryptophan pyrrolase to cortisone administration, tyrosine α-ketoglutarate transaminase appears to be "overresponsive" to the presence of adrenal steroids. Work with 5123 hepatoma (Pitot and Morris, 1963) showed that this tumor exhibited very high levels of this enzyme in the intact host. Tyrosine administration increased the tumor enzyme only slightly while bringing the host liver tyrosine transaminase to tumor levels. Cortisone administration almost doubled the tumor basal level while giving the normal response in the host liver. Adrenalectomy of the host lowered the tumor enzyme to near that of the liver, and cortisone administration caused a response in both the liver and tumor of the adrenalectomized tumor-bearing host. Later work (Pitot, 1963a) showed that the Morris hepatoma 7800 gave results identical with the 5123, and since then results obtained with the H-35 and 7793 also show similar patterns. However, the exception to this is seen in the Morris 7316 hepatoma, which has a normal (compared with host liver) tyrosine transaminase basal level. Cortisone administration induced the enzyme in both liver and tumor to the same extent; however, if cortisone was given in low doses over a 7-day period, the enzyme level in the tumor was high while that in the host liver was normal. This effect is also seen with the Morris hepatoma 7288-C and 7800 in adrenalectomized hosts (Pitot, 1963a).

D. Template Lifetimes in Liver Neoplasms

A somewhat more subtle alteration in control mechanism has been described by Pitot *et al.* (1965b) wherein enzyme induction, specifically that of serine dehydratase and tryptophan pyrrolase, has been found to be differentially sensitive to the effects of actinomycin D when compared with the same mechanisms in normal liver. An example of this for the enzyme serine dehydratase is seen in Fig. 12. Normal rats and rats bearing the Reuber H-35 or Morris 5123 hepatoma were given a mixture of amino acids at zero time, and the sensitivity of enzyme induction at that time and at hourly or 2-hourly intervals thereafter was retested to see whether enzyme induction was still sensitive or resistant to the antibiotic. In normal liver, the induction of serine dehydratase becomes insensitive to actinomycin 2 hours after the initial amino acid administration and remains so for another 6 hours. In the H-35 hepatoma, at all times tested up to 6 hours, induction of the enzyme was found to be sensitive to the antibiotic whereas in the 5123, even at

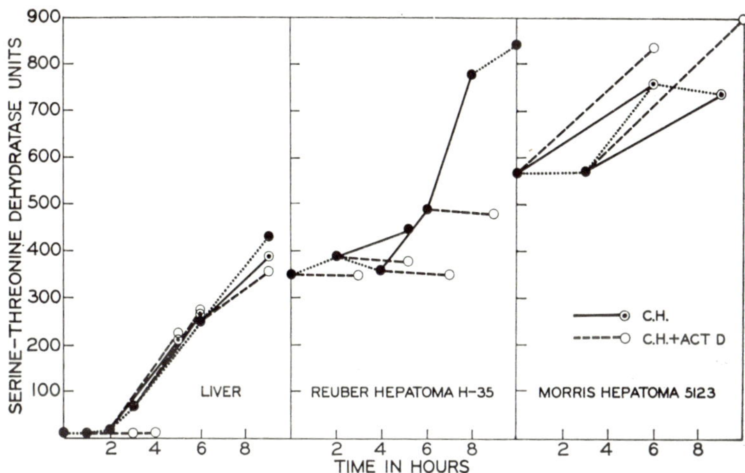

FIG. 12. The effect of actinomycin D (Act D) administration on the induction of serine dehydratase in liver, Reuber hepatoma H-35, and Morris hepatoma 5123. See Pitot (1968a) for experimental details.

zero time after the animal had been on a low protein for a week, induction of serine dehydratase was completely resistant to the antibiotic.

The altered effects of actinomycin on enzyme induction in neoplasms has its parallel in the process of differentiation. Rutter *et al.* (1963) have demonstrated that in developing pancreas, the appearance of the enzyme amylase is initially sensitive to the antibiotic during development whereas after the establishment of the acinar cell, the antibiotic has no further effect on the increase in level of this enzyme. A somewhat similar effect has been noted by Yaffe and Feldman (1964) in the case of developing skeletal muscle as well as in the eye (Kirk, 1965). Thus, it would appear that during differentiation messenger RNA templates become stabilized as an expression of the differentiation of the cell. Thus, each differentiated cell type is characterized by its own set of control mechanisms which are a reflection of its own particular set of stable messenger RNA templates. In abnormal differentiation, that of neoplasia, the stability of certain messenger RNA templates is altered, thus giving rise to a new differentiated state which is the expression of altered mechanisms for the control of enzyme synthesis.

E. Mechanisms of Template Stabilization in Liver and Its Abnormalities in Hepatomas

Pitot and co-workers (Pitot, 1932; Pitot *et al.*, 1965a) suggested that one of the major sites of biochemical abnormalities seen in hepatomas is the intracellular membrane systems. Partially on the basis of their own work (Pitot *et al.*, 1965a,b) and of other studies which have come to light in microorganisms (Aronson, 1966; Yudkin and Davis, 1965), a model was devised (Pitot, 1969a) to explain many of the alterations that are seen in template stability in liver neoplasms. Basically, the model states that messenger RNA is stabilized on intracellular membranes such as those of the endoplasmic reticulum by the association of the RNA with specific sites in the membrane. The spatial arrangement of these membrane sites is determined by the mosaic pattern of the membrane itself. If the membrane is altered in its mosaic pattern, then one would expect variations in the association of template and membrane. These associations may be either less stable or more stable than the situation in liver. A selective growth advantage results in those cell populations whose templates that are involved with enzymes important to DNA synthesis are more stable than those in normal cells. Evidence for this model has been obtained which indicates that in fact a DNA-like RNA is found in association with endoplasmic reticulum membranes in liver and this RNA is relatively tightly bound to the membrane in liver but poorly associated with the membrane in hepatomas (Lamar *et al.*, 1966). Furthermore, studies by Süss *et al.* (1966) have shown that *in vitro,* polysomes from both hepatomas and liver will bind to a similar degree on membranes of the endoplasmic reticulum of liver but much less so to membranes of the endoplasmic reticulum from highly differentiated as well as poorly differentiated hepatomas.

An interesting corollary of the mechanism suggested here is that the malignant transformation may be reversible. Such a phenomenon has been found in several different types of neoplasms but as yet no one has succeeded in the controlled reversion of malignant to relatively normal cellular populations.

It is difficult as yet to draw any general conclusions on the basis of the relatively small number of biochemical studies on control mechanisms in neoplasms. However, as has been pointed out (Pitot, 1963b), when experimental rat hepatomas are compared as a class with their cell of origin, in every neoplasm, primary or transplanted, qualitative defects in environmental mechanisms controlling enzyme levels, and probably enzyme synthesis, have been found. The pattern of control defects is virtually unique to each of the more highly differentiated hepatocellular carcinomas, thus providing each tumor with a unique phenotype. This fact appears to complicate the problem to an enormous degree; however, the study of mechanisms involved in the alteration of the control of genetic expression in systems not directly involved with cellular proliferation such as these highly differentiated or "minimal deviation neoplasms" (Potter, 1968) may take us a long way in our understanding of the neoplastic process itself.

VII. Concluding Remarks

As indicated previously, this review is not meant to be comprehensive. However, the overall format should give the reader an indication of the breadth of the field and the necessities for considerably more work.

It is of interest that perhaps some of the major clues to intermediary amino acid metabolism have resulted from various mutations, both in microorganisms and especially in the human being. A number of the degradative pathways involved in amino acid metabolism were first uncovered by the finding of specific mutations in the human being leading to diseases such as phenylketonuria, alkaptonuria, maple syrup urine disease, and others. A further interesting generalization which may be made in this area concerning the regulatory mechanisms involved in amino acid metabolism in the mammal is that most of the mutations seen in the human which involve amino acid metabolism are concerned with enzymes which are essentially absent or of extremely low activity in the fetus. Most of these enzymes do not appear until after birth and then some only slowly increase over a number of days before reaching the adult level. It is thus interesting to speculate that a number of the human mutants are in fact regulatory mutants involving the regulation of the synthesis of key enzymes in intermediary amino acid metabolism. It is for this reason that we have endeavored to at least briefly discuss a number of the aspects involved in the developmental regulatory mechanisms of enzymes of amino acid metabolism.

Although this review has been primarily designed to indicate our knowledge in the general area of the regulation of amino acid metabolism, the mechanisms of the regulation of enzyme synthesis in mammalian tissues should very briefly be indicated. While it is apparent that no single concept of the regulation of genetic expression in mammalian tissues has been forthcoming to compete with the operon concept in microorganisms, recent studies from this laboratory have suggested that such a concept is at least not impossible. The concept of a stable translating cytoplasmic messenger RNA operon, the MEMBRON (Pitot, 1969a) at least offers a working model to explain most, if not all, aspects of the regulation of genetic expression seen in mammalian tissues. The mechanism for the regulation of messenger RNA stability in mammalian tissues is incorporated into this model. The key role of regulation of the rate of degradation of enzymes may also be considered in such a model. Whether or not the MEMBRON concept will be upheld by experimentation in the future remains to be seen.

References

Adhya, S., and Echols, H. (1966). *J. Bacteriol.* **92,** 601.

Ames, B. N., and Hartman, P. E. (1963). *Cold Spring Harbor Symp. Quant. Biol.* **28,** 349.

Ansell, G. B., and Spanner, S. (1967). *J. Neurochem.* **14,** 873.

Aronson, A. (1966). *J. Mol. Biol.* **15,** 505.

Ashida, K., and Harper, A. E. (1961). *Proc. Soc. Exptl. Biol. Med.* **107,** 151.

Ballard, F. J., and Hanson, R. W. (1967). *Biochem. J.* **104,** 866.

Ballard, F. J., and Oliver, I. T. (1963). *Biochim. Biophys. Acta* **71,** 578.

Ballard, F. J., and Oliver, I. T. (1965). *Biochem. J.* **95,** 191.

Berl, S. (1966). *Biochemistry* **5,** 916.

Berlin, C. M., and Schimke, R. T. (1965). *Mol. Pharmacol.* **1,** 149.

Betheil, J. J., Feigelson, M., and Feigelson, P. (1965). *Biochim. Biophys. Acta* **104,** 92.

Blakley, R. L. (1957). *Biochem. J.* **65,** 342.

Bonner, J., and Ts'O, P., eds. (1964). "The Nucleohistones." Holden-Day, San Francisco, California.

Bottomley, R. H., Pitot, H. C., and Morris, H. P. (1963). *Cancer Res.* **23,** 392.

Bremer, J., and Greenberg, D. M. (1961). *Biochim. Biophys. Acta* **46,** 205.

Broome, J. D. (1968). *J. Exptl. Med.* **127,** 1055.

Brown, F. C., Mallady, J., and Roszell, J. A. (1966). *J. Biol. Chem.* **241,** 5220.

Budd, M. A., Tanaka, K., Holmes, L. B., Efron, M. L., Crawford, J. D., and Isselbacher, K., J. (1967). *New Engl. J. Med.* **277,** 321.

Bushnell, D., and Potter, V. R. (1969). Unpublished data.

Calissano, P., Bonsignore, D., and Cartasegna, C. (1966). *Biochem. J.* **101,** 550.

Carney, S. A., Lawrence, J. C., and Ricketts, C. R. (1966). *J. Inves. Dermatol.* **46,** 586.

Castellucci, A. (1965). *Experientia* **21,** 219.

Chari-Bitron, A. (1963). *Biochem. Biophys. Res. Commun.* **12,** 310.

Cho-Chung, Y. S., and Pitot, H. C. (1967). *J. Biol. Chem.* **242,** 1192.

Cho, Y. S., Pitot, H. C., and Morris, H. P. (1964). *Cancer Res.* **24,** 97.

Cihak, A., Pliska, V., and Sorm, F. (1966). *Collection Czech. Chem. Commun.* **31,** 4154.

Cittadini, D., Pietropaolo, C., deCristofaro, D., and d'Ayjello-Caracciolo (1964). *Nature* **203,** 643.

Civen, M., Brown, C. B., and Trimmer, B. W. (1967). *Arch. Biochem. Biophys.* **120,** 352.

Connelly, J. L., Danner, D. J., and Bowden, J. A. (1968). *J. Biol. Chem.* **243,** 1198.

Corredor, C., Mausbach, C., and Bressler, R. (1967). *Biochem. Biophys. Acta* **144,** 366.

Curry, D. M., and Beaton, G. H. (1958). *Endocrinology* **63,** 252.

Dancis, J., Hutzler, J., Tada, K., Wada, Y., Mairbawa, T., and Arahawa, T. (1967). *Pediatrics* **39,** 813.

deDuve, C., and Baudhuin, P. (1966). *Physiol. Rev.* **46,** 323.

DeGroot, L. J. (1965). *New Engl. J. Med.* **272,** 243.

Durieu-Trautmann, O., and Chatagner, F. (1966). *Bull. Soc. Chim. Biol.* **48,** 77.

Dyer, H. M., Gullino, P., and Morris, H. P. (1964). *Cancer Res.* **24,** 97.

Eisenstein, A. B. (1967). *Am. J. Clin. Nutr.* **20,** 282.

Eliasson, E. E. (1963). *Exptl. Cell Res.* **30,** 74.

Elliott, W. H. (1959). *Nature* **183,** 1051.
Exton, J. H., and Park, C. R. (1967). *J. Biol. Chem.* **242,** 2622.
Fallon, H. J. (1968). *J. Nutr.* **96,** 220.
Fallon, H. J., and Byrne, W. L. (1966). *Proc. Soc. Exptl. Biol. Med.* **123,** 378.
Farber, E. (1963). *Advan. Cancer Res.* **7,** 383.
Field, J. B. (1968). *Metabolism, Clin. Exptl.* **17,** 226.
Finkelstein, J. D. (1962). *Arch. Biochem. Biophys.* **122,** 583.
Finkelstein, J. D., and Mudd, S. H. (1967). *J. Biol. Chem.* **242,** 873.
Fleck, A., Shepherd, J., and Munro, H. N. (1965). *Science* **150,** 628.
Fleshood, L. (1969). *Federation Proc.* **28,** 727.
Fonnum, F. (1968). *Biochem. J.* **106,** 401.
Foster, D. O., Ray, P. D., and Lardy, H. A. (1966). *Biochemistry* **5,** 563.
Foster, D. O., Lardy, H. A., Ray, P. D., and Johnston, J. B. (1967). *Biochemistry* **6,** 2120.
Freedland, R. A. (1964). *Proc. Soc. Exptl. Biol. Med.* **116,** 692.
Freedland, R. A., and Avery, E. H. (1964). *J. Biol. Chem.* **239,** 3357.
Freedland, R. A., Wadzinski, I. M., and Waisman, H. A. (1961). *Biochem. Biophys. Res. Commun.* **6,** 227.
Freedland, R. A., Krakowski, M. C., and Waisman, H. A. (1964). *Am. J. Physiol.* **206,** 341.
Freedland, R A., Murad, S., and Hurvitz, A. I. (1968). *Federation Proc.* **27,** 1217.
Freedman, A. D., and Kohn, L. (1964). *Science* **145,** 58.
Freedman, M. L., Fisher, J. M., and Rabinovitz, M. (1968). *J. Mol. Biol.* **33,** 315.
Frieden, C. (1963). *In* "The Enzymes" (P. D. Boyer, H. Lardy, and K. Myrbäck. eds.), Vol. 7, pp. 3–24. Academic Press, New York.
Garcia, A., Williamson, J. R., and Cahill, G. F., Jr. (1966). *Diabetes* **15,** 188.
Gilbert, J. B. (1957). *J. Am. Chem. Soc.* **79,** 2342.
Goldstein, L. (1965). *Nature* **205,** 1330.
Goldstein, L., Knox, W. E., and Behrman, E. J. (1962). *J. Biol. Chem.* **237,** 2855.
Goodfriend, T. L., and Kaufman, S. (1961). *J. Clin. Invest.* **40,** 1743.
Gordon, L. S., and Farber, E. (1965). *Arch. Biochem. Biophys.* **112,** 233.
Goswami, M. N. D., and Knox, W. E. (1963). *J. Chronic Diseases* **16,** 363.
Granick, S. (1963). *J. Biol. Chem.* **238,** PC2247.
Granick, S., and Kappas, A. (1967). *J. Biol. Chem.* **242,** 4587.
Greenberg, D. M. (1960). *In* "The Enzymes" (P. D. Boyer, H. Lardy, and K. Myrbäck, eds.), Vol. 4, pp. 257–267. Academic Press, New York.
Greenberg, D. M. (1962). *In* "Methods in Enzymology" (S. P. Colowick and N. O. Kaplan, eds.), Vol. 5, pp. 931–951. Academic Press, New York.
Greengard, O., and Dewey, H. K. (1967). *J. Biol. Chem.* **242,** 2986.
Greengard, O., and Gordon, M. (1963). *Federation Proc.* **22,** 232.
Grillo, M. A. (1964). *Clin. Chim. Acta* **10,** 259.
Gupta, R. N., and Spenser, I. D. (1969). *J. Biol. Chem.* **243,** 88.
Hakanson, R. (1963). *Biochem. Pharmacol.* **12,** 1289.
Harper, A. (1968). *Am. J. Clin. Nutr.* **21,** 358.
Hartshorne, D., and Greenberg, D. M. (1964). *Arch. Biochem. Biophys.* **105,** 173.
Hauge, J. G., MacQuillan, A. M., Cline, A. L., and Halvorson, H. O. (1961). *Biochem. Biophys. Res. Commun.* **5,** 267
Hayaishi, O., Gefter, M., and Weissbach, H. (1963). *J. Biol. Chem.* **238,** 2040.
Hayashi, N., Yoda, B., and Kibuchi, G. (1968). *J. Biochem.* **63,** 446.

Henning, H. V., Seiffert, I., and Seubert, W. (1963). *Biochim. Biophys. Acta* **77,** 345.

Herrera, M. G., Kamm, G. D., Ruderman, N. B., and Cahill, G. F., Jr. (1966). *Advan. Enzyme Regulation* **4,** 225.

Herzfeld, A., and Knox, W. E. (1968). *J. Biol. Chem.* **243,** 3327.

Hickman, R., Saunders, S. J., Dowdle, E., and Eales, L. (1968). *Biochim. Biophys. Acta* **161,** 197.

Hirata, Y., Kawachi, T., and Sugimura, T. (1967). *Biochim. Biophys. Acta* **144,** 233.

Hope, D. B. (1955). *Biochem. J.* **59,** 497.

Horie, S., and Shimazono, N. (1961). *J. Biochem.* **49,** 768.

Horowitz, B., Madras, B. K., Meister, A., Old, L. J., Boyse, E. H., and Stockert, E. (1968) *Science* **160,** 533.

Hutzler, J., and Dancis, J. (1968). *Biochim. Biophys. Acta* **158,** 62.

Ichihara, A., and Greenberg, D. M. (1957). *J. Biol. Chem.* **224,** 331.

Ichihara, A., and Koyama, E. (1966). *J. Biochem.* **59,** 160.

Ichihara, A., and Takahashi, H. (1968). *Biochim. Biophys. Acta* **167,** 274.

Illnerova, H. (1968a). *Physiol. Bohemoslov.* **17,** 70.

Illnerova, H. (1968b). *Physiol. Bohemoslov.* **17,** 81.

Ishikawa, E., Ninagawa, T., and Suda, M. (1965). *J. Biochem.* **57,** 506.

Isselbacher, K. J. (1967). *New Engl. J. Med.* **277,** 321.

Jacob, F., and Monod, J. (1961). *J. Mol. Biol.* **3,** 318.

Jacobsen, J. G., and Smith, L. H. (1968). *Physiol. Rev.* **48,** 424.

Jervell, K. F., Christoffersen, T., and Morland, J. (1965). *Arch. Biochem. Biophys.* **111,** 15.

Jones, M. E., Anderson, A. D., Anderson, C., and Hodes, S. (1961). *Arch. Biochem. Biophys.* **95,** 499.

Jost, J.-P., and Pitot, H. C. (1969). Unpublished data.

Jost, J.-P., Khairallah, E. A., and Pitot, H. C. (1968). *J. Biol. Chem.* **243,** 3057.

Jost, J.-P., Hsie, A. W., and Rickenberg, H. V. (1969). *Biochem. Biophys. Res. Commun.* **34,** 748.

Kaplan, B. H., and Stadtman, E. R. (1968). *J. Biol. Chem.* **243,** 1787.

Kaplan, J. H., and Pitot, H. C. (1969). Unpublished data.

Kaplan, J. H., Scarbrough, E., and Pitot, H. C. (1969). *Federation Proc.* **28,** 730.

Karasek, M. A., and Greenberg, D. M. (1957). *J. Biol. Chem.* **227,** 191.

Kato, A., Matsuzawa, T., Suda, M., Nakagawa, H., and Ishizuka, J. (1964). *J. Biochem.* **55,** 401.

Kato, A., Ogura, M., Kimura, H., Kawai, T., and Suda, M. (1966a). *J. Biochem.* **59,** 34.

Kato, A., Ogura, M., and Suda, M. (1966b). *J. Biochem.* **59,** 40.

Katsunuma, T., Temma, M., and Katunuma, N. (1968). *Biochem. Biophys. Res. Commun.* **32,** 433.

Katunuma, N., Matsuda, Y., and Tomino, I. (1964). *J. Biochem.* **56,** 499.

Katunuma, N., Tomino, I., and Sanada, Y. (1968). *Biochem. Biophys. Res. Commun.* **32,** 426.

Kaufman, S. (1963). *Proc. Natl. Acad. Sci. U.S.* **50,** 1085.

Kaufman, S. (1964). *Trans. N.Y. Acad. Sci.* **26,** 977.

Kaufman, S. (1966). *Pharmacol. Rev.* **18,** 61.

Kenney, F. T., Reem, G. H., and Kretchmer, N. (1958). *Science* **127,** 86.

Khairallah, E. A., and Pitot, H. C. (1967). *Biochem. Biophys. Res. Commun.* **29**, 269.

Kim, S., and Cohen, P. P. (1965). *Arch. Biochem. Biophys.* **109**, 421.

Kirk, D. L. (1965). *Proc. Natl. Acad. Sci. U.S.* **54**, 1345.

Knox, W. E., and Greengard, O. (1965). *Advan. Enzyme Regulation* **3**, 247.

Kolenbrander, H. M., and Berg, C. P. (1967). *Arch. Biochem. Biophys.* **119**, 110.

Koszalka, T. R. (1967). *Arch. Biochem. Biophys.* **122**, 400.

Krebs, E. G. (1966). *Pharmacol. Rev.* **18**, 163.

Kretchmer, N., and McNamara, H. (1956). *J. Clin. Invest.* **35**, 1089.

Kwong, E., and Barnes, R. H. (1967). *J. Nutr.* **92**, 233.

Lamar, C., and Sellinger, O. Z. (1965). *Biochem. Pharmacol.* **14**, 489.

Lamar, C., Prival, M., and Pitot, H. C. (1966). *Cancer Res.* **26**, 1909.

Lardy, H. A., Foster, D. O., Young, J. W., Shrago, E., and Ray, P. D. (1965a). *J. Cellular Comp. Physiol.* **66**, Suppl. 1, 39.

Lardy, H. A., Paetkau, V., and Walter, P. (1965b). *Proc. Natl. Acad. Sci. U.S.* **53**, 1410.

Lee, N. D., and Baltz, B. E. (1962). *Endocrinology* **70**, 84.

Levere, R. D. (1965). *Exptl. Med. Surgery Suppl.* **172**.

Lin, E. C. C., and Knox, W. E. (1957). *Biochim. Biophys. Acta* **26**, 85.

Litwack, G., and Nemeth, A. M. (1965). *Arch. Biochem. Biophys.* **109**, 316.

Litwack, G., Al-Nejjar, Z. H., Sears, M. L., and Ostheimer, G. W. (1964). *Nature* **201**, 1028.

Loiselet, J., and Chatagner, F. (1966). *Biochim. Biophys. Acta* **130**, 180.

Loomis, W. F., and Magasanik, B. (1966). *J. Bacteriol.* **92**, 107.

Macchia, V., and Pastan, I. (1967). *J. Biol. Chem.* **242**, 1864.

McFarlane, I. G., and von Holt, C. (1966). *Abstr. 6th Intern. Congr. Biochem., New York* **5**, 499.

McKenzie, J. M., Adiga, P. R., and Murthy, P. V. N. (1968). *Endocrinology* **83**, 1132.

McLean, P., and Gurney, M. W. (1963). *Biochem. J.* **87**, 96.

Magasanik, B. (1961). *Cold Spring Harbor Symp. Quant. Biol.* **26**, 249.

Makoff, R., and Baldridge, R. C. (1964). *Biochim. Biophys. Acta* **90**, 282.

Mallette, L. E., Exton, J. H., and Park, C. R. (1969). *J. Biol. Chem.* **244**, 5713.

Marver, H. S., Collins, A., Tschudy, D. P., and Rechcigl, M., Jr. (1966). *J. Biol. Chem.* **241**, 4323.

Matsuoka, T., Yoda, B., and Kikuchi, G. (1968). *Arch. Biochem. Biophys.* **126**, 530

Meister, A. (1965). "Biochemistry of the Amino Acids," 2nd ed., Vols. I and II. Academic Press, New York.

Mider, G. B. (1951). *Cancer Res.* **11**, 821.

Miller, L. L. (1961). *Recent Progr. Hormone Res.* **17**, 539.

Mimura, T., Yamada, C., and Swendseid, M. E. (1968). *J. Nutr.* **95**, 493.

Morris, H. P. (1963). *Progr. Exptl. Tumor Res.* **3**, 370.

Mortimore, G. E. (1963). *Am. J. Physiol.* **204**, 669.

Moses, V., and Prevost, C. (1966). *Biochem. J.* **100**, 336.

Mudd, S. H., Finkelstein. J. D., Irreverre, F., and Laster, L. (1965). *Biochem. Biophys. Res. Commun.* **19**, 665.

Mulford, D. J., Longmore, W. J., and Kreye, G. M. (1959). *Arch. Biochem. Biophys.* **82**, 1.

Munro, H. N. (1968). *In* "Regulatory Mechanisms for Protein Synthesis in Mammalian Cells" (A. San Pietro, M. Lamborg, and F. T. Kenney, eds.), pp. 183–190. Academic Press, New York.

Nagabhushanam, A., and Greenberg, D. M. (1965). *J. Biol. Chem.* **240,** 3002.

Nakada, D., and Magasanik, B. (1964). *J. Mol. Biol.* **8,** 105.

Nakagawa, H., and Kimura, H. (1968). *Biochem. Biophys. Res. Commun.* **32,** 208.

Nakagawa, H., Kimura, H., and Miura, S. (1967). *Biochem. Biophys. Res. Commun.* **28,** 359.

Nemer, M. J., and Elwyn, D. (1960). *J. Biol. Chem.* **235,** 2070.

Nesheim, M. C. (1968). *Federation Proc.* **27,** 1210.

Noall, M. W., Riggs, T. R., Walker, L. M., and Christensen, H. N. (1957). *Science* **126,** 1002.

Nordlie, R. C., and Lardy, H. A. (1963). *J. Biol. Chem.* **238,** 2259.

Paik, W., and Cohen, P. P. (1960). *J. Gen. Physiol.* **43,** 683.

Park, H. S. (1969). M. S. Thesis. Univ. Wisconsin, Madison, Wisconsin.

Patterson, M. K., and Orr, G. R. (1968). *J. Biol. Chem.* **243,** 376.

Peisach, J., and Strecker, H. J. (1962). *J. Biol. Chem.* **237,** 2255.

Peraino, C. (1968). *Biochim. Biophys. Acta* **165,** 108.

Peraino, C., and Pitot, H. C. (1963). *Biochim. Biophys. Acta* **73,** 222.

Peraino, C., and Pitot, H. C. (1964). *J. Biol. Chem.* **239,** 4308.

Peraino, C., and Pitot, H. C. (1969). Unpublished data.

Peraino, C., Blake, R. L., and Pitot, H. C. (1965). *J. Biol. Chem.* **240,** 3039.

Perlman, R. L., and Pastan, I. (1968). *J. Biol. Chem.* **243,** 5420.

Peterkofsky, B., and Udenfriend, S. (1963). *Biochem. Biophys. Res. Commun.* **12,** 257.

Phillips, A. T., and Wood, W. A. (1964). *Biochem. Biophys. Res. Commun.* **6,** 530.

Pitot, H. C. (1959). Ph.D. Thesis. Tulane Univ., New Orleans, Louisiana.

Pitot, H. C. (1962). *Federation Proc.* **21,** 1124.

Pitot, H. C. (1963a). *Advan. Enzyme Regulation* **1,** 309.

Pitot, H. C. (1963b). *Cancer Res.* **23,** 1474.

Pitot, H. C. (1966). *Ann. Rev. Biochem.* **35,** 335.

Pitot, H. C. (1968a). *Cancer Res.* **28,** 1880.

Pitot, H. C. (1968b). *In* "The Control of Growth Processes by Chemical Agents," pp. 67–81. Pergamon, Oxford.

Pitot, H. C., and Jost, J.-P. (1968). *In* "Regulatory Mechanisms for Protein Synthesis in Mammalian Cells" (A. San Pietro, M. Lamborg, and F. T. Kenney, eds.), pp. 283–298. Academic Press, New York.

Pitot, H. C. (1969a). *Arch. Pathol.* **87,** 212.

Pitot, H. C. (1969b). Unpublished data.

Pitot, H. C., and Morris, H. P. (1963). *Cancer Res.* **21,** 1009.

Pitot, H. C., and Peraino, C. (1963). *J. Biol. Chem.* **238,** PC1910.

Pitot, H. C., and Peraino, C. (1964a). *J. Biol. Chem.* **239,** 1783.

Pitot, H. C., and Peraino, C. (1964b). *J. Biol. Chem.* **239,** 4308.

Pitot, H. C., Potter, V. R., and Morris, H. P. (1961). *Cancer Res.* **21,** 1001.

Pitot, H. C., Peraino, C., Bottomley, R. H., and Morris, H. P. (1963). *Cancer Res.* **23,** 135.

Pitot, H. C., Peraino, C., Pries, N., and Kennan, A. L. (1964). *Advan. Enzyme Regulation* **2,** 237.

Pitot, H. C., Peraino, C., and Lamar, C. (1965a). *In* "Developmental and Metabolic Control Mechanism and Neoplasia," pp. 413–426. Williams & Wilkins, Baltimore, Maryland.

Pitot, H. C., Peraino, C., Lamar, C., and Kennan, A. L. (1965b). *Proc. Natl. Acad. Sci. U.S.* **54**, 845.

Pizer, L. I. (1964). *J. Biol. Chem.* **239**, 4219.

Pizer, L. I. (1966). *Biochim. Biophys. Acta* **124**, 418.

Pontremoli, S., and Grazi, E. (1968). *In* "Carbohydrate Metabolism and Its Disorders" (F. Dickens, P. J. Randle, and W. J. Whelan, eds.), Vol. 1, pp. 260–295. Academic Press, New York.

Potter, V. R. (1968). *Cancer Res.* **28**, 1901.

Raiha, N. C. R., and Kekomäki, M. P. (1968). *Biochem. J.* **108**, 521.

Rao, D. R., Deodhar, A. D., and Hairharan, K. (1965). *Biochem. J.* **97**, 311.

Ray, P. D., Foster, D. O., and Lardy, H. A. (1966). *J. Biol. Chem.* **241**, 3904.

Reel, J. R., and Kenney, F. T. (1968). *Proc. Natl. Acad. Sci. U.S.* **61**, 200.

Rehbinder, D., and Greenberg, D. M. (1965). *Arch. Biochem. Biophys.* **108**, 110.

Reif, L., and Amos, H. (1966). *Biochem. Biophys. Res. Commun.* **23**, 39.

Renson, J., Weissbach, H., and Udenfriend, S. (1962). *J. Biol. Chem.* **237**, 2261.

Reuber, M. D. (1966). *Gann Monogr.* **1**, 43.

Roberts, E. (1960). *In* "The Enzymes" (P. D. Boyer, H. Lardy and K. Myrback, eds.), Vol. 4, pp. 285–300. Academic Press, New York.

Robison, G. A., Butcher, R. W.. and Sutherland, E. W. (1968). *Ann. Rev. Biochem.* **37**, 149.

Ronzio, R. A., and Meister, A. (1968). *Proc. Natl. Acad. Sci. U.S.* **59**, 164.

Rosen, F., Roberts, N. R., and Nichol, C. A. (1959). *J. Biol. Chem.* **234**, 476.

Ruderman, N. B., and Herrera, M. G. (1968). *Am. J. Physiol.* **214**, 1346.

Rudnick, D., and Waelsch, H. (1955). *J. Exptl. Zool.* **129**, 309.

Russell, D., and Snyder, S. H. (1968). *Proc. Natl. Acad. Sci. U.S.* **60**, 1420.

Russell, R. L., and Coleman, D. L. (1963). *Genetics* **48**, 1033.

Rutter, W. V., Wessels, N. K., and Grobstein, C. (1963). *Natl. Cancer Inst. Monogr.* **13**, 51.

Ryan, W. L., and Orr, W. (1966). *Arch. Biochem. Biophys.* **113**, 684.

Schayer, R. W., Janoff, A., and Zweifach, B. W. (1963). *Am. J. Physiol.* **204**, 369.

Schimke, R. T. (1963a). *Natl. Cancer Inst. Monogr.* **13**, 197.

Schimke, R. T. (1963b). *J. Biol. Chem.* **238**, 1012.

Schimke, R. T. (1964a). *J. Biol. Chem.* **239**, 136.

Schimke, R. T. (1964b). *J. Biol. Chem.* **239**, 3808.

Schirch, L., and Gross, T. (1968). *J. Biol. Chem.* **243**, 1968.

Scrutton, M. C., and Utter, M. F. (1968). *Ann. Rev. Biochem.* **37**, 249.

Seed, R. W., and Goldberg, I. H. (1963). *Proc. Natl. Acad. Sci. U.S.* **50**, 275.

Segal, H. L., Beattie, D. S., and Hopper, S. (1962). *J. Biol. Chem.* **237**, 1914.

Segal, H. L., and Kim, Y. S. (1963). *Proc. Natl. Acad. Sci. U.S.* **50**, 912.

Seubert, W., and Huth, W. (1965). *Biochem. Z.* **343**, 176.

Shanley, B. C., Zail, S. S., and Joubert, S. M. (1968). *Lancet* **i**, 70.

Shrago, E., and Lardy, H. A. (1966). *J. Biol. Chem.* **241**, 663.

Shrago, E., Lardy, H. A., Nordlie, R. C., and Foster, D. O. (1963). *J. Biol. Chem.* **238**, 3188.

Shrago, E., Young, J. W., and Lardy, H. A. (1967). *Science* **158**, 1572.

Sidransky, H., Bongiorno, M., Sarma, D. S. R., and Verney, E. (1967). *Biochem. Biophys. Res. Commun.* **27**, 242.

Smith, A. D., Benziman, M., and Strecker, H. J. (1967). *Biochem. J.* **104**, 557.

Smith, L. C., and Rossi, F. M. (1958). *Proc. Soc. Exptl. Biol. Med.* **99**, 754.

Snyder, S. H., and Epps, L. (1968). *Mol. Pharmacol.* **4**, 187.

Söling, H. D., Kaplan, J., Erbstoezer, M., and Pitot, H. C. (1968). *Advan. Enzyme Regulation* **7,** (in press).

Spector, S., Gordon, R., Sjoerdsma, A., and Udenfriend, S. (1967). *Mol. Pharmacol.* **3,** 549.

Spolter, H., and Baldridge, R. C. (1963). *J. Biol. Chem.* **238,** 2071.

Spolter, H., and Baldridge, R. C. (1964). *Biochim. Biophys. Acta* **90,** 287.

Stetten, D. Jr. (1942). *J. Biol. Chem.* **144,** 501.

Stokstad, E. L. R., and Koch, J. (1967). *Physiol. Rev.* **47,** 83.

Stone, N., and Meister, A. (1962). *Nature* **194,** 555.

Strecker, H. J. (1965). *J. Biol. Chem.* **240,** 1225.

Subrahmanyan, D. (1963). *Ind. J. Exptl. Biol.* **1,** 182.

Sung, G. P., and Johnstone, R. M. (1967). *Biochem. J.* **105,** 497.

Süss, R., Blobel, G., and Pitot, H. C. (1966). *Biochem. Biophys. Res. Commun.* **23,** 299.

Swick, R. W., Rexroth, A. K., and Stange, J. L. (1968). *J. Biol. Chem.* **243,** 3581.

Tanaka, T. (1965). *Protein, Nucleic Acid, Enzymes, Tokyo* **10,** 668.

Tepperman, H. M., and Tepperman, V. (1958). *Diabetes* **7,** 468.

Tomkins, G. M., Thompson, E. B., Hayashi, S., Gelehrter, T. D., Granner, D., and Peterkofsky, B. (1966). *Cold Springs Harbor Symp. Quant. Biol.* **31,** 349.

Trautmann, O., and Chatagner, F. (1964). *Bull. Soc. Chim. Biol.* **44,** 129.

Tschudy, D. P., Welland, F. H., Collins, A., and Hunter, G. (1964). *Metabolism, Clin. Exptl.* **13,** 396.

Tschudy, D. P., Wakman, A., and Collins, A. (1967). *Proc. Natl. Acad. Sci. U.S.* **58,** 1944.

Utter, M. F., and Keech, D. B. (1963). *J. Biol. Chem.* **238,** 2603.

Utter, M. F., and Kurahashi, K. (1954). *J. Biol. Chem.* **207,** 821.

Van den Berg, C. J., van Kempeu, G. M. J., Schade, J. P., and Veldstra, H. (1965). *J. Neurochem.* **12,** 863.

Van Pilsum, J. F., and Canfield, T. M. (1962). *J. Biol. Chem.* **237,** 2574.

Velick, S. F., and Vavra, J. (1962). *J. Biol. Chem.* **237,** 2109.

Veneziale, C M., Walter, P., Kneer, N., and Lardy, H. A. (1967). *Biochemistry* **6,** 2129.

Veneziale, C. M., Grabrielli, F., Kneer, N., and Lardy, H. A. (1969). *Federation Proc.* **28,** 411.

Villa-Treviño, S., Shull, K. H., and Farber, E. (1963). *J. Biol. Chem.* **238,** 1757.

Wada, O., Sassa, S., Takaku, F., Yano, Y., Urata, G., and Nakao, K. (1967). *Biochim. Biophys. Acta* **148,** 585.

Waldorf, M. A., and Harper, A. E. (1963). *Proc. Soc. Exptl. Biol. Med.* **112,** 955.

Waldorf, M. A., Kirk, M. C., Linkswiler, H., and Harper, A. E. (1963). *Proc. Soc. Exptl. Biol. Med.* **112,** 764.

Walker, J. B. (1959). *Biochim. Biophys. Acta* **36,** 574.

Walker, J. B. (1960). *J. Biol. Chem.* **235,** 2357.

Walker, J. B. (1961). *J. Biol. Chem.* **236,** 493.

Walker, M. S., and Walker, J. B. (1962). *J. Biol. Chem.* **237,** 473.

Walter, P., Paetkau, V., and Lardy, H. A. (1966). *J. Biol. Chem.* **241,** 2523.

Weber, G., Singhal, R. L., Stamm, N. B., Fisher, E. A., and Mantendiek, M. A. (1964). *Advan. Enzyme Regulation* **2,** 1.

Weiss, J., and Pitot, H. C. (1969). Unpublished observations.

Weissbach, A., and Sprinson, D. B. (1953). *J. Biol. Chem.* **203,** 1031.

Wells, I. C. (1963). *Federation Proc.* **22,** 652.

Wells, I. C., and Remy, C. N. (1965). *Arch. Biochem. Biophys.* **112,** 201.
Wergedal, J. E., and Harper, A. E. (1964). *Proc. Soc. Exptl. Biol. Med.* **116,** 600.
Whitely, H. R., and Tahara, M. (1966). *J. Biol. Chem.* **241,** 4881.
Williams, D. B., and Elvehjem, C. A. (1950). *J. Biol. Chem.* **183,** 539.
Williams-Ashman, H. G. (1968). *Advan. Enzyme Regulation* **7,** (in press).
Willis, J. E., and Sallach, H. J. (1962). *J. Biol. Chem.* **237,** 910.
Wilson, J. D., Gibson, K. D., and Udenfriend, S. (1960). *J. Biol. Chem.* **235,** 3539.
Wilson, J. E., and Dove, J. L. (1965). *J. Elisha Mitchell Sci. Soc.* **81,** Suppl. 1, 21.
Wu, C., Roberts, E. H., and Bauer, J. M. (1965). *Cancer Res.* **25,** 677.
Wunner, W. H., Bell, J., and Munro, H. N. (1966). *Biochem. J.* **101,** 417.
Yaffe, D., and Feldman, M. (1964). *Develop. Biol.* **9,** 347.
Yeung, D., and Oliver, I. T. (1967). *Biochem. J.* **103,** 744.
Young, J. E., Prager, M. D., and Atkins, I. C. (1967). *Proc. Soc. Exptl. Biol. Med.* **125,** 860.
Young, J. W., Shrago, E., and Lardy, H. A. (1964). *Biochemistry* **3,** 1687.
Yudkin, M. D., and Davis, B. D. (1965). *J. Mol. Biol.* **12,** 193.

CHAPTER 36

Diurnal Rhythms in Mammalian Protein Metabolism

RICHARD J. WURTMAN

Department of Nutrition and Food Science,
Massachusetts Institute of Technology,
Cambridge, Massachusetts

I. Introduction

A. *Regulation of the Extracellular Fluid*

Perhaps the most important advantage that accrues to mammalian cells by virtue of their communal relationship is the constancy of their immediate environment, the extracellular fluid. The cells of mammals have the good fortune to be bathed in a medium whose temperature and whose concentrations of perhaps forty important compounds vary over a surprisingly narrow range; hence these cells are never faced with the problem of surviving in the absence of glucose or water or amino acids, or of adapting to temperatures which differ from their normal mean by more than a degree or two.

There is good evidence that the constancy of several of these biochemical and physical functions (e.g., those which keep the temperature, osmolality, and calcium concentration of the extracellular fluid from rising too high) results from the operation of a particular type of regulatory mechanism, the closed feedback loop. Moreover, the systems which control these functions appear to utilize parallel structural components, termed sensors, set-points, comparators, and effectors. The biological sensor is thought to be a group of cells, in the brain or elsewhere, which continuously monitors the level of the regulated function in the plasma; the set-point is another group (or the same cells, as in the case of the parathyroid gland) which stores information as to the highest (or lowest) absolute value that the regulated function is allowed to attain. The cells which function as comparators presumably receive inputs from both the sensor and the set-point; they subtract the former value from the latter, and if the difference is greater than zero they issue an "error signal." This signal is transmitted via neural or hormonal channels to distant effector organs, which then act to dissipate the heat, conserve the water, or remove the excess calcium. It cannot be assumed *a priori* that all compounds whose concentrations in the internal milieu remain fairly constant do so because of the operation of closed feedback loops; processes not susceptible to internal control (such as the spillage of glucose into the urine of hyperglycemic subjects) can also contribute to the stabilization of blood levels. However, more often than not, the search for a regulatory system as the basis of the constancy of a particular compound in the extracellular fluid has proved rewarding.

B. Biological Rhythms

For the greater portion of the past century, the recognition that the levels of certain compounds in the extracellular fluid remain surprisingly constant tended to obscure the fact that most of these levels could also be shown to undergo characteristic changes, albeit within a narrow range. Perhaps the most general type of variation which is exhibited by regulated functions is diurnal rhythmicity. It now appears likely that the rates of most physiological processes and the concentrations of most biochemical compounds within the cells of mammals show at least some tendency to vary as a function of time of day. These variations are reflected in changes in the composition and temperature of the extracellular fluid. For example, body temperature in humans is regulated at all times; however, the set-point around which temperature is held constant varies predictably during each 24-hour period. Early in the morning the "normal" temperature for most males is a full degree lower than it is 12 hours later. It can be shown that if the temperature is artificially elevated from 97.5°F to 98.0°F at 7:00 AM the subject will perspire; however, if the same subject's temperature is *lowered* from 98.6°F to 98.0°F at 4:00 PM, he will shiver.

This type of demonstration provides compelling evidence that the increase in body temperature observed during the hours of daylight is not simply the result of the higher ambient temperature, or of the greater heat load generated by muscular work during these hours. Similar studies, described in greater detail below, have shown that the daily rhythm in plasma amino acid levels is not simply the consequence of man's habit of presenting a load of dietary amino acids to his gut at mealtimes. The concentrations of most amino acids in human plasma have recently been shown to undergo characteristic diurnal fluctuations. Among subjects maintained on a normal activity schedule (i.e., in bed between 11:00 PM and 7:00 AM), the levels of most of the individual amino acids are lowest between 2:00 and 4:00 AM, and rise by as much as 100% during the next 8 hours (Wurtman *et al.*, 1968a). These rhythms persist when subjects are deprived of dietary protein, and must therefore reflect endogenous changes in the metabolism of the amino acids. The concentration of any amino acid in plasma depends upon two sets of rates, the rates at which the compound enters the blood from its various sources (e.g., dietary protein, the breakdown of tissue protein), and the rates at which it is removed from the blood by entering one of its "sinks" (e.g., uptake into cellular amino acid pools, transamination). Some of these rates may be regulated in the sense described above; that is, they may be controlled by feedback loops which monitor the existing level of the amino acid, and operate to maintain this level within a narrow range. In addition, almost all these rates can be modified by hormones, food intake, and other factors which also display diurnal rhythmicity. This chapter will describe diurnal rhythms in the metabolism and plasma concentrations of individual amino acids, and will consider the mechanisms which might generate such rhythms.

C. *Mechanisms of Biological Rhythmicity*

The mechanism of a rhythm will be defined as (1) the site of origin of the oscillating signals which instruct the cells of the "target organ" [for example, the hepatic cells which display rhythmic changes in tyrosine transaminase activity (Wurtman and Axelrod, 1967)] to vary the rate of a physiological process rhythmically, and (2) the intracellular biochemical events which mediate the rhythmic behavior. Theoretically, a daily rhythm in, for example, the activity of a hepatic enzyme could be caused by signals originating within any of three possible loci:

(1) *Within the cell itself.* Evidence has been presented that diurnal rhythms which are independent of the environment do exist among certain single-celled organisms. For example, *Euglena* incorporates more phenylalanine-^{14}C into protein during the daily light period than in dark-

ness; this rhythm persists under conditions of continuous darkness for at least 2 days (Feldman, 1968). Mammalian cells maintained in synchronous tissue cultures show distinct biochemical rhythms which are thought to reflect the time-constants of the various steps necessary for protein synthesis. The periods of these rhythms are usually considerably shorter than 24 hours (Klevecz and Ruddle, 1968). If cell-free extracts of yeasts are incubated in a medium which is continuously injected with glycolytic substrates (e.g., glucose or fructose), the activity of phosphofructokinase in the medium oscillates with periodicities of 3.5–8.6 minutes (Hess *et al.,* 1969). Oscillations in hemoglobin biosynthesis with a frequency similar to the time necessary for the synthesis of one molecule have been described in rabbit reticulocytes (Tepper *et al.,* 1969). The interaction of several such high-frequency cycles could theoretically generate a rhythm whose period approximated 24 hours.

(2) *Elsewhere in the body* (*e.g., in the brain*). It is generally believed that the daily rhythms in body temperature and plasma cortisol level result from parallel changes in the set-points around which these functions are regulated (Yamamoto and Brobeck, 1965). These set-points are presumably a property of neurons localized within the brain; their oscillations are thought to be endogenous and not to depend upon the presence of environmental cycles. There is at the present time no direct evidence that any biochemical rhythm in a mammalian tissue is endogenous; there is only indirect evidence that various rhythms persist in animals deprived of one or more environmental cycles, and thus *could* be of endogenous origin. This type of proof is less than persuasive, since it is never possible to place experimental subjects in an environment which is truly devoid of cyclic inputs. Natural diurnal cycles in light, ambient temperature, food intake, and humidity can be damped without too much difficulty; however, cycles in electromagnetic field strength, gamma irradiation, and the gravitational pull of the moon persist in all terrestrial experiments, and can probably induce certain rhythms (Brown, 1965).

(3) *Outside the body.* The environment of the particular planet on which mammals happened to evolve is characterized by 24-hour cycles in a variety of physical functions. One of the cycles, the presence and absence of light, has been shown to generate daily rhythms in the biochemical activity of the pineal gland (Wurtman and Axelrod, 1965; Wurtman, 1967b). The pineal is the unique locus in mammals of an enzyme, hydroxyindole-*O*-methyl transferase (HIOMT), which synthesizes the hormone melatonin (Axelrod *et al.*, 1961). Under normal lighting conditions (i.e., 12 hours of light per day), the activity of this enzyme shows marked time-dependence; it rises several fold during the hours

of darkness and falls during the light period (Axelrod *et al.*, 1965). Rats blinded or placed in an environment of continuous darkness immediately lose the pineal enzyme rhythm; on the other hand, these treatments do not extinguish the presumably endogenous rhythms in body temperature or plasma corticosterone content (Haus *et al.*, 1967). The concentration of norepinephrine within pineal sympathetic nerve endings also varies diurnally with a rhythm which is generated by the light-dark cycle. The neural pathway through which the retinal input reaches the pineal and generates these biochemical rhythms is distinct from the visual system; it utilizes a special nerve bundle, the inferior accessory optic tract (Wurtman *et al.*, 1967a).

Even though few rhythms appear to be generated by the light-dark cycle, essentially all rhythms in mammals are influenced to some extent by this cyclic environmental input. In general, light acts as the dominant synchronizer for 24-hourly rhythms, determining the times of their maxima and minima. The light-dark cycle is thought to "entrain" endogenous rhythms, causing their period lengths to become exactly 24 hours (Aschoff, 1960; Halberg *et al.*, 1959; Haus *et al.*, 1967). The advantages to the animal of utilizing this stable environmental cycle to entrain its internal rhythms are obvious; within an individual animal, unrelated rhythmic events can be made to occur in a regular sequence, and metabolic events which generally occur at a specific time of day can be anticipated; within a species the rhythms of many individuals can be synchronized. Rhythms which are caused by the light-dark cycle and rhythms which are simply entrained by this cycle tend to show similar responses to shifts in the phase of the daily light period. If the time that subjects are exposed to light is changed from 6:00 AM—6:00 PM to noon–midnight, the times that the maxima and minima of both types of rhythms occur change until they reestablish a normal interval (termed a "phase angle"), with reference to the onset of light or darkness (Halberg *et al.*, 1959; Haus *et al.*, 1967). Rhythms generated by the light cycle can, however, be distinguished from endogenous rhythms by their response to continuous darkness: The former are immediately extinguished in the absence of light cycles (Wurtman *et al.*, 1967a), while the latter persist.

Among a small group of rhythmic functions it has been possible to demonstrate that, when animals are maintained under constant lighting conditions (i.e., continuous darkness), not only does the rhythm persist, but also the length of its period changes from exactly 24 hours to something slightly greater or less (i.e., it becomes *circadian*) (Halberg, 1959; Halberg *et al.*, 1959; Aschoff, 1960). The demonstration that a daily rhythm "free-runs" in the absence of cyclic lighting inputs is usually offered as further evidence that the rhythm is, in fact, the product of an endogenous mechanism. It is argued that a free-running period of, for example, 23 hours and 41 minutes is not isomorphic with any known environmental cycle; thus it cannot result from a one-for-one response to an oscillating environmental input. However, even the demonstration that the period of a rhythm is not exactly 24 hours does not prove that the oscillation is of endogenous origin. Environmental cycles do exist whose periods are slightly longer or shorter than 24 hours (e.g., the lunar-tidal cycle, 24.8 hours); the resultant of the lunar and solar frequencies might appear as a circadian rhythm.

It is feasible to demonstrate circadian periodicity only when the experimental design allows the rhythmic function to be sampled repeatedly and frequently in the same subjects. In general, data of this sort cannot be obtained in studies on rhythms in the activity of tissue enzymes (Wurtman, 1967a). The necessity of killing the experimental animal to obtain the tissue specimen allows each animal to be sampled only once; hence the enzyme rhythm must be studied by analyzing tissues from groups of 6–8 animals taken at multiple intervals during a period of at least 24 hours. The difficulty in maintaining perfect synchrony among such large groups of animals, and the logistics problems involved in collecting their tissues every few hours during a full 24-hour period have, to date, made it impossible for any investigator to demonstrate that the period length of any tissue enzyme rhythm becomes circadian in mammals maintained under constant environmental conditions. In the remainder of this chapter, the term "circadian" will be applied only to rhythms (e.g., in body temperature and in the concentrations of certain substances in the plasma) which have been demonstrated to free-run under controlled environmental conditions with a period significantly different from 24 hours.

In addition to the oscillating physical inputs described above, most mammals also receive cyclic inputs of chemicals (i.e., foodstuffs) because of their tendency to confine their eating behavior to a portion of the 24-hour period. Among rats exposed to light between 6:00 AM and 6:00 PM, the portal vein and the liver receive large amounts of amino acids, glucose, and other components of the diet between 2:00 PM and 2:00 AM, when the animal consumes most of its daily food intake, and little during the rest of the day. This cyclic input of dietary amino acids generates the daily rhythm in hepatic tyrosine transaminase activity (Wurtman *et al.*, 1968b); it seems likely that additional biochemical rhythms, in the liver and perhaps elsewhere, will be shown to result from the cyclic intake of foodstuffs.

II. Twenty-four-Hour Rhythms in Hepatic Enzymes

A. Tyrosine Transaminase

1. Characteristics

If Sprague-Dawley rats are maintained under light and darkness for alternating 12-hour periods and given access *ad libitum* to a diet containing protein (e.g., Purina Chow), the activity of the enzyme tyrosine transaminase shows a marked dependence upon the hour of the day that their livers are sampled (Wurtman and Axelrod, 1967; Civen *et al.*, 1967; Shambaugh *et al.*, 1967). Tissues taken from animals killed 5 hours after the onset of darkness can catalyze the *in vitro* transamination of tyrosine about four times as fast as livers taken early in the light period (Fig. 1).

2. Mechanism

Since the concentrations of corticosterone in the adrenals and blood of the rat also vary diurnally (Fig. 1), and since the administration of this steroid is known to produce an increase in tyrosine transaminase activity (Lin and Knox, 1957; Nichol and Rosen, 1964), it initially seemed likely that the nocturnal increase in transaminase activity was a consequence of the preceding rise in plasma glucocorticoid concentration. However, this hypothesis was not supported by studies on adrenalectomized or hypophysectomized rats; in both experimental preparations, the transaminase rhythm persisted with unchanged amplitude (Fig. 2). It could thus be concluded that the cyclic signal which instructs the hepatocytes to increase their tyrosine transaminase activity is not mediated by pituitary hormones (e.g., growth hormone, ACTH, TSH), by compounds secreted in response to these hormones (e.g., corticosterone, thyroxine), or by epinephrine [which is secreted in mammals only from the adrenal medulla (Wurtman, 1966)].

Studies on intact rats had shown that injections of large amounts of tyrosine, the physiological substrate for tyrosine transaminase, or of smaller quantities of tryptophan could also produce a rise in enzyme activity (Lin and Knox, 1957). In general, these studies did not include adequate controls to rule out the possibility that the enzyme changes observed after administering the amino acid were not simply a manifestation of the same transaminase rhythm as that observed in the untreated animal. However they at least suggested that the oscillating signal responsible for the tyrosine transaminase rhythm might be an amino acid, whose availability to liver cells depended upon the time of day. To

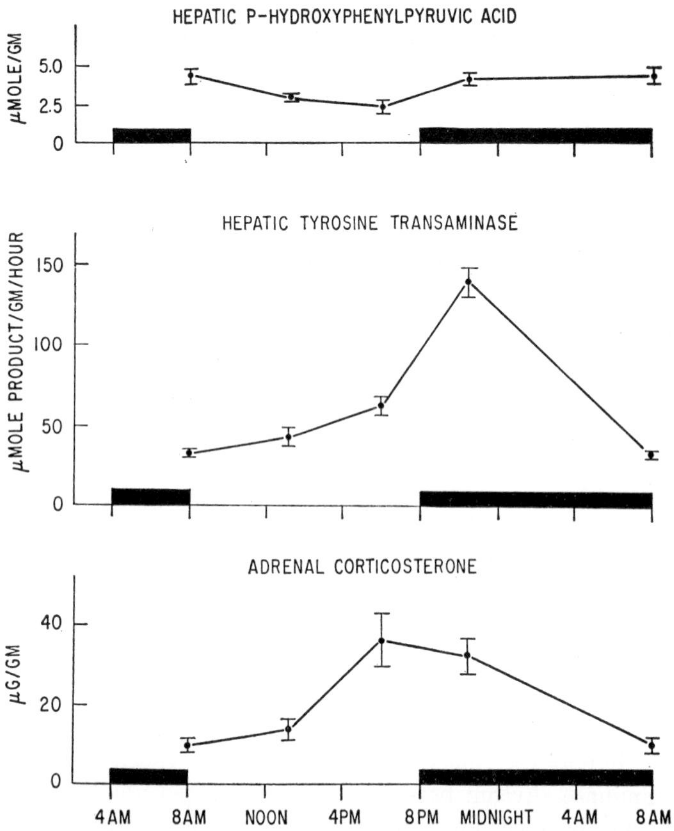

FIG. 1. Daily rhythms in hepatic tyrosine transaminase activity, hepatic content of *p*-hydroxyphenylpyruvic acid, and adrenal content of corticosterone. Rats were exposed to light from 8:00 AM to 8:00 PM for 1 week prior to assay and given access *ad libitum* to Purina Chow and water. Vertical lines in all figures refer to standard errors of the mean. (Reprinted from Wurtman and Axelrod, 1967.)

examine this possibility, livers taken at intervals throughout the day from untreated or hypophysectomized rats were assayed for tyrosine transaminase activity and for free tyrosine or tryptophan concentration; plasma samples were also assayed for free tyrosine level. It was hypothesized that, if changes in the availability of free tyrosine or tryptophan were responsible for the enzyme rhythm, it might be possible to demonstrate that the concentration of the amino acid in plasma or liver increased several hours before the daily rise in enzyme activity. No such relationship was observed between tyrosine levels and the enzyme rhythm (Fig. 3). Instead of rising several hours before the daily enzyme peak, plasma and hepatic tyrosine levels showed no change during this period but fell during the hours following the daily rise in transaminase activity (Wurtman *et al.,* 1968b). This observation was considered compatible with the catalytic function of the enzyme; i.e., as tyrosine transaminase activity increased, more of its substrate was removed in the liver, and the concentration of its product, *p*-hydroxyphenylpyruvic acid, increased (Fig. 1). In contrast to tyrosine, hepatic tryptophan levels did vary in a manner which suggested that this amino acid might be involved in producing the transaminase rhythm: Hepatic free tryptophan content increased by about 50% several hours before the daily rise in enzyme activity (Fig. 4) (Wurtman *et al.,* 1968b).

The liver differs from other organs in that the amounts of amino

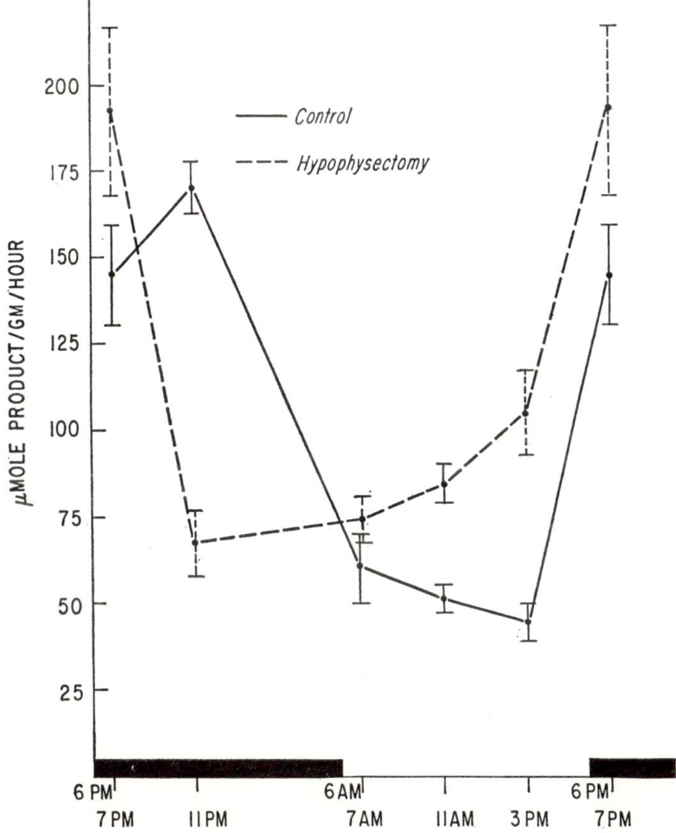

FIG. 2. Effect of hypophysectomy on the daily rhythm in hepatic tyrosine transaminase activity. Rats were hypophysectomized and kept under light from 6:00 AM to 6:00 PM for 10 days prior to assay; they had access *ad libitum* to Purina Chow and water. (Reprinted from Wurtman and Axelrod, 1967.)

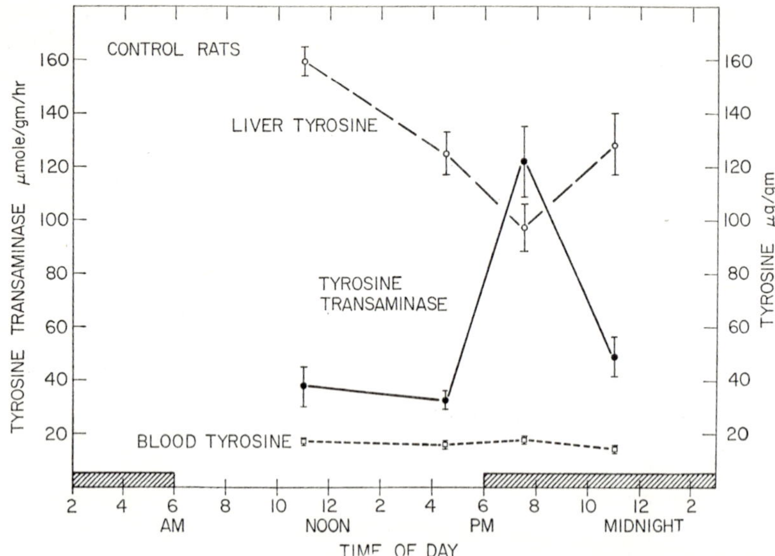

FIG. 3. Relation between blood and liver tyrosine concentrations and hepatic tyrosine transaminase activity in adult female rats given Purina Chow and water *ad libitum*. Lights were on from 6:00 AM to 6:00 PM daily. (Reprinted from Wurtman *et al.*, 1968b.)

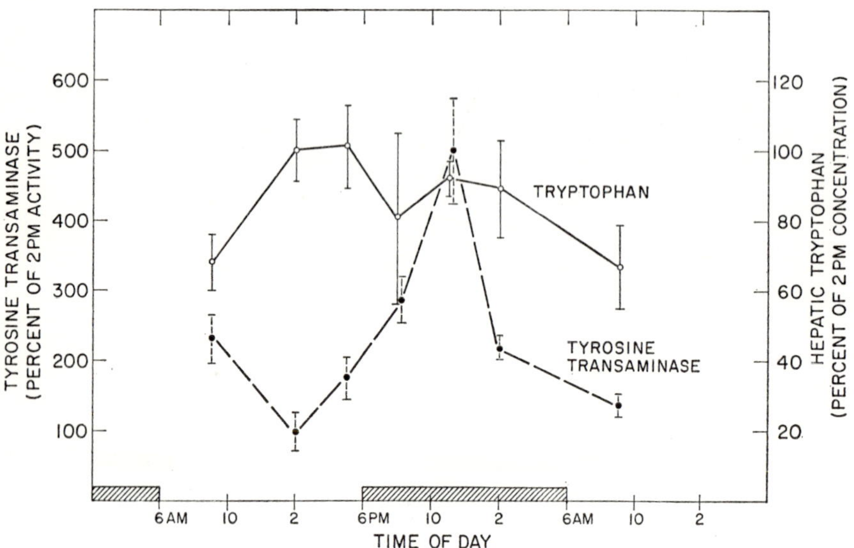

FIG. 4. Relation between hepatic tryptophan concentration and tyrosine transaminase activity in adult female rats given access *ad libitum* to Purina Chow and water. Lights were on from 6:00 AM to 6:00 PM daily. (Reprinted from Wurtman *et al.*, 1968b.)

acids with which it is perfused fluctuate markedly during the course of the day (see Chapter 38). After the animal has consumed protein, the liver receives concentrations of amino acids from the portal vascular system which are much greater than those delivered between meals from the general circulation (i.e., from the hepatic artery). Since the rat consumes its food cyclically, the quantity of tryptophan available to its liver soon after the onset of darkness, when the rate of food consumption is maximal (Wurtman *et al.*, 1968b), is far greater than the amount presented to it early in daylight period. On the basis of such reasoning it seemed likely that the oscillations in both the tryptophan content and the tyrosine transaminase activity of the liver might result from the tendency of the rat to consume dietary protein cyclically. To test this hypothesis, groups of rats were given access *ad libitum* to diets containing 18% protein (casein) or 0% protein starting soon after the onset of a daily light period; other animals were starved for 2 days. The animals were killed at intervals, and their livers were assayed for tyrosine transaminase activity. Rats having no access to protein (i.e., the 0% protein and starved groups) showed low transaminase activity throughout the first day of the experimental period (Fig. 5). On the second day, the transaminase activity in animals without protein but

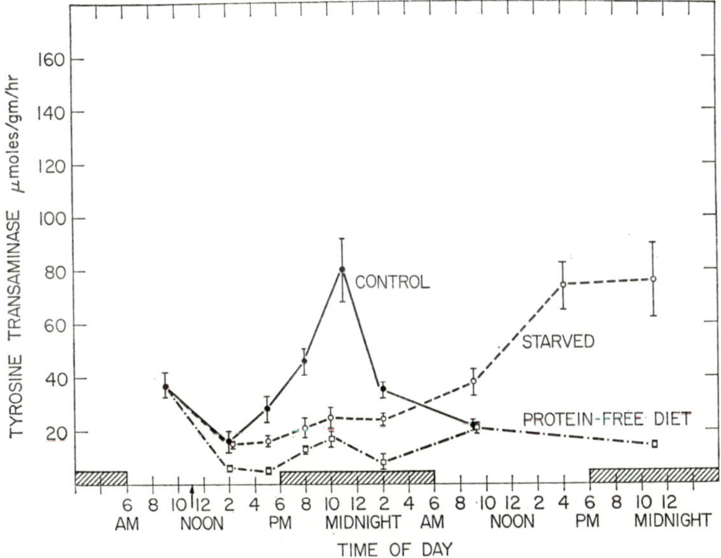

FIG. 5. Hepatic tyrosine transaminase activity in adult female rats given access to a control diet (18% protein), a 0% protein diet, or no food at all, starting at 11:00 AM. Lights were on from 6:00 AM to 6:00 PM daily. (Reprinted from Wurtman *et al.*, 1968b.)

with free access to calories (i.e., the 0% protein group) remained low, and showed little tendency to vary with time of day (Wurtman *et al.*, 1968b). At this point, enzyme activity in the starved rats rose to attain the peak evening levels observed in rats given protein (Wurtman *et al.*, 1968b). The response of the starved group presumably resulted from the increased adrenocortical secretion caused by the stress of starvation.

If rats were given access to a synthetic diet containing a complete mixture of the essential amino acids instead of protein, hepatic tyrosine transaminase activity continued to display 24-hour rhythmicity (Wurtman *et al.*, 1968b). The time of the daily peak in enzyme activity shifted slightly, possibly because amino acids from this food source did not enter the portal circulation at the same rate as amino acids derived from dietary protein. The omission of tryptophan from the amino acid mixture caused the transaminase rhythm to be extinguished (Fig. 6). It is possible that the omission of other essential amino acids might also have extinguished the enzyme rhythm; however dietary tryptophan may serve a primary function in the physiological regulation of tyrosine transaminase activity, by virtue of its special role in the aggregation of hepatic polysomes. This hypothesis is outlined below.

These studies suggested that an exogenous input, the cyclic delivery

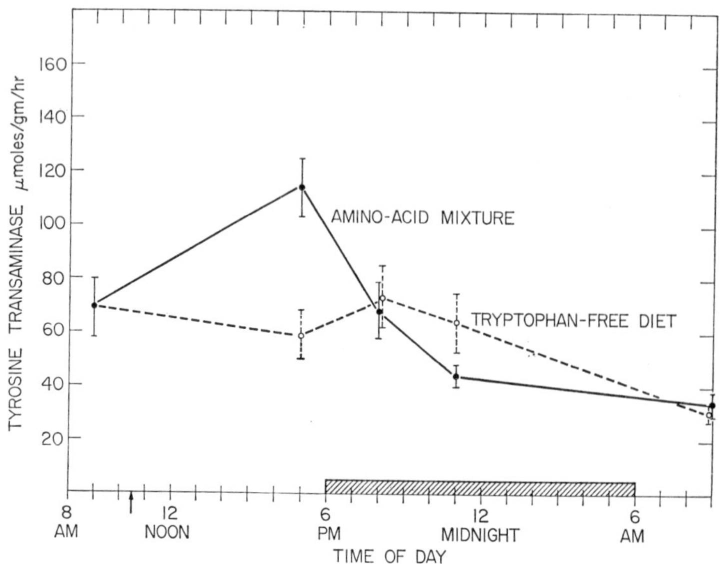

FIG. 6. Hepatic tyrosine transaminase activity in adult female rats given access to special diets starting at 10:30 AM on the day of the experiment. Lights were on from 6:00 AM to 6:00 PM daily. (Reprinted from Wurtman *et al.*, 1968b.)

to the liver of dietary amino acids, was the major factor in generating the tyrosine transaminase rhythm. This cyclic input depends upon both the composition of the diet and the tendency of the animal to confine its eating behavior to certain hours of the day. The importance of dietary composition in generating the transaminase rhythm was illustrated by the observation (Fig. 5) that the rhythm was extinguished in animals given a diet lacking in protein. Potter and his colleagues (1966) have shown that the opposite is also true; the rate at which tyrosine transaminase activity rises after an animal is given access to dietary protein is roughly proportional to the protein content of its diet. The importance of the feeding rhythm in generating the transaminase rhythm has been demonstrated by a variety of experiments: (1) The phasing of the feeding rhythm has been changed by shifting the daily onset of the light period by 12 hours. About 6 days are required for the feeding cycle to resume its normal relationship to the light cycle. During this period, the phasing of the transaminase rhythm changes at approximately the same rate as the phasing of the feeding rhythm (Zigmond *et al.*, 1969; Black and Axelrod, 1968). (2) Animals have been given access to food for only 4 hours per day (8:00 AM to noon). This constraint has shifted the phase of the tyrosine transaminase rhythm, such that the daily peak occurs several hours after the onset of eating, but almost 12 hours earlier than in animals given free access to food (Fuller and Snoddy, 1968). (3) The time of food intake has been dissociated from the time of protein ingestion by placing rats on a 0% protein diet early in the light period, and substituting an 18% protein mixture after 3, 9, or 15 hours. Under these conditions, rats continued to display a normal feeding cycle regardless of the composition of their food; however tyrosine transaminase activity did not start to rise until several hours after the animal first had access to protein (Zigmond *et al.*, 1969; Black and Axelrod, 1968). (4) Fuller has shown (1969) that mice made hyperphagic by treatment with gold thioglucose lose the tyrosine transaminase rhythm. Such animals also fail to display a normal feeding rhythm (Anliker and Mayer, 1955). If the mice are forced to eat cyclically (i.e., by being allowed access to food for only 4 hours per day), the activity of the transaminase again exhibits daily fluctuations. (5) The feeding rhythm has been extinguished by training animals to consume food in 24 equal portions during each of the 24 hours of the day. "Wheel-fed" animals prepared in this manner and exposed to light for 12 hours daily maintained a normal rhythm in corticosterone secretion; however, the transaminase rhythm was largely, but not totally, extinguished. The magnitude of the daily rise in enzyme activity was reduced to 10–15% of the increment observed among rats allowed to consume their food cyclically (Cohn *et al.*, 1970).

The persistence of a small daily rhythm in tyrosine transaminase activity among "wheel-fed" rats suggests that, in addition to the postprandial delivery of amino acids, other cyclic inputs to the liver can also generate oscillations in the activity of the enzyme. One such input might be plasma glucocorticoid level; the concentration of corticosterone in the blood of rats is known to undergo a daily rise several hours before the daily increment in enzyme activity (Guillemin *et al.*, 1959). Accordingly, the following hypothesis for the mechanism of the enzyme rhythm attributes significance to both nutritional and hormonal inputs. In the second half of the daily light period, the adrenals of the rat secrete increased amounts of glucocorticoids (Guillemin *et al.*, 1959). These agents enhance the synthesis of ribonucleic acid (RNA) within hepatic cells (Halberg *et al.*, 1958) (Fig. 7). Most of the new RNA is probably ribosomal, but some may also represent messenger coded for tyrosine transaminase (Hager and Kenney, 1968). Around this time, the animal begins to increase its rate of food consumption (Wurtman *et al.*, 1968b); this causes high concentrations of amino acids to be delivered to the liver via the portal circulation, and further stimulates RNA synthesis (Blobel and Potter, 1967; Whittle and Potter, 1968). Within the hepatic cells, dietary tryptophan (Wunner *et al.*, 1966; Sidransky *et al.*, 1967; Baliga *et al.*, 1968) and perhaps other amino acids (Mandel *et al.*, 1966) cause ribosomal and messenger RNA to

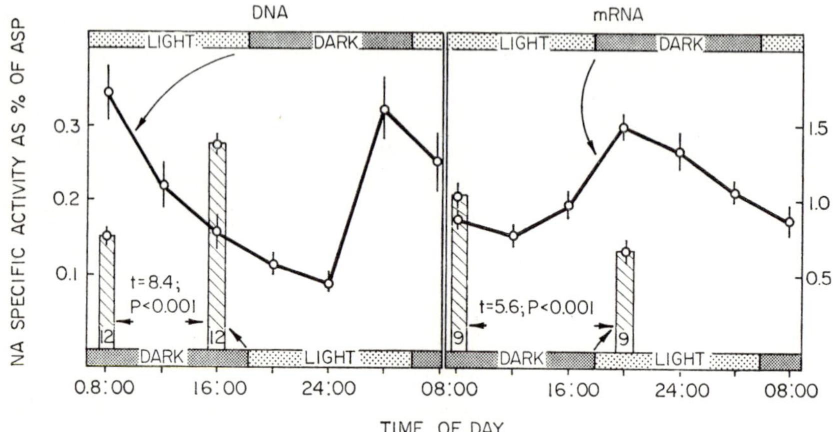

FIG. 7. Twenty-four-hour rhythms in incorporation of ^{32}P into DNA and "microsomal" RNA of mouse liver. Upper curves describe data obtained from animals kept under light from 6:00 AM to 6:00 PM; lower bars describe data obtained from animals kept under reversed lighting (i.e.. lights on from 6:00 PM to 6:00 AM) for 8–14 days. (Reprinted from Halberg *et al.*, 1959. Copyright 1959 by the American Association for the Advancement of Science.)

aggregate, forming polysomes, the units of protein synthesis. These structures then begin to synthesize tyrosine transaminase more rapidly than the protein is destroyed, hence net enzyme activity rises.

The above hypothesis rests on the assumption that the diurnal rise in transaminase activity represents an increase in the amount of enzyme protein, not simply a change in the activity of preexisting enzyme. This assumption is supported by recent evidence (D. Granner and M. Civen, unpublished observations) that the content of immunologically distinct tyrosine transaminase protein in rat liver also varies with a 24-hourly rhythm which is in phase with the rhythm in enzyme activity. Comparing the minor changes in the enzyme rhythm which follow adrenalectomy (Wurtman and Axelrod, 1967) with the near-extinction of the rhythm caused by depriving rats of dietary protein (Wurtman *et al.*, 1968b), it is apparent that the contribution of the adrenocortical cycle to the enzyme rhythm is minor. This suggests that the rate-limiting factor in the physiological regulation of tyrosine transaminase biosynthesis (i.e., in the unstressed, untreated animal) is not the amount of messenger RNA coded for this protein, but the proportion of the messenger which is bound with ribosomes in polysomal units. This proportion depends upon a nutritional input, dietary tryptophan and other amino acids, which produces its effect by acting within the cytoplasm. This hypothesis is compatible with the finding that the messenger RNA for tyrosine transaminase produced by hepatoma cells in synchronous tissue cultures is relatively long-lived (Martin *et al.*, 1968). It is also consistent with the recent observation that the proportion of hepatic RNA bound to polysomes varies diurnally in the rat (Fishman *et al.*, 1969). This proportion increases from about 50 to 73% soon after darkness among animals given free access to a diet which contains protein (Fig. 8). The rhythm is absent among animals placed on a no-protein diet.

Tyrosine transaminase activity attains a peak about 6–8 hours after the onset of darkness; it subsequently declines rapidly, even though the animal continues to consume food during the next few hours (Fig. 1). The mechanism of the daily fall in enzyme activity is entirely unknown; like the rise, it also appears to be independent of the adrenals and pituitary (Wurtman and Axelrod, 1967). The fall in tyrosine transaminase activity observed several hours after starved, adrenalectomized rats are given glucocorticoids can be blocked by inhibitors of protein synthesis (Kenney, 1967); these compounds may produce their effect by inhibiting the breakdown of the enzyme protein (see Chapter 31). The fall can also be blocked by the oral administration of casein hydrolyzates (Rosen, 1965).

The ontogenesis of the transaminase rhythm has been related to patterns of food ingestion by Honova and her colleagues (1968). An enzyme

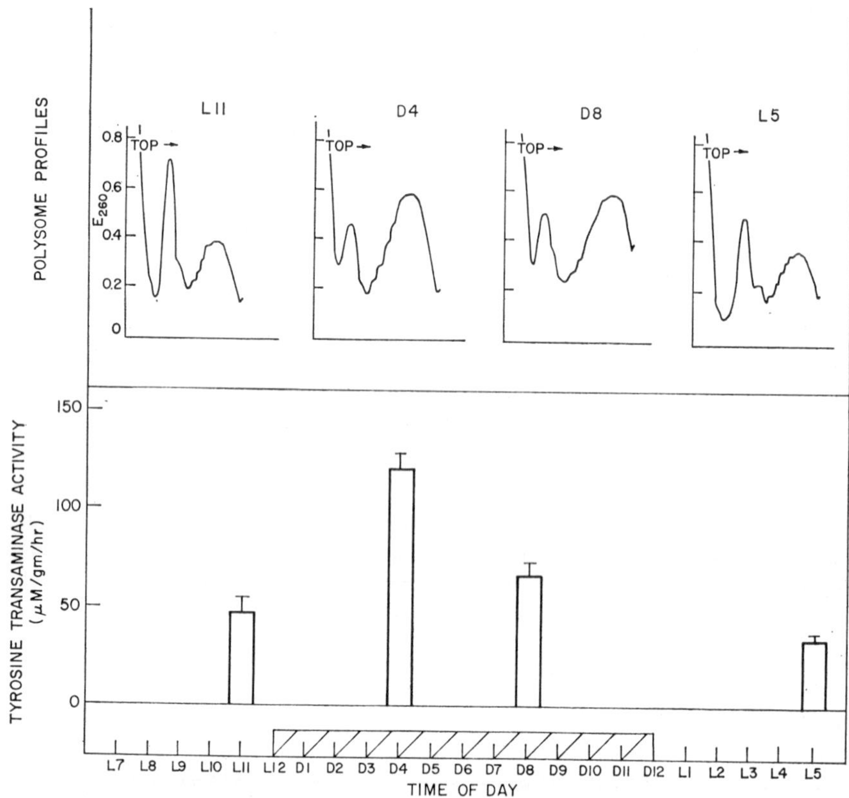

FIG. 8. Hepatic polysome profiles and tyrosine transaminase activity among groups of six rats fed Purina Chow and killed at the times indicated. Hatching along the abscissa indicates the daily dark period. (Reprinted from Fishman *et al.,* 1969.)

rhythm is present immediately after birth; however, its phasing is quite different from that seen in the adult: The daily peak in enzyme activity occurs in the morning instead of in the evening. On day 21 of life (i.e., at the time of weaning), the time of the daily peak in transaminase activity abruptly shifts to that observed in adults. The authors attribute this sudden change in phase to the fact that prior to weaning, the newborn rats ingest milk at the convenience of the mother, i.e., at times of day that she is not eating. With weaning, they suddenly assume the nocturnal feeding pattern which is characteristic of their species, and shift the time of day that they deliver dietary amino acids to their liver.

Axelrod and Black (1968) have suggested that the tyrosine transaminase rhythm is somehow related to tissue norepinephrine levels, inasmuch as rats treated with inhibitors of monoamine oxidase, an enzyme which metabolizes norepinephrine, or with L-dihydroxyphenylalanine, a precursor of the catecholamine, exhibit both an increase in brain norepinephrine levels and a partial suppression of the hepatic transaminase rhythm. These data are difficult to interpret since they include no measurements of the rates at which the treated animals consumed food; it is possible that both compounds affected the enzyme rhythm simply by suppressing the feeding cycle; both compounds are used psychologically to modify affect and behavior. Moreover, L-dihydroxyphenylalanine, an amino acid not normally present in the circulation, might have influenced hepatic protein synthesis. Studies in other laboratories have shown that a profound pharmacological depletion of brain norepinephrine has no effect on the transaminase rhythm so long as the experimental animal continues to eat rhythmically (Wurtman *et al.*, 1968c).

Baril and Potter (1968) have observed that, following the subcutaneous injection of cycloleucine-^{14}C (a nonutilizable synthetic amino acid), the steady-state ratio of its concentrations in liver and blood varies with a daily rhythm. The ratio is highest at the time of day that transaminase activity is rising. On the basis of this observation, they have suggested that the enzyme rhythm results from a rhythm in the rate at which liver cells take up amino acids from the general circulation. This hypothesis is not compatible with the observation that adrenalectomy obliterates the rhythm in cycloleucine-^{14}C uptake (Baril and Potter, 1968) but does not alter the tyrosine transaminase rhythm (Wurtman and Axelrod, 1967). Moreover, animals given 0% protein diets retain their rhythms in plasma amino acid concentrations (Wurtman *et al.*, 1968a), but fail to display rhythms in tyrosine transaminase activity (Wurtman *et al.*, 1968b). It is well established that corticoids affect the capacity of the liver to concentrate amino acids (Chapter 4, Volume I).

The administration of glucagon (Csany *et al.*, 1967) or of insulin (Nichol and Rosen, 1964) to rats elevates tyrosine transaminase activity (see Chapter 31); hence it seemed possible that enhanced secretion of the former pancreatic hormone toward the end of the daily "starvation period" (i.e, late in the light period), or of the latter hormone soon after the animal started to eat, might be responsible for the enzyme rhythm. That this is not the case has recently been demonstrated by studies on pancreatectomized animals by Fuller (1969). Removal of this organ did not extinguish the enzyme rhythm (Fig. 9). The loss of the rhythm in intact animals given a high carbohydrate, 0% protein diet (Wurtman *et al.*, 1968b) also provides evidence that neither glucagon nor insulin participates in its genesis.

The demonstration that untreated animals, left to their own devices,

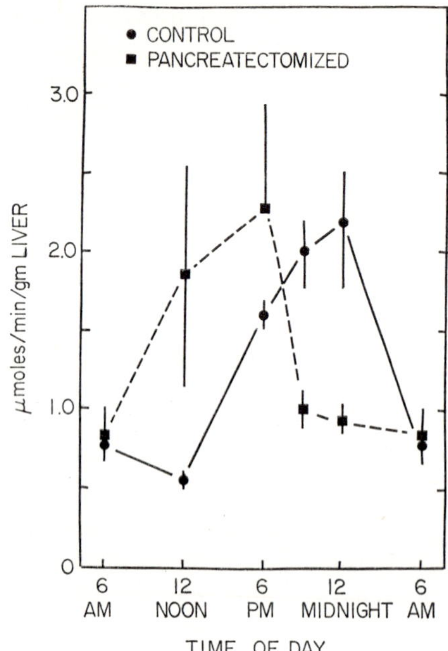

FIG. 9. Persistence of daily rhythm in hepatic tyrosine transaminase activity among pancreatectomized animals. Male rats were kept under cool-white fluorescent light from 7:30 AM to 4:30 PM and given access *ad libitum* to Purina Chow. (Figure kindly supplied by Dr. Ray Fuller, Eli Lilly Research Laboratories, Indianapolis, Indiana.)

display fourfold daily changes in tyrosine transaminase activity, and that these changes are the result of the cyclic ingestion of dietary protein, makes it mandatory that future studies on the physiological regulation of tyrosine transaminase provide considerably more information about the experimental animals than has usually been the case in the past. The environmental lighting conditions, the hour-to-hour pattern of food intake, and the composition of the diet should all be described in detail. The apparent effect of a particular experimental manipulation on transaminase activity cannot be interpreted if the reader is not told whether or not the treatment also altered the composition of the food or the time of its ingestion.

3. *Physiological Significance*

The nocturnal rise in tyrosine transaminase activity is associated with a decrease in the content of tyrosine in the liver (Fig. 3), and with an increase in the concentration of *p*-hydroxyphenylpyruvic acid (Fig. 1). If the enzyme rhythm is extinguished (by placing rats on a 0% protein diet), hepatic tyrosine content no longer declines with the onset of darkness. The rate at which intraperitoneally injected tyrosine-^{3}H disappears from the whole mouse varies diurnally with a rhythm in phase with the hepatic tyrosine transaminase cycle (Rose *et al.*, 1969). These observations indicate that the daily rhythmic changes in tyrosine transaminase activity, measured *in vitro,* are, in fact, associated with parallel changes in the rate that tyrosine is transaminated *in vivo.* This implies that the proportion of the tyrosine in a given amount of food which remains available to the body for utilization in the synthesis of endogenous proteins depends upon the hour of its ingestion. It may be possible to increase the efficiency with which protein foodstuffs are utilized in areas in which they are in short supply by arranging for them to be eaten at metabolically favorable times of day.

During the entire part of the 24-hour day when rats eat, they tend to consume their food in multiple small meals: they nibble at a rate sufficient to keep their stomachs from becoming empty. The animals can, however, be forced to adopt human eating habits. By withholding food, the investigator can train them to eat one or two individual meals each day. Under these conditions they lose protein and accumulate fat in their carcasses (Cohn, 1966); their tissues synthesize excessive amounts of lipids from isotopically labeled precursors (Cohn and Joseph, 1967). The metabolic basis of these effects of meal-eating may be related to the mechanism of the tyrosine transaminase rhythm. The ingestion of more than threshhold amounts of protein during a short interval may cause enzyme adaptations (i.e., a rise in the activity of tyrosine transaminase, and perhaps in other transaminases) which alter the fate of any protein consumed subsequently. As a consequence of meal-eating, a larger fraction of the daily protein intake is consumed after tyrosine transaminase activity has already been increased by the prior ingestion of protein. Hence more protein is converted to precursors of lipids.

It is generally held that the major function of tyrosine transaminase relates to its role in gluconeogenesis; i.e., it deaminates an amino acid to form compounds which are utilized for the production of energy. If this hypothesis were correct, one might expect to observe in normal animals a direct relationship between the number of hours since the animal had last eaten and the level of hepatic transaminase activity; enzyme activity should rise when dietary glucose is unavailable. In contrast, it is found that the daily elevation in tyrosine transaminase activity occurs not in response to the absence of dietary carbohydrates, but to the presence of dietary proteins. It thus appears that the function of tyrosine transaminase in the unstressed, unstarved animal has little to do with gluconeogenesis. More likely, the primary action of the enzyme is not to make its products (glycogenic and ketogenic energy sources) available, but to destroy its potentially toxic (Alam, 1965) substrate, tyrosine. When animals are starved for longer than 18–24 hours or are otherwise stressed, the secretion of glucocorticoids from the adrenal gland is stimulated, and these hormones act on the hepatic cells to induce the formation of tyrosine transaminase and other transaminases (see Chapter 31; see also, Nichol and Rosen, 1964). In this circumstance, these enzymes very likely do act to increase the rate of gluconeogenesis. Hence tyrosine transaminase activity can be regulated by two distinct inputs, depending upon the physiological state of the animal. It is enhanced by dietary protein in normal animals, and probably serves to destroy excessive amounts of its substrate; it is enhanced by adrenocortical hormones in stressed or starved animals, and serves in these circumstances to generate substrates for energy metabolism.

B. Other Enzymes

1. Demonstration of Enzyme Rhythms

Daily rhythmic changes have been demonstrated in the activities of several hepatic enzymes besides tyrosine transaminase; these include cytoplasmic enzymes which metabolize amino acids, and membrane-bound enzymes which metabolize drugs. Relatively few enzymes have been examined at intervals to determine whether they exhibit diurnal variations. Moreover, many of the publications which purport to provide evidence for the existence of enzyme rhythms lack critical experimental details without which their findings cannot be evaluated. In general, all reports on enzyme rhythmicity should include the following information: (1) the age, sex, and strain of the experimental animals; (2) the number of animals housed in each cage; (3) the lighting conditions at the times that the animals were sampled (i.e., nature of the light source; light intensity at the level of the animals; the hours each day that animals were exposed to darkness and light); (4) the number of days that the animals were maintained in their special lighting environment prior to sampling; (5) the composition of the food offered to the animals, and the hours that food and water were made available; (6) the actual hour-to-hour food intake of the animals; (7) the times of day at which enzyme activity was sampled, and the nature of the statistical method chosen to evaluate the data.

Since naturally occurring rhythms in plasma steroid concentrations have already been shown to be capable of generating some enzyme rhythms, the concentration of corticosterone in the plasma or the adrenals should be ascertained in samples taken at the same time as the specimens of liver. This information is especially important in justifying the conclusion that a particular enzyme does *not* show daily oscillations; i.e., if an enzyme rhythm is generated by the steroid secretory rhythm, environmental manipulations which obliterate the steroid rhythm will also extinguish the enzyme rhythm, leading the investigator to the erroneous conclusion that enzyme activity does not oscillate. The availability of data showing that the adrenal cycle is also absent will help the investigator to recognize that his particular experiment does not allow conclusions to be drawn about the existence of an enzyme rhythm.

2. Amino Acid-Metabolizing Enzymes

The activity of tryptophan pyrrolase in livers of male CD-1 mice exposed to light from 6:00 AM to 6:00 PM and given *ad libitum* access to mouse chow varies about twofold during each day; it is lowest at the start of the daily dark period and peaks about 8 hours later (Rapoport *et al.*, 1966). Bilateral adrenalectomy lowers the basal enzyme level; however, its activity continues to show some periodicity, now rising by about 50% between 8:00 PM and 2:00 AM. The rise in tryptophan pyrrolase activity is observed in intact animals 15–21 hours after plasma tryptophan levels reach their peak; hence the enzyme rhythm is probably not related to the delivery of tryptophan to the liver by the general circulation. The role of *dietary* tryptophan in producing the enzyme rhythm cannot be evaluated, inasmuch as no information was provided in the above study on the eating behavior of the experimental animals, nor are data available on changes in the rhythm among animals on a 0% protein diet.

The rhythm in hepatic tryptophan pyrrolase activity may be responsible for the observed relationship between the metabolic fate of a dietary tryptophan load and the hour of its ingestion. Rapoport and Beisel (1968) administered 3 gm of tryptophan per kilogram body weight to normal males, aged 20–42, at different times of the day; the subjects' urines were collected for the next 6 hours and assayed for kynurenine derivatives (formed through the action of tryptophan pyrrolase) and for indicans (synthesized by bacteria in the gut). The excretion of kynurenine derivatives was greatest if the amino acid was administered between 6:00 AM and 9:00 AM, or several hours after the daily peak in plasma cortisol levels; it was least if the tryptophan was administered between 6:00 PM and 9:00 PM.

Civen *et al.* (1967) assayed livers from adult male rats exposed to light for 12 hours daily and given Purina Chow *ad libitum* for both tyrosine transaminase and phenylalanine-pyruvate transaminase (PPT).

In contrast to the tyrosine-metabolizing enzyme, PPT activity showed no tendency to vary with time of day. Potter and his colleagues (1966) examined the activities of various enzymes in the livers of Buffalo rats which bore a transplantable tumor (Morris hepatoma No. 7793) for 149 days; the animals were exposed to light from 6:00 AM to 6:00 PM daily and given access to a diet containing 0–90% protein during the hours of darkness. Tyrosine transaminase activity exhibited a significant rhythm in host livers only among animals receiving a diet containing at least 30% protein; it is possible that the enzyme rhythm was not observed in rats eating customary amounts of protein (15–20%) because no animals were sampled between 6:00 PM and midnight. Serine dehydrase activity did not exhibit diurnal rhythmicity; ornithine transaminase, a mitochondrial enzyme, showed daily oscillations among rats whose diets contained 60% or 90% protein. No data were provided in this study on the actual amounts of food consumed by the experimental animals; such information is especially important in view of the observation (Szepesi and Freedland, 1968) that rats tend to decrease their consumption of diets which are extremely high in protein after about 4 days.

The activity of arginine transaminase in mouse kidney also varies diurnally. Among Bagg albino or C strain animals kept in light from 6:00 AM to 6:00 PM, enzyme activity was highest during the light period and attained its nadir at midnight (Van Pilsum and Halberg, 1964). The phase of the rhythm could be shifted by reversing the lighting conditions.

3. Drug-Metabolizing Enzymes

Radzialowski and Bosquet (1968) observed daily rhythms in the activities of several drug-metabolizing enzymes in livers of intact male Holtzman rats exposed to light from 6:30 AM to 8:00 PM daily. The oxidative enzymes which transform *p*-nitrosoanisole, aminopyrine, and hexobarbital were all 50–100% more active in livers sampled at 2:00 AM than in those removed at 2:00 PM. Bilateral adrenalectomy appeared to extinguish these rhythms; however, it had no effect on the rhythm in 4-dimethylaminoazobenzene reductase activity. No data were provided on the feeding rhythms or the protein intakes of the experimental animals. Starvation for 24 hours failed to affect the three adrenal-dependent rhythms, an observation that is difficult to reconcile with the marked effect of this stress on adrenocortical secretion.

C. Characteristics of Enzymes Displaying 24-Hour Rhythmicity

On the basis of information cited above, it can be postulated that two requirements must be met in order for a particular hepatic enzyme to display rhythmicity:

(1) The turnover time of the enzyme must be sufficiently short so that a change in the rate of enzyme production or degradation during part of a 24-hour period would generate measurable differences in enzyme activity. Thus, it is *a priori* more likely that tyrosine transaminase should display rhythmicity than arginase, an enzyme which turns over relatively slowly. All the enzymes for which rhythmicity has thus far been described have half-lives of less than 3 days (Schimke, Chapter 32).

(2) The enzyme should be induced (or activated) by a physiological input to the liver whose amplitude displays 24-hour rhythmicity. Hence enzymes which are activated or induced by components of food (e.g., protein, glucose) might be expected to vary diurnally, as might enzymes whose activities are enhanced by hormones whose plasma concentrations show similar variations: e.g., corticosterone (Guillemin *et al.,* 1959) and insulin (Freinkel *et al.,* 1968). Enzymes regulated primarily by plasma thyroxine levels would not be expected to display 24-hour rhythmicity inasmuch as the plasma concentration of this hormone shows little tendency to vary diurnally (Schatz and Volpe, 1959). The pineal organ receives a cyclic input of nerve impulses generated by the retinal response to environmental lighting (Wurtman *et al.,* 1964); these impulses drive a 24-hour rhythm in a pineal enzyme, hydroxyindole-*O*-methyl transferase (Axelrod *et al.,* 1965).

III. Twenty-four-Hour Rhythms in Plasma Concentrations of Amino Acids

A. Tyrosine

The concentration of tyrosine in human plasma is not constant but varies predictably as a function of time of day (Wurtman *et al.,* 1967b). Among healthy young men who consumed a "formula diet" (0.71 gm protein per kilogram body weight) or a "house diet" (1.50 gm protein per kilogram body weight) in four equal meals at 8:00 AM, 12:30 PM, 5:30 PM, and 10:00 PM, and who remained in bed from 11:00 PM to 7:00 AM, tyrosine levels were lowest at 2:00 AM and rose to a peak by or after 10:30 AM (Figs. 10 and 11). The magnitude of the daily rise was about 60% above the basal (2:00 AM) levels.

Subjects fed a "low-protein formula diet" (containing only 0.04 gm of protein per kilogram body weight) for 2 weeks continued to display 24-hour oscillations in plasma tyrosine levels; a sharp peak was attained at 8:00 AM (Fig. 12) (Wurtman *et al.,* 1967b, 1968a). This observation indicated that the plasma amino acid rhythm was not simply the consequence of the overflow of dietary amino acids from the portal circulation

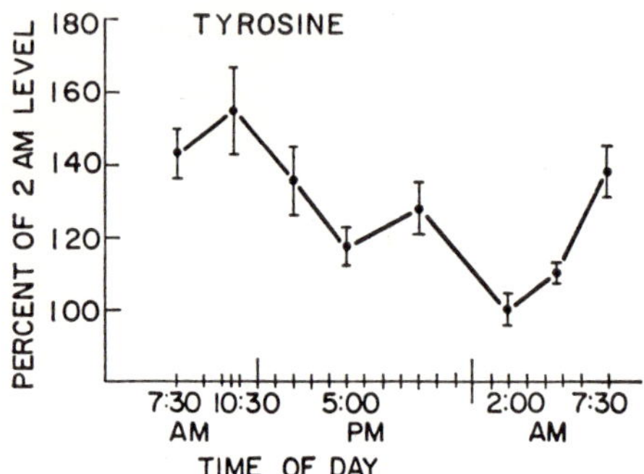

Fig. 10. Variation in plasma tyrosine concentrations in nine healthy male subjects given the "Formula Diet" (0.71 gm of egg protein per kilogram body weight). The 2:00 AM mean concentration of the amino acid was 10.04 ± 0.39 μg/ml. (Reprinted from Wurtman *et al.*, 1968a.)

to the vena cava. It also demonstrated an important difference between the rhythms in plasma tyrosine concentration and hepatic tyrosine transaminase activity. The former rhythm, but not the latter, persisted in subjects deprived of a cyclic input of dietary protein. This indicated that, although the hepatic enzyme might influence the plasma tyrosine rhythm, the primary mechanism responsible for the amino acid cycle was unrelated to alterations in the rate at which tyrosine was metabolized within the liver. Feigin and his colleagues (1968) have provided further evidence that plasma amino acid rhythms are independent of dietary protein. Subjects consumed 100 gm of protein at 9:00 AM or at 9:00 PM; this dietary load accelerated the morning rise in amino acid concentration but did not alter the expected evening decline.

To examine the possibility that the diurnal changes in plasma tyrosine level were related to man's activity rhythm, plasma samples were assayed for the amino acid before and immediately after 45 minutes of vigorous exercise. The effect of physical activity was to increase tyrosine levels only slightly (Wurtman *et al.*, 1967b); hence the daily rise in plasma tyrosine (which began at 2:00 AM, while the subjects were in bed) could not have been the result of cycles of muscular work. Changes in the sleep–wakefulness cycle can, however, entrain plasma amino acid rhythms; subjects made to sleep from 10:00 AM to 6:00 PM (instead of from 10:00 PM to 6:00 AM) showed a rapid shift in the time of peak amino acid levels. Within 48 hours, the hour of peak plasma concentrations stabilized at 4:00 AM (Feigin *et al.*, 1968). The daily rhythm in body temperature required considerably more time to resynchronize to the new activity schedule. It could not be determined whether the shift in the phase of the amino acid rhythm resulted from shifts in

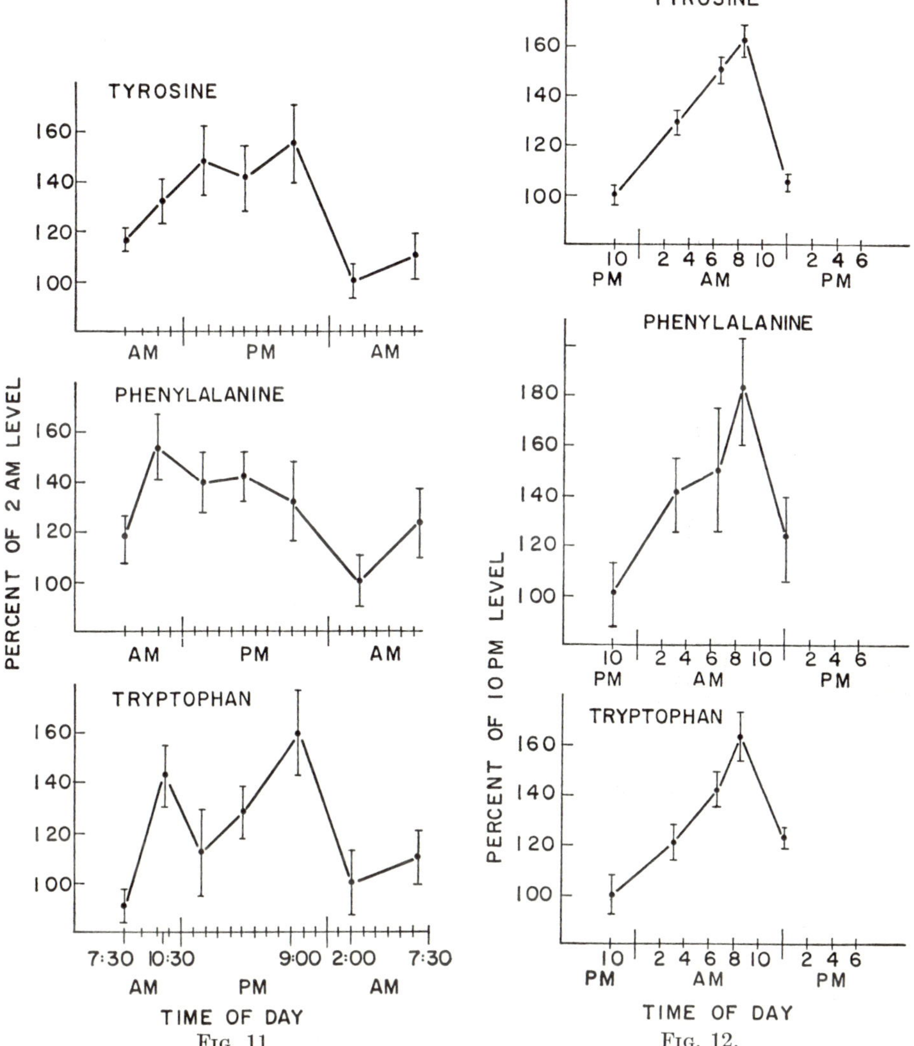

FIG. 11.

FIG. 12.

FIG. 11. Variations in plasma concentrations of tyrosine, phenylalanine, and tryptophan with time of day among six healthy male subjects given the "House Diet" (approximately 1.5 gm of protein per kilogram body weight). The 2:00 AM concentrations of the amino acids were: tyrosine: 12.20 ± 0.94 μg/ml; phenylalanine: 8.14 ± 0.85 μg/ml; tryptophan: 9.13 ± 1.22 μg/ml. (Reprinted from Wurtman *et al.*, 1968a.)

FIG. 12. Variations in plasma concentrations of tyrosine, phenylalanine, and tryptophan with time of day among eight healthy male subjects given the "Low Protein Formula Diet" (0.04 gm of protein per kilogram body weight). The 10:00 PM mean concentrations of the amino acids were: tyrosine: 7.80 ± 0.31 μg/ml; phenylalanine: 6.46 ± 0.86 μg/ml; tryptophan: 6.30 ± 0.49 μg/ml. (Reprinted from Wurtman *et al.*, 1968a.)

the sleep cycle, the times of physical activity, the hours of food ingestion, the light-dark cycle, or another cyclic behavioral or environmental input.

B. Other Amino Acids

Daily oscillations parallel to those described for tyrosine have now been observed among most other amino acids. Feigin *et al.* (1967a) measured the total concentration of 18 amino acids in plasmas of healthy young males, using a densitometric assay of spots on paper chromatograms. The total concentration was lowest (25 mg/100 gm) at 4:00 AM, and rose by 15–35% in the afternoon. Blood samples were collected at intervals of 4 hours for 6 days; there was relatively little day-to-day variability among individual subjects. Both Wurtman *et al.* (1968a) and Feigin *et al.* (1968) examined diurnal rhythms in individual amino acids. The former group observed rhythms of greatest amplitude among tyrosine, tryptophan, phenylalanine, methionine, cysteine, and isoleucine; amino acids whose plasma concentrations were highest showed the smallest percentage of variation (Table I). Feigin *et al.* (1968) found the greatest percentage differences among leucine, isoleucine, and

TABLE I

PLASMA AMINO ACID CONCENTRATIONS AT TIMES OF DAY WHEN TYROSINE LEVEL WAS HIGHEST OR LOWEST[a,b]

	"House Diet" (1.5 gm protein/kg) Amino acid concentration (μmole/ml)			"Low-Protein Formula Diet" (0.04 gm protein/kg) Amino acid concentration (μmole/ml)		
Amino acid	2:00 PM	2:00 AM	Ratio	8:30 AM	10:00 PM	Ratio
Threonine	0.17 ± 0.02	0.14 ± 0.05	1.45 ± 0.27	0.21 ± 0.10	0.15 ± 0.07	1.39 ± 0.03
Serine	0.22 ± 0.07	0.17 ± 0.05	1.39 ± 0.39	0.21 ± 0.09	0.15 ± 0.03	1.43 ± 0.19
Glutamic acid	0.28 ± 0.12	0.24 ± 0.15	1.29 ± 0.54	0.32 ± 0.10	0.26 ± 0.12	1.26 ± 0.23
Glycine	0.27 ± 0.05	0.21 ± 0.02	1.25 ± 0.17	0.41 ± 0.09	0.32 ± 0.05	1.29 ± 0.05
Alanine	0.46 ± 0.10	0.30 ± 0.07	1.59 ± 0.37	0.58 ± 0.29	0.52 ± 0.21	1.12 ± 0.33
Valine	0.36 ± 0.07	0.28 ± 0.05	1.31 ± 0.32	0.17 ± 0.03	0.11 ± 0.02	1.50 ± 0.16
Cysteine	0.08 ± 0.12	0.04 ± 0.02	1.83 ± 0.61	0.05 ± 0.02	0.03 ± 0.02	1.89 ± 0.55
Isoleucine	0.13 ± 0.02	0.09 ± 0.02	1.51 ± 0.42	0.07 ± 0.02	0.03 ± 0.01	2.16 ± 0.02
Leucine	0.23 ± 0.05	0.15 ± 0.02	1.63 ± 0.54	0.13 ± 0.03	0.06 ± 0.01	2.02 ± 0.36
Tyrosine	0.09 ± 0.02	0.05 ± 0.02	2.01 ± 1.00	0.05 ± 0.01	0.02 ± 0.02	2.13 ± 0.21
Phenylalanine	0.08 ± 0.02	0.05 ± 0.02	1.60 ± 0.44	0.06 ± 0.02	0.03 ± 0.02	1.74 ± 0.30
Lysine	0.27 ± 0.05	0.18 ± 0.05	1.58 ± 0.34	0.16 ± 0.03	0.12 ± 0.03	1.40 ± 0.09
Histidine	0.14 ± 0.02	0.11 ± 0.02	1.39 ± 0.34	0.14 ± 0.02	0.09 ± 0.02	1.60 ± 0.47
Arginine	0.13 ± 0.05	0.11 ± 0.02	1.46 ± 0.24	0.12 ± 0.02	0.08 ± 0.03	1.43 ± 0.30

[a] Reprinted from Wurtman *et al.* (1968a).

[b] Data obtained from 6 subjects on "House Diet" (1.5 gm protein/kg) and 3 subjects on "Low-Protein Formula Diet" (0.04 gm protein/kg) is presented as mean ± standard deviation.

methionine (tryptophan was not assayed); citrulline concentrations tended not to show significant diurnal oscillations.

The concentrations of tyrosine in rat plasma varies with a rhythm similar to that observed in humans, except that times of the peak and the nadir are shifted by 6–8 hours (Coburn *et al.*, 1969). Plasma phenylalanine does not display significant diurnal oscillations in this species. If rats are forced to consume all of their daily food intake between 8:00 AM and noon, the phase of the tyrosine rhythm shifts, and parallel rhythms appear in the plasma concentrations of both tyrosine and phenylalanine. Daily rhythms in free amino acid concentrations have been reported in the plasma of White Leghorn chicks (Squibb, 1966) and within the hearts of rats (Lyons *et al.*, 1967).

C. Mechanisms of Plasma Amino Acid Rhythms

A variety of factors can influence the steady-state levels of an individual amino acid in the plasma; any of these could impart rhythmicity to its net plasma concentration. Plasma tyrosine originates from a variety of "sources" and disappears by entering any of several "sinks." Its sources include the following: (1) that fraction of the dietary tyrosine which is not utilized or transformed in its passage through the portal circulation and liver; (2) tyrosine synthesized in the liver by the oxidation of phenylalanine (or, less likely, through the transamination of *p*-hydroxyphenylpyruvic acid) and secreted into the general circulation; (3) tyrosine produced in the liver and other organs from the degradation of proteins and peptides; (4) tyrosine released from intracellular depots in liver, brain, muscle, and other tissues. Its sinks include: (1) delivery to the liver via the hepatic artery, followed by deamination or incorporation into proteins or peptides; (2) uptake into tyrosine pools in peripheral tissues, followed by incorporation into proteins and such special products as catecholamines, thyroxine, or melanin; (3) excretion into the urine or feces, or decarboxylation to form tyramine (little if any tyrosine is lost via these latter pathways).

Each of the above processes depends upon many individual biochemical operations, which, if rhythmic, could impart rhythmicity to plasma tyrosine concentration. For example, the uptake of plasma tyrosine into the brain might be stimulated by an increase in the utilization of tyrosine to form proteins or catecholamines; it might also change as a result of an alteration in the steady-state ratio of intraneuronal to extracellular tyrosine, perhaps resulting from the action of a hormone.

It should be obvious that a full understanding of the mechanisms responsible for the plasma tyrosine rhythm will not be attained until considerably more information is available about the control of plasma amino acid levels in general. Indeed, the experimental analysis of the tyrosine rhythm should provide an excellent model system for studying the physiological regulation of amino acid metabolism in the whole animal. Unlike changes in plasma amino acid levels produced by experimental manipulations, the diurnal variations in tyrosine concentration observed in untreated animals are, *prima facie*, physiologically significant.

The plasma tyrosine rhythm cannot result solely from the cyclic ingestion of dietary protein or of rhythms in the intrahepatic transamination of the amino acid, inasmuch as it persists in subjects deprived of all but trace amounts of dietary protein (Wurtman *et al.*, 1968a). Although no data are available as to the time-dependence of phenylalanine hydroxylase activity, it is not likely that the plasma tyrosine rhythm results from diurnal variations in tyrosine synthesis, inasmuch as it persists indefinitely in human subjects suffering from phenylketonuria (Coburn *et al.*, 1969). Moreover the tyrosine rhythm is in phase with a plasma phenylalanine rhythm in humans (Wurtman *et al.*, 1968a), but not in rats (Coburn *et al.*, 1969). It has not yet been determined whether the rates at which endogenous tyrosine is liberated following the degradation of proteins in the liver or elsewhere vary diurnally; however, it might be expected that more protein would be catabolized toward the middle of the daily light period (i.e., when the rate of food intake is least) than soon after the onset of darkness. Similarly, no data are available on the relation between time of day and the tendency of tyrosine pools in peripheral tissues to reenter the blood stream.

The extent to which tissues can concentrate amino acids from the plasma has been shown to vary diurnally and may provide a basis for rhythmicity in plasma tyrosine levels. Baril and Potter (1968) have shown that the steady-state ratio of the concentrations of a nonutilizable amino acid, cycloleucine-^{14}C, in liver and blood varies in relation to the number of hours since the animal last ate. The ratio is depressed by feeding, and rises during fasting. These investigators initially suggested that this rhythm was responsible for the tyrosine transaminase rhythm, i.e., higher concentrations of amino acids were presented to the liver for incorporation into protein at certain times of day. The disappearance of the rhythm in intracellular amino acid-concentrating ability among adrenalectomized animals makes this hypothesis untenable. However, it does appear likely that a daily rhythm in the ability of tissues to concentrate amino acids could contribute significantly to a plasma amino acid rhythm. Baril and Potter suggested that the rhythm in the intracellular concentration of cycloleucine-^{14}C was generated by the diurnal oscillations in plasma glucocorticoid concentrations. This hypothesis is supported by the observation that human subjects who receive synthetic glucocorticoids chronically according to a schedule which disrupts their endogenous plasma cortisol cycle show parallel shifts in the phasing of the tyrosine rhythm (Wurtman *et al.*, 1968d). Plasma insulin levels vary diurnally (Freinkel *et al.*, 1968), and insulin can also modify the rates at which amino acids are transferred across cellular membranes (Roberts and Simonson, 1962); hence it is also possible that plasma amino acid rhythms are related to insulin secretion. This likelihood is especially great in subjects eating a high-carbohydrate, low-protein diet, as discussed below.

Preliminary data suggest that the rates at which individual tissues convert tyrosine-^{14}C to labeled protein-^{14}C depend upon the hour that the amino acid is administered. Adult male Sprague-Dawley rats exposed to light from 6:00 AM to 6:00 PM and given access to Purina Chow and water *ad libitum* received intraperitoneal injections of tyrosine-^{14}C at 6:00 AM, noon, 6:00 PM, and midnight. They were killed after 30 minutes, and the specific activities of tyrosine-^{14}C and ^{14}C-labeled protein were determined in liver and brain. The rates at which the labeled amino acid was incorporated into protein were highest in the middle of the light period and lowest during the dark period. The differences in the amounts of ^{14}C-labeled protein formed reflect rhythms in both the destruction of tyrosine-^{14}C and the rates of total protein synthesis (Rose, *et al.*, 1969). Marked changes in the rate at which intracellular tyrosine is utilized for protein synthesis could cause sufficient alterations in the uptake of the amino acid from the blood (or in its secretion from the tissues into the plasma) to produce rhythms in plasma tyrosine concentration.

In the case of the plasma amino acid rhythm observed in subjects receiving the low-protein diet (Fig. 12), it seems likely that this rhythm results from cyclic intake of dietary carbohydrate. In both man and other species, it has been shown that ingestion of carbohydrate causes a rapid fall in plasma free amino acid levels (see Chapter 10, Section III,B and Chapter 34, Section III,A for a complete review of the literature). Munro and Thomson (1953) examined this phenomenon in rats and in adult men, and showed that the reduction is maximal within 2 hours of feeding a dose of glucose. Feeding fat to fasting subjects or animals did not produce significant changes in free amino acid levels. The lowering of free amino acid concentrations by carbohydrate administration is not uniform for different amino acids; on the contrary, amino acids are removed from the plasma in the proportions in which they occur in tissue proteins generally.

Zinneman *et al.* (1966) have recorded the reduction in the levels of

22 circulating amino acids 2 hours after feeding glucose; with the single exception of glutamic acid, the percentage fall in individual amino acid levels correlates closely with the changes between maximal and minimal values on the "low-protein formula diet" shown in Table I. On the other hand, the relative differences in amino acid concentrations at two points on the cycle observed with the "house diet" (Table I) do not correlate with the pattern of change caused by glucose administration, thus suggesting that other factors are at work. Figure 13 shows these correlations.

The cause of the change in plasma amino acid concentrations after carbohydrate administration has been examined. Since the pattern of removal suggests deposition as tissue protein, a search was made by Munro and his colleagues for the site or sites of such deposition. Munro *et al.* (1959) showed that administration of glucose to fasting rats leads to actual *loss* of protein from the liver, rather than deposition. This was confirmed by injecting rats with ^{35}S-labeled methionine or ^{14}C-labeled glycine at the time of glucose administration. As compared with controls fed fat or water, glucose-fed animals showed no deposition of extra methionine or glycine in the tissue proteins of the liver or intestinal mucosa, but there was a significant accumulation in both the diaphram and leg muscles. It had previously been demonstrated that an intact insulin-secreting mechanism is required for the action of dietary carbo-

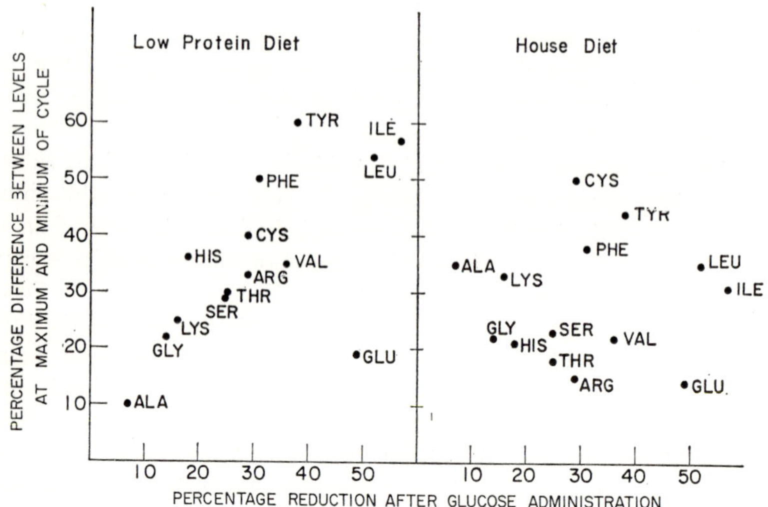

FIG. 13. Change in levels of individual amino acids after feeding glucose to fasting subjects and sampling plasma 2 hours later (Zinneman *et al.*, 1966) plotted against changes in same amino acids between daily extremes on low protein diet (left) or house diet (right).

hydrate on plasma amino acid levels (Munro, 1956). Thus it was concluded that the carbohydrate acted by stimulating the secretion of insulin; this hormone, in turn accelerates the transport of amino acids into muscle, and possibly also muscle protein synthesis. The reduction in plasma amino acid levels after giving insulin has been extensively documented (see Chapter 10, Section III,B for a review), and has been shown to occur in the eviscerated rat within 15 minutes of injecting the hormone (Ingle *et al.*, 1956). This time interval is comparable to the speed of action of insulin on carbohydrate metabolism in muscle, whereas carbohydrate metabolism in the liver responds much more sluggishly on administration of insulin to the intact animal (Renold *et al.*, 1955).

The influence of dietary carbohydrate on protein metabolism also extends to the fate of protein consumed in the same meal. As demonstrated in man by Cuthbertson and Munro (1939) and on the rat by Munro (1949), protein in the diet is better retained in the tissues if fed in a meal containing carbohydrate than if these two nutrients are fed at separate times of day. This presumably occurs because a proportion of the incoming amino acids after mixed meals is deposited in muscle as a result of the stimulus provided by the carbohydrate absorbed from the same meal. Indeed, Rabinowitz *et al.* (1966) have shown that insulin secretion is much more enhanced by a meal containing carbohydrate and protein than by carbohydrate or protein given singly. The disposal of amino acids in peripheral tissues after carbohydrate–protein meals can now be explored directly, using the arteriovenous difference technique on the arms of human subjects, as described by Foley *et al.* (1966).

As a first approximation to an explanation of the diurnal rhythms in protein metabolism disclosed in this survey, we may conclude that the main factor which causes tyrosine transaminase and perhaps other liver enzymes to undergo cyclic change in activity is variations in the levels of amino acids present in the portal vein. This hypothesis is compatible with the known sensitivity of liver protein metabolism to amino acid supply, as discussed in Chapters 10, 34, and 38. On the other hand, there is good reason to conclude that the carbohydrate absorbed from the diet exerts its major effect on the entry into, and perhaps the utilization of, amino acids within skeletal muscle. Consequently, plasma amino acid concentrations mainly reflect the ebb and flow of amino acids into and out of this large tissue mass and continue to display rhythmic change even when subjects are given diets lacking protein. When mixed meals of protein and carbohydrate are consumed, not all of the amino acids are trapped by the liver; thus a small postprandial wave of amino acids passes into the peripheral circulation, and influences the levels of the plasma amino acids. As discussed in Chapter 34, changes in plasma free amino acid levels following a mixed meal are dependent on the mixture of amino acids provided by the dietary protein in relation to the requirements of the tissues, on the interaction with dietary carbohydrate described above, and on the later postabsorptive breakdown of labile tissue proteins (notably in liver) which returns amino acids to circulation. It is thus not surprising that cyclical changes in plasma free amino acids in subjects receiving the "house diet" show complex patterns that cannot be correlated with the disappearance of amino acids from plasma after carbohydrate is given alone (Fig. 13).

D. Plasma Amino Acid Rhythms and Disease States

The tendency of plasma amino acid concentrations to vary diurnally by as much as twofold in healthy, untreated subjects makes it necessary that studies on the effects of clinical or experimental situations on "amino acid levels" be designed so that blood can be sampled at several different times of day. If blood can be taken from a given subject at only one hour, note should at least be made of when it was obtained, when and what the subject ate, and when he slept. Rivlin and Melmon (1965) reported that plasma tyrosine levels were elevated in hyperthyroid subjects; blood was taken only once from each subject, early in the morning. In the absence of data on tyrosine levels at other times of day, these data are compatible with any of the following hypothesis: (1) hyperthyroidism increases the "set-point" around which tyrosine levels vary; (2) hyperthyroidism extinguishes the tyrosine rhythm, and plasma levels remain at the 10:30 AM level at all hours of the day; (3) hyperthyroidism causes a shift in the phase of the tyrosine rhythm, so that levels present in normal subjects at 10:30 AM now occur at 7:00 AM.

Feigin has examined the effects of experimental infections with *Pasteurella tularensis* (Feigin and Dangerfield, 1967) or a live, attenuated Venezuelan equine encephalomyelitis virus vaccine (Feigin *et al.*, 1967b) on plasma amino acid rhythms. The administration of the bacteria by inhalation was followed after 4 days by a 43% fall in total plasma amino acid concentration; subsequently amino acid rhythmicity was perturbed until the subjects had recovered. Inoculation with the viral agent also damped the amino acid rhythm, but without causing the marked decline in amino acid levels. These studies included no data on the time or quantity of food ingestion.

E. Physiological Significance of Plasma Amino Acid Rhythms

There is at present little direct evidence relating changes in the plasma concentration of an amino acid to changes in its availability to the tissues. Since all tissues except the liver are dependent upon the general circulation for their supply of amino acids, it seems reasonable to speculate that a threshold plasma concentration must exist for each amino acid below which the tissues cannot extract sufficient material to meet their requirements; the rates of synthesis of protein and other products of amino acids (e.g., catecholamines, hormones) would then be slowed. If this hypothetical threshhold is higher than the plasma concentration of the amino acid at the daily nadir (i.e., at 2:00 AM), then the amino acid rhythm *will* be physiologically significant, and will limit the amount of the substance that is available to the tissues at certain times of day. On the other hand, if tissues are able even at 2:00 AM to extract from the blood as much of each amino acid as they need, then rhythmic changes in amino acid concentration will not exert a significant influence on the metabolism of the substances. The determination of the significance of amino acid rhythms awaits further studies.

ACKNOWLEDGMENT

Studies described in this review were supported in part by grants from the National Aeronautics and Space Administration (NGR-22-009-272) and the United States Public Health Service (AM-11709 and AM-11237).

REFERENCES

Alam, S. Q. (1965). Ph.D. Thesis, Department of Nutrition and Food Science, M.I.T., Cambridge, Massachusetts.

Anliker, J., and Mayer, J. (1955). *J. Appl. Physiol.* **8,** 667.

Aschoff, J. (1960). *Cold Springs Harbor Symp. Quant. Biol.* **25,** 11.

Axelrod, J., and Black, I. B. (1968). *Nature* **220,** 161.

Axelrod, J., MacLean, P. D., Albers, R. W., and Weissbach, H. (1961) *In* "Regional Neurochemistry" (S. S. Kety and J. Elkes, eds.), pp. 307–311. Pergamon, Oxford.

Axelrod, J., Wurtman, R. J., and Snyder, S. H. (1965). *J. Biol. Chem.* **240,** 949.

Baliga, B. S., Pronczuk, A. W., and Munro, H. N. (1968). *J. Mol. Biol.* **34,** 199.

Baril, E. F., and Potter, V. R. (1968). *J. Nutr.* **95,** 228.

Black, I. B., and Axelrod, J. (1968). *Proc. Natl. Acad. Sci. U.S.* **61,** 1287.

Blobel, G., and Potter, V. R. (1967). *J. Mol. Biol.* **23,** 279.

Brown, F. A., Jr. (1965). *In* "Circadian Clocks" (J. Aschoff, ed.), pp. 231–261. North-Holland Publ., Amsterdam.

Civen, M., Ulrich, R., Trimmer, B. M., and Brown, C. B. (1967). *Science* **157,** 1563.

Coburn, S. P., Seidenberg, M., and Fuller, R. W. (1969). *Proc. Soc. Exptl. Biol. Med.* (in press).

Cohn, C. (1966). *Proc. Assoc. Res. Nervous Mental Dis.* **43,** 212.

Cohn, C., and Joseph, D. (1967). *Can. J. Physiol. Pharmacol.* **45,** 610.

Cohn, C., Joseph, D., Larin, F. F., Shoemaker, W. J., and Wurtman, R. J. (1970). *Proc. Soc. Exptl. Biol. Med.* (in press).

Csany, V., Greengard, O., and Knox, W. E. (1967). *J. Biol. Chem.* **242,** 2688.

Cuthbertson, D. P., and Munro, H. N. (1939). *Biochem. J.* **31,** 694.

Feigin, R. D., and Dangerfield, H. G. (1967). *J. Infect. Diseases* **117,** 346.

Feigin, R. D., Klainer, A. S., and Beisel, W. R. (1967a). *Nature* **215,** 512.
Feigin, R. D., Jaeger, R. F., McKinney, R. W., and Alevizatos, A. C. (1967b). *Am. J. Trop. Med. Hyg.* **16,** 769.
Feigin, R. D., Klainer, A. S., and Beisel, W. R. (1968). *Metab., Clin. Exptl.* **17,** 764.
Feldman, J. F. (1968). *Science* **160,** 1454.
Fishman, B., Wurtman, R. J., and Munro, H. N. (1969). *Proc. Natl. Acad. Sci. U.S.* **64,** 677.
Foley, T. H., London, P. R., and Prenton, M. A. (1966). *J. Clin. Endocrinol. Metab.* **26,** 781.
Freinkel, N., Mager, M., and Vinnick, L. (1968). *J. Lab. Clin. Med.* **71,** 171.
Fuller, R. W. (1969). Unpublished observations.
Fuller, R. W., and Snoddy, H. D. (1968). *Science* **159,** 738.
Guillemin, R., Dear, W. E., and Liebelt, R. A. (1959). *Proc. Soc. Exptl. Biol. Med.* **101,** 394.
Hager, C. B., and Kenney, F. T. (1968). *J. Biol. Chem.* **243,** 3296.
Halberg, F. (1959). *Z. Vitamin-, Hormon-, Fermentforsch.* **10,** 225.
Halberg, F., Barnum, C. P., Silber, R. H., and Bittner, J. J. (1958). *Proc. Soc. Exptl. Biol. Med.* **97,** 897.
Halberg, F., Halberg, E., Barnum, C. P., and Bittner, J. J. (1959). *In* "Photoperiodism and Related Phenomena in Plants and Animals" (R. B. Withrow, ed.), Publ. #55, pp. 803–878. Am. Assoc. Adv. Sci., Washington, D.C.
Haus, E., Lakatua, D., and Halberg, F. (1967). *Exptl. Med. Surg.* **25,** 8.
Hess, B., Boiteux, A., and Krüger, J. (1969). *Advan. Enzymol.* **7,** 149.
Honova, E., Miller, S. A., Ehrenkranz, R. A., and Woo, A. (1968). *Science* **162,** 999.
Ingle, D. J., Torralba, G., and Flores, V. (1956). *Endocrinology* **58,** 388.
Kenney, F. T. (1967). *Science* **156,** 525.
Klevecz, R. R., and Ruddle, F. H. (1968). *Science* **159,** 634.
Lin, E. C. C., and Knox, W. E. (1957). *Biochim. Biophys. Acta* **26,** 85.
Lyons, M. M., Squibb, R. L., and Siegel, H. (1967). *Nature* **216,** 1113.
Mandel, P., Quirin, C., Bloch, M., and Jacob, M. (1966). *Life Sci.* **5,** 325.
Martin, D., Jr., Tomkins, G. M., and Granner, D. K. (1968). Unpublished observations.
Munro, H. N. (1949). *J. Nutr.* **39,** 375.
Munro, H. N. (1956). *Scot. Med. J.* **1,** 285.
Munro, H. N., and Thomson, W. S. T. (1953). *Metab., Clin. Exptl.* **2,** 354.
Munro, H. N., Black, J. G., and Thomson, W. S. T. (1959). *Brit. J. Nutr.* **13,** 475.
Nichol, C. A., and Rosen, F. (1964). *In* "Actions of Hormones on Molecular Processes" (G. Litwack and D. Kritchevsky, eds.), pp. 234–256. Wiley, New York.
Potter, V. R., Gebert, R. A., Pitot, H. C., Peraino, C., Lamar, C., Jr., Lesher, S., and Morris, H. P. (1966). *Cancer Res.* **26,** 1547.
Rabinowitz, D., Merimee, T. J., Maffezzoli, R., and Burgess, J. A. (1966). *Lancet* **ii,** 454.
Radzialowski, F. M., and Bosquet, W. F. (1968). *J. Pharmacol. Exptl. Therap.* **163,** 229.
Rapoport, M. I., and Beisel, W. R. (1968). *J. Clin. Invest.* **47,** 934.
Rapoport, M. I., Feigin, R. D., Bruton, J., and Beisel, W. R. (1966). *Science* **153,** 1642.
Renold, A. E., Hastings, A. B., Nesbett, F. B., and Ashmore, J. (1955). *J. Biol. Chem.* **213,** 135.
Rivlin, R. S., and Melmon, K. L. (1965). *J. Clin. Invest.* **44,** 1690.

Roberts, E., and Simonson, D. G. (1962). *In* "Amino Acid Pools: Distribution, Formation, and Function of Free Amino Acids" (J. T. Holden, ed.), p. 284. Elsevier, Amsterdam.

Rose, C. M., Chou, C., Zigmond, M. J., and Wurtman, R. J. (1969). *Federation Proc.* **28,** 690.

Rosen, F. (1965). *J. Cellular Comp. Physiol.* **66,** Suppl. 1, 146.

Schatz, D. L., and Volpe, R. (1959). *J. Clin. Endocrinol. Metab.* **19,** 1495.

Shambaugh, G. E., III, Warner, D. A., and Beisel, W. R. (1967). *Endocrinology* **81,** 811.

Sidransky, H., Bongiorno, M., Sarma, D. S. R., and Verney, E. (1967). *Biochem. Biophys. Res. Commun.* **27,** 242.

Squibb, R. L. (1966). *J. Nutr.* **90,** 71.

Szepesi, B., and Freedland, R. A. (1968). *J. Nutr.* **94,** 463.

Tepper, T., Hommes, F. A., Thürkow, I., and Nijhof, W. (1969). *FEBS Letters* **2,** 217.

Van Pilsum, J. F., and Halberg, F. (1964). *Ann. N.Y. Acad. Sci.* **117,** 337.

Whittle, E. D., and Potter, V. R. (1968). *J. Nutr.* **95,** 238.

Wunner, W. H., Bell, J., and Munro, H. N. (1966). *Biochem. J.* **101,** 417.

Wurtman, R. J. (1966). "Catecholamines." Little, Brown, Boston, Massachusetts.

Wurtman, R. J. (1967a). *Science* **156,** 104.

Wurtman, R. J. (1967b). *In* "Neuroendocrinology" (L. Martini and W. F. Ganong, eds.), pp. 19–59. Academic Press, New York.

Wurtman, R. J., and Axelrod, J. (1965). *Sci. Am.* **213,** 50.

Wurtman, R. J., and Axelrod, J. (1967). *Proc. Natl. Acad. Sci. U.S.* **57,** 1594.

Wurtman, R. J., Axelrod, J., and Fischer, J. E. (1964). *Science* **143,** 1328.

Wurtman, R. J., Axelrod, J., Sedvall, G., and Moore, R. Y. (1967a). *J. Pharmacol. Exptl. Therap.* **157,** 487.

Wurtman, R. J., Chou, C., and Rose, C. M. (1967b). *Science* **158,** 660.

Wurtman, R. J., Rose, C. M., Chou, C., and Larin, F. F. (1968a). *New Engl. J. Med.* **279,** 171.

Wurtman, R. J., Shoemaker, W. J., and Larin, F. (1968b). *Proc. Natl. Acad. Sci. U.S.* **59,** 800.

Wurtman, R. J., Shoemaker, W. J., Larin, F., and Zigmond, M. J. (1968c). *Nature* **219,** 1049.

Wurtman, R. J., Rose, C. M., Rose, L., Williams, G., and Lauler, D. (1968d). *Clin. Res.* **16,** 355.

Yamamoto, W. S., and Brobeck, J. R. (1965). "Physiological Controls and Regulations." Saunders, Philadelphia, Pennsylvania.

Zigmond, M. J., Shoemaker, W. J., Larin, F., and Wurtman, R. J. (1969). *J. Nutr.* **98,** 71.

Zinneman, H. H., Nuttall, F. Q., and Goetz, F. C. (1966). *Diabetes* **15,** 5.

CHAPTER 37

The Role of the Gastrointestinal Tract in the Regulation of Protein Metabolism[1]

G. Fauconneau

Laboratoire d'Études des Métabolismes, C.R.Z.V., Theix, France

AND

M. C. Michel

Laboratoire de Physio-Pathologie de la Nutrition, C.R.Z.V., Theix, France

I. Introduction

The digestive tract is made up of the digestive tube proper and its ancillary glands (salivary glands, pancreas, etc.) which secrete enzymes

[1] Translation by H. N. Munro.

required for digestion. The digestive tube itself is a complex structure consisting of a serous surface, connective tissue, muscle, blood, lymph, glands and the mucosa. Of these elements, more attention has been focused on the mucosa and on the collagenous content than on the other constituents of the wall.

The relative size of the digestive tract depends to a considerable extent on the age and state of alimentation of the animal. Thus in the case of the newborn calf, the digestive tract accounts for 3.2% of the live empty weight of the animal whereas from 2 to 12 weeks of age, it represents 4–6% of live weight (Mathieu, personal communication). Further changes parallel alterations in the nutrition of the calf; as the calf passes from being a monogastric animal to becoming a ruminant, the digestive tract increases from 7.4% of live empty weight (Mathieu, personal communication) to some 9–13% of live weight in the ruminant calf of 4 months (Stobo *et al.*, 1966). Thereafter the importance of the digestive tract diminishes, so that it represents no more than 6–6.5% of the live empty weight of the bullock of 500 kg body weight (Béranger, 1968). The relative proportions of different parts of the digestive tract of the ruminant also change during development; in the case of the monogastric calf, the small intestine forms about 50% of the total tract (Mathieu, personal communication), whereas in the case of the ruminant calf it is no more than 35–40% of the total tract (Béranger, 1968).

In the case of rats, the small intestine increases first by positive allometric growth during the first 2 weeks of life, followed by isometric growth up to 5 weeks of age, when it accounts for 6% of body weight. The wet weight continues to increase before puberty, but accounts for a diminishing proportion of the weight of the animal; thus at 10 weeks of age, when the rat weighs some 300 gm, the small intestine represents 3% of its body weight. In this respect, Durand *et al.* (1965) have demonstrated that the weight of the small intestine is directly proportional to level of food intake (that is, to ingested energy). During growth, the amount of protein in the intestinal wall increases in parallel with the wet weight of the tissue, as evidenced by the finding that the ratio of dry defatted solids to wet weight of intestine remains constant at 10.5%.

Development of the intestine of the growing rat has been studied to determine whether it increases by hypertrophy (increase in cell size) or by hyperplasia (increase in cell number) or by both processes. Using the total DNA content of the organ as a measure of cell number, Durand *et al.* (1965) showed that growth of the intestine occurs essentially by hyperplasia. The rate of cell increase is rapid up to puberty, thereafter diminishing in intensity. Between weaning and adult life, the cell population of the intestinal wall increases 3.5-fold, while the weight of the intestine increases by a factor of 3.8. Consequently, the average amount of cytoplasm per cell remains essentially constant throughout the growing period. In this respect, the intestinal wall displays greater cellularity than the liver, since the concentration of DNA per unit of dry defatted solids is 2–4 times greater in the intestine than in the liver (Durand *et al.,* 1965). Thus, the average amount of cytoplasm per cell is much less in the intestine than in the liver (see also Chapter 30, Section VII,D). Further discussion of the growth of the intestine of the rat will be found in Chapter 26, Section III,D in Volume III of this treatise.

The general regulatory role of the digestive tract in various aspects of metabolism has been analyzed by Spencer and Zamcheck (1962). The digestive tract is metabolically very active; it accounts for 15% of the oxygen requirements of a lamb fed on milk, and this proportion rises to 23% in the case of the adult sheep. The digestive tract has many functions; the principal one, as the name indicates, is the function of ensuring digestion of food. In the case of monogastric mammals, this consists of hydrolysis of large molecules into smaller ones, a process that is progressive and spread over a period of time, thanks to controlled emptying of the stomach. The products then traverse the intestinal wall, undergoing certain transformations; finally, they are transported by the blood to furnish the nutritional requirements of the tissues.

Digestion requires the formation of important quantities of nitrogenous compounds (enzymes, mucus, etc.) which, in their turn, are digested partly through the action of intestinal flora in the rear of the intestinal tract. Consequently, the digestive tract has a considerable synthetic function, not only in order to form its own constituents, some of which have a rapid turnover, but also of certain other proteins such as lipoproteins.

In summary, the digestive tract is a very active organ containing some proteins with a rapid turnover which confer on the gastrointestinal tract, in conjunction with the liver, an important role in regulation of protein metabolism. In addition to its function as a specific molecular filter, the digestive tract allows higher organisms (mammals, birds) to provide for continuous synthesis of proteins (e.g., for lactation and for growth), while receiving meals that are discontinuous (two to three meals every 24 hours). The role of the digestive tract in regulating protein metabolism is exercised at different levels:

1. Control over transport, and particularly gastric emptying, allows retardation of the time of arrival in the duodenum of food taken in brief meals.

2. Endogenous proteins, secreted as part of the digestive function of the gut, undergo relatively continuous digestion throughout the day in the middle part of the intestine.

3. The flora play an important part in the digestion of endogenous nitrogen.

4. Labile proteins in the gut, other than those associated with endogenous secreted proteins, contribute to the continuing supply of essential amino acids to the organism.

5. Changes in the relative importance of the digestive tract with physiological state and with level of food intake allow modification of the effects on the organism of variations in diet.

These various aspects of regulation of protein metabolism in relation to the gastrointestinal tract will now be considered. Digestion of protein has been dealt with in Volume I, Chapter 2, the special case of the ruminant in Chapter 3, and abnormalities of protein digestion are reviewed briefly in the Introduction to Section III in Volume II; the function of the liver in response to variations in supply of amino acids coming from the intestine is described in Chapter 38 of this volume. Finally, the reader may also wish to consult the proceedings of a conference on "The Role of the Gastro-Intestinal Tract in Protein Metabolism," edited by Munro (1964).

II. Role during Digestion of Exogenous Nitrogen

A. Gastric Emptying

The digestive tract begins to exert its effects right from the time food is taken, by impregnating the food with saliva and by mastication. Gastric emptying retards the release from each meal of a part of the food which varies in importance according to the quality and quantity of the nutrients held back by the stomach (Peraino *et al.*, 1959; Rérat and Lougnon, 1963; Auffray *et al.*, 1967). Coagulation of milk by gastric juice in the young animal causes a very significant retardation of passage of fats and proteins, whereas lactose and the minerals pass more rapidly (Mathieu, personal communication). Small meals are associated with more rapid gastric emptying in the case of the rat (Peraino *et al.*, 1959). In the case of the pig, standard meals are followed by one large emptying, followed by two or three smaller emptyings spaced 1 hour apart (Auffray *et al.*, 1967).

Gastric emptying is affected by chemical, physical, physiological, and psychological factors (Rogers *et al.*, 1960). Thanks to gastric emptying and its control, the nitrogenous constituents of a meal consumed in 15 minutes by the pig finally all arrive in the duodenum only after several hours have elapsed; thus in studies on pigs, Rérat and Lougnon (1963) found that only 60% of the ingested proteins had disappeared from the stomach in 3 hours. In the case of the rat, 60% of dietary proteins have also passed from the stomach within 3 hours, and 80% in 5 hours (Dawson and Holdsworth, 1962); Gupta *et al.* (1958) reported passage of 60% of a meal from the stomach of the rat in 2 hours.

B. Rapidity of Digestion of Proteins

Proteins make up essentially the entire nitrogenous components of foodstuffs (Pion and Fauconneau, 1968). The digestion of the majority of dietary proteins after denaturation by the HCl of the stomach appears to occur very rapidly. As pointed out in Chapter 2 (Volume I), this is related to the high concentrations in the gut contents of proteolytic enzymes which undergo only slow autodegradation (Börgström *et al.*, 1957). This agrees with the observation of Snook (1965a) that the proteolytic enzymes trypsin and chymotrypsin undergo little autodegradation in the upper half of the small intestine because of the presence of peptides arising from hydrolysis of dietary protein, which thus exert a protective effect on the enzymes.

In the small intestine, various methods have been used to demonstrate the rapid digestion of protein.

1. Some authors have studied the kinetics of disappearance after a meal, e.g., the studies of Rérat and Lougnon (1963) on pigs, of Dawson and Porter (1962) on rats and on cats eating *Chlorella* protein tagged with ^{14}C, and of Ochoa-Solano and Gitler (1968) on rats consuming ovalbumin marked with ^{75}Se-labeled selenomethionine.

2. Some have cannulated the digestive tube (e.g., Zimmerman-Nielsen and Schønheyder, 1962).

3. Studies have been made of changes in portal free amino acid levels, e.g., the studies of Pion *et al.* (1964) on pigs and of Goldberg and Guggenheim (1962) on rats.

4. Crane and Neuberger (1960) examined the excretion of urinary urea and ammonia N in human subjects receiving yeast marked with ^{15}N and found that rate of appearance of the ^{15}N was the same whether the yeast protein was intact or hydrolyzed, thus indicating very rapid hydrolysis of protein by the gut.

In general, these different approaches agree in showing very rapid breakdown of dietary protein in the gut. However, certain proteins, especially those of vegetable origin, are digested slowly (Peraino and Harper, 1963). When the portal and systemic levels of free amino acids are compared after a meal of zein and of zein hydrolyzate, the hydrolyzate causes a greater rise in the levels of certain amino acids (glutamic acid, proline, alanine, leucine) that are abundant in zein. This difference between the intact protein and its hydrolyzate is very slight in the case of casein, thus demonstrating that retardation of digestion of zein is a significant factor in the rate of absorption of its amino acids.

The amounts of digestive enzymes secreted by the pancreas increase with the concentration of protein in the diet; protein hydrolyzates given at the same level also induce considerable enzyme secretion (Snook and Meyer, 1964a). The response to a change in diet can be detected very soon, e.g., 2½ hours later. This regulation appears to occur through signals coming from the products of digestion which modify the pools of free essential amino acids in the tissues (Abdeljlil and Desnuelle, 1964; Reboud *et al.*, 1966). However, the possibility that the proteins of the diet directly regulate pancreatic secretion is not excluded (Wang and Grossmann, 1951). Inhibitors of trypsin found in raw soybean and in egg hydrolyzates increase pancreatic secretion (Guggenheim and Goldberg, 1964; Lepkovsky *et al.*, 1965, 1966; Ma'ayani and Kulka, 1968; Snook, 1968). In man, distension of the stomach increases pancreatic secretion, and in the dog there is a direct correlation between HCl secretion and pancreatic secretion (Preshaw, 1967). This is to be expected, since both secretions are controlled by the hormone gastrin which is secreted by the stomach in response to distension or to certain chemical agents (Farrar and Bower, 1967). The action of gastrin is amplified by the presence of serotonin (Haverback and Dyce, 1967).

C. Absorption of the End Products of Digestion

The mechanism of absorption is discussed in Chapter 2, Section V of Volume I and has also been reviewed by Wiseman (1964) and by Fisher (1967a,b). Free amino acids and dipeptides hydrolyzed in the intestinal epithelium cross the intestinal wall. Those not utilized by the intestinal tissue for its own needs are carried by the portal vein to the liver. The rate of absorption of compounds infused directly into the intestine through a cannula can be obtained by monitoring the free amino acid levels in the portal vein after implantation of a permanent cannula in the portal vein. Amino acids cross the intestinal wall (or more usually disappear from the lumen of the intestine) at a rate characteristic for each (Finch and Hird, 1960; Delhumeau *et al.*, 1962; Matthews and Laster, 1965a; Tasaki and Takahashi, 1966). The rate of absorption diminishes with increasing electric charge; amino acids with nonpolar side chains are rapidly absorbed (e.g., leucine, isoleucine, valine, tryptophan, and phenylalanine), whereas glycine, aspartic acid, glutamic acid, and arginine are slowly absorbed. Using everted pieces of intestine, Wiseman (1964) has shown that the L-amino acids are more actively transported than the D series. The amounts of amino acids transported vary with different factors, such as the concentration of amino acids in the medium, the amount of water transported, the location in the small intestine (maximal transport occurs between the jejunum and mid ileum), and individual characteristics of the amino acids (Benson and Rampone, 1966; Matthews and Laster, 1965a). These absorptive mechanisms are reviewed by Matthews and Laster (1965b).

The concentrations of different amino acids increase in the portal blood very rapidly after a meal (Wiseman, 1964; Fauconneau, 1966; Chapter 34, Section III,A,1). The maximal level is reached at 1 to 1½ hours after the end of a meal in the case of the young pig (Pion *et al.*, 1966). The degree of aminoacidemia in the portal vein depends on several factors, such as the amino acid content of the food eaten, the rate of gastric emptying, the speed of enzymic digestion, the rate of absorption, and finally the extent to which the amino acids undergo transformations in the intestinal wall in order to provide it with nitrogen, carbon, and energy. Among these, the most important factors are the rate of gastric evacuation, which particularly determines the time of arrival of the dietary amino acids in the portal blood, the amino acid requirements of the intestinal tissue, and, in the case of certain proteins, the speed of digestion. Proteins and amino acids introduced *directly* into the duodenum result in a rapid (half-hour) increase in amino acid concentration in the portal blood (Peraino and Harper, 1963). Similarly, casein administered by force-feeding causes an extensive increase in the free amino acid levels of portal and systemic blood (Peraino and Harper, 1963). Most of the amino acids of zein and about half of those in casein appear more rapidly in the portal blood when hydrolyzates of these proteins are administered by force-feeding in place of the intact proteins, an observation that indicates that the speed of digestion can be more limiting than absorption, notably in the case of certain vegetable proteins that are only slowly digested. Finally, there is a relationship between the rise in blood amino acid level and the removal by the tissues of amino acids coming from dietary protein. The variability of the blood amino acid curves obtained for different essential amino acids after ingestion of a protein shows the presence of factors limiting this approach (Pion *et al.*, 1964; Goldberg and Guggenheim, 1962; Morrison *et al.*, 1961; also Chapter 34, Section III,A,1). This variability in pattern of absorbed amino acids may also reflect differences in absorptive efficiency between members of the same species (Williams, 1969).

The intestine is an efficient and selective filter, since only free amino acids pass into the bloodstream. Dipeptides can be rapidly absorbed, but are then hydrolyzed in the wall before the amino acids liberated from them are transferred to the portal blood. For example, the ingestion by human subjects of diglycine (glycylglycine) causes a more rapid and more extensive (30 minutes as against 45 minutes) increment in portal free glycine levels than the ingestion of a similar amount of glycine (Craft *et al.,* 1968). Hydroxyproline peptides, however, are absorbed intact (Prockop and Sjoerdsma, 1961).

Numerous phenomena of competition have been observed in studying the absorption of different amino acids. Many studies employing the isolated intestine are unphysiological, and conclusions drawn from them are suspect if they have not been confirmed by *in vivo* studies. Thus isoleucine and valine inhibit the uptake of leucine, and leucine inhibits absorption of isoleucine and valine; however, Szmelcman and Guggenheim (1966) showed by both *in vivo* and *vitro* studies on the rat that this inhibitory effect disappears after some 30–60 minutes, probably because of absorption of the inhibitory amino acid. The inhibitory action of leucine on several amino acids could partly explain leucine-isoleucine antagonism (Chapter 13, Section III,B). However, a leucine-rich diet causes a fall in isoleucine and valine levels in the free amino acid pools of muscle and blood without a reduction in the levels in the liver and intestinal wall (Tannous *et al.,* 1966). Using everted portions of intestine, Munck (1966c) has found that absorption of tryptophan is accelerated by methionine and leucine, but is inhibited by high concentrations of methionine and by lysine in the presence of methionine. Furthermore, tryptophan inhibits the transport of leucine and lysine competitively. On the other hand, leucine increases the absorption of lysine and arginine. By the same technique, Munck (1966a,b) has identified a separate transport system for imino acids additional to that for neutral and basic amino acids. This transport system for proline and hydroxyproline is equally available to glycine, betaine, leucine, alanine, and sarcosine. Iminoacids can inhibit absorption of glycine by two transport mechanisms, namely that for neutral amino acids and that for imino acids. In the hen, Tasaki and Takahashi (1966) have shown that absorption of leucine and phenylalanine is inhibited by methionine. The opposite effect (inhibition of methionine by leucine and phenylalanine) is weaker. Methionine also inhibits to a similar extent absorption of glutamic acid, whereas glutamic acid accelerates the absorption of methionine. Interactions between the absorption of amino acids and of sugars have been observed by Newey and Smyth (1964) and are discussed by Munck (1968).

Most neutral amino acids cross the intestinal barrier by an active process, at least at the level of the duodenum and jejunum which are the sites of absorption of exogenous amino acids coming from the diet. Transport occurs against a concentration gradient, and its rate is diminished when energy reactions are inhibited (Randall and Evered, 1964). It would appear that pyridoxal phosphate is also implicated in the intermediate steps of absorption (Jacobs *et al.*, 1960). Another hypothesis formulated by Spencer and Zamcheck (1962) suggests that phosphatide peptides, similar to those isolated from the liver (Tria and Barnabei, 1963) may be involved. Using anesthetized rats with intestinal perfusion, Jacobs and Lang (1965) have demonstrated two-way transport across the intestinal wall; there is both very rapid absorption and excretion of model amino acids such as aminoisobutyric acid. However, the most slowly absorbed amino acids (diacidic amino acids) are the most rapidly excreted.

The concentration of amino acids in the portal blood is affected both by the amounts of amino acids brought by the intestine and their utilization by various tissues, notably liver, intestine and muscle. Peraino and Harper (1962) force-fed rats with various amino acids, namely glutamic acid, glutamine and alanine, and determined the fate of these amino acids. As well as transformation of glutamic acid to alanine and a less extensive formation of glutamic acid from administered alanine, they observed the disappearance of large amounts of glutamic acid and of glutamine, compounds that are rapidly metabolized and used for energy purposes.

Finally, Dawson and Porter (1962) gave *Chlorella* protein labeled with ^{14}C to rats and to cats and measured amino acids appearing in the portal and peripheral blood. During the first 6 hours after the meal, 41% of the absorbed amino acids had been recovered in the free amino acids of the portal blood. A proportion of this could have come from digestion of endogenous nitrogen labeled with ^{14}C, since Junqueira *et al.* (1957) have demonstrated in the rat that pancreatic juice contains radioactive label 1 hour after injection of glycine-^{14}C. Dawson and Porter (1962) found that maximal labeling was observed in the portal blood 45 minutes after ingestion of the ^{14}C-labeled *Chlorella*, whereas the systemic blood was maximally labeled at the end of 1½ hours. Crane and Neuberger (1960) found in man fed yeast protein labeled with ^{15}N that maximal blood labeling occurred at 35–50 minutes after ingestion.

D. Metabolic Transformations in the Intestinal Wall

This topic is dealt with also in Chapter 2, Section V,C. Metabolic alterations of amino acids in the wall of the intestine have been demonstrated by several authors. From their *in vivo* studies on the dog, the cat, and the rabbit, Neame and Wiseman (1957, 1958) hypothesized that transamination must occur actively in the intestinal wall, since absorbed aspartic and glutamic acids are not recovered among the products of absorption. Sizable amounts of alanine appear instead of glutamic acid but the corresponding keto acids (α-ketoglutaric and pyruvic acids) are not altered in concentration in the portal blood. It has been concluded that these products are metabolized by the intestinal cells (Wiseman, 1964). Peraino and Harper (1962) and Pion *et al.* (1964) have likewise obtained evidence of the transformation of glutamic acid to alanine. Furthermore, Finch and Hird (1960) and Pion *et al.* (1964) have demonstrated that a diet rich in arginine causes a sharp elevation in the concentration of ornithine in the portal blood. Alanine level in portal blood seems to rise after consumption of a diet rich in starch (Pion *et al.*, 1964). Metabolic transformations during digestion are discussed on a quantitative basis in Chapter 38, Section II,A,1, which shows that very little glutamic or aspartic acids are transferred to portal blood, their nitrogen appearing as alanine and glutathione. This is relevant to recent demonstrations (e.g. Olney and Sharpe, 1969) that *parenteral* administrations of glutamic acid damages the nervous system.

Marty and Raynaud (1965) have studied the amino acids in blood used for perfusing the cecum of the rabbit, and have observed that the concentrations of free leucine and valine rise steadily as a result of their release from the wall of the cecum. On the other hand, the concentrations of glutamic acid, aspartic acid, and alanine increase both in the perfusate and in the contents of the lumen of the cecum. Within the cecum, amino acids are partitioned in different proportions between the bacteria and the intestinal fluid (Marty and Carles, 1968).

E. Absorption of Whole Protein by the Young Mammal

At the time of birth, the intestine of the young mammal does not as yet act as a selective filter. Earlier work on this subject is cited in Chapter 16. For some time after birth the intestine permits the passage of whole protein (Payne and Marsh, 1962), mainly into the lymph (El-Nageh, 1967). Colostrum, the first milk secreted, constitutes the sole source of immunoglobulins for the piglet, the calf, and the lamb since the maternal placenta is impermeable to these proteins. This permeability of the intestine seems not to be selective for immunoglobulins, since other proteins are absorbed and excreted in the urine during the first 48 hours of life. Intestinal permeability to protein diminishes rapidly after the first 24 hours, and by 48 hours it is essentially zero in the case of the calf and the pig. In the case of the calf, absorption is important for the first 12 hours and then falls off rapidly, to become completely suppressed at 36 hours after birth (El-Nageh, 1967). The amounts of

γ-globulins absorbed do not affect this change in permeability. Absorption is very rapid, so that the immunoglobulins can be detected by polyacrylamide electrophoresis within 30 minutes after feeding, and the maximum concentration is attained after 3 hours. In the case of the calf, the proteins cross the intestinal wall by pinocytosis, mainly in the jejunum and to a small extent in the ileum. Proteins labeled with fluorescein isothiocyanate can be observed in the cytoplasm of the intestinal mucosal cell, passing in the direction of the base of the cell and then appearing in the central lymphatic channels of the villi, the lacteals (El-Nageh, 1967). The little protein globules are particularly abundant in the region of the striated border of the intestinal cell, their number increasing and diminishing in proportion as they pass into the cytoplasm. The molecules of protein are first adsorbed onto the cell membrane, then incorporated into microvesicles formed by the plasma membrane at the base of the microvilli. These vesicles migrate toward the base of the mucosal cell, where the protein globules fuse. After crossing the epithelial membrane, the γ-globulins traverse the lamina propria and then the basal membrane of the fine capillary lymphatics of the mucosa. El-Nageh (1967) stated that he had never observed direct passage of the labeled globulins into the blood stream.

In the infant pig, Payne and Marsh (1962) have made a systematic study of the absorption of fluorescein-labeled γ-globulins in the pig, using both homologous pig globulins and heterologous globulins from the cow, the horse, and man. Absorption was found to be proportional to the amount ingested; it attains a maximum when 60 ml of colostrum has been taken, thereafter declining. If the animals are fed from birth, protein absorption ceases at the end of 12 hours, but if the baby pigs receive water, protein absorption can be sustained for 106 hours after birth. Absorption of protein is an all-or-nothing phenomenon, and absorption ceases when the epithelial cells of fetal type are replaced by the adult type (Vodovar, 1967). The constituents of milk that are responsible for this transformation appear to be soluble, dialyzable, and thermostable compounds.

In the case of the rat, the period over which the intestine allows whole protein to pass is important. It is some 18–21 days (Wiseman, 1964).

III. Role of the Intestine during the Digestion of Endogenous Protein

Gitler (Chapter 2) has already provided a survey of endogenous protein secretion into the intestinal tract. The normal function of the tract involves the entry at various levels of numerous products: enzymes, mucoproteins, desquamated cells, excreted urea (some 25% of the urea formed per day is recycled through the tract of man on a normal diet, according to Walser and Bodenlos, 1959), secretion into the lumen of amino acids (Jacobs and Lang, 1965; Combe and Gordon, 1968), and serum albumin.

A. Significance of Endogenous Nitrogen

The magnitude of the endogenous secretion of nitrogen into the intestinal tract has been disputed. In man, Nasset (1965) has calculated that desquamation of the intestinal mucosal cell with a life-span of 1–3 days (Creamer, 1967) results in a loss of 77–91 gm of protein daily into the intestinal lumen, and that the total intestinal secretion of protein amounts to 64–263 gm of protein daily. These estimates correspond to some 0.75–3 times the intake of dietary protein, and are much larger than those calculated by Twombly and Meyer (1961) in their studies on rats. The latter authors used a protein-free diet, and found that the amount of endogenous protein formed under these circumstances was equivalent to consuming a diet containing 10% of protein.

This picture is amplified by study of the intestinal contents. Ingestion of a meal increases the secretion of endogenous digestive secretions. Using casein tagged with ^{14}C to indicate the amount of exogenous protein in the alimentary tract, Nasset and Ju (1961) showed that the proportion of endogenous N is 4–7 times greater than that of exogenous N in the jejunum of the rat and the dog after a meal of protein. Also, the spectrum of free amino acids in the intestine is far more constant than that of the dietary proteins fed (Nasset, 1965). Nasset draws the conclusion that the large secretion of endogenous protein stabilizes the amino acid composition of the diet. However, Gitler points out (Chapter 2) that the large proportion of endogenous N in the gut and the relatively constant free amino acid patterns are reflections of the fact that dietary protein is rapidly absorbed, leaving only the less digestible endogenous protein. Ochoa-Solano and Gitler (1968) have recently confirmed this interpretation by injecting methionine-^{35}S into rats in order to label the endogenously secreted proteins, followed 2 hours later by a meal of ovalbumin labeled with selenomethionine-^{75}Se. Table I displays some of their data. At 30 minutes after feeding, the percentage of dietary protein evacuated from the stomach and still in the intestine amounted to 15%, indicating rapid absorption of protein from the diet. The unabsorbed exogenous protein tended to accumulate in the lower part of the small intestine (Table I), where transport is sluggish (Marcus and Lengemann, 1962). Digestion of endogenous protein, which the authors could not estimate

TABLE I

Distribution of Exogenous and Endogenous Proteins and of Trypsin in the Contents of the Gastrointestinal Tract of the Rat[a]

Components	Stomach	Duodenum	Jejunum	Jejunum-ileum	Ileum	Total in intestine
Exogenous N (mg)	87.6	1.8	2.7	6.0	6.6	16.8
Endogenous N (mg)	21.0	6.7	7.1	11.9	11.4	37.1
Tryptic activity (units)	—	994	630	3025	5026	9675

[a] From Ochoa-Solano and Gitler (1968).

precisely, appeared to be much slower than that of exogenous protein; its digestion takes place in the ileum and probably also in the cecum, through the joint action of bacteria and the host's enzymes. Furthermore, the increase in amino acid levels in the portal blood during the first 6 hours after a meal gives little evidence of being affected by the amino acid composition of endogenous proteins (Goldberg and Guggenheim, 1962; Pion *et al.*, 1964).

Quantitative and qualitative analysis of endogenous nitrogen can be more easily achieved in germ-free animals (Combe and Pion, 1966). The absence of the alimentary flora further retards digestion of endogenous N, although overall digestion is not appreciably altered. The intestinal wall of the germ-free animal retains the structural features of fetal intestine (Vodovar, 1967), and the turnover of cells in the crypts and villi is reduced (Lesher *et al.*, 1964; Dubos, 1966). In contrast, the weight of the cecum is increased 3–5 times (Luckey, 1963). These morphological alterations could well involve variations in the composition of endogenous nitrogen compounds. Furthermore, Combe and Gordon (1968) have found that, per unit of mucosal surface, the cecum of the germ-free rat absorbs water 7 times slower, and the amino acids serine and glycine 130 times slower, than in the case of the conventional rat, whereas excretion through the wall (efflux) is similar in the gnotobiote and conventional animal. These results explain why the contents of the cecum of the germ-free (axenic) rat are richer in water and free amino acids than those of the conventional rat. In fact, in terms of the proportion of organic N in the form of acid-soluble compounds, the cecal contents of the axenic rat contain proportionately twice as much soluble nitrogen as the conventional rat, namely, 84% of the total N against 45% (Briend and Pion, 1967). These soluble constituents consist of free amino acids, of more or less degraded mucoproteins of molecular weights between 50,000 and 400 (these more abundant than in the conventional rat), and peptides of low molecular weights, from 350 to 400.

Recent studies of endogenous N indicate that the digestive tube constitutes an important reserve of available essential amino acids. In fact, the endogenous nitrogenous compounds are well digested, since only 10% of this N appears in the feces (Twombley and Meyer, 1961). This is achieved probably through bacterial and enzymic action. These enzymes consist of cathepsins liberated from desquamated intestinal cells, and secreted digestive enzymes that have escaped autodigestion, especially since proteolytic enzymes are unusually stable and the intestinal flora have little action on them at various levels of the intestinal tract. This is shown by the studies of Lepkovsky *et al.* (1966) on germ-free and conventional animals, as illustrated in Table II. Recently, Reddy *et al.* (1969) have challenged these conclusions; they find that the pancreatic enzymes are much better preserved in the lower small intestine of germ-free than of conventional rats, and they conclude that bacteria are responsible for inactivation in the conventional animal.

TABLE II

NITROGEN CONTENT AND PROTEOLYTIC ENZYME ACTIVITIES OF PANCREAS, INTESTINAL, CECAL, AND COLON CONTENTS OF GERM-FREE AND CONVENTIONAL RATS[a]

Site sampled	Type of rat	Nitrogen (% of dry matter)	Protease activity (per gram solids)	Trypsin activity (% of total protease activity)
Pancreas	Germ-free	8.5	6.21	78
	Conventional	8.2	4.76	80
Duodenum	Germ-free	10.6	0.75	74
	Conventional	10.5	0.92	74
Jejunum	Germ-free	7.3	1.79	78
	Conventional	7.4	1.96	76
Ileum	Germ-free	4.1	1.50	79
	Conventional	5.1	1.22	74
Cecum	Germ-free	5.2	1.15	81
	Conventional	4.4	0.77	76
Colon	Germ-free	4.3	0.93	79
	Conventional	3.9	0.82	77
Colon (feces)	Germ-free	3.6	0.78	77
voided)	Conventional	3.8	0.54	74

[a] From Lepkovsky *et al.* (1966).

The quantity of endogenous N discharged into the lumen of the digestive tract is very important but inadequately known. It varies with the diet fed, and notably with protein intake (Twombley and Meyer, 1961). The amount of endogenous N absorbed by the dog is described in Chapter 38, where Elwyn has continuously monitored the flow of amino acids from the gut to the liver over a 24-hour period, including a meal of meat and subsequent periods of absorption and fasting. Under these conditions, the contribution of endogenous amino acids is significant but not excessively large.

Finally, it should be added that similar studies have been carried out on birds. By the use of dietary cellulose (Bolton, 1964) or chromic oxide (Bird, 1968), the dilution of the diet by endogenous protein secreted into the intestine of the chick has been estimated. It has been concluded by these authors that secretion of endogenous protein into the upper small intestine of the chick is considerable. On the other hand, Crompton and Nesheim (1969) fed various proteins to ducks and found that the free amino acid patterns in the small intestine reflected those in the administered protein, in contrast to the findings on the dog by Nasset (1965). Consequently, Crompton and Nesheim have concluded that the amount of protein secreted endogenously into the gut is too small to result in a constant mixture of amino acids during digestion of dietary protein. It is also possible that the rate of digestion of endogenous protein is slower than that of exogenous protein, and this is the reason for the lack of effect of endogenous protein on the amino acid pattern during digestion after a meal.

B. Digestion of Endogenous Nitrogen

Digestion of endogenous nitrogen occurs along the length of the intestine and probably is continuous throughout the day. In contrast, the greater part of the dietary protein is digested in the upper part of the small intestine at a rapid rate, and is discontinuous, being related to meal frequency and gastric emptying. Since the digestion of endogenous N is considerable in those areas where transit is sluggish and where the flora is significant (Michel, 1961a,b), some of the essential amino acids of the protein secreted into the gut are liable to be lost through bacterial action, and there is often a net loss of essential amino acids from the body as a consequence. These amino acids cannot therefore contribute to the dietary protein, and as a result there may be little or no supplementation of amino acids among the products absorbed over short periods after each meal of dietary protein.

By synthesizing some of the endogenous protein discontinuously (e.g., pancreatic enzymes), and by continuous digestion of this endogenous protein in the intestine, the digestive tract exercises a significant regulatory action over protein metabolism. It thus continuously presents essential amino acids for absorption and distribution to the tissues and helps ensure stable amino acid concentrations in the blood of the fasting animal. This allows the animal to be relatively independent of the discontinuous nature of the food supply. Consequently, one or two meals a day ensure the daily needs of essential amino acids, even during the intervening postabsorptive periods.

Feeding can affect the endogenous N output both quantitatively and qualitatively through modifying secretion and by increasing the stability of the enzymes (Snook, 1965a). While the protein of the food stimulates output of endogenous N and particularly secretion of proteolytic enzymes, it also protects the less stable enzymes against proteolysis (Snook and Meyer, 1964a,b). This occurs because the dietary proteins subjected to denaturation in the stomach seem to provide a preferred substrate for proteolytic enzymic attack.

In this connection, soybean meal acts on protein digestion in several ways. Some soya constituents exert a specific stimulant action on pancreatic secretion (Lyman *et al.*, 1962), one preparation of trypsin inhibitor is known to retard the digestion of exogenous protein (Ochoa-Solano and Gitler, 1968), and finally another preparation of trypsin inhibitor increases the oxidation of methionine-2-^{14}C to $^{14}CO_2$ (Barnes and Kwong, 1965). The nature of the action of soybean on pancreatic function has been explored by Goldberg and Guggenheim (1962), Lepkovsky *et al.* (1966), and by Ma'ayani and Kulka (1968), and a relationship of egg-white trypsin inhibitor to pancreatic hypertrophy has also been suggested (Snook, 1968; 1969).

Among proteins formed endogenously in the wall of the intestine are some fractions of the serum proteins. Dawson and Holdsworth (1962) have shown that the intestinal mucosa synthesizes large amounts of protein, but Dawson and Porter (1962), who studied rats absorbing *Chorella* protein labeled with ^{14}C, calculated that serum proteins (with a half-life of several days) could not be responsible for transporting more than 5% of the labeled amino acids absorbed in their experiments during the 5 hours after feeding. This contrasts with the recovery from portal blood as free amino acids of 41% of the radioactivity absorbed, of which 36% was in the plasma amino acids and 5% in the red cell free amino acids, while amino acids in lymph accounted for only 0.04% of the absorbed activity, a proportion that was increased by feeding fat. However, Hartmann (1959) has claimed that a significant synthesis of serum globulins occurs in the intestine, as evidenced by administration of yeast labeled with methionine-^{35}S fed to dogs. This emphasizes that more study is required before a quantitatively satisfying picture of the partition of absorbed products can be presented. Lipoproteins are certainly made by the intestine, as discussed below.

IV. Synthesis of Various Nitrogenous Compounds

The digestive tract synthesizes many compounds other than those specifically involved in protein digestion and for its own functional activities. Thus, it is known to manufacture the lipoprotein of the chylomicrons. Scott and Winterbourn (1965) have demonstrated that low-density lipoproteins are made in the intestine and passed directly into the portal vein, a result that has been confirmed in eviscerated rats by Roheim *et al.* (1966). This synthesis is blocked by puromycin (Nestel, 1967). Using orotic acid as a specific inhibitor of β-lipoprotein synthesis by the liver, Windmueller and Levy (1968) have demonstrated synthesis of this protein by the intestine which is not inhibited. In addition, some enzymes appearing in the plasma (e.g., amylase, lipase, trypsin, phosphatase) are absorbed from the tract or its accessory organs (Spencer and Zamcheck, 1962). In addition, the gastrointestinal tract allows the absorption of endotoxin proteins made by bacteria and contributes serotonin to the circulation, probably through its adsorption to the platelets.

Finally, the gut may be the site of elimination of serum proteins, although earlier assessment of its role in this process seems to have been grossly exaggerated (Wetterfors, 1965). The site of albumin catabolism nevertheless remains obscure, as discussed in Chapter 8 of this treatise. It seems probably that the liver can account for only some 20% of albumin degradation and the kidney for 10%, leaving some 70% unaccounted for. This has led to the investigation of the alimentary tract as a major catabolic site. The careful studies of Freeman (1964) suggest that this is in fact only a minor site of degradation in normal subjects. The gut flora do not affect leakage of plasma proteins; germfree rats do not differ from conventional animals in this respect (Levenson *et al.*, 1969).

V. Role of the Intestinal Flora

In all animal species, complex relationships occur between, on the one hand, the presence and nature of the intestinal microorganisms, and on the other hand the nutrition, age, and physiological state of the host. Purely microbiological studies have attempted to provide as exact as possible a description of the flora in different animal species. Another approach consists of determining the biochemical activities of the whole flora or of single species isolated from the flora. In order to integrate the metabolism of the flora into that of the host, many studies have been made with animals that are either aseptic (gnotobiotic, axenic), contaminated with microorganisms in a controlled fashion, or conventionally contaminated. The proper means of orienting the metabolism of the intestinal microbial population in relation to the needs of the host have mainly been connected with nutritional and environmental factors.

A. Nature of the Intestinal Flora

At this time, it is still not possible to obtain a precise picture of the intestinal flora in all animals that have been studied (Dubos, 1966). The number of organisms capable of growing in the usual culture media represents hardly 10% of the total number seen by direct microscopy (Bryant and Burkey, 1953). Raibaud *et al.* (1966a) have confirmed this by showing that most of the bacteria seen on microscopy are dead. The reconstitution of the normal flora by isolating a large number of bacterial species and inoculating these into an aseptic animal is thus impossible (Evrard *et al.,* 1964; Sacquet *et al.,* 1966; Raibaud *et al.,* 1966b).

Appreciable differences occur between different groups of mammals regarding the nature of their intestinal flora.

1. Certain monogastric species, such as man, the pig, the rat, the mouse, and the preruminant calf, have a complex microbial flora. Lactobacilli generally predominate. The other large groups comprise various anaerobes, enterobacteria, streptococci, yeasts, etc., as shown by the studies of Raibaud *et al.* (1966a,b) on the pig and rat. The localization within the alimentary tract of these various species has been examined in pigs by Van Der Heyde and Henderickx (1964). Many species of high catabolic activity (streptococci, enterobacteria, clostridia, etc.) grow from the stomach to the rectum in the pig. Kenworthy and Crabb (1963) have shown that the proliferation of coliforms in the small intestine of the young pig can be very extensive (3×10^9) after a change of diet.

2. Some monogastric animals have a simplified gastrointestinal flora, made up almost exclusively of lactobacilli. This group includes the human infant raised on its mother's milk and the "specific pathogen free" (SPF) mouse of Dubos and Schaedler (1960).

3. Ruminants have a microfauna (protozoans) associated with a microflora. These are mostly made up of anaerobes (Bryant and Burkey, 1953). Although the properties of several species have been studied (e.g., by Hungate, 1950; Slyter *et al.,* 1968), the characteristics of this flora are still less well understood than in the case of the monogastric animals.

4. The rabbit and the guinea pig are semiruminants whose flora approximates to that of the preceding group.

B. Metabolic Properties of the Intestinal Flora and Their Relationship to the Physiology of the Intestine

The presence of the intestinal flora modifies the physiology and biochemistry of the intestinal tract in the following ways.

1. By comparison with aseptic rats, conventional animals have a thickened intestinal mucosa and a cecum of greatly reduced weight (Wostmann and Bruckner-Kardoss, 1959; Hudson and Luckey, 1964). As regards development of the small intestine, the same conclusion has been reached for chickens by Eyssen and de Sommer (1965). In the case of the mouse, the presence of intestinal flora results in a greatly augmented turnover of epithelial cells in the small intestine (Lesher *et al.*, 1964). In addition, it has been observed that the presence of flora stimulates gastric emptying and intestinal peristalsis (Abrams and Bishop, 1967).

The influence of microorganisms on intestinal secretion is still controversial. Lepkovsky *et al.* (1964) found no difference between the concentrations of different pancreatic enzymes in the small intestine of the aseptic or conventional chick, and they made similar observations on rats (Table II). Furthermore, indirect stimulation is not prevented. Indeed, Khayambashi and Lyman (1966) demonstrated that a preparation of soybean trypsin inhibitor produced a considerable increase in the endogenous N secreted into the intestinal tract of both aseptic and conventional rats. The presence of a microbial flora aggravates the destruction of certain amino acids (especially methionine), so that the reutilization of amino acids from the secreted endogenous protein will not necessarily be the same in aseptic and conventional groups. Similar conclusions have been reached for the hen by Miller and Coates (1966).

2. The microbial flora shows a high metabolic activity toward many nitrogenous substances. The question arises whether this action is sufficient to modify the biological value of the diet, or even whether the products formed are toxic, as Metchnikoff's (1903) unconfirmed theory of autointoxication requires. Accordingly, comparison of metabolism in aseptic and conventional animals should indicate the biochemical contribution of the flora and thus provide definitive answers to these questions. One likely action of the flora would be to change loss of N in the feces. Regarding the action of the intestinal flora on fecal N excretion, the reported observations are contradictory. Luckey (1963) observed excretion of the same order in both axenic and control animals. However, Levenson and Tennant (1963) found more fecal N in the feces of germ-free rats; furthermore, soluble nitrogenous compounds accounted for a higher proportion of the fecal N of the axenic animal. On the other hand, Hoskins and Zamcheck (1968) fed rats on glucose only

and noted a higher excretion of undigested protein and mucin carbohydrate by aseptic animals. Miller (1966) too reports greater output of endogenous N by germ-free chicks. In contrast to these various observations, Harmon *et al.* (1968) observed with a sterile liquid diet that axenic rats had a very much *lower* output of fecal N at zero protein intake. These authors pointed out that the weight of small intestine (Gordon and Wostmann, 1960), the rate of renewal of the mucosa of the small intestine (Abrams *et al.*, 1963), the surface area of the small intestine (Gordon and Bruckner-Kardoss, 1961a), and the thickness of the intestinal lamina propria (Gordon and Bruckner-Kardoss, 1961b) are all reduced in the axenic animals, and they drew the conclusion that this causes reduced intestinal secretion and thus less fecal loss of endogenous N. Harmon *et al.* (1968) present evidence that these differences in fecal output in the axenic versus control groups may reflect the amount and nature of the diet ingested. Although emphasis has generally been placed on changes in morphology of the small intestine in germ-free animals, it is nevertheless in the cecum that the action of the flora appears to be most important. In the cecal contents one finds the greatest increase in free amino acids in the axenic rat, reaching an average of 500 μM concentration compared with 20 μM in the conventional rat (Michel and Sacquet, unpublished data). The level of urea, which is negligible in the conventional animal, approximates to the blood level in the case of the axenic rat (Combe and Sacquet, 1966). In chicks, it is claimed that the bacteria of the cecum play a role in proteolysis of poorly digestible proteins, such as raw soybean (Nitsan, 1965) and deteriorated fish concentrates (Payne *et al.*, 1968). Loesche (1968) fed germ-free rats on a protein-free diet and found that the protein content of the cecum was similar to that found in germ-free rats receiving dietary protein. This may indicate that the excessive amount of organic nitrogen in the cecum of the germ-free animal comes from endogenous protein that is not digested in the lower small intestine in the absence of a bacterial flora.

The action of the intestinal flora on protein utilization and excretion could result from several factors. First, stimulation of intestinal peristalsis, which would increase absorption rate, could arise from hypotensive compounds formed in the tissue and destroyed by the bacteria (Dubos, 1966). Second, increase of endogenous N may come from the intestinal wall, as demonstrated by indirect methods (e.g., Holtzman and Visek, 1965). Third, the intestinal flora can degrade amino acids, amides, urea, etc. Several studies allow this action of the flora to be more precisely identified (see Lewis, 1955, and Lewis and Emery, 1962, for the sheep; Borchers, 1965, for the cow; Doft, 1965, for the rat). Michel (1966)

TABLE III

Degradation of Amino Acids by the Flora of the Cecum of the Pig[a,b]

Amino acid	Percentage of α-amino N released	Percentage amino acid disappearing	Nitrogenous substances formed other than NH_3
Arginine	188	72	Agmatine, ornithine, citrulline, putrescine
Glutamic acid	100	100	—
Serine	100	100	—
Aspartic acid	79	78	—
Threonine	70	62	—
Alanine	47	50	—
Citrulline	46	38	Ornithine, putrescine
Leucine	41	52	—
Isoleucine	33	42	—
Histidine	32	98	Urocanic acid, histamine
Cystine	31	—	—
Glycine	28	28	—
Valine	25	34	—
Ornithine	15	28	Putrescine
Tyrosine	14	22	Tyramine
Lysine	12	21	Cadaverine
Phenylalanine	11	30	—
Tryptophan	10	15	Indole, tryptamine
Methionine	10	20	—
Proline	6		

[a] From Michel (1968).

[b] The substrate was added at a concentration of 0.02 *M* to a 1:4 dilution of the flora, and incubation was continued for 16 hours. The data are based on studies made with 80 animals killed in the winter time.

has shown that the cecal flora of the pig is capable of degrading all the amino acids by decarboxylation or deamination (Table III). The intestinal flora provides considerable complexity in its enzymic equipment. Thus degradation of arginine by the flora results in activation of several metabolic pathways other than those of the arginine–ornithine cycle in the liver; thus the flora provides enzymes decarboxylating arginine and ornithine, agmatinase, urease, and there is also degradation by enzymes of putrescine (Michel, 1968). These enzymes are potentially very active and undergo seasonal variations from unknown causes (Michel, 1966). The amines produced *in vitro* by the intestinal flora have also been identified in the whole animal in corresponding relative abundances. This demonstrates that they are of microbial origin, as shown also by comparing their formation in germ-free and conventional animals (Michel and

Sacquet, 1966). Similar results have been obtained with calves by measuring fecal output of amines (Michel and Mathieu, 1967). These studies show that, in the healthy calf, few amino acids are degraded, whereas in the animal with diarrhea, fecal excretion of free amino acids and their degradation products is much increased (Table IV). As shown in the table, in the case of arginine and lysine, fecal excretion of the corresponding amines (putrescine and cadaverine) represent, respectively, 34% and 11% of the ingested amino acids. From various observations, it would appear that this nutritional effect may be more important than the presumed toxic action of the amines formed. Indeed, administration to healthy calves of amines at ten times the dose level compared with the amount excreted during diarrhea has no influence at all on the metabolism of the animal. Indeed, protein malnutrition has been observed in a patient with stasis and extensive bacterial infection of the small intestine (Jones *et al.*, 1968).

3. *Urea metabolism.* In species that break down urea extensively, a source of intestinal nitrogen of particular importance is the urea recycled through the intestinal secretions. Using urea marked with ^{14}C and ^{15}N, Regoeczi *et al.* (1965) have demonstrated that the excretion of urea by the rabbit represents less than 50% of the amount synthesized. Similar conclusions have been drawn by Packett and Groves (1965)

TABLE IV

COMPARISON OF NITROGEN METABOLISM IN NORMAL CALVES AND CALVES WITH DIARRHEA[a]

Observation	Normal animals[b]	Animals with diarrhea[b]
Milk consumed per day (kg)	7.2	4.9
Apparent digestibility (%)	94(18)	86(3)
Retention of nitrogen (%)	65(16)	44(6)
Fecal nitrogenous compounds (mg/day)		
Total N	1900(25)	7100(25)
Amino N	21	840
Amine N	7	440
Urea N	7	200
Fecal excretion of amines (mg/day)		
NH_3 and volatile amines	1.2(25)	454(25)
Tyramine	Traces	94
Diamines	130	1211
Fecal cadaverine as % lysine eaten	0.4	11
Fecal putrescine as % arginine eaten	0.8	34

[a] Data from Michel (1968).

[b] Numbers in parentheses are numbers of observations for either a week (first two figures in each column) or a day (groups of 25).

for the sheep. During periods when the animal is taking a diet rich in starch and low in protein, urea recycled at the level of the rumen exceeds the amount provided by hepatic synthesis by 260 mg N per hour. Houpt (1963) considers the cecum to be main site of attack in the case of the rabbit.

Urea recycled in the intestine is degraded by intestinal urease to ammonia which can be used for anabolic purposes. It has been established that urea, on addition to a diet containing only essential amino acids, allows the mammal to obtain its nonessential amino acid requirement (Rose *et al.*, 1949; Rose and Dekker, 1956; Adkins *et al.*, 1967; Rechcigl *et al.*, 1959). In spite of contradictory evidence obtained with organism-bearing animals, it would appear that intestinal urease is entirely microbial in origin. The level of urea in the cecum of the germ-free rat is close to that of the blood, whereas it is essentially zero in the animal bearing bacterial flora (Evrard *et al.*, 1964; Combe and Sacquet, 1966; Michel and Sacquet, 1966). Furthermore, urea labeled with ^{14}C and administered to germ-free rats does not give rise to $^{14}CO_2$, whereas the conventional animal makes this conversion (Levenson *et al.*, 1959). Recent studies by Delluva *et al.* (1968) have similarly demonstrated that the urease thought to be secreted by the gastric mucosa of nonruminant mammals is in fact absent from the stomach of germ-free animals of various species.

The metabolic utilization of the urea nitrogen depends in the first instance on the action of microbial urease, but its utilization is also determined by the energy content of the diet and the availability of dietary amino acids. In the case of man, tissue incorporation of ^{15}N from urea is especially demonstrable in the case of uremic subjects receiving a diet of essential amino acids rather than in normal subjects receiving proteins that already contain an adequate quota of nonessential amino acids (Giordano *et al.*, 1966).

It should, however, be pointed out that, under pathological conditions, production of ammonia in the intestine can prove toxic (McDermott, 1966) and necessitate regulation of its formation (Summerskill *et al.*, 1967). This occurs in subjects with gross liver damage. The role of ammonia in the production of hepatic coma in such cases is complex, and is discussed in the introductory chapter to Part III in Volume II of this treatise.

C. Factors Capable of Influencing the Metabolism of Intestinal Organisms

Several factors influence the role of the intestinal flora in the metabolism of the host.

1. Both nutritional and environmental factors determine the type of flora present in the intestine and in consequence its metabolic role. In contrast to the infant raised on cow's milk, whose flora is similar to that of an adult human subject, the human baby receiving mother's milk presents flora confined exclusively to *Lactobacillus bifidus* (György, 1957; Haenel, 1965). The absence of reducing organisms, and particularly of enterobacteria, allows a rise in pH and of redox potential of the intestinal contents, which approach the values found in blood (Grütte, 1965; Gätzsche, 1965). Similar observations have been made on the germ-free rat by Wostmann and Bruckner-Kardoss (1966). In the case of the young pig fed by its mother, fecal excretion of amines arising from microbial degradation of amino acids does not occur. However, after weaning, excretion of amines increases and is maximal when diarrhea occurs (Michel *et al.,* 1964). The nature of the substances present in mammalian milk and responsible for the establishment of a simplified intestinal flora is still no more than hypothetical.

In addition to nutritional factors, environmental conditions can play a role in determining the intestinal organisms; these, in turn, influence the digestion of food and speed of growth. The chickens used by Coates *et al.* (1951) showed greater growth rate when placed in new hen houses than in previously occupied quarters. The "specific pathogen free" mice of Dubos and Schaedler (1960), which were obtained by cesarean section and maintained under rigorously hygienic conditions, showed distinct metabolic differences from mice reared under conventional conditions. In particular, their growth was much better on a diet containing gluten as protein. Their intestinal flora was simplified and in particular was free from enterobacteria. Since gluten is a protein particularly deficient in lysine, it is interesting to note that Stoewsand *et al.* (1968) have found that maximal growth of germ-free mice can occur on an amino acid diet containing only 0.4% lysine, whereas mice inoculated with the wild mouse flora achieved equivalent growth only at 0.6% lysine intake.

2. Action of antibiotics has provided some interesting effects both on the bacterial flora and on the host. Modifications in the metabolism of intestinal organisms can be obtained by the addition to the diet of small doses of antibiotics (10 to 20 parts per millon). In this way, chlortetracycline (Larson and Hill, 1955), and oxytetracycline and sodium acrylate (Michel *et al.,* 1964) inhibit the formation of nitrogenous bases in the intestine of the young pig. *In vitro* studies performed over several years have shown that chlortetracycline inhibits the overall metabolism of the flora, especially that of nitrogenous substances (Michel 1961a) and of carbohydrates (Michel, 1961b). The same effect is pro-

duced by a series of other antibiotics, by copper salts, by unsaturated acids, nitrofuranes, etc.

It is interesting to observe that those substances having a marked inhibitory action on the flora have an equally strong stimulating action on the growth of the host (Michel and François, 1955). It can be concluded that this stimulation of growth is mediated by changes in the intestinal flora (François, 1962; François and Michel, 1968). Even if this mechanism is not entirely proved, the following correlations have nevertheless been established. (a) Ingestion of the antibiotic is followed by a reduction in the level of ammonia in the portal vein, as shown by François and Michel (1960) for the pig, and by Warren and Newton (1959) for the guinea pig. (b) The antibiotic causes an increase in nitrogen retention in rats (Henry and Rérat, 1962; Michel and Durand, unpublished work), in calves (Hogue *et al.*, 1956), and in the pig (Delort-Laval *et al.*, 1963). Food intake is generally less in relation to the weight gain and nitrogen retention. (c) There is a reduced fecal excretion of nitrogen. It is difficult to decide whether this effect is due to a reduction in endogenous nitrogen output (Delort-Laval *et al.*, 1963) or to superior absorption of dietary amino acids (Carrol *et al.*, 1953). (d) The characteristics of animals receiving antibiotics approximate to those of germ-free animals, particularly the reduction in thickness of the small intestine and the increase in volume of the cecum.

In summary, the influence of the intestinal flora on the nitrogen metabolism of the host is the resultant of a series of metabolic interactions. Unfavorable effects consist of degradation of nitrogenous and energy-yielding nutrients to form products not available to the host, such as CO_2, amines, etc. In addition, the flora probably increase endogenous nitrogen output and utilize these and dietary products for the synthesis of bacterial protein that is partly excreted. On the other hand, there are several advantages derived from the flora. They provide protection against foreign bacteria, some of which can be pathogenic, they increase intestinal tone, and they destroy some toxic materials of endogenous origin (Dubos, 1966). They may also digest endogenous protein.

Even if, in the case of normal digestion, the overall function of the flora on metabolism of dietary nitrogen seems rather limited, and is more important in the case of the metabolism of endogenous nitrogen, nevertheless, the picture can be quite different in the case of malabsorption from various causes. Under these circumstances it is possible to have increased catabolism by the microorganisms which can aggravate metabolic changes in the host. Whether in man or in animals, the economic results of digestive illnesses are very high (Grossman, 1967; Amstutz, 1965; Almy and Sleisenger, 1967). The relationship of infection to malnutrition in the case of the human infant is discussed in detail in Chapter 23 in Volume II of this treatise. It is accordingly most desirable that some effort should be made to understand and to control the action of the intestinal flora under these pathological conditions.

VI. Role of the Metabolic Activity of the Digestive Tract

The epithelium and the accessory glands of the digestive tract are in a highly dynamic state. The daily turnover of protein involved in the secretion of digestive juices and the shedding of epithelial cells is extensive. Table V gives some estimates of the amounts of protein secreted daily into the gastrointestinal tract by an adult man and compares them with intake of dietary protein and with protein synthesis in the body as a whole and in certain selected organs and systems. Although many of these estimates, including the amount of protein secretion into the alimentary tract, are quite crude, it is apparent that the magnitude of endogenous nitrogen output into the gut is large, especially the fraction provided by epithelial shedding, and that variations in the amount of protein secreted in this way will have a considerable influence on the protein metabolism of the body as a whole.

This dynamic state results in a more rapid turnover of proteins in the organs associated with digestion than in most other tissues, which is further evidence of the importance of this area in total protein synthesis in the body. After administration of leucine-^{14}C to rats, the specific activities of the proteins of the intestinal mucosa (Munro *et al.*, 1959) and of the pancreas (Vandermeers *et al.*, 1967) are much greater than that of the liver proteins. This process of protein turnover is, however, due to different processes in the case of the accessory glands (salivary, pancreas) than in the case of the gastrointestinal mucosa. The glands secrete protein without much turnover of cells, whereas the mucosae discharge cells continuously into the lumen of the gut and therefore undergo extensive cell renewal. Each type of turnover will now be briefly considered.

The renewal of proteins in the pancreas has been considered by several authors. In the rat of 200 gm body weight, the average renewal time of the pancreatic proteins is 34 hours, that is, they have a half-life of 24 hours (Vandermeers *et al.*, 1967), which can be compared with another estimate of 48 hours by Junqueira *et al.* (1957). However, these figures for mixed tissue proteins may obscure more rapid synthesis of enzymes. In the cow, the renewal time of the enzymes trypsinogen, chymotrypsinogen, and ribonuclease has been found with cystine-^{35}S to be about 2–3 hours (Keller *et al.*, 1961). Rapid renewal is also suggested

TABLE V

Daily Nitrogen Exchange in the Digestive Tract of a 70-kg Man Compared with His Turnover of Body Proteins

Component	Amount of protein (gm/day)	Reference
Protein secreted into alimentary tract		
Saliva	3	Chapter 2
Gastric juice	5	Chapter 2
Bile	1	Russell *et al.* (1964)
Pancreatic juice	8	Howard *et al.* (1951)
Mucosal shedding	50[a]	Leblond and Walker (1956)
Balance sheet of protein entering and leaving tract:		
Average dietary intake	90	Part 2, Introductory Chapter, Table V
Total secreted into tract	67	Sum of data given above
Fecal output	10	Part 3, Introductory Chapter, Table II
Amount absorbed (157 − 10)	147	—
Amount of protein synthesized		
Total body protein synthesis	285–340	San Pietro and Rittenberg (1953); Crispell *et al.* (1956); Kassenaar *et al.* (1960)
Skin protein replacement	5	Mitchell and Edman (1962)
Blood protein replacement		
Hemoglobin	8[b]	—
Albumin	12	Chapter 8
γ-Globulin	3	Chapter 8
Fibrinogen	2	Chapter 8

[a] Leblond and Walker (1956) calculated that, if the data they cited for the rat apply to man, the mass of cells released daily from the surface epithelia into the digestive tract would weigh half a pound (250 gm). Assuming 20% protein, this would represent a loss of 50 gm of protein daily. From the estimate of 167×10^6 cells shed per minute from the mucosa of the small intestine of man (Crosby, 1961), it is possible to calculate that some 36 gm of protein are lost from this site alone in 24 hours.

[b] By calculation based on 1% replacement of Hb per day, with a Hb concentration of 16 gm/100 ml blood and a blood volume of 5 liters.

by the autoradiographic studies of Caro and Palade (1964) on the formation of enzyme-rich granules in the pancreas.

Turnover of the cells lining the digestive tract is also rapid. The details have been the subject of two recent short reviews (Lipkin, 1965; Creamer, 1967). The cells are formed in the crypts which specialize in cell division and from this site the epithelial cells pass up the villi and are extruded from the tip, where they are lost in the intestinal contents. The process of cell turnover can be followed by counting mitoses in the epithelial cells or by labeling with thymidine-^{3}H which is incorporated into newly formed DNA. Uptake of thymidine is generally equated with cell division, though the recent recognition of "metabolic" DNA not connected with cell division (e.g., Pelc, 1968) suggests caution in interpretation. Nevertheless, Lipkin and Quastler (1962) have successfully used the thymidine labeling technique not only to analyze the frequency of cell division in the intestinal mucosa, but also to quantitate the mitotic cycle and its timing. After a pulse dose of labeled thymidine, cells that are making DNA (S phase) incorporate the label and a peak in labeled mitoses (M phase) can be identified some 6–8 hours later, followed by a decline at 16–24 hours and thereafter a second smaller peak. From this pattern, the duration of the synthetic, mitotic, and resting phases can be computed (Fig. 1).

In the small intestine, cell division is limited to the crypts of Lieberkühn. According to Cairnie *et al.* (1965), one-third of these cells are in the DNA-synthesizing (S) phase at any one time in rat intestinal mucosa. In man, Lipkin (1965) states that new epithelial cells are produced in the epithelium of the stomach, jejunum, ileum, colon, and rectum at the rate of 1–2 cells per 100 epithelial cells per hour and in consequence the whole population of epithelial cells is replaced in 3–6 days. The S phase (DNA synthesis) lasts 10–15 hours, the G_2 (pre-

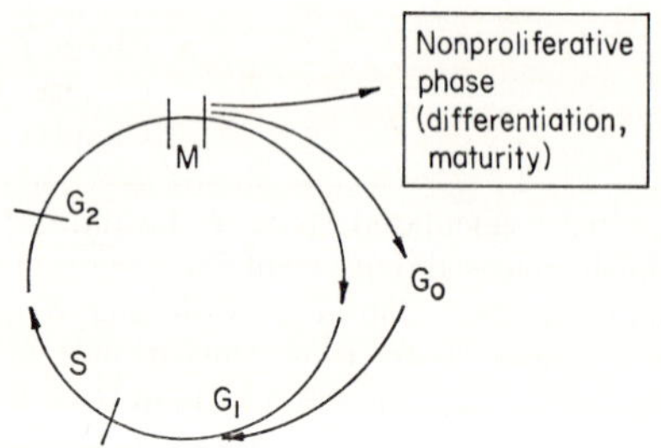

G_1 Interval between mitosis and DNA synthesis
S DNA synthesis
G_2 Interval between DNA synthesis and mitosis
M Mitosis
G_0 Prolonged interphase

FIG. 1. Diagram of phases of cell renewal cycle. Phase G_1 (postmitotic, presynthetic), S (DNA synthesis), G_2 (postsynthetic, premitotic), and M (mitotic) constitute the cycle. Phase G_0: prolonged interphase, contains fertile cells, not proliferating. Cells enter a nonproliferative phase during which differentiation and maturation take place. (From Lipkin, 1965.)

mitotic) phase 1 to 2 hours, mitosis (M) is less than 1 hour, while the G_1 (postmitotic) phase can be as short as 7–12 hours, so that the whole cycle can sometimes be accomplished in 24 hours.

The mechanism of control of cell replication is not understood, but must be rather strict since the intestinal mucosa maintains a constant appearance. The usual shape of the villus is maintained by the normal balance of cell turnover, so that flattening of villi follows loss of cell number; this is due either to diminished rate of cell production or increased rate of cell loss without compensatory changes in other phases of cell turnover and can happen in certain abnormal conditions. Normally, in conditions under which rate of cell loss increases, the crypts lengthen and mitosis becomes more frequent (Creamer, 1967). This implies a feedback of information from the villi to the crypts. When a small part of the intestine is removed, the remainder shows lengthening of villi and increase in rate of cell turnover (Loran and Althausen, 1960). This is due to some form of hormonal mechanism, since resection of the intestine of one parabiotic rat causes this response in the intestine of the other.

A. *Stabilizing Role of the Intestinal Tract throughout the Day*

Each meal has a direct action on the functions of the digestive tract, and furthermore the increase in blood amino acids after each meal has a direct and positive effect on protein synthesis in different organs (Hanking and Roberts, 1965; see also Chapter 34). The cells of the digestive tract, certain components of which have a short half-life, have first choice of free amino acids among the absorbed products and discharge them again into the blood several hours later when the endogenous protein is digested back to free amino acids. Evidence for such a stabilizing effect is suggested by the response of the liver polysomes to free amino acid mixtures deficient in tryptophan, which cause polysome disaggregation within an hour of feeding (Wunner *et al.*, 1966). Wunner (1967) has subsequently observed a tendency for the polysomes to reaggregate when the period of observation after the amino acid meal is prolonged to 7 hours (Fig. 2), suggesting that tryptophan liberated from other parts of the animal (especially the digestive tract) has made up for its deficiency in the mixture fed. Indeed, Arnal *et al.* (1968) have observed such exchanges of lysine-^{14}C between body compartments (Table VI). Six hours after injection of this labeled amino acid, muscle obtained significant quantities of lysine for protein synthesis from labeled lysine released from tissues containing proteins with a short half-life (liver, digestive tract) and from digestive enzymes labeled in the pancreas and later broken down in the gut.

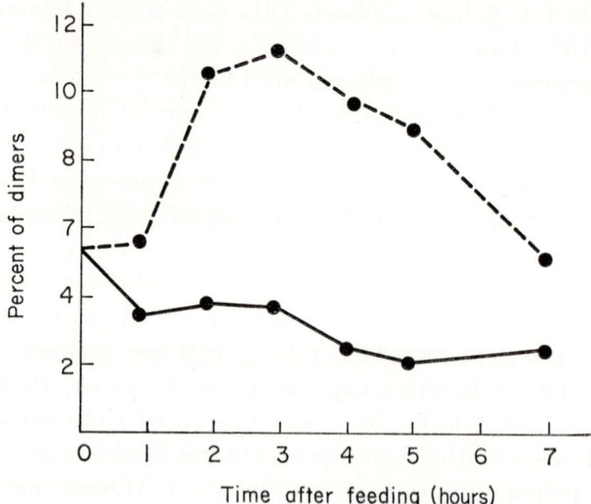

FIG. 2. Effect of amino acid mixture tube-fed to rats on polysome aggregation in the liver. The curves show the percentage of dimers in the liver polysome profiles at various times after feeding either a complete mixture of amino acids (——) or a mixture lacking tryptophan (- - - -) to fasting rats. An increase in dimer population indicates disaggregation of polysomes. (From Wunner, 1967.)

TABLE VI

PERCENTAGE OF INJECTED DOSE OF LYSINE-^{14}C IN DIFFERENT BODY COMPARTMENTS AT DIFFERENT TIME INTERVALS AFTER INJECTION[a]

Compartment[b]	Short time intervals				Long time intervals			
	30 min	60 min	120 min	240 min	480 min	5640 min	7000 min	9000 min
Blood free amino acids	4.1	1.8	1.5	1.2	0.6	0.4	0.4	0.45
Muscle free amino acids	29.2	26.6	17.4	3.1	1.9	1.1	0.6	0.5
Protein-bound amino acids of muscle	4.3	7.4	11.6	15.2	23.1	29.2	26.4	32.3
Sum of free and protein-bound amino acids in muscle	33.5	34.0	29.0	18.3	25.0	30.3	27.0	32.8

[a] Data of Arnal *et al.* (1968).

[b] The blood represents 10% of the weight of the rat; muscle represents 40% of live weight.

B. Role of the Intestinal Tract following Dietary Alterations

The effect of complete starvation and of protein deficiency have been studied by Ju and Nasset (1959), and also the response to the specific deficiency of lysine (Vandermeers *et al.*, 1966; Durand *et al.*, 1966). Under these conditions, the capacity of the intestinal tract to lose protein rapidly is considerable. Ju and Nasset (1959) found that, in rats subjected to starvation for 8 days, the stomach loses only 10% of its protein, whereas the small intestine, pancreas, and liver lost half; the rate of loss over the 8-day period was more rapid in the pancreas than for the intestine. When the rats were given a protein-free diet, the stomach and pancreas lost protein more rapidly than during total starvation, and during realimentation these organs lost a further small amount of protein before beginning slowly to recover their original protein content. On the other hand, the small intestine showed rapid restoration and recovered in 3 or 4 days all the protein it had lost in 8 days. Vandermeers *et al.* (1966, 1967) have extended these observations, using rats fed on a diet deficient in lysine for 20 days. Under these conditions, the pancreas lost 19% of its RNA and 38% of its protein, notably in the form of hydrolases, amylase, chymotrypsin, ribonuclease, trypsinogen, lipase, and procarboxypeptidases. Following 3 days' realimentation, there was a marked rise in mitotic index and nuclear uptake of thymidine-^{14}C by the pancreas, and the total DNA content of the organ increased. There was also a change in acinar cell morphology with reappearance of hydrolase granules. Synthesis of pancreatic protein reflected these changes; compared with a renewal time of 34 hours in rats receiving an adequate diet throughout the experiments, animals undergoing realimentation with protein showed a renewal time of 8 hours.

The small intestine reacts to protein deficiency with a reduction in cell number and a loss of ribosome population, whereas the liver loses only the latter (see Chapter 10 in Volume I). Starvation has an extensive retarding effect on intestinal cell division (Hooper and Blair, 1958), and protein deficiency has a similar action on DNA metabolism in the small intestine as judged by uptake of ^{32}P into DNA (Munro and Goldberg, 1964). The last-named authors also showed that the mean cell composition (protein and RNA per unit of DNA) was unaffected by protein deficiency. This indicates that, in contrast to the liver, the intestinal mucosa responds to a change in protein intake with an alteration in rate of cell production but not in cell composition. This has been confirmed by Durand *et al.* (1966), who fed a gluten-containing diet deficient in lysine, tryptophan, and threonine to rats and observed a diminution of 10% in the amount of intestinal DNA without a change in the RNA or protein content per unit of DNA. The nature of the response in intestinal cell population to starvation has been analyzed by Brown *et al.* (1963), who found slower cell renewal and reduced migration of the cells. Hopper *et al.* (1968) examined the small intestine of rats that were starved or fed on a protein-free diet; they found fewer cells per crypt, but the number of crypt cells in mitosis or labeled with thymidine-^{14}C was unaffected. However, both starvation and protein deficiency appeared to reduce the migration of cells from the crypts to the villi, and it was concluded that the time of generation of each cell was lengthened. Eventually, protein deficiency leads to a flattening of the villi and consequent reduction in surface area (Platt *et al.*, 1964). Creamer (1967) points out, however, that there are species differences in response of the mucosa to protein deficiency; in the monkey, the whole mucosa becomes thinner with a slower turnover of cells, whereas in the pig the villi alone become shorter, while the crypts remain unaffected. Creamer (1967) suggested that the Paneth cells at the bottom of the crypts provide a secretion nourishing the other crypt cells, and that the absence of Paneth cells from pig mucosa makes the cells of this species notably sensitive to undernutrition. Using whole intestinal wall, Imondi *et al.* (1968) demonstrated decreased synthesis of DNA when rats were stressed.

Nutritional state affects the capacity of the intestinal mucosa to absorb amino acids. Two groups of investigators studying protein-depleted animals have reported that everted sacs of small intestine prepared from the depleted animals transported amino acids into the sac more rapidly than did preparations made from normally nourished animals (Suda and Shimomura, 1964; Herskovic, 1969). Kirsch *et al.* (1968) failed to obtain this response with sacs prepared from protein-depleted rats, but did observe increased uptake of amino acids into the mucosal cells. Starvation has been found to increase the transport of amino acids *in vivo* across the gut wall (Kershaw *et al.*, 1960) and also *in vitro* using everted sac preparations made from rats or guinea pigs (Neame and Wiseman, 1959; Hindmarsh *et al.*, 1967). This effect appears to be species-specific since starvation does not accelerate absorption into everted sacs prepared from hamsters (Hindmarsh *et al.*, 1967). It is therefore important to determine what effects nutritional factors may have in modifying amino acid absorption in man, especially under conditions of malnutrition in infancy. Steiner and his colleagues have used human intestinal mucosa obtained by biopsy and have examined its chemical composition and its capacity to concentrate amino acids when incubated *in vitro* with radioactive amino acids. In the case of obese patients, Steiner *et al.* (1969a) found that the mucosal cells had rather more cytoplasm and that their *in vitro* capacity for active accumulation of amino acids was greater than that of mucosal cells obtained from normal subjects. They suggest that this represents hypertrophic adaptation of the mucosa of the obese subject to the increased intake of food. When moderately obese patients were fasted for 2 weeks, the capacity of the mucosal cells to concentrate amino acids fell considerably, but not below the levels found in normal subjects (Steiner *et al.*, 1969b). On the other hand, when three healthy men were deprived of protein for 2 weeks, mucosal cells obtained by biopsy showed no change in their capacity for amino acid concentration. It is not possible to decide whether the effect of starvation on amino acid uptake by the mucosa of the obese subjects represents a reduction to normal function or an effect of starvation that would be evident in nonobese human subjects deprived of food.

Changes in the intestinal mucosa relevant to the absorptive mechanism may differ in starvation from protein depletion. It is known in the rat that starvation reduces the average size of the intestinal cell (Thaysen and Thaysen, 1949) and its content of protein and of RNA (Steiner *et al.*, 1968), whereas a period of protein deficiency diminishes mucosal cell population without altering cell composition or structure, especially in the jejunum (Munro and Goldberg, 1964; Hill *et al.*, 1968). It is possible that the mucosal cells can maintain their normal composition during protein depletion through utilizing proteins secreted endogenously into the intestine. Such endogenous secretions still continue to be formed even when the diet contains no protein since the nonprotein food continues to be digested (see Section III,A, above). Amino acids released on digestion of this endogenous protein are subsequently reabsorbed. Hirschfield and Kern (1968) have found that labeled amino acids given orally to protein-depleted rats are much better utilized by the mucosal cells than are amino acids given by injection. This difference was not observed in normally nourished animals, and was taken to imply that the endogenous protein is preferentially utilized for mucosal cell renewal on a protein-deficient diet. Eventually, prolonged protein deficiency causes extensive changes in the mucosa which undergoes atrophy (Platt *et al.*, 1964) that is fully reversible (Cook and Lee, 1966). The atrophic mucosa frequently shows loss of disaccharidases, particularly lactase (Bowie *et al.*, 1965), thus representing impaired function of the mucosal cells. The mucosal atrophy does not appear to diminish absorption of protein given to the malnourished child to promote recovery, but the lactose intolerance may persist for many years (Cook and Lee, 1966).

C. Role of the Intestinal Tract in Different Physiological States

Studies on the sow (Salmon-Legagneur, 1965, 1968) and on the rat (Fell and Campbell, 1964; Campbell and Fell, 1964; Cairnie and Bent-

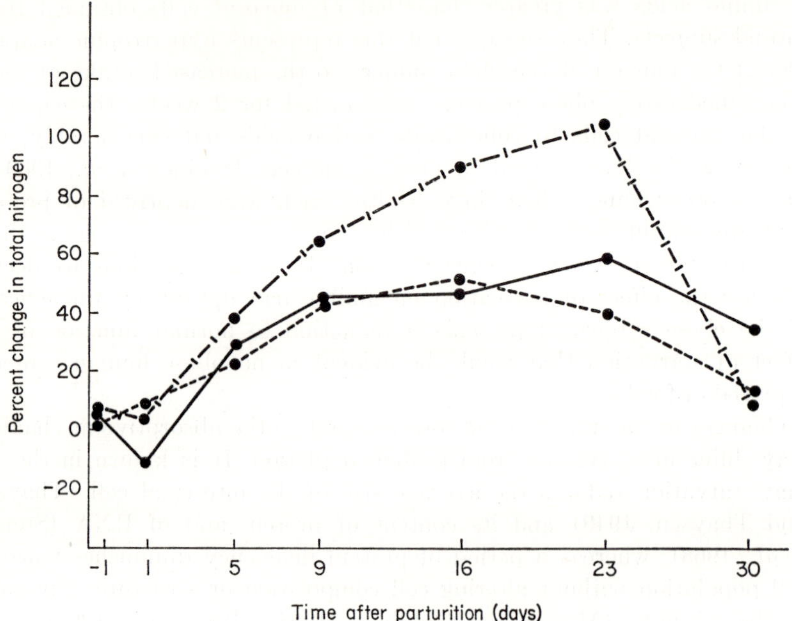

FIG. 3. Percentage change in the total nitrogen of the wall of digestive tract in rats during lactation and subsequent weaning. Small intestine (— —), cecum (——), stomach (- - - -). (From Fell and Campbell, 1964.)

ley, 1967) show that there is a considerable hypertrophy of the digestive tract during lactation (Fig. 3). By day 23 of lactation, the small intestine of the rat has increased 100%, and the stomach and cecum, 60%. Pregnant ewes show gastric hypertrophy, which is accentuated during lactation and is accompanied by hyperplasia of the parietal cells and gastric epithelium (Fell and Campbell, 1964). The cause of the hypertrophy in lactating rats has been explored by Fell and Campbell (1964), who found that it was especially marked when food was abundant; this was, however, considered to be only a partial reason for the hypertrophy, which was also attributed by these authors to the action of the various hormones regulating lactation. These interpretations were supported, first, by injecting sheep prolactin into a nonlactating mouse and obtaining mucosal hypertrophy; and second, by restricting the food intake of lactating sheep to that of virgin females and observing a reduction in alimentary hypertrophy except for the stomach, which weighed more because of water retention without an accompanying deposition of protein. Cairnie and Bentley (1967) have found that the population of cells both in the crypts and on the intestinal villi of lactating rats is increased, but the turnover time for the cells of the villi is somewhat shorter than in nonlactating rats. Consequently, the rate of loss of cell protein into the intestine must be considerably increased during lactation. Cairnie and Bentley were unable to say whether these changes in cell population and dynamics are due entirely to the increased food intake, but they suggested that this is not likely to be so.

The role of food intake in determining the development of the small intestine has been examined by Durand *et al.* (1965, 1967). During growth of the rat, there is a parallel increase in food intake and in intestinal weight, both of which augment vigorously until puberty. Thereafter, while the weight of the rat increases from 200 to 400 gm, food intake rises only from 18 to 20 gm per day and intestinal weight from 8.4 to 9.7 gm. During the same period, liver weight goes up from 9.4 to 15.6 gm. This large increase in liver weight is proportional to the increment in general body metabolism; that is, it is proportional to the increase in surface area. It can be concluded that the size of the digestive tract is particularly influenced by level of food intake. Similarly, hyperphagia induced by interference with appetite regulation in the brain causes a considerable increase in the weights of different portions of the alimentary tract (Auffray, 1968).

Finally, it should be pointed out that the shedding of intestinal epithelium is subject to change from disease, and in consequence the amount of protein involved in the process can be altered. Croft *et al.* (1968a) have devised a method for washing out a segment of the small intestine of human subjects and measuring the number of shed cells present from DNA estimations performed on the washings. In celiac disease loss of mucosal cells was found by Croft *et al.* (1968b) to be considerably increased, and they concluded that excessive loss of endogenous compounds coupled with poor reabsorption through the abnormal mucosa in such cases may be a factor in the malnutrition of these patients. To this must be added the leakage of plasma proteins common to intestinal diseases.

VII. Conclusion

The gastrointestinal tract plays a significant role in the protein metabolism of the whole body. The magnitude of this role is shown in Table V where the daily exchange of protein within the tract is compared with the total amount of protein synthesized per day in the body. Nevertheless, much more precise quantitative detail still needs to be provided in order to assess how each aspect of alimentary function affects the regulation of protein metabolism.

First, gastric emptying plays an important part in determining the arrival of amino acids in the bloodstream over a period after each meal, but there is insufficient information about the rate of discharge of individual constituents of the meal. Second, the magnitude of the endogenous secretion of proteins is still a matter of opinion. We do not know how far the continuing digestion of this protein source maintains the supply of available amino acids throughout the periods of fasting between meals. Third, it is not possible as yet to quantitate the relative importance of the intestinal flora and of the enzymes of the host in making available the essential amino acids present in endogenous protein. That this nitrogenous material does become reabsorbed is evident from the fact that only some 10 gm of protein are excreted daily in the feces of man (Table V). Four, the digestive tract provides a pool of labile protein, but we do not know the metabolic importance of deposition of amino acids from the diet in these temporary locations within the alimentary tract. Five, metabolic transformations occur within the lumen due to bacterial action and also in the wall, and these affect the available supply of amino acids. Finally, we have the biologically interesting response mechanisms within the cells of the mucosa and accessory glands. The mucosa is one of the areas of the body where cell division occurs continuously throughout life. This process is subject to regulation through changes in diet and in response to other stimuli. The synthesis and secretion of enzymes by the pancreas is also regulated by requirements for digestive purposes, and again provides a challenging area for future research into control mechanisms operating at the subcellular level. In conclusion, it can be stated that the digestive tract still provides many areas for quantitative evaluation and research before it can be said that we really understand the role of the gastrointestinal tract in the regulation of protein metabolism.

References

Abdeljlil, A. B., and Desnuelle, P. (1964). *Biochim. Biophys. Acta* **81**, 136.

Abrams, G. D., and Bishop, J. E. (1967). *Proc. Soc. Exptl. Biol. Med.* **126**, 301.

Abrams, G. D., Bauer, H., and Sprinz, H. (1963). *Lab. Invest.* **12**, 355.

Adkins, J. S., Wertz, J. M., Boffman, R. H., and Hove, E. L. (1967). *Proc. Soc. Exptl. Biol. Med.* **126**, 500.

Almy, T. P., and Sleisenger, M. H. (1967). *Gastroenterology* **52**, 917.

Amstutz, H. E. (1965). *J. Am. Vet. Med. Assoc.* **147**, 1360.

Arnal, M., Fauconneau, G., and Pech, R. (1968). *Compt. Rend. Acad. Sci.* **267**, 1016.

Auffray, P. (1968). Private communication.

Auffray, P., Martinet, J., and Rérat, A. (1967). *Ann. Biol. Animale Biochim. Biophys.* **7**, 261.

Barnes, R. H., and Kwong, E. (1965). *J. Nutr.* **86**, 245.

Benson, J. A., and Rampone, A. J. (1966). *Ann. Rev. Physiol.* **28**, 201.

Béranger, C. (1968). Private communication.

Bird, F. H. (1968). *Federation Proc.* **27**, 1194.

Bolton, W. (1964). *In* "Role of the Gastrointestinal Tract in Protein Metabolism" (H. N. Munro, ed.), p. 117. Blackwell, Oxford.

Borchers, R. (1965). *J. Animal Sci.* **24,** 1033.
Borgström, B., Dahlquist, A., Lundh, G., and Sjövall, J. (1957). *J. Clin. Invest.* **36,** 1521.
Bowie, M. D., Brinkman, G. L., and Hansen, J. D. L. (1965). *J. Pediat.* **66,** 1083.
Briend, G., and Pion, R. (1967). Private communication.
Brown, H. O., Levine, M. L., and Lipkin, M. (1963). *Am. J. Physiol.* **205,** 868.
Bryant, M. P., and Burkey, L. A. (1953). *J. Dairy Sci.* **36,** 205.
Cairnie, A. B., and Bentley, R. E. (1967). *Exptl. Cell Res.* **46,** 428.
Cairnie, A. B., Lamerton, L. F., and Steel, G. G. (1965). *Exptl. Cell Res.* **39,** 528.
Campbell, R. M., and Fell, B. F. (1964). *J. Physiol. (London)* **171,** 90.
Caro, L. G., and Palade, G. (1964). *J. Cell Biol.* **20,** 473.
Carrol, R. W., Hensley, G. W., Sittler, C. L., Wilcox, E. L., and Graham, W. R. (1953). *Arch. Biochem. Biophys.* **45,** 260.
Coates, M. E., Dickinson, C. D., Harrison, G. F., Kon, S. K., Cummins, S. H., and Cuthbertson, W. F. J. (1951). *Nature* **168,** 332.
Combe, E., and Gordon, H. A. (1968). *Congr. Gnotobiotic Assoc. Buffalo.* Plenum Press, New York.
Combe, E., and Pion, R. (1966). *Ann. Biol. Animale Biochim. Biophys.* **6,** 255.
Combe, E., and Sacquet, E. (1966). *Compt. Rend. Acad. Sci.* **262,** Ser. D, 685.
Cook, G. C., and Lee, F. D. (1966). *Lancet* **II,** 1263.
Craft, I. L., Geddes, D., and Matthews, D. W. (1968). *J. Physiol. (London)* **196,** 31P.
Crane, C. W., and Neuberger, A. (1960). *Biochem. J.* **74,** 313.
Creamer, B. (1967). *Brit. Med. Bull.* **23,** 226.
Crispell, K. R., Parson, W., and Hollifield, G. W. (1956). *J. Clin. Invest.* **35,** 164.
Croft, D. N., Loehry, C. A., Taylor, J. F. N., and Cole, J. (1968a). *Lancet* **II,** 70.
Croft, D. N., Loehry, C. A., and Creamer, B. (1968b). *Lancet* **II,** 68.
Crompton, D. W. T., and Nesheim, M. C. (1969). *J. Nutr.* **99,** 43.
Crosby, W. H. (1961). *Am. J. Digest. Diseases* [N.S.] **6,** 492.
Dawson, R., and Holdsworth, E. S. (1962). *Brit. J. Nutr.* **16,** 13.
Dawson, R., and Porter, J. W. G. (1962). *Brit. J. Nutr.* **16,** 27.
Delhumeau, G., Pratt, V., and Gitler, C. (1962). *J. Nutr.* **77,** 52.
Delluva, A. M., Markley, K., and Davies, R. E. (1968). *Biochim. Biophys. Acta* **151,** 646.
Delort-Laval, J., Charlet-Lery, G., and Zelter, S. Z. (1963). *Ann. Biol. Animale Biochim. Biophys.* **3,** 369.
Doft, F. S. (1965). "The Germfree Animal in Research," Abstr. NATO, Advanced Study, p. 40. NATO Advanced Study Inst.
Dubos, R. J. (1966). *Gastroenterology* **51,** 868.
Dubos, R. J., and Schaedler, R. W. (1960). *J. Exptl. Med.* **111,** 407.
Durand, G., Fauconneau, G., and Pénot, E. (1965). *Ann. Biol. Animale Biochim. Biophys.* **5,** 163.
Durand, G., Fauconneau, G., and Pénot, E. (1966). *Ann. Biol. Animale Biochim. Biophys.* **6,** 389.
Durand, G., Fauconneau, G., and Pénot, E. (1967). *Ann. Biol. Animale Biochim. Biophys.* **7,** 145.
El-Nageh, M. M. (1967). *Ann. Med. Vet.* **6,** 370, 380, and 384.
Evrard, E., Hoet, P. P., Eyssen, H., Charlier, H., and Sacquet, E. (1964). *Brit. J. Exptl. Pathol.* **45,** 409.
Eyssen, H., and de Sommer, P. (1965). *Ernaehrungsforschung* **10,** 264.
Farrar, G. E., and Bower, R. J. (1967). *Ann. Rev. Physiol.* **29,** 141.

Fauconneau, G. (1966). *9th Intern. Congr. Animal Prod. Edinburgh,* p. 47. Oliver & Boyd, Edinburgh and London.

Fell, B. F., and Campbell, R. M. (1964). *In* "The Role of Gastrointestinal Tract in Protein Metabolism" (H. N. Munro, ed.), p. 199. Blackwell, Oxford.

Finch, L. R., and Hird, F. J. R. (1960). *Biochim. Biophys. Acta* **43,** 268.

Fisher, R. B. (1967a). *Brit. Med. Bull.* **23,** 241.

Fisher, R. B. (1967b). *Proc. Nutr. Soc.* **26,** 23.

François, A. C. (1962). *World Rev. Nutr. Diet.* **3,** 1.

François, A. C., and Michel, M. C. (1960). *Cahiers Colloq. Hôp.* **12,** 949.

François, A. C., and Michel, M. C. (1968). *Nutr. Dieta* **10,** 35.

Freeman, T. (1964). *In* "The Role of Gastrointestinal Tract in Protein Metabolism" (H. N. Munro, ed.), p. 125. Blackwell, Oxford.

Gätzsche, L. (1965). *Ernaehrungsforschung* **10,** 641.

Giordano, G., de Pascale, C., Balestrieri, C., Cittadini, D., and Crescenzi, A. (1966). *J. Clin. Invest.* **45,** 1013.

Goldberg, A., and Guggenheim, K. (1962). *Biochem. J.* **83,** 129.

Gordon, H. A., and Bruckner-Kardoss, E. (1961a). *Acta Anat.* **44,** 210.

Gordon, H. A., and Bruckner-Kardoss, E. (1961b). *Am. J. Physiol.* **201,** 175.

Gordon, H. A., and Wostmann, B. S. (1960). *Anat. Record* **137,** 65.

Grossman, M. I. (1967). *Gastroenterology* **53,** 821.

Grütte, F. K. (1965). *Ernaehrungsforschung* **10,** 633.

Guggenheim, K., and Goldberg, A. (1964). *In* "The Role of Gastrointestinal Tract in Protein Metabolism" (H. N. Munro, ed.), p. 63. Blackwell, Oxford.

Gupta, J. D., Dakroury, A. M., and Harper, A. E. (1958). *J. Nutr.* **64,** 447.

György, P. (1957). *Extr. Ann. Nutr. Alim.* **11,** 189.

Haenel, H. (1965). *Ernaehrungsforschung* **10,** 289.

Hanking, B. M., and Roberts, S. (1965). *Biochim. Biophys. Acta* **104,** 427.

Harmon, B. G., Becker, D. E., Jensen, A. H., and Baker, D. H. (1968). *J. Nutr.* **96,** 391.

Hartmann, F. (1959). *In* "Protides in Biological Fluids" (H. Peeters, ed.), p. 204. Elsevier, Amsterdam.

Haverback, B. J., and Dyce, B. J. (1967). *Gastroenterology* **53,** 326.

Henry, Y., and Rérat, A. (1962). *Ann. Biol. Animale Biochim. Biophys.* **2,** 267.

Herskovic, T. (1969). *Am. J. Clin. Nutr.* **22,** 300.

Hill, R. B., Prosper, J., Hirschfield, J. S., and Kern, F. (1968). *Exptl. Mol. Pathol.* **8,** 66.

Hindmarsh, J. T., Kilby, D., Ross, B., and Wiseman, G. (1967). *J. Physiol. (London)* **188,** 207.

Hirschfield, J. S., and Kern, F. (1968). *Clin. Res.* **14,** 298.

Hogue, D. E., Warner, R. G., Loosli, J. K., and Grippin, C. A. (1956). *J. Animal Sci.* **15,** 788.

Holtzman, J. L., and Visek, W. J. (1965). *J. Nutr.* **87,** 101.

Hooper, C. S., and Blair, M. (1958). *Exptl. Cell Res.* **14,** 175.

Hopper, A. F., Wannemacher, R. W., and McGovern, P. A. (1968). *Proc. Soc. Exptl. Biol. Med.* **128,** 695.

Hoskins, L. C., and Zamcheck, N. (1968). *Gastroenterology* **54,** 210.

Houpt, T. R. (1963). *Am. J. Physiol.* **205,** 1144.

Howard, J. M., James, C. L., and Evans, S. S. (1951). *Surg. Forum.* **2,** 577.

Hudson, J. A., and Luckey, T. D. (1964). *Proc. Soc. Exptl. Biol. Med.* **116,** 628.

Hungate, R. E. (1950). *Bacteriol. Rev.* **14,** 1.

Imondi, A. K., Balis, M. E., and Lipkin, M. (1968). *Exptl. Mol. Pathol.* **9,** 339.

Jacobs, F. A., and Lang, A. H. (1965). *Proc. Soc. Exptl. Biol. Med.* **118,** 772.
Jacobs, F. A., Flaa, R. C., and Belk, W. F. (1960). *Federation Proc.* **19,** 183.
Jones, E. A., Craigie, A., Tavill, A. S., Franglen, G., and Rosenoer, V. M. (1968). *Gut.* **9,** 466.
Ju, J. S., and Nasset, E. S. (1959). *J. Nutr.* **68,** 633.
Junqueira, L. C. U., Rothschild, H. A., and Fajer, A. (1957). *Exptl. Cell. Res.* **12,** 338.
Kassenaar, J., deGraeff, J., and Kouwenhoven, A. T. (1960). *Metab. Clin. Exptl.* **9,** 831.
Keller, P. J., Cohen, E., and Neurath, H. (1961). *J. Biol. Chem.* **236,** 1404.
Kenworthy, R., and Crabb, W. E. (1963). *J. Comp. Pathol.* **73,** 215.
Kershaw, T. G., Neame, K. D., and Wiseman, G. (1960). *J. Physiol.* (*London*) **152,** 182.
Khayambashi, H., and Lyman, R. L. (1966). *J. Nutr.* **89,** 455.
Kirsch, R. E., Saunders, S. J., and Brock, J. F. (1968). *Am. J. Clin. Nutr.* **21,** 1302.
Larson, N. L., and Hill, E. G. (1955). *J. Animal Sci.* **14,** 674.
Leblond, C. P., and Walker, C. E. (1956). *Physiol. Rev.* **36,** 225.
Lepkovsky, S., Wagner, M., Furuta, F., Ozone, K., and Koike, T. (1964). *Poultry Sci.* **43,** 722.
Lepkovsky, S., Furuta, F., Koike, T., Hasegawa, N., Dimick, M. K., Krause, K., and Barnes, F. J. (1965). *Brit. J. Nutr.* **19,** 41.
Lepkovsky, S., Furuta, F., Ozone, K., Koike, T., and Wagner, M. (1966). *Brit. J. Nutr.* **20,** 257.
Lesher, S., Walburg, H. E., and Sacker, G. A. (1964). *Nature* **202,** 884.
Levenson, S. M., and Tennant, B. (1963). *Federation Proc.* **22,** 109.
Levenson, S. M., Crowley, L. V., Morowitz, R. E., and Malm, O. J. (1959). *J. Biol. Chem.* **234,** 2061.
Levenson, S. M., Gruber, C., and Kan, D. (1969). *J. Nutr.* **98,** 99.
Lewis, D. (1955). *Brit. J. Nutr.* **9,** 519.
Lewis, T. R., and Emery, R. S. (1962). *J. Dairy Sci.* **45,** 765.
Lipkin, M. (1965). *Federation Proc.* **24,** 10.
Lipkin, M., and Quastler, H. (1962). *J. Clin. Invest.* **41,** 141.
Loesche, W. J. (1968). *Proc. Soc. Exptl. Biol. Med.* **129,** 380.
Loran, M. R., and Althausan, T. L. (1960). *J. Biophys. Biochem. Cytol.* **7,** 667.
Luckey, T. D. (1963). "Germfree Life and Gnotobiology." Academic Press, New York.
Lyman, R. L., Wilcox, S. S., and Monsen, E. R. (1962). *Science* **136,** 155.
Ma'ayani, S., and Kulka, R. G. (1968). *J. Nutr.* **96,** 363.
McDermott, W. V. (1966). *Gastroenterology* **51,** 721.
Marcus, C. S., and Lengemann, F. W. (1962). *J. Nutr.* **76,** 179.
Marty, J., and Carles, J. (1968). *Compt. Rend. Acad. Sci.* **267,** 638.
Marty, J., and Raynaud, P. (1965). *Arch. Sci. Physiol.* **19,** 321.
Matthews, D. M., and Laster, L. (1965a). *Am. J. Physiol.* **208,** 593.
Matthews, D. M., and Laster, L. (1965b). *Gut.* **6,** 411.
Metchnikoff, E. (1903). *Bull. Inst. Pasteur* **7,** 265.
Michel, M. C. (1961a). *Ann. Biol. Animale Biochim. Biophys.* **1,** 16.
Michel. M. C. (1961b). *Ann. Biol. Animale Biochim. Biophys.* **1,** 213.
Michel, M. C. (1966). *Ann. Biol. Animale Biochim. Biophys.* **6,** 33.
Michel, M. C. (1968). *Ann. Biol. Animale Biochim. Biophys.* (in press).
Michel, M. C., and François, A. C. (1955). *Compt. Rend. Acad. Sci.* **240,** 808.
Michel, M. C., and Mathieu, C. M. (1967). Personal communication.

Michel, M. C., and Sacquet, E. (1966). *Ernaehrungsforschung* **11,** 68.
Michel, M. C., Jouandet, C., Salmon-Legagneur, E., Aumaitre, A., and François, A. C. (1964). *Ann. Zootech.* **13,** 341.
Miller, W. S. (1966). *Proc. Nutr. Soc.* (*Engl. Scot.*) **26,** x.
Miller, W. S., and Coates, M. E. (1966). *Proc. Nutr. Soc.* (*Engl. Scot.*) **25,** 1.
Mitchell, H. H., and Edman, M. (1962). *Am. J. Clin. Nutr.* **10,** 163.
Morrison, A. B., McLaughlan, J. M., Noel, F. J., and Campbell, J. A. (1961). *Can. J. Biochem. Physiol.* **39,** 1681.
Munck, B. G. (1966a). *Biochim. Biophys. Acta* **120,** 97.
Munck, B. G. (1966b). *Biochim. Biophys. Acta* **120,** 282.
Munck, B. G. (1966c). *Biochim. Biophys. Acta* **126,** 299.
Munck, B. G. (1968). *Biochim. Biophys. Acta* **156,** 192.
Munro, H. N., ed. (1964). "The Role of the Gastrointestinal Tract in Protein Metabolism." Blackwell, Oxford.
Munro, H. N., and Goldberg, D. M. (1964). *In* "The Role of Gastrointestinal Tract in Protein Metabolism" (H. N. Munro, ed.), p. 189. Blackwell, Oxford.
Munro, H. N., Black, J. G., and Thomson, W. S. T. (1959). *Brit. J. Nutr.* **13,** 475.
Nasset, E. S. (1965). *Federation Proc.* **24,** 953.
Nasset, E. S., and Ju, J. S. (1961). *J. Nutr.* **74,** 461.
Neame, K. D., and Wiseman, G. (1957). *J. Physiol.* (*London*) **135,** 442.
Neame, K. D., and Wiseman, G. (1958). *J. Physiol.* (*London*) **140,** 148.
Neame, K. D., and Wiseman, G. (1959). *J. Physiol.* (*London*) **146,** 10P.
Nestel, P. J. (1967). *In* "Newer Methods of Nutritional Biochemistry" (A. A. Albanese, ed.), Vol. 3, p. 243. Academic Press, New York.
Newey, H., and Smyth, D. H. (1964). *Nature* **202,** 400.
Nitsan, Z. (1965). *Poultry Sci.* **44,** 1036.
Ochoa-Solano, A., and Gitler, C. (1968). *J. Nutr.* **94,** 249.
Olney, J. W., and Sharpe, L. G. (1969). *Science* **166,** 386.
Packett, L. V., and Groves, T. D. D. (1965). *J. Animal Sci.* **24,** 341.
Payne, L. C., and Marsh, C. L. (1962). *J. Nutr.* **76,** 151.
Payne, W. L., Combs, G. F., Kifer, R. R., and Snyder, D. G. (1968). *Federation Proc.* **27,** 1199.
Pelc, S. R. (1968). *Nature* **219,** 162.
Peraino, C., and Harper, A. E. (1962). *Arch. Biochem. Biophys.* **97,** 442.
Peraino, C., and Harper, A. E. (1963). *J. Nutr.* **80,** 270.
Peraino, C., Rogers, Q. R., Yoshida, M., Chen, M. L., and Harper, A. E. (1959). *Can. J. Biochem. Physiol.* **37,** 1475.
Pion, R., and Fauconneau, G. (1968). *In* "Isotope Studies on the Nitrogen Chain." Intern. Atomic Energy Agency, Vienna (in press).
Pion, R., Fauconneau, G., and Rérat, A. (1964). *Ann. Biol. Animale Biochim. Biophys.* **4,** 383.
Pion, R., Fauconneau, G., and Rérat, A. (1966). *Cahier* **6,** 328. Private collection Soc. Org. Biol. Chem., Paris.
Platt, B. S., Heard, C. R. C., and Stewart, R. J. C. (1964). *In* "The Role of Gastrointestinal Tract in Protein Metabolism" (H. N. Munro, ed.), p. 227. Blackwell, Oxford.
Preshaw, R. M. (1967). *Federation Proc.* **25,** 1454.
Prockop, D. J., and Sjoerdsma, A. (1961). *J. Clin. Invest.* **40,** 843.
Raibaud, P., Dickinson, A. B., Sacquet, E., Charlier, H., and Mocquot, G. (1966a). *Ann. Inst. Pasteur* **110,** 861.

Raibaud, P., Dickinson, A. B., Sacquet, E., Charlier, H., and Mocquot, G. (1966b). *Ann. Inst. Pasteur* **110,** 193.

Randall, H. G., and Evered, D. F. (1964). *Biochim. Biophys. Acta* **93,** 98.

Reboud, J. P., Marchis-Mouren, G., Pasero, L., Cozzone, A., and Desnuelle, P. (1966). *Biochim. Biophys. Acta* **117,** 351.

Rechcigl, M., Kwong, E., Barnes, R. H., and Williams, H. H. (1959). *Proc. Soc. Exptl. Biol. Med.* **101,** 342.

Reddy, B. S., Pleasants, J. R., and Wostmann, B. S. (1969). *J. Nutr.* **97,** 327.

Regoeczi, E., Irons, L., Koj, A., and McFarlane, A. S. (1965). *Biochem. J.* **95,** 521.

Rérat, A., and Lougnon, J. (1963). *Ann. Biol. Animale Biochim. Biophys.* **3,** 21.

Rogers, Q. R., Chen, M. L., Peraino, C., and Harper, A. E. (1960). *J. Nutr.* **72,** 331.

Roheim, P. S., Gidez, L. L., and Eder, H. A. (1966). *J. Clin. Invest.* **45,** 297.

Rose, W. C., and Dekker, E. E. (1956). *J. Biol. Chem.* **223,** 107.

Rose, W. C., Smith, L. C., Womack, M., and Shane, M. (1949). *J. Biol. Chem.* **181,** 307.

Russell, I. S., Fleck, A., and Burnett, W. (1964). *Clin. Chim. Acta* **10,** 210.

Sacquet, E., Charlier, H., Raibaud, P., Dickinson, A. B., Evrard, E., and Eyssen, H. (1966). *Compt. Rend. Acad. Sci.* **262,** 786.

Salmon-Legagneur, E. (1965). *Ann. Zootech.* **14,** 137.

Salmon-Legagneur, E. (1968). Private communication.

San Pietro, A., and Rittenberg, D. (1953). *J. Biol. Chem.* **201,** 457.

Scott, P. J., and Winterbourn, C. C. (1965). *Nature* **208,** 494.

Slyter, L. L., Oltjen, R. R., Kern, D. L., and Weaver, J. M. (1968). *J. Nutr.* **94,** 185.

Snook, J. T. (1965a). *J. Nutr.* **87,** 297.

Snook, J. T. (1965b). *Federation Proc.* **24,** 941.

Snook, J. T. (1968). *J. Nutr.* **94,** 351.

Snook, J. T. (1969). *J. Nutr.* **97,** 286.

Snook, J. T., and Meyer, J. H. (1964a). *J. Nutr.* **82,** 409.

Snook, J. T., and Meyer, J. H. (1964b). *J. Nutr.* **83,** 94.

Spencer, R. P., and Zamcheck, N. (1962). *Gastroenterology* **42,** 732.

Steiner, M., Bourges, H. R., Friedman, L. S., and Gray, S. J. (1968). *Am. J. Physiol.* **215,** 75.

Steiner, M., Bourges, H. R., Farrish, G. C. M., Ross, J. R., and Gray, S. J. (1969a). *Am. J. Med. Sci.* **257,** 234.

Steiner, M., Farrish, G. C. M., and Gray, S. J. (1969b). *Am. J. Clin. Nutr.* **22,** 871.

Stobo, I. J. F., Roy, J. H. B., and Gaston, H. J. (1966). *Brit. J. Nutr.* **20,** 189.

Stoewsand, G. S., Dymsza, H., Ament, D., and Trexler, P. C. (1968). *Life Sci.* **7,** 689.

Suda, M., and Shimomura, A. (1964). *Osaka Univ. Med. J.* **16,** 11.

Summerskill W. H. J., Faythorsell, B. S., Feinberg, J. H., and Aldrete, J. S. (1967). *Gastroenterology* **54,** 20.

Szmelcman, S., and Guggenheim, K. (1966). *Biochem. J.* **100,** 7.

Tannous, R. I., Rogers, Q. R., and Harper, A. E. (1966). *Arch. Biochem. Biophys.* **113,** 356.

Tasaki, I., and Takahashi, N. (1966). *J. Nutr.* **89,** 359.

Thaysen, E. H., and Thaysen, J. H. (1949). *Acta Pathol. Microbiol. Scand.* **26,** 370.

Tria, E., and Barnabei, O. (1963). *Nature* **197,** 598.

Twombly, J., and Meyer, J. H. (1961). *J. Nutr.* **74,** 453.

Van Der Heyde, H., and Henderickx, H. (1964). *Zentr. Bakteriol. Parasitenk. Abt. I. Orig.* **195,** 215.

Vandermeers, A., Robberecht, P., Rathe, J., and Christophe, J. (1966). *Bull. Soc. Chim. Biol.* **48,** 133.

Vandermeers, A., Vandermeers-Piret, M. C., and Christophe, J. (1967). *Bull. Soc. Chim. Biol.* **49,** 759.

Vodovar, N. (1967). Thesis, Faculty of Sciences of Paris.

Walser, M., and Bodenlos, L. J. (1959). *J. Clin. Invest.* **38,** 1617.

Wang, C. C., and Grossmann, M. I. (1951). *Am. J. Physiol.* **164,** 527.

Warren, K. S., and Newton, W. L. (1959). *Am. J. Physiol.* **197,** 717.

Wetterfors, J. (1965). *Acta Med. Scand.* **177,** Suppl. 430, p. 72.

Williams, V. J. (1969). *Comp. Biochem. Physiol.* **29,** 865.

Windmuller, H. G., and Levy, R. I. (1968). *J. Biol. Chem.* **243,** 4878.

Wiseman, G. (1964). "Absorption from the Intestine." Academic Press, New York.

Wostmann, B. S., and Bruckner-Kardoss, E. (1959). *Am. J. Physiol.* **197,** 1345.

Wostmann, B. S., and Bruckner-Kardoss, E. (1966). *Proc. Soc. Exptl. Biol. Med.* **121,** 1111.

Wunner, W. H. (1967). *Proc. Nutr. Soc.* (*Engl. Scot.*) **26,** 153.

Wunner, W. H., Bell, J., and Munro, H. N. (1966). *Biochem. J.* **101,** 417.

Zimmermann-Nielsen, C., and Schønheyder, F. (1962). *Biochim. Biophys. Acta* **63,** 201.

CHAPTER 38

The Role of the Liver in Regulation of Amino Acid and Protein Metabolism

DAVID H. ELWYN

Department of Surgery,
Mt. Sinai School of Medicine,
New York, New York

I. Introduction

Each cell or tissue is competent to synthesize and degrade its own constituent proteins; the extent of these reactions is regulated largely by processes taking place within the cell. These intracellular or intraorgan processes are in turn influenced by external factors of hormonal, neural, or metabolic nature. Thus, intercellular or interorgan regulation is superimposed upon the more general and more primitive intracellular regulatory processes.

The metabolic aspects of interorgan regulation are of major concern in assessing the role of the liver in regulation of protein metabolism. These center around the regulation of the supply of amino acids to tissues: the nature of this supply, whether as protein or free amino acids; the extent to which tissue needs vary or remain constant; and the influence of the postprandial amino acid influx on tissue requirements.

There are many considerations which point to the liver as potentially the key organ in regulation of amino acid supply. The anatomical position of the liver gives it access to dietary amino acids ahead of other tissues. Most amino acid catabolism and urea synthesis takes place in the liver. Liver proteins, together with the plasma proteins synthesized by the liver, comprise a major part of labile protein reserves. Deposition or loss of protein is faster in the liver than in any other organ except the pancreas. Of great significance is the fact that catabolic hormones which cause protein depletion in most other tissues, cause protein deposition in the liver. Conversely, under special circumstances, anabolic hormones cause protein deposition in some tissues, but protein depletion in the liver (Chapter 10).

This ability of the liver to supply or utilize amino acids rapidly, in a fashion complementary to other tissues, is subject to at least two quite different explanations. Perhaps the simplest is that the primary effects of hormone action are on protein breakdown or synthesis in nonhepatic tissues. This produces a rise or fall in amino acid concentrations in blood. Changing circulatory levels of amino acids in turn act on the liver to cause synthesis or breakdown of protein. An alternate explanation is that hormone action on the liver is in opposite direction to that of nonhepatic tissues. Protein synthesis or breakdown in the liver would then be the cause rather than the effect of increased hepatic uptake or output of amino acids. (This subject is discussed more fully in Chapter 34.)

The choice between these and other possible explanations requires observation of actual movements of amino acids between organs. This can best be accomplished by measuring the net metabolism of amino acids in organs through observation of venous-arterial (V-A) concentration changes and measurement of blood flow. Net organ output or uptake of each compound can then be calculated as the product of V-A concentration difference multiplied by blood flow. Although these methods are not new conceptually, their practical application to the study of amino acid metabolism has been made possible only recently, through development of techniques for catheterization and measurement of blood flow and modern methods of amino acid analysis. Several investigators have measured portal-arterial concentration differences in rats (Peraino and Harper, 1963) and dogs (Ganapathy and Nassett, 1962; Nassett *et al.*, 1963) and hepatic venous-arterial differences in man (Onen *et al.*, 1956; Carlsten *et al.*, 1967). As yet, however, there have been very few studies, particularly of the splanchnic region, in which both venous-arterial concentration differences of amino acids and blood flow have been measured.

To a large extent, therefore, the following description of net metabolism of the liver and the gut in the intact animal rests on a single study of dogs on a horsemeat diet (Elwyn, 1966; Elwyn *et al.*, 1968).

II. Diurnal Patterns of Amino Acid Movements

Unanesthetized dogs were studied 4 or 5 days after implantation of catheters in the portal vein, hepatic vein, and the splenic artery (Shoemaker *et al.*, 1959). Animals weighed 16.0, 20.4, and 28.3 kg; they were fed, respectively, 400, 454, and 565 gm ground horsemeat at noon. They were kept in a Pavlov stand or restraining cage. Blood was sampled continuously over 24-hour periods. Total hepatic plasma flow was measured by bromosulfophthalein clearance (Shoemaker, 1960). Concentrations of urea, ammonia, amino acids, and related compounds in whole blood, plasma, or erythrocytes were determined by ion-exchange chromatography (Elwyn, 1966).

The nonhepatic splanchnic (NHS) output for each compound was obtained as the product of portal vein-arterial (P-A) concentration differences and portal blood flow. The latter was taken to be 80% of total hepatic blood flow (Elwyn *et al.*, 1968). The organs comprising the NHS region include most of the stomach, the small intestine, some of the large intestine, and the pancreas (the spleen having been removed). As this approximates the total gut, the term gut will be used interchangeably with NHS region.

The liver output of each compound was obtained by multiplying total hepatic blood flow by the difference between hepatic venous concentration and a weighted average of portal and arterial concentrations. The output of the total splanchnic region, which includes both gut and liver, was obtained as the product of hepatic vein-arterial (H-A) concentration differences and total hepatic blood flow. By definition, total splanchnic output is the algebraic sum of gut and liver outputs. Outputs are positive if there is a net transfer from tissue to blood, negative if transfer is from blood to tissue. Uptake is defined as negative output.

The interrelations between blood concentrations, organ outputs of amino acids, and blood flow are illustrated graphically for leucine in Fig. 1. The concentration changes in blood in all three animals are illustrated for phenylalanine in Fig. 2. The patterns for portal and arterial blood are qualitatively similar to those of previous studies (Dent and Schilling, 1949; Christensen, 1949; Parshin and Rubel, 1951; Denton and Elvehjem, 1954; Wheeler and Morgan, 1958; Frame, 1958; Levenson *et al.*, 1959; Guggenheim *et al.*, 1960; Ganapathy and Nassett, 1962; Goldberg and Guggenheim, 1962; Neame and Wiseman, 1957; Peraino

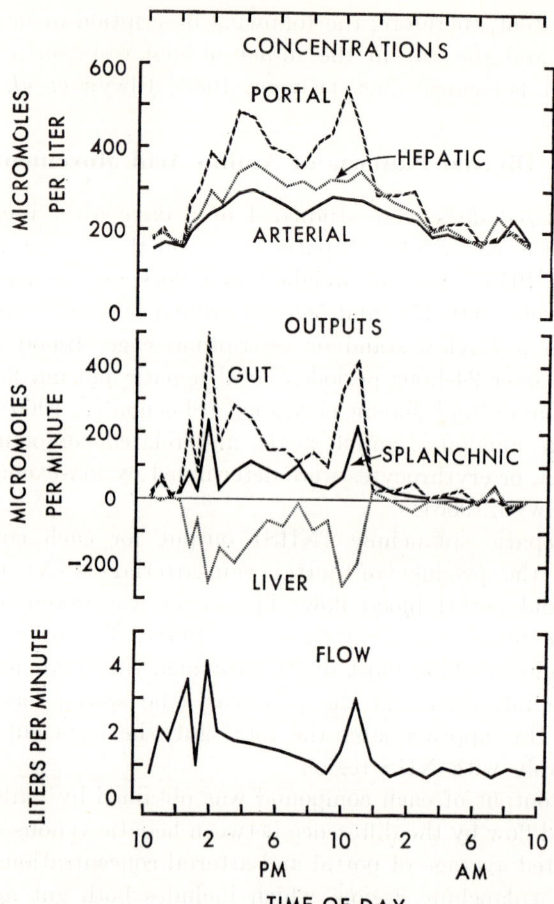

FIG. 1. Leucine concentrations in whole blood from the splenic artery and portal and hepatic veins; rates of output into blood of leucine from gut, liver, and total splanchnic region; and hepatic blood flow in dog 100. Flow and outputs have been normalized to 1 kg of liver or 35 kg of body weight.

and Harper, 1963; Pion *et al.*, 1964). Within 30 minutes after the single meal, given daily at noon, there is an abrupt rise in portal concentrations, which remain high for 8–11 hours and then return to preprandial levels, thus dividing the day into an absorptive and a nonabsorptive period. Arterial and hepatic venous concentrations show smaller, less abrupt increases and return sooner than portal values to nonabsorptive levels. The size and duration of these increases show considerable variation from animal to animal. There is also considerable variation between compounds in each animal (Elwyn *et al.*, 1968).

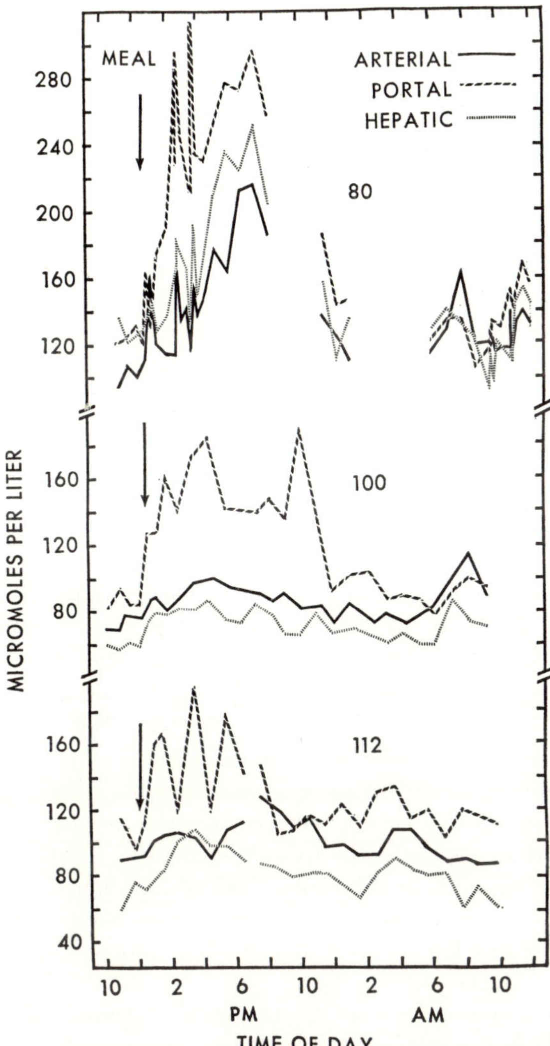

FIG. 2. Phenylalanine concentrations in whole blood from the splenic artery and portal and hepatic veins of 3 dogs sampled over 24-hour periods. The experiment with dog **112** started at **7:01** PM and continued until **6:51** PM the next day. The experiments with the other animals started in the morning and continued until the following morning, except that sampling of dog **80** was interrupted for two periods. From Elwyn *et al.* (1968). (Published with permission of *Am. J. Physiol.*)

A. *Gut Output of Amino Acids*

A detailed analysis of the role of the liver must include some description of the composition, amount, and time course of gut amino acid output. A comprehensive consideration of the role of the gut in protein metabolism is presented in Chapter 37.

1. *Composition as Compared to Ingested Protein*

Gut outputs of amino acids in dogs on horsemeat diets are shown in Fig. 3 as cumulative outputs for the absorptive and nonabsorptive periods. Outputs in the absorptive period are greater than the amount ingested for 2 animals, and somewhat smaller in the third. Nevertheless, for most compounds, including all the essential amino acids, the composition of gut output is very similar to that of the meal. Exceptions are discussed in Sections a and b below.

a. Aspartic and Glutamic Acids, Glutamine, Alanine, and Ammonia. These are the compounds most actively involved in nitrogen exchange reactions in all tissues. Considerable interchanges of nitrogen between them has occurred in gut, so that very little glutamic or aspartic acid and disproportionately high amounts of alanine and ammonia appear in portal blood. This low absorption of glutamic and aspartic acids is in agreement with earlier observations in the dog (Neame and Wiseman, 1957; Dent and Schilling, 1949), the pig (Pion *et al.*, 1964), and the isolated rat intestine (Finch and Hird, 1960).

b. Glutathione. This compound is not present in the diet but is put out in large quantities by the gut. It accounts for more glutamic acid and cysteine than appear as the free amino acids.

Gut outputs in the nonabsorptive period are much smaller than in the absorptive period. Nevertheless, they are positive and similar in composition, indicating that net hydrolysis of protein to free amino acids takes place during the entire day.

The detailed time course of gut output, illlustrated for phenylalanine in Fig. 4, shows considerable short-term (hour to hour) fluctuations, particularly in the absorptive period. Only rarely throughout the entire day do negative gut outputs occur.

2. *Influence of Endogenous Protein on Composition of Gut Amino Acid Output*

Considerable amounts of endogenous protein daily enter the lumen of the gut, where they are hydrolyzed. Nassett and Ju (1961) have observed that ingested amino acids may be diluted as much as sixfold by amino acids derived from endogenous protein. This has led to the

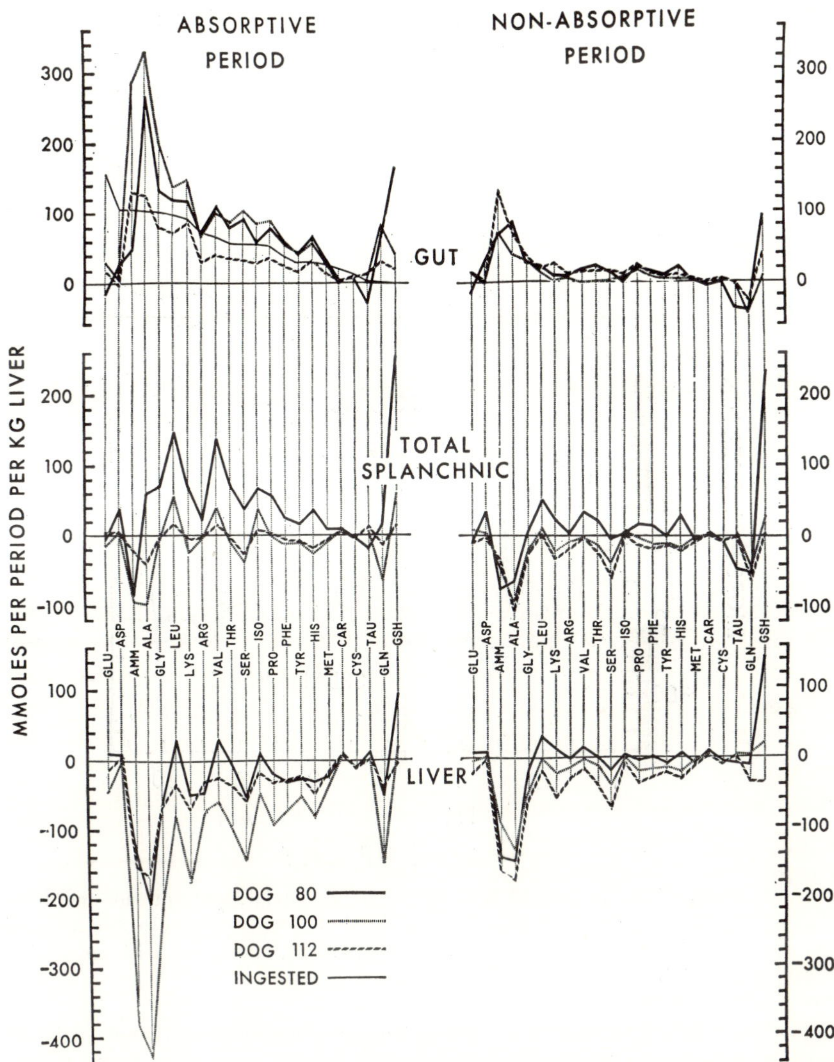

FIG. 3. Outputs of amino acids into whole blood from gut, liver, and total splanchnic region in the absorptive and nonabsorptive periods for 3 dogs. The amount ingested is shown for dog 112. It was 9% higher for dog 80, and 8% lower for dog 100. From Elwyn *et al.* (1968). (Published with permission of *Am. J. Physiol.*)

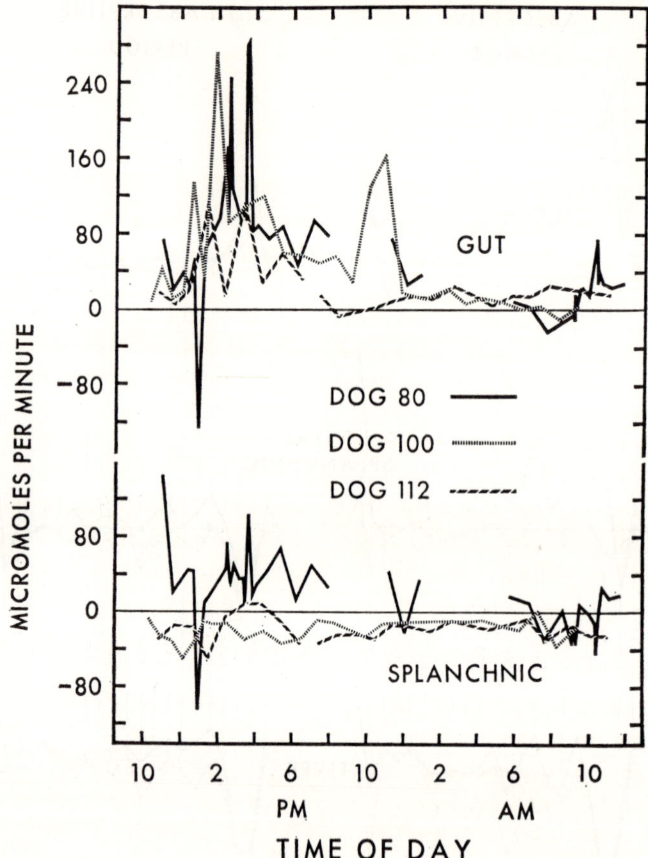

FIG. 4. Rates of output into whole blood of phenylalanine from gut and total splanchnic region over 24-hour periods in 3 dogs. Rates are normalized to 1 kg of liver or 35 kg of body weight. From Elwyn *et al.* (1968). (Published with permission of *Am. J. Physiol.*)

suggestion (Nassett and Ju, 1961; Nassett, 1965) that hydrolysis of endogenous protein in the gut may serve to protect the rest of the organism from the immediate effects of ingestion of imbalanced protein. Two hours after feeding dogs zein, which is deficient in tryptophan, lysine, and glycine, Nassett *et al.* (1963) observed appreciable quantities of these three amino acids in mesenteric vein blood and concluded that they were probably derived from the hydrolysis of endogenous protein. This comparison of the amino acid composition of ingested protein and mesenteric vein blood is misleading, since the quantity of free amino acids present in mesenteric vein blood is equal to the quantity present in arterial blood plus the amount added by the gut. Amino acid concentrations in arterial blood were also measured in these dogs (Nassett *et al.*, 1963), and the data show negative mesenteric vein-arterial differences for tryptophan, lysine, and glycine and positive differences for all other amino acids. This indicates little or no effect of the hydrolysis of endogenous protein on the composition of gut output of amino acids 2 hours after the meal. Denton and Elvehjem (1954) in dogs, and Peraino and Harper (1963) in rats, have also shown that the composition of ingested zein or zein hydrolyzates is immediately reflected in the relative concentrations of lysine in portal vein and systemic blood. As presented in Chapter 34, many experiments have shown that amino acid concentrations in arterial or peripheral venous blood rapidly reflect imbalances in dietary protein. Examples are the experiments of Longenecker and Hause (1959) in which imbalanced protein meals containing wheat gluten or gelatin were fed to dogs. Arterial concentrations of lysine or tryptophan, the most limiting amino acids, declined relative to other amino acids within 1–2 hours after feeding, again indicating that imbalances in ingested protein are quickly reflected in gut output of amino acids. Wunner *et al.* (1966) studied the effects of tryptophan-deficient and complete diets on the degree of aggregation of rat liver ribosomes. Marked effects of tryptophan deficiency could be demonstrated 1–2 hours after feeding.

From these data, and in the absence of contrary evidence, it would appear that the gut has little if any modifying role on the amino acid composition of the essential amino acids of the diet. Rather, gut output of amino acids quickly reflects imbalances in the diet, passing on to the liver and other organs the metabolic problems relating to their assimilation.

3. Quantity of Gut Amino Acid Output

Figure 3 indicates that, in 2 animals, during the absorptive period, the quantity of gut output of total amino acids and of each of the essential amino acids was greater than that ingested (1700 mmoles N per kilogram of liver). In the third animal it was less. The exact amount of gut output is in doubt in these experiments since portal blood flow rates were not independently determined. However, it seems reasonable to assume that portal flow accounts for 60–90% of total hepatic blood flow (Elwyn *et al.*, 1968). For any values of portal flow between these limits, the relative relations shown in Fig. 3 hold true. For the three animals the average 24-hour gut output was somewhat greater than that ingested. This supplies additional evidence, to that discussed in Chapter 2, that the bulk of ingested protein appears in portal blood as free amino acids rather than as peptides or proteins.

These experiments indicate also that a part of 24-hour gut amino acid output, 150–1000 mmoles N, depending on flow assumptions, was derived from hydrolysis of endogenous protein. Estimates of total daily protein secretion into the gut in man are given by Nassett (1965) as 141–354 gm, and by Munro (1966) as 70 gm. A 35-kg dog may be expected to produce one-half that amount, or, using Munro's figure, 35 gm of protein equal to 400 mmoles N.[1] This corresponds in order of magnitude to the amounts of gut output in excess of diet noted above. However, in a normal animal, under steady-state conditions, there should be no net addition to gut amino acid output over 24-hour periods from hydrolysis of proteins synthesized in organs drained by the portal vein. The mass and composition of these organs should remain constant if measured at 24-hour intervals. As a consequence, the amounts of amino acids derived from hydrolysis of endogenous protein from these organs must exactly equal the amounts of amino acids required to resynthesize protein. The extensive synthesis and breakdown of protein in pancreas and gastrointestinal tract constitutes an intraregional cycle with little net effect on the rest of the organism under normal conditions. Under abnormal conditions this will not be true. Extensive protein depletion and repletion of both the small intestine and pancreas can be effected by dietary means (Chapter 10; Ju and Nassett, 1959). This would cause nonhepatic splanchnic output of amino acids to be considerably less or considerably greater than the amount of protein ingested. The animals studied by Elwyn *et al.* (1968) were recovering from an operation 4–5 days previously and were additionally subjected to the stress of sampling and replacement of large amounts of blood. It is probable that net changes in protein content of the nonhepatic splanchnic region partially accounted for the differences observed between the amounts of ingested protein and gut amino acid output.

In addition to those synthesized in the nonhepatic splanchnic region, some of the endogenous proteins hydrolyzed in the gut are derived from organs not drained by the portal vein. These include salivary, bile, and plasma proteins. Since the amino acids required for their synthesis are not reflected in portal blood, hydrolysis of these proteins would at all times make a positive addition to gut amino acid output. In a 35-kg dog, secretion of bile and salivary proteins is of the order of 2 gm, or 23 mmoles, N per day (Altman and Dittmer, 1961; Munro, 1966; Nassett, 1965). This is approximately 5% of total protein secretion into the gut. Hydrolysis of plasma proteins in the gut appears to be quite small under normal conditions, of the order of 5–10% of plasma protein

[1] In converting weight of protein to millimoles of nitrogen, it has been assumed that 1 mmole N corresponds to 87.5 mg protein.

turnover (Freeman, 1964; Waldman *et al.*, 1967), roughly equal to the contribution of bile and salivary proteins.

B. Liver Uptake and Metabolism of Amino Acids

1. Comparison with Gut Output and Total Splanchnic Output

The pattern of liver output of amino acids is almost the mirror image of that from the gut (Fig. 3). Liver outputs, which are usually negative, are approximately equal to, although opposite in sign to, gut outputs. Since total splanchnic output is the sum of liver and gut outputs, the pattern of splanchnic amino acid outputs indicates the net balance between liver and gut. If liver uptake is greater than gut output, splanchnic output will be negative; if gut output is greater than liver uptake, splanchnic output will be positive.

Splanchnic outputs shown in Fig. 3 indicate marked differences between the animals studied. Two of the animals had quite similar patterns. For most amino acids of protein origin, there was net splanchnic uptake in both the absorptive and nonabsorptive periods. Marked exceptions were the branched-chain amino acids leucine, isoleucine, and valine. The rate of splanchnic output as a function of time is illustrated for phenylalanine in Fig. 4. In dogs 100 and 112 splanchnic output is almost always negative. In contrast to gut output of phenylalanine, it shows little hourly fluctuation and no significant difference between the absorptive and nonabsorptive periods. This indicates a close correspondence, on an hourly basis, between liver uptake and gut output, with the former always slightly larger than the latter. In these animals the liver acts as a buffer between the digestive tract and the rest of the organism with respect to most amino acids, preventing almost all effects of the diurnal events associated with ingestion from reaching peripheral tissues.

The time course of leucine outputs in one animal (dog 100) are shown in Fig. 1. In contrast to phenylalanine, splanchnic output of leucine was always positive and much larger in the absorptive than in the nonabsorptive period; it also reflected the hourly fluctuations observed in gut output of leucine. With respect to leucine and the other branched-chain amino acids, the buffering effect of the liver is largely absent.

In a third animal, there were positive liver outputs of glutamic and aspartic acids, leucine, lysine, valine, isoleucine, and histidine for either the nonabsorptive or the absorptive periods, or both. Liver uptake of the other amino acids was considerably less than gut output. As a result splanchnic outputs for most compounds, including all the essential amino acids, were positive in both periods (Fig. 3). The rate of splanchnic output of phenylalanine (Fig. 4) showed marked hourly fluctuation and considerable diurnal variation. The mean rate in the nonabsorptive period, 17 μmoles per minute per kilogram of liver was one-half that in the absorptive period. However, the diurnal variation of splanchnic output was much less than that of gut output. Although less effective than in the other animals, the liver still served to buffer the impact of the meal on peripheral tissues.

2. *Metabolic Processes in Liver Related to Amino Acid Uptake*

The two major metabolic fates of amino acids in liver are catabolism to urea and CO_2 and incorporation into proteins. Both of these have been estimated in these experiments. The pattern of diurnal changes in the rates of these processes should throw light on their possible roles in regulation of amino acid uptake by the liver.

a. Urea Synthesis. Cumulative liver outputs of urea are shown for 3 dogs in Fig. 5. Apart from short-term fluctuations,[2] the diurnal patterns are quite similar in all animals. Maximal rates of hepatic urea synthesis, which were achieved within 1 hour after the meal, were about 160 mmoles

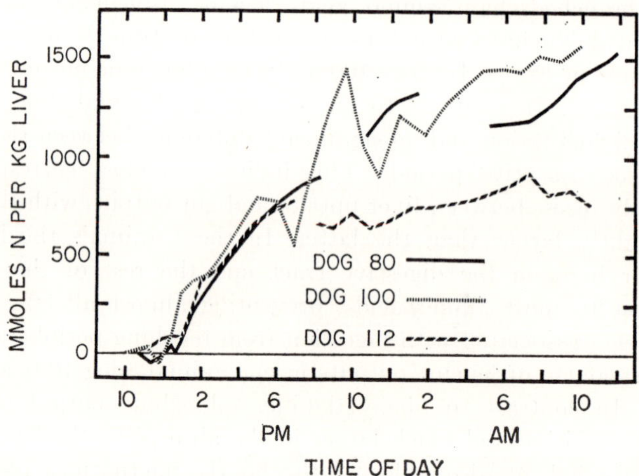

FIG. 5. Cumulative liver outputs of urea over 24-hour periods in 3 dogs.

[2] The considerable fluctuations shown for liver urea output, particularly in dog 100, do not necessarily imply that breakdown of urea occurs in liver tissue. A number of factors may contribute to these fluctuations, of which the most important are errors in determination of urea concentrations in blood. The standard deviation of replicate urea determinations was found to be 4% of the mean value (Elwyn, 1966). At average urea concentrations and blood flow, a 4% concentration error would introduce an error of 30 mmoles N per hour per kilogram of liver in the estimated rate of urea output. Most errors would be smaller than this, but many would be larger. Although these errors would tend to cancel out over the total course of the day, they may account for most of the short-term fluctuations.

TABLE I

MAXIMUM RATES OF UREA PRODUCTION FROM INFUSION OF AMINO ACIDS INTO THE RADIAL VEIN OF DOGS[a]

Amino acids infused	Urea production, increase above control values (mmoles N per kg liver per hour)
L-Methionine	8
L-Glutamic acid	17
L-Aspartic acid	17
L-Alanine	30
L-Lysine	30
L-Leucine	36
L-Tyrosine	53
L-Histidine	98
Glycine	120
L-Asparagine	120
Casein hydrolyzate	270
L-Arginine	285
L-Glutamine	390
L-Glutamine + casein hydrolyzate	375
L-Arginine + casein hydrolyzate	525
L-Arginine + L-glutamine	540

[a] Adapted from Kamin and Handler (1951).

N per hour per kilogram of liver. The rates remained maximal for 4–6 hours and then declined. During the nonabsorptive period, urea synthesis in dog 112 was substantially less than for the other animals and resulted in a much smaller total daily urea output. Nitrogen intake, which was quite similar in the 3 dogs, was not well correlated with 24-hour hepatic urea production. In dogs 80, 100, and 112, respectively, urea synthesis comprised 84, 100, and 45% of ingested nitrogen. There was much better correlation of urea synthesis with gut amino acid output. Urea synthesis comprised 52, 57, and 45%, respectively, of 24-hour gut output of total amino acid nitrogen.[3] This indicates that rates of hepatic urea synthesis are controlled, in part, by the amounts of amino acids supplied by the gut.

Despite the high protein intake and still higher gut output of amino acids in these dogs, the rates of urea synthesis attained were much below the liver's capabilities. Kamin and Handler (1951) infused amino acids into the radial vein of lightly anesthetized dogs and measured the maximum increases in rates of urea formation (Table I). With casein hy-

[3] Total amino acid nitrogen is taken to be the sum of all ninhydrin-positive compounds determined in blood (less urea). This includes ammonia, amino acids, and related compounds, such as glutathione, carnosine, phosphoethanolamine, and taurine (Elwyn *et al.*, 1968).

drolyzates, arginine, glutamine, or mixtures of these, rates of urea synthesis ranged from 1.7 to 3.4 times those found in the experiments of Elwyn *et al.* In keeping with its role in the Krebs-Henseleit ornithine cycle, stimulation of urea synthesis by arginine was additive to that of either glutamine or casein hydrolyzate. Similar effects of arginine have been observed in the intact rat (du Ruisseau *et al.*, 1956; Winitz *et al.*, 1956). These data suggest that, in animals fed by way of the gastrointestinal tract, liver urea synthesis will be affected by gut outputs of arginine and glutamine as well as of total amino acids. It is of interest in this respect that in the experiments of Elwyn *et al.* (1) gut outputs of arginine were smaller, relative to the amounts ingested, than were outputs of most other amino acids (Fig. 3), and (2) gut outputs of glutamine were positive during the absorptive period when rates of urea synthesis were high, but negative during the nonabsorptive period when there was little synthesis of urea (Fig. 3).

This role of the gut in regulation of the rates of hepatic urea synthesis suggests a partial explanation for the slow response of the total amount of daily urea excretion to abrupt changes in protein intake. As discussed more fully in Chapter 10, an abrupt decrease in nitrogen intake results in a period of negative nitrogen balance lasting several days, until urea excretion, which decreases slowly during this period, falls sufficiently to restore N balance. As noted in Section II,A,3, a reduction in protein intake causes a rapid loss of protein from the organs of the nonhepatic splanchnic region, resulting in a gut output of amino acids which is greater than the amount of protein ingested. This would keep urea synthesis higher than would be expected if gut amino acid output equaled the quantity of protein ingested. After several days the protein content of these organs would level off, and gut amino acid output and liver urea synthesis would again correspond to the amount of ingested proteins. Conversely, an abrupt increase in protein intake would cause protein deposition in the nonhepatic splanchnic region. Gut amino acid output would be less than the amount of ingested protein, thus tending to keep urea synthesis at a low rate. As the protein content of these organs reached a higher plateau, gut amino acid output and liver urea synthesis would again correspond to the amount of ingested protein.

Superimposed on the role played by substrate availability in regulation of diurnal and longer-term patterns of urea synthesis, are changes in activity of urea cycle enzymes in the liver. Schimke (1962) has shown that activities of these enzymes increase with increasing protein in the diet, taking 4–8 days to adjust to new dietary levels. In addition, starvation or administration of corticosteroids increases the level of these enzymes over control levels.

The quantitative significance in the intact animal of these and other possible regulatory mechanisms of urea synthesis remains to be assessed.

b. Net Protein Synthesis or Breakdown in the Liver. The differences between liver uptake of total amino acid nitrogen[3] and output of urea may be called *net noncatabolic metabolism of amino acids.* This can be positive or negative depending on whether amino acid uptake is greater or less than urea output. The products of this metabolism will be, in part, nonprotein compounds, such as phospholipids, nucleotides, creatine, and heme. However, the bulk of the products may be expected to be protein. The measurement of *net noncatabolic amino acid metabolism,* therefore, is an approximate measure of the rate of net protein synthesis or breakdown in the liver.

Cumulative net synthesis of protein, as so defined, is shown for 3 animals in Fig. 6. In contrast to the very similar patterns observed for urea synthesis, there is wide variation between animals in the pattern of net protein synthesis. In two animals there was extensive net synthesis throughout the day. In a third animal there was extensive protein break-

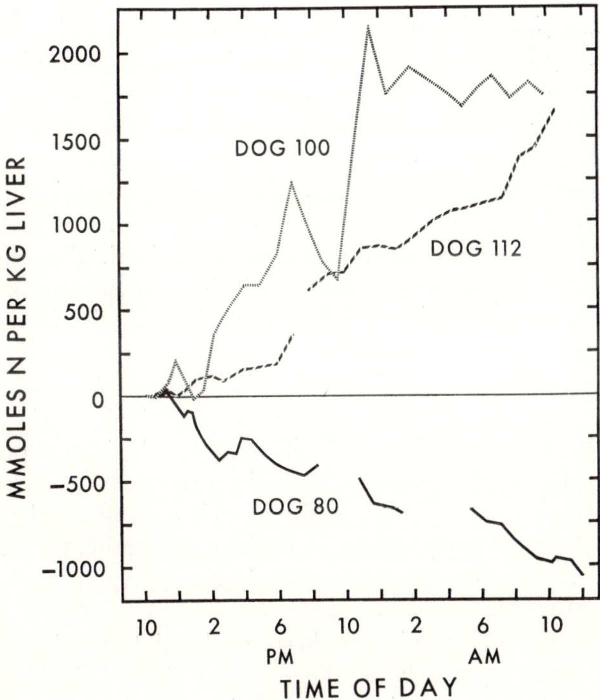

FIG. 6. Cumulative synthesis or breakdown of protein in liver over 24-hour periods in 3 dogs. From Elwyn *et al.* (1968). (Published with permission of *Am. J. Physiol.*)

down. For the average of the three animals, synthesis was more rapid after the meal than during the nonabsorptive period. The absolute values for protein synthesis and breakdown are uncertain since they are subject to the errors of measurement of both liver amino acid uptake and urea output. The errors in urea output have been noted.[2] The major source of error in liver amino acid uptake, as with gut amino acid output (Section II,A,3), is the uncertainty in estimating the distribution of total hepatic blood flow between the hepatic artery and portal vein. For limiting values for portal flow of 60% and 90% of total hepatic flow, the corresponding values for 24-hour protein synthesis in two animals are 1100–2100 mmoles N, and for breakdown in the third animal the values are 600–1900 mmoles N (Elwyn *et al.*, 1968). Thus, the qualitative pattern shown in Fig. 6 appears to be valid, although the exact extent of protein synthesis and breakdown is in doubt.

Net synthesis of protein has also been observed in the perfused canine liver in response to an amino acid load (McMenamy *et al.*, 1962). Considerable net protein synthesis (noncatabolic amino acid utilization) followed amino acid addition at the start of perfusion (Fig. 7). However, when the livers were first perfused for 90 minutes, the subsequent administration of amino acids caused no net protein synthesis. By contrast, urea synthesis was stimulated by amino acid addition under both circumstances. Furthermore, the rates of turnover of liver protein and of synthesis of plasma albumin and α-glycoprotein were not measurably different in the two situations (Richmond *et al.*, 1963).

Increased net synthesis of protein, although it may result from an increased supply of amino acids, is not an obligatory response of either the perfused liver or the liver of the intact animal. The physiological state of the liver, determined in part by the level of circulating hormones, plays a permissive role. The loss of net synthetic ability in the perfused liver after 90 minutes may reflect loss of hormones in this system.

In both preparations, blood amino acid concentrations were lower in experiments in which there was net protein synthesis in the liver than they were in experiments in which either no synthesis or protein breakdown took place. This suggests that the extent and direction of net protein synthesis or breakdown in the liver is a factor in the regulation of blood amino acid concentrations. High concentrations of amino acids in arterial blood do not appear to stimulate net protein synthesis in the liver (cf. Section I).

The extent of net protein synthesis or breakdown observed in the intact animals was quite large. The smaller figures cited above, 1100 mmoles N or 96 gm of protein for synthesis and 600 mmoles N or 52 gm protein for breakdown, are equivalent to 48% and 26%, respectively,

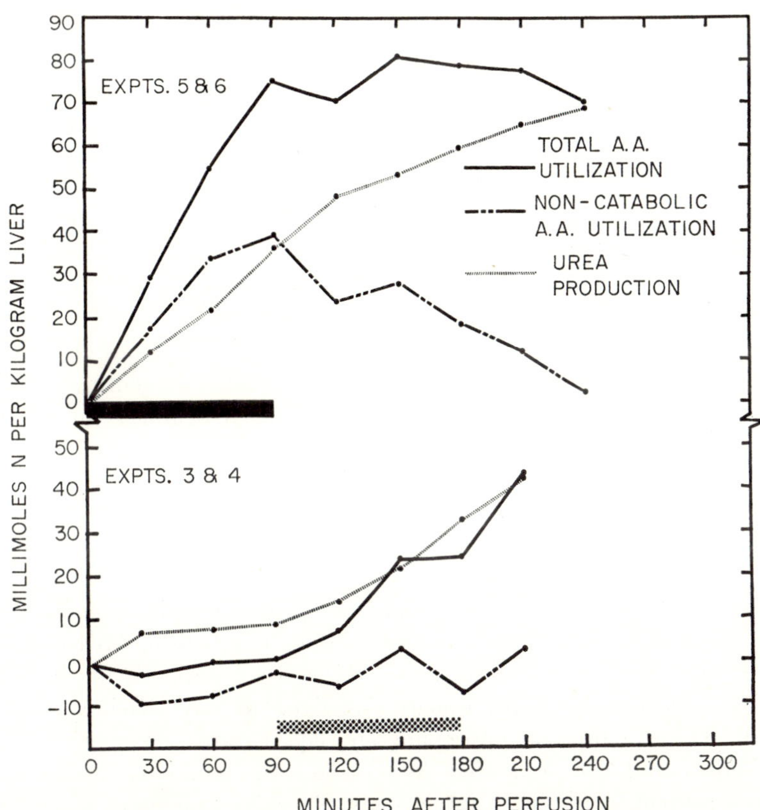

FIG. 7. Urea production and total and noncatabolic utilization of amino acids by perfused liver. Horizontal bars refer to periods of amino acid addition. From McMenamy *et al.* (1962). (Published with permission of *Am. J. Physiol.*)

of total protein in a 1 kg liver. In part this may be accounted for by net changes in liver content of protein. These animals were studied 4–5 days after operation and may have been in either a catabolic state or a subsequent state of repletion of liver protein. Gains or losses of protein of these magnitudes have been observed in livers of animals changing from protein-free diets to high-protein diets and vice versa (Chapter 10) (Addis *et al.*, 1936).

A part of net protein synthesis or breakdown may also be accounted for by plasma proteins. The liver is the major, if not the sole, source of plasma albumin, fibrinogen, α-globulins, and much of the β-globulins (Miller and Bale, 1954; Miller *et al.*, 1964; Kukral *et al.*, 1963). An estimation of the total amounts of plasma protein replaced daily in the normal dog may be made from the amounts in the intravascular pool

(Allison and Wannemacher, 1957) and the ratio of extravascular to intravascular pool size and the turnover rate (Anker, 1960). The amount of protein replaced in a 35-kg dog amounts to 7–13 gm of albumin, 9 gm of globulins, and 1.2 gm of fibrinogen, or a total of about 20 gm per day. Abnormal or nonsteady-state conditions may cause large changes in these rates of synthesis. Protein depletion over many days results in extensive loss of plasma albumin (Allison and Wannemacher, 1957). This may be effected either by a decreased rate of synthesis, an increased rate of breakdown or both. Jeffay *et al.* (1958) have shown, in rats that steady-state turnover rates of albumin increase as much as two- to threefold with increasing protein in the diet. Plasma concentrations of proteins may be an important factor regulating their rates of synthesis. Miller *et al.* (1964) have demonstrated that fibrinogen synthesis in perfused rat liver was low, 0.2 mg per gram of liver per hour, when the perfusing blood contained normal amounts of fibrinogen. If defibrinated blood was used, net rates of fibrinogen synthesis of 0.77 mg per liver per hour were attained. These are about four times the rate calculated for normal turnover of fibrinogen in the rat.

Changes in amount or direction of net protein synthesis could result from changes in rates either of synthesis or catabolism. Measurements in perfused dog liver indicate rates of protein turnover in the range of 2700-5200 mmoles N per kilogram of liver per day (Richmond *et al.*, 1963). These represent approximately equal rates of synthesis and catabolism and are somewhat larger than the estimates made above for net protein synthesis in intact dogs. Therefore, increases in net synthesis could result either from increases in synthetic rates or decreases in catabolic rates. Conversely, decreases in net synthesis could result from increased catabolism or decreased synthesis. There is some evidence that regulation of both synthesis and catabolism of proteins is effected by dietary means:

1. Wunner *et al.* (1966) have shown that the degree of aggregation of the polyribosomes of rat liver microsomes is increased by feeding complete amino acid mixtures. This is associated with an increased ability to incorporate radioactive amino acids *in vitro,* indicating an increased rate of protein synthesis *in vivo.* (This subject is discussed in more detail in Chapter 34.)

2. Gan and Jeffay (1967) infused lysine-^{14}C at a constant rate into rats and measured the specific activities of plasma, liver, and muscle free lysine for periods up to 20 hours. After 2 hours the ratio of lysine specific activities in liver and plasma remained constant. However, the absolute value of this ratio was dependent on the dietary state of the animals. As shown in Fig. 8, the ratio was 0.5 on a normal diet. This

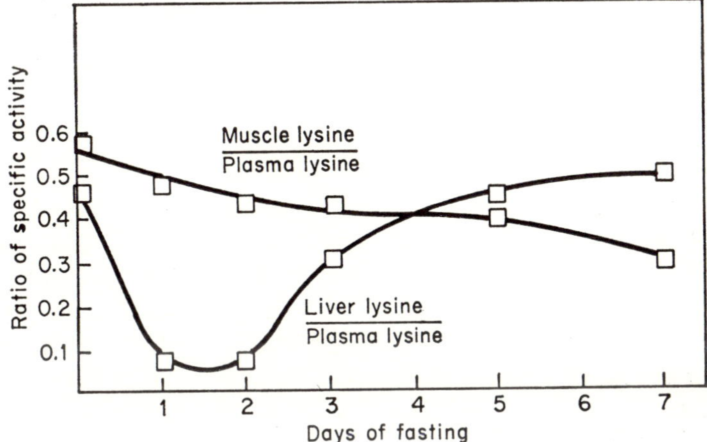

FIG. 8. Effect of fasting on the sources of intracellular lysine. No priming dose was given, and the animals were infused with 1.1×10^6 cpm per 0.65 μg of lysine-^{14}C every hour for 10 hours. Each point is an average from two animals. From Gan and Jeffay (1967). (Published with permission of *Biochim. Biophys. Acta.*)

was reduced to 0.1 on the first and second days of fasting and subsequently, with continued fasting, returned to fed levels. These data suggest a marked increase in the rate of release of unlabeled lysine from protein breakdown during the first 2 days of fasting when the rate of net loss of proteins from the liver was maximal. Net loss of protein is seen in this case to result in large part from an increase in the rate of protein catabolism. After 3 days there was no further net loss of protein from the liver, and the rate of protein catabolism appears to have returned to fed values. Muscle, by contrast, showed much smaller changes which continued for a longer time.

3. Liver concentrations of the enzyme arginase change markedly with the amount of protein in the diet. Schimke (1964) has shown that changes in the rates both of synthesis and degradation are involved in regulation of arginase concentrations. (This subject is discussed in Chapter 35.)

The foregoing discussion illustrates the extent and flexibility of response of the rates of net protein synthesis in the liver to stresses of dietary and of traumatic origin. As yet, however, there is little quantitative information as to the part played by individual proteins and particularly the relative importance of liver and plasma proteins in contributing to net synthesis or breakdown by the liver under stressed conditions.

The extent and variability of protein synthesis must be much smaller in normal animals in nitrogen balance. For such animals, on constant daily diets, there can be no net synthesis or degradation of liver proteins over 24-hour intervals. It may be surmised, however, that during absorptive periods there would be a net synthesis, and during nonabsorptive periods a net breakdown, of liver proteins. The duration of these periods and the extent of protein synthesis and breakdown would depend on the frequency of feeding and the amount of protein in the diet.

The situation is somewhat different with respect to plasma proteins. Although it is well established that almost all plasma protein synthesis takes place in the liver, the sites of catabolism are not well known. This subject is discussed at some length in Chapter 8, Sections VII and VIII. From this discussion it seems likely that the liver accounts for less than one-half of plasma protein catabolism. In addition, kidney, intestine, and phagocytes have been implicated. Data of Katz *et al.* (1961) show that rates of degradation of albumin in eviscerated rats are 50% of rates in normal animals. To the extent that plasma proteins are broken down elsewhere, their synthesis by the liver constitutes net protein synthesis. Thus, it appears that, in the normal animal, the daily rate of net protein synthesis in the liver is more than one-half the rate of plasma protein turnover.

A tentative scheme of the quantitative interrelations between ingested protein, gut, and splanchnic amino acid outputs, and urea and net protein synthesis in the liver is shown in Fig. 9. The quantities are shown in order to give an indication of the magnitude of these processes. Rates of plasma protein synthesis and secretion of protein into the intestinal tract were estimated as described above. The other values represent an average of the 3 dogs of Elwyn *et al.* (1968), which were not in N balance, and are not meant to represent the normal animal. The combination of data from all 3 dogs obscures the wide variation in the values for 2 of the animals (100 and 112) as compared with the other (80). Nevertheless, the diagram serves to show that more than half the amino N absorbed after the meal of meat is transformed to urea without passing through the liver, thus demonstrating the extensive buffering action of the liver, and furthermore that a considerable proportion of the absorbed amino N passed into the systemic circulation is in the form of plasma proteins. In terms of net balances, 57% of the absorbed amino N is converted to urea as it passes through the liver, 23% enters the general circulation in the form of free amino acids, and 6% as plasma proteins during the 12-hour period. The residual 14% of the absorbed amino acids not accounted for by these three routes are retained in the liver in the form of a temporary increment in the protein content of this organ. Finally, it should be pointed out that this picture would undoubtedly vary with the amino acid composition of the meal fed to

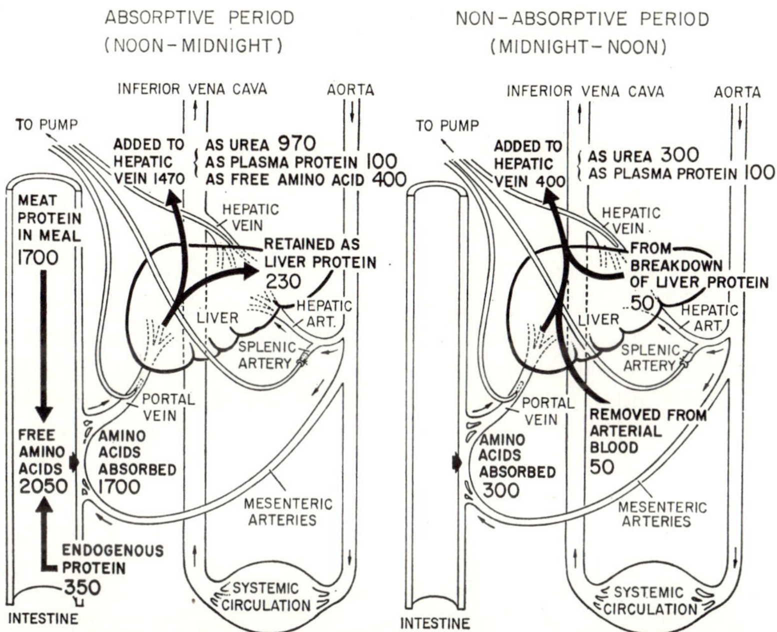

FIG. 9. Movement of amino acids into, out of, and within the splanchnic region during the absorptive period (12 hours after eating) and nonabsorptive period (12 hours before eating). All values are in mmoles N per kilogram of liver. Quantities, except as noted in text, are mean values for 3 dogs fed daily at noon a diet of 780 gm of horsemeat per kilogram of liver. In converting weight of protein to mmoles N, it has been assumed that 1 mmole N corresponds to 87.5 mg of protein.

the animal and that the buffering action of the liver does not protect the peripheral tissues from amino acid deficiencies in the meal fed, as shown by the fall in systemic plasma lysine levels observed when a meal deficient in lysine is fed to dogs (Longenecker and Hause, 1959) and by other similar evidence reviewed in Chapter 34. Although the data shown in Fig. 9 appear to represent an enormous wastage of ingested amino acids, it should be recollected that the daily needs for amino N are not large by comparison with the capacity of the dog to consume meat. From Table V of Chapter 25, it can be calculated by interpolation that a 35 kg dog receiving a protein-free diet has an endogenous urinary N output of about 60 mg per kilogram of body weight, equivalent to a loss of 150 mmoles of amino acids per animal daily. Allowing for other sources of obligatory N loss, it is likely that a 35 kg dog can be maintained in nitrogen equilibrium on an intake of about 200 to 250 mmoles of an amino acid mixture of high biological value. Thus the intake of

1700 mmoles eaten by the dog as meat is some seven to eight times greater than it needs. It is very likely that the proportion of the incoming amino acids disposed of by way of catabolism would be much reduced if the protein content of the meal was small instead of large. Studies on dogs fed various levels of protein will be needed in order to obtain a complete picture of the role of the liver in regulating amino acid supply to the remainder of the body.

3. *Differential Utilization and Synthesis of Specific Compounds in the Liver*

The extent of liver uptake of amino acids in the experiments of Elwyn *et al.* (1968), varied from animal to animal, as shown in Fig. 3. In addition, in each animal there were considerable differences in the amounts of each amino acid extracted by the liver. In Table II, the essential and nonessential amino acids are listed in order of decreasing uptake by the liver expressed as percent of gut amino acid output. A very similar

TABLE II

RELATIVE RATES OF AMINO ACID UPTAKE BY LIVER DURING THE ABSORPTIVE PERIOD IN THE INTACT DOG AND BY PERFUSED DOG LIVER

Amino acid	Hepatic uptake in intact dog as percent of gut output[a]	Uptake by perfused dog liver as percent of amount added[b]
Essential		
Tryptophan	—	90
Histidine	118	69
Methionine	114	79
Phenylalanine	104	90
Lysine	90	44
Threonine	77	24
Isoleucine	38	17 (Isoleucine and Leucine)
Leucine	34	
Valine	33	22
Nonessential		
Cystine	156	—
Glutamine	134	66
Serine	128	81
Tyrosine	114	75
Alanine	114	57
Arginine	95	76
Glutamic acid	94	—
Glycine	86	47
Proline	74	44

[a] Calculated from data of Elwyn *et al.* (1968). Average for three dogs.
[b] Data of McMenamy *et al.* (1962).

order was observed when liver uptake was measured in perfused dog liver subjected to an amino acid load (McMenamy *et al.*, 1962). These differences may reflect in part differences in composition of protein synthesized in the liver and of that ingested; in part they may be due to differences in the rate of catabolism of amino acids by the liver. The latter appears to account for the remarkably low uptakes of the branched-chain amino acids observed in both preparations.

a. Valine, Leucine, and Isoleucine. In the intact dog, for most essential amino acids, either liver uptake was greater than that required for protein synthesis, or liver output was less than that required to account for protein breakdown, indicating that each is catabolized in the liver (Figs. 3 and 6). This was not true of valine, leucine, or isoleucine. The amount and time course of liver uptake or output of these amino acids could be entirely accounted for by the amount of protein synthesized or broken down (Elwyn *et al.*, 1968).

When dog liver (McMenamy *et al.*, 1962) or rat liver (Miller, 1962; Fisher and Kerly, 1964; Schimassek and Gerok, 1965) was perfused without addition of amino acids, concentrations of most amino acids in the perfusing fluid remained quite constant and in the range of normal plasma values. Concentrations of valine, leucine, and isoleucine increased markedly under these conditions indicating (1) that net breakdown of protein was occurring; (2) that amino acid catabolism to urea serves as a very sensitive regulatory mechanism in maintaining concentrations of most amino acids; and (3) that if any catabolism of the branched chain amino acids occurred, it was much less than of other amino acids. In eviscerated rats (Miller, 1962) or hepatectomized dogs (Freeman and Svec, 1951) plasma concentrations of most amino acids were higher than in normal animals, but concentrations of valine, leucine, and isoleucine were the same or lower. This indicates that catabolic rates in nonhepatic tissues are greater for the branched-chain compounds than for most other essential amino acids.

Miller (1962) has studied oxidation of labeled amino acids by perfused rat liver, eviscerated rats, and eviscerated rats with the liver retained. These data, shown in Table III, indicate that the nonessential amino acids, with the exception of arginine, are readily oxidized by both the liver and nonhepatic tissues.[4] Oxidation is greater when the liver is left in than when it is removed from the eviscerated rat, indicating that these processes in liver and in other tissues are additive. For most of the essential amino acids catabolism in the eviscerated rat was negligible

[4] In addition to the data shown in Table III, it has been found that alanine, serine, cysteine and pyrollidone carboxylate are as effectively oxidized by the nonhepatic tissues as by the liver (L. L. Miller, personal communication).

TABLE III

OXIDATION OF ^{14}C-LABELED AMINO ACIDS BY THE PERFUSED LIVER AND BY NONHEPATIC TISSUES IN 5-H EXPERIMENTS[a]

		$^{14}CO_3$ as percent dose		
			Eviscerated rat	
Amino acid[b]	Dose (mg)	Liver perfusion	Liver out	Liver in[c]
L-Phenylalanine	14	25.8	0.13	27.7
DL-Tryptophan-2-^{14}C	4.4	16.8	1.6	21.2
L-Arginine · HCl	156	16.1	3.0[d]	—
L-Threonine	8	66	1.2	—
DL-Methionine-2-^{14}C	1.2 + 9.6 L-	47.7	2.0	22.0
L-Histidine-2-^{14}C HCl	10	36.7	0.35	—
DL-Lysine-6-^{14}C HCl	0.6 + 21 L-	25	1.1	—
L-Valine	11	—	19.3	31
L-Leucine	34	7.4	13.3	14.3
L-Isoleucine	15	4.3	27.5[e]	17.7
L-Glutamic acid	43	47.8	40.2	62.5[f]
L-Glutamine	1.0	43.0	20.3	—
Glycine-1-^{14}C	—	—	8.4	19.0
L-Aspartic acid	2	—	53[g]	63
L-Proline	2	—	17.2[g]	—

[a] From Miller (1962). Reprinted with permission of Elsevier.
[b] All amino acids uniformly labeled unless otherwise indicated.
[c] Liver left in supplied with hepatic artery.
[d] Dose, 2 mg.
[e] Duration, 4 hours.
[f] Dose, 5.5 mg.
[g] Kidneys also removed.

compared either to the perfused liver or to the eviscerated rat with the liver retained. However, catabolism of the branched-chain amino acids is low in perfused liver when compared to the other essential amino acids, but very high in the eviscerated rat. These data are consistent with the conclusions from the nonisotopic experiments, discussed above, that a major part of leucine, isoleucine, and valine catabolism occurs in nonhepatic tissues. In addition, they indicate that some oxidation takes place in the liver. In the case of leucine and isoleucine, oxidation in the eviscerated rat with the liver still *in situ* was no greater than with the liver out. This suggests that the rate of oxidation by the liver is affected by its interrelation with other tissues. The higher concentrations present in perfused liver may result in more rapid oxidation of leucine and isoleucine than occurs in the liver of the whole animal.

The low catabolic rate of these compounds in liver may be useful, in some circumstances, in estimating liver metabolism. Swendseid *et al.* (1967), for instance, have interpreted the differential rise in plasma concentrations of leucine, isoleucine, and valine in fasting as an indication of liver gluconeogenesis.

b. Glutathione. The liver appears able either to utilize or produce this tripeptide. Production is high when protein breakdown is occurring in the liver, low or negative when protein synthesis is taking place. Production of glutathione by the gut seems also to be roughly proportional to protein hydrolysis in that organ. This association between protein hydrolysis and glutathione synthesis coupled with the extensive transport of glutathione between gut, liver, and peripheral tissues suggests that glutathione may serve as a readily transportable and available storage form of its constituent amino acids (Elwyn *et al.,* 1968). Since glycine and glutamic acid are readily synthesized in most tissues, this is probably of greatest significance in the case of cysteine.

c. Carnosine. There is a consistent output of carnosine by the liver at all times of day resulting in a continuous splanchnic output of this peptide. Dietary carnosine is completely hydrolyzed in the gut and therefore makes no contribution to splanchnic output.

C. Differential Roles of Plasma and Erythrocytes in Interorgan Transport of Amino Acids

Plasma and erythrocytes appear to play quite distinct roles in interorgan transport of amino acids. Concentrations in red blood cells and plasma are illustrated for phenylalanine in Fig. 10. Plasma concentrations show very abrupt changes with time, particularly in the portal vein, and large differences between sampling sites. Erythrocyte concentrations change more gradually and differences between arterial, portal, and hepatic venous blood are relatively much smaller. It is particularly interesting that concentration changes may be in opposite direction in plasma and erythrocytes as blood circulates through either gut or liver. The absolute concentrations and the pattern of concentration changes in plasma show little variation between animals. By contrast, absolute concentrations in erythrocytes vary markedly between animals. As a result, since relative changes in erythrocyte concentration are similar in different animals, the magnitude of the contribution of red blood cells to interorgan transport is roughly proportional to the absolute concentrations of amino acid within them.

These relations are summarized for all amino acids in Figs. 11 and 12, which show gut, liver, and splanchnic outputs into plasma and erythrocytes, respectively. Patterns for plasma are remarkably constant. Except for glutamine and carnosine, gut output into plasma is always

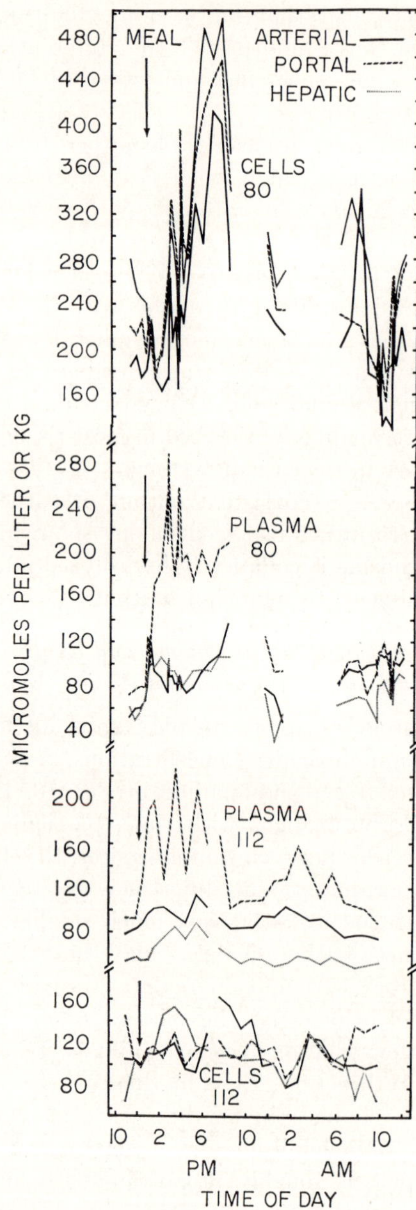

FIG. 10. Phenylalanine concentrations in plasma and erythrocytes of arterial, portal, and hepatic venous blood of two dogs sampled over 24-hour periods. Conditions are as described for Fig. 2. From Elwyn *et al.* (1968). (Published with permission of *Am. J. Physiol.*)

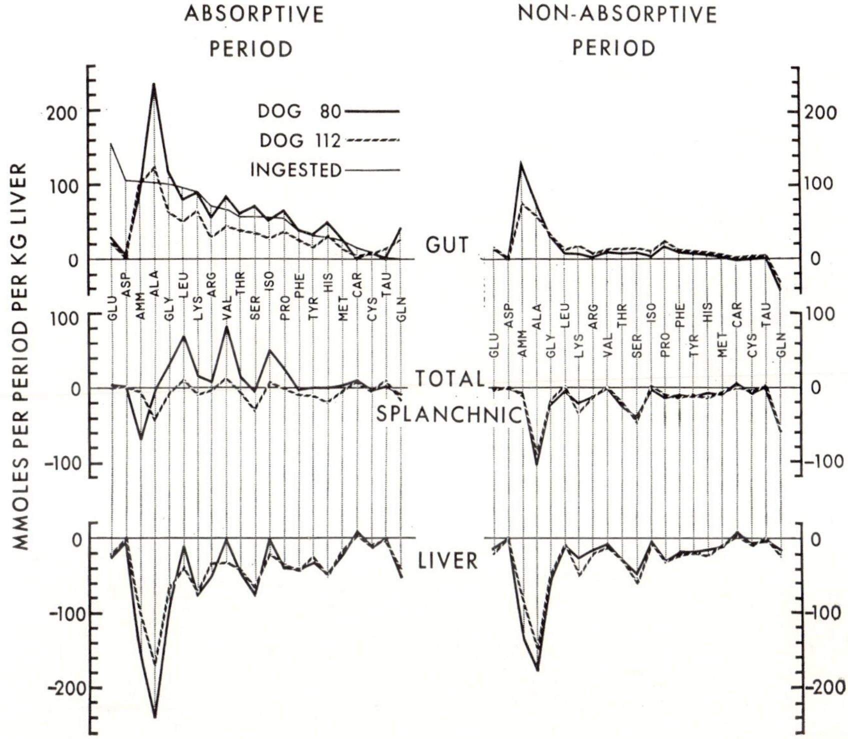

FIG. 11. Outputs of amino acids into plasma from gut, liver, and total splanchnic region in the absorptive and nonabsorptive periods for 2 dogs. The amount ingested is shown for dog 112. It was 9% higher for dog 80. From Elwyn *et al.* (1968). (Published with permission of *Am. J. Physiol.*)

positive, liver output is always negative. Splanchnic output is generally negative except for the branched-chain amino acids.

In contrast to plasma, amino acid transport in erythrocytes is quite variable (Fig. 12). Depending on the animal or the period, gut, liver, or splanchnic outputs can be either positive or negative and in either the same or opposite direction to plasma outputs. However, when erythrocyte concentrations of amino acids are low, as in the case of dog 112, they make little contribution. Whole blood output patterns are, therefore, almost identical to plasma patterns. When amino concentrations in erythrocytes are high, as in dog 80, transport in erythrocytes may be equal to or greater than that in plasma. The differences in behavior of red blood cells in these 2 dogs were evident throughout the day despite the fact that more than the total blood volume was transfused to replace sampling losses in each animal. These differences must

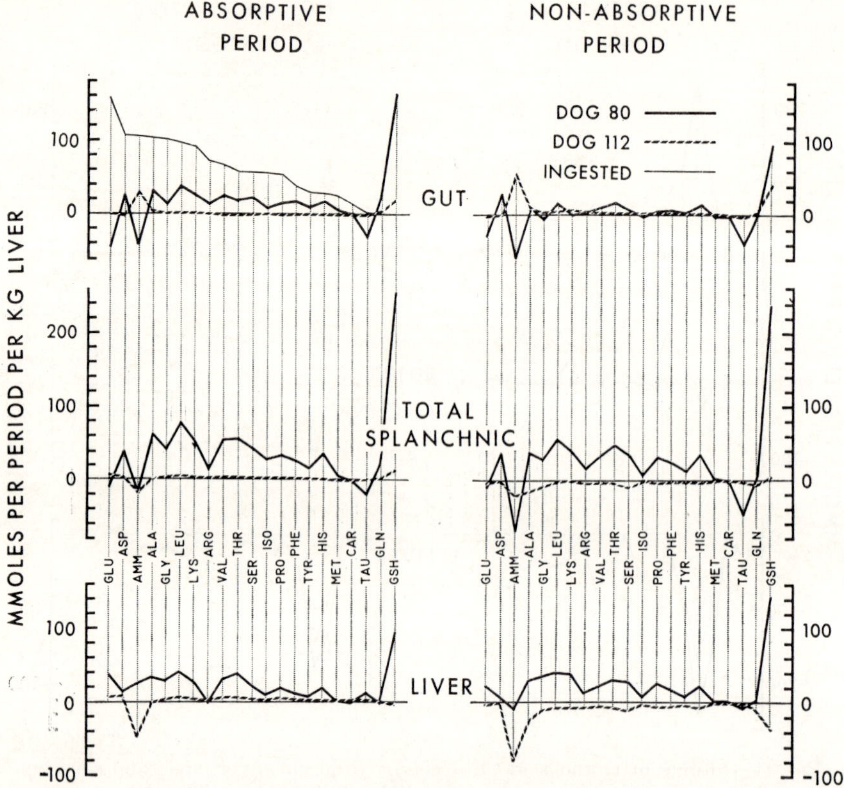

FIG. 12. Outputs of amino acids into erythrocytes from gut, liver, and total splanchnic region in the absorptive and nonabsorptive periods for 2 dogs. The amount ingested is shown for dog 112. It was 9% higher than dog 80. From Elwyn *et al.* (1968). (Published with permission of *Am. J. Physiol.*)

be attributed, therefore, to the physiological states of the animals rather than to the previous history of the erythrocytes themselves.

D. Factors Affecting Amount and Direction of Total Splanchnic Output of Amino Acids

Studies of human subjects in the nonabsorptive state show that total splanchnic outputs of all amino acids except glutamic acid are consistently negative (Onen *et al.*, 1956; Carlsten *et al.*, 1967). Consistent with this are studies in the human forearm which show amino acid outputs of this region to be consistently positive (Foley *et al.*, 1966). Studies of anesthetized dogs show positive outputs from the head 2 hours after eating zein or beef (Ganapathy and Nassett, 1962; Nassett *et al.*, 1963). Studies on the 3 dogs described above indicate that splanchnic amino acid outputs are the same in both the absorptive and nonabsorptive periods (Figs. 3 and 4). Taken together these studies suggest that negative splanchnic outputs of amino acids normally continue throughout the day.

Two of the dogs described here had negative splanchnic outputs of amino acids accompanied by net protein synthesis in the liver and an insignificant contribution by red blood cells to interorgan transport of amino acids. These two animals, based on the information cited above appear to approach the normal situation. The third animal, dog 80, had a high positive splanchnic output of amino acids associated with net protein breakdown in the liver, high concentrations of amino acids in erythrocytes, and a major contribution by red blood cells to interorgan amino acid transport. This pattern, which is quite different from the normal, probably represents a stage in recovery from the operation to implant catheters.

The major factor associated with the duration and amount of total splanchnic amino acid output appears to be the net rate of synthesis or breakdown of proteins in the liver. Insofar as these two processes are directly related, it seems likely that variations in splanchnic amino acid output are the result rather than the cause of variations in hepatic protein synthesis. If the reverse were true, one would expect that a positive splanchnic output, or its associated peripheral uptake, of amino acids would cause lowered arterial concentrations, which in turn would tend to lower liver amino acid concentrations and thus induce hepatic protein breakdown. Conversely a positive peripheral output of amino acids would increase arterial concentrations, causing increased liver amino acid concentrations and net hepatic protein synthesis. However, in the animals studied this was not true. High arterial amino acid concentrations were associated with a positive splanchnic output and protein loss from the liver. Low concentrations were associated with splanchnic uptake of amino acids and protein synthesis in the liver (Figs. 2, 6, 7, and 10) (Elwyn *et al.*, 1968). The relationship of free amino acid levels to regulation of tissue protein synthesis is discussed in Chapter 34.

E. The Role of Plasma Proteins in Interorgan Transport of Amino Acids

Since the balance sheet of amino acid interchange between the viscera and the peripheral tissues shows a net gain of amino acids by the viscera over a 24-hour period in the steady-state animal, then some other form of amino N must reach the peripheral tissues so that the carcass does not become progressively depleted of its protein and fail to maintain the steady state. The candidates for such a rebalancing of the flow between carcass and viscera are presumably amino acids bound by peptide linkage. Since the quantity of nonprotein peptides in blood is small, it is likely that these bound amino acids are in the form of plasma proteins. This would imply that some but not necessarily all peripheral tissues, such as kidney, muscle, skin, bone marrow, or connective tissue, are able to hydrolyze plasma proteins to their constituent amino acids, which subsequently are returned to the splanchnic region. Evidence that the eviscerated rat catabolizes plasma proteins at one-half the normal rate (Section II,B,2,b) supports this thesis. Since there will also be net consumption of amino acids, in these tissues, for such requirements as synthesis of hair and skin proteins and urinary excretion of ammonia and amino acids, the transport of bound amino acids from viscera to periphery will be greater than transport of free amino acids from periphery to viscera. The magnitude of the deficit to be made up can be gauged from the following considerations. From the data on different species given in Table V of Chapter 25, a 35 kg dog will incorporate daily some 300 mg amino N per kilogram body weight into its tissues as protein, that is, 750 mmoles of amino acids per 35 kg animal. This is much larger than the amino acid requirements of such an animal, computed in Section II,B,2 to be 200–250 mmoles per day. The larger figure for body protein synthesis is due to recycling of amino acids liberated from tissue proteins as a consequence of turnover (see Chapter 28). Much of this reutilization takes place within the same tissue, and furthermore a good deal of the organ interchange of recirculated acids occurs within the splanchnic area. Consequently, the magnitude of the daily exchange of amino acids between carcass and viscera must be less than 500 mmoles. In order to compensate for this, the daily secretion of 20 gm (230 mmoles) of plasma proteins by the liver would provide about half the needed amount, and the remaining 250 mmoles of amino acids fall within the level of excess amino acids passing through the liver after the daily meal of meat (Fig. 9).

A dual system of interorgan transport of amino acids in either free or bound form has potential advantages over one involving free amino acids only. Because of the low concentrations of free amino acids in blood, a change in uptake of amino acids by one tissue must be matched within seconds by a corresponding change in output from some other tissue. Blood concentrations of plasma proteins are 50–100 times the concentration of total free amino acids. Sizable quantities of plasma proteins could thus be utilized by one tissue without requiring immediate synthesis by some other tissue.

F. Factors Affecting Arterial Concentrations of Amino Acids

Numerous studies have demonstrated increased concentrations of amino acids in arterial or peripheral venous blood or plasma after meals (e.g., Dent and Schilling, 1949; Christensen, 1949; Parshin and Rubel, 1951; Denton and Elvehjem, 1945; Wheeler and Morgan, 1958; Frame, 1958; Levenson *et al.,* 1959; Ganapathy and Nassett, 1962; Peraino and Harper, 1963). This is also evident as shown in Figs. 2 and 10 for the experiments described above. In two of these dogs, however, there was a net splanchnic uptake of most amino acids in the absorptive period roughly equal to that in the nonabsorptive period. Arterial amino acid increases in these instances do not represent a part of ingested amino acids which get past the liver. Rather they reflect a change in the balance between liver uptake and peripheral output of amino acids. To what extent this is due to decreased liver uptake, to increased peripheral output, or to both is not clear at present.

Positive splanchnic outputs are normal for the branched-chain amino acids. Under abnormal conditions they may occur for most other amino acids. The postprandial increase and the absolute values of arterial amino acid concentrations are much greater when splanchnic outputs are positive than when they are negative. These effects are more pronounced in erythrocytes than in plasma (Elwyn *et al.,* 1968). These relations suggest that measurement of arterial whole blood or erythrocyte concentrations of amino acids might be, under some circumstances, a useful indicator of amino acid metabolism in the liver.

G. Implications for Parenteral Administration of Amino Acids

Two aspects of concern in parenteral administration of amino acids are the route of administration and the composition of the amino acid mixture. Although present information is too fragmentary to warrant any firm conclusion, examination of the patterns of interorgan movements of amino acids suggest some avenues for further exploration.

From consideration of gut action on the meal, desirable compositional changes from a balanced protein hydrolyzate would be replacement of glutamic and aspartic acids by alanine and glutamine, and addition of glutathione. Such a mixture might be suitable for infusion via the portal vein. If, however, splanchnic amino acid output is normally negative, as discussed in Section II,D, such a mixture might be highly inappropriate for administration in a peripheral vein. The same amino acids introduced into the systemic circulation might have effects quite different from their introduction into the portal vein. The possibility that plasma proteins are a major source of amino acids for peripheral tissues suggests that they might be a more suitable source for administration into the peripheral circulation than are free amino acids. This is consistent with much clinical experience with protein deficiencies requiring parenteral alimentation. Although more costly, larger amounts of amino acids may be administered as plasma or plasma proteins with fewer complications than as amino acid mixtures. However, some combination of free amino acid mixtures with the appropriate plasma proteins might be most satisfactory. Examination of some of these questions through study of net organ metabolism of amino acids might be a fruitful avenue of approach.

III. Interrelations of Urea and Protein Synthesis in the Liver with Regulation of Protein Metabolism in the Whole Organism

The regulatory role of the liver may be defined in terms of two interrelated functions. The first is the response to both diurnal and long-term variations in protein ingestion which maintains a relatively constant supply of amino acids to the rest of the organism. The second is the response to changing tissue requirements for amino acids which increases or decreases the available supply. The two processes in the liver primarily involved in this regulation are catabolism of amino acids to urea and net synthesis or breakdown of protein. It is evident that the rates of both these processes can show marked diurnal as well as long-term changes in response to dietary and hormonal factors, although information on the mechanism and patterns of these responses is still fragmentary.

A. Influence of Dietary Factors on Urea and Protein Synthesis in Liver

Classical studies of nitrogen balance in humans and other mammals have shown that changes in amount of ingested nitrogen are not immediately reflected in nitrogen excretion. After an abrupt increase in nitrogen intake, it may take 1–2 weeks before urea excretion reaches a higher steady-state level. During this period the subject is in positive nitrogen balance, the excess nitrogen being deposited as labile protein. With a decrease in intake, there is a loss of protein which continues until urea synthesis has fallen to a new lower level (Chapter 10). In the whole organism, total daily output of urea adapts only slowly to changing diets. This slow adaptation of urea excretion has been shown by Schimke (1962) to be closely correlated with changes in liver concentrations of all the urea cycle enzymes.

In contrast to the slow changes in daily urea production there are very marked diurnal changes in rates of hepatic urea synthesis (Fig. 5), which are correlated *in vivo* and *in vitro* (Fig. 7) with the amino acid supply. Studies in both dogs (Kamin and Handler, 1951) and rats (du Ruisseau *et al.*, 1956; Winitz *et al.*, 1956) demonstrate that the rate of urea synthesis can be rapidly increased *in vivo* by injection of a variety of different amino acids and amino acid mixtures. Thus, rates of urea synthesis in the intact mammal are readily affected by the availability of substrates. The diurnal pattern of urea synthesis, which is closely related to the direction and course of digestion and absorption of protein (Fig. 5), appears to be regulated in large part by substrate availability.

In contrast to the relative stability of the pattern and daily quantity of urea synthesis, the rate of net protein synthesis in liver is quite variable. Although it appears to respond to changes in amino acid supply from the diet, the limits of this response are largely dependent on other factors. Taken together, the rates of urea and protein synthesis in the liver respond rapidly to changes in gut amino acid output and maintain a splanchnic output of most amino acids which is almost constant from hour to hour throughout the day.

B. Influence of Tissue Requirements on Liver Amino Acid Metabolism

Free amino acids, plasma proteins, and peptides such as glutathione and carnosine are implicated in interorgan transport of amino acids. The regulation of net protein synthesis in the liver seems directed toward maintenance of a splanchnic output of most free amino acids which is quite constant over 24-hour periods, but which can vary with the physiological state of the animal. The rate and direction of splanchnic output are presumably set in relation to tissue requirements. In the adult organism these may vary with stress, trauma, disease, and dietary conditions. This regulation requires some kind of controls, external to the liver, acting on hepatic protein synthesis. Such controls are probably largely hormonal in character. Certainly, as discussed in Chapters 9 and 10, several hormones have quite marked effects on protein depletion and deposition in the liver and on net rates of synthesis and breakdown of various plasma proteins. Not presently clear is the extent to which synthesis and breakdown of individual liver or plasma proteins determine the amount and direction of total net synthesis of protein in the liver.

In summary, the liver plays an active role in regulation of the supply of amino acids in free and bound form to the rest of the organism. It serves as a buffer between diurnal variations in protein ingestion and the requirements of tissues for a constant supply of amino acids. The two processes primarily responsible for the liver regulatory functions are net protein synthesis or breakdown, and catabolism of amino acids to urea. These processes respond rapidly to changes in gut output and tissue requirements of amino acids. However, they do not seem to respond directly to increases or decreases in arterial levels of amino acids. Rather, hepatic protein synthesis and amino acid catabolism play a major role in determining arterial amino acid concentrations and, as a result, the supply of amino acids to the rest of the organism. Further examples of differences in the response of liver and peripheral tissues to the exogenous supply of amino acids from the diet will be found in Chapter 34, which emphasizes that the free amino acid pool of the liver reflects mainly dietary variations, whereas the peripheral concentrations of free amino acids are determined by other factors.

References

Addis, T., Poo, L. J., and Lew, W. (1936). *J. Biol. Chem.* **116,** 343.

Allison, J. B., and Wannemacher, R. W., Jr. (1957). *In* "Amino Acid Malnutrition" (W. H. Cole, ed.), p. 1. Rutgers Univ. Press. New Brunswick, New Jersey.

Altman, P. L., and Dittmer, D. S. (1961). "Blood and Other Body Fluids." Federation Am. Soc. Exptl. Biol., Washington, D.C.

Anker, H. S. (1960). *In* "The Plasma Proteins" (F. W. Putnam, ed.), Vol II. p. 267. Academic Press, New York.

Carlsten, A., Hallgren, B., Jagenburg, R., Svanborg, A., and Werkö, L. (1967). *Acta Med. Scand.* **181,** 199.

Christensen, H. N. (1949). *Biochem. J.* **44,** 333.

Dent, C. E., and Schilling, J. A. (1949). *Biochem. J.* **44,** 318.

Denton, A. E., and Elvehjem, C. A. (1954). *J. Biol. Chem.* **206,** 449.

du Ruisseau, J. P., Greenstein, J. P., Winitz, M., and Birnbaum, S. M. (1956). *Arch. Biochem. Biophys.* **64,** 355.

Elwyn, D. H. (1966). *Federation Proc.* **25,** 854.

Elwyn, D. H., Parikh, H. C., and Shoemaker, W. C. (1968). *Am. J. Physiol.* **215,** 1260.

Finch, L. R., and Hird, F. J. R. (1960). *Biochim. Biophys. Acta* **43,** 268.

Fisher, M. M., and Kerly, M. (1964). *J. Physiol.* (*London*) **174,** 273.

Foley, T. H., London, D. R., and Prenton, M. A. (1966). *J. Clin. Endocrinol. Metab.* **26,** 781.

Frame, E. G. (1958). *J. Clin. Invest.* **37,** 1710.

Freeman, S., and Svec, M. (1951). *Am. J. Physiol.* **167,** 201.

Freeman, T. (1964). *In* "The Role of the Gastro-Intestinal Tract in Protein Metabolism" (H. N. Munro, ed.), p. 125. Blackwell, Oxford.

Gan, J. C., and Jeffay, H. (1967). *Biochim. Biophys. Acta* **148,** 448.

Ganapathy, S. N., and Nassett, E. S. (1962). *J. Nutr.* **78,** 241.

Goldberg, A., and Guggenheim, K. (1962). *Biochem. J.* **83,** 129.

Guggenheim, K., Halevy, A., and Friedmann, N. (1960). *Arch. Biochem. Biophys.* **91,** 6.

Jeffay, H., Winzler, R. J., and Donnelly, J. S. (1958). *J. Biol. Chem.* **231,** 101; 111.

Ju, J. S., and Nassett, E. S. (1959). *J. Nutr.* **68,** 633.

Kamin, H., and Handler, P. (1951). *J. Biol. Chem.* **188,** 193.

Katz, J., Rosenfeld, S., and Sellers, A. L. (1961). *Am. J. Physiol.* **200,** 1301.

Kukral, J. C., Sporn, J., Louch, J., and Winzler, R. J. (1963). *Am. J. Physiol.* **204,** 262.

Levenson, S. M., Rosen, H., and Upjohn, H. L. (1959). *Proc. Soc. Exptl. Biol. Med.* **101,** 178.

Longenecker, J. B., and Hause, N. L. (1959). *Arch. Biochem. Biophys.* **84,** 46.

McMenamy, R. H., Shoemaker, W. C., Richmond, J. E., and Elwyn, D. H. (1962). *Am. J. Physiol.* **202,** 407.

Miller, L. L. (1962). *In* "Amino Acid Pools" (J. T. Holden, ed.), pp. 708–721. Elsevier, Amsterdam.

Miller, L. L., and Bale, W. F. (1954). *J. Exptl. Med.* **99,** 125.

Miller, L. L., Hanavan, H. R., Titthasiri, N., and Chowdhury, A. (1964). *Advan. Chem. Ser.* **44,** 17.

Munro, H. N. (1966). *In* "Postgraduate Gastroenterology" (T. J. Thompson and I. E. Gillespie, eds.), pp. 58–66. Ballière, London.

Nassett, E. S. (1965). *Federation Proc.* **24,** 955.

Nassett, E. S., and Ju, J. S. (1961). *J. Nutr.* **74,** 461.

Nassett, E. S., Ganapathy, S. N., and Goldsmith, D. P. J. (1963). *J. Nutr.* **81,** 343.

Neame, K. D., and Wiseman, G. (1957). *J. Physiol.* (*London*) **135,** 442.

Onen, K. H., Wade O. L., and Blainey, J. D. (1956). *Lancet* **271,** 1075.

Parshin, A. N., and Rubel, L. N. (1951). *Dokl. Akad. Nauk. SSSR* **77,** 313.

Peraino, C., and Harper, A. E. (1963). *J. Nutr.* **80,** 270.

Pion, R., Fauconneau, G., and Rerat, A. (1964). *In* "The Role of the Gastrointestinal Tract in Protein Metabolism" (H. N. Munro, ed.), pp. 309–321. Blackwell, Oxford.

Richmond, J. E., Shoemaker, W. C., and Elwyn, D. H. (1963). *Am. J. Physiol.* **205,** 848.

Schimassek, H., and Gerok, W. (1965). *Biochem. Z.* **343,** 407.

Schimke, R. T. (1962). *J. Biol. Chem.* **237,** 459.

Schimke, R. T. (1964). *J. Biol. Chem.* **239,** 3808.

Shoemaker, W. C. (1960). *J. Appl. Physiol.* **15,** 473.

Shoemaker, W. C., Walker, W. F., VanItallie, T. B., and Moore, F. D. (1959). *Am. J. Physiol.* **196,** 311.

Swendseid, M E., Yamada, C., Vinyard, E., Figuera, W. G., and Drenick, E. J. (1967). *Am. J. Clin. Nutr.* **20,** 52.

Waldman, T. A., Morell, A. G., Wochner, R. D., Strober, W., and Sternleib, I. (1967). *J. Clin. Invest.* **46,** 10.

Wheeler, P., and Morgan, A. F. (1958). *J. Nutr.* **64,** 137.

Winitz, M., du Ruisseau, J. P., Otey, N. C., Birnbaum, S. M., and Greenstein, J. P. (1956). *Arch. Biochem. Biophys.* **64,** 368.

Wunner, W. H., Bell, J., and Munro, H. N. (1966). *Biochem. J.* **101,** 417.

CHAPTER 39

The Role of the Kidney in the Regulation of Protein Metabolism

GEORGE F. CAHILL, JR.,

Elliott P. Joslin Research Laboratory, Department of Medicine, Harvard Medical School, Boston, Massachusetts

AND

OLIVER E. OWEN

Temple University School of Medicine, Philadelphia, Pennsylvania

I. Introduction

The kidney plays an important part in overall nitrogen homeostasis by virtue of its role as a primary excretory organ. This role is more or less passive with regard to urea excretion, which is mainly a function of the concentration of this substrate in the blood. In contrast, ammonia excretion is actively regulated by the kidney, and under certain conditions it may equal or exceed urea excretion. Ammonia production, control, and its correlation with renal gluconeogenesis is emphasized in this review.

The kidney has been implicated as a primary site of proteolysis, particularly for proteins in circulatory fluids. The weight of evidence, as presented, suggests that this proteolysis under normal conditions is probably limited to the very small amount of protein with low molecular weight which is filtered through the glomerular sieve.

Control of the kidney's own mass, as a reflection of renal protein homeostasis, continues to be poorly defined, and, as will be discussed, both excretory load and circulating hormonal factors appear to be involved.

Finally, there is an enormous amount of data indicating that with renal failure, particularly with proteinuria, many metabolic alterations occur throughout the body, particularly in the liver; these last-mentioned phenomena will not be discussed in this brief review.

II. Control of Kidney Mass

How the mass and function of an organ or tissue are integrated into total body needs is a problem shared by the kidney, as well as by many other systems, particularly liver, heart, and bone marrow. That such a regulation does definitely exist has been known for over half a century by the anatomists, who frequently noted that a solitary kidney, as long as it is not diseased, is larger than it would be were the animal to possess the contralateral organ. Other evidence that such a regulation does exist is the very close correlation within a given species between renal function and body surface area (Smith, 1955), which, in itself, is closely correlated to overall metabolic mass.

A part, but not all, of renal growth and its integrated relationship with overall mass is a function of growth hormone, as reflected by the increased renal mass and function noted in animals or man treated with growth hormone or with spontaneous acromegaly. Measures of renal function augmented by growth hormone include an increased glomerular filtration rate, increased renal blood flow, as well as an increased tubular transport for PAH, glucose, and other actively absorbed or excreted metabolites. It is obvious, however, that growth hormone cannot control the size of a specific organ in relation to the rest of the body. The specific control mechanism for kidney has been searched for, but remains ill-defined. There has been evidence for several decades that the excretory load placed on the kidney is a stimulus to its growth. For example, MacKay *et al.* (1928) found a linear correlation between dietary protein and renal size in rats and dogs. No experimental evidence has been given in man.

A. Compensatory Renal Hypertrophy

Compensatory renal hypertrophy after unilateral nephrectomy, as mentioned above, has been well known for decades and has served as the experimental model for investigation of renal growth mechanisms. A recent review by Bucher (1967) described in detail the known biochemical processes in liver regeneration following subtotal hepatectomy, and these processes are probably applicable to kidney. In addition, this author discussed two mechanisms whereby an organ's mass can be autoregulated. First, the tissue may release, per cell, a material which can be detected and which can control directly or indirectly the tissue's overall mass by inducing hyperplasia and/or hypertrophy. Such a mechanism is well illustrated by the thyroid-pituitary axis of the endocrine system. Second, the presentation of a substrate or accumulation of a product can serve as a monitor of organ mass, particularly in an excretory organ. For liver, the choice betweeen the latter, substrate control, or the former, a type of hormonal control, is still not clear, and many experimental data support each hypothesis, as reviewed by Bucher (1967). Both may be involved.

In man the kidney is different from liver: the newborn has his lifelong number of nephrons. There is no evidence that an increase in the number of nephrons ever occurs after completion of fetal development. Any increase in function must be accomplished by increased filtration per glomerulus in addition to increased number and perhaps size of tubular cells in each nephron. After unilateral nephrectomy there is an upper limit of renal compensatory function in the remaining kidney to about 70% of that of two normal kidneys. In contrast, the liver can form new lobes and return to 100% of normal function after a major subtotal resection with a tenfold increase over the starting size.

Rollason (1949) noted a marked increase in the number of mitoses in tubular cells 48 hours after unilateral nephrectomy in young rats. Schaffenburg *et al.* (1954) found a significant increase in renal tissue weight as early as 24 hours after surgery and noted this to occur in hypophysectomized, as well as normal animals, suggesting nonpituitary factors to be responsible. A very interesting observation was made by Paschkis *et al.* (1957), who noted that if a subtotal hepatectomy was performed coincident with the unilateral nephrectomy, the degree of compensatory renal hypertrophy was accentuated. These data suggest some overlap between renal and hepatic stimuli, but another possibility would be more available substrate, such as an increased level of amino acids. This briefly described observation of Paschkis *et al.* has not been corroborated or refuted.

That the excretory load presented to the kidney was not responsible for the hypertrophy was suggested by Simpson (1961), who performed ureteroperitoneostomies in rats, thus returning the excreted load to the animal, and failed to observe any evidence of compensatory growth. This observation was countered by Royce (1963), who demonstrated that peritonitis or food or water restriction could each prevent compensatory renal hypertrophy, thus suggesting the observation of Simpson to be due to nonspecific stress effects inhibiting the compensatory hypertrophy. In agreement, Goss (1965), performing ureteral ligation, did observe compensatory hypertrophy, suggesting the metabolic load as being a most important stimulus. He also observed that the adrenal gland was necessary for this process and that salt or deoxycorticosterone administration was able to correct for the lack of the adrenal gland, thereby eliminating glucocorticoid as an essential factor.

That no humoral agent served as the signal was suggested by Williams (1962), who injected serum from nephrectomized rats into other rats and failed to find an increase in mitoses. In contrast, Lowenstein and Stern (1963) described a significant increase in tritiated thymidine incorporation into kidneys of rats receiving serum obtained from unilaterally nephrectomized animals. As control, no increase in labeled thymidine incorporation into liver was noted in these same animals.

Royce (1967, 1968) has partially characterized an α-globulin with a molecular weight of approximately 18,500. This protein in rat serum is in very low concentration (14 μg/ml) and rises within one hour to 76 μg/ml after nephrectomy due, persumably, to cessation fo renal removal. Further evidence suggests that of this protein which is filtered by the glomerulus, approximately one-half is absorbed from tubular urine, the remainder being excreted. It is suggested that this protein or a similar material of low molecular weight which is removed mainly by kidney may serve as a signal for compensatory hypertrophy after reduction in renal mass. Although yet hypothetical, the rapidity and degree of the change in the concentration of this material are timewise in accord with the earliest chemical changes in RNA, as will be discussed below.

B. Growth Hormone

The role of growth hormone in compensatory renal hypertrophy was most thoroughly studied by Astarabadi (1963b), who noted that growth hormone augmented compensatory renal hypertrophy in hypophysectomized rats, but that the hypertrophy "factor" and growth hormone effects were independent and additive. Nakamura *et al.* (1964) reported an increase in tritiated thymidine incorporation into the pituitary and thyroid glands of animals following unilateral nephrectomy, suggesting a pituitary- or thyroid-mediated servomechanism, but this isolated observation remains to be corroborated or further clarified.

In man, compensatory renal hypertrophy is well documented, and the recent use of normal donors for renal transplantation has provided several excellent studies of the effects of unilateral nephrectomy in non-diseased subjects. These studies (Ogden, 1967; Krohn *et al.*, 1966; Donadio *et al.*, 1967) have shown the remaining kidney to assume 70% of the total renal function that existed prior to the unilateral nephrectomy. Increase in function of the remaining kidney, although measurable within weeks, may take months and years (Mitchell and Valk, 1962) to achieve a maximum. It is also more striking in younger than in older subjects.

C. Excretory Load as Stimulus

Some of the previous discussion has suggested that renal mass may be partly a function of excretory demands. Lotspeich (1965) found that administration of an acid load to rats caused an increase in renal weight. After 25 hours, protein and RNA contents were significantly increased. However, unilateral nephrectomy, combined with the acid load, caused an even greater increase in this response, so one cannot assume that acid loading, *per se,* is the sole stimulus. The acid-induced hypertrophy was blocked by low doses of actinomycin D, as was the increase in glutaminase activity (Lotspeich, 1967). Actinomycin D also, as might be expected, blocked the compensatory hypertrophy after unilateral nephrectomy. Finally, with the acid loading, as mentioned above, glutaminase activity was increased, as was glutamine extraction; however, with compensatory renal hypertrophy after unilateral nephrectomy, there is no increase in these moieties, clearly distinguishing the two growth-promoting stimuli as working through separate mechanisms. Katz and Epstein (1967a,b) found that the increase in kidney weight which occurs in the first 24 hours after unilateral nephrectomy is not accompanied by a demonstrable change in glomerular filtration rate. After 3 days, filtration and sodium reabsorption, expressed per gram of kidney, are 27% above normal. The data are interpreted as showing that increased reabsorptive work, at least as represented by sodium loading to the tubules, is not the initial stimulus leading to compensatory renal hypertrophy. Subsequently, Hayslett *et al.* (1968), using micropuncture techniques to study fuel and electrolyte exchange, found proximal and distal tubules to respond differently to compensatory hypertrophy. Both sites showed increases in tubular volume, but the proximal increase was far greater. Reabsorptive half-time in the distal tubule, on the other hand, was shortened markedly. These data suggest that even within renal substructures, such as tubules, hypertrophy (or hyperplasia), during compensatory hypertrophy is not uniform as evidenced by functional alterations.

Many other physiological stimuli also produce renal hypertrophy. Androgen administration to several species (man, dogs, rats, mice, to name a few) has been shown to increase renal mass and function (Dorfman and Shipley, 1956). Similar effects have been noted with thyroid administration. However, the decreased renal size in hypophysectomized animals is not returned to normal with prolactin or ACTH administration (Astarabadi, 1963a,b).

Much attention has recently been directed to the initial biochemical events in compensatory hypertrophy, and a recent volume (Nowinski and Goss, 1970) has been devoted to this topic. Malt and collaborators, as described in a section in the aforementioned volume and in a separate extensive review (Malt, 1969), have studied the early changes in RNA isolated from kidney cortex after contralateral nephrectomy. Within 10 minutes a decrease in a fraction of nucleoplasmic RNA (Hn RNA $>$ 28S), previously labeled with ^{3}H-uridine, was noted suggesting this event to precede synthesis of renal cytoplasmic RNA (Willems *et al.*, 1969). Within the next 2 days, as renal mass increases by 10% or less, the RNA/DNA ratio increases 20–33%, and the high ratio of free/bound microsomes of 3/1 in kidney (Priestley and Malt, 1969) remains unchanged during this active phase of growth. It is enticing to correlate the very rapid increase in the protein factor described by Royce (1968), its increased concentration in filtered urine, and the concomitant alteration in nucleoplasmic RNA metabolism noted by Malt (1969).

In summary, normal kidney has the capacity to increase its size and weight by hypertrophy and hyperplasia of the glomeruli (Ungvary and Faller, 1967) and tubular cells, mainly proximal (Threlfall *et al.*, 1967), and its functional work by increasing glomerular filtration (Katz and Epstein, 1967a) and some enzymic specific activities involved in transport as expressed on the basis of protein weight (Katz and Epstein, 1967b). As discussed, the responsible servomechanisms for these responses remain to be defined.

III. Kidney as a Proteolytic Organ

A. Glomerulus as a Molecular Sieve for Proteins

In general terms, protein catabolism by kidney is a function of glomerular filtration of plasma proteins. Studies using model molecules which are metabolically inert have shown that the glomerular membrane is essentially impervious to substances larger than 50,000–60,000 molecular weight (Wallenius, 1954). On the other hand, substances with a molecular weight less than 7000 permeate freely. Between these weights there are variable permeabilities (Hardwicke and Soothill, 1961). To explain this selectivity, Pappenheimer (1953) has suggested pores with a diameter of 60–90 Å which occupy 1–2% of the available filtration area. Chinard *et al.* (1955), on the other hand, postulated a nonporous gel-like membrane through which all components of plasma may diffuse; and the extremely low diffusion rates of macromolecules prevent their permeation through this gel. Govaerts and Lambert (1954) calculated that both mechanisms may exist, filtration through pores and diffusion through a gel, with diffusion being three times larger than filtration in the case of hemoglobin transfer. Morphologically demonstrable pores in the glomerular basement membrane, although originally reported (Spiro, 1959), have not been observed by more detailed electron microscopy even in abnormal states associated with marked urinary loss of many macromolecules (Farquhar and Palade, 1961). Nevertheless, steroid-induced decreases in proteinuria are not accompanied by alterations in the pattern of proteinuria, suggesting a decreased number of "pores" (Vere and Walduck, 1966). The older study of Dock (1942), whereby active tubular metabolic processes were blocked by perfusion of kidneys with media at 0°C and no significant proteinuria was observed, heavily favors the nonporous hypothesis of Chinard.

Since glomerular filtration approximates 180 liters of plasma per day, and since only 10–80 mg of protein appear daily in the urine of a normal man (Tidstrom, 1963a,b), an overall renal efficiency of 99.99% is achieved. The actual efficiency is even greater, since protein from desquamated urothelium and protein (such as one or more of the uromucoids) added directly by other nonglomerular processes are not taken into account.

B. Protein Reabsorption

Addis (1948) had originally postulated that as much as 18 gm of protein per day might be filtered and reabsorbed, and if catabolized would have accounted for the major portion of plasma protein catabolism. Whether a significant quantity of protein is filtered and actively reabsorbed in an intact form as a normal physiological process continues to be queried, and conclusive information is lacking. Current opinion is now inclined to the hypothesis that some small proteins (or peptides) may be filtered, reabsorbed, and catabolized; but quantitatively, in terms of overall protein catabolism, this amount is relatively insignificant. In certain renal diseases involving primarily the tubule, such as tubular acidosis, Fanconi's syndrome, or cadmium poisoning, a slightly different spectrum of protein may appear in the urine when compared to the proteinuria associated with glomerular disease, suggesting tubular protein reabsorption to be quantitatively a significant process in the normal state (Butler *et al.*, 1962; Creeth *et al.*, 1963). It is dangerous to assume, however, that any disease process is specific for the tubule; and furthermore, the difference in the pattern of proteinuria between supposedly tubular and glomerular disease is actually very slight.

One must be guarded in accepting the so-called "normal" level of protein (40 mg/100 ml) found by numerous investigators in glomerular fluid obtained by micropuncture, since many minimal manipulations to even the normal intact animal induce proteinuria, as does slight exercise or even changes in posture in normal man (Strauss and Welt, 1963). For example, simple stop-flow procedures (Lathem *et al.*, 1960) may result in the increased appearance of protein in the glomerular filtrate. Likewise, most of the experimental data supporting this alleged 40 mg/100 ml in normal filtrate were obtained in the rat, an animal noted for its "normal" proteinuria, particularly in the male.

Using immunochemical methods for determining protein concentration, Dirks *et al.* (1964) found proximal tubular fluid in the dog to contain less than 2 mg/100 ml albumin in half of micropuncture fluids examined. γ-Globulin was essentially excluded. This study adds credence to the hypothesis that the normal glomerular sieve is relatively impermeable to molecules of this dimension.

C. Protein Catabolism

Early studies on mice, using impure protein preparations, seemed to indicate that bilateral nephrectomy prolonged the half-life of radioactive iodoalbumin from 0.7 to 1.4 days (Gitlin *et al.*, 1958). It has been recognized that iodination causes serious denaturation of proteins (see Chapter 8). Damaged molecules are more rapidly catabolized than their intact counterparts and are actively catabolized by the perfused kidney (Katz *et al.*, 1960a). The labeled and metabolically unstable molecules can be selectively removed from a preparation by biological screening (McFarlane, 1956; see also Chapter 8). With appropriately screened homologous ^{131}I-labeled albumin, no significant decrease in the catabolic rate in rats was observed (Katz *et al.*, 1960a). Concurrent experiments with kidney slices or homogenates either failed to show any significant albumin catabolism or were inconclusive (Armstrong and Tarver, 1960; Katz *et al.*, 1960b). Rosenfeld *et al.* (1962) soon confirmed their observed lack of renal catabolism of albumin noted in rats by similar studies in dogs. However, when rats were made nephrotic by amino nucleoside or nephrotoxic serum, nephrectomy halved the catabolic rate (Katz *et al.*, 1964), an effect far beyond that expected by elimination of the proteinuria. Thus, a clear relationship between escape of protein into the glomerular filtrate and catabolism of protein was demonstrated.

Recently, three lines of evidence have supported the hypothesis that very little protein is filtered in the normal kidney and that of the small amount that is filtered a portion may appear in the urine and another portion is catabolized but *not* reabsorbed intact by the tubule. First, Rosenfeld *et al.* (1967) perfused the rabbit kidney *in vitro* and found essentially no albumin catabolism. When glomerular permeability was increased by either addition of renin to the system or by perfusing with an excessive concentration of plasma protein, both proteinuria and catabolism appeared simultaneously. Likewise, Cora *et al.* (1962) found far greater arteriovenous differences of protein in man with diseased kidneys than could be explained by the magnitude of proteinuria. A second line of evidence was obtained by studies of low molecular weight proteins such as the Bence-Jones proteins of plasma cell dyscrasias. These polypeptides have molecular weights of approximately 22,000 and are sufficiently small to pass through the nondiseased or otherwise unaltered glomerulus. The observation of Solomon *et al.* (1964) in man that 50–90% of the removal rate of Bence-Jones protein is endogenous, and not due to proteinuria, suggests that renal protein catabolism may be coupled with filtration, absorption, and proteolysis. Wochner *et al.* (1967) found that, in mice, renal catabolism of Bence-Jones protein and isolated light chains accounted for over 75% of their total catabolism as evidenced by decreased removal rates after nephrectomy. However, among the less common forms of plasma cell dyscrasias are those described by Franklin *et al.* (1964) and Osserman and Takatsuki (1964) which are associated with the elaboration of excessive quantities of polypeptides with molecular weights of approximately 55,000 and related to the heavy chains (and, more specifically, to the F_c fragment) of IgG immunoglobulins. Proteinuria (4–15 gm/24 hours) was the initial indication of a protein abnormality in these patients. Wochner *et al.* (1967) were unable to show in mice that the kidney catabolized heavy-chain immunoglobulin fragments. Thus, the mere presence of protein in renal tubular filtrate cannot be simply equated to protein catabolism. Furthermore, Wochner and his associates (1967) showed that ureter-severed and ureter-ligated mice were identical in their ability to catabolize Bence-Jones protein. Ureteral ligation is known to decrease glomerular filtration. Thus, whether renal catabolism of light-chain (Bence-Jones) protein occurs in the course of resorption of filtered protein by the tubular epithelium or after entry of light-chain protein into the renal parenchyma via the peritubular capillaries has not been definitely determined. Both processes may be operative and additive in catabolism of such polypeptides.

The third line of evidence is derived from studies on specific proteins. Chamberlain and Stimmler (1967) found an arteriovenous insulin difference of 29% across the kidneys in man, this fraction being independent of the insulin concentration. However, only 1.5% of the filtered load, if this 29% had been cleared by filtration, appeared in the urine; thus, the remainder was degraded at least to a sufficient degree to render the molecule unreactive in the immunoassay. With progressive tubular disease, the amount of insulin in urine approached the amount filtered. Similar data were obtained by Rubenstein *et al.* (1967), who, measuring insulin in plasma and urine (mol. wt. 6000 as the monomer or 12,000 as the dimer), found it to be cleared into urine at a rate of only 0.42 ml of plasma per minute and considered it to be either filtered and reabsorbed or else to exist as a higher molecular weight aggregate and therefore not filtered. In view of the previous reference, the first hypothesis is the more probable. Similarly, many studies have failed to demonstrate insulin in physiological solutions as other than a single, low molecular weight, unbound monomer or dimer (Rasio *et al.,* 1967). More recent data by McCormick *et al.* (1969) have shown that in the dog, insulin clearance by the kidney may be greater than that amount passing through the glomerulus, supporting prior studies by Rubenstein and Spitz (1968). Thus, renal protein extraction is not limited to that amount present in tubular urine but also may occur at other sites, presumably by uptake directly from peritubular capillaries.

Again using insulin as the model (thanks to the interest of other disciplines in its metabolic fates), the effective total clearance of the amount filtered through the glomerulus serves as an excellent mechanism to clear blood of it and other small molecular weight materials. Although this process may appear wasteful at first consideration, it does provide a means for relatively rapid removal rates of components in biological feedback loops with brief time constants; thence, it is not surprising that protein hormones with short half-lives, such as insulin, thyrotropin (Bakke *et al.,* 1962; Odell *et al.,* 1967), ACTH (Syndor and Sayers, 1953) and growth hormone (Parker *et al.,* 1962), may all be rapidly catabolized by filtration, absorption, and proteolysis, thereby permitting a more variable and rapid hormonal control mechanism with a brief time constant.

D. Pinocytosis and Emiocytosis

It is well known that droplets of protein-like material may appear in the tubule in subjects with renal disease and proteinuria (Oliver and

MacDowell, 1958). These have been interpreted, and probably correctly so, as evidence of protein reabsorption. In general, there appears to be a close correlation between these droplets and the severity of proteinuria (Mostofi and Smith, 1966; Strauss and Welt, 1963). A reasonable hypothesis is that these droplets represent a temporary saturation of the proteolytic process rather than bulk transport of protein through the tubular cell, since no observer has found any evidence of discharge (emiocytosis) of the vacuole contents on the vascular side. In contradiction to this hypothesis, however, is the fact that each patient with renal glomerular disease appears to excrete an almost unique pattern of urinary protein (Joachim *et al.*, 1964). Since this suggests more selective processes of glomerular leakage and tubular reabsorption, processes which are difficult to explain by bulk pinocytosis, there are still some doubts as to the overall nonspecificity of protein reabsorption by the tubule.

In summary, renal protein catabolism appears to be related to the capability of the molecule to be filtered through the glomerulus. Normally, this is limited to the trace quantities of proteins which have small molecular weights. With increased permeability due to diseased glomeruli or to diseases with accumulation of large quantities of small molecular weight protein, such as Bence-Jones protein, renal protein catabolism may become extremely important and may surpass the remainder of endogenous catabolism plus that amount excreted in the urine (Wochner *et al.*, 1967). In support of the minimal proteolytic role of the kidney in the normal state, studies by the authors in man during prolonged fasting have shown that arteriovenous balances of amino acids and various other metabolites such as glucose, lactate, and pyruvate across the kidney fail to reflect any significant proteolysis (Owen *et al.*, 1969). These data are discussed Section IV.

IV. Nitrogen Excretion and Renal Metabolism

In this section, certain aspects of amino acid and protein metabolism in kidney will be discussed with special emphasis on the metabolic role of renal cortex as it relates to overall nitrogen economy. By necessity, enzyme data or data referrable to substrate concentrations in tissues are derived from animal experimentation; however, those relating to nitrogen balance and arteriovenous differences of substrate will be primarily from experiments on man. This emphasis on human data is due to the availability of test subjects, the well-developed techniques for renal catheterization and flow determinations, and the obvious clinical relevance of such data. Previous sections (Chapter 11 by Allison and Bird, and Chapter 18 by Holt and Snyderman) in this treatise have dealt in further detail with specific amino acid reabsorption and nitrogenous products appearing in urine.

A. Body Nitrogen Stores

In adult man (and other animals in overall steady state, comparing one day to the next), food consumption equals caloric need. In contrast, protein consumption must be equal to or greater than physiological requirements. This necessity stems from the lack of a specific depot of protein or amino acids, in contrast to the massive store of fat as adipose tissue triglyceride and the relatively smaller, but yet significant, store of carbohydrate as glucose and glycogen. Surplus amino acid is either metabolized to carbohydrate or to lipid, or oxidized as fuel, and the nitrogen is excreted in the urine. Accumulation of catabolic nitrogen is incompatible with life.

In order to place body protein into perspective in man, Table I lists the stores of the above-mentioned fuels in a normal 70-kg man, and, in contrast, compares similar data in an extremely obese subject of approximately 150 kg.

Although obese individuals usually have a slightly greater protein mass (necessary to carry about this excessive carcass), the major difference between normal and obese is the fivefold or more increase in adipose tissue. The relatively insignificant carbohydrate reserve is apparent. The daily net flux through these pools is obviously a function of diet. Caloric intake in all humans, obese and thin, whether in the form of protein, carbohydrate, or fat, if in excess of caloric expenditure, is mainly converted into lipid and stored in adipose tissue. Accordingly, surplus amino acids must be metabolized and excreted, and the corresponding residue oxidized for its caloric value or converted into carbohydrate or fat for storage.

B. Nitrogen Flux

The average sedentary man living in an affluent society eats approximately 300 gm of carbohydrate, 100 gm of protein, and 100 gm of fat

TABLE I

Body Fuel Stores in Normal and Obese Subjects

Store	Normal	Obese
Fat (adipose triglyceride) (kg)	15.00	80.00
Protein (kg)	6.00	8.00
Glycogen, muscle (kg)	0.12	0.16
Glycogen, liver (kg)	0.07	0.07
Glucose, extracellular fluid (kg)	0.02	0.03

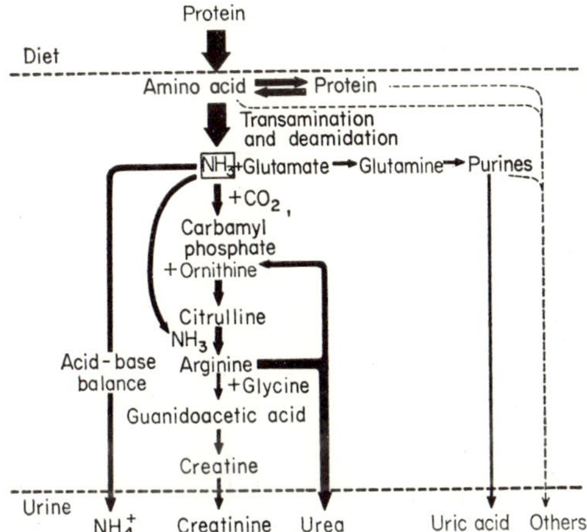

FIG. 1. Schematic representation of nitrogen metabolism, tracing dietary protein intake to urinary nitrogen products.

daily. Thus, approximately 16.5 gm of nitrogen must be excreted, and about 95% of this process is a function of the kidney, the remainder being excreted as stool nitrogen in addition to trace quantities in other excretions and exuviae. As shown in Fig. 1, the bulk of urinary nitrogen is excreted as four components, urea and ammonia predominating with lesser amounts as uric acid and creatinine. Nitrogen in various other forms such as proteins, polypeptides, glycoproteins, amines, amino acids, purines, and pyrimidines appear as normal trace constituents in urine.

The primary pathway for surplus nitrogen elimination in man is urea synthesis from ammonia and carbon dioxide in liver, release of urea into the blood, and clearance by the kidney. In man consuming the usual occidental diet, urea constitutes 80–90% of nitrogen excreted (Fig. 2).

1. *Uric Acid*

A lesser alternative pathway for nitrogen excretion in man is provided by diversion of amino acid nitrogen into purines with subsequent uric acid excretion (Gutman, 1965). In the fed state, urate nitrogen accounts for only 2–4% of urinary nitrogen in man and reflects primarily that moiety of purine turnover which escapes reutilization.

Glutamine synthetase catalyzes the formation of glutamine from glutamic acid and ammonia. Nitrogen from glutamine, glycine, and aspartic acid not utilized for synthesis of various needed compounds or ammonia production can be converted *de novo* into inosinic acid and subsequently into adenine and guanine nucleotides. The catabolism of adenine and guanine bases results in formation of xanthine and hypoxanthine, respec-

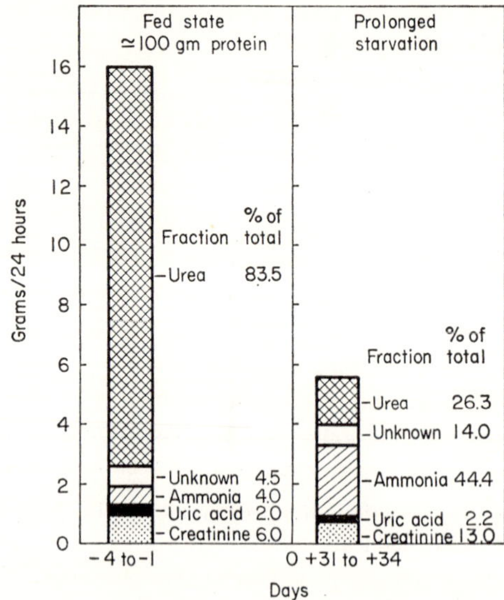

FIG. 2. Urinary nitrogen components in the 4 male obese adults pre and post starvation.

tively. The latter purines can be reutilized (Rosenbloom *et al.*, 1967) by conversion back into inosinic acid via the action of the widely distributed enzyme phosphoribosyltransferase or further metabolized by the hepatic enzyme xanthine oxidase producing uric acid for renal excretion. In mammals uric acid formation occcurs primarily in liver (Dietrich and Borries, 1954), and xanthine-oxidase activity disappears from rat liver after 2 weeks on a protein-free regime (White *et al.*, 1964). Starvation with its complete protein restriction results in ketonuria and a striking reduction in renal clearance of urate (Goldfinger *et al.*, 1965; Cahill *et al.*, 1966). The resulting hyperuricemia and xanthine-oxidase atrophy may promote conversion of hypoxanthine and xanthine back to inosinic acid. Elevated oxypurine levels have been related to congenital brain dysfunction, and prolonged hyperoxypurinemia probably cannot be tolerated (Lesch and Nyhan, 1964; Engelman *et al.*, 1964; Rosenbloom *et al.*, 1967).

2. Creatinine

Another important component, which, like uric acid, represents only a few percent of total excreted nitrogen, is creatinine, the anhydrous metabolite of creatine. As shown by numerous investigators, the quantity excreted is mainly a function of muscle mass, with a lesser component contributed by the diet if much meat is eaten. The initial step in endogenous creatinine formation takes place in the kidney, where arginine and glycine react through a process of transamidation to guanidoacetic acid (see Fig. 1). Most of the guanidoacetic acid is carried to the liver, where transmethylation with methionine converts it to creatine, and subsequent deposition in muscle as creatine phosphate. For details, see Chapter 11 of this treatise by Allison and Bird.

3. Ammonia and Urea

Ammonia nitrogen ordinarily constitutes 5–10% of daily urinary nitrogen and with urea serves as the principal form of nitrogen lost in the urine. Ammonia and urea differ in one important aspect; ammonia is excreted primarily as a cation, whereas urea is excreted without a net charge. The excretion of ammonia is of quantitative significance in relation to maintenance of acid–base balance but not in relation to elimination of waste nitrogen *per se* in the state of adequate nourishment (Pitts, 1964). However, the normal elimination of approximately 50 mEq of metabolic acid per day as ammonium salts can be increased severalfold in respiratory and starvation acidosis. In diabetic ketoacidosis, ammonium ion excretion may exceed 750 mEq/day (Pitts, 1948). In severe acidosis, NH_4^+ may even supplant urea as the principal urinary nitrogenous component. Thus, there is a reciprocity between these two moieties, their relative proportions in urine being dictated by the subject's need to maintain acid–base balance and their total sum being a function of nitrogen balance (see Fig. 2). Therefore, kidney and liver must mutually adapt to both total nitrogen flux and to synthesis of adequate amounts of ammonia to preserve acid–base homeostasis. In the absence of adequate nitrogen in the diet, acid–base homeostasis assumes first priority, as reflected by the fact that ammonia excretion in urine continues at the expense of body protein stores (Owen *et al.*, 1969). In fact, the largest expendable cation pool in man is protein nitrogen, which is potentially convertible to tens of thousands of milliequivalents of NH_4^+ in comparison to the limited and unexpendable sodium and potassium pools needed for the preservation of fluid volume.

The amino acids derived from peripheral protein breakdown not only supply ammonia, but also comprise the largest noncarbohydrate source of gluconeogenic precursors. Recently, renal ammoniagenesis and gluconeogenesis have been integrated; they will be discussed in the next section.

C. Ammoniagenesis

Nash and Benedict (1921) concluded that ammonia is synthesized in the kidney from precursors in arterial blood. Since then, numbers of investigators (Van Slyke *et al.*, 1943; Owen and Robinson, 1963; Shalhoub *et al.*, 1963; Pitts *et al.*, 1963) have shown glutamine to be the

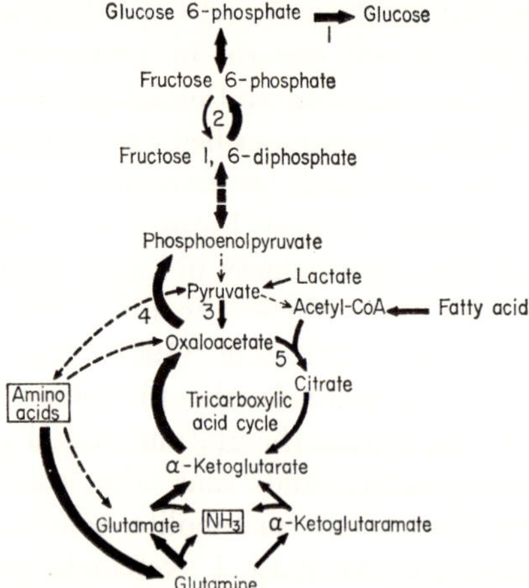

FIG. 3. Coupling of amino acid consumption in the acidotic kidney with ammonia and glucose production.

rats rendered acidotic by ammonium chloride administration, slices of kidney cortex were more active in producing glucose not only from glutamine, but also from glutamate, α-ketoglutarate, and oxaloacetate, but not from glycerol or fructose. Conversely, rats made alkalotic by sodium bicarbonate feeding showed decreased glucose synthesis from glutamine, glutamate, and α-ketoglutarate. Kidneys from rats relatively depleted of potassium, a maneuver associated with increased ammonia production, also exhibited increased glucose synthesis. Adrenalectomy was without effect. Goorno *et al.* (1967) reported a similarly accelerated rate of gluconeogenesis in kidney cortex of dogs rendered acidotic by ammonium chloride administration.

Kamm *et al.* (1967) evaluated the effect of altering pH and bicarbonate concentration on the rate of gluconeogenesis in rat kidney and found a fall in pH to be the most potent stimulus; however, for any given pH, a decrease in bicarbonate concentration also accelerated glucose production. These data all suggest that phosphoenolpyruvate carboxykinase (step 4 in Fig. 3) may be the rate-controlling step in gluconeogenesis from the aforementioned substrates, and, as such, is the step altered by acid–base manipulations. Goorno *et al.* (1967) also came to this same conclusion. In accord, Alleyne and Scullard (1969) demonstrated an increase in renal phosphoenolpyruvate carboxykinase activity within 6 hours after administration of ammonium chloride to rats. No change in glutaminase I activity was observed at this short time interval.

To assemble the above discussion and to interrelate gluconeogenesis and ammoniagenesis into a plausible theory, the following can be proposed: accelerated phosphorylation of oxaloacetate depletes tricarboxylic acid cycle intermediates, glutamate is decreased, glutaminase inhibition is removed, and ammonia is produced.

E. Extrarenal Glutamine Metabolism

Since glutamine is the primary donor of nitrogen for ammonia synthesis, it must be made readily available as a substrate for ammoniagenesis by the kidney. It is not surprising, therefore, that glutamine is plentiful in plasma, comprising one-quarter of the total plasma amino acids (Soupart, 1962; Dickinson *et al.*, 1965). Braunstein and Kritzman (1937) demonstrated wide distribution of tissue transaminases for amino acids in general, but the liver is undoubtedly the principal producer of glutamine. In fact, Lotspeich (1967) has found an increased rate of glutamine synthesis in livers of acidotic animals, again illustrating the close metabolic cooperation between the liver and the kidney.

F. Kidney's Energy Metabolism

The above discussion has correlated gluconeogenesis with ammoniagenesis leaving open the question of the source of the kidney's own metabolic needs. Acetyl-CoA for tricarboxylic acid cycle activity can be derived from pyruvate or from free fatty acids. Hohenleitner and Spitzer (1961) have found by renal vein catheterization that the uptake of free fatty acids provides the major proportion of fuel for the kidney's oxygen consumption. Uptake of fatty acids is associated with high levels of acetyl-CoA and their acyl-CoA derivates. These intermediates inhibit pyruvate dehydrogenation (Randle *et al.*, 1963; Garland and Randle, 1964) and enhance pyruvate carboxylation (Utter and Keech, 1960). Krebs *et al.* (1965) showed *in vitro* that fatty acids, acetoacetate, or β-hydroxybutyrate served as excellent metabolic fuels for kidney cortex slices, and all stimulate renal gluconeogenesis from glycerol, lactate, pyruvate, tricarboxylic acid cycle intermediates, and amino acids. Thus, fat or fat-derived products provide fuel for the kidney, the oxidation of other substrates being kept to a minimum.

Of interest are other metabolic changes noted in accelerated ammonia production. For example, Dies and Lotspeich (1967) noted increased activities of hexose monophosphate enzyme activities and in a subsequent publication found increased incorporation of glutamine-^{14}C carbon into

ferences of metabolic fuels. The mean renal blood flow in the 5 subjects was 1.207 liters per minute. By combining the measurement of renal blood flow with the arteriovenous concentration differences, the rate of utilization or production of specific metabolites can be quantitatively estimated. The calculated glucose production is 38 gm/24 hours. Although this is small compared to the average hepatic glucose production after an overnight fast, data after prolonged starvation show a marked attenuation of hepatic glucose production (Owen *et al.,* 1969). Of importance, however, is the stoichiometric correlation between observed glucose production and the theoretical glucose which could be made from the various glucogenic precursors (lactate, pyruvate, α-amino nitrogen and glycerol). This close approximation strongly supports the importance of fatty acids (or keto acids) as fuel for kidney and the close correlation between ammonia production and gluconeogenesis as observed in experimental animals. It also supports the concept that renal proteolysis is minimal under normal circumstances; otherwise the substrate balance would not have been observed.

In summary, these studies on substrate balance across the human kidney demonstrate that the deaminated and deamidated residues of ammonia production are quantitatively incorporated into glucose. Free fatty acids serve as the principal substrate for the kidneys' energy needs, the entire process participating in the body's overall effort to conserve carbohydrate or its precursors from combustion. Since amino acid uptake, ammonia production, and glucose synthesis are quantitatively in proportion to each other, little, if any, proteolysis can be taking place in kidney under these conditions.

V. Summary

The kidney participates in several mechanisms relating to total nitrogen metabolism. It serves to excrete urea, normally the primary excreted nitrogen metabolite. It actively regulates ammonia excretion, which, under certain conditions, may quantitatively approach or even surpass urea as the primary nitrogen loss. The control mechanisms involved in kidney's own protein mass and turnover are, at best, only poorly described. Finally, kidney serves normally as a proteolytic organ, but probably only to small protein or peptide molecules capable of passing through the normal glomerulus. As such, it plays an important role in the metabolism of certain peptide hormones, thereby providing a short half-life. Under abnormal conditions where there is an excessive accumulation of protein molecules of small molecular weight, or where there is increased glomerular permeability to large molecules, the kidney may be a major site for protein catabolism.

References

Addis, T. (1948). "Glomerular Nephritis, Diagnosis and Treatment." Macmillan, New York.

Alleyne, G. A. O., and Scullard, G. H. (1969). *J. Clin. Invest.* **48,** 364.

Armstrong, F. B., and Tarver, H. (1960). *Proc. Soc. Exptl. Biol. Med.* **103,** 323.

Astarabadi, T. (1963a). *Quart. J. Exptl. Physiol.* **48,** 80.

Astarabadi, T. (1963b). *Quart. J. Exptl. Physiol.* **48,** 85.

Bakke, J., Lawrence, H., and Roy, S. (1962). *J. Clin. Endocrinol.* **22,** 352.

Benoy, M. P., and Eliott, K. A. C. (1937). *Biochem. J.* **31,** 1268.

Braunstein, A. E., and Kritzmann, M. G. (1937). *Enzymologia* **2,** 129.

Bucher, N. L. R. (1967). *New Engl. J. Med.* **277,** 686 and 738.

Butler, E. A., Flynn, F. V., Harris, H., and Robson, E. B. (1962). *Clin. Chim. Acta* **7,** 34.

Cahill, G. F., Jr., Herrera, M. G., Morgan, A. P., Soeldner, J. S., Steinke, J., Levy, P. L., Reichard, G. A., Jr., and Kipnis, D. M. (1966). *J. Clin. Invest.* **45,** 1751.

Chamberlain, M. J., and Stimmler, L. (1967). *J. Clin. Invest.* **46,** 411.

Chinard, F. P., Vosburgh, G. J., and Enns, T. (1955). *Am. J. Physiol.* **183,** 221.

Cora, D., Debiasi, S., Dalla Rosa, C., Maggia, A., and Tonello, A. (1962). *Clin. Sci.* **23,** 331.

Creeth, J. M., Kekwick, R. A., Flynn, F. V., Harris, H., and Robson, E. B. (1963). *Clin. Chim. Acta* **8,** 406.

Davies, B. M. A., and Yudkin, J. (1952). *Biochem. J.* **52,** 407.

Dickinson, J. C., Rosenblum, H., and Hamilton, P. B. (1965). *Pediatrics* **36,** 2.

Dies, F., and Lotspeich, W. D. (1967). *Am. J. Physiol.* **212,** 61.

Dietrich, L. S., and Borries, E. (1954). *J. Biol. Chem.* **208,** 287.

Dirks, J. H., Clapp, J. R., and Berliner, R. W. (1964). *J. Clin. Invest.* **43,** 916.

Dock, W. (1942). *New Engl. J. Med.* **227,** 633.

Donadio, J. V., Jr., Farmer, C. D., Hunt, J. C., Tauxe, W. N., Hallenback, G. A., and Shouter, R. G. (1967). *Ann. Internal Med.* **66,** 105.

Dorfman, R. I., and Shipley, R. A. (1956). *In* "Androgens: Biochemistry, Physiology, and Clinical Significance," p. 506. Wiley, New York.

Engelman, K., Watts, R. E. W., Klinenberg, J. R., Sjoerdsma, A., and Siegmiller, J. E. (1964). *Am. J. Med.* **37,** 839.

Farquhar, M. G., and Palade, G. E. (1961). *J. Exptl. Med.* **114,** 699.

Franklin, E. C., Lowenstein, J., Bigelow, B., and Meltzer, M. (1964). *Am. J. Med.* **37,** 332.

Gabuzda, G. J., and Hall, P. W. (1966). *Medicine* **45,** 481.

Garland, P. B., and Randle, P. J. (1964). *Biochem. J.* **91,** 6c.

Gitlin, D., Klinenberg, J. R., and Hughes, W. L. (1958). *Nature* **181,** 1064.

Goldfinger, S., Klinenberg, J. R., and Seegmiller, J. E. (1965). *New Engl. J. Med.* **272,** 351.

Goldstein, L. (1965). *Nature* **205,** 1330.

Goldstein, L. (1966). *Am. J. Physiol.* **210,** 661.

Goldstein, L. (1967). *Am. J. Physiol.* **213,** 983.

Goldstein, L., and Copenhaver, J. H., Jr. (1960). *Am. J. Physiol.* **198,** 227.

Goodman, A. D., Fuisz, R. E., and Cahill, G. F., Jr. (1966). *J. Clin. Invest.* **45,** 612.

Goorno, W. E., Rector, F. C., Jr., and Seldin, D. W. (1967). *Am. J. Physiol.* **213,** 969.

Goss, R. J. (1965). *Proc. Soc. Exptl. Biol. Med.* **118,** 342.

Smith, H. W. (1955). "The Kidney." Oxford Univer. Press, London and New York.

Solomon, A., Waldmann, T. A., Fahey, J. L., and McFarlane, A. J. (1964). *J. Clin. Invest.* **43,** 103.

Soupart, P. (1962). *In* "Amino Acid Pools" (J. T. Holden, ed.), p. 220, Elsevier, Amsterdam.

Spiro, D. (1959). *Am. J. Pathol.* **35,** 47.

Strauss, M. B., and Welt, L. G. (1963). "Diseases of the Kidney." Little, Brown, Boston, Massachusetts.

Syndor, K. L., and Sayers, G. (1953). *Proc. Soc. Exptl. Biol. Med.* **83,** 719.

Threlfall, G., Taylor, D. M., and Buck, T. (1967). *Am. J. Pathol.* **50,** 1.

Tidstrøm, B. (1963a). *Scand. J. Clin. Lab. Invest.* **15,** 167.

Tidstrøm, B. (1963b). *Scand. J. Clin. Lab. Invest.* **15,** 259.

Ungvary, G., and Faller, V. (1967). *Acta Morphol. Acad. Sci. Hung.* **15,** 39.

Utter, M. F., and Keech, D. B. (1960). *J. Biol. Chem.* **235,** PC 17.

Van Slyke, D. D., Phillips, R. A., Hamilton, P. B., Archibald, R. M., Futcher, P. H., and Hiller, A. (1943). *J. Biol. Chem.* **150,** 481.

Vere, D. W., and Walduck, A. (1966). *Clin. Sci.* **30,** 315.

Wallenius, G. (1954). *Acta Soc. Med. Upsalien. Suppl.* **4,** 1.

Weber, G., Lea, M. A., Convery, H. J. H., and Stamm, N. B. (1967). *Advan. Enzyme Regulation* **5,** 257.

White, A., Handler, P., and Smith, E. L., eds. (1964). *In* "Principles of Biochemistry," p. 937. McGraw-Hill, New York.

Willems, M., Musilova, H. A., and Malt, R. A. (1969). *Proc. Natl. Acad. Sci. U.S.* **62,** 1189.

Williams, G. E. C. (1962). *Lab. Invest.* **11,** 1295.

Wochner, R. D., Strober, W., and Waldmann, T. A. (1967). *J. Exptl. Med.* **127,** 207.

CHAPTER 40

The Role of Skeletal and Cardiac Muscle in the Regulation of Protein Metabolism

V. R. Young

Department of Nutrition and Food Science, Massachusetts Institute of Technology, Cambridge, Massachusetts

I. Introduction

Skeletal muscle is the largest single tissue in the body of the mammal, and it can be assumed from this observation alone that it must play a significant role in metabolism. In the past, research has been directed largely toward the most obvious biological function of muscle, namely to transformation of energy from nutrients into mechanical work. The functions of skeletal muscle are, however, more complex and subtle than such a statement would imply, since the musculature of the body participates in many of the adaptations of the animal to environmental changes (Hensel and Hildebrandt, 1964). For example, many animals react to a moderate reduction in environmental temperature by nonshivering thermogenesis, a process in which the resting energy metabolism of skeletal muscle is increased without concomitant muscle contraction (Davis and Mayer, 1955; Hart, 1958). Since skeletal muscle forms such a large proportion of body mass, it is not surprising to find that it participates in other regulatory reactions in metabolism. Thus it is involved in general adaptive mechanisms to environmental change, such as those of water and mineral metabolism (Widdowson *et al.,* 1960), and also in such specialized adaptations as the provision of energy from muscle carbohydrate during arousal from hibernation (Lyman and Leduc, 1953; Zimny and Gregory, 1958; Hannon and Vaughan, 1961).

This chapter emphasizes the role of skeletal and cardiac muscle in protein metabolism, and tries to identify regulatory mechanisms in which these tissues play a part. First the anatomical features of skeletal muscle are described with an account of those chemical components of muscle relevant to protein metabolism. At the outset it needs to be emphasized that skeletal muscle consists not only of the contractile cells but also of various other cell types associated with the vascular and nervous supply, as well as the supporting connective tissue. The presence of these various cell types has not been emphasized sufficiently in previous biochemical studies concerned with muscle protein metabolism, and in the discussion which follows, attention is focused on those observations that are complicated by the occurrence of the various noncontractile cells in muscle tissue. The next section deals with changes occurring in skeletal muscle during growth and development. This is followed by a description of the free amino acid pools of muscle, with a discussion of muscle enzymes that are involved in protein metabolism, and then with a description of protein synthesis in muscle. This is succeeded by descriptions of the responses of skeletal muscle metabolism to nutritional changes, to hormones, and to exercise. A short section is devoted to protein metabolism in muscular dystrophy and in the denervated muscle. Finally, our present knowledge of protein metabolism in cardiac muscle is reviewed briefly.

II. General Features of Skeletal and Cardiac Muscle

A. *Proportion of Skeletal and Cardiac Muscle in Mammals*

Skeletal muscle is normally the largest tissue in the body of the adult mammal, being about 45% of body weight irrespective of species size (Chapter 25, Section IV,A). In this respect, it contrasts with the viscera, which form a much smaller proportion of body weight in large mammals than in smaller species. Possibly because of this, energy metabolism of muscle accounts for 25% of the total resting oxygen consumption of the mouse, 33% in the case of the rat, and 62% in the case of the dog (Martin and Fuhrman, 1955). The progression upward with increasing weight of the species is probably exaggerated by these data, however, since Havel (1966) has computed that, in resting man, skeletal muscle accounts for only 30% of his basal oxygen consumption and cardiac muscle for a further 11%. During severe exercise, energy expenditure in these two tissues can rise to as much as 90% of the total energy metabolism of the body. The relationship of the skeletal musculature to the viscera also changes during development. During growth of the rat, the proportion of skeletal muscle in the body increases, and since it has a lower Q_{O_2} than the viscera (Krebs, 1950; Martin and Fuhrman, 1955), the total metabolic rate per unit of body weight decreases (Chinn and Hannon, 1966). In old age the oxygen consumption of the rat per unit of body weight rises again (Ring *et al.*, 1964), and this has been attributed to the reduction in the proportion of muscle in the body with advancing age (Andrew *et al.*, 1959; Yiengst *et al.*, 1959; Barrows *et al.*, 1962; Neumaster and Ring, 1965). These various observations indicate that, in assessing the role of skeletal muscle in whole-body metabolism, the age and state of muscular development of the animal should be specified.

The contribution of skeletal muscle to the total body protein content of man can be seen at different stages of development in Table I. In the fetus of 20–22 weeks and at birth, skeletal muscle accounts for some

TABLE I

COMPARISON OF THE NITROGEN CONTENT AND PERCENTAGE CONTRIBUTION OF SKELETAL MUSCLE WITH THAT OF OTHER ORGANS OF MAN AT DIFFERENT AGES

Organ	Fetus, 20–22 weeks		Newborn		Adult	
	% Body weight[a]	% Body N	% Body weight[a]	% Body N	% Body weight[a]	% Body N
Skeletal muscle[b]	25	—	25	—	43	—
Total N	—	28.6	—	23.0	—	38.6
Noncollagen N	—	26.2	—	20.0	—	37.3
Heart[c]	0.6	0.6	0.5	0.43	0.4	0.3
Liver[c]	4	6.7	5	5	2	1.7
Kidney[c]	0.7	0.7	1	0.8	0.5	0.4
Brain[c]	13	8.2	13	5.3	2	1.0
Body weight (kg)[d]	0.271	—	3.5	—	65	—
Body N (gm N/kg whole body)[d]	—	13.3	—	22.6	—	34.0

[a] From Widdowson (1968).
[b] Calculated from Dickerson and Widdowson (1960).
[c] Calculated from Widdowson and Dickerson (1960).
[d] From Widdowson and Dickerson (1964).

25% of the weight and total N content of the body, whereas the liver represents 7% of body N, and the heart 0.6% of body N. During postnatal growth, there are two phenomena that increase the relative proportion of total body protein present in muscle (Widdowson and Dickerson, 1964). First, the weight of muscle increases out of proportion to the rest of the body, so that in adult man it forms about 43% of body weight (Table I). Second, the increase in the percentage of protein in the tissues as the infant grows (chemical maturation, as described in Chapter 26, Section II) is more marked in muscle and in skin than in most other tissues (Widdowson and Dickerson, 1960; Dickerson and Widdowson, 1960). In consequence, the skeletal musculature of the adult human subject accounts for about 40% of the total protein of his body (Table I). The increase in the percentage of body protein found in muscle from about 20% at birth to about 45% in the adult has also been noted in other species, including the rat (Moulton, 1923; Spray and Widdowson, 1950; Chinn and Hannon, 1966; Cheek, 1968). This increase in muscle mass relative to other tissues probably reflects the need for muscular activity as growth proceeds. Between birth and weaning, there is presumably a limited need for muscular movement, but thereafter the growing animal becomes more dependent on movement for obtaining food and there is a coincident acceleration of muscle development. This aspect is considered further in Section III,B.

B. Cell Types in Skeletal and Cardiac Muscle

As in other tissues, the cell population in both skeletal and cardiac muscle is made up of a mixture of cell types (Table II). In the skeletal muscle of the adult mouse or rat, about 60–80% of the nuclei are present in contractile cells, the remaining nuclei belonging to endothelial cells of capillaries, fibrous tissue cells, pericytes (perivascular cells), and satellite cells first described by Mauro (1961) and confirmed in several species (Midsukami, 1964; Flood, 1964; Muir *et al.*, 1965; Venable, 1966a,b). In general, the percentage of nuclei belonging to contractile cells in the skeletal muscles of the rat is not very different from the early postnatal period to maturity (Enesco and Puddy, 1964). However, during development the contribution of contractile cells to the muscle mass increases owing to a thickening of the muscle fibers. In suckling rats, muscle fibers constitute some 70–77% of total muscle weight, but from the prepubertal stage onward, they account for about 85% of muscle mass (Enesco and Puddy, 1964). It is possible that the satellite cell nuclei in the muscle of newborn animals is high [20% according to Muir *et al.* (1965) for newborn mice] and decreases with normal growth and development [2–5% according to Venable (1966a,b) for adult mouse levator ani muscle].

In the case of the heart, there are in contrast considerable changes in the ratio of contractile to interstitial cells as growth progresses (Sasaki *et al.*, 1968a). At birth, there are more nuclei belonging to contractile muscle cells than to interstitial cells, but by 2 weeks of age the number of these two classes of nuclear types are about equal. As growth progresses the interstitial cells outstrip the contractile elements, so that in the mature 2-year-old rat, only 23% of the nuclei in the heart belong to contractile cells. There is some evidence (Linzbach, 1947; Naeye, 1965, 1967) that the proportion of contractile cell nuclei in human heart may be even lower than in the rat. The relative volume occupied by contractile cells per nucleus is about 115% of that contributed by the interstitial cells in the human infant heart (Naeye, 1967); this compares with the heart of young mice, where the contractile cells occupy about five times the volume relative to the interstitial cells (Naeye, 1966).

The fine structure of mammalian skeletal muscle has been studied extensively. The function of muscle proteins is given by reversible changes of their molecular arrangement. The detailed structure of the contractile elements in both skeletal muscle (Huxley, 1960; Bennett,

TABLE II

Cell Types in Skeletal and Cardiac Muscle of Mice and Rats

Species	Age	Muscle	Percent of total cells					References
			Contractile	Endothelial	Fibroblast	Pericyte	Satellite	
Mouse	Adult	Levator ani	57[a]	21	14	6	2	Venable (1966a)
Rat	Adult	Various	65	25	—	10 (perimysial)	—	Enesco and Puddy (1964)
Rat	Adult	Various	82–85[b]	—	—	—	—	Enesco and Puddy (1964)
Rat	Adult	Heart	27	73 (interstitial)				Morkin and Ashford (1968)
Rat	Adult	Heart	24	76 (interstitial)				Sasaki *et al.* (1968a)
Rat	Day-old	Heart	58	42 (interstitial)				Sasaki *et al.* (1968a)

[a] Percentages on the basis of nuclear types.

[b] Values on a total muscle weight basis.

1960; Price, 1963) and cardiac muscle (James and Sherf, 1968) have been the subject of recent reviews. The contractile cell of skeletal muscle (Fig. 1) consists of interdigitating thick and thin fibrils (filaments) and two systems of intermyofibrillar membranes, namely, the sarcoplasmic reticulum (reviewed by Porter, 1961; Peachey, 1965a,b; Smith, 1966) and the transverse tubular systems. The sarcoplasmic reticulum has been considered to be analogous to the endoplasmic reticulum in other tissues (Muscatello *et al.*, 1965; Margreth *et al.*, 1966; Ezerman and Ishikawa, 1967); its major role, however, may be concerned with the uptake and release of calcium during relaxation and contraction (Peachey, 1968), calcium being an ATPase activator. The relative amounts of

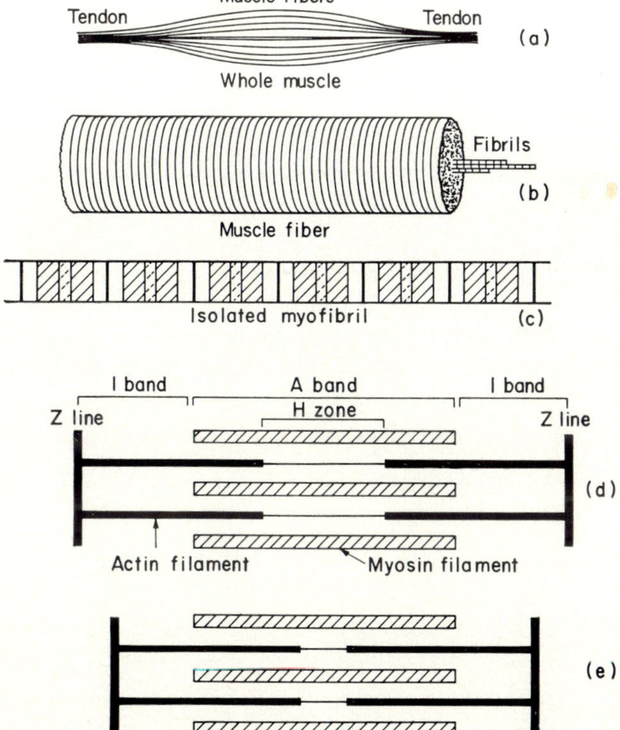

FIG. 1. A schematic representation of the organization of skeletal muscle. The muscle (a) consists of individual muscle fibers (b) each of which constitutes the contractile cell. The fibers are made up of individual myofibrils (c), the striations of which form repeating light and dark bands. An individual unit of this pattern [(d) extended, (e) resting] is bounded by the Z lines, and the repeating pattern is due to the overlapping of thick (myosin) and thin (actin) filaments. (Taken from Huxley, 1960.)

transverse system and sarcoplasmic reticulum have been broadly correlated with speed of muscular contraction, there being more of these structures in white, rapidly contracting muscles, particularly in lower vertebrates (Peachey and Huxley, 1962; Hess, 1965; Page, 1965) and in some mammals (Pellegrino and Franzini, 1963; Shafiq *et al.*, 1969). However, the difference in fine structure of red and white fibers in human vastus lateralis muscle is less striking than that observed for the rodent species (Shafiq *et al.*, 1966) so indicating that the organization and complexity of the sarcoplasmic reticulum varies with the species, as well as fiber type. It is also clear that the relationship between fine structure and function of muscle fibers is more complex than has been previously considered. This point is particularly well demonstrated by the studies of Gauthier and Padykula (1966), who observed that the red fibers within the diaphragm of a number of small mammals have elaborate networks of tubules of the sarcoplasmic reticulum. This aspect of the problem will be considered further in Section II,D.

C. Chemical Composition of Skeletal Muscle

About one-fifth of the wet weight of muscle consists of protein, and Table III shows that 60–65% of this is associated with the contractile fibrils, and that 20–25% is due to sarcoplasmic protein. The myofibril consists mainly of the three structural proteins, myosin, actin, and tropomyosin B (Hanson and Lowy, 1963, 1964; Perry, 1960a, 1965). Myosin, which shows ATPase activity, associated with the heavy meromyosin

TABLE III

DISTRIBUTION OF NITROGEN (EXPRESSED AS A PERCENTAGE OF TOTAL NITROGEN) IN SKELETAL MUSCLE OF ADULT MAMMALIAN SPECIES AND THE HEN

Muscle	Rat[a] Quadriceps	Pig[b] Thigh	Man[b] Thigh	Hen[c] Pectoral
Component				
Nonprotein N	10.5	12.9	9.7	14.8
Sarcoplasmic protein N	19.0	24.3	22.0	30.0
Fibrillar protein N	65.7	60.3	65.6	53.2
Extracellular protein N	2.5	3.3	4.6	3.3
Total N	31.5[d]	31.1[e]	30.8[e]	37.0[e]

[a] Calculated from data of Cabak *et al.* (1963).

[b] Taken from Dickerson and Widdowson (1960). Similar to values for myofibrillar protein N for porcine longissimus dorsi and gastrocnemius muscles reported by McLoughlin (1968).

[c] Taken from Dickerson (1960).

[d] Grams of total N per kilogram fat-free fresh muscle.

[e] Grams of total N per kilogram fresh muscle.

fragments which are released together with the light meromyosin fragments during tryptic digestion of the parent molecule, is the most abundant and accounts for about 54% of the total fibrillar structural protein compared with 21% for actin (Hanson and Lowy, 1964) and for one-third of the total protein in skeletal muscle (Perry, 1967). Consequently, myosin represents about 15% of the total protein content of the adult mammalian body, and must in consequence have some importance in protein metabolism. It may be noted that myosin differs in ATPase activity when the same muscle is compared from different species (Bailey, 1942; Perry, 1960a; Quass and Briskey, 1968), when different muscles in the same animal are compared (Seidel *et al.*, 1964; Sreter *et al.*, 1966; Barany *et al.*, 1965) and in the same muscle at different stages of growth (Trayer and Perry, 1966; Khyl'ko, 1965).

In skeletal muscle, extracellular protein is distributed as a sheath of endomysium around individual muscle fibers, perimysium around the groups (fasciculi) of fibers, and epimysium around the large bundles of fasciculi which make up the entire muscle. This extracellular protein is mainly collagen and differs in abundance from muscle to muscle within a species (Pettengill and Martin, 1947; Beatty *et al.*, 1966) and for homologous muscles between species (Pettengill and Martin, 1947). In general, however, the connective tissue content of both skeletal and cardiac muscle is low, in the region of 5% of total nitrogen or 1–2% of fresh weight (Harkness, 1961; Cheek *et al.*, 1965a).

The sarcoplasmic protein fraction can be obtained by extraction with buffers of low ionic strength. It contains many hundreds of different species of proteins, probably most of which have enzymic activity (Czok and Bucher, 1960). Though analogous to the cytoplasmic fraction of other tissues (cell sap), the spectrum of proteins in the sarcoplasmic fraction differs from those of cell sap in other tissues. A wide variation in the types and distribution of various sarcoplasmic proteins has been observed within a given muscle in different species and between different muscles in the same species (Perry, 1960a; Johnson and Stephen, 1964) which reflects the large differences in the organization of various metabolic pathways in this tissue.

D. Fiber Types in Skeletal Muscles

There are considerable differences between skeletal muscles, as far as their functional, morphological, biochemical, and other properties are concerned (see Hnik, 1962, for review). From both a functional and biochemical point of view, skeletal muscle fibers have long been divided into two types: the "red" or "dark" muscle fibers, functionally recognized as "slow" and "tonic," and the "white" fibers, which are "fast" and "phasic" in function (Needham, 1926). The red fibers are smaller, rich in large mitochondria and possess the widest Z lines. The white fibers possess fewer and smaller mitochondria and Z lines of about half the width as in the red fibers. Recent studies (Padykula and Gauthier, 1967; Gauthier, 1969) suggest, however, that a new category, the intermediate fiber, should be added to the earlier terminology. These three

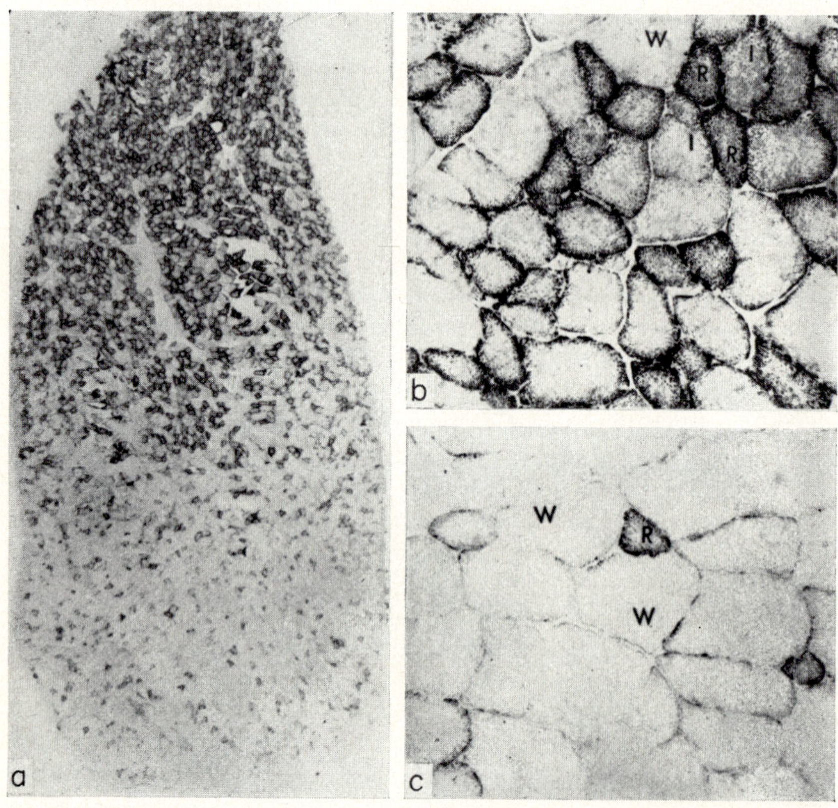

FIG. 2. (a) Transverse section through the entire width at the center of the rat semitendinosus muscle. The red (upper) portion consists primarily of fibers rich in the mitochondrial enzyme succinic dehydrogenase (red and intermediate fibers). The white (lower) portion consists almost entirely of fibers with low mitochondrial content (white fibers). ×18. (b) Red band. In this region, there are numerous small red fibers (*R*), intermediate fibers (*I*), and a few large white fibers (*W*). Cross-sectional dimensions are smallest in red and largest in white fibers (see Table IV). ×150. (c) White band. Fibers are similar to corresponding fiber types in the red band (*R*, *W*), except that large white fibers predominate. Cross-sectional dimensions are equivalent to those of the red region, but white fibers are larger in the white fibers located in the red band region. ×150. (Taken from Gauthier, 1969.)

TABLE IV

COMPARATIVE DIMENSIONS AND AREA OF FIBER TYPES IN RAT SEMITENDINOSUS[a]

Region	Fiber type	Mean fiber diameter (μ)	Percentage of fiber types	Percentage of area contributed by each fiber type
Red band	Red	45	52	44
	Intermediate	52	40	44
	White	59	9	13
White band	Red	47	4	2
	Intermediate	47	14	7
	White	69	82	91

[a] From Gauthier (1969), obtained with adult male rats.

fiber types (Fig. 2) have different morphological and ultrastructural features (Gauthier, 1969). The intermediate fibers are similar to red fibers but their diameters are larger. Although some muscles are made up of predominantly red or of white fibers, most muscles in mammals consist of a mixture of types of fiber, especially in the primates (Adams *et al.*, 1962). Also red and white regions can be clearly distinguished by the naked eye in some muscles, for example, the semitendinosus of the rat (Gauthier, 1969) and the pig (Beecher *et al.*, 1968). Within each region of the rat semitendinosus the relative proportions of the three fiber types, red, intermediate, and white, differs (Table IV). The red band consists primarily of small red fibers (52%) and intermediate fibers (40%) and only a few (9%) large white fibers. In comparison, the white band consists almost entirely (82%) of large white fibers and of 14% intermediate and 4% red fibers (Gauthier, 1969). The relationship between structure and function is further complicated in comparing a specific muscle of various mammalian species since it cannot be assumed that the muscle is representative of either red or white muscle type. This is true for the diaphragm, where, in small mammals such as the shrew or bat, this muscle consists almost entirely of small red fibers whereas in large mammals (the pig or steer) large white fibers account for the bulk of the muscle mass (Gauthier and Padykula, 1966). Along with differences in the contractile properties of the two muscle types ("slow" and "fast" contraction), there are differences in innervation (Denny-Brown, 1929; Buller *et al.*, 1960a,b; Hess and Pilar, 1963; Guth and Watson, 1967). It has been stated that at birth all muscles in the rat and cat show slow contraction, but after several weeks the contraction period shortens to varying extents in all muscles, and there is a differentiation into fast and slow species (Buller *et al.*, 1960a; Buller and Lewis, 1965a; Close, 1964, 1967; Guth, 1968).

Attempts have been made to correlate the biochemical and functional features of red and white muscle fibers with their fine structure. The relationships, however, have been made largely on the basis of studies with whole muscles and therefore a precise correlation between structure and function of fiber types cannot yet be drawn. It should also be pointed out that the traditional view which has been to consider "red" muscles as slow acting and "white" muscles as fast acting is not entirely valid when different specific muscles are considered or interspecies comparisons are made. Cytological studies of the cricothyroid muscle of the bat (Revel, 1962) and the diaphragm of a number of small mammalian species (Gauthier and Padykula, 1966; Padykula and Gauthier, 1967) show that red fibers are predominant in these muscles, but they tend to be fast acting. Red muscle is highly vascular; the fibers contain numerous mitochondria, and it is rich in the hemoglobin-related oxygen-carrying pigment myoglobin (Millikan, 1937; Lawrie, 1952, 1953a; Beecher *et al.*, 1968), whereas white muscle contains little myoglobin and has fewer mitochondria. It is thus not surprising to find that red muscle has a greater capacity for oxidative metabolism of energy-yielding substrates (Domonkos and Latzkovits, 1961; Dubowitz and Pearse, 1960a,b; Stein and Padykula, 1962), and shows higher activities of oxidative enzymes (Beatty *et al.*, 1966) and of the enzymes for fatty acid degradation (George and Talesara, 1961) than are found in white muscle. In contrast, white muscle is principally composed of large, pale fibers with a higher glycogen content than is found in red muscle (Beatty *et al.*, 1963; Ogata, 1960; Pellegrino and Franzini, 1963; Bocek *et al.*, 1966), has higher activities of phosphorylase (Dubowitz and Pearse, 1960a; Stubbs and Blanchaer, 1965) and other glycolytic enzymes (Lawrie, 1952; Dawson and Romanul, 1964; Blanchaer *et al.*, 1963; Stubbs and Blanchaer, 1965), and, although lower in hexokinase activity (Peter *et al.*, 1968), contains larger stores of the high-energy phosphate reserve creatine phosphate (Beatty *et al.*, 1963), so that it can continue to contract vigorously until these reserves of glycogen and high-energy phosphate are exhausted (Szent-Györgyi, 1953). White muscle is thus oriented to obtain more of its energy requirements from glycolysis. It also has a higher concentration of protein in its sarcoplasm (Barany *et al.*, 1965) presumably representing the higher concentration of soluble glycolytic enzymes; in contrast, the myofibrillar content of red and white muscle appears to be similar (Barany *et al.*, 1965; Beecher *et al.*, 1968). Although this discussion suggests that muscles of varying pigmentation can be broadly differentiated on a biochemical basis, the work of Burleigh and Schimke (1969) indicates that there are difficulties encountered in correlating the histochemical and biochemical observations made on various muscles. They systematically surveyed the relative activities of hexokinase and glycogen phosphorylase in mammalian muscles and found that there was no clear relationship between hexokinase activity and visible redness of muscles.

The color of red muscle is partly due to the cytochromes present in the mitochondria, but is due mainly to the presence of myoglobin, a respiratory pigment structurally related to hemoglobin (Chapter 24, Section V) but with a higher affinity for oxygen than hemoglobin (Korner, 1959). Comparison of different muscles suggests that those with a high myoglobin content have a greater capacity for oxidative metabolism and less for glycolysis (Lawrie, 1953a). However, this relationship does not hold when mammals of different body size are compared, since there is a general tendency for myoglobin content of skeletal muscle to be higher in the larger species (see Table V in Chapter 25, Volume III), whereas oxidative metabolism in muscle is less in mammals of larger size and the mitochondrial population in their cells is reduced (Kunkel and Campbell, 1952; Drabkin, 1950). Thus in the case of the blue whale, the myoglobin content of its psoas muscle is as great as that of the horse, although oxidative enzyme preparations made from whale psoas muscle are only one-third as active as those obtained from the horse. From these observations, Perry (1960a) has concluded that the relatively high myoglobin content is advantageous to the whale during diving. Myoglobin content also seems to vary with exercise and activity of muscles (Whipple, 1926; Lawrie, 1953b; Pattengale and Holloszy, 1967). In response to the hypoxia of high altitude, the myoglobin content of muscle is increased (Hurtado *et al.*, 1937; Stickney and Van Liere, 1953; Tappan and Reynafarje, 1957; Vaughan and Pace, 1956; Anthony *et al.*, 1959; Reynafarje and Morrison, 1962); this increment occurs slowly and is completed some time after hemoglobin has attained a new level as a result of hypoxia (see Hensel and Hildebrandt, 1964). Finally, the myoglobin content of muscle is known to be subject to seasonal variation in the case of the Alaskan snowshoe hare (Rosenmann and Morrison, 1965) and the red-backed vole (Morrison *et al.*, 1966); the same species show a more modest seasonal variation in the myoglobin content of cardiac muscle. In adult man, Akeson *et al.* (1968) have estimated that the total skeletal musculature contains about 150 gm of myoglobin. Human skeletal muscle in general has about twice the myoglobin content of heart muscle (Biorck, 1949; Akeson *et al.*, 1968). Among the skeletal muscles of man, abdominal muscle has less myoglobin than thigh muscles, while diaphragm has a slightly higher content than

TABLE V

RNA Content and Protein Metabolism of "Red" and "White" Muscles of Rats[a]

Muscle	Type[b]	AIB Distribution ratio[c]	RNA (μg/mg wet muscle)	Incorporation of Leucine-^{14}C (cpm/mg fresh muscle)	
				Soluble fraction	Myofibrillar fraction
Diaphragm	Predominantly red	2.4	5.6	4.4	6.3
Soleus	Predominantly red	3.4	7.1	3.7	5.6
Medial gastrocnemius	Predominantly red	2.1	5.8	3.3	5.3
Extensor digitorium longus	Predominantly white	1.4	4.0	3.0	4.5
Lateral gastrocnemius	Predominantly white	1.2	3.8	2.3	3.7

[a] Combined from data of Goldberg (1967): Hypophysectomized rats (100–120 gm body weight) given subcutaneous injection leucine-^{14}C and killed 24 hours later. Similar results were obtained by Goldberg and by Dreyfus (1967) for amino acid incorporation with intact rats, by Margreth and Novello (1964) for RNA concentration of different muscle types in rats, rabbits, and pigeons.

[b] Type of muscle was based upon classification given by Sreter and Woo (1963).

[c] Values for the ratios of concentration of ^{14}C-labeled α-aminoisobutyric acid in total muscle water to that of plasma 4 hours after intravenous injection.

the heart. It has been noted by Reynafarje and Morrison (1962) that animals with habitats at high altitude usually have more myoglobin in their hearts and diaphragms than in their thigh muscles.

The concentration of RNA in skeletal muscle is much less than in the viscera and can be correlated with the lower uptake of injected labeled amino acids into muscle protein (Chapter 25, Table VI). In respect to RNA content and uptake of labeled amino acids, there are differences between "red" and "white" skeletal muscles. Muscles capable of continuous tonic work, presumably predominantly those with red fibers, show a higher concentration of RNA than do muscles that perform rapid, phasic work (Margreth and Novello, 1964; Goldberg, 1967). This can be correlated with the finding of Biron *et al.* (1964) that administered labeled amino acids are incorporated more extensively into the myosin, aldolase, and lactic dehydrogenase of red than of white muscle. Similar observations have been made by Dreyfus (1967) on red and white muscles of the normal rabbit, a ^{14}C-labeled *Chlorella* hydrolyzate being used as the labeled protein precursor. In a study of amino acid incorporation into myofibrillar and sarcoplasmic proteins of different muscles, Goldberg (1967) has demonstrated a relationship of the intensity of uptake into protein with the RNA content of the muscles (Table V). He suggested on this basis that the turnover of protein differs in red and white muscle. However, the *in vitro* capacity for protein synthesis by ribosome preparations from various muscle types is quite similar (Wool *et al.*, 1968) and the greater incorporation into the proteins of red muscle may to some extent reflect more rapid penetration of the labeled amino acid into the free amino acid pool of this type of muscle. It has been shown that the nonmetabolizable amino acid AIBA is also taken up more rapidly by red muscle (Dreyfus, 1967; Goldberg, 1967). More extensive capillary density (Romanul, 1965) and a higher rate of blood flow (Reis *et al.*, 1967) in red than in white muscle may account for a greater penetration, and therefore a higher uptake of labeled amino acids by the proteins of rat diaphragm as compared with other muscles (Altman *et al.*, 1949; Munro *et al.*, 1959). Differences in amino acid uptake cannot be explained entirely on differences in blood supply, however, because Short (1969) observed a higher rate of ^{14}C-labeled methionine incorporation into protein of red muscle fibers than in white muscle fibers during *in vitro* incubation.

III. Changes in Skeletal Muscle during Growth and Development

Changes in skeletal muscle metabolism during growth and development are discussed in detail in Chapter 26, Section III,D. Here the main features in the development of muscle cell structure and composition will be recapitulated.

A. *Changes in Muscle Cells during Growth*

Changes in the size of organs during growth occur through an increase in the number of cells (hyperplasia) and by an increase in the size of the constituent cells (hypertrophy). Table VI illustrates the contribution of both these factors to the growth of the extensor carpi radialis longus muscle of the rat. Between 2 weeks and 5 weeks of age, this muscle grows 4.5 times in mass and there is simultaneously a 3-fold increase in the total amounts of DNA in the muscle and in its population of nuclei. Consequently, growth during this period occurs mainly by hyperplasia with some hypertrophy. From 5 to 12 weeks of age, there is a further increase of DNA and nuclear population by about 50%, whereas muscle weight continues to increase by a factor of 2.5. This phase of growth is consequently achieved mainly by hypertrophy. Similar patterns have been observed in other rat muscles, except that the increases in weight and in nuclear number change by different factors during the growing period (Enesco and Puddy, 1964). Data on the DNA content of rat skeletal muscle shows that increase in nuclear number ceases at about 60–90 days of age (Gordon *et al.*, 1966; Cheek *et al.*, 1968). Histological studies demonstrate that the increments in cell nuclei after the early postnatal period represents a proportional increase in both the nuclei of the contractile cells and of the endomysium and

TABLE VI

THE GROWTH PATTERN OF A SKELETAL MUSCLE (EXTENSOR CARPI RADIALIS LONGUS) IN THE RAT[a]

Datum	Suckling, 2 weeks	Prepubertal, 5 weeks	Young adult, 12 weeks
Weight of muscle (mg)	6.5	30.0	81.0
Area of muscle cross section (mm^2)	0.38	1.18	2.52
Number of fibers/muscle	2030	—	2445
Weight of fiber (μg)	2.1	11.6	29.9
Number of nuclei $\times 10^6$	1.6	4.9	6.8
Number of nuclei per fiber	474	1452	2122
DNA content per muscle (mg)	10.0	30.8	42.4
Weight of muscle (mg/μg DNA)	0.65	0.98	1.95

[a] Data taken from Enesco and Puddy (1964) obtained with Sherman rats. Muscle area from Chiakulus and Pauly (1965) for Sprague-Dawley rats. The muscle area for suckling and 5-week-old rats are extrapolated values. The mean number of fibers for the E. carpi radalis longus reported by Chiakulus and Pauly was three to four times greater than those given in this table.

perimysium (Enesco and Leblond, 1962; Enesco and Puddy, 1964). Similar findings have been reported for the chick (Robinson, 1952; Moss *et al.*, 1964; Moss, 1968a,b; Herrmann *et al.*, 1957; Marchok and Herrmann, 1967), except that the mean cross-sectional area of chick muscles increases in proportion to nuclear number throughout growth (Moss, 1968a). It should be pointed out, however, that the terms hyperplasia and hypertrophy, as commonly conceived in reference to the amount of DNA in an organ, may not be applicable to muscle growth. As will be shown below, growth of muscle cytoplasm occurs mainly by continued enlargement of fibers laid down before birth, not by the addition of new fibers. Increase in nuclear number occurs through fusion of mononucleate cells with a developed fiber (Stockdale and Holtzer, 1961; Konigsberg, *et al.*, 1960; Betz *et al.*, 1966). This evidence is discussed below.

While the nuclear population of skeletal muscle continues to increase for a large part of the growing period, the number of muscle fibers does not (Table VI). In the published literature (MacCallum, 1898; McMeekan, 1940; LeGros Clark, 1958; Ham and Leeson, 1961; Adams *et al.*, 1962), some authors have concluded that muscle cells grow exclusively by increase in size of muscle fibers. However, there is some evidence for the rat and mouse (Morpurgo, 1895; Chiakulas and Pauly, 1965; Goldspink, 1962a) and for the human infant (Montgomery, 1962a) that a modest increase in fiber number occurs for a short period after birth, and then the population of fibers becomes stabilized at a fixed level. Thereafter, growth in muscle mass occurs through increase in the size of the individual muscle fibers (Table VI). Bowden and Goyer (1960) noted that the size of the fibers of different human muscles closely approximates the size of other muscles in the fetus, but after birth the size differs in various muscles and is related to function. This also seems to apply to rat (Chiakulas and Pauly, 1965) and mouse muscle (Rowe and Goldspink, 1969a). In an extensive study of the effect of growth on fiber dimensions in several species, Joubert (1956) has demonstrated that there is no correlation between the size of the muscle fiber at birth and at maturity and that the diameter of the fiber in the adult is unrelated to its final body size. On the other hand, Gauthier and Padykula (1966) made a survey of fiber sizes in the diaphragms of different species and observed that red fibers were characteristic of small mammals and large white fibers predominant in the diaphragm of large mammals with relatively low metabolic activity and breathing rate. They also concluded that large animals have somewhat larger fibers. This cannot, however, greatly contribute to the difference in size of the muscles of small and large mammals; thus, Black-Schaffer *et al.* (1965) have computed that the rat heart of 1.4 gm contains 9×10^7 fibers and the heart of the blue whale (weighing 2.9×10^5 gm) contains 2×10^{13} fibers, an increase in fiber number roughly proportional to the weight difference.

Since nuclear number continues to rise after fiber number has ceased to increase (see Table VI), it follows that the number of nuclei per fiber becomes progressively greater throughout growth. Mitotic activity has been noted in developing muscle (MacConnachie and co-workers, 1964; Overy and Priest, 1966) and has been attributed to mitotic activity in the satellite cells of skeletal muscle (Shafiq *et al.*, 1968; Mauro, 1961) and in the case of cardiac muscle, which does not have satellite cells, to mitosis in undifferentiated myoblasts and in other "free" cells. Formation of myofibrils in the satellite cells of normal (Muir *et al.*, 1965) and regenerating muscle (Shafiq and Gorycki, 1965; Shafiq *et al.*, 1967) has also been described; this supports the view that the development of muscle cell bears a close relationship to satellite cell activity. Mitosis has not been observed within the muscle fibers or in myoblasts already showing fibril formation, and the presence of numerous nuclei in the striated muscle fibers has therefore been attributed to fusion of mononucleate myoblasts and/or multinucleate myostrap with a developed fiber (Message, 1968). The length of the fiber increases with growth, which is accomplished by a modest increase in the sarcomere length but mainly by the addition of new sarcomeres (Goldspink, 1968). Polyploidy has not been observed in the skeletal or cardiac muscle of the rat or mouse (Enesco, 1957; Enesco and Puddy, 1964; MacConnachie *et al.*, 1964; Petersen and Baserga, 1965), although it has been reported in the heart muscle of man (Sandritter and Scomazzoni, 1964; Kompmann *et al.*, 1966). Finally, there appears to be a relationship between nuclear density in skeletal muscle fibers and the size of the species of mammal. Munro and Gray (1969) examined the ratio of muscle mass to DNA for the posterior leg muscles in a range of mammals of widely different body weight and found that DNA per unit of muscle weight decreased by a factor of two as body size increased from the mouse to the horse. This implies that larger species have somewhat fewer nuclei per unit of muscle cell mass.

B. Changes in Muscle Composition during Growth

During growth, muscle undergoes a reduction in concentration of water and extracellular electrolytes whereas protein concentration increases. These are changes shared with other tissues showing chemical maturation during the early stages of growth, except that muscle and skin undergo larger increases in protein concentration than have been observed in other tissues (Table I). Blaxter (Chapter 15, Section III,D,1) describes changes occurring in individual fractions of muscle protein during growth, and similar changes have been noted in muscles of growing chicks (Robinson, 1952; Dickerson, 1960). The relative proportions of fibrillar and sarcoplasmic proteins differ according to the stage of development, the rate of change being related to the time of functional development of the muscle. Thus in the pig, which is mobile shortly after birth, the fibrillar fraction of skeletal muscle attains adult proportions by about 4–6 weeks of age, whereas in man this fraction undergoes only a small increase in relative proportion to other fiber proteins for the first 5–7 months of postnatal life (Widdowson and Dickerson, 1960). In the chick, which is active as soon as it leaves the egg, fibrillar proteins are laid down even more quickly than in the case of the pig (Robinson, 1952; Dickerson, 1960; Baril and Herrmann, 1967). The concentration of extracellular N reaches a maximum level at birth and thereafter decreases to a low level in the adult pig and man (Widdowson and Dickerson, 1960) and in growing steers (Bendall and Voyle, 1967). The collagen content of rat thigh muscle reaches its adult level of 1% of muscle weight within the first month of life (Herrmann and Nicholas, 1948; Stewart, 1955) and possibly by about 1 week of age (Cheek *et al.*, 1965a).

In addition to the marked changes in the concentration of the myofibrillar protein fraction in developing muscle, small but functionally significant alterations occur in the composition of the sarcoplasmic protein fraction (Hartshorne and Perry, 1962; Kendrick-Jones and Perry, 1966, 1967; Singh and Kanungo, 1968; Burleigh and Schimke, 1969). The pattern of change in the enzyme 5′-adenosine monophosphate deaminase, which is similar to that for creatine phosphokinase, adenylate kinase, and aldolase, varies among species (Fig. 3) and also among muscles within a species (Fig. 4). The changes in enzyme activity correlate with the increased intensity of muscular activity. The guinea pig moves about shortly after birth and the adult enzyme levels of the leg muscles are more rapidly attained than in the rat, which is born at a less advanced stage. Similarly, in the case of the diaphragm, which must function efficiently at birth, the adult enzyme level is achieved very soon after birth; the leg muscles, however, reveal a slower change, and adult enzyme levels are reached at about the time of weaning, when the demand for movement is greatly increased. The presence of two glycolytic dehydrogenases within the sarcoplasmic reticulum (Fahimi and Karvosky, 1966), and possibly other enzymes (Margreth *et al.*, 1967), suggests that increased activities of some enzymes after birth may be linked with the elaboration of the sarcoplasmic reticulum which occurs postnatally (Shiaffino and Margreth, 1969).

The nucleic acid content of skeletal muscle also changes during growth. The concentration of DNA is high in the muscles of the rat just before birth (Table VII), and decreases rapidly immediately after birth, attaining a relatively constant concentration only at about 200 gm body weight. A parallel change occurs in the DNA content of the liver, except

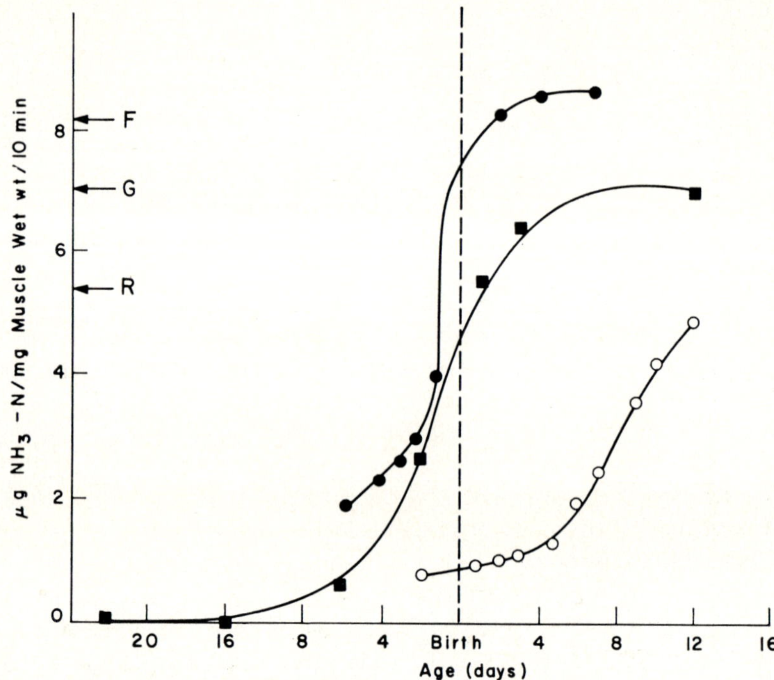

FIG. 3. A comparison of the developmental changes in 5′-AMP deaminase activities in the leg muscles of different species. Adult values are shown by the arrows; ■, guinea pig (*G*); ○, rat (*R*), ●, chick (*F*). (From Kendrick-Jones and Perry, 1967.)

TABLE VII

CONCENTRATIONS OF DNA AND RNA IN LIVER AND SKELETAL MUSCLES OF THE RAT AT DIFFERENT STAGES OF GROWTH[a]

Body weight (gm)	DNA (mg/100 g tissue)		RNA (mg/100 g tissue)		RNA:DNA Ratio	
	Liver	Muscle	Liver	Muscle	Liver	Muscle
2.4 (2 days before birth)	750	228	1007	307	1.3	1.3
25	500	100	1180	290	2.4	2.9
45	380	85	1210	170	3.2	2.0
115	270	75	960	140	3.6	1.9
225	270	50	870	130	3.2	2.6
400	270	55	820	90	3.0	1.6

[a] The data for RNA and DNA concentration for the fetus were taken from the tabulated data published by Devi *et al.* (1963); for older rats, the data have been read off their graphs.

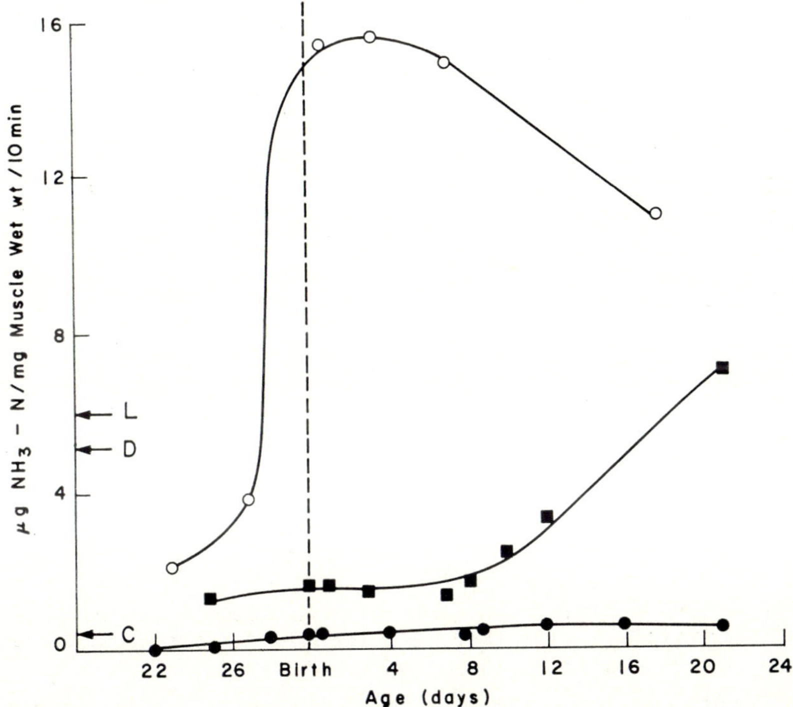

FIG. 4. Developmental changes in 5′-AMP deaminase activities in three muscles of the rabbit. Adult values are shown by arrows; ■, leg (*L*); ○, diaphragm (*D*); ●, heart muscle (*C*). (Taken from Kendrick-Jones and Perry, 1967).

that the plateau comes earlier. The concentration of RNA also decreases during postnatal life (Table VII). Calculations based on these data and on the relative amounts of liver and skeletal muscle mass in the body at different stages of growth (Table I) lead to the conclusion that the total amount of RNA in the skeletal musculature relative to the amount in the liver increases steadily throughout growth. This implies that protein synthesis in muscle may represent an increasing proportion of total body protein synthesis as growth progresses.

IV. Free Amino Acid Pools of Muscle

Free amino acids in mammalian tissues are discussed in Chapter 4 of Volume I and in Chapter 34 of this volume. Table VIII shows some representative published values for free amino acid concentrations in the plasma, liver, and skeletal and cardiac muscles of the rat. Similar data for the former three tissues, from a different publication, are shown

TABLE VIII

CONCENTRATIONS AND TOTAL AMOUNTS OF AMINO ACIDS IN PLASMA, LIVER, SKELETAL AND HEART MUSCLE OF FASTING RATS[a]

	Concentration (μ moles/100 gm or 100 ml)				Total amount in organ[b] (μ moles/kg body wt.)			
Amino acid	Plasma[a]	Liver[a]	Skeletal muscle[a]	Heart	Plasma	Liver	Skeletal muscle	Heart
Threonine	43	45	148	76	14	23	622	3
Methionine	3	7	6	—	1	4	25	—
Histidine	5	72	167	12	17	36	701	0.5
Leucine	10	69	13	36	3	35	55	1.4
Isoleucine	6	37	8	15	2	19	34	0.6
Valine	13	40	15	24	4	20	63	1
Phenylalanine	4	19	8	28	1	10	34	1.1
Lysine	30	59	52	97	10	30	218	3.9
Arginine	9	Trace	16	19	3	Trace	67	1.2
Tyrosine	6	37	12	5	2	19	50	0.2
Serine	35	189	156	113	12	95	655	4.5
Glycine	29	481	336	129	10	241	1411	5.2
Glutamic acid	11	1052	122	614	4	526	512	25
Alanine	45	361	230	500	15	181	966	20
Aspartic acid	1	189	32	275	0.3	95	134	11

[a] Values for various tissues taken from data of Leung (1965), except for heart, which were calculated from the graphs given by Wannemacher and McCoy (1969).

[b] Weight of rats was assumed to be 200 gm; plasma, liver, muscle, and heart were taken to be 3.3%, 5%, 42%, and 0.4% of body weight, respectively (Table I, Chapter 25).

in Table VII of Chapter 30, in Volume III. The concentrations shown in Table VIII illustrate the general observation that the intracellular levels of some amino acids are higher than those of the plasma. In skeletal muscle, threonine and histidine are highly concentrated as compared to plasma, but the remaining essential amino acids show little difference from the levels found in the plasma. On the other hand, all the nonessential amino acids other than tyrosine are 5–30 times more concentrated in muscle than in plasma, a phenomenon common to most tissues (Chapter 34, Section III,A). The synthesis of the nonessential amino acids in muscle has been examined. Using intraperitoneal injections of amino acids and then determining the amino acid content in tissues, Coulson and Hernandez (1968) have concluded that skeletal muscle of rats cannot synthesize glutamic and aspartic acids. However, studies with ^{14}C-labeled carbonate indicate that in rat diaphragm there is synthesis of aspartic and glutamic acids (Manchester and Young, 1959), but that skeletal muscle is capable of only limited synthesis of the other dietary nonessential acids (Swick and Handa, 1956; Manchester and Young, 1959). Taurine, not recorded in Table VIII, is present in exceptionally high concentration in cardiac and skeletal muscle of various mammalian species including man (Scharff and Wool, 1964, 1965a,b, 1966; Nyhan *et al.*, 1968; and for review see Jacobsen and Smith, 1968). The high concentration of taurine in the heart and skeletal muscles of the rat accounts for about 75% of total body taurine (Stern and Stim, 1959), the latter being estimated to about 600–1200 μmoles/100 gm body weight (Schram, 1960; Boquet and Fromogeot, 1965). The significance of the high level of taurine in muscle is not known, except to note that it appears to have an effect on excitability of cardiac muscle (see Jacobsen and Smith, 1968). Paradoxically, the excretion of taurine in the urine is increased in malnourished children that have lost a significant proportion of their muscle mass (Vis *et al.*, 1958). However, this may be related to an increased taurine content of muscle, because Allison *et al.* (1963) found high levels of taurine in muscle of rats after 3 days of dietary protein restriction and also after starvation.

Two dipeptides occur in relatively high concentration in skeletal muscle. These are carnosine and its methyl derivative, anserine (Davey, 1957). Carnosine is synthesized from β-alanine and histidine, or in the case of anserine, 1-methylhistidine (Martignoni and Winnick, 1954; Winnick and Winnick, 1957). The functional importance of the imidazole dipeptides in skeletal muscle is probably related to a role in the transmission of nerve impulses (Severin *et al.*, 1965), they are also effective myosin ATase activators (Yun and Parker, 1965; Avena and Bowen, 1969). Low concentrations of free 1-methylhistidine and 3-methylhistidine are also present in skeletal muscle (Tallan *et al.*, 1954; Block *et al.*, 1965). The presence of the former histidine derivative is presumably functionally related to anserine and 3-methylhistidine is related to the contractile proteins actin and myosin which have been found to contain this amino acid (Asatoor and Armstrong, 1967; Johnson *et al.*, 1967; Trayer *et al.*, 1968). There is circumstantial evidence that the methylation occurs after activation of histidine and that 3-methylhistidine cannot be used for protein synthesis. Cowgill and Freeburg (1957) have shown that the radioactivity in the muscles of chicks, rabbits, rats, and frogs, injected with histidine-3-$^{14}CH_3$ was present in muscle as the free amino acid and was not incorporated into dipeptides or protein. Reporter (1969) incubated ^{14}C-methyl-labeled methionine or ^{14}C-labeled histidine with tissue cultures of rat leg muscle and observed radioactivity in the 3-methylhistidine of muscle protein derived from both these donors. Finally, Hardy and Perry (1969) found that S-adenosylmethionine could serve as a methyl donor for the formation of protein-bound 3-methylhistidine by crude muscle homogenates. Using amino-acyl ligases prepared from rat skeletal muscle, we have compared the *in vitro* binding of various radioactive amino acids to tRNA with the binding to tRNA of 3-methylhistidine-^{3}H (Young *et al.*, 1969a). The results gave no evidence of attachment of the 3-methylhistidine to tRNA. These observations are analogous to the proline and hydroxyproline relationship in collagen metabolism (Gould 1968), and free 3-methylhistidine in muscle probably originates from the catabolism of actin and myosin. Furthermore, it is likely that the major proportion of urinary 3-methylhistidine in adult man (Block *et al.*, 1965, 1967) arises from the breakdown of these fibrillar proteins in skeletal muscle.

The somewhat higher concentrations of many free amino acids in liver than in skeletal muscle (Table VIII) gives a false impression that liver is a more important depot for free amino acids. Since liver contributes only some 4–5% of the body weight of the adult, whereas muscle represents 45% of its weight, the total amounts of free amino acids in muscle are quantitatively much more important than in the liver (Table VIII). The relationship between the size of the muscle pool of free essential amino acids and the requirements of the animal for these essential amino acids has been explored in Chapter 25, where it was demonstrated that, as the body size of the adult mammal increases, its requirements for essential amino acids per kilogram of body weight becomes less, whereas the free amino acid pools do not change so markedly. Consequently, the pool of free amino acids in the muscles of large animals can represent a large part of their daily requirements, thus acting in a buffering capacity in relation to amino acid needs. This idea has been explored in Table IX in relation to young and adult human subjects.

Table IX shows that the pool of free threonine in muscle is equivalent to the threonine needs of adult man for about 5 days; in the case of muscle free phenylalanine, the pool represents the requirements for nearly 1 day. In the younger subjects shown in Table IX, the free amino acid pool of muscle provides a smaller proportion of the daily need for the essential amino acids. This suggests that skeletal muscle may contribute more to buffering amino acid needs in the adult than in the young human subject. The buffering function of muscle is possibly greater than the comparisons shown in Table IX so indicate because mean minimum requirements for the individual essential amino acids are probably lower than suggested in this table. The presence of such pools of free amino acids may be significant in determining how closely amino acids have to be consumed in order to be utilized for protein synthesis. The question of delayed supplementation of amino acids is reviewed in Chapter 34, Section III,A; it may be said in summary here that tryptophan is generally agreed to require simultaneous administration along with the other essential amino acids in order to be utilized for protein synthesis, whereas lysine probably can be fed several hours apart from the other amino acids of the diet and still allow full utilization of the amino acid mixture. This would be compatible with the demonstration (Tables VIII and IX) that free lysine is present in muscle in high concentrations, and thus provides a significant proportion of the daily needs. The importance of the free amino acid pool in skeletal muscle in the economy of total body amino acids is also clearly illustrated in the study of Pawlack and Pion (1969). They gave rats, for 2 weeks, graded levels of dietary lysine, varying from about one-third to twice the amount required for maximal growth. The free lysine concentration in skeletal muscle increased nearly 27-fold between the low and high levels of lysine intake, whereas the increase was only 7-fold in the plasma of the rats. This again demonstrates the capacity of muscle to act as a store for free lysine. The values summarized in Chapter 34, Table I, also show that the proportion of free lysine in muscle is higher than most other dietary essential amino acids. This is consistent with the observations of Pawlack and Pion and together with their experiment shows that muscle acts as a variable reserve for this amino acid. On the basis of the concentrations of other amino acids in muscle (Table VIII and in Chapter 34, Table I), it can be predicted that both threonine and histidine are similarly deposited in muscle when dietary intake exceeds their requirement. Hider *et al.* (1969) also believe from their data that the greater proportion of amino acids in muscle functions

TABLE IX

CONCENTRATIONS AND TOTAL AMOUNTS OF ESSENTIAL AMINO ACIDS IN PLASMA AND SKELETAL MUSCLE OF MAN AT TWO AGES COMPARED WITH REQUIREMENTS[a]

Amino acid	Concentration (mg/100 ml or 100 gm tissue)				Amount (mg/kg body wt.)				Requirement (mg/kg body wt. per day)	
	Young		Adult		Young		Adult			
	Plasma	Muscle	Plasma	Muscle	Plasma	Muscle	Plasma	Muscle	Young[b]	Adult[c]
Threonine	1.42	3.26	1.18	5.34	0.68	11.7	0.57	24.0	35	5.0
Valine	2.23	1.78	2.03	2.49	1.07	6.4	0.97	11.2	33	9.0
Methionine	0.33	0.62	0.37	0.89	0.15	2.2	0.18	4.0	27	3.9
Isoleucine	0.80	0.66	0.73	1.09	0.38	2.4	0.35	4.9	30	7.8
Leucine	1.48	1.20	1.30	1.95	0.71	4.3	0.62	8.8	45	8.5
Phenylalanine	1.00	0.67	0.74	1.34	0.48	2.4	0.35	6.0	27	3.7
Lysine	2.44	6.83	—	—	1.17	24.6	—	—	60	6.1

[a] Plasma and muscle data are taken from Nyhan *et al.* (1968) for boys and girls (age range 5–17 years, mean weight 30 kg). The values for adults were calculated from Zachmann *et al.* (1966) for 5 subjects (mean weight 69 kg). Plasma volume for both young and adults is taken to be 4.8% of body weight (Table V, Chapter 25), and muscle weight is assumed to be 36% and 45% of body weight for young and adults, respectively.

[b] From Nakagawa *et al.* (1964, 1965).

[c] Mean value calculated from data for men and women (see Rose *et al.*, 1955 and NRC, 1959),

as a body storage pool and that the extracellular pool in this tissue supplies the amino acids for muscle protein synthesis.

The role of muscle in accepting amino acids during absorption after a meal of protein has also been examined. In 1913 Van Slyke and Meyer infused dogs with protein hydrolyzates and showed that within a few minutes of administration most of the amino N had disappeared from the blood and that a major proportion of the missing amino N could be accounted for in muscle. With the advent of labeled amino acids, the uptake of amino acids into muscle was re-examined. Borsook *et al.* (1950) injected several ^{14}C-labeled amino acids intravenously into mice and showed that 5 minutes later much of the label could be found in the acid-soluble pool of muscle. Some 30 minutes later, most of the radioactivity had left muscle and was now in the viscera, where it was soon partly incorporated into protein or transformed to degradation products including CO_2. This is not quite in accord with results obtained by other investigators. Using labeled glycine (Henriques *et al.*, 1955) and lysine (Gan and Jeffay, 1967; Waterlow and Stephen, 1967, 1968), penetration of each free amino acid into muscle has been found to be slower than into liver. However, the data of Henriques *et al.* (1955) do exhibit a small and transient initial spike of penetration of glycine, which may correspond to the temporary deposition of amino acids in muscle observed by Borsook *et al.* (1950).

The concentrations of free amino acids in tissues arise from a balance between a number of factors additional to the rate of penetration of amino acids into the tissue. Amino acids are added to the pool through synthesis of nonessential amino acids from precursors within the tissue, and through release of amino acids from breakdown of proteins within the cells of the tissue; amino acids are removed by exit from the tissue to the plasma, by utilization for synthesis of proteins and other substances, and by amino acid catabolism. In some tissues the recycling of amino acids released within the cell from tissue proteins can be a major source of amino acids. By infusing labeled amino acids intravenously into rats and comparing the specific activity of the same amino acid in plasma and in the free pool of the tissue, Gan and Jeffay (1967) have shown that recycling provides about 60% of the free lysine and tyrosine in the liver of the fed rat and about 90% in the rat during the first few hours of fasting, when tissue protein breakdown is accelerated. In the case of muscle, the same technique has shown that only some 30% of the free amino acid pool represents recycling of muscle protein, and that fasting does not greatly disturb this proportion until several days have elapsed (see Fig. 10 in Chapter 34 for their data). These data on turnover are largely confirmed by the studies of Henshaw *et al.* (1968), and those of Waterlow and Stephen (1968), although they obtained lower values for each tissue. These latter workers also note that free lysine in the liver of the rat has a turnover time of about 30 minutes and in muscle of about 60 minutes. However, since muscle provides much larger proportions of the total free amino acid pool of the body than does liver (Table VIII), the total amount of amino acid turned over per hour is much greater for muscle than for liver.

V. Enzymes of Amino Acid Metabolism in Muscle

The muscle enzymes to be considered here are the transaminases (aminotransferases), the amino acid-activating enzymes, and the proteolytic enzymes. The amounts of some of these in the liver, heart, skeletal muscle, and kidney of the rat are presented in Table X as activities per gram of tissue and also as total activity in the tissue per kilogram body weight of the whole animal.

A. Transaminases

The degradation of many of the essential amino acids is dependent on enzymes present exclusively in the liver, whereas the breakdown of the nonessential amino acids is effectively carried out by other tissues. For example, Miller (1962) administered labeled amino acids individually to intact rats and to hepatectomized rats and also in the perfused liver and observed that the hepatectomized animals were unable to convert arginine, histidine, phenylalanine, tryptophan, lysine, threonine, and methionine to $^{14}CO_2$. On the other hand, hepatectomy did little to reduce the catabolism of the nonessential amino acids and also of isoleucine, leucine, and valine. This can be correlated with studies of transaminase distribution in different tissues (Table X). The enzymes associated with catabolism of aromatic amino acids are confined to the liver (Lin and Knox, 1958), whereas the transaminase that attacks the branched-chain amino acids is present in highest activity in kidney, followed by skeletal and cardiac muscle, and is present in low concentration in liver cells (Ichihara and Koyama, 1966; Mimura *et al.*, 1968). From the point of view of total enzyme activity, the large amount of skeletal muscle in the body outweighs its somewhat lower enzyme activity than in kidney; consequently, the whole skeletal musculature of the rat contains three to ten times the amount of enzyme in the two kidneys, and about 100 times the amount of enzyme in liver, so it can be concluded that catabolism of branched-chain amino acids is primarily effected in muscle (Table X). This is confirmed by Elwyn's studies on the fate of amino acids absorbed from the gut after a meal of protein (Chapter 38). Elwyn has shown that much of the load of

TABLE X: ACTIVITY AND DISTRIBUTION OF VARIOUS ENZYMES ASSOCIATED WITH PROTEIN METABOLISM IN LIVER, HEART, SKELETAL MUSCLE, AND KIDNEY OF RATS

Enzyme	Liver		Heart		Skeletal Muscle		Kidney	
	Per gm tissue	Per kg body wt.[a]	Per gm tissue	Per kg body wt.[a]	Per gm tissue	Per kg body wt.[a]	Per gm tissue	Per kg body wt.[a]
Branched-chain transaminase[b]								
Valine	75	3,225	4,140	14,490	833	374,850	14,500	116,000
Leucine	145	6,235	14,490	50,715	2,367	1,065,150	16,000	128,000
Isoleucine	220	9,460	10,557	36,949	1,960	882,000	18,500	148,000
Glutamic-oxaloacetic transaminase[c]	32	1,380	32	109	17.0	7,650	49	390
Glutamic-pyruvic transaminase[c]	35	1,513	19	67	17	7,595	20	163
Amino acid-activating enzymes[d]	190	8,170	46	161	58	26,100	—	—
RNA Content (mg)[e]	8.2	—	2.0	—	0.9	—	—	—
Acid cathepsin[f]	1.4	49	—	—	0.8	360	—	—
Creatine phosphokinase[g]	0.04	1.7	2.6	9.1	15.0	6,750	0.08	6.4

[a] For total body activity (expressed per kilogram of body weight) calculations are made for adult 400-gm rats from data summarized in Table I of Chapter 26. As a percentage of body weight, liver, heart, skeletal muscle, and kidney are taken to be 4.3%, 0.35%, 45%, and 0.8%, respectively. The assumption is made that age-related changes in enzyme activity of the tissues does not substantially alter the relative contributions of each organ to total body activity.

[b] From Ichihara and Koyama (1966). Activity is given as millimicromoles of keto acid formed in 10 minutes incubation per gram of fresh tissue. Weight of rats is not given. Comparable distributions are reported by Awapara and Seale (1952) and for 250–300 gm rats by Mimura *et al.* (1968).

[c] Obtained with two rats (weight not given) by Laferté *et al.* (1963).

[d] Calculated from data of Gaetani *et al*, (1964) as micro moles of pyrophosphate-^{32}P exchange with ATP per hour of incubation per gram of tissue. The calculations were made assuming the liver and muscle DNA concentrations given for 400 gm rats in Table VII and 1.49 mg of DNA per gram of heart (Wool *et al.*, 1968). Comparable enzyme data are reported by Stephen (1968).

[e] RNA content for liver and muscle from Table VII and for heart from Korecky and French (1967).

[f] From Bird *et al.* (1968). The data given by these authors are for rats of about 90 gm body weight assuming protein concentration in liver and muscle (noncollagen protein) to be 20% and 16%, respectively.

[g] From Oliver (1955). Activity is given as micromoles of ATP per hour of incubation per gram of tissue. The distribution is similar to that reported by Tanzer and Gilvarg (1959) for 1 rat weighing 150 gm.

amino acids absorbed into the portal vein is metabolized within the liver so that the amounts transferred to the peripheral blood are small; the branched-chain amino acids are exceptions in that a larger proportion of these pass into the general circulation, presumably being transaminated in muscle and kidney. Another consequence of the high activity of this transaminase in muscle is that the levels of the branched-chain amino acids in heart muscle (Scharff and Wool, 1965a,b) and in skeletal muscle (Table VIII) are relatively low.

The distribution of transaminases for the nonessential amino acids is widespread. Although the pattern of distribution of glutamate–oxaloacetate transaminase (GOT) and of glutamate–pyruvate transaminase (GPT) is somewhat species-specific, some general observations can be made. Liver and myocardium are particularly rich in GOT (Cornelius *et al.*, 1959; Wroblewski and LaDue, 1956; Cohen and Hekhius, 1941; Laferté *et al.*, 1963; Zimmerman, *et al.*, 1968; Howarth *et al.*, 1968). However, appreciable GOT activity is found in the skeletal muscle of all species of mammal examined (reviewed by Cohen, 1942); in the case of the rat, the ox, and the horse, the activity per gram of tissue is equal to or higher than is found in the liver (Cohen and Hekhuis, 1941; Freedland *et al.*, 1965; Cardinet *et al.*, 1967; Cornelius *et al.*, 1959; Howarth *et al.*, 1968). The distribution of GPT activity is somewhat different, most species exhibiting a higher activity per gram of tissue in the case of liver than in other tissues (Cornelius *et al.*, 1959; Zimmerman *et al.*, 1968; Laferté *et al.*, 1963) with the exception of the skeletal and cardiac muscle of the horse (Freedland *et al.*, 1965), ox, and pig (Cornelius *et al.*, 1959). Comparison of human red muscles (pectoral and abdominal) with white muscles (deltoid and semimembranosus) demonstrates higher GOT and GPT activities in the former (Kleine and Chlond, 1967).

Since the total weight of the liver relative to that of skeletal musculature changes in different species (Chapter 25, Fig. 3), these activities per gram of tissue should be correlated with the mass of liver and other tissues in order to arrive at a picture of the intensity of transamination in different organs. Such calculations have been shown in Table XI for GOT and GPT in the case of the liver and the skeletal muscle of species of mammals ranging in size from the mouse to the horse. The table shows total amounts of enzyme activity in each of these tissues expressed per kilogram of body weight of the animal. It is assumed that the subjects analyzed by the various authors were adults and that age of animals within the range studied is not a significant factor in enzyme activity. The work of Burleigh and Schimke (1969) showing that hexokinase and glycogen phosphorylase activities in rat, mouse, and

TABLE XI

TOTAL ACTIVITY (PER KILOGRAM OF BODY WEIGHT) OF GLUTAMIC–OXALOACETIC TRANSAMINASE AND GLUTAMIC–PYRUVIC TRANSAMINASE IN THE LIVER, HEART, AND SKELETAL MUSCLE OF VARIOUS MAMMALIAN SPECIES

Species	Weight (kg)	GOT (values/kg body wt.)			Muscle: liver	GPT (values/kg body wt.)			Muscle: liver	Reference
		Liver	Heart	Muscle		Liver	Heart	Muscle		
Mouse	0.025	1460	186	4,940	3.4	18,680	10	3645	0.2	*a*
Rat	0.250	3680	648	18,000	4.9	800	18	1575	2.0	*b*
Guinea pig	0.404	2490	450	22,500	9.0	340	39	1125	3.3	*b*
Rabbit	2	1500	377	3,600	2.4	150	52	4	2.1	*b*
Man	70	2700	239	44,550	18.5	840	39	2250	2.7	*c*
Horse	600	700	858	17,100	24.0	220	5	630	2.8	*b*

[a] Calculated from Laferte *et al.* (1963).

[b] Read from graphs of Zimmerman *et al.* (1968).

[c] Calculated from Wrobelewski and LaDue (1956).

[d] Values for body weight and organ weights, expressed as a percentage of adult body weight, have been taken from Tables II and III in Chapter 25.

rabbit diaphragm stablizes at about 6 weeks lends support for this assumption. In the case of GOT, there is a general tendency for muscle to make an increasing contribution to enzyme activity as compared with liver in mammals of increasing body size. This tendency is not so definite in the case of the relative amounts of GPT in liver and muscle but emphasize that muscle is a major location for GPT activity in most species. Thus, because of its size, muscle is a major site of transamination reactions involving glutamic and aspartic acids and alanine which may be correlated with the high concentration of these amino acids in tissues generally and with the large proportion of the body pool of the three amino acids located in skeletal muscle.

A final point regarding the significance of the large pool of alanine in skeletal muscle is that this amino acid provides a means of transport of amino-nitrogen from the carcass. In fasting man, the arteriovenous differences of amino acids in the forearm muscles are greatest for alanine (London *et al.*, 1965; Pozefsy *et al.*, 1969). Therefore, this amino acid provides the other organs, notably the kidney and liver, with amino-nitrogen, as well as a quantitatively important carbon source for the synthesis of glucose. The transport of alanine from skeletal muscle during fasting is discussed again in Section VIII,C.

B. Other Enzymes

In terms of activity per unit of tissue weight, Table X shows that amino acid activation in muscle is less intense than in many other tissues (Gaetani *et al.*, 1964; Stephen, 1968). This may be correlated with the less active synthesis of proteins in muscle (Gaetani *et al.*, 1964; Mariani *et al.*, 1963; Waterlow and Stephen, 1966; Stephen, 1968; see also Table VI in Chapter 25). Once more, however, the larger mass of muscle than of liver results in a greater *total* enzyme activity in the musculature than in the liver. An interesting comparison is between the activating enzyme content of different tissues and their content of RNA, also concerned with protein synthesis. Table X shows that the total amino acid-activating capacity of liver and heart are in proportion to the total amount of RNA in these two organs, but that the total activating enzyme content of skeletal muscle is about twice that expected on the basis of the RNA content of skeletal muscle. This discrepancy will be discussed later in connection with the efficiency of protein synthesis in muscle.

Proteolytic activity has been observed in various mammalian tissues. Cathepsins are known to occur in lysosomes (de Duve, 1959, 1963) which show low enzymic activity in both cardiac and skeletal muscle of several species (Zender *et al.*, 1958; Tappel *et al.*, 1962; Sottocasa *et al.*, 1962; Shibko *et al.*, 1963; Buchanan and Schwartz, 1967; Park and Pennington,

1967). Cathepsins are much more active in fish muscle (Siebert, 1958). The term cathepsin is a collective one and at least two different enzyme systems with differing pH optima and substrate specificity have been identified in the muscles of the mouse (Pennington, 1963), the rat (Koszalka and Miller, 1960; Bouma and Gruber, 1964), and the bovine (Landmann, 1963; Randall and MacRae, 1967). The cathepsins are frequently implicated as the main agents for degradation of body protein, and calculation of the total cathepsin content in the liver and in the skeletal musculature of the rat (Table X) shows that muscle has seven times the total activity of liver. Consequently, changes in the proteolytic activity of muscle could cause significant alterations in the protein metabolism of the whole body. It is also interesting to note, in passing, that the muscles of 2-year-old rats show greater proteolytic activity than do the muscles of 3-month-old rats (Hájek *et al.*, 1965) suggesting that the catabolism of muscle protein accounts for a reduction in the skeletal muscle mass in old age (Section II,A). Although it is assumed that proteolysis in muscle is a result of lysosomal cathepsin activity, electron microscopic and histochemical studies generally have not confirmed the presence of lysosomes in skeletal muscle contractile cells (Price *et al.*, 1964; Pellegrino and Franzini, 1963). Furthermore, Youhotsky-Gore and Pathmanathan (1968) have observed that the increased acid phosphatase activity with age in mouse gastrocnemius is associated with the interfibrillary connective tissue and not within the contractile cell. Smith (1964) concluded from an examination of rabbit muscle that acid phosphatase, a typical lysosomal enzyme, is confined to the blood-vessel cells of the muscle. On the other hand, the presence of lysosomes in cardiac muscle cells has been demonstrated both by electron microscopy and by histochemical tests (Wheat, 1965). It is thus possible that proteolysis in skeletal muscle is accomplished by enzymes that are not present in lysosomes (Joseph and Sanders, 1966). In this connection, it is interesting to note that the skeletal muscle of the rat and the hamster contains a moderately active protease that attacks muscle proteins at an optimum pH of nine, and is thus different from the acid protease of lysosomes (Koszalka and Miller, 1960; Koszalka *et al.*, 1961). Furthermore, Kohn (1969) has recently described the presence of a soluble factor in normal rat muscle which reacts, at neutral pH, with isolated myofibrils to yield peptides and amino acids. Finally, mammalian muscle shows several different types of ribonuclease activity (Tappel *et al.*, 1962; Epshtein, 1964; Abdullah and Pennington, 1968) and some of these are likely to be associated with lysosomes.

Table X also shows the relative activities of the enzyme creatine phosphokinase in four tissues of the rat and illustrates that the highest activities are found in muscle (Kuby *et al.*, 1954; Oliver, 1955; Tanzer and Gilvarg, 1959; Colombo *et al.*, 1962; Hess *et al.*, 1964). This enzyme plays an important role in the release of energy-rich phosphate for use in muscle contraction and also in muscle protein synthesis.

VI. Protein Synthesis and Degradation in Muscle

General features of the synthesis of proteins in tissues are dealt with in Chapter 6 of Volume I and in the Introductory Chapter to this volume. The degradation (turnover) of proteins is discussed in Chapter 7 (Volume I) and also in Chapter 32 of this volume. Here we must consider features of protein synthesis and degradation that are specific to muscle.

A. Muscle Protein Synthesis

Systems for incorporating amino acids into muscle proteins have been described by several investigators (Earl and Korner, 1965; Earl and Morgan, 1968; Breuer *et al.*, 1964; Rampersad *et al.*, 1965; Strohman, 1966; Chen and Young, 1968; Wool *et al.*, 1968). A special feature of protein synthesis in muscle cells is the presence of unusually large polysomes consisting of over 50 ribosomes aggregated on a messenger strand whose length presumably accounts for the size of the polysome aggregate. This unusual messenger length corresponds to the large size of the subunit of the myosin molecule. Myosin has a molecular weight of about 500,000, and its subunit is about 200,000 (Kielley and Harrington, 1960; Driezen *et al.*, 1966; Perry, 1967); the latter could be synthesized by a messenger RNA capable of carrying some 50–60 ribosomes (Heywood *et al.*, 1967), and Herrmann (1968) has estimated the time of synthesis of each myosin molecule by such a large messenger to be 2.4 minutes. The presence in muscle of polysomes of such a size has been confirmed both by electron microscopy and by their behavior on sucrose gradients (Heywood *et al.*, 1967, 1968; Heywood and Rich, 1968; Heywood and Nwagwu, 1968). Such large polysomes have been identified by these techniques in chick embryo skeletal muscle (Heywood *et al.*, 1967; Allen and Pepe, 1965; Fischman, 1967), human skeletal muscle (Nihei, 1969) and in rat skeletal and cardiac muscle (Breuer *et al.*, 1964; Bergman, 1962; Heuson-Stiennon, 1964; Chen and Young, 1968; Rabinowitz *et al.*, 1964; Cedergren and Harary, 1967). However, Larson *et al.* (1969) observed, with the electron microscope, that polyribosomes were in close association with the myosin filaments in mouse triceps and the quadriceps muscles of a normal human fetus but they did not find ribosome aggregates of more than 15 ribosomes. Other polysomes are also present in muscle. In sucrose-gradient studies of chick embryo muscle ribosomes, Heywood and Rich (1968) showed that polysome aggregates containing 15–25 ribosomes are able to synthesize actin, while aggregates of 5–9 ribosomes synthesize tropomyosin. These aggregate sizes correspond closely to expected messenger lengths for actin, which has a molecular weight of 60,000–70,000 (Hanson and Lowy, 1963, 1964; Perry, 1960a) and tropomyosin, with a molecular weight of 30,000–35,000 (Woods, 1966).

These proteins and their attendant polysomes do not appear simultaneously during development of muscle. In chicks the thin filaments of actin appear before the thicker myosin (Allen and Pepe, 1965; Fischman, 1967; Obinata *et al.*, 1966), and myosin synthesis at this time is relatively low (Coleman and Coleman, 1968; Herrmann, 1968). The appearance of myosin in the muscle of the 10-day-old chick embryo coincides with the identification for the first time of polysome aggregates containing 70–73 ribosomes (Allen and Pepe, 1965). Heywood and Rich (1968) have confirmed by sucrose gradients this sequential appearance of polysomes associated with synthesis of muscle proteins. As shown in Fig. 5, in the 10-day chick embryo there is a sharp peak on the gradient

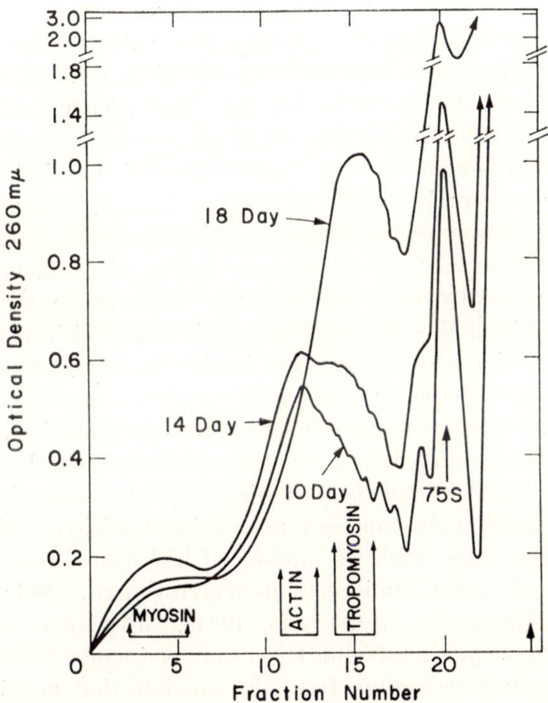

FIG. 5. Sucrose gradient analyses of ribosomes prepared from cytoplasmic extracts of chick embryo leg muscle at 10, 14, and 18 days of embryological development. The position of the gradient associated with myosin, actin, or tropomyosin synthesis is identified in the figure. (From Heywood and Rich, 1968.)

corresponding to polysome aggregates for actin synthesis, whereas at 14 days the profile indicates polysomes compatible with myosin and tropomyosin synthesis. Less complete information is available for mammalian embryos. Heuson-Stiennon (1964) observed the large polysomes in the muscles of rat embryos 16–17 days old, some of which were associated with a reticulum. Finally, the rate of development of the myofibrils and presumably the rate of protein synthesis appears to differ between typical slow-tonic and fast-phasic muscles in the chicken. Thus, during embryological development in the chick, Gutmann *et al.* (1969) have observed with the electronmicroscope that the myofibrils in the "slow" anterior latissimus dorsi accumulate at a markedly greater rate than in the "fast" posterior latissimus dorsi.

In connection with muscle protein synthesis, it is relevant to examine the distribution of RNA in muscle cells. Table XII shows data gathered on the skeletal and cardiac muscle of several species. By comparison with the subcellular distribution of RNA in liver, also shown in the table, the distribution of muscle RNA appears to differ in two respects. First, there is a lower content of soluble RNA; however, technical differences in cell sap preparation may account for this finding. Second, the distribution of ribosomes between those bound to membranes and those free in the cytoplasm differs in the two tissues. There are many more free ribosomes in the cells of skeletal muscles than in those of the liver. This observation is confirmed by studies on cardiac and skeletal muscle by means of electron miscroscopy (Porter and Palade, 1957; Larson *et al.*, 1969) and by sucrose gradients following the use of detergents (Earl and Korner, 1965; Earl and Morgan, 1968). Since membrane-bound ribosomes are usually found in secreting cells where they obviously discharge peptide chains through the membranes and into the extracellular space, it is important to try to identify the function of such membrane-attached ribosomes in muscle cells. The occurrence of filaments in cells containing free ribosomes but only a sparse endoplasmic reticulum has led to the suggestion that actin and myosin are synthesized by free ribosomes (Lentz, 1969). By electron microscopy, Heuson-Stiennon (1964) has observed large polysomes often associated with a reticulum, and this may be correlated with the finding of high concentrations of RNA in fragments of sarcoplasmic reticulum (Hulsmans, 1961; Muscatello *et al.*, 1961; Margreth and Novello, 1964). Muscatello *et al.* (1965) suggest that the membrane-bound RNA may be involved in the synthesis of the sarcoplasmic reticulum. It is also possible that membrane-bound polysomes may be involved in increasing the numbers of fibrils during growth which are added at the peripheral region of the fiber (Venable, 1969). It has long been known that RNA is commonly present in

TABLE XII

COMPARATIVE ESTIMATES OF THE DISTRIBUTION OF RNA IN SKELETAL AND CARDIAC MUSCLE OF SOME SPECIES AND IN RAT LIVER

		Percent RNA in homogenate				
			Ribosomal			
Species	Tissue or organ	Nuclear	Free	Bound	Total	"Transfer"[a]
Rat	Liver[b]	4	20	62	82	15
Rat	Diaphragm[c]	—	—	—	94	5
Rat	Mixed thigh muscles[d]	3	50	34	84	13
Pigeon	Breast muscle[c]	—	—	—	83	6
Rat	Heart[e]	—	—	—	—	5
Chick	Heart[f]	About 5	—	—	85	About 11

[a] Nonsedimentable at 105,000 *g* for 60 minutes.

[b] From Blobel and Potter (1967).

[c] Calculated from Margreth and Novello (1964). The total ribosomal RNA includes nuclear RNA and RNA in unbroken tissue.

[d] Unpublished results of Young and Alexis. Procedure used was a modification of the method described by Blobel and Potter (1967).

[e] From Hulsmans (1961).

[f] Calculated from Zak *et al.* (1967) for 16-day-old embryos. These authors found that 15% of total RNA was 4 S type. Nuclear RNA estimated from their DNA data and DNA:RNA ratio of 6.6:1 for purified nuclei of chick embryo heart.

significant amounts along with myosin when the latter protein is extracted from muscle (Perry, 1952; 1960b; Perry and Zydowo, 1959a,b; Baril *et al.*, 1964, 1966; Zak *et al.*, 1967) and that this RNA is ribosomal in type (Mihalyi *et al.*, 1957; Zak *et al.*, 1967), but recent studies suggest that much of this adherent RNA results from coprecipitation of ribosomes and myosin when extraction of the muscle is carried out with buffers too low in ionic strength (Heywood *et al.*, 1967, 1968; Chen and Young, 1968; Earl and Morgan, 1968).

Knowledge of the distribution of total muscle RNA between the various cell types of skeletal and cardiac muscle is not available. However, in biochemical studies of the nature and function of the RNA in skeletal and cardiac muscle, it is most frequently assumed that the RNA material extracted from these tissues is derived mainly from the contractile cells. A large proportion of the DNA in these tissues is present in the noncontractile cells (Section II,B); similarly it is likely that a significant proportion of the total tissue RNA is associated with the various cell types located outside of the muscle fibers. Electron microscope observations on the psoas muscle of newborn rats reveal an abundance of polyribosomes and ribosomes (Shiaffino and Margreth 1969) but similar observations on the semitendinosus muscle of 90–110 day-old rats (Gauthier, 1969) fail to show the presence of a significant ribosome population, suggesting the need for a detailed examination of the cellular origin of the ribosomes extracted from the tissue preparations of skeletal and cardiac muscle. The proportion of total muscle RNA which is derived from the contractile cell may, therefore, depend upon the age of the animal used for study.

B. Muscle Protein Turnover and Degradation

The relative stability of muscle protein has been commented on in Chapter 7 and is supported by a body of published data (e.g., Sprinson and Rittenberg, 1950; Thompson and Ballou, 1954, 1956; Davison, 1961; Zak, 1962; Funabiki and Kandatsu, 1965; Dreyfus *et al.*, 1960). From a study of isotope incorporation into isolated muscle proteins, Velick (1956) concluded that individual muscle proteins turn over at different rates. Thus he computed that the half-life of muscle glyceraldehyde 3-phosphate in the rat is 100 days, heavy meromyosin gives 80 days, actin is 67 days, aldolase and phosphorylase are both 50 days, tropomyosin is 27 days, and light meromyosin is 20 days. However, McManus and Mueller (1966) have concluded that the turnover times of light and heavy meromyosin are both 29 days, thus supporting the contention that the light and heavy meromyosins derived from myosin are synthesized and degraded at similar rates. Nevertheless, these figures are all much longer than the half-lives of proteins made in the liver of the rat.

The problem of measuring protein degradation rates in tissues is complicated by reutilization of amino acids liberated from degradation of protein within the cell. This phenomenon is discussed in Chapter 28, Section VI,A, in Volume III. After administration of a labeled amino acid, the rate of loss of radioactivity from the tissue proteins is the resultant of two processes: first, the rate of breakdown of tissue protein, and second, the extent to which radioactive amino acids liberated as a result of this breakdown are utilized again in making new tissue protein. In liver, reutilization of endogenous amino acids can vary according to nutritional circumstances from 50 to 90% (Henshaw *et al.*, 1968; Gan and Jeffay, 1967) of the total amount incorporated into protein, whereas in muscle reutilization is about 10–30% (Gan and Jeffay, 1967; Waterlow and Stephen, 1968). The lower reutilization of amino acids in muscle as compared with the liver may be related to the view expressed by Hider *et al.* (1969) that the extracellular amino acid pool of muscle provides the amino acids which are directly used for tissue protein synthesis. Nevertheless, reutilization gives rise to considerable errors in estimates of turnover times for tissue proteins, since the loss of radioactivity is less rapid than true rate of turnover would require. However, this source of error can be considerably reduced by using amino acids labeled in groups that are rapidly renewed when the amino acid is in the free state. In the case of the liver, labeling of arginine in the guanidino group results in rapid removal of the label from the free amino acid as urea synthesis proceeds (Stephen and Waterlow, 1966). This strategy is not valid for muscle, but a similar removal of readily exchangeable groups from the endogenous free amino acids of muscle can be achieved if $^{14}CO_2$ is used as the labeling source; in this case, the label passes into the carboxylic groups of aspartate and glutamate that undergo rapid transamination, and when the labeled amino acids are released again from the tissue protein, the label is once more eliminated from the amino acid by transamination. Millward (1969) has utilized this technique to compare the fractional rates of synthesis and catabolism of proteins in the liver and muscle of the rat, using the mathematical treatment of Koch (1962) to analyze the data (Table XIII). He computes the half-life of mixed muscle proteins to be 6.3 days, a period much shorter than the estimates given above, but in agreement with the findings of Waterlow and Stephen (1968), who in-

TABLE XIII

COMPARISON OF PROTEIN TURNOVER IN LIVER AND SKELETAL MUSCLE OF YOUNG FEMALE RATS[a]

		Muscle protein fraction		
	Liver	Mixed	Sarcoplasmic	Fibrillar
Fractional rate constants				
Synthesis	0.25	0.11	0.14	0.098
Catabolism	0.20	0.087	0.12	0.079
Half-life (days)				
Synthesis	2.9	6.3	5.0	7.1
Catabolism	3.4	8.0	5.8	8.8
Pool size				
Organ (gm)	5	6[b]	—	—
Protein (gm)	1.41[c]	6.48	2.16	4.32
Protein synthesis (mg/day)	353	713	302	423
Protein catabolism (mg/day)	282	564	259	341

[a] From data of Millward (1969) obtained with female rats 30–35 days old. They were fasted overnight prior to injection of ^{14}C-labeled carbonate and sacrificed at 2 and 7 days. Data calculated for a 100 gm rat.

[b] Muscle assumed to be 36% of body weight (Table I, Chapter 26).

[c] Includes an allowance of 0.6 gm of serum protein per 100 gm rat.

fused labeled lysine at a constant rate into the rat, and Garlick (1969), who did a similar experiment using labeled glycine. Millward used female rats 30–35 days old; assuming a body weight of 100 gm, it can be estimated that 713 mg of muscle protein and 353 mg of protein are synthesized by the liver each day. Also assuming a total body protein synthesis rate of 28 gm/kg per day (Waterlow and Stephen, 1967), synthesis in muscle may thus account for about 25% of the total in the young female rat. In a previous section of this chapter (Section IV), the presence of 3-methylhistidine in myosin and actin was discussed in relation to the lack of binding of this amino acid to tRNA and the synthesis of fibrillar proteins. It seems likely, therefore, that measurement of the loss of radioactivity from the 3-methylhistidine residues in myosin and actin, after injection of isotopic histidine or $^{14}CH_3$-methionine, would enable an accurate estimate of the true rates of degradation of these two muscle proteins and allow a further comparison with Millward's estimates based on the use of ^{14}C-labeled carbonate.

According to Waterlow and Stephen (1968), the turnover of muscle protein is higher in males, although the renewal of liver proteins is similar in the two sexes. From the data of Waterlow and Stephen (1967, 1968), it can be calculated that muscle protein turnover in the young male rat accounts for 23% of total body protein turnover, and in the adult it represents 30% (Table XIV); on the other hand, the proportion of total body protein synthesis contributed by liver tends to decline during normal growth. These differential effects of growth are largely reflections of the changes in relative size of muscle and liver as the animal attains its adult proportions. There are also changes in relative tissue weights as the rat grows old. With advancing age, the turnover of body protein decreases (Waterlow and Stephen, 1967) and it has been shown that this affects both muscle and liver protein (Waterlow and Stephen, 1968). Their data on total body protein turnover and muscle protein turnover are shown in Table XV, in which the decline in muscle RNA concentration with increasing age of the rat is also recorded. The reduction in RNA content thus parallels the slower rate of muscle protein formation, but it is claimed that additionally there is a reduction in the activity of muscle ribosomes *in vitro* with increasing age (Breuer and Florini, 1965; Srivastava and Chaudhary, 1969) and a decrease in the proportion of polyribosomes (Srivastava, 1969). Finally, it may be noted that turnover of muscle RNA is reputed to be slower than that of liver RNA. Thus for the rabbit the half-life of ribosomal RNA is 11.5 days in muscle, 7.5 days in heart, and 5.3 days in liver (Erdos and Bessada, 1966). In the case of the rat, half-life of muscle RNA is 11.7 days and 5.0 days for liver (Gerber *et al.*, 1960).

TABLE XIV

AN ESTIMATE OF THE CONTRIBUTION OF PROTEIN SYNTHESIS BY THE LIVER AND SKELETAL MUSCLES TO THE DAILY TOTAL BODY PROTEIN TURNOVER IN YOUNG AND ADULT MALE RATS

Age	Young	Adult
Body weight (gm)	100	300
	Per kilogram body weight	
Lysine flux[a] (mmoles/day)	26	18
Total protein turnover[b] (gm/day)	38	25
Liver[c]		
Protein turnover (gm/day)	13.7	8.3
Percent body turnover	36	33
Muscle[c]		
Protein turnover (gm/day)	8.6	7.2
Percent body protein turnover	23	29
Muscle:Liver turnover	0.63	0.87

[a] Read from graph of Waterlow and Stephen (1967).

[b] Calculated assuming a lysine content of body protein to be 4.5 μmoles per milligram of body N (Waterlow and Stephen, 1966).

[c] Calculated from Waterlow and Stephen (1968). Data for muscle read from graphs given in their paper. Liver values calculated using data for <160 gm and >160 gm rats for the young and adult groups, respectively. Included in the liver figures is an allowance for albumin synthesis by assuming the body albumin pool to be 0.6 gm/100 gm rat (intra- plus extravascular albumin pool) and a turnover rate equal to that of liver protein. Size of muscle and liver plus albumin protein pools were estimated from organ size shown in Table 1, Chapter 26, and protein concentrations of 16% (noncollagen protein) and 20% for muscle and liver, respectively.

TABLE XV

EFFECT OF BODY WEIGHT ON TOTAL LYSINE FLUX, TURNOVER RATE OF MUSCLE PROTEIN, AND RNA CONTENT OF MUSCLE IN RATS

Body weight (gm)	Total body lysine flux[a] (μmoles 100 $gm^{-1}hr^{-1}$)	Muscle protein[b] turnover rate ($days^{-1}$)	Muscle[c] RNA (mg/gm)
100	110 (100)[d]	0.15 (100)	1.9 (100)
200	90 (82)	0.12 (80)	1.7 (90)
300	75 (68)	0.10 (67)	1.2 (63)

[a,b] Read from graphs published by Waterlow and Stephen (1967) and Waterlow and Stephen (1968), respectively.

[c] From Table VIII.

[d] Figures in parentheses are percentages of values obtained in 100 gm rats.

VII. Response of Muscle Protein Metabolism to Nutritional Change

A number of nutritional deficiency conditions are recognized to influence protein metabolism in muscle. For example, potassium deficiency affects the levels of the basic amino acids in muscle (Muntwyler *et al.*, 1958; Iacobellis *et al.*, 1956; Eckel *et al.*, 1954); it could be postulated that the reduced concentration of potassium in skeletal muscle of malnourished children (Frenk *et al.*, 1957; Waterlow and Mendes, 1957; Vis *et al.*, 1965; Nichols *et al.*, 1969) may affect the status of protein synthesis in this tissue. A potassium-deficient diet, when given to chicks, reduces the incorporation of ^{14}C-labeled leucine into mixed muscle proteins (Rinehart, Featherston, and Rogler, 1968). The postulate is further supported by the observations of Lubin and co-workers (Ennis and Lubin, 1961, 1965; Lubin, 1967) demonstrating an *in vivo* requirement for potassium in cellular protein synthesis and also the requirement for potassium in the transfer of, or binding of, aminoacyl-sRNA to the polysome (Schlessinger, 1964; Conway, 1964; Spyrides, 1964). Ascorbic acid deficiency also causes changes in muscle free amino acid pools that are not explainable through change in appetite (Schønheyder and Lyngbye, 1962) and alterations in muscle free amino acid levels have also been noted in rats suffering from vitamin B_6 deficiency (Swendseid *et al.*, 1964; Runyan and Gershoff, 1969). However, these diverse observations have not been pursued to the point at which they can be integrated into a general picture of the action on protein metabolism of the specific nutrient in short supply. In contrast, we now have a considerable number of observations on the effects of protein deficiency and of caloric deficiency in the diet on protein metabolism in skeletal muscle, many of the important studies being the work of Waterlow and his group. We shall therefore limit this review to the effects of protein-calorie malnutrition on muscle protein content and on muscle protein synthesis and turnover, the effect of carbohydrate administration and of amino acid imbalance and excess on muscle protein metabolism.

A. *Effect of Protein-Calorie Malnutrition on Muscle Protein Content*

When an animal or human subject is starved or is fed on a diet inadequate in protein content, the tissues lose both weight (Aron, 1911; Jackson, 1925; Widdowson *et al.*, 1960) and protein content (Chapters 10 and 28) at very different rates. This is clearly illustrated by the studies of Addis *et al.* (1936a,b,c), who showed rapid and extensive changes in the total amount of protein in the liver of the rat under conditions of starvation or protein deficiency, but much slower changes in the case of the carcass, a large part of which is muscle. A representative experiment by this group is illustrated in Fig. 6, Chapter 10, Volume I, of this treatise. In Chapter 10, Section II,C, Munro concluded from a survey of the literature that the liver of the rat has completed its adaptation of protein deficiency within 3 days of consuming a protein-free diet, whereas muscle does not begin to lose protein significantly until after this time. This deduction has been confirmed through direct experimentation by Waterlow and Stephen (1966), who showed that the livers of rats have lost 30% protein after 3 days on a low-protein diet, and 50% protein after 6 weeks on a low-protein diet, thus indicating that adaptation is nearly complete within the first 3 days; on the other hand, they demonstrated no loss of muscle protein during the first 3 days on the diet, but a loss of 58% of the original muscle protein content after 6 weeks on this regimen. The full data for this experiment are given in Table II of Chapter 28. Thus during prolonged protein-calorie malnutrition, muscle eventually loses a greater percentage of its initial protein content than does liver and in terms of total protein lost from the body, skeletal muscle contributes much more protein than does liver (Waterlow, 1956; Allison *et al.*, 1962; Hagan and Scow, 1957; Mendes and Waterlow, 1958; Widdowson *et al.*, 1960). In man, a similar picture probably occurs; Kerpel-Fronius and Frank (1949) showed in malnourished infants that the loss of muscle mass was considerably greater than the reduction in body weight. It is generally argued teleologically by workers in this field that muscle protein is catabolized during protein-calorie malnutrition in order to provide amino acids for the liver and other vital organs (e.g., Widdowson *et al.*, 1960; Viteri *et al.*, in Chapter 22, Section II). It would be more instructive to know how the balance between different organs is achieved.

The weight and amount of protein lost as a result of malnutrition varies in different muscles. In the case of the fowl, the pectoral muscles develop later and are used less than are the sartorius muscles, and malnutrition has more effect on the weight of the pectorals (Dickerson and McCance, 1960). At the cellular level, protein-calorie malnutrition causes a decrease in mean fiber diameter in the muscle of man (Vincent and Rademacker, 1959; Montgomery, 1962b), cattle (Robertson and Baker, 1933), sheep (Joubert, 1956), pigs (McMeekan, 1940), and the mouse (Goldspink, 1964, 1965; Rowe, 1968a). In the mouse, it has been proposed that there is a bimodal distribution of fibers into large phase and small phase, and inadequate intake of food decreases the relative proportion of large phase fibers (Goldspink, 1964; Rowe, 1968a). Starvation decreases the number and diameter of myofibrils in the fibers (Goldspink, 1965), and in the rat undernutrition reduces the number of myofilaments at the outer periphery of the myofibril (Wechsler, 1964).

Whether there is an actual diminution in the number of muscle fibers during undernutrition is not certain. Many investigators have concluded that only the mean diameter of the fibers changes, without an alteration in fiber number. In contrast, Elliott and Cheek (1968) observed a loss of muscle DNA from food-restricted weanling rats and concluded that there must have been a decrease in fiber number. However, Mendes and Waterlow (1958) fed a low-protein diet to rats at a level that caused even greater restriction on growth than that imposed in Elliott and Cheek's study, but failed to obtain a loss of muscle DNA. It may be that caloric restriction is more severe than protein restriction alone. In human infants, there is histological evidence that protein-calorie malnutrition leads to loss of muscle fibers (Frankl and Freund, 1884; Montgomery, 1962b). The effect of a period of undernutrition on subsequent muscle development has been investigated by Winick and Noble (1966) as part of their survey of tissue responses to malnutrition. If the period of malnutrition is applied for a short period while muscle cell population is still increasing, the amount of DNA in the muscle reaches a final plateau at a lower level than it would in normally nourished animals, even if some time elapses between the time of terminating the malnutrition and the end of the growing period.

Malnutrition has a differential effect on the constituent cells of skeletal muscle. The proteins of the contractile fibers (fibrillar and sarcoplasmic) diminish, whereas the extracellular proteins (chiefly collagen) are not reduced during a period of malnutrition. Indeed there is evidence from studies on animals (Mendes and Waterlow, 1958; Cabak *et al.*, 1963; Widdowson *et al.*, 1960; Hagan and Scow, 1957) and on man (Picou *et al.*, 1966) that muscle collagen can continue to accumulate during severe malnutrition, a phenomenon that also occurs with body collagen generally (Harkness *et al.*, 1958). Part of the collagen increment is contributed by a reduction in collagen catabolism during malnutrition (Picou *et al.*, 1965). Within the contractile cells, it has been concluded from studies on pigs (Widdowson *et al.*, 1960) and rats (Waterlow and Stephen, 1966) that prolonged malnutrition has little effect on the ratio of fibrillar to sarcoplasmic proteins. However, other studies on rats (Yamatami and Kandatsu, 1967; Hagan and Scow, 1957) and on cocks (Dickerson and McCance, 1960; Montgomery *et al.*, 1964) have shown a greater proportional loss of sarcoplasmic proteins.

B. Effect of Protein-Calorie Malnutrition on Muscle Protein Synthesis

When rats are given a diet deficient in protein or in amino acid content or balance, the distribution of radioactivity following injection of a labeled amino acid differs from that in well-nourished rats. Liver protein is more highly labeled whereas muscle protein has less label than is found in the well-fed controls (Solomon and Tarver, 1952; Waterlow, 1959; Schreier and Kazassis, 1960; Muramatsu *et al.*, 1963; Sidransky and Verney, 1968). Several of these experiments are summarized in Table II of the Introductory Chapter to Part V in Volume III of this treatise. Many of these studies suffer from the obvious criticism that, if the free amino acid pools of liver and muscle change to different extents as a result of protein deficiency, dilution of the labeled amino acid in the precursor pools will be altered and in consequence labeling of the tissue protein could be accounted for in this way. However, Waterlow and Stephen (1966) have taken account of this in studying uptake of radioactivity from lysine-^{14}C into the muscle proteins of rats receiving a low-protein diet. At 3 days after labeled lysine injection, uptake into muscle protein was lower than that of control animals, but there was no difference in either the extracellular:intracellular lysine concentration ratio or the specific activity of muscle free lysine; consequently, Waterlow and Stephen concluded that there was a real decrease in muscle protein synthesis due to protein malnutrition. This view was further supported by a later series of experiments by these workers (Waterlow and Stephen, 1968) in which a continuous lysine infusion technique was used (Waterlow and Stephen, 1967). After 3 days on a protein-free diet or 10 days on a low-protein diet, muscle protein synthesis was greatly reduced, whereas the rate of liver protein synthesis was at first increased and later was reduced, but less so than for muscle, as protein depletion progressed. The greater reduction in the capacity of muscle to incorporate amino acids into protein would be a sufficient reason for the redistribution of amino acids toward liver.

The mechanism responsible for reduced muscle protein synthesis is not yet clear. Waterlow and Stephen (1966, 1968) suggest that free amino acid concentrations in muscle are not likely to be the causal factor during protein depletion, since they could not detect alterations in free lysine level, at least during the first 3 days of protein-free feeding. This is not compatible with some other data in the literature. Although Gan and Jeffay (1967) found no change in rat plasma and muscle lysine levels during a 7-day fast, Thompson *et al.* (1950) observed a 20% reduction in muscle free lysine concentration after adult rats had been given a protein-free diet for 3 weeks, and Allison *et al.* (1963) also found changes in the free lysine content of rat liver and muscle after short periods of protein deficiency. More cogent, however, is the argument put forward by Munro in Chapter 34, Section III,A,1, that the concentration of one free amino acid in a tissue can limit the rate of protein synthesis in that tissue. This limiting amino acid need not be lysine, which can consequently be present in normal concentrations even when amino acid supply is responsible for reducing the rate of protein synthesis. Whatever the cause, there is a demonstrable effect of protein deficiency on the capacity of muscle ribosomes to make proteins. Using a diet deficient in threonine that was force-fed to rats, Farber, Sidransky and co-workers (Sidransky *et al.*, 1964; Sidransky and Farber, 1958; Sidransky and Verney, 1965, 1968) observed an increased hepatic protein synthesis both *in vivo* and *in vitro* and a decreased incorporation of labeled amino acids by skeletal muscle ribosomes (Sidransky and Verney, 1967). Similarly, Young and Alexis (1968) observed that amino acid incorporation by ribosomes in a cell-free system prepared from rats fed on a low-protein diet was about half that obtained when ribosomes prepared from well-nourished rats were used. These changes were observed within 6 days of starting to feed the low-protein diet and persisted for the 4 weeks of the experiment. Their findings can be correlated with a decrease in the proportion of large polysome aggregates harvested from the muscles of rats on low-protein or protein-free diets (Young *et al.*, 1968a; Young and Alexis, 1968). However, the changes in polysome profiles were not as marked as might be expected from the observed *in vivo* and *in vitro* alterations in muscle protein synthesis. This is not likely to be due to differences in the charging of tRNA with amino acids. In our laboratory, we have observed that the tRNA content of muscle decreases in proportion to the change in total and ribosomal RNA during protein depletion, and Stephen (1968) has observed that amino acid activating enzymes in muscle do not decrease in activity during protein depletion, although she confirmed the observations of Gaetani *et al.* (1964) that the activating enzymes of liver increase in activity during prolonged protein depletion. The latter group (Mariani *et al.*, 1963; Gaetani *et al.*, 1964) claimed that activating enzyme levels in muscle decrease during depletion when the activities are expressed on the basis of muscle DNA content. It is not known whether changes in hormonal secretion due to protein deficiency can account for the change in muscle protein synthesis. In protein-calorie malnutrition, the levels of cortisol (Alleyne and Young, 1966; 1967) and of growth hormone (Pimstone *et al.*, 1967, 1968) are increased and insulin level is reduced (Baig and Edozien, 1965; Edozien and Young, 1969). As will be discussed later, increased cortisol level and reduced insulin secretion each have effects on muscle protein synthesis and polysome patterns similar to those observed in protein-calorie malnutrition.

Finally, some observations have been made during protein depletion on muscle enzymes involved in protein metabolism. When adult cocks were given a protein-free diet, there was a reduction in activity of GOT and GPT in muscle (Ashley and Fisher, 1967), whereas rats on a protein-free diet show reduced GOT activity in muscle (Yamaguchi and Kandatsu, 1967) and a significantly *increased* activity of the branched-chain amino acid transaminases (Mimura *et al.*, 1968). This latter observation may partially explain the marked reduction in the plasma levels of the branched-chain amino acids in children suffering with kwashiorkor (Arroyave *et al.*, 1962; Snyderman *et al.*, 1963; Edozien, 1966). However, in protein-calorie malnourished rats, McFarlane and Holt (1969a,b) have observed a decreased oxidative degradation of leucine. They also found that the urinary excretion of 4-methyl-2-oxopentanoate (2-oxoisocaproate) was increased; it appears that in the liver the decarboxylation of the α-keto acid of leucine may be reduced under conditions of a deficient dietary protein supply. The acid cathepsin of rat muscle is also increased in activity by 5 days of food restriction (Bird *et al.*, 1968). Enzymes connected with energy metabolism in skeletal muscle also undergo changes during malnutrition. Red and white muscle fibers may respond to starvation differently. Taskar and Tulpule (1964) found a marked decrease in the activities of the enzymes of aerobic oxidation with starvation in rats. In comparison, the glycolytic enzymes of white muscle did not change in activity with starvation. However, they (Taskar and Tulpule, 1964) did observe that a reduction in the activities of adenosine triphosphatase and creatine phosphokinase occurred in both red and white muscle on feeding a low-protein diet. Metcoff and co-workers (Metcoff *et al.*, 1966; Metcoff, 1967) have noted in muscles from malnourished children that the activities of pyruvic kinase, malic dehydrogenase, and isocitric dehydrogenase were lower, but lactic dehydrogenase was unchanged. This may account for the decreased basal oxygen consumption (Monckeberg *et al.*, 1964) and reduced oxidative metabolism of skeletal muscle (Nichols *et al.*, 1968) observed in malnourished children.

C. *Muscle Protein Metabolism after Carbohydrate Administration*

As discussed in Chapter 10, Section III,B,2, ingestion of carbohydrate by fasting subjects or animals results in a reduction in urinary N output, and in liver protein content, and a decrease in plasma amino acid concentrations lasting several hours. In the case of man and other species studied, the different plasma amino acids are not affected proportionately, the pattern of decrease resembling the pattern of amino acid requirements of man (Munro and Thomson, 1953; Swendseid *et al.*, 1967), the amino acid composition of skeletal muscle in the case of the dog (Lotspeich, 1949) and the amino acid pattern of carcass protein in the case of the rat (Rao and McLaughlan, 1967). These patterns all show essentially the same proportions of amino acids and imply that the amino acids have probably been used for protein synthesis in some tissue. The decrease in plasma amino acids is accompanied by an increase in uptake of amino acids into muscle, as evidenced by greater labeling of muscle protein with glycine-^{14}C, leucine-^{14}C and methionine-^{35}S when glucose is given to rats (Munro *et al.*, 1959). Liver and intestinal mucosal proteins did not participate in this increased deposition. Recently, Knipfel *et al.* (1969) confirmed the change in muscle protein labeling, but also observed some increase in liver protein uptake. Even if this is so, the large mass of muscle in the body must make it the major repository of the amino acids removed from the blood. Munro (1956) has demonstrated that the action of glucose on blood amino acid levels is dependent on insulin secretion; in view of evidence to be discussed later, the greater labeling of muscle protein after giving glucose is due to a real increase in muscle protein synthesis and not simply to increased labeling of the precursor pool. The relationship between deposition of plasma amino acids in muscle after each meal containing carbohydrate and diurnal rhythms in plasma amino acid levels is discussed in Chapter 36, Section III,C.

D. Effect of Amino Acid Imbalance and Excess on Muscle Protein Metabolism

Within a short time in rats fed an amino acid imbalanced diet, the plasma amino acid pattern changes (Sanahuja and Harper, 1963; Harper *et al.*, 1964), and in particular the most limiting amino acid falls to a low level in the plasma. This decrease is paralleled by a similar but often a more marked change in the muscle (Harper and Rogers, 1965; Rogers and Harper, 1968; Leung *et al.*, 1968). It is not clear whether muscle protein synthesis is altered after ingestion of a threonine-imbalanced diet. Harper and Rogers (1965) conclude that muscle protein synthesis continues at a normal rate, but their data suggest the possibility of an increased rate of amino acid incorporation despite the marked decrease in the size of the free threonine pool in muscle. More extensive studies on the metabolic effect of feeding threonine- and histidine-imbalanced diets to rats have been reported from their laboratory (Yoshida *et al.*, 1966; Benevenga *et al.*, 1968) with the similar observation that the *in vivo* incorporation of ^{14}C from the limiting amino acid into muscle proteins is not decreased, whereas the activity in acid-soluble fraction of muscle is depressed. On the other hand, after force-feeding rats a threonine-devoid diet, there is a reduction in the *in vitro* protein synthetic capacity of skeletal muscle ribosomes (Sidransky and Verney,

1967). However, the reduction in the plasma level of the limiting amino acid after feeding an imbalanced diet would not appear to be due to an alteration in muscle protein synthesis (Yoshida *et al.*, 1966; Benevenga *et al.*, 1968). If muscle protein metabolism plays a role in the marked plasma amino acid response, it may be that the rate of protein catabolism in muscle is decreased, but as yet evidence is lacking to support this point.

Changes in the free amino acid content of muscle also occur after giving rats an excess of a specific amino acid (Alam *et al.*, 1966; Clark *et al.*, 1968; Pawlack and Pion, 1969). The increased plasma levels are paralleled by similar changes in liver and muscle, but the latter tissue shows a much greater increase in its free amino acid content than does the liver, and it may be concluded that deposition of the amino acid given in excess occurs mainly in the muscle shortly after administration.

VIII. Hormones and Muscle Protein Metabolism

As discussed in Chapter 10, Section IV, hormones can cause either anabolic changes in skeletal muscle with gain in muscle protein (androgens, growth hormones, insulin) or they can cause a catabolic change, with a net loss of muscle protein (corticosteroids, thyroid hormones, estrogens). Since muscle contributes such a large proportion of the total protein of the body, these hormonal effects on muscle protein content are usually paralleled by similar changes in nitrogen balance, but frequently they are the opposite of changes in the viscera caused by the same hormonal treatment. The subcellular sites of action of hormones are discussed in detail in Chapter 33 in this volume.

A. Androgens

The influence of androgens in muscle growth has long been recognized. This is evidenced by the reduced gain in muscle weight, total protein content, and total DNA content in female animals (Cheek *et al.*, 1968) and in the castrated male (Kochakian *et al.*, 1956; Scow and Roe, 1953; Cheek *et al.*, 1968; Swartz, 1962), and by the deposition of protein in muscle when androgens are administered to castrated animals (Kochakian *et al.*, 1950). Although both the pituitary and thyroid glands are necessary for normal muscle maturation (Cheek *et al.*, 1965a,b, 1966, 1968), androgens act on muscle independently of other endocrine secretions (Scow and Hagan, 1957, 1965; Kochakian *et al.*, 1956). The interrelationship between diet and androgenic action is discussed in Chapter 10, Section IV.

The nature of the anabolic action of androgens on muscle has been examined. After administration of testosterone, the amount of DNA in muscle does not change (Kochakian *et al.*, 1964; Venable, 1966a), nor are there changes in the number of nuclei or muscle cells (Venable, 1966a). The increase in muscle weight caused by the androgen has been attributed to an increase in thickness of the muscle fibers without a change in fiber number (Venable, 1966a); in agreement with this, castration causes loss of muscle weight without a decrease in the number of muscle fibers (Rowe, 1968b; Venable, 1966b). The effect of androgen supply on muscle composition varies somewhat from muscle to muscle. The temporal and masseter muscles in the guinea pig and the perineal muscles in the rat are particularly sensitive to atrophy following castration and to enlargement on treatment with androgens (Scow and Roe, 1953; Scow and Hagan, 1955; Kochakian *et al.*, 1948; Wainman and Shipounoff, 1941). Scow and Hagan (1955) observed that castration reduced the myosin, actomyosin, and sarcoplasmic protein fractions of the temporal muscle of guinea pigs, but the rectus femoris muscle showed a loss of only alkali-soluble stroma and collagen. Different muscles also vary in their loss of myoglobin following castration (Scow and Roe, 1953). The decrease in RNA:DNA ratio caused by castration is also confined to some muscles (Saunders *et al.*, 1962).

Muscle protein synthesis appears to be stimulated by androgen administration. Following androgen administration to the intact and castrated animal, there is an increased ^{14}C incorporation *in vivo* into muscle protein after injection of ^{14}C-labeled amino acids (Novak, 1957; DeLoecker, 1965) and of amino acids added to excised muscle strips (DeLoecker, 1965). This may be related to muscle ribosome function. Breuer and Florini (1965) observed that skeletal muscle ribosomes obtained from castrated animals are less active in cell-free protein synthesis than are ribosomes prepared from normal animals or from castrated animals receiving androgens. These investigators correlated this finding with a reduction in the proportion of ribosomes present in muscle extracts as polysomes. On the other hand, they found unexpectedly that administration of testosterone to normal rats reduced the capacity of muscle ribosomes for *in vitro* protein synthesis. In this series of studies, Breuer and Florini (1965) observed that actinomycin D pretreatment prevented the stimulant action of androgens on the formation of polyribosomes and ribosome activity, and in subsequent studies (Breuer and Florini, 1966) they demonstrated an increase in RNA polymerase activity in the aggregate enzyme prepared from the muscles of androgen-treated animals; they conclude that chromosome priming activity is increased by hormonal treatment.

Combination of androgens with other hormones appear to produce the expected changes in muscle metabolism. Growth hormone, in combination with testosterone, has an additive effect on the weight increase and nitrogen retention of growing rats (Kochakian, 1960) and on the *in vitro* activity of muscle ribosomes (Florini and Breuer, 1966) and of muscle RNA polymerase (Breuer and Florini, 1966). Bullock *et al.* (1968) have shown that cortisone administration reduces the *in vitro* activity of muscle ribosomes and that the giving of a synthetic anabolic steroid related to the androgens can prevent this effect.

B. Growth Hormone

The effects of growth hormone on protein metabolism have been surveyed from a number of different points of view (de Bodo and Altszuler, 1957; Russell and Wilhelmi, 1960; Knobil and Hotchkiss, 1964; Engel and Kostyo, 1964; Root, 1965; Korner, 1967; Kostyo, 1968; Snipes, 1968) and is also dealt with in Chapter 10, Section IV,A in Chapter 33, Section V,C of this treatise. As in the case of the androgens, a preferential effect of growth hormone on muscle has been well documented. After administration of the hormone, liver weight increases less than total body weight (Fraenkel-Conrat *et al.*, 1941), whereas muscle grows faster than body weight (Greenbaum and Young, 1953). The increase in muscle bulk appears not to affect the average composition of muscle (Scow and Hagan, 1965). However, although the DNA *concentration* of muscle does not increase with growth hormone injection to hypophysectomized rats, the *total* amount of DNA in the skeletal muscles is increased (Beach and Kostyo, 1968).

The studies of Friedberg and Greenberg (1948) with rats injected with methionine-^{35}S showed that treatment with growth hormone increased uptake of the amino acid into muscle protein but diminished incorporation into liver protein, observations confirmed by Lee and Williams (1952). The stimulus of growth hormone on muscle protein uptake of labeled amino acids is confirmed by studies of excised muscle samples, which show an enhanced capacity for incorporation of labeled amino acids *in vitro* when taken from animals pretreated with growth hormone and a decreased incorporation when prepared from hypophysectomized animals (Kostyo and Knobil, 1959; Manchester *et al.*, 1959; Reiss and Kipnis, 1959). This appears to be a direct hormonal effect on muscle since increased transport of amino acids into the muscle cell and increased labeling of muscle protein have been observed when growth hormone was added to the fluid bathing portions of diaphragm excised from hypophysectomized rats (Kostyo, 1968).

It is difficult to disentangle the effect of the hormone on transport of amino acids from its action on muscle protein labeling; thus growth hormone treatment stimulates not only passage of natural amino acids

(Snipes and Kostyo, 1962; Snipes, 1967), but also of the metabolically inert AIBA (Riggs and Walker, 1960; Kipnis and Reiss, 1960; Eichhorn *et al.,* 1961). Consequently, the action of growth hormone on muscle protein synthesis, as evidenced by the increased capacity of muscle ribosomes for cell-free protein synthesis (Florini and Breuer, 1966), could be the result of the increased intracellular concentration of amino acids. It does not, however, appear to involve enhanced RNA synthesis (Martin and Young, 1965; Dawson *et al.,* 1966). Thus Kostyo (1966) incubated diaphragms from hypophysectomized rats with labeled uridine and observed no change in RNA uptake of the label, although amino acid uptake into muscle protein was increased. This is confirmed by the observation of Florini and Breuer (1966) that a single injection of growth hormone into hypophysectomized rats caused increased ribosome activity for protein synthesis before it stimulated RNA polymerase activity in muscle. The subcellular site of growth hormone action is discussed in detail in Chapter 33, Section V,C.

Changes in growth hormone secretion resulting from nutritional change are not readily interpreted. Plasma growth hormone level in man is said to increase within 2–5 hours after consuming protein, whereas glucose administration causes a decrease in the level of the hormone in plasma (Rabinowitz *et al.,* 1966, 1968), an observation which suggests that the anabolic action of this hormone may fluctuate in relation to meal consumption. The large increase in the growth hormone content of plasma obtained from protein-calorie malnourished children (Pimstone *et al.,* 1968) suggests that, in this condition, muscle protein formation may be maintained to some extent even though this nutritional condition is associated with low levels of plasma insulin (Baig and Edozien, 1965) and increased levels of plasma cortisol (Alleyne and Young, 1966, 1967), both circumstances that would be inimical to retaining protein in skeletal muscle.

C. Insulin

The actions of insulin on protein metabolism and on protein synthesis are discussed in Chapter 10, Section III,B and in Chapter 33, Section V,A. After injection of insulin, plasma levels of all amino acids are swiftly reduced, an effect that also occurs with eviscerated animals (Mirsky, 1938; Frame and Russell, 1946; Ingle *et al.,* 1947), thus indicating that viscera are not the site of action of the hormone. Direct isotopic evidence indicates the main site of amino acid deposition to be muscle (Munro, 1956, and reviewed in Chapter 10, Section III,B).

Consumption of a meal containing protein, carbohydrate, or both results in secretion of insulin (Rabinowitz *et al.,* 1966, 1968; Rabinowitz and Merimee, 1968), and this release of insulin appears to account for the lowering of plasma amino acid levels after feeding glucose, rather than through a direct action of glucose taken up by the muscle cell (Manchester and Young, 1958). This response to dietary carbohydrate is discussed further in Chapter 10, Section III,B, in Chapter 34, Section III,A,4, and in Chapter 36, Section III,C. In man, uptake of amino acids from plasma into muscle can be studied by measuring the arteriovenous difference in amino acid concentration in the forearm, since about 85% of the blood flow in this area passes through muscle (Andres *et al.*, 1956; Baltzan *et al.*, 1962). Using this technique, Pozefsky *et al.* (1969) have observed that, after an overnight fast, the arteriovenous difference in amino acid concentration indicates an extensive negative balance of amino acids passing out from muscle. This passage of amino acid N out of muscle could be reduced by 74% after insulin infusion into the brachial artery, but the 17% reduction in alanine release was not significant. Pozefsky *et al.* (1969) have computed from these data that the musculature of the body provides enough amino acids during a 24-hour fast to permit synthesis of 43 gm of glucose, a quantity agreeing with their estimates of the daily extent of gluconeogensis during fasting, which they based on urinary N excretion (Cahill *et al.*, 1966). These data also confirm the view expressed in Chapter 10 that insulin secretion regulates the passage of amino acids between plasma and muscle in relation to meals. For example, we have observed that adult male subjects fed four mixed meals a day show high plasma concentrations of tryptophan in the early morning, but these levels decrease after meals are started; this diurnal fluctuation in tryptophan level is not observed if the subjects receive the same diet as six meals spaced throughout the 24 hours (Young *et al.*, 1969b).

The effect of insulin on *in vitro* uptake of labeled amino acids by isolated muscle cells confirms the observations made on the intact animal. There is both an increase in labeling of muscle protein during the period of incubation and in the concentrations of about ten of the free amino acids within the muscle cell (Kipnis and Noall, 1958; Manchester and Young, 1960; Akedo and Christensen, 1962; Manchester and Wool, 1963; Wool, 1964). Scharff and Wool (1965b) found that if protein synthesis was suppressed in the perfused rat heart by puromycin treatment, then all free amino acids in muscle showed an increase in concentration level following insulin treatment, and they conclude that the reason for only ten amino acids being affected by insulin treatment in the absence of puromycin is that protein synthesis keeps the others from accumulating even though their transport is enhanced by insulin. These experiments are discussed critically by Manchester in Chapter 33, Section V,A, where some objections to this interpretation are made. Using protein-depleted rats, Waterlow and Stephen (1968) found no change in free lysine concentration in skeletal muscle following insulin administration. The position of free amino acids in cardiac muscle is confused. Although the concentrations of these in the perfused heart do not change when insulin is added to the perfusate (Scharff and Wool, 1965a), nevertheless, alloxan diabetes causes significant alterations in the free amino acid levels of cardiac muscle (Scharff and Wool, 1966).

The relationship of changes in muscle free amino acid levels to the stimulant action of insulin on muscle protein synthesis is also subject to dispute. Some contend that changes in free amino acid level are responsible for the effect of insulin on protein synthesis (Kipnis and Noall, 1958; Goldstein and Reddy, 1967), but Wool (Wool, 1965; Stirewalt and Wool, 1966) considers that the evidence favors a direct action of insulin on protein synthesis. Although muscle RNA labeling from precursors increases after insulin administration (Wool, 1963; Wool and Munro, 1963), insulin can still stimulate uptake of amino acids into muscle protein after giving doses of actinomycin D sufficient to block RNA synthesis (Wool and Moyer, 1964; Eboué-Bonis *et al.*, 1963). Wool has recently demonstrated changes in muscle ribosome capacity for *in vitro* protein synthesis following administration of insulin. Ribosomes prepared from the muscles of diabetic rats show decreased *in vitro* capacity for protein synthesis, but their protein synthetic activity is restored 5 minutes after giving a physiological dose of insulin to the diabetic rats (Wool and Cavicchi, 1967). This effect is not reproduced when insulin is added directly to a cell-free system (Wool *et al.*, 1968). Stimulation of ribosome activity following injection of insulin into diabetic rats is accompanied by a shift of ribosome aggregates seen on sucrose gradients toward accumulation of larger polysomes, a response that can be prevented by administering cycloheximide (Stirewalt *et al.*, 1967). Injection of insulin into *normal* rats appears not to change the protein synthetic capacity of muscle ribosomes (Wool *et al.*, 1968) but increases the proportion of large polysomes (Young *et al.*, 1968a). The response of polysome aggregation to insulin is diminished by fasting for 2 days (Young, *et al.*, 1968a). This latter may occur because injection of insulin into the starved animal activates the pituitary-adrenal axis and results in release of adrenocortical hormones which tend to suppress muscle protein synthesis (Chapter 10, Section IV,B).

Wool has made several studies of the defect in muscle ribosomes obtained from diabetic animals and its correction by insulin administration. In diabetic rats, only a small proportion of the ribosome population (about 9%) actively synthesize protein *in vitro*, whereas ribosomes derived from the muscles of normal rats show some 25% of active ribosomes (Wool and Kurihara, 1967). Martin and Wool (1968) have identified the defect in ribosomes from diabetic animals to lie in the 60 S subunit. This defect may interfere with polysome aggregation since it has been shown that the largest polysome aggregates in muscle have similar activities whether prepared from diabetic or insulin-treated rats (Wool *et al.*, 1968). Other changes contribute to the decreased incorporation of amino acids into muscle protein in diabetes and the enhanced rate after insulin administration. In particular, the amount of RNA per cell diminishes in diabetes (Saunders *et al.*, 1962), an effect that appears to be equally distributed over the various major species of RNA within the muscle cell (Wool *et al.*, 1968). Insulin also increases the binding of radioactive amino acids to tRNA *in vivo* (Davey and Manchester 1969), which may also lead to increased incorporation of amino acids into muscle proteins.

In conclusion, it can be said that insulin appears to play a significant role in regulating muscle protein metabolism over short time intervals. Insulin secretion is decreased in starvation (Roth *et al.*, 1963; Berson and Yalow, 1965), and during protein depletion (Baig and Edozien, 1965), whereas it increases transiently after meals containing carbohydrate or protein (Rabinowitz *et al.*, 1966; Rabinowitz and Merimee, 1968). Consequently, short-term diurnal variations in muscle protein synthesis must result from these diet-related changes in the release of insulin from the pancreas. This aspect of diurnal rhythms in protein metabolism is discussed in Chapter 36.

D. Corticosteroids

Release of adrenocortical hormones causes different changes in the protein metabolism of the peripheral tissues and of the viscera. As discussed in Chapter 10, Section IV,B, administration of glucocorticoid adrenal hormones causes loss of protein from the carcass but accumulation of protein in the liver and in some of the other viscera. Since the carcass constitutes most of the mass of the body, amino acid loss from the carcass predominates, and the overall action of corticosteroids is consequently catabolic, as evidenced by negative nitrogen balance following their administration. However, the anabolic action on the liver and other viscera probably has evolutionary advantages. Secretion of glucocorticoids is part of the reaction to stresses, such as physical trauma, infection, and starvation, and thus results in mobilization of amino acids from the peripheral tissues, making them available for protein synthesis and gluconeogenesis in the visceral organs during the period of emergency. The major contributor to the loss of carcass protein in response to corticosteroid secretion is skeletal muscle. We shall therefore consider, first, the cause of protein loss from muscle following treatment with these hormones, and second the significance of this response in various forms of stress.

A few hours after adminstration of glucocorticoids to animals, there is an increase in the levels of all free amino acids within the muscle cell (Kaplan and Shimizu 1963; Betheil *et al.*, 1965), accompanied by an increase in many of the amino acid levels in plasma (Friedberg and Greenberg, 1947; Ryan and Carver, 1963; Kaplan and Shimizu, 1963; Betheil *et al.*, 1965). In spite of this enrichment of the intracellular concentration of free amino acids, there are reductions in the transport of amino acids into the cell and incorporation of labeled amino acids into the muscle protein on treatment *in vivo* or *in vitro* with corticosteroids (Manchester *et al.*, 1959; Wool and Weinshelbaum, 1959, 1960; Wool, 1960b; Shimizu and Kaplan, 1964; Kostyo and Schmidt, 1963; Kostyo and Redmond, 1966). Kostyo and Redmond (1966) have suggested that the inhibition of protein synthesis may occur because of interference with amino acid transport into muscle, which is known to occur from the inhibitory action of corticosteroids on uptake of the model amino acid AIBA by muscle (Wool, 1960a,b). This theory of action is, however, incompatible with the accumulation of free amino acids in the muscle cells of hormone-treated animals. It seems inherently much more probable that free amino acids accumulate in the muscle cell and pass into the blood from either a reduction in rate of muscle protein synthesis or an increase in rate of muscle protein breakdown. The former seems likely, since corticosteroid injection into rats and rabbits causes muscle ribosomes to have reduced protein-synthesizing capacity *in vitro* (Bullock *et al.*, 1968) and also reduces the proportion of large polysome aggregates in skeletal muscle (Young *et al.*, 1968a). Nevertheless, some investigators favor the view that catabolism may be accelerated by the hormone. It is claimed (DeLoecker, 1966; DeLoecker and de Wever, 1967) that small doses of corticosteroid stimulate incorporation of precursors into both protein and RNA of skeletal muscle, whereas larger doses inhibit incorporation; from this it has been deduced by DeLoecker that the effect of large doses of hormone represent increased catabolism rather than reduced synthesis. It has also been observed that cortisone administration to animals increases the activity in muscle of dipeptidase and leucine aminopeptidase (Rose *et al.*, 1959). These enzymes are not lysosomal in origin (DeDuve *et al.*, 1962), and indeed it has been shown that true lysosomal enzymes, such as acid phosphatase and β-glucuronidase, do not increase in activity in muscle when cortisone is administered (Buchanan and Schwartz, 1967). However, it has already been pointed out in Section V,B that the evidence on muscle protein turnover shows that lysosomal enzymes do not account for the phenomena of normal muscle protein degradation.

In the context of adrenocortical effects on muscle, it is relevant to discuss the changes in muscle protein metabolism that follow stresses of various kinds. After physical injury (see review in Chapter 19 by Cuthbertson) or a general systemic infection (Beisel, 1966; Beisel *et al.*, 1967; Scrimshaw *et al.*, 1968), urinary nitrogen output increases, and there is a negative nitrogen balance that may persist for several weeks and can deplete the body of up to 10% of its normal protein content. While injury and infection undoubtedly cause increased adrenocortical activity, a review of the evidence given in Chapter 19 shows that this lasts only a few days after even quite severe injuries and does not seem likely to account for the prolonged and extensive nitrogen depletion. It should also be acknowledged that the mechanisms involved in metabolic responses to different forms of stress may not be identical; fractures, burns, and infections are each likely to present individual components to the metabolic response, such as the extensive loss of protein that occurs in the exudate from a burned area.

Because the loss of body nitrogen following injury is diminished or abolished by feeding a diet low in protein before the insult (Munro and Chalmers, 1945; Calloway *et al.*, 1955; Fleck and Munro, 1963), it has been earlier suggested that the nitrogen lost by the well-nourished animal that is injured includes labile or deposit protein. However, the magnitude of the nitrogen loss after injury is usually much greater than the amount of labile protein available, which does not exceed some 3% in the well-nourished dog or cat (Chapter 10, Section II) and is as little as 1% in the normal human subject (Young *et al.*, 1968c; Chan, 1968). Furthermore, the distribution of body nitrogen is not consistent with the dissipation of all labile protein sources in the body, since liver weight, protein content, and RNA content are not decreased by injury (Calloway *et al.*, 1955; Fleck and Munro, 1963) and also liver protein synthesis is increased in response to infection (Williams *et al.*, 1963; 1965; Lust, 1966) and physical trauma (Chandler and Neuhaus, 1968; Liu and Neuhaus, 1968; Neuhaus *et al.*, 1966). Hence, the extra N comes from sites other than the viscera. Although the source of the increased N loss has not been identified, the magnitude of the loss and the increased creatinine excretion which follow injury and infection (Chapter 19, Section IV; Beisel *et al.*, 1967) imply that a large proportion of the N arises from skeletal muscle and Levenson *et al.* (1959) calculated that all of the nitrogen which appears in the urine of burned rats comes entirely from carcass protein. Lawrie (1966) has also pointed out the metabolic responsiveness of muscle to various stressful stimuli.

The mechanisms associated with the loss of muscle protein have received limited investigation. Infection in mice reduces the *in vivo* uptake of leucine-^{14}C by muscle microsomes (Lust, 1966) and the uptake of the labeled amino acids by muscle ribosomes in infected rats (Young *et al.*, 1968b). These studies, however, are complicated by possible changes in the size of the free amino acid pools, but Young *et al.* (1968b) did observe a reduced capacity for *in vitro* protein synthesis by muscle ribosomes in infected rats, supporting the conclusion that muscle protein synthesis is reduced by infection. Total cellular RNA also decreases in the skeletal muscle (Young *et al.*, 1968b), which would presumably reduce the capacity for protein synthesis in the contractile cells. The changes in muscle protein catabolism in response to a stressful stimulus have not received detailed investigation. Levenson *et al.* (1959) found that the uptake of glycine-^{15}N given at the height of the negative nitrogen balance by tissues of burned rats was at least equal to that of uninjured controls and concluded that increased protein catabolism could account entirely for the loss of muscle protein. The *in vivo* uptake of radioactivity from leucine-^{14}C by muscle ribosomes in rats at different times after injury was examined by Young and Huang (1969), and they observed a reduced incorporation of radioactivity at 3 days post-fracture in rats fed a 25% casein diet. In contrast, there was little or no change in the case of rats given a 5% casein diet. These data suggest that muscle protein synthesis may be reduced in response to physical injury and that the response seems to depend upon the nutritional state of the animal, thus correlating with the observations noted above of a loss of response of nitrogen balance to injury among animals given low-protein diets.

Glucocorticoids have been implicated in the loss of body nitrogen which occurs after injury or generalized infection. Although the catabolic nitrogen response to physical trauma is prevented by adrenalectomy (Campbell *et al.*, 1954), the increased N output occurs in adrenalectomized animals maintained on low levels of cortisone (Campbell *et al.*, 1954). Furthermore, the time sequence of changes in glucocorticoid metabolism are not consistent with the protein catabolic response observed to occur in man following infection and surgical trauma (Moore *et al.*, 1965; Beisel and Rapoport, 1969), and the alterations in muscle polysome profiles after a relatively high dose of glucocorticoids are not as marked as the changes in muscle polysomes that occur after infection in rats (Young *et al.*, 1968a,b). In this connection, it should be pointed out that insulin has effects on muscle protein metabolism which are the reverse of those obtained after corticosteroid administration. It is thus possible that changes relative to insulin may be of more fundamental importance in the response of muscle protein metabolism to stressful stimuli. Infection and physical trauma are associated with a prolonged diabetic glucose tolerance curve and insulin resistance (Shambaugh and Beisel, 1967; Allison *et al.,* 1967). Consequently, a relative limitation of insulin secretion and action in the presence of normal glucocorticoid metabolism would result in a reduced rate of muscle protein synthesis. This may explain the observation that the catabolic nitrogen response is blocked by adrenalectomy but occurs on fixed doses of adrenal corticoids (Campbell *et al.,* 1954).

IX. Effect of Exercise on Muscle Protein Metabolism

In 1897, Morpurgo demonstrated that within 1 day strenuous exercise could cause the cross-sectional area of the sartorius muscle to increase appreciably without altering the total number of muscle fibers. This early demonstration of the hypertrophic effect of exercise has been repeatedly confirmed (Siebert, 1928; Thorner, 1939; Hoffmann, 1947; Walker, 1966; Carrow *et al.,* 1967; Goldberg, 1967). Although Morpurgo (1897) and more recently Holmes and Rasch (1958) considered that increase of myofibrils within the fiber was excluded, an increase in the myofibril number and thickness is shown by biochemical studies (Helander, 1961) and histological studies on skeletal muscle (Goldspink, 1964, 1965) and on cardiac muscle (Molbert and Jijima, 1959; Richter and Kellner, 1963).

All the fibers within a muscle may not respond to the same degree following muscular exercise. Using a pulley and weights system of training in mice, Goldspink (1964) observed for the m. biceps brachii that the increase in mean fiber size was due to conversion of "small phase" fibers (about 20 μ diameter) into "large phase" fibers (Goldspink, 1962b). Using forced and voluntary exercise programs, Carrow *et al.* (1967) observed in rats that there was a greater increase in the cross-sectional area of the smaller red than the larger white fibers in the gastrocnemius muscle, possibly due to differences in the innervation of the different muscle fiber types. There is also an enlargement of the vascular bed of exercised muscle (Vannotti and Magiday, 1934), an increase in the number of capillaries per muscle fiber (Carrow *et al.,* 1967), and blood flow in the exercised muscle may increase by as much as four to twenty times (Zierler, 1966); the resting blood flow in trained muscles does not increase (Vanderhoof *et al.,* 1961).

Changes in the composition of trained muscle occur, but the observations are not entirely consistent. In guinea pigs, Helander (1961) observed a decrease in the water content of muscle, whereas in rats (Pattengale and Holloszy, 1967) no change was observed. The myofibrillar nitrogen fraction of muscle increased in exercised guinea pigs (Helander,

1961), but in the hypertrophied soleus muscle of rats 4 days after tenotomy of the gastrocnemius and plantaris muscle, Hamosh *et al.* (1967) reported a decrease in *concentration* of myofibrillar protein, although the absolute amount increased. Whipple (1926) appears to have been the first to suggest that the concentration of myoglobin increases with exercise in muscle. Comparative studies have revealed that exercised muscle contain higher concentrations of myoglobin than nonexercised muscle. Lawrie (1953a) observed a lower concentration of myoglobin in the psoas muscle of the sedentary laboratory rabbit than in the same muscle of the active wild hare. Also Shenk *et al.* (1934) have observed that the concentration of myoglobin in skeletal muscle is higher in grazing than in stall-fed cattle. These observations are supported by controlled studies which show that muscular exercise increases the amount of myoglobin in skeletal muscle (Lawrie, 1953b; Pattengale and Holloszy, 1967). The increase in response to exercise occurs also in rats living at high altitude (Vaughan and Pace, 1956). However, a single exhausting exercise of short duration has little effect on myoglobin content of muscle.

Changes in enzymes in muscular exercise have been reviewed by Hensel and Hildebrandt (1964). A correlation exists between the content of respiratory enzymes in muscle and its ability to perform prolonged exercise (Lawrie, 1953a). Moderate exercise appears to have little effect on the activity of respiratory enzymes in muscle (Hearn and Wainio, 1956; Gould and Rawlinson, 1959) but may increase enzymes associated with anaerobic oxidative metabolism (Hearn and Wainio, 1957). With more vigorous and prolonged exercise, there is, however, an increase in the mitochondrial protein content of muscle and the activity of enzymes associated with the electron transport chain (Holloszy, 1967). Although a low-protein diet decreases the amount of myoglobin, cytochrome, and respiratory enzyme activity, the increase in respiratory enzyme activity and of myoglobin content in skeletal muscle still occurs in response to vigorous exercise (Fuge *et al.*, 1968).

The changes in the cellular events of protein metabolism of muscle which occur in response to exercise have received very little attention. Goldberg (1968) carried out tenotomy of the gastrocnemius muscle of rats so that the soleus muscle was consequently subjected to increased activity. After repeated injections of leucine-^{14}C, incorporation of radioactivity in both the sarcoplasmic and myofibrillar protein fractions of the hypertrophied soleus muscle was enhanced. Goldberg (1969a) also suggests that the degradation of myofibrillar and sarcoplasmic proteins is decreased during work hypertrophy. Increased AIB accumulation also parallels the enhanced incorporation of radioactive amino acids into the work-induced hypertrophying muscle (Goldberg and Goodman,

1969a). Also the RNA concentration of hypertrophied muscle following tenotomy of the synergistic muscle is increased and the activity of skeletal muscle microsomes is enhanced (Hamosh *et al.,* 1967). This is probably related to an increased ribosome content of the microsome fraction. Growth hormone does not appear to be involved in the mechanism since Goldberg (1967) found that the increase in amino acid incorporation into the soleus and plantaris muscles, after tenotomy of the gastrocnemius, was similar in hypophysectomized and intact rats.

X. Effect of Muscular Dystrophy and of Denervation on Skeletal Muscle Protein Metabolism

In this section some of the characteristic features of protein metabolism in muscle wasting due to dystrophy and to denervation will be considered briefly. A number of reviews have appeared on the varieties of muscular dystrophy in man (Walton, 1965), on their histopathology (Pearson, 1965), and on the biochemical features of various myopathies (Hughes, 1965; Weinstock, 1966). Also Milhorat (1967) has edited an excellent series of papers on current advances in knowledge of muscular dystrophy. The morphological (Gutmann and Zelena, 1962) protein metabolic (Zak, 1962) and comparative aspects of response of various muscles to denervation atrophy (Hnik, 1962) have also been well summarized in recent reviews. Accordingly, this section will cover only certain aspects of protein metabolism under situations of muscle wasting.

A. Changes in Muscular Dystrophy

Structural changes occur in the skeletal muscle of the genetically dystrophic animal (Ross *et al.,* 1960; Pearce, 1966; Banker, 1967; Shafiq *et al.,* 1969) showing a reduction in fiber diameter and in the amount of sarcoplasmic reticulum (Shafiq *et al.,* 1969) as well as a reduced fiber number (Rowe and Goldspink, 1969b). In human muscular dystrophy (Vignos and Lefkowitz, 1959) and in hereditary dystrophy in experimental animals, the content of myosin and possibly even the structure of the myosin molecule are altered (Schapira *et al.,* 1954; Oppenheimer *et al.,* 1964; Smoller and Fineberg, 1965; Samaha and Gergely, 1969) although the latter is not always changed (Morey *et al.,* 1967, 1968). The cause of the loss of muscle protein in dystrophy and following denervation has been extensively examined. Increased cathepsin activity is associated with the loss of muscle protein in nutritional muscular dystrophy resulting from vitamin E deficiency (Weinstock *et al.,* 1955; Koszalka *et al.,* 1961; Zalkin *et al.,* 1962; Weinstock, 1966) or hereditary muscular dystrophy in mice (Weinstock *et al.,* 1958) and in the chick

(Tappel *et al.*, 1962; Weinstock and Lukas, 1965a). A similar increase in the activities of cathepsin (Weinstock and Lukas, 1965b; Hajek *et al.*, 1964) and other acid hydrolases (Pollack and Bird, 1968) follows denervation of skeletal muscle. The reason for increased proteolysis is not known. The origin of the increased activity may be extracellular, as Zalkin *et al.* (1962) suggested on the basis of studies with vitamin E-deficient rabbits, or intracellular, as Weinstock (1966) claimed in mouse muscular dystrophy. Although infiltration of tissue by macrophage cells is a recognized feature of nutritional or genetic muscular dystrophy, macrophages are not reported to increase during the early stage of denervation atrophy (Gutmann and Zelena, 1962). Pollack and Bird (1968) observed an increase in hydrolase activity in denervated rat leg muscles which was associated with lysosome-like particles and concluded that these particles play an important role in the process of muscle tissue breakdown due to denervation. It thus seems improbable that a common enzymatic mechanism can be responsible for the loss of muscle constituents in these different types of muscle wasting.

The synthesis of muscle protein is maintained or even increased in dystrophic muscles. In genetically dystrophic mice, various investigators (Kruh *et al.*, 1960; Simon *et al.*, 1962; Coleman and Ashworth, 1959; Srivastava and Berlinguet, 1966; Weinstock, 1966) have observed a normal or increased incorporation of radioactivity into muscle protein following *in vivo* injection of labeled amino acids. Weinstock and associates (1969) also found enhanced incorporation of ^{14}C-labeled valine into protein during incubation of minced breast muscle obtained from genetically dystrophic chicks. Nichoalds *et al.* (1967) favor the view that, in the muscle of vitamin E-deficient rabbits, the increased labeling of protein is a secondary response since histological changes precede the change. Conclusions on protein synthesis based upon *in vivo* incorporation data should be carefully evaluated because of the possibility of changes in the free amino acid pools in skeletal muscle. However, the increased incorporation of labeled amino acids occurs in spite of an increased size of the free amino acid pool in muscle from dystropic animals (Weinstock 1966; Tallan, 1955; Peterson *et al.*, 1963), which supports the view that a normal or increased incorporation of radioactivity reflects a real increment in the rate of protein synthesis. The enhanced levels of free amino acids in dystrophic muscle may be due to an increased permeability of the cell to some amino acids (Diehl, 1963; Diehl and Jones, 1966). In support of this conclusion, Srivastava and Berlinguet (1966) observed higher cell-free activity of ribosomes obtained from skeletal muscle of dystrophic mice. Although Monckton and Nihei (1966) observed from sucrose gradient analysis that there was a decrease in the proportion of heavy ribosome aggregates in a ribosome fraction from the deltoid muscle of Duchenne dystrophy patients, they (Monckton and Nihei, 1969; Nihei and Monckton, 1969) found that the activity of these ribosomes in a cell-free system was three times higher than that obtained with ribosomes from healthy persons. They also reported that the RNA content of dystrophic human muscle is within the normal range. On the other hand, the RNA content of dystrophic mouse muscle is increased two- to threefold above normal (Coleman and Ashworth, 1959; Girkin *et al.*, 1962; Srivastava *et al.*, 1963). Histological (Walton and Adams, 1956; Pearce and Walton, 1963) and cytochemical (Kitiyakara, 1961) observations also suggest an altered RNA metabolism. Srivastava (1967) observed increased incorporation of uridine-2-^{14}C into RNA of dystrophic muscle at 30 days. At a later stage in the disease, the difference had disappeared, and by 90 days incorporation was less in the muscles of the dystrophic animals. Girkin *et al.* (1962) also found no difference in the incorporation of formate-^{14}C into the RNA of the muscle of dystrophic mice at 60–70 days of age. It thus appears that the degradation of RNA may be increased in dystrophic muscle. Kinetic studies to test this have not been reported, but there is increased RNase activity (Krupnick *et al.*, 1964; Tappel *et al.*, 1962; Epshtein, 1964; Abdullah and Pennington, 1968) in myopathies in experimental animals and man. The increased ribonuclease activities conceivably may cause an accelerated breakdown of RNA in the muscle fibers and contribute to the degeneration of the muscle in this disease.

In summary, although nutritional and muscular dystrophies may well differ in several respects, and although a variety of muscular dystrophies are known to occur in man, it would appear that there is a similarity in the metabolic picture for those dystrophies that have been studied. Loss of muscle protein appears to be due to a greater increase in protein catabolism associated with an increase in protein synthesis in these conditions. The shift toward catabolism is probably also reflected in the increased activities of GOT and GPT in the muscular dystrophy of rabbits and mice (Laferté *et al.*, 1963) and in the augmented activity of muscle glutamic dehydrogenase in mice (Tsuji, 1967). Similar studies on branch-chain transaminases would be of interest in muscular dystrophy, in view of the important role of these muscle enzymes in amino acid catabolism, but no such data appear to have been reported.

B. Changes Following Denervation

As mentioned in Section II,D, innervation plays a role in the functional and metabolic activity of skeletal muscle. For example, it has been found that reversing the nerve supply of tonic and phasic muscles results in a corresponding reversal of both the functional and biochemical characteristics of the two muscle types (Drahota and Gutmann, 1963; Buller *et al.,* 1960a,b; Buller and Lewis, 1965b; Prewitt and Salofsky, 1967; McPherson and Tokunaga, 1967; Guth *et al.,* 1968). The trophic influence of innervation is also indicated by changes in muscle metabolism, a subject that has been reviewed by Gutmann (1962).

When the nerve supply to a skeletal muscle is severed, there is loss of weight of the paralyzed muscle. During the early stages of atrophy, the rate of weight loss differs in various muscles (Zelena and Hnik, 1957), but in the later stages of atrophy this distinction between muscles is less obvious (Hnik, 1962). The rapidity with which atrophy occurs is known to vary in different species, being more rapid in the rat (Hnik, 1962) than in man (Bowden and Gutmann, 1944). Indeed, it seems that the process follows the general relationship of metabolic processes to body size of species (Chapter 25, Section IV,B), since the rate of weight loss from the denervated gastrocnemius muscle of different mammals can be correlated with the intensity of their metabolic rates (Knowlton and Hines, 1936). In other words, the larger the species, the slower is its metabolic rate and also the rate of atrophy of its denervated muscle. In the young animal, denervated muscle may continue to grow, although at a slower rate than other muscles (Schapira *et al.,* 1950; Zelena and Hnik, 1957; Leonard, 1957; Stewart, 1968). Consequently, muscle atrophy can only be assessed by comparison with a corresponding muscle that has retained its innervation. Under these circumstances, it would appear that the percentage weight difference between denervated and normal muscle is greater in young than in adult animals (Schapira *et al.,* 1950; Stewart, 1968). Denervation of the diaphragm through section of the phrenic nerve provides a specific early response not seen in other skeletal muscles. In several species, it has been shown that phrenic nerve section causes a transient hypertrophy of that half of the diaphragm, followed by atrophy (Sola and Martin, 1953; Stewart, 1955; Stewart and Martin, 1956; Manchester and Harris, 1968). This hypertrophy affects both the structural and sarcoplasmic protein in the muscle fiber of the denervated rat diaphragm (Zak *et al.,* 1969). A similar transitory hypertrophy has been observed in the denervated anterior tibialis muscle of the rat (Gutmann, 1960), but other skeletal muscles do not appear to undergo this preliminary hypertrophy following denervation (Guth, 1968).

Changes in muscle structure following denervation have also been described. There is dispute about the loss of the basic units of muscle, that is the muscle fibers. No reduction in muscle fiber population was observed by Sunderland and Ray (1950), whereas Gutmann and Zelena

(1962) concluded that the muscle loses fibers following denervation. A decrease in the diameter of the muscle fiber has been described (Gutmann, 1948) and also a reduction in the number of myofibrils within the fiber (Gutmann and Zelena, 1962), along with a loss of myofilaments from the outer surface of the myofibrils so that their thickness is reduced (Muscatello *et al.*, 1965; Pellegrino and Franzini, 1963; Wechsler and Hager, 1960). These demonstrable changes in the contractile elements of muscle are consistent with the rapid and extensive loss of myosin described below.

After denervation, loss of muscle weight is accompanied by loss of muscle protein which affects some muscle constituents sooner and more extensively than others. Among the earliest to undergo loss in myosin, so that the concentration of this contractile protein decreases soon after nerve section (Fischer and Ramsey, 1946; Fischer, 1948). The sarcoplasmic proteins also decrease in concentration, but not so extensively as does myosin (Schmidt and Schlief, 1956; Helander, 1957). In contrast, collagen is lost more slowly and less extensively than other muscle proteins (Zak, 1962), and in consequence, the *concentration* of collagen in the denervated muscle increases as atrophy progresses (Helander, 1957; Fischer and Ramsey, 1946).

Changes in muscle protein metabolism following denervation have been reviewed by Gutmann (1962). It is difficult to determine the nature of changes in muscle protein synthesis. Injection of labeled amino acids into the intact animal has demonstrated decreased incorporation into denervated muscle (Schapiro *et al.*, 1953; Padieu, 1959). This is not consistent with the conclusion of other investigators that there is increased incorporation of radioactivity from labeled amino acids into mixed muscle proteins or muscle protein fractions by denervation (Sheves *et al.*, 1956; Slack, 1954; Pearlstein and Kohn, 1966; Pater and Kohn, 1967). Studies of protein synthesis in whole animals are limited by differences in rate of uptake of the labeled precursor into the tissue free amino acid pool and by subsequent dilution of the label in the pool. In the case of denervated muscle, Padieu (1959) observed a decreased level of glycine-^{14}C in the free amino acid pool of muscle 3 days after infusion of glycine-^{14}C, which may represent reduced penetration of muscle. On the other hand, *in vivo* uptake of the model amino acid AIBA into muscle is decreased initially (Goldberg and Goodman, 1969b) and later accelerated by denervation (Diehl and Jones, 1966; Bombara and Bergamini, 1968). With regard to the free amino acid pool, it would appear that the levels of individual amino acids tend to increase or to remain unchanged (Telepneva, 1961; Harman *et al.*, 1960; Yudaev *et al.*, 1953). In view of these changes, the interpretation of *in vivo* data on muscle protein synthesis after denervation becomes hazardous. Studies of protein synthesis in denervated muscle have also been made *in vitro,* using frog muscle (Muscatello *et al.,* 1965). After incubation of the muscle with valine-^{14}C, there was increased incorporation of this amino acid into the sarcoplasmic fraction of the denervated muscle. This could be correlated with an increase in the RNA and polysome content of the denervated muscle. On the other hand, studies with incubated frog muscle showed that valine-^{14}C uptake by myofibril protein decreased in the denervated preparation (Margreth *et al.,* 1966). From these results, it appears possible that denervation specifically decreases myofibril protein synthesis while accelerating the synthesis of the sarcoplasmic proteins in the denervated muscle. Finally, denervation impairs the capacity of muscle to increase its protein synthetic capacity in response to an anabolic stimulus. This is demonstrated by the failure of the bulbocavernosus muscle of the rat to respond to testosterone after denervation (Kare *et al.,* 1955); this muscle is normally very sensitive to injection of androgens.

A special case of altered protein synthesis following denervation is provided by the transitory hypertrophy of the diaphragm after section of the phrenic nerve. The increase in the amount of diaphragm protein is accompanied by elevated incorporation of amino acids into protein (Buse *et al.,* 1965; Harris and Manchester, 1966) and by an increment in RNA content which may be due to decreased RNA breakdown (Manchester and Harris, 1968). Although Manchester and Harris (1968) did not observe any change in the DNA content of the hypertrophied diaphragm, Zak *et al.* (1969) found a significant increment in DNA and in DNA polymerase, which they attributed to changes in connective tissue cells rather than to contractile cell DNA. This increased DNA content may represent activity in satellite cells since these cells proliferate in denervated muscle (Lee, 1965) and in regenerating muscle (Shafiq *et al.,* 1967).

There is evidence that protein breakdown is increased in denervated muscle (Goldberg, 1969b). Following denervation, there is an increase in the size of the muscle lysosomes (Pellegrino and Franzini, 1963), and in the activities of proteolytic enzymes of the cathepsin type and of other acid hydrolases (Hajek *et al.,* 1964; Weinstock and Lucas, 1965b; Weinstock, 1966). As discussed earlier, the type of enzyme responsible for proteolysis in muscle is uncertain. Pollack and Bird (1968) consider the lysosomal enzymes to be responsible for muscle protein disintegration following denervation, but Kohn (1969) has recently described a soluble enzyme in muscle that attacks isolated myofibrils at neutral pH, and may play a significant role in myofibril degradation during the atrophy of denervation.

XI. Protein Metabolism in Cardiac Muscle

In the preceding sections, sporadic reference has been made to studies on cardiac muscle in order to correlate them with results obtained from skeletal muscle (see Sections II–VI,VIII). Here, a short account of protein metabolism in cardiac muscle will be assembled. Protein accounts for some 80% of the dry weight of the adult heart in all mammals studied (Widdowson and Dickerson, 1964). In terms of total body protein, the contribution of the heart is small, since this organ represents only 0.6% of body weight in the rat, 0.5% in man, and 0.4% in the largest of mammals, the blue whale (Chapter 25, Table V).

A. Cardiac Muscle Structure and Composition

The heart consists of four chambers (two auricles and two ventricles). Entry of blood into the chambers and its passage into the pulmonary artery and the aorta is regulated by valves. The two sides of the heart are separated by a muscular wall in which are embedded the tissues generating and transmitting the heart impulse, namely the nodal tissue and the conducting bundle. Consequently, the heart is made up of a variety of tissues and cell types (Bloom and Fawcett, 1964; James and Sherf, 1968), and the various parts of the heart differ in gross chemical composition (see Widdowson and Dickerson, 1964, for references).

In the normal human heart, the right ventricle is only half the weight of the left ventricle, but each ventricle has approximately the same number of fibers and nuclei (Linzbach, 1960). This implies that, in man, the right ventricular fibers are smaller than the left ventricular fibers and in consequence the amount of fiber per nucleus or protein per unit of DNA should differ in the two ventricles. In the rat, the right ventricle is also smaller than the left (Krames and Van Liere, 1966), but the DNA concentrations of each ventricle are quite similar (Matsumoto and Krasnow, 1968). However, this does not necessarily invalidate the data obtained for man, since about 75% of myocardial DNA is associated with connective tissue and endothelial cells (Morkin and Ashford, 1968) and therefore the DNA concentration does not accurately reflect the true myocardial cell number.

Earlier it was believed that the myocardial cell was syncytial in structure, but electron micrographic studies have revealed that heart cells of higher animals are discrete and that overlapping of the cells gives a syncytial appearance that is only apparent (Stenger and Spiro, 1961;

Spiro and Sonnenblick, 1964). Most nuclei contain a diploid compliment of DNA (Capers, 1964), but polyploidy has been observed to occur (Sandritter and Scomazzoni, 1964) and tends to increase with increased fiber width (Kompmann *et al.*, 1966).

The ultrastructural aspects of heart muscle cells have received a good deal of recent attention, and it is obvious that there are marked differences in the fine structure of atrial and ventricular muscle cells; the absence of T-tubules in the atrial fibers and a poorly organized sarcoplasmic reticulum (Hibbs and Ferrans, 1969) differentiates these cells from the ventricular cells. Furthermore, various biochemical differences have been observed between these cells. The content of succinic dehydrogenase and other oxidative enzymes (Cooper, 1955; Herbener, 1966; Pratt, 1967; Toth and Schiebler, 1967) is lower in the atrial cell, and this correlates with the lower level of mechanical work required of atrial muscle. There are relatively few myofibrils that are peripherally located in the atrial muscle cells. However, in the suckling rat the ventricular cells are similar in appearance to the atrial cells of the adult rat (Schiebler and Wolff, 1966).

Changes in myocardial structure during growth and development are discussed in Chapter 26, Section III,D. During the early phase of normal growth of myocardial tissue, the heart increases mainly by cell division and later by an increase in cell size. Petersen and Baserga (1965) observed that cell division accounted for a major proportion of heart growth in mice before 5 weeks of age, but that after this time the total amount of DNA increased only slightly, suggesting hypertrophy of existing muscle cells. Other data are consistent with this. The DNA concentration, as well as RNA concentration of heart muscle, decreases with continued growth and development (Devi *et al.*, 1963), presumably reflecting dilution by the increased fiber size. In man the diameter of the myocardial fibers doubles with age from about 7 μ at birth to 14 μ in adult life (Wearn, 1941). In contrast to this picture, Matsumato and Krasnow (1968) observed a continued linear increase in the total DNA content of the right and left ventricles during growth of rats from 3 to 12 months and suggested that hyperplasia accounted for all the increased heart weight. It must be pointed out that although Sasaki *et al.* (1968b) observed a similar continuous increase in cardiac DNA content, the elevation was observed to be due to a greater increase in the contribution of the interstitial cells to total DNA during the later stages of growth (Sasaki *et al.*, 1968a). This is also discussed earlier in Section II,B of this chapter. In parallel with these changes, the collagen concentration of the ventricles in the rat doubles during the first year of life and continues to increase so that in the 3-year-old rat the concentration has reached three times that in the suckling rat (Schaub, 1964, 1965). For the pig heart, the collagen concentration doubles between birth and the adult animal, but in man the change appears not to be as marked (Widdowson and Dickerson, 1960).

Finally, we can ask how the cardiac muscle of mammals is affected by differences in body size of the species. The diameter of heart muscle fibers appears to be similar over a wide range of mammals from the rat to the blue whale (Black-Schaffer *et al.*, 1965). Consequently, the differences in the weight of the heart in various mammals are largely due to differences in total cell number (Section III,A).

B. Protein Synthesis in Cardiac Muscle

The developmental aspects of protein synthesis in cardiac muscle are covered in Chapter 26 of this treatise. The basic protein synthetic machinery in this muscular tissue is broadly similar to that in skeletal muscle. Heywood *et al.* (1968) have observed that, in the case of chick embryo heart muscle, the polyribosome profiles obtained on sucrose gradients are similar to those observed with skeletal muscle ribosomes. Biochemical studies (Heywood *et al.*, 1968; Earl and Korner, 1965; Earl and Morgan, 1968) and cytological studies (Porter and Palade, 1957) reveal that the ribosomes in mammalian heart muscle are free and not attached to a reticulum. Rapidly sedimenting polysomes have also been observed (Rabinowitz *et al.*, 1964; Heywood *et al.*, 1968) suggesting the presence of large myosin-synthesizing polysomes in this tissue similar to those established for skeletal muscle (Heywood and Rich, 1968).

C. Hypertrophy of Cardiac Muscle

A characteristic response of the heart to a chronic load is an increase in myocardial mass. It is well established that, in rats subjected to severe bouts of exercise, there is myocardial hypertrophy with the left ventricle hypertrophied relatively more than the right. On the other hand, when rats are subjected to effective degrees of hypoxia, the opposite seems to be the case, namely that the right ventricle hypertrophies relatively more than the left (Van Liere *et al.*, 1964).

Hypertrophy can be induced by a variety of other means, including thyroxine administration (Beznak, 1962), aortic constriction (Nowy and Frings, 1960; Gluck *et al.*, 1964; Nair *et al.*, 1968), diet-induced anemia (Widdowson and McCance, 1955; Hoyle *et al.*, 1963; Korecky and French, 1967), exercise (Van Liere *et al.*, 1964; Van Liere and Northup, 1957; Bloor *et al.*, 1968), and hypertension due to unilateral nephrectomy (Wannemacher and McCoy, 1969). Hypertrophy is considered to be an important adaptive mechanism whereby the heart is able to sustain the increased circulatory load (Norman, 1962).

Hypertrophy is usually associated with an increase in fiber diameter (Norman, 1962; Linzbach, 1960) without an increase in fiber number. In anemic rats, Korecky and French (1967) found no change in the RNA:DNA ratio of the hypertrophied hearts and suggested that hyperplasia is the most likely explanation for the increase in heart weight. Similarly from RNA-DNA relationships, Widdowson and McCance (1955) have concluded that the hearts of anemic piglets were larger because the cells were more numerous, but neither their size nor their composition had changed. However, myocardial DNA synthesis has been observed to increase markedly between the second and seventh days after aortic constriction in rats, but the increase seems to be due to increased DNA synthesis by interstitial cells (Morkin and Ashford, 1968). Hence it seems hazardous to draw conclusions concerning hyperplasia in the heart which is increasing in size in response to an increased work load. However, Linzbach (1960) considers that there is also a stage of hypertrophy beyond the normal physiological response which represents a pathological hypertrophy, and here an increase in fiber number without an alteration in fiber size is suggested to be the basis for the increased heart weight.

Because the total protein content of the hypertrophic heart increases, it is not surprising that alterations in the nucleic acid and protein metabolism of this tissue have been observed (Posner and Fanburg, 1968). However, the precise nature of the change and the underlying mechanisms remain to be more clearly defined. The *total* RNA content has been found to increase, but changes in the *concentration* of RNA have been variable. Nair *et al.* (1968) summarize some of the published values for the RNA content of hypertrophic hearts and suggest that the differences in published values reflect differences in the degree of hypertrophy and the time of measurement in relation to development of the hypertrophy. The increase in RNA in the hypertrophic myocardial tissue appears to be mainly ribosomal in type (Fanburg and Posner, 1968), and this probably accounts for the increased *in vitro* protein synthetic activity of microsomes obtained from the hypertrophied ventricle (Moroz, 1967). Schreiber *et al.* (1968) and Nair *et al.* (1968) consider that there may be increased mRNA synthesis in rat heart in response to aortic constriction. Nair *et al.* (1968) have observed increased Mg^{2+}-activated RNA polymerase activity in the nuclei isolated from hypertrophic rat heart; this increase occurred within 12 hours of applying the aortic constriction and reached peak values on the second day, thereafter returning to initial levels at the end of one week.

The *in vivo* incorporation of labeled amino acids into myocardial protein is also increased in the hypertrophied heart (Cohen *et al.*, 1966; Wannemacher and McCoy, 1969), leading to the suggestion that the rate of cardiac protein synthesis is increased. This appears to be a reasonable conclusion in view of changes in RNA metabolism noted above and also because feeding a diet low in protein reduces the hypertrophic response to unilateral nephrectomy (Allison *et al.*, 1961; Wannemacher and McCoy, 1969). However, the primary mechanism(s) behind the alterations in protein and nucleic acid metabolism in the hypertrophic heart remain to be identified. Wannemacher and McCoy (1969) suggested that an increased cellular free amino acid content might be responsible for the increased rate of amino acid incorporation into the protein of the hypertrophic heart.

D. Nutritional Effects on Cardiac Protein Metabolism

The alterations that occur in skeletal muscle protein metabolism during the feeding of a diet inadequate in protein and/or calories are paralleled by changes in the protein and amino acid content of cardiac muscle. In 1866 Voit observed that starvation in cats reduced the weight of the heart, and later Addis *et al.* (1936a,b) found that during a 7-day fast in rats the heart lost relatively more of its protein content than the carcass (skin, muscle, skeleton). However, in comparison with the liver, which lost 40% of its initial protein, the heart lost 18% (Addis *et al.*, 1936a) and on refeeding gained protein more slowly than the liver (Addis *et al.*, 1936c). During the initial stages of protein depletion, as well as starvation, Allison *et al.* (1963) also observed a marked reduction in the RNA:DNA ratio and the protein:DNA ratio in the heart of rats, thus confirming the findings of Addis. However, after a period of prolonged undernutrition, the heart appears to show less marked changes in chemical composition than the skeletal muscles (Widdowson *et al.*, 1960). Furthermore, histological study of protein-depleted monkeys reveal that the myocardium is less severely affected than skeletal muscle (Chauhan *et al.*, 1965) and although the absolute weight of the heart is decreased in children with kwashiorkor, gross and histological changes have been found to be minimal (Béhar *et al.*, 1956). Moreover, the decrease in heart weight in the severely undernourished pig is less than the overall loss in body weight, and consequently the heart as a percentage of body weight is increased (Widdowson *et al.*, 1960).

The decreased RNA concentration in the heart of protein-depleted rats (Allison *et al.*, 1963) is also accompanied by a decrease in the free amino acid levels of this organ (Wannemacher and McCoy, 1969). These changes would presumably reduce the rate of protein synthesis in the myocardial cell, but there is a lack of information on the alterations in the subcellular aspects of protein metabolism of this tissue as a result of dietary protein and calorie restriction. An interesting observation, however, is that a dietary tryptophan deficiency results in the production of both endocardial and myocardial lesions in the guinea pig (Spatz, 1969), and this is significant in view of the relationship between dietary tryptophan intake and protein synthesis in tissues (see Chapter 34, Section III,A,1). Reid and Berjak (1966) have found the production of myocardial lesions in rats given a maize diet, which also suggests a possible relationship to the relatively low tryptophan content of the experimental diet utilized in their studies. Naeye (1965) considers that, in the young human, the size and number of myocardial fibers is altered by nutritional status. He has observed in some monovular twins in which a placental transfusion syndrome develops that the heart of the recipient twin consists of a greater number of fibers, which also are larger in size. However, the significant nutritional factors associated with this syndrome remain to be established.

XII. Conclusion

The capacity for independent and coordinated movement provided by the skeletal muscles has been partially responsible for the evolution of amino acid requirements and protein metabolism in the animal kingdom. It has also emerged during the course of this review that, although there are important variations in the protein metabolism of different skeletal muscles and even between individual muscle cells, significant qualitative and quantitative relationships exist between the protein metabolism of the muscular tissue, the liver and the body as a whole.

During the course of growth in the normal mammal, the skeletal muscles account for an increasing proportion of total body protein metabolism, the net effect being that skeletal muscle draws more heavily upon the daily essential amino acid needs of the organism during the course of growth and development. It is also clear that in adult mammals of increasing size the protein metabolism of the visceral tissues decreases in importance relative to skeletal muscles, and it follows that changes in muscle protein metabolism closely reflect changes in overall body protein status. Under unfavorable nutrition conditions and in diseases associated with stress, the protein metabolism of skeletal muscle is profoundly altered, frequently to a greater extent and in an opposite direction to that of other organs, such as the liver, and often in a way best suited for the continued survival of the organism. In consideration of the circulatory function of the heart, which provides the body tissues with essential substrates required for maintenance of function, the protein metabolism of cardiac cells is similarly well designed to meet the demands imposed upon the animal following environmental change. Evidence for this is shown by the rapid and marked alterations in protein metabolism of cardiac muscle which occurs during myocardial hypertrophy. It is unfortunate that the importance of protein metabolism of skeletal muscle and heart and the relatively high efficiency whereby protein is fabricated in the muscle cell has not received sufficient recognition in the past. Furthermore, protein metabolism in the skeletal muscles is considerably more labile than hitherto appreciated, and to a measurable extent this provides the mammal with a significant capacity for adaptation to environmental change.

Acknowledgments

Financial support for the author's studies has been provided by Grant No. 386 from The Nutrition Foundation Inc., New York, a contract, No. DA-49-193-MD-2560 with the U.S. Army Medical Research and Development Command, and recently by a U.S. Public Health Service Grant No. AM13432. The author is extremely grateful to Professor Hamish N. Munro for his encouragement and thoughtful suggestions given during the preparation of this chapter, to Dr. D. J. Millward and Dr. T. Nihei for making their work available prior to publication, and to Dr. Helen Padykula and Dr. Geraldine Gauthier for their instruction and time spent in helping him to gain further insight into the ultrastructural aspects of skeletal muscle. This is Publication No. 1490 from the Department of Nutrition and Food Science, Massachusetts Institute of Technology.

References

Abdullah, F., and Pennington, R. J. (1968). *Clin. Chim. Acta* **20,** 365.

Adams, R. D., Denny-Brown, D., and Pearson, C. M. (1962). "Diseases of Muscle," 2nd ed. Harper, New York.

Addis, T., Poo, L. J., and Lew, W. (1936a). *J. Biol. Chem.* **115,** 111.

Addis, T., Poo, L. J., and Lew, W. (1936b). *J. Biol. Chem.* **115,** 117.

Addis, T., Poo, L. J., and Lew, W. (1936c). *J. Biol. Chem.* **116,** 343.

Akedo, H., and Christensen, H. N. (1962). *J. Biol. Chem.* **237,** 118.

Akeson, A., Biorck, G., and Simon, R. (1968). *Acta Med. Scand.* **183,** 307.

Alam, S. Q., Rogers, Q. R., and Harper, A. E. (1966). *J. Nutr.* **89,** 97.

Allen, E. R., and Pepe, F. A. (1965). *Am. J. Anat.* **116,** 115.

Alleyne, G. A. O., and Young, V. H. (1966). *Lancet* **i,** 911.

Alleyne, G. A. O., and Young, V. H. (1967). *Clin. Sci.* **33,** 189.

Allison, J. B., Brande, P. F., Wannemacher, R. W., Parmer, L. P., and Leathem, J. H. (1961). *J. Nutr.* **73,** 321.

Allison, J. B., Wannemacher, R. W., Banks, W. L., Wunner, W. H., and Gomez-Brenes, R. A. (1962). *J. Nutr.* **78,** 333.

Allison, J. B., Wannemacher, R. W., and Banks, W. L. (1963). *Federation Proc.* **22,** 1126.

Allison, S. P., Prowse, K., and Chamberlain, M. J. (1967). *Lancet* **i,** 478.

Altman, K. I., Cassarett, G. W., Noonan, T. R., and Salomon, K. (1949). *Arch. Biochem.* **23,** 131.

Andres, R., Cader, G., and Zierler, K. L. (1956). *J. Clin. Invest.* **35,** 671.
Andrew, W., Shock, N. W., Barrows, C. H., and Yiengst, M. J. (1959). *J. Gerontol.* **14,** 405.
Anthony, A., Ackerman, E., and Strother, G. K. (1959). *Am. J. Physiol.* **196,** 512.
Aron, H. (1911). *Philippine J. Sci.* (*B, Med.*) **6,** 1.
Arroyave, G., Wilson, D., Funes, C. de, and Béhar, M. (1962). *Am. J. Clin. Nutr.* **11,** 517.
Asatoor, A. M., and Armstrong, M. D. (1967). *Biochem. Biophys. Res. Commun.* **26,** 168.
Ashley, J. H., and Fisher, H. (1967). *Brit. J. Nutr.* **21,** 661.
Avena, R. M., and Bowen, W. J. (1969). *J. Biol. Chem.* **244,** 1600.
Awapara, J., and Seale, B. (1952). *J. Biol. Chem.* **194,** 497.
Baig, H. A., and Edozien, J. C. (1965). *Lancet* **ii,** 662.
Bailey, K. (1942). *Biochem. J.* **36,** 121.
Baltzan, M. A., Andres, R., Cader, G., and Zierler, K. L. (1962). *J. Clin. Invest.* **41,** 116.
Banker, B. Q. (1967). *J. Neuropath. Exptl. Neurol.* **26,** 259.
Barany, M., Barany, K., Reckard, R., and Volpe, A. (1965). *Arch. Biochem. Biophys.* **109,** 185.
Baril, E. F., and Herrmann, H. (1967). *Develop. Biol.* **15,** 318.
Baril, E. F., Love, D. S., and Herrmann, H. (1964). *Science* **146,** 413.
Baril, E. F., Love, D. S., and Herrmann, H. (1966). *J. Biol. Chem.* **241,** 822.
Barrows, C. H., Roeder, L. M., and Falzone, J. A. (1962). *J. Gerontol.* **17,** 144.
Beach, R. K., and Kostyo, J. L. (1968). *Endocrinology* **82,** 882.
Beatty, C. H., Peterson, R. D., and Bocek, R. M. (1963). *Am. J. Physiol.* **204,** 939.
Beatty, C. H., Basinger, G. M., Dully, C. C., and Bocek, R. M. (1966). *J. Histochem. Cytochem.* **14,** 590.
Beecher, G. R., Kastenschmidt, L. L., Cassens, R. G., Hoekstra, W. G., and Briskey, E. J. (1968). *J. Food Sci.* **33,** 84.
Béhar, M., Arroyave, G., Tejada, C., Viteri, F., and Scrimshaw, N. S. (1956). *Rev. Col. Med. Guatemala* **7,** 221.
Beisel, W. R. (1966). *Federation Proc.* **25,** 1682.
Beisel, W. R., and Rapoport, M. I. (1969). *New Engl. J. Med.* **280,** 541 and 596.
Beisel, W. R., Sawyer, W. D., Ryll, E. D., and Crozier, D. (1967). *Ann. Internal Med.* **67,** 744.
Bendall, J. R., and Voyle, C. A. (1967). *J. Food Technol.* **2,** 259.
Benevenga, N. J., Harper, A. E., and Rogers, Q. R. (1968). *J. Nutr.* **95,** 434.
Bennett, H. S. (1960). *In* "Structure and Function of Muscle" (G. H. Bourne, ed.), Vol. I, p. 137. Academic Press, New York.
Bergman, R. A. (1962). *Bull. Johns Hopkins Hosp.* **110,** 187.
Berson, S. A., and Yalow, R. S. (1965). *Diabetes* **14,** 549.
Betheil, J. J. Feigelson, M., and Feigelson, P. (1965). *Biochim. Biophys. Acta* **104,** 92.
Betz, E., Firket, H., and Reznik, M. (1966). *Intern. Rev. Cytol.* **19,** 203.
Beznak, M. (1962). *Can. J. Biochem. Physiol.* **40,** 1647.
Biorck, G. (1949). *Acta Med. Scand. Suppl.* **226.**
Bird, J. W. C., Berg, T., and Leathem, J. H. (1968). *Proc. Soc. Exptl. Biol. Med.* **127,** 182.
Biron, P., Dreyfus, T. C., and Schapira, G. (1964). *Compt. Rend. Soc. Biol.* **158,** 1841.

Black-Schaffer, B., Grinstead, C. E., and Braunstein, J. N. (1965). *Circulation Res.* **16,** 383.

Blanchear, M. C., Van Wijhe, M., and Mosersky, D. (1963). *J. Histochem. Cytochem.* **11,** 500.

Blobel, G., and Potter, V. R. (1967). *J. Mol. Biol.* **26,** 279.

Block, W. D., Hubbard, R. W., and Steele, B. F. (1965). *J. Nutr.* **85,** 419.

Block, W. D., Westhoff, M. H., and Steele, B. F. (1967). *J. Nutr.* **91,** 189.

Bloom, W., and Fawcett, D. W. (1964). "A Textbook of Histology," Chapter 8. Saunders Co., Philadelphia, Pennsylvania.

Bloor, C. M., Leon, A. S., and Pasyk, S. (1968). *Lab. Invest.* **19,** 675.

Bocek, R. M., Peterson, R. D., and Beatty, C. H. (1966). *Am. J. Physiol.* **210,** 1101.

Bombara, G., and Bergamini, E. (1968). *Biochim. Biophys. Acta* **150,** 226.

Boquet, P. L., and Fromogeot, P. (1965). *Biochim. Biophys. Acta* **97,** 222.

Borsook, H., Deasy, C. L., Haagen-Smit, A. J., Keighley, G., and Lowy, P. H. (1950). *J. Biol. Chem.* **187,** 839.

Bouma, J. M. W., and Gruber, M. (1964). *Biochim. Biophys. Acta* **89,** 545.

Bowden, D. H., and Goyer, R. A. (1960). *Arch. Pathol.* **69,** 188.

Bowden, R. M., and Gutmann, E. (1944). *Brain* **67,** 273.

Breuer, C. B., and Florini, J. R. (1965). *Biochemistry* **4,** 1544.

Breuer, C. B., and Florini, J. R. (1966). *Biochemistry* **5,** 3857.

Breuer, C. B., Davies, M. C., and Florini, J. R. (1964). *Biochemistry* **3,** 1713.

Buchanan, W. E., and Schwartz, T. B. (1967). *Am. J. Physiol.* **212,** 732.

Buller, A. J., and Lewis, D. M. (1965a). *J. Physiol. (London)* **176,** 355.

Buller, A. J., and Lewis, D. M. (1965b). *J. Physiol. (London)* **178,** 343.

Buller, A. J., Eccles, J. C., and Eccles, R. M. (1960a). *J. Physiol. (London)* **150,** 399.

Buller, A. J., Eccles, J. C., and Eccles, R. M. (1960b). *J. Physiol. (London)* **150,** 417.

Bullock, G., White, A. M., and Worthington, J. (1968). *Biochem. J.* **108,** 417.

Burleigh, I. G., and Schimke, R. T. (1969). *Biochem. J.* **113,** 157.

Buse, M. G., McMaster, J., and Buse, J. (1965). *Metab. Clin. Exptl.* **14,** 1220.

Cabak, V., Dickerson, J. W. T., and Widdowson, E. M. (1963). *Brit. J. Nutr.* **17,** 601.

Cahill, G. F., Jr., Herrera, M. G., Morgan, A. P., Soeldner, J. S., Steinke, J., Levy, P. L., Reichard, G. A., Jr., and Kipnis, D. M. (1966). *J. Clin. Invest.* **45,** 1751.

Calloway, D. H., Grossman, M. I., Bowman, J., and Calhoun, W. K. (1955). *Surgery* **37,** 935.

Campbell, R. M., Sharp, G., Boyne, A. W., and Cuthbertson, D. P. (1954). *Brit. J. Exptl. Pathol.* **35,** 566.

Capers, T. (1964). *Am. Heart J.* **68,** 102.

Cardinet, G. H., Littrell, J. F., and Freedland, R. A. (1967). *Res. Vet. Sci.* **8,** 219.

Carrow, R. E., Brown, R. E., and Van Huss, W. D. (1967). *Anat. Record* **159,** 33.

Cedergren, B., and Harary, I. (1964). *J. Ultrastruct. Res.* **11,** 428.

Chan, H. (1968). *Brit. J. Nutr.* **22,** 315.

Chandler, A. M., and Neuhaus, O. W. (1968). *Biochim. Biophys. Acta* **166,** 186.

Chauhan, S., Nayak, N. C., and Ramalingaswami, V. (1965). *J. Pathol. Bacteriol.* **90,** 301.

Cheek, D. B., ed. (1968). "Human Growth." Lea & Febiger, Philadelphia, Pennsylvania.

Cheek, D. B., Powell, G. K., and Scott, R. E. (1965a). *Bull. Johns Hopkins Hosp.* **116,** 378.

Cheek, D. B., Powell, G. K., and Scott, R. E. (1965b). *Bull. Johns Hopkins Hosp.* **117,** 306.

Cheek, D. B., Brasel, J. A., Elliott, D., and Scott, R. E. (1966). *Bull. Johns Hopkins Hosp.* **119,** 46.

Cheek, D. B., Brasel, J. A., and Graystone, J. E. (1968). *In* "Human Growth" (D. B. Cheek, ed.), p. 306. Lea & Febiger, Philadelphia, Pennsylvania.

Chen, S. C., and Young, V. R. (1968). *Biochem. J.* **106,** 61.

Chiakulas, J. J., and Pauly, J. E. (1965). *Anat. Record* **152,** 55.

Chinn, K. S. K., and Hannon, J. P. (1966). *Am. J. Physiol.* **211,** 993.

Clark, A. J., Yamada, C., and Swendseid, M. E. (1968). *Am. J. Physiol.* **215,** 1324.

Close, R. (1964). *J. Physiol.* (*London*) **173,** 74.

Close, R. (1967). *Excerpta Med. Found. Intern. Congr. Ser.* **147,** 142.

Cohen, J., Aroesty, J. M., and Rosenfeld, M. G. (1966). *Circulation Res.* **18,** 388.

Cohen, P. P. (1942). *Federation Proc.* **1,** 273.

Cohen, P. P., and Hekhuis, G. L. (1941). *J. Biol. Chem.* **140,** 711.

Coleman, D. L., and Ashworth, M. E. (1959). *Am. J. Physiol.* **197,** 839.

Coleman, J. R., and Coleman, A. W. (1968). *J. Cell Physiol.* **72,** Suppl. 1, 19.

Colombo, J. P., Richterich, R., and Rossi, E. (1962). *Klin. Wochschr.* **40,** 37.

Conway, T. W. (1964). *Proc. Natl. Acad. Sci. U.S.* **51,** 1216.

Cooper, W. G. (1955). *Anat. Record* **123,** 103.

Cornelius, C. E., Bishop, J., Switzer, J., and Rhode, E. A. (1959). *Cornell Vet.* **49,** 116.

Coulson, R. A., and Hernandez, T. (1968). *Am. J. Physiol.* **215,** 741.

Cowgill, R. W., and Freeburg, B. (1957). *Arch. Biochem. Biophys.* **71,** 466.

Czok, R., and Bucher, Th. (1960). *Advan. Protein Chem.* **15,** 315.

Davey, C. L. (1957). *Nature* **179,** 209.

Davey, P. J., and Manchester, K. L. (1969). *Biochim. Biophys. Acta* **182,** 85.

Davis, T. R. A., and Mayer, J. (1955). *Am. J. Physiol.* **181,** 675.

Davison, A. N. (1961). *Biochem. J.* **78,** 272.

Dawson, D. M., and Romanul, F. C. A. (1964). *Arch. Neurol.* **11,** 369.

Dawson, K. G., Patey, P., Rubenstein, D., and Beck, J. C. (1966). *Mol. Pharmacol.* **2,** 269.

deBodo, R. C., and Altszuler, N. (1957). *Vitamins Hormones* **15,** 205.

de Duve, C. (1959). *In* "Subcellular Particles" (T. Hayashi, ed.), p. 128. Ronald Press, New York.

de Duve, C. (1963). *In* "Lysosomes" (A. V. S. de Reuck, and M. P. Cameron, eds.), p. 1. Little, Brown, Boston, Massachusetts.

de Duve, C., Wattiaux, R., and Baudhuin, P. (1962). *Advan. Enzymol.* **24,** 291.

DeLoecker, W. (1965). *Arch. Intern. Pharmacodyn.* **153,** 69.

DeLoecker, W. (1966). *Acta Endocrinol.* **52,** 416.

DeLoecker, W., and de Wever, F. (1967). *Arch. Intern. Pharmacodyn.* **165,** 127.

Denny-Brown, D. (1929). *Proc. Roy. Soc.* **B104,** 371.

Devi, A., Mukundan, M. A., Srivastava, U., and Sarkar, N. K. (1963). *Exptl. Cell Res.* **32,** 242.

Dickerson, J. W. T. (1960). *Biochem. J.* **75,** 33.

Dickerson, J. W. T., and McCance, R. A. (1960). *Brit. J. Nutr.* **14,** 331.

Dickerson, J. W. T., and Widdowson, E. M. (1960). *Biochem. J.* **74,** 247.

Diehl, J. F. (1963). *Biochem. Z.* **337,** 333.

Diehl, J. F., and Jones, R. R. (1966). *Am. J. Physiol.* **210,** 1080.

Domonkos, J., and Latzkovits, L. (1961). *Arch. Biochem. Biophys.* **95,** 144.

Drabkin, D. L. (1950). *J. Biol. Chem.* **182,** 317.
Drahota, Z., and Gutmann, E. (1963). *Physiol. Bohemoslov.* **12,** 339.
Dreizen, P., Hartshorne, D. J., and Stracher, A. (1966). *J. Biol. Chem.* **241,** 443.
Dreyfus, J. C. (1967). *Rev. Franc. Etudes Clin. Biol.* **12,** 343.
Dreyfus, J. C., Kruh, J., and Schapira, G. (1960). *Biochem. J.* **75,** 574.
Dubowitz, V., and Pearse, A. G. (1960a). *Nature* **185,** 701.
Dubowitz, V., and Pearse, A. G. (1960b). *Histochemie* **2,** 105.
Earl, D. C. N., and Korner, A. (1965). *Biochem. J.* **94,** 721.
Earl, D. C. N., and Morgan, H. E. (1968). *Arch. Biochem. Biophys.* **128,** 460.
Eboué-Bonis, D., Chambaut, A. M., Volfin, P., and Clauser, H. (1963). *Nature* **199,** 1183.
Eckel, R. E., Pope, C. E., and Norris, J. E. C. (1954). *Arch. Biochem. Biophys.* **52,** 293.
Edozien, J. C. (1966). *Clin. Sci.* **31,** 153.
Edozien, J. C., and Young, V. R. (1969). Unpublished results.
Eichhorn, J., Feinstein, M., Halkerston, I. D. K., and Hechter, O. (1961). *Proc. Soc. Exptl. Biol. Med.* **106,** 781.
Elliott, D. A., and Cheek, D. B. (1968). *In* "Human Growth" (D. B. Cheek, ed.), p. 326. Lea & Febiger, Philadelphia, Pennsylvania.
Enesco, M. (1957). Ph.D. Thesis, McGill University, Montreal.
Enesco, M., and Leblond, C. P. (1962). *J. Embryol. Exptl. Morphol.* **10,** 530.
Enesco, M., and Puddy, D. (1964). *Am. J. Anat.* **114,** 235.
Engel, F. L., and Kostyo, J. L. (1964). *In* "The Hormones" (G. Pincus, K. V. Thimann, and E. B. Astwood, eds.), Vol. 5, p. 69. Academic Press, New York.
Ennis, H. L., and Lubin, M. (1961). *Biochim. Biophys. Acta* **50,** 399.
Ennis, H. L., and Lubin, M. (1965). *Biochim. Biophys. Acta* **95,** 624.
Epshtein, S. F. (1964). *Ukranian Biochem. J.* **36,** 855.
Erdos, T., and Bessada, R. (1966). *Biochim. Biophys. Acta* **129,** 628.
Ezerman, E. B., and Ishikawa, H. (1967). *J. Cell Biol.* **35,** 405.
Fahimi, H. D., and Karnovsky, M. (1966). *J. Cell. Biol.* **29,** 113.
Fanburg, B. L., and Posner, B. I. (1968). *Circulation Res.* **23,** 123.
Fischer, E. (1948). *Arch. Phys. Med. Rehabil.* **29,** 291.
Fischer, E., and Ramsey, V. W. (1946). *Am. J. Physiol.* **145,** 571.
Fischman, D. A. (1967). *J. Cell Biol.* **32,** 557.
Fleck, A., and Munro, H. N. (1963). *Metab. Clin. Exptl.* **12,** 783.
Flood, P. R. (1964). *Proc. 3rd. European Regional Conf. Electron Microscopy,* p. 575.
Florini, J. R., and Breuer, C. B. (1966). *Biochemistry* **5,** 1870.
Fraenkel-Conrat, H. L., Simpson, M. E., and Evans, H. M. (1941). *Am. J. Physiol.* **135,** 398.
Frame, E. G., and Russell, J. A. (1946). *Endocrinology* **39,** 420.
Frankl, L., and Freund, E. (1884). Quoted by Jackson (1925).
Freedland, R. A., Hjerpe, C. A., and Cornelius, C. E. (1965). *Res. Vet. Sci.* **6,** 18.
Frenk, S., Metcoff, J., Gomez, F., Ramos-Galvan, R., Cravioto, J., and Antonowicz, I. (1957). *Pediatrics* **20,** 105.
Friedberg, F., and Greenberg, D. M. (1947). *J. Biol. Chem.* **168,** 405.
Friedberg, F., and Greenberg, D. M. (1948). *Arch. Biochem.* **17,** 193.
Fuge, K. W., Crews, E. L., Pattengale, P. K., Holloszy, J. O., and Shank, R. E. (1968). *Am. J. Physiol.* **215,** 660.
Funabiki, R., and Kandatsu, M. (1965). *J. Agr. Chem. Soc. Japan* **39,** 109.

Gaetani, S., Paolucci, A. M., Spadoni, M. A., and Tomassi, G. (1964). *J. Nutr.* **84**, 173.
Gan, J. C., and Jeffay, H. (1967). *Biochim. Biophys. Acta* **148**, 448.
Garlick, P. J. (1969). *Nature* **233**, 61.
Gauthier, G. F. (1969). *Z. Zellforsch.* **95**, 462.
Gauthier, G. F., and Padykula, H. A. (1966). *J. Cell Biol.* **28**, 333.
George, J. C., and Talesara, C. L. (1961). *J. Cellular Comp. Physiol.* **58**, 253.
Gerber, G., Gerber, G., and Altman, K. I. (1960). *J. Biol. Chem.* **235**, 2682.
Girkin, G., Fitch, C. D., and Dinning, J. S. (1962). *Arch. Biochem. Biophys.* **98**, 224.
Gluck, L., Talner, N. S., Stern, H., Gardner, T. H., and Kulovich, M. V. (1964). *Science* **144**, 1244.
Goldberg, A. L. (1967). *Am. J. Physiol.* **213**, 1193.
Goldberg, A. L. (1968). *J. Cell Biol.* **36**, 653.
Goldberg, A. L. (1969a). *J. Biol. Chem.* **244**, 3217.
Goldberg, A. L. (1969b). *J. Biol. Chem.* **244**, 3223.
Goldberg, A. L., and Goodman, H. M. (1969a). *Am. J. Physiol.* **216**, 1111.
Goldberg, A. L., and Goodman, H. M. (1969b). *Am. J. Physiol.* **216**, 1116.
Goldspink, G. (1962a). *Proc. Roy. Irish Acad. Sect. B* **62**, 135.
Goldspink, G. (1962b). *Comp. Biochem. Physiol.* **7**, 157.
Goldspink, G. (1964). *J. Cellular Comp. Physiol.* **63**, 209.
Goldspink, G. (1965). *Am. J. Physiol.* **209**, 100.
Goldspink, G. (1968). *J. Cell Sci.* **3**, 539.
Goldstein, S., and Reddy, W. J. (1967). *Biochim. Biophys. Acta* **141**, 310.
Gordon, E. E., Kowalski, K., and Fritts, M. (1966). *Am. J. Physiol.* **210**, 1033.
Gould, B. S. (1968). *In* "Collagen" (B. S. Gould, ed.), Vol. 2, Part A, Chapter 3. Academic Press, New York.
Gould, M. K., and Rawlinson, W. A. (1959). *Biochem. J.* **73**, 41.
Greenbaum, A. L., and Young, F. G. (1953). *J. Endocrinol.* **9**, 127.
Guth, L. (1968). *Physiol. Rev.* **48**, 645.
Guth, L., and Watson, P. K. (1967). *Exptl. Neurol.* **17**, 107.
Guth, L., Watson, P. K. and Brown, W. C. (1968). *Exptl. Neurol.* **20**, 52.
Gutmann, E. (1948). *J. Neurophysiol.* **11**, 279.
Gutmann, E. (1960). *Physiol. Bohemoslov.* **9**, 382.
Gutmann, E., ed. (1962). "The Denervated Muscle." Czechoslovak Acad. Sci., Prague.
Gutmann, E., and Zelena, J. (1962). *In* "The Denervated Muscle" (E. Gutmann, ed.), p. 57. Czechoslovak Acad. Sci., Prague.
Gutmann, E., Hanzliková, V., and Holečková, E. (1969). *Exptl. Cell. Res.* **56**, 33.
Hagan, S. N., and Scow, R. O. (1957). *Am. J. Physiol.* **188**, 91.
Hajek, I., Gutmann, E., and Syrovy, I. (1964). *Physiol. Bohemoslov.* **13**, 32.
Hajek, I., Gutmann, E., and Syrovy, I. (1965). *Physiol. Bohemoslov.* **14**, 481.
Ham, A. W., and Leeson, T. W. (1961). "Histology," 4th ed. Lippincott, Philadelphia, Pennsylvania.
Hamosh, M., Lesch, M., Baron, J., and Kaufman, S. (1967). *Science* **157**, 935.
Hannon, J. P., and Vaughan, D. A. (1961). *Am. J. Physiol.* **201**, 217.
Hanson, J., and Lowy, J. (1963). *J. Mol. Biol.* **6**, 46.
Hanson, J., and Lowy, J. (1964). *In* "Biochemistry of Muscle Contraction" (J. Gergely, ed.), p. 141. Little, Brown, Boston, Massachusetts.
Hardy, M. F., and Perry, S. V. (1969). *Nature* **223**, 300.

Harkness, M. L. R., Harkness, R. D., and James, D. W. (1958). *J. Physiol.* (*London*) **144**, 307.

Harkness, R. D. (1961). *Biol. Rev. Cambridge Phil. Soc.* **36,** 399.

Harman, P. J., Magnani, J. L., Sanwal, S. D., and Thompson, W. A. L. (1960). *J. Bone Joint Surg.* **42A,** 1008.

Harper, A. E., and Rogers, Q. R. (1965). *Proc. Nutr. Soc.* **24,** 173.

Harper, A. E., Leung, P., Yoshida, A., and Rogers, Q. R. (1964). *Federation Proc.* **23,** 1087.

Harris, E. J., and Manchester, K. L. (1966). *Biochem. J.* **101,** 135.

Hart, J. S. (1958). *Federation Proc.* **17,** 1045.

Hartshorne, D. J., and Perry, S. V. (1962). *Biochem. J.* **85,** 171.

Havel, R. (1966). *In* "Proceedings of a Conference on Energy Metabolism and Body Fuel Utilization" (E. A. Morgan, ed.), p. 20. Harvard Univ. Press, Cambridge, Massachusetts.

Hearn, G. R., and Wainio, W. W. (1956). *Am. J. Physiol.* **185,** 348.

Hearn, G. R., and Wainio, W. W. (1957). *Am. J Physiol.* **190,** 206.

Helander, E. (1957). *Acta Physiol. Scand.* **41,** Suppl. 141.

Helander, E. A. S. (1961). *Biochem. J.* **78,** 478.

Henriques, O. B., Henriques, S. B., and Neuberger, A. (1955). *Biochem. J.* **60,** 409.

Hensel, H., and Hildebrandt, G. (1964). *In* "Handbook of Physiology," Sect. 4: Adaptation to the Environment," p. 73. Am. Physiol. Soc., Washington, D.C.

Henshaw, E. C., Hirsch, C. A., Milosevic, P., and Hiatt, H. H. (1968). *Trans. Assoc. Am. Physicians* **81,** 116.

Herbener, G. H. (1966). *Anat. Record* **154,** 357.

Herrmann, H. (1968). *J. Cell Physiol.* **72,** Suppl. 1, 30.

Herrmann, H., and Nicholas, J. S. (1948). *J. Exptl. Zool.* **107,** 165.

Herrmann, H., White, B. N., and Cooper, M. (1957). *J. Cellular Comp. Physiol.* **49,** 227.

Hess, A. (1965). *J. Cell Biol.* **26,** 467.

Hess, A., and Pilar, G. (1963). *J. Physiol.* (*London*) **169,** 780.

Hess, J. W., MacDonald, R. P., Frederick, R. J., Jones, R. N., Neely, J., and Gross, D. (1964). *Ann. Internal. Med.* **61,** 1015.

Heuson-Stiennon, J. A. (1964). *J. Microscop.* **3,** 229.

Heywood, S. M., and Nwagwu, M. (1968). *Proc. Natl. Acad. Sci. U.S.* **60,** 229.

Heywood, S. M., and Rich, A. (1968). *Proc. Natl. Acad. Sci. U.S.* **59,** 590.

Heywood, S. M., Dowben, R. M., and Rich, A. (1967). *Proc. Natl. Acad. Sci. U.S.* **57,** 1002.

Heywood, S. M., Dowben, R. M., and Rich, A. (1968). *Biochemistry* **7,** 3289.

Hibbs, R. G., and Ferrans, V. J. (1969). *Am. J. Anat.* **124,** 251.

Hider, R. C., Fern, E. B., and London, D. R. (1969). *Biochem. J.* **114,** 171.

Hnik, P. (1962). *In* "The Denervated Muscle" (E. Gutmann, ed.), p. 341. Czechoslovak Acad. Sci., Prague.

Hoffmann, A. (1947). *Anat. Anz.* **96,** 191.

Holloszy, J. O. (1967). *J. Biol. Chem.* **242,** 2278.

Holmes, R., and Rasch, P. J. (1958). *Am. J. Physiol.* **195,** 50.

Howarth, R. E., Baldwin, R. L., and Ronning, M. (1968). *J. Dairy Sci.* **51,** 1270.

Hoyle, T. C., Sumner, R. G., and McIntosh, H. D. (1963). *J. Lab. Clin. Med.* **62,** 632.

Hughes, B. P. (1965). *Postgrad. Med.* **41,** 313.

Hulsmans, H. A. M. (1961). *Biochim. Biophys. Acta* **54,** 1.

Hurtado, A., Rotta, A., Merino, C., and Pons, J. (1937). *Am. J. Med. Sci.* **194,** 708.
Huxley, H. E. (1960). *In* "The Cell" (J. Brachet and A. E. Mirsky, eds.), Vol. IV, p. 365. Academic Press, New York.
Iacobellis, M., Muntwyler, E., and Dodgen, C. L. (1956). *Am. J. Physiol.* **185,** 275.
Ichihara, A., and Koyama, E. (1966). *J. Biochem.* (*Japan*) **59,** 160.
Ingle, D. J., Prestrud, M. C., and Nezamis, J. E. (1947). *Am. J. Physiol.* **150,** 682
Jackson, C. M. (1925). "The Effects of Inanition and Malnutrition upon Growth and Structure." Churchill, London.
Jacobsen, J. G., and Smith, L. H. (1968). *Physiol. Rev.* **48,** 424.
James, T. N., and Sherf, L. (1968). *Am. J. Cardiol.* **22,** 389.
Johnson, O. W., and Stephen, W. P. (1964). *Nature* **203,** 207.
Johnson, P., Harris, C. I., and Perry, S. V. (1967). *Biochem. J.* **105,** 361.
Joseph, R. L., and Sanders, W. J. (1966). *Biochem. J.* **100,** 827.
Joubert, D. M. (1956). *J. Agr. Sci.* **47,** 59.
Kaplan, S. A., and Shimizu, C. S. N. (1963). *Endocrinology* **72,** 267.
Kare, M. R., Medway, W., and Brandt, C. S. (1955). *Am. J. Physiol.* **183,** 416.
Kendrick-Jones, J., and Perry, S. V. (1967). *Biochem. J.* **103,** 207.
Kendrick-Jones, J., and Perry, S. V. (1967). *Excerpta Med. Found. Intern. Congr. Ser.* **147,** 64.
Kerpel-Fronius, E., and Frank, K. (1949). *Ann. Paediat.* **173,** 321.
Khyl'ko, O. K. (1965). *Ukr. Biokhim. Zh.* **37,** 8.
Kielley, W. W., and Harrington, W. F. (1960). *Biochim. Biophys. Acta* **41,** 401.
Kipnis, D. M., and Noall, M. W. (1958). *Biochim. Biophys. Acta* **28,** 226.
Kipnis, D. M., and Reiss, E. (1960). *J. Clin. Invest.* **39,** 1002.
Kitiyakara, A. (1961). *Arch. Pathol.* **71,** 579.
Kleine, T. O., and Chlond, H. (1967). *Clin. Chim. Acta* **15,** 19.
Knipfel, J. E., Botting, H. G., Noel, F. J., and McLaughlin, J. M. (1969). *Can. J. Biochem.* **47,** 323.
Knobil, E., and Hotchkiss, J. (1964). *Ann. Rev. Physiol.* **26,** 47.
Knowlton, G. C., and Hines, H. M. (1936). *Proc. Soc. Exptl. Biol. Med.* **35,** 394.
Kochakian, C. D. (1960). *Proc. Soc. Exptl. Biol. Med.* **103,** 196.
Kochakian, C. D., Humm, J. H., and Bartlett, M. N. (1948). *Am. J. Physiol.* **155,** 242.
Kochakian, C. D., Robertson, E., and Bartlett, M. N. (1950). *Am. J. Physiol.* **163,** 332.
Kochakian, C. D., Tillotson, C., and Endahl, G. L. (1956). *Endocrinology* **58,** 226.
Kochakian, C. D., Hill, J., and Harrison, D. G. (1964). *Endocrinology* **74,** 635.
Koch, A. L. (1962). *J. Theoret. Biol.* **3,** 283.
Kohn, R. R. (1969). *Lab. Invest.* **20,** 202.
Kompmann, M., Paddags, I., and Sandritter, W. (1966). *Arch. Pathol.* **82,** 303.
Konigsberg, I. R., McElvain, N., Tootle, M., and Herrmann, H. (1960). *J. Cell Biol.* **8,** 333.
Korecky, B., and French, I. W. (1967). *Circulation Res.* **21,** 635.
Korner, A. (1967). *Progr. Biophys. Mol. Biol.* **17,** 63.
Korner, P. I. (1959). *Physiol. Rev.* **39,** 687.
Kostyo, J. L. (1966). *Biochim. Biophys. Acta* **129,** 294.
Kostyo, J. L. (1968). *Excerpta Med. Found. Intern. Congr. Ser.* **158,** 175.
Kostyo, J. L., and Knobil, E. (1959). *Endocrinology* **65,** 395.
Kostyo, J. L., and Redmond, A. F. (1966). *Endocrinology* **79,** 531.
Kostyo, J. L., and Schmidt, J. E. (1963). *Am. J. Physiol.* **204,** 1031.
Koszalka, T. R., and Miller, L. L. (1960). *J. Biol. Chem.* **235,** 665.

Koszalka, T. R., Mason, K. E., and Krol, G. (1961). *J. Nutr.* **73,** 78.
Krames, B. B., and Van Liere, E. J. (1966). *Anat. Record* **156,** 461.
Krebs, H. A. (1950). *Biochim. Biophys. Acta* **4,** 249.
Kruh, J., Dreyfus, J. C., Schapira, G., and Gey, G. O. (1960). *J. Clin. Invest.* **39,** 1180.
Krupnick, A. B., Casa, C. M., and Rosenkrantz, H. (1964). *Arch. Biochem. Biophys.* **106,** 89.
Kuby, S. A., Noda, L., and Lardy, H. A. (1954). *J. Biol. Chem.* **209,** 191.
Kunkel, H. O., and Campbell, J. E. (1952). *J. Biol Chem.* **198,** 229.
Laferté, R. O., Rosenkrantz, H., and Berlinguet, L. (1963). *Can. J. Biochem. Physiol.* **41,** 1423.
Landmann, W. A. (1963). *Proc. The Meat Tenderness Symp.* p. 87. Campbell Soup, Camden, New Jersey.
Larson, P. F., Hudgson, P., and Walton, J. N. (1969). *Nature* **222,** 1168.
Lawrie, R. A. (1952). *Nature* **170,** 122.
Lawrie, R. A. (1953a). *Biochem. J.* **55,** 298.
Lawrie, R. A. (1953b). *Nature* **171,** 1069.
Lawrie, R. A. (1966). *In* "The Physiology and Biochemistry of Muscle as a Food" (E. J. Briskey, R. G. Cassens, and J. C. Trautman, eds.), p. 137. Univ. of Wisconsin Press, Madison, Wisconsin.
Lee, J. C. (1965). *Exptl. Neurol.* **12,** 123.
Lee, N. D., and Williams, R. H. (1952). *Endocrinology* **51,** 451.
Le Gros Clark, W. E. (1958). "The Tissues of the Body." Oxford Univ. Press (Clarendon), London and New York.
Lentz, T. L. (1969). *Am. J. Anat.* **124,** 447.
Leonard, S. L. (1957). *Proc. Soc. Exptl. Biol. Med.* **94,** 458.
Leung, P. M. B. (1965). Ph.D. Thesis, Massachusetts Inst. Technol., Cambridge, Massachusetts.
Leung, P. M. B., Rogers, Q. R., and Harper, A. E. (1968). *J. Nutr.* **96,** 303.
Levenson, S. M., and Watkin, D. M. (1959). *Federation Proc.* **18,** 1155.
Levenson, S. M., Braasch, J. W., Mueller, H., and Crowley, L. (1959). Quoted by Levenson and Watkin (1959).
Lin, E. C. C., and Knox, W. E. (1958). *J. Biol. Chem.* **233,** 1186.
Linzbach, A. J. (1947). *Arch. Pathol. Anat.* **314,** 534.
Linzbach, A. J. (1960). *Am. J. Cardiol.* **5,** 370.
Liu, A. Y., and Neuhaus, O. W. (1968). *Biochim. Biophys. Acta* **166,** 195.
London, D. R., Foley, T. H., and Webb, C. G. (1965). *Nature* **208,** 588.
Lotspeich, W. D. (1949). *J. Biol. Chem.* **179,** 175.
Lubin, M. (1967). *Nature* **213,** 451.
Lust, G. (1966). *Federation Proc.* **25,** 1688.
Lyman, C. P., and Leduc, E. H. (1953). *J. Cellular Comp. Physiol.* **41,** 471.
MacCallum, J. B. (1898). *Bull. Johns Hopkins Hosp.* **9,** 208.
MacConnachie, H. F., Enesco, M., and Leblond, C. P. (1964). *Am. J. Anat.* **114,** 245.
McFarlane, I. G., and van Holt, C. (1969a). *Biochem. J.* **111,** 557.
McFarlane, I. G., and van Holt, C. (1969b). *Biochem. J.* **111,** 565.
McLoughlin, J. V. (1968). *J. Food Sci.* **33,** 383.
McManus, I. R., and Mueller, H. (1966). *J. Biol. Chem.* **241,** 5967.
McMeekan, C. P. (1940). *J. Agr. Sci.* **30,** 276.
McPherson, A., and Tokunaga, J. (1967). *J. Physiol.* (*London*) **188,** 121.
Manchester, K. L., and Harris, E. J. (1968). *Biochem. J.* **108,** 177.

Manchester, K. L., and Wool, I. G. (1963). *Biochem. J.* **89,** 202.
Manchester, K. L., and Young, F. G. (1958). *Biochem. J.* **70,** 353.
Manchester, K. L., and Young, F. G. (1959). *Biochem. J.* **72,** 136.
Manchester, K. L., and Young, F. G. (1960). *Biochem. J.* **75,** 487.
Manchester, K. L., Randle, P. J., and Young, F. G. (1959). *J. Endocrinol.* **18,** 395.
Marchok, A. C., and Herrmann, H. (1967). *Develop. Biol.* **15,** 129.
Margreth, A., and Novello, F. (1964). *Exptl. Cell Res.* **35,** 38.
Margreth, A., Catani, C., and Shiaffino, S. (1967). *Biochem. J.* **102,** 35c.
Margreth, A., Novello, F., and Aloisi, M. (1966). *Exptl. Cell Res.* **41,** 666.
Mariani, A., Spadoni, M. A., and Tomassi, G. (1963). *Nature* **199,** 378.
Martignoni, P., and Winnick, T. (1954). *J. Biol. Chem.* **208,** 251.
Martin, A. W., and Fuhrman, F. A. (1955). *Physiol. Zool.* **28,** 18.
Martin, T. E., and Wool, I. G. (1968). *Proc. Natl. Acad. Sci. U.S.* **60,** 569.
Martin, T. E., and Young, F. G. (1965). *Nature* **208,** 684.
Matsumoto, S., and Krasnow, N. (1968). *Am. J. Physiol.* **214,** 620.
Mauro, A. (1961). *J. Biophys. Biochem. Cytol.* **9,** 493.
Mendes, C. B., and Waterlow, J. C. (1958). *Brit. J. Nutr.* **12,** 74.
Message, M. A. (1968). *In* "Growth and Development of Mammals" (G. A. Lodge and G. E. Lamming, eds.), pp. 26–50. Plenum Press, New York.
Metcoff, J. (1967). *Ann. Rev. Med.* **18,** 377.
Metcoff, J., Frenk, S., Yoshida, T., Pinedo, R. T., Kaiser, E., and Hansen, J. D. L. (1966). *Medicine* **45,** 365.
Midsukami, M. (1964). *Okajimas Folia Anat. Japan* **40,** 173.
Mihalyi, E., Laki, K., and Knoller, M. I. (1957). *Arch. Biochem. Biophys.* **68,** 130.
Milhorat, A. T., ed. (1967). *Excerpta Med. Found. Intern. Congr. Ser.* **147.**
Miller, L. L. (1962). *In* "Amino Acid Pools" (J. T. Holden, ed.), p. 708. Elsevier, Amsterdam.
Millikan, G. A. (1937). *Proc. Roy. Soc.* **B123,** 218.
Millward, D. J. (1969). Unpublished results.
Mimura, T., Yamada, C., and Swendseid, M. E. (1968). *J. Nutr.* **95,** 493.
Mirsky, I. J. (1938). *Am. J. Physiol.* **124,** 569.
Molbert, E., and Jijima, S. (1959). *Verhandl. Deut. Ges. Pathol.* **42,** 349.
Monckeberg, F., Beas, F., Horwitz, I., Debancens, A., and Gonzalez, M. (1964) *Pediatrics* **33,** 554.
Monckton, G., and Nihei, T. (1966). *Ann. N.Y. Acad. Sci.* **138,** Art. 1, 329.
Monckton, G., and Nihei, T. (1969). *Neurology* **19,** 415.
Montgomery, R. D. (1962a). *Nature* **195,** 194.
Montgomery, R. D. (1962b). *J. Clin. Pathol.* **15,** 511.
Montgomery, R. D., Dickerson, J. W. T., and McCance, R. A. (1964). *Brit. J. Nutr.* **18,** 587.
Moore, F. D., Steenberg, R. W., Ball, M. R., Wilson, G. M., and Myrden, J. A. (1955). *Ann. Surg.* **141,** 145.
Morey, K. S., Tarczy-Hornoch, K., Richards, E. G., and Brown, W. D. (1967). *Arch. Biochem. Biophys.* **119,** 491.
Morey, K. S., Tarczy-Hornoch, K., and Brown, W. D. (1968). *Arch. Biochem. Biophys.* **124,** 521.
Morkin, E., and Ashford, T. P. (1968). *Am. J. Physiol.* **215,** 1409.
Moroz, L. A. (1967). *Circulation Res.* **21,** 449.
Morpurgo, B. (1895). *Arch. Sci. Med.* **19,** 327.
Morpurgo, B. (1897). *Arch. Pathol. Anat. Physiol. Klin. Med.* **150,** 522.

Morrison, P., Rosenmann, M., and Sealander, J. A. (1966). *Am. J. Physiol.* **211,** 1305.
Moss, F. P. (1968a). *Am. J. Anat.* **122,** 555.
Moss, F. P. (1968b). *Am. J. Anat.* **122,** 565.
Moss, F. P., Simmonds, R. A., and McNary, H. W. (1964). *Poultry Sci.* **43,** 1086.
Moulton, C. R. (1923). *J. Biol. Chem.* **57,** 79.
Muir, A. R., Kanji, A. H. M., and Allbrook, D. (1965). *J. Anat.* **99,** 435.
Munro, H. N. (1956). *Scot. Med. J.* **1,** 285.
Munro, H. N., and Chalmers, M. I. (1945). *Brit. J. Exptl. Pathol.* **26,** 396.
Munro, H. N., and Gray, J. A. M. (1969). *Comp. Biochem. Physiol.* **28,** 897.
Munro, H. N., and Thomson, W. S. T. (1953). *Metab. Clin. Exptl.* **2,** 354.
Munro, H. N., Black, J. G., and Thomson, W. S. T. (1959). *Brit. J. Nutr.* **13,** 475.
Muntwyler, E., Griffin, G. E., and Iacobellis, M. (1958). *Am. J. Physiol.* **195,** 347.
Muramatsu, K., Sato, T., and Ashida, K. (1963). *J. Nutr.* **81,** 427.
Muscatello, U., Andersson-Cedergren, E., Azzone, G. F., and Von Der Decker, A. (1961). *J. Biophysiol. Biochem. Cytol.* **10,** 201.
Muscatello, U., Margreth, A., and Aloisi, M. (1965). *J. Cell Biol.* **27,** 1.
Naeye, R. L. (1965). *Am. J. Pathol.* **46,** 829.
Naeye, R. L. (1966). *Lab. Invest.* **15,** 700.
Naeye, R. L. (1967). *J. Am. Med. Assoc.* **200,** 281.
Nair, K. G., Cutilletta, A. F., Zak, R., Koide, T., and Rabinowitz, M. (1968). *Circulation Res.* **23,** 451.
Nakagawa, I., Takahashi, T., Suzuki, T., and Kobayashi, K. (1964). *J. Nutr.* **83,** 115.
Nakagawa, I., Takahashi, T., Suzuki, T., and Kobayashi, K. (1965). *J. Nutr.* **86,** 333.
Needham, D. M. (1926). *Physiol. Rev.* **6,** 1.
Neuhaus, O. W., Balegno, H. F., and Chandler, A. M. (1966). *Am. J. Physiol.* **211,** 151.
Neumaster, T. D., and Ring, G. C. (1965). *J. Gerontol.* **20,** 379.
Nichoalds, G. E., Diehl, J. F., and Fitch, C. D. (1967). *Am. J. Physiol.* **213,** 759.
Nichols, B. L., Alleyne, G. A. O., Barnes, D. J., and Hazlewood, C. F. (1969). *J. Pediat.* **74,** 49.
Nichols, B. L., Barnes, D. J., Ashworth, A., Alleyne, G. A. O., Hazelwood, C. F., and Waterlow, J. C. (1968). *Nature* **217,** 475.
Nihei, T. (1969). Personal communication.
Nihei, T., and Monckton, G. (1969). *2nd Intern. Congr. Neuro-Genetics* (in press).
Norman, T. D. (1962). *Progr. Cardiovascular Diseases* **4,** 439.
Novak, A. (1957). *Am. J. Physiol.* **191,** 306.
Nowy, H., and Frings, H. D. (1960). *Z. Ges. Exptl. Med.* **132,** 538.
NRC (1959). *Natl. Acad. Sci.—Natl. Res. Council Publ.* **711.**
Nyhan, W. L., Yujnavsky, A. O., and Wehr, R. F. (1968). *In* "Human Growth" (D. B. Cheek, ed.), p. 396. Lea & Febiger, Philadelphia, Pennsylvania.
Obinata, T., Yamamoto, M., and Maruyama, K. (1966). *Devel. Biol.* **14,** 192.
Ogata, T. (1960). *J. Biochem. (Tokyo)* **6,** 726.
Oliver, I. T. (1955). *Biochem. J.* **61,** 116.
Oppenheimer, H., Barany, K., and Milhorat, A. T. (1964). *Proc. Soc. Exptl. Biol. Med.* **116,** 877.
Overy, H. R., and Priest, R. E. (1966). *Lab. Invest.* **15,** 1100.
Padieu, P. (1959). *Bull. Soc. Chim. Biol.* **41,** 57.
Padykula, H. A., and Gauthier, G. F. (1967). *Excerpta Med. Found. Intern. Congr. Ser.* **147,** p. 117.

Page, S. (1965). *J. Cell Biol.* **26,** 477.
Park, D. C., and Pennington, R. J. (1967). *Enzymol. Biol. Clin.* **8,** 149.
Pater, J. L., and Kohn, R. R. (1967). *Proc. Soc. Exptl. Biol. Med.* **125,** 476.
Pattengale, P. K., and Holloszy, J. O. (1967). *Am. J. Physiol.* **213,** 783.
Pawlak, M., and Pion, R. (1968). *Ann. Biol. Anim. Biochim. Biophys.* **8,** 517.
Peachey, L. D. (1965a). *J. Cell Biol.* **25,** 209.
Peachey, L. D. (1965b). *Federation Proc.* **24,** 1124.
Peachey, L. D. (1968). *Ann. Rev. Physiol.* **30,** 401.
Peachey, L. D., and Huxley, A. F. (1962). *J. Cell Biol.* **13,** 177.
Pearce, G. W. (1966). *Ann. N.Y. Acad. Sci.* **138,** 138.
Pearce, G. W., and Walton, J. N. (1963). *J. Pathol. Bacteriol.* **86,** 25.
Pearlstein, R., and Kohn, R. R. (1966). *Am. J. Pathol.* **48,** 823.
Pearson, C. M. (1965). *In* "Muscle" (W. M. Paul, E. E. Daniel, C. M. Kay, and G. Monckton, eds.), p. 423. Pergamon, Oxford.
Pellegrino, C., and Franzini, C. (1963). *J. Cell Biol.* **17,** 327.
Pennington, R. J. (1963). *Biochem. J.* **88,** 64.
Perry, S. V. (1952). *Biochim. Biophys. Acta* **8,** 499.
Perry, S. V. (1960a). *In* "Comparative Biochemistry" (M. Florkin and H. S. Mason, eds.), Vol. II. Academic Press, New York.
Perry, S. V. (1960b). *Biochem. J.* **74,** 94.
Perry, S. V. (1965). *In* "Muscle" (W. M. Paul, E. E. Daniel, C. M. Kay, and G. Monckton, eds.), p. 29. Pergamon, Oxford.
Perry, S. V. (1967). *Progr. Biophys. Mol. Biol.* **17,** 325.
Perry, S. V., and Zydowo, M. (1959a). *Biochem. J.* **71,** 220.
Perry, S. V., and Zydowo, M. (1959b). *Biochem. J.* **72,** 682.
Peter, J. B., Jeffress, R. N., and Lamb, D. R. (1968). *Science* **160,** 200.
Petersen, R. O., and Baserga, R. (1965). *Exptl. Cell Res.* **40,** 340.
Peterson, D. W., Lilyblade, A. L., and Lyon, J. (1963). *Proc. Soc. Exptl. Biol. Med.* **113,** 798.
Pettengill, N. E., and Martin, A. W. (1947). *Federation Proc.* **6,** 179.
Picou, D., Alleyne, G. A. O., and Seakins, A. (1965). *Clin. Sci.* **29,** 517.
Picou, D., Halliday, D., and Garrow, J. S. (1966). *Clin. Sci.* **30,** 345.
Pimstone, B. L., Barbezat, G., Hansen, J. D. L., and Murray, P. (1967). *Lancet* **ii,** 1333.
Pimstone, B. L., Barbezat, G., Hansen, J. D. L., and Murray, P. (1968). *Am. J. Clin. Nutr.* **21,** 482.
Pollack, M. S., and Bird, J. W. C. (1968). *Am. J. Physiol.* **215,** 716.
Porter, K. R. (1961). *J. Biophys. Biochem. Cytol.* **10,** 219.
Porter, K. R., and Palade, G. E. (1957). *J. Biochem. Biophys. Cytol.* **3,** 269.
Posner, B. I., and Fanburg, B. L. (1968). *Circulation Res.* **23,** 137.
Pozefsky, T., Felig, P., Tobin, J. D., Soeldner, J. S., and Cahill, G. F. (1969). *J. Clin. Invest.* **48,** 2273.
Pratt, N. E. (1967). *Anat. Record* **157,** 303.
Prewitt, M. A., and Salefsky, B. (1967). *Am. J. Physiol.* **213,** 295.
Price, H. M. (1963). *Am. J. Med.* **35,** 589.
Price, H. M., Howes, E. L., and Blumberg, J. M. (1964). *Lab. Invest.* **13,** 1279.
Quass, D. W., and Briskey, E. J. (1968). *J. Food Sci.* **33,** 180.
Rabinowitz, D., and Merimee, T. J. (1968). *In* "Human Growth" (D. B. Cheek, ed.), p. 207. Lea & Febiger, Philadelphia, Pennsylvania.
Rabinowitz, D., Merimee, T. J., Maffezzoli, R., and Burgess, J. A. (1966). *Lancet* **ii,** 454.

Rabinowitz, D., Merimee, T. J., Nelson, J. K., Schultz, R. B., and Burgess, J. A. (1968). *Excerpta Med. Found. Intern. Congr. Ser.* **158.**

Rabinowitz, M., Zak, R., Beller, B., Rampersad, O., and Wool, I. G. (1964). *Proc. Natl. Acad. Sci. U.S.* **52,** 1353.

Rampersad, O. R., Zak, R., Rabinowitz, M., Wool, I. G., and DeSalle, L. (1965). *Biochim. Biophys. Acta* **108,** 95.

Randall, C. J., and MacRae, H. F. (1967). *J. Food Sci.* **32,** 182.

Rao, M. N., and McLaughlan, J. M. (1967). *Can. J. Biochem.* **45,** 1653.

Reid, J. V. O., and Berjak, P. (1966). *Am. Heart J.* **71,** 240.

Reis, D. J., Wooten, G. F., and Hollenberg, M. (1967). *Am. J. Physiol.* **213,** 592.

Reis, D. J., Moorhead, D., and Wooten, G. F. (1969). *Am. J. Physiol.* **217,** 541.

Reiss, E., and Kipnis, D. M. (1959). *J. Lab. Clin. Med.* **54,** 937.

Reporter, M. (1969). *Biochemistry* **8,** 3489.

Revel, J. P. (1962). *J. Cell Biol.* **12,** 571.

Reynafarje, B., and Morrison, P. (1962). *J. Biol. Chem.* **237,** 2861.

Richter, G. W., and Kellner, A. (1963). *J. Cell Biol.* **18,** 195.

Riggs, T. R., and Walker, L. M. (1960). *J. Biol. Chem.* **235,** 3603.

Rinehart, K. E., Featherston, W. R., and Rogler, J. C. (1968). *J Nutr.* **95,** 627.

Ring, G. C., Dupuch, G. H., and Emeric, D. (1964). *J. Gerontol.* **19,** 215.

Robertson, D. D., and Baker, D. D. (1933). *Missouri, Univ. Agr. Expt. Sta. Res. Bull.* **200.**

Robinson, D. S. (1952). *Biochem. J.* **52,** 628.

Rogers, Q. R., and Harper, A. E. (1968). *In* "Protein Nutrition and Free Amino Acid Patterns" (J. H. Leathem, ed.), p. 107. Rutgers Univ. Press, New Brunswick, New Jersey.

Romanul, F. C. A. (1965). *Arch. Neurol.* **12,** 497.

Root, A. (1965). *Pediatrics* **36,** 940.

Rose, H. G., Robertson, M. C., and Schwartz, T. B. (1959). *Am. J. Physiol.* **197,** 1063.

Rose, W. C., Wixom, R. L., Lockhart, H. B., and Lambert, G. F. (1955). *J. Biol. Chem.* **217,** 987.

Rosenmann, M., and Morrison, P. (1965). *J. Biol. Chem.* **240,** 3353.

Ross, M. H., Pappas, G. D., and Harman, P. J. (1960). *Lab. Invest.* **9,** 388.

Roth, J., Glick, S. M., Yalow, R. S., and Berson, S. A. (1963). *Science* **140,** 987.

Rowe, R. W. D. (1968a). *J. Exptl. Zool.* **167,** 353.

Rowe, R. W. D. (1968b). *J. Exptl. Zool.* **169,** 59.

Rowe, R. W. D., and Goldspink, G. (1969a). *J. Anat.* **104,** 519.

Rowe, R. W. D., and Goldspink, G (1969b). *J. Anat.* **104,** 531.

Runyan, T. J., and Gershoff, S. N. (1969). *J. Nutr.* **98,** 113.

Russell, J. A., and Wilhelmi, A. E. (1960). *In* "Structure and Function of Muscle" (G. H. Bourne, ed.), Vol. II, p. 141. Academic Press, New York.

Ryan, W. L., and Carver, M. J. (1963). *Proc. Soc. Exptl. Biol. Med.* **114,** 816.

Samaha, F. J., and Gergely, J. (1969). *New Engl. J. Med.* **280,** 184.

Sanahuja, J. C., and Harper, A. E. (1963). *Am. J. Physiol.* **204,** 686.

Sandritter, W., and Scomazzoni, G. (1964). *Nature* **202,** 100.

Saunders, H. L., Steciw, B., and Kline, A. (1962). *Endocrinology* **71,** 314.

Sasaki, R., Watanabe, Y., Morishita, T., and Yamagata, S. (1968a). *Tohoku J. Exptl. Med.* **95,** 177.

Sasaki, R., Wanantabe, Y., Morishita, T., and Yamagata, S. (1968b). *Tohoku J. Exptl. Med.* **95,** 185.

Schapira, G., Dreyfus, J., and Schapira, F. (1950). *Compt. Rend. Soc. Biol.* **144,** 829.
Schapira, G., Coursaget, J., Dreyfus, J. C., and Schapira, F. (1953). *Bull. Soc. Chim. Biol.* **35,** 1309.
Schapira, G., Joly, M., and Dreyfus, J. C. (1954). *Compt. Rend. Soc. Biol.* **148,** 1056.
Scharff, R., and Wool, I. G. (1964). *Nature* **202,** 603.
Scharff, R., and Wool, I. G. (1965a). *Biochem. J.* **97,** 257.
Scharff, R., and Wool, I. G. (1965b). *Biochem. J.* **97,** 272.
Scharff, R., and Wool, I. G. (1966). *Biochem. J.* **99,** 173.
Schaub, M. C. (1964–1965). *Gerontologia* **10,** 38.
Schiebler, T. H., and Wolff, H. H. (1966). *Z. Zellforsch* **69,** 22.
Schlessinger, D. (1964). *Biochim. Biophys. Acta* **80,** 473.
Schmidt, C. G., and Schlief, H. (1956). *Z. Ges. Exptl. Med.* **127,** 53.
Schønheyder, F., and Lyngbye, J. (1962). *Brit. J. Nutr.* **16,** 75.
Schram, E. (1960). *Arch. Intern. Physiol. Biochim.* **68,** 698.
Schreiber, S. S., Oratz, M., Evans, C. D., Silver, E., and Rothschild, M. A. (1968). *Am. J. Physiol.* **215,** 1250.
Schreier, K., and Kazassis, C. (1960). *Nature* **187,** 1117.
Scow, R. O., and Hagan, S. N. (1955). *Am. J. Physiol.* **180,** 31.
Scow, R. O., and Hagan, S. N. (1957). *Endocrinology* **60,** 273.
Scow, R. O., and Hagan, S. N. (1965). *Endocrinology* **77,** 852.
Scow, R. O., and Roe, J. H. (1953). *Am. J. Physiol.* **173,** 22.
Scrimshaw, N. S., Taylor, C. E., and Gordon, J. E. (1968). *World Health Organ. Monogr. Ser.* **57.**
Seidel, J. C., Sreter, F. A., Thompson, M. M., and Gergely, J. (1964). *Biochem. Biophys. Res. Commun.* **17,** 662.
Severin, S. E., Bocharnikova, I. M., Vulfson, P. L., Grigorovich, J. A., and Solovieva, G. A. (1963). *Biokhimiya* **28,** 510.
Shafiq, S. A., and Gorycki, M. A. (1965). *J. Pathol. Bacteirol.* **90,** 123.
Shafiq, S. A., Gorycki, M. A., Goldstone, L., and Milhorat, A. T. (1966). *Anat. Record* **156,** 283.
Shafiq, S. A., Gorycki, M. A., and Milhorat, A. T. (1967). *Neurology* **17,** 567.
Shafiq, S. A., Gorycki, M. A., and Mauro, A. (1968). *J. Anat.* **103,** 135.
Shafiq, S. A., Gorycki, M. A., and Milhorat, A. T. (1969). *J. Anat.* **104,** 281.
Shambaugh, G. E., III, and Beisel, W. R. (1967). *Diabetes* **16,** 369.
Shenk, J. H., Hall, J. L., and King, H. H. (1934). *J. Biol. Chem.* **105,** 741.
Sheves, G. S., Epelbaum, S. E., and Riumina, W. I. (1956). *Biokhimiya* **21,** 71.
Shiaffino, S., and Margreth, A. (1969). *J. Cell. Biol.* **41,** 855.
Shibko, S., Caldwell, K. A., Sawant, P. L., and Tappel, A. L. (1963). *J. Cellular Comp. Physiol.* **61,** 85.
Shimizu, C. S. N., and Kaplan, S. A. (1964). *Endocrinology* **74,** 709.
Short, F. A. (1969). *Am. J. Physiol.* **217,** 307.
Siebert, G. (1958). *Experientia* **14,** 65.
Siebert, W. W. (1928). *Z. Klin. Med.* **109,** 360.
Sidransky, H., and Farber, E. (1958). *Arch. Pathol.* **66,** 135.
Sidransky, H., and Verney, E. (1965). *J. Nutr.* **86,** 73.
Sidransky, H., and Verney, E. (1967). *Biochim. Biophys. Acta* **138,** 426.
Sidransky, H., and Verney, E. (1968). *J. Nutr.* **96,** 349.
Sidransky, H., Staehelin, T., and Verney, E. (1964). *Science* **146,** 766.
Simon, E. J., Gross, C. S., and Lessell, I. M. (1962). *Arch. Biochem. Biophys.* **96,** 41.
Singh, S. N., and Kanungo, M. S. (1968). *J. Biol. Chem.* **243,** 4526.

Slack, H. G. B. (1954). *Clin. Sci.* **13,** 155.
Smith, B. (1964). *Neurology* **14,** 857.
Smith, D. S. (1966). *Progr. Biophys. Mol. Biol.* **16,** 107.
Smoller, M., and Fineberg, R. A. (1965). *J. Clin. Invest.* **44,** 615.
Snipes, C. A. (1967). *Am. J. Physiol.* **212,** 279.
Snipes, C. A. (1968). *Quart. Rev. Biol.* **43,** 127.
Snipes, C. A., and Kostyo, J. L. (1962). *Am. J. Physiol.* **203,** 933.
Snyderman, S. E., Holt, L. E., Norton, P. M., Roitman, E., and Finch, J. (1963). *Am. J. Clin. Nutr.* **12,** 333.
Sola, O. M., and Martin, A. W. (1953). *Am. J. Physiol.* **172,** 324.
Solomon, G., and Tarver, H. (1952). *J. Biol. Chem.* **195,** 447.
Sottocasa, G. L., Romeo, D., Stagni, N., and Bernard, B. de (1962). *Arch. Intern. Physiol. Biochim.* **70,** 581.
Spatz, M. (1969). *Ann. N.Y. Acad. Sci.* **156,** 152.
Spiro, D., and Sonnenblick, E. H. (1964). *Circulation Res.* **15** (Suppl. II), II-14.
Spray, C. M., and Widdowson, E. M. (1950). *Brit. J. Nutr.* **4,** 332.
Sprinson, D. B., and Rittenberg, D. (1950). *J. Biol. Chem.* **184,** 405.
Spyrides, G. J. (1964). *Proc. Natl. Acad. Sci. U.S.* **51,** 1220.
Squibb, R. L., Lyons, M. M., and Beisel, W. R. (1968). *J. Nutr.* **96,** 509.
Sreter, F. A., and Woo, G. (1963). *Am. J. Physiol.* **205,** 1290.
Sreter, F. A., Seidel, J. C., and Gergely, J. (1966). *J. Biol. Chem.* **241,** 5772.
Srivastava, U. (1967). *Can. J. Biochem.* **45,** 1419.
Srivastava, U. (1969). *Arch. Biochem. Biophys.* **130,** 129.
Srivastava, U., and Berlinquet, L. (1966). *Arch. Biochem. Biophys.* **114,** 320.
Srivastava, U., and Chaudhary, K. D. (1969). *Can. J. Biochem.* **47,** 231.
Srivastava, U., Devi, A., and Sarkar, N. K. (1963). *Exptl. Cell Res.* **29,** 289.
Stein, J. M., and Padykula, H. A. (1962). *Am. J. Anat.* **110,** 103.
Stenger, R. J., and Spiro, D. (1961). *Am. J. Med.* **30,** 653.
Stephen, J. M. L. (1968). *Brit. J. Nutr.* **22,** 153.
Stephen, J. M. L., and Waterlow, J. C. (1966). *Nature* **211,** 978.
Stern, D. N., and Stim, E. M. (1959). *Proc. Soc. Exptl. Biol. Med.* **101,** 125.
Stewart, D. M. (1955). *Biochem. J.* **59,** 553.
Stewart, D. M. (1968). *Am. J. Physiol.* **214,** 1139.
Stewart, D. M., and Martin, A. W. (1956). *Am. J. Physiol.* **186,** 497.
Stickney, J. C., and Van Liere, E. J. (1953). *Physiol. Rev.* **33,** 13.
Stirewalt, W. S., and Wool, I. G. (1966). *Science* **154,** 284.
Stirewalt, W. S., Wool, I. G., and Cavicchi, P. (1967). *Proc. Natl. Acad. Sci. U.S.* **57,** 1885.
Stockdale, F. E., and Holtzer, H. (1961). *Exptl. Cell Res.* **24,** 508.
Strohman, R. C. (1966). *Biochem. Z.* **345,** 148.
Stubbs, S. St. G., and Blanchaer, M. C. (1965). *Can. J. Biochem.* **43,** 463.
Sunderland, S., and Ray, L. J. (1950). *J. Neurol. Neurosurg. Psychiatr.* **13,** 159.
Swartz, F. J. (1962). *Growth* **26,** 167.
Swendseid, M. E., Villalobos, J., and Friedrich, B. (1964). *J. Nutr.* **82,** 206.
Swendseid, M. E., Tuttle, S. G., Drenick, E. J., Joven, C. B., and Massey, F. J. (1967). *Am. J. Clin. Nutr.* **20,** 243.
Swick, R. W., and Handa, D. T. (1956). *J. Biol. Chem.* **218,** 577.
Szent-Györgyi, A. (1953). "Chemical Physiology of Contraction in Body and Heart Muscle," p. 104. Academic Press, New York.
Tallan, H. H. (1955). *Proc. Soc. Exptl. Biol. Med.* **89,** 553.

Tallan, H. H., Moore, S., and Stein, W. H. (1954). *J. Biol. Chem.* **211,** 927.
Tanzer, M. L., and Gilvarg, C. (1959). *J. Biol. Chem.* **234,** 3201.
Tappan, D. V., and Reynafarje, B. (1957). *Am. J. Physiol.* **190,** 99.
Tappel, A. L., Zalkin, H., Caldwell, K. A., Desai, I. D., and Shibko, S. (1962). *Arch. Biochem. Biophys.* **96,** 340.
Taskar, K., and Tulpule, P. G. (1964). *Biochem. J.* **92,** 391.
Telepneva, V. I. (1961). *Vopr. Med. Khim.* **7,** 409.
Thompson, H. T., Schurr, P. E., Henderson, L. M., and Elvehjem, C. A. (1950). *J. Biol. Chem.* **182,** 47.
Thompson, R. C., and Ballou, J. E. (1954). *J. Biol. Chem.* **208,** 883.
Thompson, R. C., and Ballou, J. E. (1956). *J. Biol. Chem.* **223,** 795.
Thorner, S. H. (1939). *Arb. Physiol.* **8,** 359.
Toth, A., and Schiebler, T. H. (1967). *Z. Zellforsch.* **76,** 543.
Trayer, I. P., and Perry, S. V. (1966). *Biochem. Z.* **345,** 87.
Trayer, I. P., Harris, C. I., and Perry, S. V. (1968). *Nature* **217,** 452.
Tsuji, S. (1967). *Experientia* **23,** 629.
Vanderhoof, E. R., Imig, G. J., and Hines, H. M. (1961). *J. Appl. Physiol.* **16,** 873.
Van Liere, E. J., and Northup, D. W. (1957). *J. Appl. Physiol.* **11,** 91.
Van Liere, E. J., Krames, B. B., and Northup, D. W. (1964). *Circulation Res.* **16,** 244.
Vannotti, A., and Magiday, M. (1934). *Arbeitphysiologie* **7,** 615.
Van Slyke, D. D., and Meyer, G. M. (1913a). *J. Biol. Chem.* **16,** 197.
Van Slyke, D. D., and Meyer, G. M. (1913b). *J. Biol. Chem.* **16,** 213.
Vaughan, B. E., and Pace, N. (1956). *Am. J. Physiol.* **185,** 549.
Velick, S. F. (1956). *Biochim. Biophys. Acta* **20,** 228.
Venable, J. H. (1966a). *Am. J. Anat.* **119,** 263.
Venable, J. H. (1966b). *Am. J. Anat.* **119,** 271.
Venable, J. H. (1969). *Anat. Record* **163,** 279.
Vignos, P. J., and Lefkowitz, M. (1959). *J. Clin. Invest.* **38,** 873.
Vincent, M., and Radermecker, M. A. (1959). *Am. J. Trop. Med. Hyg.* **8,** 511.
Vis, H., DuBois, R., Loeb, H., Vincent, M., and Bigwood, E. J. (1958). *Ann. Soc. Belge Med. Trop.* **38,** 991.
Vis, H., DuBois, R., Vanderborght, H., and DeMaeyer, E. (1965). *Rev. Franc. Etudes Clin. Biol.* **10,** 729.
Voit, C. (1886). *Z. Biol.* **2,** 307.
Wainman, P., and Shipounoff, G. C. (1941). *Endocrinology* **29,** 975.
Walker, M. G. (1966). *Comp. Biochem. Physiol.* **19,** 791.
Walton, J. N. (1965). *In* "Muscle" (W. M. Paul, E. E. Daniel, C. M. Kay, and G. Monckton, eds.), p. 501. Pergamon, Oxford.
Walton, J. N., and Adams, R. D. (1956). *J. Pathol. Bacteriol.* **72,** 273.
Wannemacher, R. W., and McCoy, J. R. (1969). *Am. J. Physiol.* **216,** 781.
Waterlow, J. C. (1956). *West Indian Med. J.* **5,** 167.
Waterlow, J. C. (1959). *Nature* **184,** 1875.
Waterlow, J. C., and Mendes, C. B. (1957). *Nature* **180,** 1361.
Waterlow, J. C., and Stephen, J. M. L. (1966). *Brit. J. Nutr.* **20,** 461.
Waterlow, J. C., and Stephen, J. M. L. (1967). *Clin. Sci.* **33,** 489.
Waterlow, J. C., and Stephen, J. M. L. (1968). *Clin. Sci.* **35,** 287.
Wearn, J. T. (1941). *Bull. Johns Hopkins Hosp.* **68,** 363.
Wechsler, W. (1964). *Naturwissenschaften* **51,** 91.
Wechsler, W., and Hager, H. (1960). *Naturwissenschaften* **47,** 185.

Weinstock, I. M. (1966). *Ann. N.Y. Acad. Sci.* **138,** Art. 1, 199.
Weinstock, I. M., and Lukacs, M. (1965a). *Enzymol. Biol. Clin.* **5,** 103.
Weinstock, I. M., and Lukacs, M. (1965b). *Enzymol. Biol. Clin.* **5,** 89.
Weinstock, I. M., Goldrich, A. D., and Milhorat, A. T. (1955). *Proc. Soc. Exptl. Biol. Med.* **88,** 257.
Weinstock, I. M., Epstein, S., and Milhorat, A. T. (1958). *Proc. Soc. Exptl. Biol. Med.* **99,** 272.
Weinstock, I. M., Soh, T. S., Freedman, H. A., and Cutler, M. E. (1969). *Biochem. Med.* **2,** 345.
Wheat, M. W. (1965). *J. Mt. Sinai Hosp. N.Y.* **32,** 107.
Whipple, G. H. (1926). *Am. J. Physiol.* **76,** 693.
Widdowson, E. M. (1968). *In* "Growth and Development of Mammals" (G. A. Lodge and G. M. Lamming, eds.), p. 224. Plenum Press, New York.
Widdowson, E. M., and Dickerson, J. W. T. (1960). *Biochem. J.* **77,** 30.
Widdowson, E. M., and Dickerson, J. W. T. (1964). *In* "Mineral Metabolism" (C. L. Comar and F. Bronner, eds.), Vol. II, Part A, Chapter 17. Academic Press, New York.
Widdowson, E. M., and McCance, R. A. (1955). *Brit. J. Exptl. Pathol.* **36,** 175.
Widdowson, E. M., Dickerson, J. W. T., and McCance, R. A. (1960). *Brit. J. Nutr.* **14,** 457.
Williams, C. A., Asofsky, R., and Thorbecke, G. J. (1963). *J. Exptl. Med.* **118,** 315.
Williams, C. A., Gonoza, M. C. and Lipmann, F. (1965). *Proc. Natl. Acad. Sci. U.S.* **53,** 622.
Winick, M., and Noble, A. (1966). *J. Nutr.* **89,** 300.
Winnick, T., and Winnick, R. E. (1957). *Biochim. Biophys. Acta* **23,** 649.
Woods, E. F. (1966). *J. Mol. Biol.* **16,** 581.
Wool, I. G. (1960a). *Am. J. Physiol.* **199,** 715.
Wool, I. G. (1960b). *Am. J. Physiol.* **199,** 719.
Wool, I. G. (1963). *Biochim. Biophys. Acta* **68,** 28.
Wool, I. G. (1964). *Nature* **202,** 196.
Wool, I. G. (1965). *Federation Proc.* **24,** 1060.
Wool, I. G., and Cavicchi, P. (1967). *Biochemistry* **6,** 1231.
Wool, I. G., and Kurihara, K. (1967). *Proc. Natl. Acad. Sci. U.S.* **58,** 2401.
Wool, I. G., and Moyer, A. N. (1964) *Biochim. Biophys. Acta* **91,** 248.
Wool, I. G., and Munro, A. J. (1963). *Proc. Natl. Acad. Sci. U.S.* **50,** 918.
Wool, I. G., and Weinshelbaum, E. I. (1959). *Am. J. Physiol.* **197,** 1089.
Wool, I. G., and Weinshelbaum, E. I. (1960). *Am. J. Physiol.* **198,** 1111.
Wool, I. G., Stirewalt, W. S., Kurihara, K., Low, R. B., Bailey, P., and Oyer, D. (1968). *Recent Progr. Hormone Res.* **24,** 139.
Wroblewski, F., and LaDue, J. S. (1956). *Proc. Soc. Exptl. Biol. Med.* **91,** 569.
Yamaguchi, M., and Kandatsu, M. (1967). *Agr. Biol. Chem.* (*Japan*) **31,** 776.
Yamatami, Y., and Kandatsu, M. (1967). *Agr. Biol. Chem.* (*Japan*) **31,** 705.
Yiengst, M. J., Barrows, C. H., and Shock, N. W. (1959). *J. Gerontol.* **14,** 400.
Yoshida, A., Leung, P. M-B., Rogers, Q. R., and Harper, A. E. (1966). *J. Nutr.* **89,** 80.
Youhotsky-Gore, I., and Pathmanathan, K. (1968). *Exptl. Gerontol.* **3,** 281.
Young, V. R., and Alexis, S. D. (1968). *J. Nutr.* **96,** 255.
Young, V. R., Huang, P. C. (1969). *Brit. J. Nutr.* **23,** 271.
Young, V. R., Chen, S. C., and MacDonald, J. (1968a). *Biochem. J.* **106,** 913.
Young, V. R., Chen, S. C., and Newberne, P. M. (1968b). *J. Nutr.* **94,** 361.

Young, V. R., Hussein, M. A., Scrimshaw, N. S. (1968c). *Nature* **218,** 568.
Young, V. R., Baliga, B. S., Alexis, S. D., and Munro, H. N. (1969a). *Biochim. Biophys. Acta* **199,** 298.
Young, V. R., Hussein, M. A., Murray, E., and Scrimshaw, N. S. (1969b). *Am. J. Clin. Nutr.* **22,** 1563.
Yudaev, N. A., Smirnov, M. I., Razina, P. G., and Dobbert, N. N. (1953). *Biokhamiia* **18,** 732.
Yun, J., and Parker, C. J. (1965). *Biochim. Biophys. Acta* **110,** 212.
Zachmann, M., Cleveland, W. W., Sandberg, D. H., and Nyhan, W. L. (1966). *Am. J. Diseases Children* **112,** 283.
Zak, R. (1962). *In* "The Denervated Muscle" (E. Gutmann, ed.), pp. 273–340. Czechoslovak Acad. Sci., Prague.
Zak, R., Rabinowitz, M., and Platt, C. (1967). *Biochemistry* **6,** 2493.
Zak, R., Grove, D., and Rabinowitz, M. (1969). *Am. J. Physiol.* **216,** 647.
Zalkin, H., Tappel, A. L., Caldwell, K. A., Shibko, S., Desai, I. D., and Holliday, T. A. (1962). *J. Biol. Chem.* **237,** 2678.
Zelena, J., and Hnik, P. (1957). *Physiol. Bohemoslov.* **6,** 193.
Zender, R., Lataste-Dorolle, C., Collet, R. A., Rowinski, P., and Mouton, R. F. (1958). *Food Res.* **23,** 305.
Zierler, K. (1966). *In* "Proceedings of the Conference on Energy Metabolism and Body Fuel Utilization" (A. P. Morgan, ed.), p. 78. Harvard Univ. Press, Cambridge, Massachusetts.
Zimmerman, H. J., Dujovne, C. A., and Levy, R. (1968). *Comp. Biochem. Physiol.* **25,** 1081.
Zimny, M. L., and Gregory, R. (1958). *Am. J. Physiol.* **195,** 233.
Zinkin, L. N., and Andreeva, L. F. (1963). *J. Embryol. Exptl. Morphol.* **11,** 353.

Author Index

Numbers in italics refer to the pages on which the complete references are listed.

B

F

G

I

J

M

Q

R

S

T

W

Subject Index

F

O

P

U